LIST OF SYMBOLS

D0078434

E_s	modulus of elasticity of steel
f_c	stress on concrete at extreme compression fiber
f_c'	compressive strength of concrete, measured at 2~~~~~ ~~~~~ casting
f_{c1}	unit stress on concrete at the compression steel location
f_{ci}'	concrete strength developed at time of prestress transfer to the concrete
f_{ct}	split-cylinder tensile strength
f_d	stress due to unfactored dead load at extreme fiber of section where tensile stress is caused by externally applied loads (Chap. 21)
f_{pc}	compressive stress in concrete (after allowance for all prestress losses) at centroid of cross section resisting externally applied loads or at junction of web and flange when the centroid lies within the flange (Chap. 21)
f_{pe}	compressive stress in concrete due to effective prestress forces only (after allowance for all prestress losses) at extreme fiber of section where tensile stress is caused by externally applied loads (Chap. 21)
f_{ps}	stress in prestressed reinforcement at nominal strength (Chap. 21)
f_{pu}	specified ASTM minimum tensile strength of prestressing tendons (Chap. 21)
f_r	modulus of rupture (tensile strength in bending)
f_s	unit stress on steel at service load conditions (Chap. 4)
f_s'	unit stress on compression steel
f_{si}	initial tensile stress in prestressing reinforcement before transfer of stress to the concrete (Chap. 21)
f_{sy}	yield stress of spiral reinforcement
f_t	tensile stress on transformed area of steel in equivalent concrete
f_y	yield stress of steel
F_{ti}	$1/K_{ti}$ (Sec. 17.4)
F_{tj}	$1/K_{tj}$ (Sec. 17.4)
G	shear modulus of elasticity
h	overall depth of section
	diameter of section
h_s	story height, or overall height of all stories (Chap. 15)
H	humidity, percent
I_b	moment of inertia of the gross section of the beam as defined in ACI-13.2.4
I_c	moment of inertia of a standard section (Chap. 14)
I_{cr}	moment of inertia of the cracked transformed section
I_e	effective moment of inertia
$(I_e)_D$	effective moment of inertia with dead load acting
$(I_e)_{D+L}$	effective moment of inertia with dead plus live load acting
I_g	gross moment of inertia (uncracked section)
I_m	effective moment of inertia at midspan for a simply supported or continuous span, and at the support section for a cantilever
I_0	moment of inertia of gross section about centroidal axis (Chaps. 21 and 22)
I_s	$t^3/12$ times the width of slab bounded laterally by the centerline of the adjacent panel, if any (Sec. 16.4)
	moment of inertia of steel reinforcement, taken about plastic centroid of concrete section (Chap. 15)
I_{sb}	moment of inertia of slab-beam combination, including the slab used for I_s, and also including the contribution of that portion of the beam stem extending above or below the slab (Chap. 16)
I_x	moment of inertia about x axis
j	ratio of moment arm of internal couple to effective depth d for an ideally reinforced beam (Sec. 3.10)
J_c	section property analogous to polar moment of inertia (Sec. 16.18)
k	effective length factor (Sec. 13.7; Chap. 15)
	ratio of neutral axis distance x to effective depth d for ideally reinforced beam (Sec. 4.10); general ratio of neutral axis distance measured from compression face to the effective depth d (Sec. 5.2)
k_b	r^2/y_t, bottom kern distance from centroid of a cross section
k_r	deflection multiplier to account for effect of compression steel on creep
k_t	r^2/y_b, top kern distance from centroid of a cross section
K	flexural stiffness ($4EI/L$ for prismatic members)
	wobble coefficient for curved prestressing tendon (Sec. 21.7)
K_b	flexural stiffness of beam; moment per unit rotation
K_c	flexural stiffness of column; moment per unit rotation
K_{ec}	flexural stiffness of *equivalent* column, $\Sigma K_c/(1 + \Sigma K_c/K_t)$ (Sec. 17.5)
K_s	flexural stiffness of slab; moment per unit rotation

K_t	torsional stiffness of member; torsional moment per unit rotation about longitudinal axis of member (Sec. 16.20; Chaps. 17 and 19)
K_{ti}	stiffness of rotational spring at end i of member ij, causing slip angle ψ_i (Chap. 17)
K_{tj}	stiffness of rotational spring at end j of member ij, causing slip angle ψ_j (Chap. 17)
L_1	span measured in the direction of the equivalent frame for two-way systems, measured center-to-center of supports (Chap. 16)
L_2	span transverse to L_1, measured center-to-center of supports (Chap. 16)
L_a	length of straight embedment provided beyond inflection point or beyond center of support at simple supports (cannot be taken greater than the larger of d or $12d_b$ for ACI-12.2.3)
L_d	development length for a straight bar; distance bar must be embedded to develop its nominal tensile strength
L_{db}	basic development length for a straight bar (before any modifications for favorable or unfavorable factors)
L_{dh}	development length of standard hooked bar in tension (Fig. 6.11.2), measured from critical section to outside end of hook (straight embedment length between critical section and start of hook [point of tangency] plus radius of bend and one bar diameter) = L_{hb} times applicable modification factors
L_{hb}	basic development length of a standard hooked bar in tension; basic value of L_{dh} defined by Fig. 6.11.2
L_n	clear span
L_u	unsupported length of a compression member (Sec. 13.7; Chap. 15)
m	ratio of the volume of longitudinal bars to volume of closed hoops, $A_\ell s/[2A_c\,(x_1 + y_1)]$ (Sec. 19.6) $f_y/(0.85f_c')$, stress ratio number of tension bars; for several sizes, $m = A_s/(A_b$ for largest bar) (Sec. 4.12)
M_{1b}	smaller of primary moments at ends of member in a *braced* frame
M_{1s}	smaller of primary moments at ends of member in an *unbraced* frame
M_{2b}	larger of primary moments at ends of member in a *braced* frame
M_{2s}	larger of primary moments at ends of member in an *unbraced* frame
M_b	portion transferred by flexure out of the total moment M to be transferred between column and slab at a joint (Sec. 16.18)
M_c	$\delta_b M_{2b} + \delta_s M_{2s}$, magnified factored moment to use in design of compression member (Chap. 15)
M_{cr}	cracking moment, $f_r I_g/y_t$
M_D	dead load moment (Chaps. 14 and 21)
M_{D+L}	dead load plus live load moment (Chap. 14)
M_{equiv}	$C_m M_{max}$, equivalent primary uniform moment along a braced member (Chap. 15)
M_i	primary bending moment (function of z) (Chap. 15)
M_m	maximum primary bending moment
M_{max}	maximum service moment for condition under which deflection is computed
M_n	nominal moment strength
M_{n0}	nominal moment strength when member is subject to flexure alone (Chap. 19)
M_{nb}	nominal moment strength per unit distance along a yield line (Chap. 18)
M_{nc}	nominal moment strength for beam having no compression steel, $C_c(d - a/2)$
M_{net}	$w_{net}L^2/8$; $M_D - M_{prestress}$ (Chap. 21)
M_{ns}	nominal moment strength added by compression steel, $C_s(d - d')$
M_{nx}	nominal moment strength about the x axis, $P_n e_y$ (Sec. 13.21)
M_{ny}	nominal moment strength about the y axis, $P_n e_x$ (Sec. 13.21)
M_0	moment at midspan on a simply supported beam nominal moment strength M_n for member subject to flexure alone (Chap. 13) full statical moment on a two-way slab system (Sec. 16.3)
M_{0i}	fixed end moment at end i of a flexural member ij (Sec. 16.10 and Chap. 17)
M_{0i}'	sum of M_{0i} and moment necessary to cancel unwanted slope, $[(1 + F_{tj}S_{ij})M_{0i} - (F_{tj}S_{ij})M_{0j}]/T_2$
M_{0j}	fixed end moment at end j of a flexural member ij (Sec. 16.10 and Chap. 17)
M_{0j}'	sum of M_{0j} and moment necessary to cancel unwanted slope, $[(1 + F_{ti}S_{ii})M_{0j} - (F_{ti}S_{ij})M_{0i}]/T_2$ (Sec. 17.4)
M_{0x}	nominal moment strength for bending about the x axis when axial compression is zero (Sec. 13.21)
M_{0y}	nominal moment strength for bending about the y axis when axial compression is zero (Sec. 13.21)
M_s	net positive midspan moment
M_u	factored moment
M_v	portion transferred by shear out of the total moment M to be transferred between

Reinforced Concrete Design

Fourth Edition

Chu-Kia Wang
Charles G. Salmon
University of Wisconsin—Madison

1817

HARPER & ROW, PUBLISHERS, New York
Cambridge, Philadelphia, San Francisco,
London, Mexico City, São Paulo, Singapore, Sydney

Sponsoring Editor: Cliff Robichaud
Project Editor: David Nickol
Text Art: Fineline Illustrations, Inc.
Production: Delia Tedoff
Compositor: Waldman Graphics, Inc.
Printer and Binder: The Maple Press

Reinforced Concrete Design, Fourth Edition

Library of Congress Cataloging in Publication Data

Wang, Chu-kia, 1917–
   Reinforced concrete design.

   Includes bibliographies.
   1. Reinforced concrete construction.   I. Salmon,
Charles G.   II. Title.
TA683.2.W3   1985   624.1′8341   84-12973
ISBN 0-06-046896-3

84 85 86 87 9 8 7 6 5 4 3 2 1

CONTENTS

PREFACE

The publication of this fourth edition reflects the continuing change occurring in design procedures relating to reinforced concrete structures. The strength design philosophy is now well established. Current design recognizes that the "limit of usefulness" or "limit state" is reached when the *strength* of the member is fully utilized, or it may be reached when the member is no longer *serviceable* (such as when it deflects too much). Either *strength* or *serviceability* may control the design and provision must be made for both.

The fourth edition reflects the changes in design rules arising from the publication of the 1983 American Concrete Institute (ACI) Building Code and Commentary. Changes in the 1983 ACI Code include those appearing in the 1980 Supplement as well as those approved by ACI Committee 318 (Standard Building Code) since 1980.

Also included is additional extension toward SI units. The 1983 ACI Code has an SI version (known as ACI 318-83M) and the SI version of each Code equation appears in this book as a footnote equation having the same equation number. According to the ACI Code, the designer must use in its entirety either the Inch-Pound units version (ACI 318-83) or the SI version (ACI 318-83M). The authors believe that sufficient metrication should be included in a textbook so that some familiarity may be gained with SI units. The text provides data on reinforcing bars in accordance with ASTM (American Society for Testing and Materials) Inch-Pound units (the commonly available bar sizes and strengths) and also ASTM SI units (the bar sizes and strengths used in Canada but not yet actually used in the United States).

Some design tables are provided for bars and material strengths in SI units, a few numerical examples are given in SI units, and many problems at the ends of chapters are given with an SI alternate in parenthesis at the end of the problem statement.

Regarding the choice between the Standard Metric unit of force (kilogram force, kgf) or the SI unit of force (newton = kilogram meter per second), the authors have concluded that the use of the newton in accordance with ASTM Standard E380 is likely to become the accepted metric approach in the United States. Thus, in all parts of the book where metric versions are used, the newton (N) or kilonewton (kN) is used to measure force. The SI unit of stress is the pascal (Pa), or newton per meter squared, which because of its typically large numerical value is usually expressed in megapascals (MPa); that is, 10^6 pascals. A few of the diagrams have along the stress axis the kilogram force per centimeter squared (kgf/cm^2) in addition to Inch-Pound and SI units. Throughout the book the term *US Customary* used previously to refer to the units commonly used in the United States has been changed to *Inch-Pound* in accordance with the latest recommendation of ASTM E380. For the convenience of the reader, some conversion factors for forces, stresses, uniform loading, and moments are provided on a separate page following this preface.

This new edition follows the same philosophical approach that has gained the wide acceptance of users since the first edition was published in 1965. Herein, as previously, strength and behavior of concrete elements are treated with the primary objective of explaining and justifying the ACI Code rules and formulas.* Then numerous examples are presented illustrating the general approach to design and analysis. Considering the limited scope of most examples, attempts to reach practical results are made insofar as possible.

Considerable emphasis is placed on presenting for the student, as well as the practicing engineer, the basic concepts deemed essential to understand and apply properly the ACI Code rules and formulas. The treatment is incorporated into the chapters in such a way that the reader may either study in detail the concepts in logical sequence, or merely accept a qualitative explanation and proceed directly to the design process using the ACI Code.

Depending on the proficiency required of the student, this book may provide material for two courses of three or four semester-hours each. It is suggested that the beginning course in concrete structures for undergraduate students might contain the material of Chapters 1 through 9, 13, and the spread footing portion of Chapter 20, excepting Sections 5.13 through 5.16 and 13.18 through 13.21. In addition, the first portion of Chapter 21, "Introduction to Prestressed Concrete," is recommended for the first course. The second course may start with the continuous beam in Chapter 10, utilizing that to review many of the topics in Chapters 1 through 9. The remaining chapters—particularly Chapter 14 on deflections; Chapter 15 on length effects on compression members; Chapter 16 on two-way slab sys-

---

*Since nearly continuous reference is made to the 1983 ACI Code, the reader will find it desirable to secure a copy of it and the Commentary from the American Concrete Institute, Box 19150, Detroit, Michigan 48219.

tems; the remaining sections of Chapter 5 on shear strength affected by axial force, deep beams, and brackets; Chapter 19 on torsion; and Chapter 21 on prestressed concrete—are suggested for inclusion.

Special features of the fourth edition are (a) complete revision and simplification of the two chapters in the third edition on two-way slab systems into one design chapter (Chapter 16); (b) a new chapter (Chapter 17) on equivalent frame analysis of unbraced frames having two-way floor systems; (c) completely revised and expanded treatment of length effects on compression members including introductory treatment of frames (Chapter 15); (d) revised treatment of development length for hooked bars (Chapter 6); (e) reduction in length and simplification of the basic chapter (Chapter 13) on compression members to consolidate the material relating to rectangular sections into successive chapter sections and remove most of the detail relating to circular sections; and (f) completely revised material on brackets and corbels and shear-friction in Chapter 5.

This complete revision has retained important features of the third edition, including (a) detailed treatment of beam deflections (Chapter 14); (b) comprehensive treatment of design for torsion (Chapter 19); (c) treatment of monolithic beam-to-column joints (Chapter 11); (d) introductory treatment of yield line theory for slabs; (e) important extensive treatment of bar development length, cutoff, and anchorage requirements using the moment capacity diagram (Chapter 6); (f) treatment of shear requirements for effect of axial loads, deep beams, and brackets (corbels); (g) prestressed concrete treatment necessary for understanding the basic concept; and (h) introductory treatment of composite concrete-on-concrete construction.

The authors continue to be indebted to students, colleagues, and other users of the first three editions who have suggested improvements of wording, identified errors, and recommended items for inclusion or omission. These suggestions have been carefully considered and the result is reflected in this complete revision. The detailed suggestions and comments from Professor J. C. Smith of North Carolina State are particularly appreciated.

Users of this fourth edition are urged to communicate with the authors regarding all aspects of this book, particularly on identification of errors and suggestions for improvement.

The authors again gratefully acknowledge the continuing patience and encouragement of their wives, Vera Wang and Bette Salmon, and to them affectionately dedicate this book.

Chu-Kia Wang
Charles G. Salmon

Conversion Factors

Some Conversion Factors, Between Inch-Pound Units and SI Units, Useful in Reinforced Concrete Design

	To Convert	*To*	*Multiply by*
Forces	kip force	kN	4.448
	N	lb	0.2248
Stresses	ksi	MPa	6.895
	MPa	ksi	0.1450
Uniform Loading	kip/ft	kN/m	14.59
	kN/m	lb/ft	68.52
Moments	ft-kip	kN·m	1.356
	kN·m	ft-kip	0.7376

Basis (ASTM E380): 1 in. = 25.4 mm; 1 lb force = 4.448 221 615 260 5 newtons; 1 kg force = 9.80665 newtons.

1

Introduction, Materials, and Properties

1.1 REINFORCED CONCRETE STRUCTURES

The three most common materials from which most structures are built are wood, steel, and reinforced (including prestressed) concrete. Lightweight materials such as aluminum and plastics are also becoming more common in use. Reinforced concrete is unique in that two materials, reinforcing steel and concrete, are used together; thus the principles governing the structural design in reinforced concrete differ in many ways from those involving design in one material.

Many structures are built of reinforced concrete: bridges, viaducts, buildings, retaining walls, tunnels, tanks, conduits, and others. This text deals primarily with fundamental principles in the design and investigation of reinforced concrete members subjected to axial force, bending moment, shear, torsion, or combinations of these. Thus these principles are basically applicable to the design of any structure, so long as information is known about the variation of axial force, shear, moment, etc., along the length of each member. Although *analysis* and *design* may be treated separately, they are inseparable in practice, especially in the case of reinforced concrete structures, which are usually statically indeterminate. In such cases relative sizes of members are needed in the preliminary analysis that must precede the final design; so the final conciliation between analysis and design is largely a matter of trial, judgment, and experience.

Water Tower Place, 74 stories, 859 ft high, hotel–condominium–shopping complex, Chicago, Ill. Tallest concrete structure other than a tower, completed 1976. (Courtesy of Portland Cement Association.)

Reinforced concrete is a logical union of two materials: plain concrete, which possesses high compressive strength but little tensile strength, and steel bars embedded in the concrete which can provide the needed strength in tension. For instance, the strength of the beam shown in Fig. 1.1.1 is

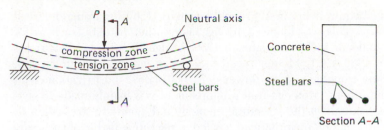

Figure 1.1.1   Position of bars in a reinforced concrete beam.

greatly increased by placing steel bars in the tension zone. However, since reinforcement steel is capable of resisting compression as well as tension, it is also used to provide part of the carrying capacity in reinforced concrete columns, and frequently in the compression zone of beams.

Steel and concrete work readily in combination for several reasons: (1) bond (interaction between bars and surrounding hardened concrete) prevents slip of the bars relative to the concrete; (2) proper concrete mixes provide adequate impermeability of the concrete against bar corrosion; and (3) sufficiently similar rates of thermal expansion—that is, 0.0000055 to 0.0000075 for concrete and 0.0000065 for steel per degree Fahrenheit (°F), or 0.000010 to 0.000013 for concrete and 0.000012 for steel per degree Celsius (°C)—introduce negligible stresses between steel and concrete under atmospheric changes of temperature.

Transverse cracks of small width may appear near the steel bars placed in the tension regions of ordinary reinforced concrete (unless prestressed); such cracks are expected and do not interfere with the performance of the member.

1.2 HISTORICAL BACKGROUND

Joseph Monier, the owner of an important nursery in Paris, generally deserves the credit for making the first practical use of reinforced concrete in 1867. He recognized many of its potential uses, and successfully undertook to expand the application of the new method [1].* Prior to Monier's work, however, the method of reinforcing concrete with iron was known and in some cases even protected by patents. Ancient Grecian structures have been found which show that builders knew something about the reinforcing of stonework for added strength [2].

In the mid-1800s, Lambot in France constructed a small boat which he exhibited at the Paris Exposition of 1854 and on which he received a patent in 1855. In Lambot's patent was shown a reinforced concrete beam and a column reinforced with four round iron bars. Another Frenchman, Francois Coignet, published a book in 1861 describing many applications and uses of reinforced concrete. In 1854 W. B. Wilkinson of England took out a patent for a reinforced concrete floor.

*Numbers in brackets refer to the Selected References at the end of the chapter.

Monier acquired his first French patent in 1867 for iron reinforced concrete tubs. This was followed by his many other patents, such as for pipes and tanks in 1868, flat plates in 1869, bridges in 1873, and stairways in 1875. In 1880–1881, Monier received German patents for railroad ties, water feeding troughs, circular flower pots, flat plates, and irrigation channels, among others. Monier's iron reinforcement was mainly made to conform to the contour of the structural element and generally strengthen it. He apparently had no quantitative knowledge regarding its behavior or any method of making design calculations [1].

In the United States the pioneering efforts were made by Thaddeus Hyatt, originally a lawyer, who conducted experiments on reinforced concrete beams in the 1850s. In a perfectly correct manner the iron bars in Hyatt's beams were located in the tension zone, bent up near the supports, and anchored in the compression zone. Additionally, transverse reinforcement (known as vertical stirrups) was used near the supports. However, Hyatt's experiments were unknown until 1877 when he published his work privately.

The first cast-in-place reinforced concrete structure in the United States is generally credited to the William Ward house in Port Chester, New York, built in 1870 [3]. E. L. Ransome, head of the Concrete-Steel Company of San Francisco, apparently used some form of reinforced concrete in the early 1870s. He continued to increase the application of wire rope and hoop iron to many structures and was the first to use and have patented in 1884 the deformed (twisted) bar.

The Monier German patents were sold to G. A. Wayss and Company of Germany in 1880. Tests of structural strength were conducted by German engineers during the 1880s. Theories and computational methods were published by Koenen and Wayss in 1886. Test results of Wayss and J. Bauschinger were published in 1887.

In 1890, Ransome built the Leland Stanford Jr. Museum in San Francisco, a reinforced concrete building two stories high and 312 ft (95 m) long. Since that time, development of reinforced concrete in the United States has been rapid. During the period 1891–1894, various investigators in Europe published theories and test results; among them were Moeller (Germany), Wunsch (Hungary), Melan (Austria), Hennebique (France), and Emperger (Hungary), but practical use was less extensive than in the United States.

Throughout the entire period 1850–1900, relatively little was published, as the engineers working in the reinforced concrete field considered construction and computational methods as trade secrets. One of the first publications that might be classified as a textbook was that of Considère in 1899. By the turn of the century there was a multiplicity of systems and methods with little uniformity in design procedures, allowable stresses, and systems of reinforcing. In 1903, with the formation in the United States of a joint committee of representatives of all organizations interested in reinforced concrete, uniform application of knowledge to design was initiated. The development of standard specifications is discussed in Chap. 2.

In the first decade of the twentieth century, progress in reinforced concrete was rapid. Extensive testing to determine beam behavior, compressive strength of concrete, and modulus of elasticity was conducted by Arthur N. Talbot at the University of Illinois, by Frederick E. Turneaure and Morton O. Withey at the University of Wisconsin, and by Bach in Germany, among others. From about 1916 to the mid-1930s, research centered on axially loaded column behavior. In the late 1930s and 1940s, eccentrically loaded columns, footings, and the ultimate strength of beams received special attention.

Since the mid-1950s, reinforced concrete design practice has made the transition from that based on elastic methods to one using strength as its basis. The use of prestressed concrete (Chap. 21), wherein the steel reinforcement is stressed in tension and the concrete is in compression even before external loads are applied, has advanced from an experimental technique to a major category in structural material. There has been a transition from cast-in-place reinforced concrete to elements precast at a manufacturer's plant and shipped to the job site for assembly.

Recent summaries of concrete building construction in the United States are given by Cohen and Heun [3] and by the American Concrete Institute [4].

Understanding of reinforced concrete behavior is still far from complete; building codes and specifications that give design procedures are continually changing to reflect latest knowledge.

1.3 CONCRETE

Plain concrete is made by mixing cement, fine aggregate, coarse aggregate, water, and frequently admixtures. When reinforcing steel is placed in the forms and wet concrete mix is placed around it, the final solidified mass becomes reinforced concrete (see Fig. 1.3.1). The strength of concrete de-

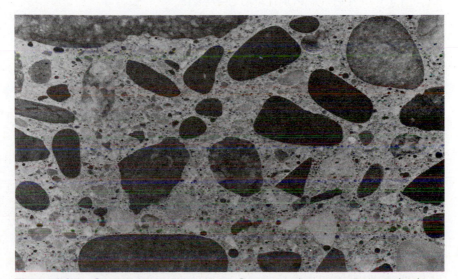

Figure 1.3.1 Cross section of concrete. Cement and water paste completely coats each aggregate particle and fills all of the space between particles. (Courtesy of Portland Cement Association.)

pends on many factors: notably the proportion of the ingredients and the conditions of temperature and moisture under which it is placed and cured.

Contained in subsequent sections are brief discussions of the materials in and the properties of plain concrete. The treatment is intended to be only introductory; an interested reader should consult standard references devoted entirely to the subject of plain concrete [5,6,7].

1.4 CEMENT

Cement is a material that has adhesive and cohesive properties enabling it to bond mineral fragments into a solid mass. Although this definition can apply to many materials, the cements of interest for reinforced concrete construction are those that can set and harden in the presence of water—the so-called *hydraulic cements*. These consist primarily of silicates and aluminates of lime made from limestone and clay (or shale) which is ground, blended, fused in a kiln, and crushed to a powder. Such cements chemically combine with water (hydration) to form a hardened mass. The usual hydraulic cement used for reinforced concrete is known as *portland cement*, because of its resemblance when hardened to Portland stone found near Dorset, England. The name was originated in a patent obtained by Joseph Aspdin of Leeds, England, in 1824.

Concrete made with portland cement ordinarily requires about 14 days to attain adequate strength so that forms can be removed and construction and dead loads carried. The design strength of such concrete is reached at about 28 days. This ordinary portland cement is identified by ASTM (American Society for Testing and Materials) C150 [8] as Type I. Other types of portland cement and their intended uses are given in Table 1.4.1.

Table 1.4.1 TYPES OF PORTLAND CEMENT[a]

TYPE	USAGE
I	Ordinary construction where special properties are not required
II	Ordinary construction when moderate sulfate resistance or moderate heat of hydration is desired
III	When high early strength is desired
IV	When low heat of hydration is desired
V	When high sulfate resistance is desired

[a]According to ASTM C150 [8].

There are also several categories of blended hydraulic cements (ASTM C595), such as *portland blast-furnace slag cement, portland-pozzolan cement, slag cement, pozzolan-modified portland cement*, and *slag-modified portland cement. Air-entraining portland cement* may be referred to for Types I, II, and III given in Table 1.4.1 by using the designation IA, IIA, or IIIA.

Air-entraining portland cement contains a chemical admixture finely ground with the cement to produce intentionally air bubbles on the order of 0.02 in. (0.5 mm) diameter uniformly distributed throughout the concrete. Such air entrainment will give the concrete improved durability against frost action, as well as better workability. Air-entraining agents may be added to the first three types of cement in ASTM C150 or to the blended hydraulic cements in ASTM C595 at the time the concrete is mixed.

Portland blast-furnace-slag cement has lower heat of hydration than ordinary Type I cement and is useful for mass concrete structures such as dams; and because of its high sulfate resistance, it is used in seawater construction. *Portland-pozzolan cement* is a blended mixture of ordinary Type I cement with pozzolan. *Pozzolan* is a finely divided siliceous or siliceous and aluminous material which possesses little or no inherent cementitious property, but in the powdery form and in the presence of moisture, will chemically react with calcium hydroxide at ordinary temperatures to form compounds possessing cementitious properties. Blended cements with pozzolan gain strength more slowly than cements without pozzolan, hence they produce less heat during hydration, and thus are widely used in mass concrete construction.

1.5 AGGREGATES

Since aggregate usually occupies about 75% of the total volume of concrete, its properties have a definite influence on the behavior of hardened concrete. Not only does the strength of the aggregate affect the strength of the concrete, its properties also greatly affect durability (resistance to deterioration under freeze-thaw cycles). Since aggregate is less expensive than cement, it is logical to try to use the largest percentage feasible. In general, for maximum strength, durability, and best economy, the aggregate should be packed and cemented as densely as possible. Hence aggregates are usually graded by size and a proper mix has specified percentages of both *fine* and *coarse* aggregates.

Fine aggregate (sand) is any material passing through a No. 4 sieve[†] [i.e., less than about $\frac{3}{16}$ in. (5 mm) diameter]. Coarse aggregate (gravel) is any material of larger size. The nominal maximum size of coarse aggregate permitted (ACI-3.3.3)[‡] is governed by the clearances between sides of forms and between adjacent bars and may not exceed "(a) $\frac{1}{5}$ the narrowest dimension between sides of forms, nor (b) $\frac{1}{3}$ the depth of slabs, nor (c) $\frac{3}{4}$ the minimum clear spacing between individual reinforcing bars. . . ." Additional information concerning aggregate selection and use is to be found in a report of ACI Committee 621 [10].

Natural stone aggregates conforming to ASTM C33 [11] are used in the majority of concrete construction, giving a unit weight for such concrete of about 145 pcf (pounds per cubic foot) or 2320 kg/m^3 (kilograms per cubic meter). When steel reinforcement is added, the unit weight of *normal weight* reinforced concrete is taken for calculation purposes as 150 pcf, or 2400 kg/m^3. Actual weights for concrete and steel are rarely, if ever, computed separately. For special purposes, lightweight or extra heavy aggregates are used.

[†]4.75 mm according to ASTM Standard E11.
[‡]Numbers refer to sections in the "ACI Code," officially ACI 318–83, *Building Code Requirements for Reinforced Concrete* [9].

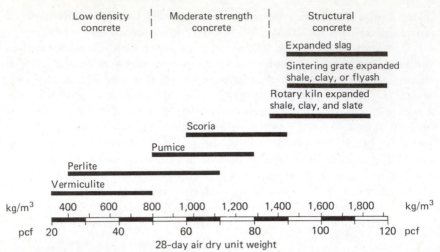

Figure 1.5.1 Approximate unit weight and use classification of lightweight aggregate concretes (from Ref. 13).

Structural lightweight concretes [12,13] are usually made from aggregates conforming to ASTM C330 [12] which are usually produced artificially in a kiln, such as expanded clays and shales. The unit weight of such concretes typically ranges from 70 to 115 pcf (see Fig. 1.5.1). Lightweight concretes ranging down to 30 pcf (often known as cellular concretes) are also used for insulating purposes and for masonry units. When lightweight materials are used for both coarse and fine aggregates in structural concrete, it is termed *all-lightweight* concrete. When only the coarse aggregate is of lightweight material but normal weight sand is used for the fine aggregate, it is said to be *sand-lightweight* concrete. Often the term "sand replacement" is used in connection with lightweight concrete. This refers to replacing all or part of the lightweight aggregate fines with natural sand. Additional information relating to lightweight concretes is to be found in guides [13,14,15] prepared by ACI Committees 213 and 523.

Heavyweight, high-density concrete is used for shielding against gamma and x radiation in nuclear reactor containers and other structures [16]. Naturally occurring iron ores, titaniferous iron ores, "hydrous iron ores" (i.e., containing bound and adsorbed water), and barites are crushed to suitable size for use as aggregates. Heavyweight concretes weigh typically from 200 to 350 pcf (3200 to 5600 kg/m^3).

1.6 ADMIXTURES

In addition to cement, coarse and fine aggregates, and water, other materials known as *admixtures* may be added to the concrete mix immediately before or during the mixing. Admixtures may be used to modify the properties of the concrete to make it better serve its intended use or for better economy.

Some of the important purposes of admixtures are as follows:

1. To increase resistance to deterioration resulting from freeze-thaw

cycles and the use of ice-removal salts (air-entraining admixtures, under ASTM C260).

2. To increase workability without increasing water content, or to decrease the water content at the same workability (finely divided materials including pozzolans, such as fly ash, are generally used for this purpose, under ASTM C618).

3. To accelerate the rate of strength development at early ages (calcium chloride is the best known and most widely used accelerator).

4. To retard the setting and thereby reduce heat evolution (ASTM C494 admixtures).

5. To increase the strength (water-reducing and set-controlling admixtures, ASTM C494, Chemical Admixtures for Concrete).

Air-entraining admixtures are probably the most widely used type. Air entrainment provides a high degree of resistance to the disruptive action of freezing and thawing and of deicing chemicals. Plasticity and workability are also improved, permitting a reduction in water content. Uniformity of placement with little segregation can be achieved. In addition, air-entrained concrete is more water tight and has greater resistance to sulfate action. For exposed concrete, the possible reduced strength (less than 15%) is far less important than the improved resistance to frost action.

ACI Committee 212 [17,18] provides the essential guidance for use of admixtures.

1.7 COMPRESSIVE STRENGTH

The strength of concrete is controlled by the proportioning of cement, coarse and fine aggregates, water, and various admixtures. The ratio of water to cement is the chief factor for determining concrete strength, as shown in Fig. 1.7.1. The lower the water–cement ratio, the higher the compressive strength. A certain minimum amount of water is necessary for the proper chemical action in the hardening of concrete; extra water increases the workability (how easily the concrete will flow) but reduces strength. A measure of the workability is obtained by a *slump test*. A truncated cone-shaped metal mold 12 in. (300 mm) high is filled with fresh concrete, the mold is then lifted off, and a measurement is made of the distance the top of the wet mass "slumps" from its original position before the mold was removed. The smaller the slump, the stiffer and less workable is the mix. In building construction, a 3 to 4 in. (75 to 100 mm) slump is common. Vibration of the concrete mix will greatly improve workability and even very stiff no-slump concrete can be placed [19].

Information regarding proportioning of concrete mixes is available in ACI Standard 211.1 for normal weight, heavyweight, and mass concrete [21], ACI Standard 211.2 for structural lightweight concrete [20], and ACI Standard 211.3 for no-slump concrete [19]. Actual strength of concrete in place in the structure is also greatly affected by quality control procedures for placement and inspection. Details regarding good practice are to be found in ACI Standard 304 [22], and in the *ACI Manual of Concrete Inspection* [23].

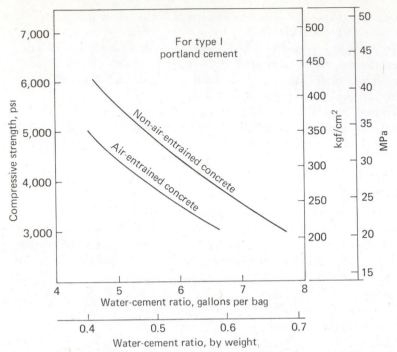

Figure 1.7.1 Effect of water–cement ratio on 28-day compressive strength. Average values for concrete containing 1.5 to 2% trapped air for non-air-entrained concrete, and no more than 5 to 6% air for air-entrained concrete (curves drawn from data in Ref. 21, Tables 5.3.4a and 5.3.3).

The strength of concrete is denoted in the United States by f_c', which is the compressive strength in psi of test cylinders 6 in. (150 mm) in diameter by 12 in. (300 mm) high measured on the 28th day after they are made. In many other parts of the world, the standard test unit is the cube, frequently of 200 mm to a side.

Since nearly all the behavior of reinforced concrete is related to the 28-day compressive strength, it is important to realize that such strength differs depending on the size and shape of the standard test specimen and the manner of testing. Since the cylinder strength test does not exhibit exactly the same properties as the cube strength test, it has been difficult to define a constant relationship between them. Factors such as tensile strength of concrete and size of the contact area of the testing machine have more effect on the cube strength than on the cylinder strength. As an average, one may assume that for ordinary weight concrete, the 6 × 12 in. (150 × 300 mm) cylinder strength is 80% of the 150-mm cube strength and 83% of the 200-mm cube strength [24]. For lightweight concrete, cylinder strength and cube strength are nearly equal.

Even in the United States tests of concrete in existing structures are often made using other than the standard 6 × 12 in. cylinder. For such tests, cores of other than 6 in. diameter are usually cut from the concrete. Tests of these are usually on cylinders having a height-to-diameter ratio of

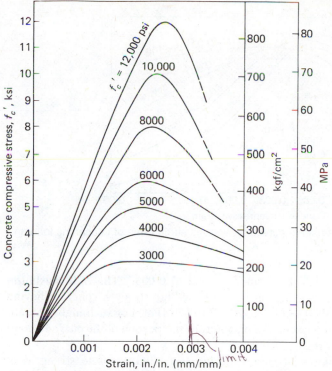

Figure 1.7.2 Typical stress-strain curves for concrete in compression under short-time loading. (Curves represent a compromise adapted from curves and results given by Wang, Shah, and Naaman [29], Bertero [30], Naaman [31], and Nilson [32]).

2 to 1 but smaller than the standard cylinder. To draw proper conclusions, it is necessary to know the effect of test specimen size.

An interesting discussion of the cylinder test is given by Shilstone [25]. When an assessment of the strength of in-place concrete is desired, there are various procedures ranging from cutting cylindrical cores from the structure to using nondestructive tests [26,27].

The stress-strain behavior of concrete is dependent on its strength, age at loading, rate of loading, aggregates and cement properties, and type and size of specimens [28,29]. Typical curves for specimens loaded in compression at 28 days using normal testing speeds are shown in Fig. 1.7.2.

The rate of applying strain during testing has a great influence on the result as shown by Fig. 1.7.3. Because of this strain rate, there is not agreement on what constitutes "typical curves." Particularly the portion of the curve after the maximum stress is reached varies considerably in test results.

One may note in Fig. 1.7.2 that lower-strength concrete has greater deformability (ductility) than higher-strength concrete, and the maximum stress is reached at a compressive strain between 0.002 and 0.0025. Ultimate strain at crushing of concrete varies from 0.003 to as high as 0.008; however, the maximum usable strain for practical cases is 0.003–0.004. The ACI Building Code [9] states (ACI-10.2.3) "Maximum usable strain at extreme concrete

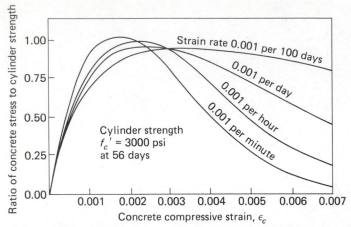

Figure 1.7.3 Stress-strain curves for various strain rates of concentric loading (from Rüsch [34]).

compression fiber shall be assumed equal to 0.003." The ACI maximum strain of 0.003 may not be conservative for high-strength concrete having f'_c in the range of 8000 to 12,000 psi (55 to 83 MPa). For such high-strength concrete there is not agreement on whether the portion of the curve beyond maximum stress is really usable.

In usual reinforced concrete design, specified concrete strengths of 3000 to 4000 psi (roughly 20 to 30 MPa) are used for nonprestressed structures, and 5000 to 6000 psi (roughly 35 to 42 MPa) for prestressed concrete. For special situations, concrete ranging from 6000 to 12,000 psi (42 to 69 MPa) has been used [35]. Experiments are being conducted on high-strength concrete in the range of 10,000 to 15,000 psi (roughly 70 to 100 MPa) [36]. Ready-mix concrete is being used with strengths to 9000 psi (62 MPa) for columns on lower stories of high-rise buildings.

1.8 TENSILE STRENGTH

The strength of concrete in tension is also an important property that greatly affects the extent and size of cracking in structures. Tensile strength is usually determined by using the *split-cylinder* test in accordance with ASTM C496 [37] in which the same size cylinder used for the compression test is placed in the testing machine lying on its side so that the compression load P is applied uniformly along the length of the cylinder in the direction of the diameter. The cylinder will split in half when the tensile strength is reached. The stress is computed by $2P/[\pi(\text{diameter})(\text{length})]$ based on theory of elasticity for a homogeneous material in a biaxial state of stress.* Tensile strength is a more variable property than compressive strength, and

*See, for example, S. Timoshenko and J. N. Goodier, *Theory of Elasticity*, 2nd ed., McGraw-Hill, 1951, pp. 85, 107.

is about 10 to 15% of it. The split-cylinder tensile strength f_{ct} has been found to be proportional to $\sqrt{f'_c}$,[†] such that

$$f_{ct} = 6\sqrt{f'_c} \quad \text{to} \quad 7\sqrt{f'_c} \text{ psi} \qquad \text{for normal-weight concrete}[‡]$$

$$f_{ct} = 5\sqrt{f'_c} \quad \text{to} \quad 6\sqrt{f'_c} \text{ psi} \qquad \text{for lightweight concrete}[‡]$$

The ACI Code has indirectly used $f_{ct} = 6.7\sqrt{f'_c}$ psi for normal-weight concrete, $f_{ct} = 5.7\sqrt{f'_c}$ for sand-lightweight concrete, and $f_{ct} = 5\sqrt{f'_c}$ for all-lightweight concrete (ACI-11.2).

Tensile strength in flexure, known as *modulus of rupture*, measured in accordance with ASTM C78 [38] is also important when considering cracking and deflection of beams. The modulus of rupture f_r, computed from the flexure formula $f = Mc/I$, gives higher values for tensile strength than the split-cylinder test, primarily because the concrete compressive stress distribution is not linear when tensile failure is imminent as is assumed in the computation of the nominal Mc/I stress. It is generally accepted (ACI-9.5.2.3) that an average value for the modulus of rupture f_r may be taken as $7.5\sqrt{f'_c}$ ($0.62\sqrt{f'_c}$ MPa) for normal-weight concrete and 75% of that value for all-lightweight concrete. Because of the large variability in modulus of rupture, as shown in Fig. 1.8.1, the selection of the coefficient 7.5, or even the entire expression $7.5\sqrt{f'_c}$, should be viewed as a practical choice for design purposes.

One may note that neither the split-cylinder nor the modulus of rupture tensile strength is correctly a measure of the strength under uniform axial tension. However, uniform axial tension is difficult to measure accurately and when compared with the modulus of rupture or split-cylinder strength it does *not* give better correlation with tension-related failure behavior such as flexural cracking in beams, inclined cracking from shear and torsion, and splitting from interaction of reinforcing bars with surrounding concrete.

1.9 MODULUS OF ELASTICITY

The modulus of elasticity of concrete varies, unlike that of steel, with strength. It also depends, though to a much lesser extent, on the age of concrete, properties of the aggregates and cement, rate of loading, and the type and size of specimen. Furthermore, since concrete exhibits some permanent set even under small loads, there are various definitions of the modulus of elasticity.

Referring to Fig. 1.9.1, representing a typical stress-strain curve for

---

[†]$\sqrt{f'_c}$ is in psi units; thus $f'_c = 3000$ psi, $\sqrt{f'_c} = 54.8$ psi. When f'_c is in kilogram-force per square centimeter (kgf/cm²), the constant in front of $\sqrt{f'_c}$ is to be multiplied by 0.265; when f'_c is in newtons per square millimeter, that is, megapascals (MPa), the constant in front of $\sqrt{f'_c}$ is to be multiplied by 0.083.

[‡]In SI, with f'_c and f_{ct} in MPa,

$$f_{ct} = 0.5\sqrt{f'_c} \quad \text{to} \quad 0.6\sqrt{f'_c} \qquad \text{for normal-weight concrete}$$

$$f_{ct} = 0.4\sqrt{f'_c} \quad \text{to} \quad 0.5\sqrt{f'_c} \qquad \text{for lightweight concrete}$$

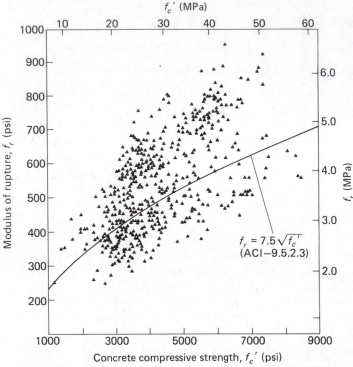

Figure 1.8.1 Comparison of test results for modulus of rupture with ACI Code expression (adapted from Mirza, Hatzinikolas, and MacGregor [39]).

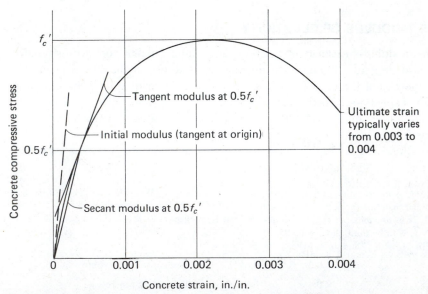

Figure 1.9.1 Stress-strain curve for concrete in compression.

concrete [33], the initial modulus, the tangent modulus, and the secant modulus are noted. Usually the secant modulus at from 25 to 50% of the compressive strength f_c' is considered to be the modulus of elasticity. For many years the modulus was approximated adequately as 1000 f_c' by the ACI Code; but with the rapidly increasing use of lightweight concrete, the variable of density needed to be included. As a result of a statistical analysis of available data, the empirical formula given by ACI-8.5.1

$$E_c = 33w_c^{1.5}\sqrt{f_c'} \qquad (1.9.1)*$$

was developed [40] for values of w_c between 90 and 155 lb/cu ft. Equation (1.9.1) may be considered as the secant modulus for a compressive stress at service load level. For normal-weight concrete weighing 145 pcf, the formula gives $E_c = 57,600\sqrt{f_c'}$. For normal-weight concrete ACI-8.5.1 suggests

$$E_c = 57,000\sqrt{f_c'} \qquad (1.9.2)^\dagger$$

Values of modulus of elasticity for various concrete strengths appear in Table 1.9.1.

Table 1.9.1 VALUES OF MODULUS OF ELASTICITY (USING $E_c = 33w_c^{1.5}\sqrt{f_c'}$ FOR NORMAL-WEIGHT CONCRETE WEIGHING 145 pcf)

INCH-POUND UNITS		SI UNITS[b]	
f_c' (psi)	E_c (psi)	f_c' (MPa)	$E_c^\dagger$ (MPa)
3000	3,150,000	21[a]	22,900
3500	3,400,000	24	24,500
4000	3,640,000	28	26,500
4500	3,860,000	31	27,800
5000	4,070,000	35	29,600

[a]These metric values are rounded values approximating concrete strengths in Inch-Pound units; actual equivalents for 3000, 3500, 4000, 4500, and 5000 psi are 20.7, 24.1, 27.6, 31.0, and 34.5 MPa, respectively.
[b]Multiply MPa values by 10.2 to obtain kgf/cm².
[†]Using $E_c = 5000\sqrt{f_c'}$ as per ACI 318–83M.

1.10 CREEP AND SHRINKAGE

Creep and shrinkage are time-dependent deformations, that along with cracking provide the greatest concern for the designer because of the in-

*For SI, with w_c in kg/m³ and E_c and f_c' in MPa,

$$E_c = 0.043w_c^{1.5}\sqrt{f_c'} \quad \text{(ACI 318–83M)} \qquad (1.9.1)$$

†For SI, with E_c and f_c' in MPa,

$$E_c = 5000\sqrt{f_c'} \quad \text{(ACI 318–83M)} \qquad (1.9.2)$$

and with E_c and f_c' in kgf/cm²,

$$E_c = 15,000\sqrt{f_c'} \quad \text{(approximate)} \qquad (1.9.2)$$

accuracies and unknowns that surround them. Concrete is elastic only under loads of short duration, and because of additional deformation with time, the effective behavior is that of an inelastic material. Deflection after a long period of time is therefore difficult to predict, but its control is needed to assure serviceability during the life of the structure.

Creep. Creep is the property of concrete (and other material) by which it continues to deform with time under sustained loads at unit stresses within the accepted elastic range (say, below $0.5f'_c$). This inelastic deformation increases at a decreasing rate during the time of loading, and its total magnitude may be several times as large as the short-time elastic deformation. Frequently creep is associated with shrinkage, since both are occurring simultaneously and often provide the same net effect: increased deformation with time. As may be noted by the general relationship of deformation versus time in Fig. 1.10.1, the "true elastic strain" decreases since the modulus of elasticity E_c is a function of concrete strength f'_c which increases with time.

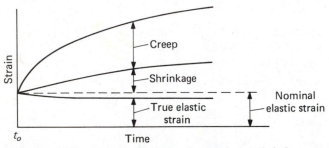

Figure 1.10.1 Change in strain of a loaded and drying specimen; t_0 is the time of application of load (from Ref. 39).

Although creep is separate from shrinkage, it is related to it. Detailed information is available for predicting creep [41,42]. The internal mechanism of creep, or "plastic flow" as it is sometimes called, may be due to any one or a combination of the following: (1) crystalline flow in the aggregate and hardened cement paste; (2) plastic flow of the cement paste surrounding the aggregate; (3) closing of internal voids; and (4) the flow of water out of the cement gel due to external load and drying. None of these alone seems to entirely account for creep. The gel theory seems to predominate at low-stress levels, whereas the crystalline theory predominates at high-stress levels [43, see also Ref. 7, pp. 511–514 for a more esoteric discussion of creep mechanisms].

Factors affecting the magnitude of creep are (1) the constituents—such as the composition and fineness of the cement, the admixtures, and the size, grading, and mineral content of the aggregates; (2) proportions such as water content and water–cement ratio; (3) curing temperature and humidity; (4) relative humidity during period of use; (5) age at loading; (6) duration of loading; (7) magnitude of stress; (8) surface–volume ratio of the member; and (9) slump.

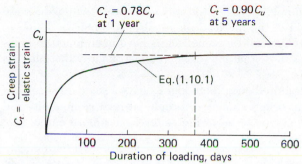

Figure 1.10.2 Standard creep coefficient variation with duration of loading (for 4 in. or less slump, 40% relative humidity, moist cured, and loading age of 7 days).

Accurate prediction of creep is complicated because of the variables involved; however, a general prediction method developed by Branson [42] gives a standard creep coefficient equation (4 in. or less slump, 40% relative humidity, moist cured, and loading age of 7 days)

$$C_t = \frac{\text{creep strain}}{\text{initial elastic strain}}$$

$$= \frac{t^{0.60}}{10 + t^{0.60}} C_u \qquad (1.10.1)$$

shown in Fig. 1.10.2, where t is duration of loading (days) and C_u is ultimate creep coefficient. (Branson [42] suggests using an average of 2.35 for standard conditions but the range is shown to be from 1.3 to 4.15.) Correction factors are given for relative humidity, loading age, minimum thickness of member, slump, percent fines, and air content. For practical purposes, the only factors significant enough to require correction are humidity and age at loading.

The effect of unloading may be seen from Fig. 1.10.3 where at a certain time t_1 the load is removed. There is an immediate elastic recovery and a long-time creep recovery, but a residual deformation remains.

Shrinkage. Shrinkage, broadly defined, is volume change that is unrelated to load application. It is possible for concrete cured continuously under

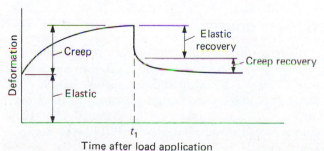

Figure 1.10.3 Typical creep and recovery relationship (from Ref. 39).

water to increase in volume; however, the usual concern is with a decrease in volume. A discussion of the mechanisms of shrinkage may be found in Ref. 7 (pp. 481–500). In general the same factors have been found to influence shrinkage strain as those that influence creep—primarily those factors related to moisture loss.

The Branson general prediction method [42] gives a standard shrinkage strain equation (for 4 in. or less slump, 40% ambient relative humidity and minimum thickness of member 6 in. or less, after 7 days moist cured)

$$\epsilon_{sh} = \left(\frac{t}{35 + t}\right)(\epsilon_{sh})_u \qquad (1.10.2)$$

shown in Fig. 1.10.4, where t is time (days) after moist curing, and $(\epsilon_{sh})_u$ is ultimate shrinkage strain. (Branson [42] suggests using 800×10^{-6} in./in. for average conditions but the range is from 415 to over 1000×10^{-6}.) Correction factors are given with the primary one relating to humidity H,

$$\text{correction factor} = 1.40 - 0.01H \qquad \text{for} \quad 40\% \leq H \leq 80\%$$
$$\text{correction factor} = 3.00 - 0.03H \qquad \text{for} \quad 80\% \leq H \leq 100\%$$

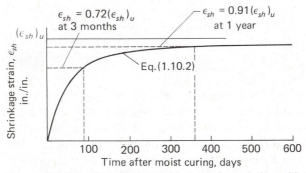

Figure 1.10.4   Standard shrinkage strain variation with time after moist curing (for 4 in. or less slump, 40% ambient relative humidity and minimum thickness of member 6 in. or less, after 7 days moist cured).

Shrinkage, particularly when restrained unsymmetrically by reinforcement, causes deformations generally additive to those of creep. For proper serviceability, it is desirable to predict or compensate for shrinkage in the structure.

1.11 CONCRETE QUALITY CONTROL

In reinforced concrete design the concrete sections are proportioned and reinforced using a *specified compressive strength* f'_c (28-day compressive strength). The strength f'_c for which each part of a structure has been designed should be clearly indicated on the design drawings. In the United States, as indicated in Sec. 1.7, f'_c is based on cylinder strength (6×12 in. cylinders), evaluated in accordance with ACI Standard 214 [44].

✳ Concrete is a material whose strength and other properties are not precisely predictable, so that test cylinders from a mix designed to provide, say, 3000 psi (roughly 20 MPa) concrete will show considerable variability. Because of this, mixes must be designed to provide an average compressive strength greater than the specified value f'_c.

ACI-4.7.2.3 indicates that adequate control of strength occurs when both of the following requirements are met:

1. Average of all sets of three consecutive strength tests* equal or exceed f'_c.
2. No individual strength test (average of two cylinders) falls below f'_c by more than 500 psi (3.4 MPa).

Statistical variations are to be expected and strength tests failing to meet the above criteria will occur perhaps once in 100 tests even though all proper procedures have been followed.

When the ready-mix plant or other concrete production facility has a record based on at least 30 consecutive strength tests for materials and conditions similar to those expected, the *standard deviation s* can be computed based on those tests to establish how variable is the concrete strength.

Standard deviation is calculated by first computing the simple average of the tests results, taking the absolute value of the difference (deviation) between each test value and the average, then obtaining the square root of the average of the squares of the deviations. The smaller the standard deviation, the more consistent the results.

ACI-4.3.2 indicates that the strength f'_{cr} used for proportioning the mix must exceed the specified f'_c by increasing amounts for increasing values of the standard deviation s, using the *larger* of the following:

$$f'_{cr} = f'_c + 1.34s \tag{1.11.1}$$

or

$$f'_{cr} = f'_c + 2.33s - 500 \tag{1.11.2}$$

when at least 30 consecutive strength tests are the basis for computing the standard deviation s. When fewer than 30 tests are available a modification in accordance with ACI-4.3.1.2 will require using a higher value of f'_{cr}. Thus if the designer has used a specified strength f'_c of 4000 psi, and the concrete producer has shown a standard deviation of 450 psi, the mix should be designed for an average strength of 4600 psi [i.e., the larger of 4000 + 1.34 (450) or 4000 + 2.33 (450) − 500]. Further, when data are not available to establish a standard deviation, the mix must be designed to produce an average strength 1200 psi (8.3 MPa) above the specified strength f'_c for f'_c of 3000 to 5000 psi (21 to 35 MPa).

With regard to evaluation and acceptance of concrete, ACI-4.7 provides requirements for the numbers of samples that must be taken. A discussion on the risks inherent in the consideration of limited test data is provided by Tait [45].

---

*According to ACI-4.7.1.4, a strength test is the "average of the strengths of two cylinders made from the same sample of concrete and tested at 28 days or at test age designated for determination of f'_c."

It should be noted that the term "quality control" entails much more than designing the concrete mix and evaluating the cylinder strength tests. The foregoing discussion of concrete strength variation should merely give an awareness of the fact that concrete having a specified compressive strength f'_c cannot be expected to provide precisely known actual strength and other properties.

Quality control in the broader sense for reinforced concrete construction is a subject of great importance, but generally lies outside the scope of this text. A series of symposium papers [46] and the paper by Peters [47] provide an excellent overall discussion of this subject.

1.12 STEEL REINFORCEMENT

Steel reinforcement may consist of bars, welded wire fabric, or wires. For usual construction, bars (called *deformed bars*) having lugs or protrusions (*deformations*) are used (Fig. 1.12.1). Such deformations inhibit longitudinal movement of the bar relative to the concrete that surrounds it. These deformed bars are available in the United States in sizes $\frac{3}{8}$ to $2\frac{1}{4}$ in. (9.5 to 57 mm) nominal diameter.

Sizes of ASTM bars (in Inch-Pound units) are indicated by numbers (see Table 1.12.1). For sizes #3 through #8, they are based on the number of eighths of an inch included in the nominal diameter of the bars. Bars of designation #9 through #11 are round bars corresponding to the former 1 in. square, $1\frac{1}{8}$ in. square, and $1\frac{1}{4}$ in. square sizes, and bars designated #14 and #18 are round bars having cross-sectional areas equal to those of $1\frac{1}{2}$ and 2 in. square sizes, respectively. The nominal diameter of a deformed bar is equivalent to the diameter of a plain bar having the same weight per foot as the deformed bar.

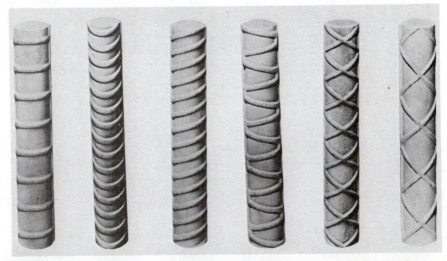

Figure 1.12.1 Deformed reinforcing bars. (Courtesy of Concrete Reinforcing Steel Institute.)

Table 1.12.1   STANDARD REINFORCING BAR DIMENSIONS AND WEIGHTS (BARS IN INCH-POUND UNITS ACCORDING TO ASTM A615 [48], A616 [49], A617 [50], and A706 [51])

BAR NUMBER	NOMINAL DIMENSIONS					
	DIAMETER		AREA		WEIGHT	
	(in.)	(mm)	(sq in.)	(cm^2)	(lb/ft)	(kg/m)
3	0.375	9.5	0.11	0.71	0.376	0.559
4	0.500	12.7	0.20	1.29	0.668	0.994
5	0.625	15.9	0.31	2.00	1.043	1.552
6	0.750	19.1	0.44	2.84	1.502	2.235
7	0.875	22.2	0.60	3.87	2.044	3.041
8	1.000	25.4	0.79	5.10	2.670	3.973
9	1.128	28.7	1.00	6.45	3.400	5.059
10	1.270	32.3	1.27	8.19	4.303	6.403
11	1.410	35.8	1.56	10.06	5.313	7.906
14	1.693	43.0	2.25	14.52	7.65	11.38
18	2.257	57.3	4.00	25.81	13.60	20.24

Metric bars in SI units according to ASTM A615M [52], which are in agreement with the Canadian Standard, are given in Table 1.12.2.

Reinforcing bar steel may be of several types as shown in Table 1.12.3: (1) ASTM A615 [48], billet steel of Grades 40 and 60 having minimum specified yield stresses* of 40,000 and 60,000 psi, respectively (276 and 414 MPa); (2) ASTM A616 [49], rail steel of Grades 50 and 60; (3) ASTM A617 [50], axle steel of Grades 40 and 60; and (4) ASTM A706 [51], low-alloy steel of Grade 60 intended for applications where welding or bending, or both, are important.

The billet steel of A615 is newly made steel having its chemical content sufficiently controlled to provide necessary ductility. Axle and rail steel bars

Table 1.12.2   CANADIAN STANDARD AND ASTM 615M [52] REINFORCING BAR DIMENSIONS AND WEIGHTS IN SI UNITS

BAR NUMBER	DIAMETER (mm)	AREA (cm^2)	WEIGHT (kg/m)	COMPARISON TO U.S. BARS BAR NUMBER (AREA)
10M	11.3	1.0	0.784	− 3 (0.71 cm^2) − 4 (1.29 cm^2)
15M	16.0	2.0	1.568	− 5 (2.00 cm^2)
20M	19.5	3.0	2.352	− 6 (2.84 cm^2) − 7 (3.87 cm^2)
25M	25.2	5.0	3.920	− 8 (5.10 cm^2)
30M	29.9	7.0	5.488	− 9 (6.45 cm^2)
35M	35.7	10.0	7.840	−10 (8.19 cm^2) −11 (10.06 cm^2)
45M	43.7	15.0	11.760	−14 (14.52 cm^2)
55M	56.4	25.0	19.600	−18 (25.81 cm^2)

*The term "yield stress" refers to either *yield point*, the well-defined deviation from perfect elasticity, or *yield strength*, the value obtained by a specified offset strain for material having no well-defined yield point.

Table 1.12.3 REINFORCING BAR STEELS

ASTM DESIGNATION	GRADE	BAR SIZES	MINIMUM YIELD STRESS,[a] f_y		MINIMUM TENSILE STRENGTH	
			ksi	MPa	ksi	MPa
A615-81 (SI)[b]	40	#3–#11	40		70	
(Billet steel)	60	#3–#18	60		90	
A615M-80[c]	300	10–35		300		500
(Billet steel)	400	10–55		400		600
A616-79[d]	50	#3–#11	50	345	80	550
(Rail steel)	60	#3–#11	60	415	90	620
A617-81	40	#3–#11	40	275	70	480
(Axle steel)	60	#3–#11	60	415	90	620
A706-81		#3–#18	60	415	80	550
(Low-alloy steel)						

[a]The term "yield stress" refers to either *yield point*, the well-defined deviation from perfect elasticity, or *yield strength*, the value obtained by a specified offset strain for material having no well-defined yield point.
[b]Bars acceptable according to the 1983 ACI Code must satisfy the more severe bend test requirements contained in the Supplementary Requirements S1. Bars supplied meeting these requirements will be marked with a letter S.
[c]Metric (SI) specification applies to bars designated number 10 through 55, as given in Table 1.12.2.
[d]Although rail steel is no longer considered a practical source of bar reinforcement, its use is still permitted provided the bend test requirements for axle steel, ASTM A617, Grade 60, can be satisfied.

are made from steel that is rerolled from old axles and rails and are in general less ductile than those of billet steel. In 1983 axle steel is rarely used for reinforcing bars and rail steel is no longer considered a practical source of bar reinforcement. Axle and rail steels are, however, still recognized by the 1983 ACI Code.

Because of relatively little extra cost, the Grade 60 steel has essentially replaced the formerly predominately used Grade 40 steel. All of the bar sizes in Table 1.12.1 are available in Grade 60 billet steel and low-alloy steel; however, only #3 through #11 bars are included under ASTM 615, Grade 40 billet steel, or under ASTM axle or rail steel specifications. The Grade 40 reinforcing bars generally exhibit the well-defined yield point and the ideal elastic-plastic stress-strain behavior in tension as shown in Fig. 1.12.2(a). Grade 60 steel also has the well-defined yield point but the plastic range (constant stress with increasing strain) is shorter than for Grade 40. The total strain capability before rupture is less for Grade 60 than for Grade 40, as shown in Fig. 1.12.2(b).

Welded wire fabric is used in thin slabs, thin shells, and other locations where available space would not permit the placement of deformed bars with proper cover and clearance. Welded wire fabric (WWF) covered under ASTM A185 [53] and A497 [54], consists of cold drawn wire in orthogonal patterns, square or rectangular, resistance welded at all intersections. The wires may be smooth (ASTM A82 [55]) or deformed (ASTM A496 [56]). The wire is specified by the symbol W (for smooth wires) or D (for deformed wires) followed by a number representing the cross-sectional area in hundredths of a square inch, varying from 1.5 to 31. On design drawings it is

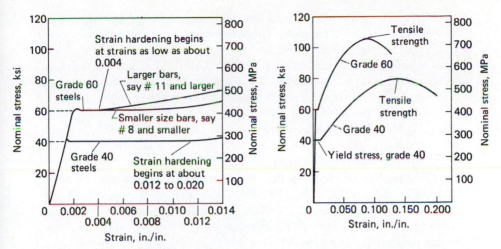

Figure 1.12.2 Typical stress-strain curves for reinforcing bar steels in tension.

(a) Enlarged portion used in design

(b) Entire curve to rupture

usually indicated by the symbol WWF followed by spacings of the wires in the two 90° directions. Thus WWF6 × 8 − W7 × W7 indicates welded wire fabric with 6-in. longitudinal wire spacing, 8-in. transverse wire spacing, and both sets of wires smooth and having a cross-sectional area of 0.07 sq in. Properties for welded wire fabric steels are given in Table 1.12.4. Additional information about welded wire fabric may be obtained from the Wire Reinforcement Institute [57].

Table 1.12.4 WIRE AND WELDED WIRE FABRIC STEELS

ASTM DESIGNATION	WIRE SIZE DESIGNATION	MINIMUM YIELD STRESS,[a] f_y		MINIMUM TENSILE STRENGTH	
		ksi	MPa	ksi	MPa
A82-79 (cold-drawn wire) (properties apply when material is to be used for fabric)	W1.2 and larger[b]	65	450	75	520
	Smaller than W1.2	56	385	70	480
A185-79 (welded wire fabric)	Same as A82; this is A82 material fabricated into sheet (so-called "mesh") by the process of electric welding.				
A496-78 (deformed steel wire) (properties apply when material is to be used for fabric)	D1–D31[c]	70	480	80	550
A497-79	Same as A82 or A496; this specification applies for fabric made from A496, or from a combination of A496 and A82 wires.				

[a]The term "yield stress" refers to either *yield point*, the well-defined deviation from perfect elasticity, or *yield strength*, the value obtained by a specified offset strain for material having no well-defined yield point.

[b]The W number represents the nominal cross sectional area in square inches multiplied by 100, for smooth wires.

[c]The D number represents the nominal cross sectional area in square inches multiplied by 100, for deformed wires.

Wires in the form of individual wires or groups of wires forming *strands* are used for prestressed concrete. There is a great variety of wire and strands of different strengths and properties, the most prevalent being the 7-wire stress relieved strand conforming to ASTM A416 [58]. These strands have a center wire enclosed by six helically wound outside wires. Usual nominal diameters for 7-wire strands are $\frac{1}{4}$, $\frac{3}{8}$, and $\frac{1}{2}$ in. The minimum ultimate tensile strength for Grade 250 strands is 250,000 psi (1720 MPa), and for Grade 270 is 270,000 psi (1860 MPa); there is no well-defined yield point. Typically under service conditions these prestressed strands have a stress of 150,000 to 160,000 psi (1030 to 1100 MPa).

The modulus of elasticity for all nonprestressed steel may be taken (ACI-8.5.2) as 29,000,000 psi (200,000 MPa). For prestressing steel it is lower and somewhat variable, therefore it must be obtained from the manufacturer. A value of 27,000,000 psi (186,000 MPa) is often used for 7-wire strands conforming to ASTM A416.

ACI Committee 439 [59] provides an excellent review of steel reinforcement properties and availability.

1.13 SI UNITS

The use of SI units has made little advance in the construction industry in the United States. In Canada, formal conversion to SI has been accomplished. The National Standard of Canada [60] for design of concrete structures for buildings was adopted in SI in 1977. The American Concrete Institute has adopted an SI version of the 1983 ACI Code (see Sec. 2.1). It is not anticipated, however, that wide use of SI units will occur in the United States for some time.

The metric system, although not SI, is used in nearly all western hemisphere countries (other than the USA and Canada). In these countries the MKS (metre-kilogram-second) system is used, where instead of the kilogram (kg) as a unit of mass, as in SI, the kilogram (kgf) is used as a unit of force. Thus, stresses are given in kilograms per square centimeter (kgf/cm²). Throughout this text, curves involving stresses have three parallel axes: pounds per square inch (psi) or kips per square inch (ksi) for Inch-Pound units, megapascals (MPa) for SI units, and kilogram-force per square centimeter (kgf/cm²) for the other metric (MKS) system. The authors believe that some familiarity with metric units is essential; on the other hand, overemphasis on units in design calculations and procedures detracts from concepts. In general, throughout the remaining chapters, design equations that involve units will have an SI version (usually ACI 318–83M) given in a footnote. The reader interested in proper use of SI units should consult the ASTM *Standard for Metric Practice* [61] and the *Standard Practice for the Use of Metric (SI) Units in Building Design and Construction* [62].

SELECTED REFERENCES

1. Hans Straub. *A History of Civil Engineering*. London: Leonard Hill, 1952 (pp. 208–211).

2. R. S. Kirby, Sidney Withington, A. B. Darling, and F. G. Kilgour. *Engineering in History*. New York: McGraw-Hill, 1956.

3. Edward Cohen and Raymond C. Heun. "100 Years of Concrete Building Construction in the United States," *Concrete International*, 1, March 1979, 38–46.

4. "Memorable Miscellany," *Concrete International*, 1, October 1979, 46–62.

5. George E. Troxell, Harmer E. Davis, and Joe W. Kelly. *Composition and Properties of Concrete* (2nd ed.). New York: McGraw-Hill, 1968.

6. A. M. Neville. *Properties of Concrete* (2nd ed.). New York: Halstead Press, 1973.

7. Sidney Mindness and J. Francis Young. *Concrete*. Englewood Cliffs, New Jersey: Prentice-Hall, Inc., 1981.

8. *Standard Specification for Portland Cement* (ASTM C150–81). Philadelphia: American Society for Testing and Materials, 1981.

9. *Building Code Requirements for Reinforced Concrete* (ACI 318–83). Detroit: American Concrete Institute, 1983.

10. ACI Committee 621. "Selection and Use of Aggregates for Concrete," *ACI Journal, Proceedings*, 58, November 1961, 513–542. (See also *ACI Manual of Standard Practice*, Part 1.)

11. *Standard Specification for Concrete Aggregates* (ASTM C33–82). Philadelphia: American Society for Testing and Materials, 1982.

12. *Standard Specification for Lightweight Aggregates for Structural Concrete* (ASTM C330–80). Philadelphia: American Society for Testing and Materials, 1980.

13. ACI Committee 213. "Guide for Structural Lightweight Aggregate Concrete," *Concrete International*, 1, February 1979, 33–62. Disc. August 1979, 66.

14. ACI Committee 523. "Guide for Cast-in-Place Low-Density Concrete," *ACI Journal, Proceedings*, 64, September 1967, 529–535.

15. ACI Committee 523. "Guide for Cellular Concretes Above 50 pcf, and for Aggregate Concretes Above 50 pcf with Compressive Strengths Less Than 2500 psi," *ACI Journal, Proceedings*, 72, February 1975, 50–66.

16. ACI Committee 304. "High Density Concrete: Measuring, Mixing, Transporting, and Placing," *ACI Journal, Proceedings*, 72, August 1975, 407–414. (See also *ACI Manual of Standard Practice*.)

17. ACI Committee 212. "Admixtures for Concrete," *Concrete International*, 3, May 1981, 24–52.

18. ACI Committee 212. "Guide for Use of Admixtures in Concrete," *Concrete International*, 3, May 1981, 53–65.

19. *Standard Practice for Selecting Proportions for No-Slump Concrete*, (ACI 211.3–75, Revised 1980). Detroit: American Concrete Institute, 1980.

20. ACI Committee 211. *Standard Practice for Selecting Proportions for Structural Lightweight Concrete* (ACI 211.2–81). Detroit: American Concrete Institute, 1981. (Published as "Proposal Revision of ACI 211.2–69: Standard Practice for Selecting Proportions for Structural Lightweight Concrete," *Concrete International*, 2, August 1980, 47–64. Disc. 3, February 1981, 65–66.)

21. ACI Committee 211. *Standard Practice for Selecting Proportions for Normal, Heavyweight, and Mass Concrete* (ACI 211.1–81). Detroit: American Concrete Institute, 1981.

22. *Recommended Practice for Measuring, Mixing, Transporting, and Placing Concrete* (ACI 304–73). Detroit: American Concrete Institute, 1973. (Reaffirmed 1978.)

23. *ACI Manual of Concrete Inspection* (SP-2, 5th ed.). Detroit: American Concrete Institute, 1967.

24. UNESCO. *Reinforced Concrete: An International Manual*. London: Butterworth, 1971 (pp. 19–22).

25. James M. Shilstone, Jr. "The Cylinder Test—Reliable Informer or False Prophet," *Concrete International,* **2,** July 1980, 63–68.
26. A. J. Chabowski and D. Brydon-Smith. "A Simple Pull-Out Test to Assess the In Situ Strength of Concrete," *Concrete International,* **1,** December 1979, 35–40.
27. R. F. Feldman. "Non-Destructive Testing of Concrete," *Canadian Building Digest,* Division of Building Research, National Research Council of Canada, Ottawa, May 1977 (Paper 187).
28. Hjalmar Granholm. *A General Flexural Theory of Reinforced Concrete.* New York: Wiley, 1965 (pp. 23–36).
29. P. T. Wang, S. P. Shah, and A. E. Naaman. "Stress-Strain Curves of Normal and Lightweight Concrete in Compression," *ACI Journal, Proceedings,* **75,** November 1975, 603–611.
30. Vitelmo V. Bertero. "Inelastic Behavior of Structural Elements and Structures," *High Strength Concrete,* Proceedings of a Workshop, University of Illinois at Chicago Circle, December 2–4, 1979, pp. 102–103, 153.
31. Antoine E. Naaman. *Prestressed Concrete Analysis and Design Fundamentals.* New York: McGraw-Hill, 1982 (p. 54).
32. Arthur H. Nilson. *Design of Prestressed Concrete.* New York: Wiley, 1978 (p. 45).
33. E. Hognestad, N. W. Hanson, and D. McHenry. "Concrete Stress Distribution in Ultimate Strength Design," *ACI Journal, Proceedings,* **52,** December 1955, 455–479.
34. Hubert Rüsch. "Researches Toward a General Flexural Theory for Structural Concrete," *ACI Journal, Proceedings,* **57,** July 1960, 1–28.
35. Kenneth L. Saucier. "High-Strength Concrete, Past, Present, Future," *Concrete International,* **2,** June 1980, 46–50.
36. S. P. Shah, editor. *High Strength Concrete,* Proceedings of a Workshop Held at the University of Illinois at Chicago Circle, December 2–4, 1979 (sponsored by National Science Foundation), 226 pp.
37. *Standard Method of Test for Splitting Tensile Strength of Cylindrical Concrete Specimens* (ASTM C496–71 RA 79). Philadelphia: American Society for Testing and Materials, 1979.
38. *Standard Test Method for Flexural Strength of Concrete (Using Simple Beam with Third-Point Loading)* (ASTM C78–75) (Reapproved 1982). Philadelphia: American Society for Testing and Materials, 1975.
39. Sher Ali Mirza, Michael Hatzinikolas, and James G. MacGregor. "Statistical Descriptions of Strength of Concrete," *Journal of Structural Division,* ASCE, **105,** No. ST6 (June 1979), 1021–1037.
40. Adrian Pauw. "Static Modulus of Elasticity of Concrete as Affected by Density," *ACI Journal, Proceedings,* **57,** December 1960, 679–687.
41. Adam M. Neville and Bernard L. Meyers, "Creep of Concrete: Influencing Factors and Prediction," *Symposium on Creep of Concrete* (SP–9). Detroit: American Concrete Institute, 1964 (pp. 1–33).
42. Dan E. Branson. *Deformations of Concrete Structures.* New York: McGraw-Hill, 1977 (pp. 11–27, 44–55).
43. Iqbal Ali and Clyde E. Kesler. "Mechanisms of Creep in Concrete," *Symposium on Creep of Concrete* (SP–9). Detroit: American Concrete Institute, 1964 (pp. 35–63).
44. *Recommended Practice for Evaluation of Strength Test Results of Concrete* (ACI 214–77). Detroit: American Concrete Institute, 1977.
45. J. Bruce Tait. "Concrete Quality Assurance Based on Strength Tests," *Concrete International,* **3,** September 1981, 79–87.

46. "Inspection and Quality Control of Concrete," Symposium series of papers, *ACI Journal, Proceedings,* **65,** August 1968, 639–658. Disc., **66,** February 1969, 154–157 (includes other references).

47. Donald J. Peters. "Concrete Quality—A Producer Looks at the Building Code Requirements," *ACI Journal, Proceedings,* **74,** October 1977, 501–505.

48. *Standard Specification for Deformed and Plain Billet-Steel Bars for Concrete Reinforcement (including Supplementary Requirements)* (ASTM A615–82). Philadelphia: American Society for Testing and Materials, 1982.

49. *Standard Specification for Rail-Steel Deformed and Plain Bars for Concrete Reinforcement* (ASTM A616–82a). Philadelphia: American Society for Testing and Materials, 1982.

50. *Standard Specification for Axle Steel Deformed and Plain Bars for Concrete Reinforcement* (ASTM A617–82a). Philadelphia: American Society for Testing and Materials, 1982.

51. *Standard Specification for Low-Alloy Steel Deformed Bars for Concrete Reinforcement* (ASTM A706–82a). Philadelphia: American Society for Testing and Materials, 1982.

52. *Standard Specification for Deformed and Plain Billet-Steel for Concrete Reinforcement (Metric)* (ASTM A615M–82). Philadelphia: American Society for Testing and Materials, 1982.

53. *Standard Specification for Welded Steel Wire Fabric for Concrete Reinforcement* (ASTM A185–79). Philadelphia: American Society for Testing and Materials, 1979.

54. *Standard Specification for Welded Deformed Steel Wire Fabric for Concrete Reinforcement* (ASTM A497–79). Philadelphia: American Society for Testing and Materials, 1979.

55. *Standard Specification for Cold-Drawn Steel Wire for Concrete Reinforcement* (ASTM A82–79). Philadelphia: American Society for Testing and Materials, 1979.

56. *Standard Specification for Deformed Steel Wire for Concrete Reinforcement* (ASTM A496–78). Philadelphia: American Society for Testing and Materials, 1978.

57. *Manual of Standard Practice, Welded Wire Fabric* (3rd ed.). McLean, Virginia: Wire Reinforcement Institute, 1979 (7900 Westpark Drive, McLean, Virginia 22102).

58. *Standard Specification for Uncoated Seven-Wire Stress-Relieved Steel Strand for Prestressed Concrete* (ASTM A416–80). Philadelphia: American Society for Testing and Materials, 1980.

59. ACI Committee 439. "Steel Reinforcement Properties and Availability," *ACI Journal, Proceedings,* **74,** October 1977, 481–492.

60. *Code for the Design of Concrete Structures for Buildings* (CAN3–A23.3–M77). Rexdale, Ontario, Canada: Canadian Standards Association, 1977.

61. *Standard for Metric Practice* (ASTM E380–82). Philadelphia: American Society for Testing and Materials, 1982.

62. *Standard Practice for the Use of Metric (SI) Units in Building Design and Construction (Committee E-6 Supplement to E-380)* (ASTM E621–78). Philadelphia: American Society for Testing and Materials, 1978.

2

Design Methods
and Requirements

2.1 ACI BUILDING CODE

When two different materials, such as steel and concrete, act together, it is understandable that the analysis for strength of a reinforced concrete member has to be partly empirical, although mostly rational. These semirational principles and methods are being constantly revised and improved as results of theoretical and experimental research accumulate. The American Concrete Institute, serving as a clearinghouse for these changes, issues building code requirements, the most recent of which is the *Building Code Requirements for Reinforced Concrete* (ACI 318–83), hereafter referred to as the ACI Code [Ref. 9, Chap. 1].*

The ACI Code is a Standard of the American Concrete Institute. In order to achieve legal status, it must be adopted by a governing body as a part of its general building code. The ACI Code is partly a specification-type code, which gives acceptable design and construction methods in detail, and partly a performance code, which states desired results rather than details of how such results are to be obtained. A building code, legally adopted, is intended to prevent people from being harmed; therefore it specifies minimum requirements consistent with good safety. It is important to realize that a building code is not a recommended practice, nor is it a design handbook, nor is it intended to replace engineering knowledge, judgment, or experience. It does *not* relieve the designer of the responsibility for having a safe economical structure.

*The reader is advised to have the ACI Code as a ready reference while using this text.

Lake Point Tower, Chicago, apartment building, 70 stories high. (Courtesy of Portland Cement Association.)

2.2 STRENGTH DESIGN AND WORKING STRESS DESIGN METHODS

Two philosophies of design have long been prevalent. The *working stress method,* focusing on conditions at service load (that is, when the structure is being used), was the principal one used from the early 1900s until the early 1960s. By 1983, the transition has been made to the *strength design method,* focusing on conditions at loads greater than service loads when failure may be imminent. The strength design method is deemed conceptually more realistic to establish structural safety.

2.3 WORKING STRESS METHOD

In the working stress method (referred to by the present ACI Code as the "alternate design method"), a structural element is so designed that the stresses resulting from the action of *service loads* (also called *working loads*) and computed by the mechanics of elastic members do not exceed some predesignated allowable values. Service load is the load, such as dead, live, snow, wind, and earthquake, which is assumed actually to occur when the structure is in service.

The working stress method may be expressed by the following:

$$f \leq [\text{allowable stress}, f_{\text{allow}}] \qquad (2.3.1)$$

where

$f =$ an elastically computed stress, such as by using the flexure formula $f = Mc/I$ for a beam

and

$f_{\text{allow}} =$ a limiting stress prescribed by a building code as a percentage of the compressive strength f'_c for concrete, or of the yield stress f_y for the steel reinforcing bars

Some of the obstacles to the working stress method are as follows.

1. Since the limitation is on the total stress under service load, there is no simple way to account for different degrees of uncertainty of various kinds of load. Generally the dead load (gravity load due to weight of structural elements and permanent attachments) is known more accurately than the live load, which is code prescribed and may have unknown and variable distribution.
2. Creep and shrinkage, which contribute major time-dependent effects on a structure, are not easily accounted for by elastic calculation of stresses.
3. Concrete stress is not proportional to strain up to its crushing strength, so that the inherent safety provided is unknown when a percentage of f'_c is used as the allowable stress.

2.4 STRENGTH DESIGN METHOD

In the *strength design method* (formerly called *ultimate strength method*), the service loads are increased sufficiently by factors to obtain the load at which failure is considered to be "imminent." This load is called the *factored load* or *factored service load.* The structure or structural element is then proportioned such that the strength is reached when the factored load is acting. The computation of this strength takes into account the nonlinear stress-strain behavior of concrete.

The strength design method may be expressed by the following,

$$\text{strength provided} \geq \begin{bmatrix} \text{strength required to} \\ \text{carry factored loads} \end{bmatrix} \qquad (2.4.1)$$

where the "strength provided" (such as moment strength) is computed in accordance with rules and assumptions of behavior prescribed by a building code, and the "strength required" is that obtained by performing a structural analysis using factored loads.

The "strength provided" has been commonly referred to by practitioners as "ultimate strength." However, it is a code-defined value for strength, and is not necessarily "ultimate" in the sense of being a value above which it is impossible to reach. The ACI Code uses a conservative definition of strength; thus the modifier "ultimate" is not appropriate.

When the strength design method is used, the comparison of provided strength with required strength (that is, axial force, shear, or bending moment, caused by factored loads) *does not* imply that any material "yields" or "fails" under service load conditions. In fact, at service loads, the behavior of the structure is essentially elastic. The use of the term "imminent failure" under factored loads is only a device for establishing adequate safety.

2.5 COMMENTS ON DESIGN METHODS

Historically, "ultimate" strength was the earliest approach to design since the failure load could be measured by test without a knowledge of the magnitude or distribution of internal stresses. With the interest in and understanding of the elastic methods of analysis in the early 1900s, the elastic working stress method was adopted almost universally by codes as the best for design. As more detailed understanding of the actual behavior of reinforced concrete structures subjected to loads in excess of the service loads developed, adjustments in the theory and in the design procedures were made.

The first modification of the elastic working stress method resulted from the study of axially loaded columns in the early 1930s. By 1940 the design of axially loaded columns was based on ultimate strength. Next, the working stress method was modified to account for creep of concrete in beams with compression steel and in eccentrically loaded columns. The early history of the ACI Code has been summarized by Kerekes and Reid [1].

The 1956 ACI Code was the first that officially recognized and permitted the strength design method, the result of work by ACI–ASCE Committee 327 [2]. The 1963 ACI Code treated the working stress method and the strength design method on an equal basis; but actually the major portion of the working stress method was based on strength. With the relegation of the working stress method to a small section referred to as the "alternate method," the 1971 ACI Code entirely accepted the strength design method. The 1977 and 1983 ACI Codes have the "alternate design method" in Appendix B. It is likely that in the near future the "alternate design method" will disappear and the only use for the elastic concepts of the working stress method will be in the computation of deflections, which are of interest under service loads rather than at factored loads.

No matter which of the above philosophies is employed in a design, *serviceability* must also be considered. Serviceability factors that may be as important as strength are excessive deflection, detrimental cracking, excessive amplitude or undesirable frequency of vibration, and excessive noise transmission. The designer must consider both *strength* and *serviceability*. Any one, or a combination, of the strength and serviceability factors may provide a criterion for the limit of structural usefulness.

The term "limit state" has come into use in recent years. Strength and a serviceability consideration such as a limit on deflection may be thought of as "limit states." MacGregor [3] has provided an excellent treatment of limit states design as applied to reinforced concrete.

2.6 SAFETY PROVISIONS

Structures and structural members must always be designed to carry some reserve load above what is expected under normal use. Such reserve capacity is provided to account for a variety of factors, which may be grouped in two general categories; factors relating to overload and factors relating to understrength (that is, less strength than computed by acceptable calculation procedures). Overloads may arise from changing the use for which the structure was designed, from underestimation of the effect of loads by oversimplification in calculation procedures, and from effects of construction sequence and methods. Understrength may result from adverse variations in material strengths, workmanship, dimensions, control, and degree of supervision, even though individually these items may be within required tolerances.

Conventionally, the term "safety factor" has been used in working stress design to designate nominally the ratio between the yield stress (real, as for steel; nominally defined, as for concrete) and the allowable working stress. Such use has resulted in structures and structural elements with the same "safety factor" but considerably variant in their strength to service load ratio. Thus the term "safety factor" as conventionally applied has little meaning so far as the prediction of strength is concerned.

The variability in the ratio of the strength to service load under the working stress method has been a major factor in the transition to the use of the strength design method. To distinguish between the conventional term "safety factor" and the strength to service load ratio, the term "load factor" has traditionally been adopted for the latter.

The ACI Code has separated the safety provision into factors U for overload and factors ϕ for understrength. The basic overload equation (ACI-9.2.1) for structures in such locations and of such proportions that the effects of wind and earthquake may be neglected is

$$U = 1.4D + 1.7L \qquad (2.6.1)$$

where

U = required strength (based on possible overload)
D = dead load under service conditions
L = live load under service conditions

The factors ϕ for understrength, called *strength reduction factors* according to ACI-9.3, are prescribed as follows:

	ϕ Factors
1. Flexure, with or without axial tension	0.90
2. Axial tension	0.90
3. Shear and torsion	0.85

	ϕ Factors
4. Compression members, spirally reinforced	0.75*
5. Compression members, tied	0.70*
6. Bearing on concrete	0.70
7. Bending in plain concrete	0.65†

Even though the overload factors U and the strength reduction factors ϕ are itemized separately in the ACI Code, they are in fact complementary parts of the provision to ensure adequate safety. The total required nominal (theoretical) strength to be provided in design is U/ϕ. Because the ϕ factors vary for different types of member action, the designer generally will use them at various stages in the calculations. The overload provision is best accounted for at the beginning of the design, where the service loads are multiplied by their respective factors in U giving what are called *factored loads*. The term "design load" is to be avoided since in common use it is unclear whether service load or factored load is referred to.

The purpose of a safety provision is to limit the probability of failure and yet to permit economical structures. Obviously if cost is no object, it is easy to design a structure whose probability of failure is nil. To arrive properly at suitable factors for safety, the relative importance of various items must be established. Some of those items are

1. Seriousness of a failure, either to humans or goods.
2. Reliability of workmanship and inspection.
3. Expectation of overload and to what magnitude.
4. Importance of the member in the structure.
5. Chance of warning prior to a failure.

By assigning percentages to the above items and evaluating the circumstances for any given situation, proper factors for safety may be determined for each case. The ACI Code committee has combined experience with historical precedent to arrive at the prescribed overload and strength reduction factors. The normal range of the factor for safety U/ϕ is from 1.55 to 2.4 for reinforced concrete structural elements. The subject of safety provisions for reinforced concrete has been summarized by Winter [4].

In recent years considerable attention has been focused on using theory of probability as a basis for a design code, thus providing a more rational basis for the various components comprising the U and ϕ factors. A series of ACI Committee 348 (structural safety) papers [5–9] gives an excellent overview of this approach, and an interesting collection of discussion and opinion is also available [8]. Benjamin and Lind [6] have stated that the following five safety conditions form the basis for the current practice in structural analysis and design.

---

*For combined compression and flexure, *both* axial load and bending moment are subject to the same ϕ factor, which may be variable and increase to 0.90 as the axial compression decreases to zero.
†From *Building Code Requirements for Structural Plain Concrete* (ACI 318.1–83) *and Commentary*. Detroit: American Concrete Institute, 1983 (Sec. 6.2.2). Formerly this was in ACI-9.3.

1. The probability of a real loading in excess of the nominal service load $D + L$ must be satisfactorily small.
2. The probability of a real loading in excess of the factored loading, say $U = 1.4D + 1.7L$, must be very small or near to zero during the life of the structure.
3. The probability of unsatisfactory performance at the factored load U must be satisfactorily small.
4. The probability of unsatisfactory performance under a load test, say $0.85U$ as given in ACI-20.4.3, must be very small or near to zero.
5. The probability of unsatisfactory performance at the service load $D + L$ must be practically zero.

It is noted that conditions 1 and 2 relate to the overload provision U along with analysis methods, whereas conditions 3, 4, and 5 relate to the factor ϕ for understrength as well as the methods for computing strength.

Currently under review by the ACI Code committee is a proposal to adopt the probability-based factors for overload and the combinations of factored loads given by the 1982 American National Standards Institute (ANSI) Code [10]. These ANSI factors and loading combinations are based on the report, *Development of a Probability Based Load Criterion for American National Standard A58* [11], which recommends using lower factors U, as well as different combinations than discussed in Sec. 2.7, to account for overload. Since the overall objective is to use the *same* factored loads and factored load combinations no matter what material is used, the use of lower factors for overload in reinforced concrete design must be accompanied by the use of lower strength reduction factors ϕ (so-called *resistance factors* according to ANSI), because the overall safety in present concrete structures design is considered appropriate. Detailed discussion of the implications of using design reliability and probability-based load criteria is given for reinforced concrete by Ellingwood [20,21] and MacGregor [24], and is treated in general by Galambos, Ellingwood, MacGregor, and Cornell [22,23].

2.7 OVERLOAD PROVISIONS FOR VARIOUS LOAD COMBINATIONS

In addition to the basic provision for load factors on dead load plus live load given by Eq. (2.6.1), other service loads may also act.

For wind load W acting in combination with other loads, ACI-9.2.2 provides

$$U = 0.75(1.4D + 1.7L + 1.7W) \tag{2.7.1}$$

Because wind is of a transient nature and acts with its maximum magnitude for a short duration, it has been traditional to allow an overstress of $33\frac{1}{3}\%$ under the working stress method. The same effect is accomplished by using three-quarters of the factored load when wind effect is included.

Frequently a more severe situation arises with wind loading if live load is absent; this possibility must be considered. When live load is absent, Eq. (2.7.1) becomes

never used

$$U = 1.05D + 1.275W \tag{2.7.2}$$

Furthermore for situations in which dead load is a gravity stabilizing effect in combination with wind (such as a tower or wall), the possibility of a reduced dead load must be considered rather than overload; thus ACI-9.2.2 also gives

$$U = 0.9D + 1.3W \tag{2.7.3}$$

When lateral earth pressure H is involved, it is treated as live load; thus

$$U = 1.4D + 1.7L + 1.7H \tag{2.7.4}$$

but when dead or live load (or both) reduces the effect of earth pressure, then

$$U = 0.9D + 1.7H \tag{2.7.5}$$

For liquid pressure F having a controllable maximum height, Eqs. (2.7.4) and (2.7.5) are used except $1.7H$ is replaced with $1.4F$. Since liquid density is generally accurately known, its pressure is treated as dead load using the 1.4 factor. On the other hand soil properties are more variable, thus earth pressure is treated as live load using the 1.7 factor.

Where the structural effects T of differential settlement, creep, shrinkage, or temperature change may be significant, they are to be included with dead load (ACI-9.2.7),

$$U = 0.75(1.4D + 1.4T + 1.7L) \tag{2.7.6}$$

but not less than

$$U = 1.4(D + T) \tag{2.7.7}$$

The 0.75 factor is to recognize the low probability of having these effects occur simultaneously with full dead and live load.

Any structure or structural element *must be designed for the most severe of any of the load combinations* given by Eq. (2.6.1) and Eqs. (2.7.1) to (2.7.7).

2.8 HANDBOOKS

It may be pointed out that this textbook does not appreciably consider office practices or the use of design aids. Once concepts and principles are thoroughly understood, the use of curves and tables can greatly speed up design. There are several handbooks in common use, the *Strength Design Handbook,* Vols. 1 and 2 [12,13], published by ACI, the *CRSI Handbook* [14], published by the Concrete Reinforcing Steel Institute, the *PCI Design Handbook* [15], published by the Prestressed Concrete Institute, and the *Handbook of Concrete Engineering* [16]. These publications contain many useful tables and charts that can speed up design for the experienced designer.

The importance of correct and clear detailing work cannot be overemphasized. For this the reader is referred to the *ACI Detailing Manual* (SP-66) [17] which contains typical detailing of steel reinforcement, engineering and placing drawings, and other reference data regarding materials and sizes.

2.9 DIMENSIONS AND TOLERANCES

Although the designer may tend to think of dimensions, clearances, and bar locations as exact, practical considerations require that there be accepted tolerances. These tolerances are the permissible variations from dimensions given on drawings.

Overall dimensions of reinforced concrete members are usually specified by the engineer in whole inches for beams, columns, and walls; sometimes half inches for thin slabs; and often 3-in. increments for more massive elements such as plan dimensions for footings. Formwork for the placing of these members must be built carefully so that it does not deform excessively under the action of workmen, construction machinery loads, and wet concrete. Accepted tolerances for variation in cross-sectional dimensions of columns and beams and in the thickness of slabs and walls are $+\frac{1}{2}$ in. and $-\frac{1}{4}$ in. [18]. For concrete footings accepted variations in plan dimensions are $+2$ in. and $-\frac{1}{2}$ in. [18], whereas the thickness has an accepted tolerance of -5% of specified thickness [18]. The strength reduction factor ϕ is intended to account for the situation in which several acceptable tolerances may combine to reduce the strength from that computed using specified dimensions.

Reinforcing bars are normally specified in 3-in. length increments and the placement tolerances are given in the ACI Code (ACI-7.5.2). For minimum clear concrete protection and for the effective depth d (distance from compression face of concrete to center of tension steel) in flexural members, walls, and compression members, the specified tolerances are as follows:

| EFFECTIVE DEPTH, d | | TOLERANCES | | | |
| | | ON EFFECTIVE DEPTH | | ON MINIMUM CONCRETE COVER | |
(in.)	(mm)	(in.)	(mm)	(in.)	(mm)
$d \le 8$	200^a	$\pm\frac{3}{8}$	± 10	$-\frac{3}{8}$	-10
$d > 8$	200	$\pm\frac{1}{2}$	± 12	$-\frac{1}{2}$	-12

^{a}These values are from ACI 318–83M, and are not strictly conversions from ACI 318–83.

Notwithstanding the stated tolerances on cover, the resulting cover shall not be less than two-thirds of the minimum cover specified on structural drawings or in specifications. Since the effective depth and the clear concrete cover are both components of total depth, the tolerances on those dimensions are directly related. When the tolerances on bar placement and cover accumulate, the overall dimension tolerance may be exceeded; thus field adjustment may have to be made. This may be particularly important on very thin sections such as in precast and shell structures.

For location of bars along the longitudinal dimension, and of bar bends, the tolerance is ± 2 in. (± 50 mm*) except at discontinuous ends where tolerance shall be $\pm\frac{1}{2}$ in. (± 12 mm*) (ACI-7.5.2).

*According to ACI 318–83M, Sec. 7.5.2.

A summary of all tolerances for concrete construction and materials has been provided by ACI Committee 117 [19].

2.10 ACCURACY OF COMPUTATIONS

When one understands that variations exist in material strength for both steel and concrete and that variations in dimensions are inevitable (and acceptable), it becomes clear that design of reinforced concrete structures does not require a high degree of precision.

The designer should place highest priority on determining proper location and length of steel reinforcement to carry the tension forces, thus making up for that capacity which is deficient in the concrete. Failures, when they occur, generally result from gross underestimating of tensile forces or lack of identification of how the structure or element will behave under loads. Failures are rarely the result of carrying too few significant figures in the design computations. However, significant figures may be lost in arithmetic operations, and gross errors may sometimes result from sloppiness. *It is recommended, therefore, more for systematic control of calculations and for ease in checking than for any improved effect on the final structure, that three significant figures (not three decimal places) be used in computations.* Recording of results from the display of an electronic calculator should not exceed four significant figures. Since rarely does the recorded result indicate a precision of even three significant figures, the recording of many figures in design computations makes scanning of results difficult and may hinder detection of errors made in design.

SELECTED REFERENCES

1. Frank Kerekes and Harold B. Reid, Jr. "Fifty Years of Development in Building Code Requirements for Reinforced Concrete," *ACI Journal, Proceedings*, **50**, February 1954, 441.
2. ACI-ASCE Committee 327. "Ultimate Strength Design," *ACI Journal, Proceedings*, **52**, January 1956, 505–524.
3. J. G. MacGregor. "Safety and Limit States Design for Reinforced Concrete," *Canadian Journal of Civil Engineering*, **3**, December 1976, 484–513.
4. George Winter. "Safety and Serviceability Provisions in the ACI Building Code," *Concrete Design: U.S. and European Practices* (SP-59). Detroit, Michigan: American Concrete Institute, 1979.
5. Robert G. Sexsmith and Mark F. Nelson. "Limitations in Application of Probabilistic Concepts," *ACI Journal, Proceedings*, **66**, October 1969, 823–828.
6. Jack R. Benjamin and N. C. Lind. "A Probabilistic Basis for a Deterministic Code," *ACI Journal, Proceedings*, **66**, November 1969, 857–865.
7. C. Allin Cornell. "A Probability-Based Structural Code," *ACI Journal, Proceedings*, **66**, December 1969, 974–985.
8. R. C. Reese, D. E. Allen, C. A. Cornell, Luis Esteva, R. N. White, R. G. Sexsmith, and George Winter. "Probabilistic Approaches to Structural Safety," *ACI Journal, Proceedings*, **73**, January 1976, 37–49.
9. Haresh C. Shah and Robert G. Sexsmith. "A Probabilistic Basis for the ACI Code," *ACI Journal, Proceedings*, **74**, December 1977, 610–611.
10. *American National Standard Building Code Requirements for Minimum Design Loads for Buildings and Other Structures* (ANSI-A58.1). Washington, D.C.: American National Standards Institute, 1982.

11. Bruce Ellingwood, Theodore V. Galambos, James G. MacGregor, and C. Allin Cornell. *Development of a Probability Based Load Criterion for American National Standard A58* (NBS Special Publication 577). Washington, D.C.: U.S. Department of Commerce, National Bureau of Standards, June 1980.
12. ACI Committee 340 (C. G. Salmon, Chairman). *Design Handbook In Accordance with the Strength Design Method of ACI 318-77. Vol. 1—Beams, Slabs, Brackets, Footings, and Pile Caps* [SP-17(81)] (3rd ed.). Detroit, Michigan: American Concrete Institute, 1981.
13. ACI Committee 340 (C. G. Salmon, Chairman). *Design Handbook In Accordance with the Strength Design Method of ACI 318-77. Vol. 2—Columns* (SP-17A) (2nd ed.). Detroit, Michigan: American Concrete Institute, 1978.
14. *CRSI Handbook* (3rd ed.). Chicago: Concrete Reinforcing Steel Institute, 1978.
15. *PCI Design Handbook—Precast and Prestressed Concrete* (2nd ed.). Chicago: Prestressed Concrete Institute, 1980 (20 North Wacker Drive, Chicago, Ill. 60606).
16. Mark Fintel, editor. *Handbook of Concrete Engineering* (2nd ed.). New York: Van Nostrand Reinhold, 1984.
17. ACI Committee 315. *ACI Detailing Manual—1980* (SP-66). Detroit, Michigan: American Concrete Institute, 1980.
18. ACI Committee 347. *Recommended Practice for Concrete Formwork* (ACI 347-78). Detroit: American Concrete Institute, 1978.
19. ACI Committee 117. "Proposed ACI Standard: Tolerances for Concrete Construction and Materials," *Concrete International*, **2**, August 1980, 38–46.
20. Bruce Ellingwood. "Reliability of Current Reinforced Concrete Designs," *Journal of the Structural Division*, ASCE, **105**, April 1979 (ST4), 699–712.
21. Bruce Ellingwood. "Reliability Based Criteria for Reinforced Concrete Design," *Journal of the Structural Division*, ASCE, **105**, April 1979 (ST4), 713–727.
22. Theodore V. Galambos, Bruce Ellingwood, James G. MacGregor, and C. Allin Cornell. "Probability Based Load Criteria: Assessment of Current Design Practice," *Journal of the Structural Division*, ASCE, **108**, May 1982 (ST5), 959–977.
23. Bruce Ellingwood, James G. MacGregor, Theodore V. Galambos, and C. Allin Cornell. "Probability Based Load Criteria: Load Factors and Load Combinations," *Journal of the Structural Division*, ASCE, **108**, May 1982 (ST5), 978–997.
24. James G. MacGregor. "Load and Resistance Factors for Concrete Design," *ACI Journal, Proceedings*, **80**, July-August 1983, 279–287.

3

Strength
of Rectangular
Sections
in Bending

3.1 GENERAL INTRODUCTION

Until 1956 the ACI Code method for the determination of whether a reinforced concrete member had adequate strength was made solely on the basis
of allowable working stresses, service (or working) loads, and the straightline
theory of flexure. Since 1971, the concepts involved in the working stress
method have been used only for establishing serviceability at service loads.
The working stress method and serviceability considerations are therefore
dealt with in Chap. 4.

The strength design method, in which factored loads and the computed
strength of sections at imminent failure are used, was first permitted as an
alternate method of design in the 1956 ACI Code and later became a substantial part of the 1963 Code. The 1971 ACI Code relegated the working
stress method to a small section of the Code and utilized primarily the
strength method of design. The adjective "ultimate" was dropped since
there was little chance of confusion over the meaning of the word "strength."
The 1977 ACI Code removed the alternate design (working stress) method
from the body of the Code and placed it in Appendix B, where it still remains
in the 1983 ACI Code.

"Strength" is used in the general sense to include various modes of
failure other than that of the concrete reaching its expected crushing strain
(as is the criterion for bending strength). Another reason for deletion of the
term "ultimate" is that the strength considered for design is a code-defined
strength, and may in fact not be "ultimate" in its literal sense.

Rectangular tapered beams, cantilevers, and rigid frame, University of Wisconsin Stadium, Madison, Wis. (Photo by C. G. Salmon.)

In the strength method the *factored loads* (including moments, shears, axial forces, etc.) are obtained by multiplying the service loads by factors U to cover possible overloads and variations in design assumptions. The *design strength* of a section is obtained by multiplying the *nominal strength* (based on statical equilibrium and compatibility of stress and strain) by a strength reduction factor ϕ to account for adverse variations in material strengths, workmanship, dimensions, control, and degree of supervision, even though all are within accepted tolerances.

In discussing the ACI strength design method for reinforced concrete structures, attention must be called to the difference between loads on the structure as a whole and loads on the cross sections of individual members. The elastic methods of structural analysis are used first to compute the service loads in the individual members due to the action of service loads on the entire structure. Only then are the overload factors applied to the service loads acting on the individual cross sections. This emphasis is more conceptual than practical. In practice one may apply the factored loads on the entire structure and still make an elastic analysis (which theoretically is no longer valid) to obtain the factored loads (such as shears and bending moments) in the individual members.

Research is continuing on the feasibility of using the inelastic, or limit, method of structural analysis in which factored loads in the individual members are determined correctly from the factored loads acting on the whole structure. This is called "limit design" and is introduced briefly in Sec. 10.12.

3.2 BASIS OF NOMINAL FLEXURAL STRENGTH

The modern analytical approach to reinforced concrete beam strength was originated by F. Stüssi in 1932 [1]. The general stress-strain behavior for concrete as presented in Sec. 1.7 shows the nonlinearity of stress and strain at stress levels above about $0.5f'_c$. Since the compression zone of a beam should be expected to have the same general variation of stress and strain as the test specimen (standard cylinder or cube), the compressive stress distribution for a beam that has achieved its theoretical (nominal) strength [2,3,4] should be as shown in Fig. 3.2.1(c).

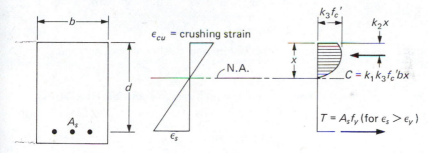

d = effective depth; distance from compression face to centroid of tension steel

(a) Singly reinforced beam

(b) Strain condition when M_n is reached

(c) Stress condition at strain condition when M_n is reached

Figure 3.2.1 Conditions when nominal strength M_n in flexure is reached.

The nominal strength is assumed to be reached when the strain in the extreme compression fiber is equal to the crushing strain ϵ_{cu} of concrete (see Fig. 3.2.2). When crushing occurs (usually a somewhat sudden occurrence), the strain in the tension steel A_s *could* be either larger or smaller than the strain $\epsilon_y = f_y/E_s$ at first yield, depending on the relative proportion of steel to concrete. If the steel amount were low enough, it would yield prior to crushing of the concrete, resulting in a ductile failure mode in which there is large deformation. On the other hand a large quantity of steel would allow the steel to remain elastic at the time of crushing of the concrete, causing a brittle or sudden mode of failure. The ACI Code has provisions which, by limiting the amount of tension steel, are intended to ensure the ductile mode of failure when the nominal strength is reached.

Although the compressive stress distribution in a beam has the same general shape as for a test cylinder, the maximum stress is less than f'_c, say $k_3f'_c$ [see Fig. 3.2.1(c)]. The average stress over a beam of constant width is $k_1k_3f'_c$; and the centroid location of the roughly parabolic distribution is k_2x measured from the compressive face, where x is the neutral axis distance. Thus the compressive force C is the summation of the compressive stresses

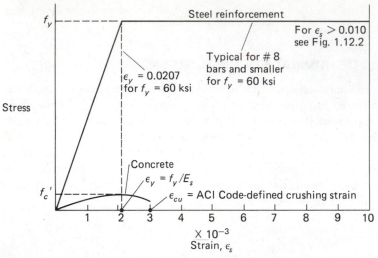

Figure 3.2.2 Stress-strain curves for concrete and steel reinforcement, and definitions of ϵ_y and ϵ_{cu}.

acting on the compression concrete area, which may be considered a "stress solid" volume, as

$$C = k_1 k_3 f_c' x b \qquad (3.2.1)$$

and for the ductile failure condition, the tensile force T is

$$T = A_s f_y \qquad (3.2.2)$$

Equilibrium requires $C = T$, from which

$$x = \frac{A_s f_y}{k_1 k_3 f_c' b} \qquad (3.2.3)$$

The nominal flexural strength then may be expressed as

$$M_n = T(\text{arm}) = T(d - k_2 x)$$

$$= A_s f_y (d - k_2 x) \qquad (3.2.4)$$

Substituting Eq. (3.2.3) for x into Eq. (3.2.4) gives

$$M_n = A_s f_y \left(d - \frac{k_2}{k_1 k_3} \frac{A_s f_y}{f_c' b} \right) \qquad (3.2.5)$$

One may note that if strength is the quantity that is of interest, it is readily obtainable from Eq. (3.2.5) *if the quantity $k_2/(k_1 k_3)$ is known.* It is not necessary to have values for k_1, k_2, or k_3 individually if the value for the combined term is known. Experimental results [2,3] have established values for the combined term, as well as the individual k values, with some of the results shown in Fig. 3.2.3. From that figure, $k_2/(k_1 k_3)$ ranges from about 0.55 to 0.63.

The values experimentally determined when crushing of the concrete occurred at the compression face necessarily involved variation in the compressive strain ϵ_{cu} for the various tests. The ACI Code (ACI-10.2.3) consid-

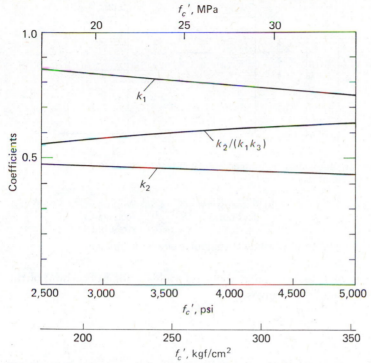

Figure 3.2.3 "Stress solid" parameters (adapted from Ref. 2).

ers the maximum usable strain ϵ_{cu} to be 0.003. Some countries use a value of 0.0035, which makes little difference in the computed flexural strength.

3.3 WHITNEY RECTANGULAR STRESS DISTRIBUTION

The computation of flexural strength M_n based on the approximately parabolic stress distribution of Fig. 3.2.1(c) may be done using Eq. (3.2.5) with given values of $k_2/(k_1k_3)$. However, it is desirable for the designer to have a simple method in which basic static equilibrium is used.

In the 1930s Whitney [5,6] proposed the use of a rectangular compressive stress distribution to replace that of Fig. 3.2.1(c). As shown in Fig. 3.3.1(c), an average stress of $0.85f_c'$ is used with a rectangle of depth $a = \beta_1 x$, determined so that $a/2 = k_2 x$. Whitney determined that β_1 should be 0.85 for concrete with $f_c' \leq 4000$ psi, and 0.05 less for each 1000 psi of f_c' in excess of 4000 psi.* The value of β_1 may not be taken less than 0.65 (ACI-10.2.7.3).

The flexural strength M_n, using the equivalent rectangle, is obtained from Fig. 3.3.1 as follows:

$$C = 0.85f_c'ba \qquad (3.3.1)$$

$$T = A_sf_y \qquad (3.3.2)$$

*For SI, ACI 318–83M gives $\beta_1 = 0.85$ for $f_c' \leq 30$ MPa and reduces by 0.08 for each 10 MPa of f_c' in excess of 30 MPa, but not less than 0.65.

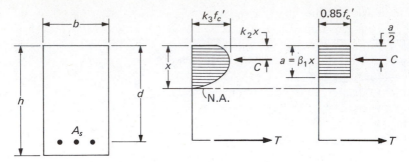

d = effective depth; distance from compression face of concrete
   to centroid of tension steel

(a) Beam   (b) Actual stress   (c) Whitney
                        distribution   rectangular
                                                   stress block

Figure 3.3.1 Definition of Whitney rectangular stress distribution.

where the use of f_y assumes that the steel yields prior to crushing of the concrete. Equating $C = T$ gives

$$a = \frac{A_s f_y}{0.85 f_c' b} \tag{3.3.3}$$

$$M_n = A_s f_y (d - a/2) \tag{3.3.4}$$

which on substituting Eq. (3.3.3) into Eq. (3.3.4) gives

$$M_n = A_s f_y \left(d - 0.59 \frac{A_s f_y}{f_c' b} \right) \tag{3.3.5}$$

Note that 0.59 corresponds to $k_2/(k_1 k_3)$ of Eq. (3.2.5). The ACI Code explicitly accepts the Whitney rectangle (ACI-10.2.7). One may also note that β_1 is only needed to establish the neutral-axis location for determining the steel strain. So long as the steel strain exceeds $\epsilon_y = f_y/E_s$, as shown in Fig. 3.2.2, the flexural strength M_n is not affected by the value of β_1.

3.4 INVESTIGATION OF RECTANGULAR SECTIONS IN BENDING WITH TENSION REINFORCEMENT ONLY

The quantities defining a rectangular section with tension reinforcement only are b, d, and A_s [Fig. 3.3.1(a)]. Such a section is said to be *singly reinforced*. The steel area A_s is, of course, furnished by the combined area of an actual number of reinforcing bars. Protective concrete cover is necessary around the bars in order to make the steel and concrete act together and, also very importantly, to provide fire protection. At high temperatures (say above 600°F) the yield strength and modulus of elasticity begin to reduce markedly, so that concrete cover is needed for insulation. Minimum cover requirements are generally prescribed by code (see ACI-7.7).

   Since the tensile strength of concrete is normally neglected in flexure calculations, the cross-sectional shape of the beam on the tension side of the neutral axis and the amount of concrete cover do not affect the flexural strength. Thus the important depth dimension for computing strength is the *effective depth d* rather than the overall depth *h*. The effective depth is defined as the distance from the extreme fiber in compression to the centroid of the tension steel area. When the tension steel is comprised of bars in several layers satisfying the minimum spacing requirement between layers, the centroid of the combined area is usually used, with all bars assumed to have the same strain.

Investigation of a rectangular section in bending is made to establish whether or not the section has adequate *strength* for given applied service loads (or service load moments). The system must also be serviceable; that is, it must perform satisfactorily under service loads without detrimental effects, such as excessive deflection, cracking, or vibration. *Serviceability* generally is treated in Chap. 4, and the detailed discussion of deflection appears in Chap. 14.

ACI Code Strength Method. According to the ACI Code, the required strength *U* for a section in bending is obtained by applying load factors (i.e., overload provision) to the moment due to dead load *D*, live load *L*, wind load *W*, earthquake load *E*, lateral earth pressure *H*, fluid pressure *F*, and structural effects *T*, as discussed in Sec. 2.6 (ACI-9.2). The design flexural strength ϕM_n is the product of the strength reduction factor ϕ and the nominal flexural strength M_n.

The basic load factors *U* in the overload provision are 1.4 and 1.7, used as follows:

$$U = 1.4D + 1.7L$$

and the ϕ factor for flexure is 0.90. The terms *U*, *D*, and *L* of the overload provision may also represent quantities that are functions of load, such as moment, shear, and axial force. If M_u is defined as the moment under factored load and M_D and M_L are the service dead load and live load moments, respectively, the basic overload provision for flexure may be expressed as

$$M_u = 1.4M_D + 1.7M_L \tag{3.4.1}$$

The *strength requirement* for flexure may then be expressed as

$$\phi M_n \geq M_u \tag{3.4.2}$$

For computation of the nominal flexural strength M_n, the following assumptions (ACI-10.2) are made:

1. The strength of members shall be based on satisfying the applicable conditions of equilibrium and compatibility of strains.
2. Strain in the steel reinforcement and in the concrete shall be assumed directly proportional to the distance from the neutral axis (except for deep members covered under ACI-10.7).
3. The maximum usable strain ϵ_{cu} at the extreme concrete compression fiber shall be assumed equal to 0.003.

4. The tensile strength of the concrete is to be neglected (except for certain prestressed concrete conditions).
5. The modulus of elasticity of nonprestressed steel reinforcement may be taken as 29,000,000 psi (200,000 MPa or 2,040,000 kgf/cm²).
6. For practical purposes the relationship between the concrete compressive stress distribution and the concrete strain when nominal strength is reached may be taken as (ACI-10.2.7) an equivalent rectangular stress distribution, wherein a concrete stress intensity of $0.85f_c'$ is assumed to be uniformly distributed over an equivalent compressive zone bounded by the edges of the cross section and a straight line located parallel to the neutral axis at a distance $a = \beta_1 x$ from the fiber of maximum compressive strain. The distance x from the fiber of maximum strain to the neutral axis is measured in a direction perpendicular to that axis. The value of β_1 is given by the following equations:

For f_c' less than 4000 psi,

$$\beta_1 = 0.85 \tag{3.4.3}$$

For f_c' greater than 4000 psi,

$$\beta_1 = 0.85 - 0.05\left(\frac{f_c' - 4000}{1000}\right) \geq 0.65 \tag{3.4.4}$$

It should be noted that assumption (6) describes the Whitney rectangular compressive stress distribution (see Sec. 3.3), but other shapes of stress solids, such as the trapezoid and the parabola, have been used [7] and are acceptable for use according to ACI-10.2.6.

EXAMPLE 3.4.1 Determine the nominal flexural strength M_n of the rectangular section shown in Fig. 3.4.1, given $f_c' = 5000$ psi, $f_y = 50,000$ psi, $b = 14$ in., $d = 21.5$ in., and $A_s = 4$-#10 bars.

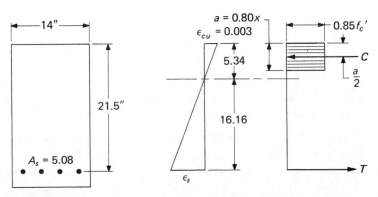

(a) Cross section   (b) Actual condition

Figure 3.4.1 Singly reinforced beam of Example 3.4.1.

Solution: Assume the steel has already yielded when the strength is reached. From Fig. 3.4.1 the internal forces are

$$C = 0.85f'_c ba = 0.85(5)(14)a = 59.5a$$

$$T = A_s f_y = 5.08(50) = 254 \text{ kips}$$

For equilibrium, $C = T$; therefore

$$a = \frac{254}{59.5} = 4.27 \text{ in.}$$

$$\beta_1 = 0.80 \qquad \text{for } f'_c = 5000 \text{ psi}$$

The neutral axis position is

$$x = \frac{a}{\beta_1} = \frac{4.27}{0.80} = 5.34 \text{ in.}$$

The strain in the tension steel when the strain 0.003 is reached in the concrete is, by straightline proportion,

$$\epsilon_s = \frac{d-x}{x}(0.003) = \frac{16.16}{5.34}(0.003) = 0.0091$$

$$\epsilon_y = \frac{f_y}{E_s} = \frac{50}{29,000} = 0.00172$$

When the nominal flexural strength is reached, ϵ_s is 5.3 times ϵ_y, which means that large and gradual deflection is presumed to occur before the crushing of concrete. The assumption that the steel yields has been shown valid.

The nominal flexural strength is

$$M_n = C\left(d - \frac{a}{2}\right) \qquad \text{or} \qquad T\left(d - \frac{a}{2}\right)$$

$$= 254(21.5 - 2.13)\tfrac{1}{12} = 410 \text{ ft-kips } (556 \text{ kN·m})$$

EXAMPLE 3.4.2 For the beam of Example 3.4.1, determine the safe service moment M_w that may be applied according to the ACI Code, if 60% of the total moment is dead load and 40% is live load.

Solution: From Example 3.4.1,

$$M_n = \text{nominal strength} = 410 \text{ ft-kips}$$

$$M_u = \text{factored moment} = 1.4M_D + 1.7M_L$$

$$\phi = 0.90 \text{ for flexure}$$

For safety, Eq. (3.4.2) requires that

$$\phi M_n \geq M_u$$

$$\geq [1.4M_D + 1.7M_L]$$

which if $M_D = 0.60M_w$ and $M_L = 0.40\ M_w$,

$$\text{required } M_n = \frac{M_u}{\phi} = \frac{1.4(0.60)M_w + 1.7(0.40)M_w}{0.90} = 1.69M_w$$

Thus the safe service moment is

$$M_w = \frac{M_n}{1.69} = \frac{410}{1.69} = 243 \text{ ft-kips (329 kN·m)}$$

The safety factor between nominal strength and service load is 1.69 for this case.

3.5 DEFINITION OF BALANCED STRAIN CONDITION

At the *balanced strain* condition [Fig. 3.5.1(b)] the maximum strain ϵ_{cu} at the extreme concrete compression fiber just reaches 0.003 simultaneously with the tension steel reaching a strain $\epsilon_y = f_y/E_s$ (Fig. 3.2.2 shows the definition of ϵ_{cu} and ϵ_y). An amount of tension steel A_{sb} would provide the neutral axis distance x_b for this balanced strain condition. If the actual A_s were greater than A_{sb} equilibrium of internal forces ($C = T$) would mean an increase in the depth a of the compression stress block (and thereby also make x exceed x_b), so that the strain ϵ_s would be less than ϵ_y when $\epsilon_{cu} = 0.003$. The failure of this beam will be sudden when the concrete reaches the strain 0.003 although the beam will exhibit little deformation (steel does not yield) to warn of impending failure.

On the other hand when the actual A_s is less than A_{sb}, the tensile force reduces so that internal force equilibrium reduces the depth a of the compression stress block (and thereby makes x less than x_b) giving a strain ϵ_s greater than ϵ_y. In this case, with the steel having yielded, the beam will have noticeable deflection prior to the concrete reaching the crushing strain of 0.003.

Thus the relative amount of tension steel compared to that in the balanced strain condition will affect significantly the mode of failure.

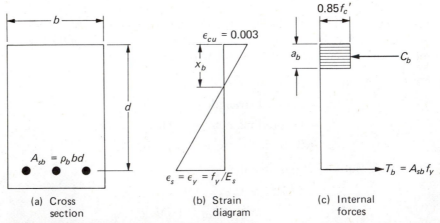

Figure 3.5.1 Balanced strain condition.

(a) Cross section   (b) Strain diagram   (c) Internal forces

Reinforcement Ratio at Balanced Strain Condition for Rectangular Beam Having Tension Reinforcement Only. The symbol ρ, the *reinforcement ratio* (commonly called *reinforcement percentage*), may be conveniently used to represent the relative amount of tension reinforcement in a beam. Thus using the dimensions of Fig. 3.5.1,

$$\rho = \frac{A_s}{bd} \tag{3.5.1}$$

The reinforcement ratio ρ_b in the balanced strain condition may be obtained by applying the equilibrium and compatibility conditions. From the linear strain condition, Fig. 3.5.1(b),

$$\frac{x_b}{d} = \frac{\epsilon_{cu}}{\epsilon_{cu} + \epsilon_y} = \frac{0.003}{0.003 + f_y/29,000,000} = \frac{87,000}{87,000 + f_y} \tag{3.5.2}$$

The compressive force C_b is

$$C_b = 0.85 f_c' b \beta_1 x_b$$

The tensile force T_b is

$$T_b = f_y A_{sb} = \rho_b b d f_y$$

Equating C_b to T_b gives

$$0.85 f_c' b \beta_1 x_b = \rho_b b d f_y$$

$$\rho_b = \frac{0.85 f_c'}{f_y} \beta_1 \left(\frac{x_b}{d}\right) \tag{3.5.3}$$

which on substitution of Eq. (3.5.2) gives

$$\rho_b = \frac{0.85 f_c'}{f_y} \beta_1 \left(\frac{87,000}{87,000 + f_y}\right) \tag{3.5.4*}$$

where the stresses f_y and f_c' are in psi.

3.6 MAXIMUM REINFORCEMENT RATIO ρ

In order to have reasonable assurance for a ductile mode of failure in flexure, ACI-10.3.3 limits the amount of tension steel to not more than 75% of the amount in the balanced strain condition, that is,

$$\max \rho = 0.75 \rho_b \tag{3.6.1}$$

The limitation on reinforcement ratio ρ is an indirect way of controlling the strain diagram when "failure" (i.e., crushing of the concrete at the extreme compression fiber) is imminent.

---

*For SI, ACI 318–83M gives

$$\rho_b = \frac{0.85 f_c'}{f_y} \beta_1 \left(\frac{600}{600 + f_y}\right) \tag{3.5.4}$$

with f_c' and f_y in MPa, and β_1 as given in footnote on p. 43.

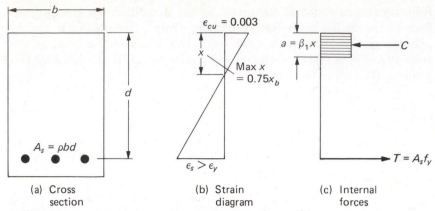

(a) Cross section   (b) Strain diagram   (c) Internal forces

Figure 3.6.1   Maximum reinforcement ratio (ACI-10.3.3); max $\rho \leq 0.75\rho_b$ or max $x \leq 0.75x_b$.

Table 3.6.1   MAXIMUM REINFORCEMENT RATIO ρ FOR SINGLY REINFORCED RECTANGULAR BEAMS (CORRESPONDING TO $0.75\rho_b$)

f_y	$f'_c = 3000$ psi $\beta_1 = 0.85$	$f'_c = 3500$ psi $\beta_1 = 0.85$	$f'_c = 4000$ psi $\beta_1 = 0.85$	$f'_c = 5000$ psi $\beta_1 = 0.80$	$f'_c = 6000$ psi $\beta_1 = 0.75$
40,000 psi	0.0278	0.0325	0.0371	0.0437	0.0491
50,000 psi	0.0206	0.0241	0.0275	0.0324	0.0364
60,000 psi	0.0160	0.0187	0.0214	0.0252	0.0283
f_y	$f'_c = 20$ MPa $\beta_1 = 0.85$	$f'_c = 25$ MPa $\beta_1 = 0.85$	$f'_c = 30$ MPa $\beta_1 = 0.85$	$f'_c = 35$ MPa $\beta_1 = 0.81$	$f'_c = 40$ MPa $\beta_1 = 0.77$
300 MPa	0.0241	0.0301	0.0361	0.0402	0.0436
350 MPa	0.0196	0.0244	0.0293	0.0326	0.0354
400 MPa	0.0163	0.0203	0.0244	0.0271	0.0295
f_y	$f'_c = 200$ kgf/cm² $\beta_1 = 0.85$	$f'_c = 240$ kgf/cm² $\beta_1 = 0.85$	$f'_c = 280$ kgf/cm² $\beta_1 = 0.85$	$f'_c = 320$ kgf/cm² $\beta_1 = 0.82$	$f'_c = 360$ kgf/cm² $\beta_1 = 0.79$
2800 kgf/cm²	0.0266	0.0319	0.0372	0.0410	0.0444
3500 kgf/cm²	0.0197	0.0236	0.0276	0.0304	0.0330
4200 kgf/cm²	0.0153	0.0184	0.0214	0.0236	0.0256

A more direct way of controlling ductility is to prescribe a maximum value for the neutral axis distance x at the "failure" imminent condition. It can be shown that for singly reinforced rectangular sections, ACI-10.3.3 is equivalent to (see Fig. 3.6.1)

$$\text{max } x = 0.75x_b \tag{3.6.2}$$

Thus the direct and indirect concepts lead to identical results of ductile failure for singly reinforced rectangular sections.

For beams containing steel in the compression zone (see Sec. 3.10) or

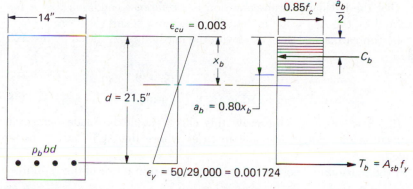

(a) Cross section   **(b) Balanced condition**

Figure 3.6.2   Balanced strain condition for Example 3.6.1 (actual condition shown in Fig. 3.4.1).

having a nonrectangular compression zone, the literal wording of ACI-10.3.3 as expressed by Eq. (3.6.1) may give somewhat different results than limiting x to $0.75x_b$, as expressed by Eq. (3.6.2). The actual code wording was *intended* to make computation of maximum ρ easier for beams having compression steel, rather than to vary the required ductility at imminent "failure." The limitation of maximum x to not exceed $0.75x_b$ has the logical (and the authors believe intended) result of requiring the same ductility of all beams.

In actual structures the provided design strength ϕM_n, being equal to or slightly greater than the factored moment M_u, is always much larger than the actual service moment acting on the beam. Thus the actual attainment of M_n in a beam is an imaginary situation or, put in another way, a philosophy of design. In this philosophy it is considered appropriate to restrict the "failure" imminent condition to that of Fig. 3.6.1(b), wherein the ductile failure mode can be expected.

EXAMPLE 3.6.1 Determine whether or not the amount of steel used in the beam ($b = 14$; $d = 21.5$; $A_s = 5.08$; $f_c' = 5000$; $f_y = 50,000$) of Example 3.4.1 (Fig. 3.4.1) is acceptable according to the ACI Code.

Solution: (a) Determine A_{sb} in the balanced strain condition using basic principles, according to the ACI Code. From Fig. 3.6.2,

$$x_b = \frac{0.003(21.5)}{0.003 + 50/29,000} = 13.65 \text{ in.}$$

$$\beta_1 = 0.80 \text{ [for } f_c' = 5000 \text{ psi]}$$

$$a_b = 0.80(13.65) = 10.92 \text{ in.}$$

$$C_b = 0.85(5)(14)(10.92) = 650 \text{ kips}$$

$$A_{sb} = \frac{650}{50} = 13.0 \text{ sq in.}$$

Note that the balanced strain condition is a *reference* condition that is not permitted by the ACI Code to actually occur in a beam.

(b) Compare actual steel used with the maximum permitted by the ACI Code.

$$\max A_s = 0.75A_{sb} = 0.75(13.0) = 9.75 \text{ sq in.}$$

$$\text{actual } A_s = 5.08 \text{ sq in.} < 9.75 \text{ sq in.} \qquad \text{OK}$$

ACI-10.3.3 is satisfied. Ordinarily this check should be made *before* an investigation of strength is made in cases where the ACI Code is to be satisfied.

If the maximum steel permitted is expressed in terms of the reinforcement ratio ρ,

$$\max \rho = 0.75\rho_b = \frac{9.75}{14(21.5)} = 0.0324$$

Values of maximum ρ for beams having a rectangular compression zone and various material strengths are given in Table 3.6.1.

3.7 MINIMUM REINFORCEMENT RATIO ρ

When the steel reinforcement in a flexural member is only a small amount because the factored moment M_u is low, the beam may perform in service in its uncracked state. The method used for computing flexural strength, however, is based on the assumption that the tension concrete is cracked. Thus, it would be possible for the computed nominal strength M_n for a section having a small amount of reinforcement, based on cracked tension concrete, to be less than the strength M_n (called M_{cr}) of the same section as a plain concrete beam (i.e., having no steel reinforcement). Since a ductile failure mode is desired, the lowest amount of steel permitted should be the amount that would equal the strength of an unreinforced beam. The desired relationship may be developed as follows:

$$\left[\begin{array}{c}\text{strength of reinforced} \\ \text{concrete beam, } M_n\end{array}\right] \geq \left[\begin{array}{c}\text{strength of plain} \\ \text{concrete beam, } M_{cr}\end{array}\right] \qquad (3.7.1)$$

The strength of a plain concrete beam is referred to as the *cracking moment* M_{cr} and is achieved when the extreme fiber in tension reaches the modulus of rupture f_r (see Sec. 1.8). The ACI Code uses for normal weight concrete,

$$f_r = 7.5\sqrt{f_c'} \qquad (3.7.2)$$

as given by ACI-9.5.2.3. Considering plain (nonreinforced) concrete as an elastic homogeneous material, the flexure formula gives M_{cr} as

$$M_{cr} = f_r \frac{I_g}{y_t} \qquad (3.7.3)$$

where

I_g = moment of inertia of the gross concrete cross section

y_t = distance from the neutral axis to the extreme fiber in tension

Thus, for a rectangular section,

$$M_{cr} = 7.5\sqrt{f'_c}\,\frac{bh^3/12}{h/2} = \frac{7.5\sqrt{f'_c}\,bh^2}{6} \tag{3.7.4}$$

For a reinforced concrete rectangular beam, using Eq. (3.3.4),

$$M_n = A_s f_y(d - a/2) \tag{3.3.4}$$

Substituting Eqs. (3.7.4) and (3.3.4) into Eq. (3.7.1) gives

$$A_s f_y\left(d - \frac{a}{2}\right) \geq \frac{7.5\sqrt{f'_c}\,bh^2}{6} \tag{3.7.5}$$

Letting $A_s = \rho bd$ and estimating $a/2$ as $\approx 0.05d$ for small values of ρ, Eq. (3.7.5) gives

$$\rho bd f_y(0.95d) \geq 1.25\sqrt{f'_c}\,bh^2$$

or

$$\rho \geq \frac{1.25\sqrt{f'_c}}{0.95 f_y}\left(\frac{h}{d}\right)^2 \tag{3.7.6}$$

Then, if $d \approx 0.9h$,

$$\min \rho \geq \frac{1.62\sqrt{f'_c}}{f_y} \tag{3.7.7}$$

For various values of f'_c, the min ρ becomes

$$f'_c = 3000 \text{ psi}; \qquad \min \rho = \frac{89}{f_y}$$

$$f'_c = 4000 \text{ psi}; \qquad \min \rho = \frac{102}{f_y}$$

$$f'_c = 5000 \text{ psi}; \qquad \min \rho = \frac{115}{f_y}$$

The requirement of ACI-10.5.1 states

$$\min \rho \geq \frac{200}{f_y} \tag{3.7.8}$$

Equation (3.7.8) is conservative for rectangular sections. Even if h/d becomes 1.25 as it might for a thin slab, the coefficient would only go as high as 145. Because it is recognized that Eq. (3.7.8) is extremely conservative for slabs, ACI-10.5.3 directs the designer to ACI-7.12, because the reinforcement indicated for temperature and shrinkage is approximately the amount indicated by Eq. (3.7.7). Temperature and shrinkage behavior is not involved but the amounts in ACI-7.12 are the approximate minimums.

Since beams in monolithic concrete construction are commonly T-shaped (see Chap. 9) rather than rectangular, the same analysis used to obtain Eq. (3.7.7) should be made for, say, the T-section with the top (the flange or slab) of the T in tension, and also with the bottom (the stem of width b_w) of the T in tension. Such an analysis will show that with the *slab in compression*, Eq. (3.7.7) using $\rho = \rho_w = A_s/b_w d$ will give

$$\min \rho \geq \frac{(150 \text{ to } 250)^*}{f_y} \tag{3.7.9}$$

For the *slab in tension* on the T-section,

$$\min \rho \geq \frac{(300 \text{ to } 600)^\dagger}{f_y} \tag{3.7.10}$$

Thus, ACI-10.5.1 is a reasonable requirement for isolated rectangular beams and the usual positive moment orientation for T-sections. It is unconservative for T-sections with the slab in tension. The authors believe that minimum reinforcement should be used in both positive and negative moment regions, and particularly on cantilevers.

Note that ACI-10.5.2 permits a waiver of Eq. (3.7.8) whenever ϕM_n equals or exceeds $1.33 M_u$. The philosophy is that extra safety is a satisfactory substitute for having a ductile failure mode.

3.8 DESIGN OF RECTANGULAR SECTIONS IN BENDING WITH TENSION REINFORCEMENT ONLY

In the design of rectangular sections in bending with tension reinforcement only, the problem is to determine b, d, and A_s from the required value of $M_n = M_u/\phi$, and the given material properties f'_c and f_y.

The two conditions of equilibrium are

$$C = T \tag{3.8.1a}$$

and

$$M_n = (C \text{ or } T)\left(d - \frac{a}{2}\right) \tag{3.8.1b}$$

Since there are three unknowns but only two conditions, there are many possible solutions. If the reinforcement ratio ρ is preset, then from Eq. (3.8.1a)

$$0.85 f'_c ba = \rho bd f_y$$

$$a = \rho\left(\frac{f_y}{0.85 f'_c}\right)d \tag{3.8.2}$$

Substituting Eq. (3.8.2) into Eq. (3.8.1b),

$$M_n = \rho bd f_y\left[d - \frac{\rho}{2}\left(\frac{f_y}{0.85 f'_c}\right)d\right] \tag{3.8.3}$$

---

*195 for flange width equal to four times web width and flange thickness equal to 20% of total gross depth.
†362 for the proportions stated above.

A strength *coefficient of resistance* R_n may be obtained by dividing Eq. (3.8.3) by bd^2 and letting

$$m = \frac{f_y}{0.85 f_c'} \tag{3.8.4a}$$

Thus

$$R_n = \frac{M_n}{bd^2} = \rho f_y (1 - \tfrac{1}{2}\rho m) \tag{3.8.4b}$$

The relationship between ρ and R_n for various values of f_c' and f_y is shown in Fig. 3.8.1.

In some situations the values of b and d may be preset, which is equivalent to having R_n preset; then ρ may be determined by solving the quadratic equation (3.8.4). Thus

$$R_n = \rho f_y (1 - \tfrac{1}{2}\rho m)$$

from which

$$\rho = \frac{1}{m}\left(1 - \sqrt{1 - \frac{2 m R_n}{f_y}}\right) \tag{3.8.5}$$

The procedure (without being concerned with certain practical decisions) to be used in the strength design of rectangular sections with tension reinforcement only involves the following steps:

1. Assume a value of ρ equal to or less than $0.75\rho_b$, but greater than the minimum ρ of $200/f_y$ (ACI-10.5). The balanced value ρ_b may be obtained from basic principles or from Eq. (3.5.4),

$$\rho_b = \frac{0.85\beta_1 f_c'}{f_y}\left(\frac{87{,}000}{87{,}000 + f_y}\right)$$

and

for $f_c' \leq 4000$ psi $\beta_1 = 0.85$

for $f_c' > 4000$ psi $\beta_1 = 0.85 - 0.05\left(\dfrac{f_c' - 4000}{1000}\right) \geq 0.65$

2. Determine the required bd^2 from

$$\text{required } bd^2 = \frac{\text{required } M_n}{R_n}$$

in which

$$R_n = \rho f_y (1 - \tfrac{1}{2}\rho m)$$

with

$$m = \frac{f_y}{0.85 f_c'}$$

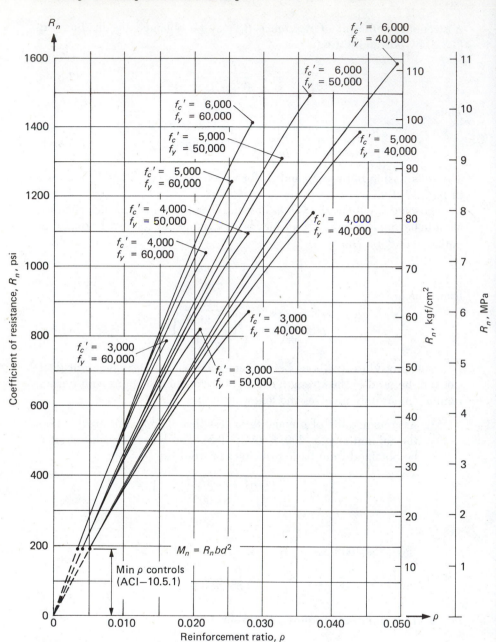

Figure 3.8.1 Strength curves (R_n vs ρ) for singly reinforced rectangular sections. Upper limit of curves is at $0.75\rho_b$.

3. Choose a suitable set of values of b and d so that the provided bd^2 is approximately equal to the required bd^2. (Note: Actually d is not chosen, rather the overall depth h is chosen and d is computed from h while maintaining the desired minimum protective cover.)

4. Determine the revised value of ρ after computing $R_n = M_n/bd^2$ for the selected section using one of the following methods.

a. By formula (most exact),

$$\rho = \frac{1}{m}\left(1 - \sqrt{1 - \frac{2mR_n}{f_y}}\right)$$

$$m = \frac{f_y}{0.85 f_c'}$$

$$R_n = \frac{M_n}{bd^2}$$

$$M_n = M_u / \phi$$

b. By curves (Fig. 3.8.1)

c. By approximate proportion (on the safe side if revised R_n is smaller than original R_n), the revised ρ is

$$\rho \approx (\text{original } \rho)\frac{(\text{revised } R_n)}{(\text{original } R_n)}$$

It may be noted from Fig. 3.8.1 that the relationship between R_n and ρ is approximately linear over short distances, even though the actual equation is a quadratic one.

5. Compute A_s from

$$A_s = (\text{revised } \rho)(\text{actual } bd)$$

6. Select reinforcement and check the strength of the section to be certain that

$$\phi M_n \geq M_u \qquad \text{or} \qquad M_n \geq \frac{M_u}{\phi}$$

EXAMPLE 3.8.1 Determine a set of values of b, d, and A_s that will carry a factored bending moment M_u of 400 ft-kips. Use $f_c' = 4000$ psi and $f_y = 40,000$ psi.

Solution: (a) Establish the limits within which the reinforcement ratio ρ can be chosen. From Eq. (3.5.4),

$$\rho_b = \frac{0.85\beta_1 f_c'}{f_y}\left(\frac{87,000}{87,000 + f_y}\right) = \frac{0.85(0.85)(4)}{40}\left(\frac{87}{87 + 40}\right) = 0.0495$$

$$\max \rho = 0.75\,\rho_b = 0.0371$$

From ACI-10.5.1,

$$\min \rho = \frac{200}{f_y} = \frac{200}{40,000} = 0.005$$

(b) Choose reinforcement ratio ρ, determine corresponding required beam size, and select actual beam size. Arbitrarily assume that $\rho = 0.03$ (any choice within prescribed limits is acceptable here),

$$m = \frac{f_y}{0.85 f_c'} = \frac{40}{0.85(4)} = 11.76$$

$$R_n = \rho f_y(1 - \tfrac{1}{2}\rho m) = 0.03(40,000)[1 - \tfrac{1}{2}(0.03)(11.76)] = 988 \text{ psi}$$

$$\text{required } M_n = \frac{M_u}{\phi} = \frac{400}{0.90} = 444 \text{ ft-kips}$$

$$\text{required } bd^2 = \frac{\text{required } M_n}{R_n} = \frac{444(12,000)}{988} = 5400 \text{ in.}^3$$

Try $b = 14$ in. and $d = \sqrt{5400/14} = 19.64$ in. Use $b = 14$ in. and compute d to be 21.5 in. after selecting overall depth (d is usually $2\frac{3}{8}$ to $2\frac{5}{8}$ in. less than overall depth if one layer of steel can be used; this is illustrated in Example 3.9.2).

(c) Compute required A_s and select actual steel bars, using actual beam size. The effective depth d is now greater than required for $\rho = 0.03$ in order to get a convenient depth, or perhaps to reduce deflection.

$$\text{required } R_n = \frac{\text{required } M_n}{\text{provided } bd^2} = \frac{444(12{,}000)}{14(21.5)^2} = 824 \text{ psi}$$

$$\rho = \frac{1}{m}\left(1 - \sqrt{1 - \frac{2mR_n}{f_y}}\right)$$

$$= \frac{1}{11.76}\left[1 - \sqrt{1 - \frac{2(11.76)(824)}{40{,}000}}\right] = 0.0240$$

$$A_s = \rho bd = 0.0240(14)(21.5) = 7.22 \text{ sq in.}$$

Two layers of reinforcement will be required. Select 3-#11 and 2-#10, $A_s = 7.22$ sq in.

(d) Check design. A review of the correctness of the above computations in which formulas are used should be made using the basic statics shown in Fig. 3.8.2.

$$T = A_s f_y = 7.22(40) = 289 \text{ kips}$$

$$a = \frac{C \text{ or } T}{0.85 f'_c b} = \frac{289}{0.85(4)(14)} = 6.07 \text{ in.}$$

$$M_n = (C \text{ or } T)\left(d - \frac{a}{2}\right) = 289[21.5 - 0.5(6.07)]\tfrac{1}{12}$$

$$= 444 \text{ ft-kips } (602 \text{ kN·m}) \qquad \text{OK}$$

(e) Final decision and design sketch. Every design requires a clear decision as its conclusion, along with an appropriate design sketch.

<u>Use</u> 3-#11 and 2-#10 in the 14 by 25 in. beam shown in Fig. 3.8.2. (The

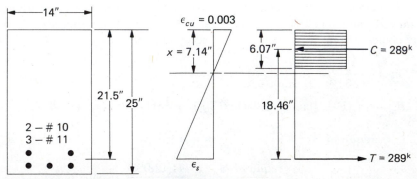

Figure 3.8.2 Section for Example 3.8.1.

d was computed from the overall 25-in. depth as discussed in succeeding examples.)

It should be noted that the beam of Example 3.8.1 with 2.4% reinforcement, which requires placement in two layers, is significantly smaller in concrete cross-sectional area than traditional designs under the working stress method would have made it. Deflection thus becomes an essential consideration in designing beams by the strength method. The subject of deflection is treated in Chap. 14.

3.9 PRACTICAL SELECTION FOR BEAM SIZES, BAR SIZES, AND BAR PLACEMENT

In the previous section the procedure and example for the design of rectangular sections in bending with tension reinforcement only have been treated on the assumption that the factored moment M_u is known. This is rarely the case, however, because the factored moment must include the effect of the weight of the beam itself which has not yet been designed. In reality, then, the dead weight of the beam has to be assumed at the outset; a trial beam size is then obtained and may be readjusted if its effect on the factored moment is significantly different from the assumed value.

The choice of the steel reinforcement ratio ρ is very much dependent on the limitation on the deflection of the beam. Years of experience with the working stress method showed that deflection problems were rarely encountered with beams having a steel reinforcement ratio ρ not more than one-half the maximum permissible value. The use of this amount (one-half of $0.75\rho_b = 0.375\rho_b$) may provide a suitable *guide* for the preliminary choice of the reinforcement ratio.

For the selection of an actual number of bars to meet a total steel area requirement, it is desirable to tabulate the combined area of several bars at a time. Table 3.9.1 gives bar areas for up to 10 bars of the different sizes. Note that once the area of one bar, say of #8, is set at 0.79 sq in., the area of 10 bars becomes 7.90, not 7.85; this practice has been carried on by tradition.

Table 3.9.1 TOTAL AREAS FOR VARIOUS NUMBERS OF REINFORCING BARS

BAR SIZE	NOMINAL DIAMETER (in.)	WEIGHT (lb/ft)	NUMBER OF BARS									
			1	2	3	4	5	6	7	8	9	10
#3	0.375	0.376	0.11	0.22	0.33	0.44	0.55	0.66	0.77	0.88	0.99	1.10
#4	0.500	0.668	0.20	0.40	0.60	0.80	1.00	1.20	1.40	1.60	1.80	2.00
#5	0.625	1.043	0.31	0.62	0.93	1.24	1.55	1.86	2.17	2.48	2.79	3.10
#6	0.750	1.502	0.44	0.88	1.32	1.76	2.20	2.64	3.08	3.52	3.96	4.40
#7	0.875	2.044	0.60	1.20	1.80	2.40	3.00	3.60	4.20	4.80	5.40	6.00
#8	1.000	2.670	0.79	1.58	2.37	3.16	3.95	4.74	5.53	6.32	7.11	7.90
#9	1.128	3.400	1.00	2.00	3.00	4.00	5.00	6.00	7.00	8.00	9.00	10.00
#10	1.270	4.303	1.27	2.54	3.81	5.08	6.35	7.62	8.89	10.16	11.43	12.70
#11	1.410	5.313	1.56	3.12	4.68	6.24	7.80	9.36	10.92	12.48	14.04	15.60
#14[a]	1.693	7.65	2.25	4.50	6.75	9.00	11.25	13.50	15.75	18.00	20.25	22.50
#18[a]	2.257	13.60	4.00	8.00	12.00	16.00	20.00	24.00	28.00	32.00	36.00	40.00

[a] #14 and #18 bars are used primarily as column reinforcement and are rarely used in beams.

Table 3.9.2 MINIMUM BEAM WIDTH (INCHES) ACCORDING
TO THE ACI CODE[a]

SIZE OF BARS	NUMBER OF BARS IN SINGLE LAYER OF REINFORCEMENT							ADD FOR EACH ADDED BAR
	2	3	4	5	6	7	8	
#4	6.8	8.3	9.8	11.3	12.8	14.3	15.8	1.50
#5	6.9	8.5	10.2	11.8	13.4	15.0	16.7	1.63
#6	7.0	8.8	10.5	12.3	14.0	15.8	17.5	1.75
#7	7.2	9.0	10.9	12.8	14.7	16.5	18.4	1.88
#8	7.3	9.3	11.3	13.3	15.3	17.3	19.3	2.00
#9	7.6	9.8	12.2	14.3	16.6	18.8	21.1	2.26
#10	7.8	10.4	12.9	15.5	18.0	20.5	23.1	2.54
#11	8.1	10.9	13.8	16.6	19.4	22.2	25.0	2.82
#14	8.9	12.3	15.7	19.1	22.5	25.9	29.3	3.40
#18	10.6	15.1	19.6	24.1	28.6	33.1	37.6	4.51

Table shows minimum beam widths when #3 stirrups are used.
For additional bars, add dimension in last column for each added bar.
For bars of different size, determine from table the beam width for smaller size bars and then add last column figure for each larger bar used.
[a]Assumes maximum aggregate size does not exceed three-fourths of the clear space between bars (ACI-3.3.3). Table computation procedure is in agreement with the ACI Code interpretation of ACI Committee 340, as used in the *Strength Design Handbook* [Ref. 12, Chap. 2].

A = $1\frac{1}{2}$-in. clear cover to stirrup
B = $\frac{3}{8}$-in. stirrup bar diameter
C = For #11 and smaller bars, use twice the diameter of #3 stirrups (i.e., C = 0.75 in.).
     For #14 and #18 bars, use $C = 0.5d_b$.
D = clear distance between bars = d_b or 1 in., whichever is greater (where d_b is the diameter of the larger adjacent longitudinal bar)

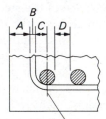

Diameter of corner bar is assumed to be located to intersect the horizontal tangent to stirrup bend

For the placement of bars within the beam width, ACI-7.6.1 specifies the clearance needed between bars to permit proper concrete placement around them. This clearance for bars in a layer is 1 in. or the nominal diameter of the bar, whichever is greater. When two or more layers of bars are required, the minimum clearance between layers is 1 in. (ACI-7.6.2). Table 3.9.2 gives minimum beam widths for various numbers of equal-sized bars, computed in the manner described above.

Bars are supported from the bottom of the forms, and layers of bars are separated by various types of bar supports, known as *bolsters* and *chairs*, some of which are shown in Table 3.9.3. Bar supports may be made of concrete, metal, or other approved materials. Usually they are standard factory-made wire bar supports. They remain in place after the concrete is cast and must have special rust protection on the portions nearest the face of the concrete.

Table 3.9.3 STANDARD TYPES AND SIZES OF BAR SUPPORTS
(ADAPTED FROM REF. 17 OF CHAP. 2)

SYMBOL	BAR SUPPORT ILLUSTRATION	TYPE OF SUPPORT	STANDARD SIZES
SB		Slab Bolster	$\frac{3}{4}$, 1, $1\frac{1}{2}$, and 2 in. heights in 5-ft and 10-ft lengths
SBU		Slab Bolster Upper	Same as SB
BB		Beam Bolster	1, $1\frac{1}{2}$, 2, over 2 to 5 in. heights in increments of $\frac{1}{4}$ in. in lengths of 5 ft
BBU		Beam Bolster Upper	Same as BB
BC		Individual Bar Chair	$\frac{3}{4}$, 1, $1\frac{1}{2}$, and $1\frac{3}{4}$ in. heights
HC		Individual High Chair	2 to 15 in. heights in increments of $\frac{1}{4}$ in.
CHC		Continuous High Chair	Same as HC in 5-ft and 10-ft lengths
CHCU		Continuous High Chair Upper	Same as CHC

In order to assist the designer further in making choices for beam sizes, bar sizes, and bar placement, the following guidelines are suggested. These may be regarded as accepted practice, and are *not* ACI Code requirements. Undoubtedly situations will arise in which the experienced designer will, for good and proper reasons, make a selection not conforming to the guidelines.

For Beam Sizes
1. Use whole inches for overall dimensions; except slabs may be in $\frac{1}{2}$-in. increments.
2. Beam stem widths are most often in multiples of 2 or 3 in.; such as 9, 10, 12, 14, 15, 16, and 18.

3. Minimum specified clear cover is measured from outside the stirrup or tie to the face of the concrete. (Thus beam effective depth d has rarely, if ever, a dimension to the whole inch.)
4. An economical rectangular beam proportion is one in which the overall depth-to-width ratio is between about 1.5 to 2.0.
5. For T-shaped beams, typically the flange thickness represents about 20% of overall depth (see Chap. 9 for treatment of T-shaped sections).

For Reinforcing Bars
6. Maintain bar symmetry about the centroidal axis which lies at right angles to the bending axis (i.e., symmetry about the vertical axis in usual situations).
7. Use at least two bars wherever flexural reinforcement is required.
8. Use bars #11 and smaller for usual sized beams.
9. Use no more than two bar sizes and no more than two standard sizes apart for steel in one face at a given location in the span (i.e., #7 and #9 bars may be acceptable, but #9 and #4 bars would not).
10. Place bars in one layer if practicable. Try to select bar size so that no less than two and no more than five or six bars are put in one layer.
11. Follow requirements of ACI-7.6.1 and 7.6.2 for clear distance between bars and between layers.
12. When different sizes of bars are used in several layers at a location, place the largest bars in the layer nearest the face of beam.

EXAMPLE 3.9.1 Select an economical rectangular beam size and select bars using the ACI strength method. The beam is a simply supported span of 40 ft and it is to carry a live load of 1.3 kips/ft and a dead load of 0.8 kip/ft (not including beam weight). Without actually checking deflection, use a reinforcement ratio ρ such that excessive deflection is unlikely. Use $f'_c = 4000$ psi, and $f_y = 60,000$ psi.

Solution: (a) Decide on a reinforcement ratio ρ to use. To have reasonable expectation that deflection will not be excessive, choose ρ at about half the maximum permitted ($0.75 \rho_b = 0.0214$). Use $\rho = 0.011$.

(b) Determine the desired R_n (corresponding to the desired ρ) using Eq. (3.8.4).

$$m = \frac{f_y}{0.85f'_c} = \frac{60}{0.85(4)} = 17.65$$

$$R_n = \rho f_y(1 - \tfrac{1}{2}\rho m)$$
$$= 0.011(60,000)[1 - 0.5(0.011)(17.65)] = 596 \text{ psi}$$

(c) Determine factored moments.

$$M_u = 1.4M_D + 1.7M_L$$

$$M_L = \frac{1.3(40)^2}{8} = 260 \text{ ft-kips}$$

$$M_D = \frac{(0.8 + 0.4)(40)^2}{8} = 240 \text{ ft-kips}$$

using a beam weight estimated at 0.4 kip/ft, based on unit weight of rein-forced concrete at 150 pcf.

$$M_u = 1.4(240) + 1.7(260) = 336 + 442 = 778 \text{ ft-kips}$$

$$\text{required } M_n = \frac{M_u}{\phi} = \frac{778}{0.90} = 864 \text{ ft-kips}$$

(d) Determine required bd^2 from desired R_n.

$$\text{required } bd^2 = \frac{M_n}{R_n} = \frac{864(12,000)}{596} = 17,400 \text{ in.}^3$$

(e) Establish beam size. Select width b and determine the correspond-ing required value for effective depth d. Make a table of possibilities.

CHOSEN b	REQUIRED d
12	38.1
15	34.1
18	31.1←Try
20	29.5

Selecting the 18-in. width will give a beam whose overall depth is between $1\frac{1}{2}$ and 2 times its width (suggested guideline).

Determine overall depth in order to verify the assumed weight. As-suming that the bars to be selected will fit in one layer, the minimum overall depth may be computed (see Fig. 3.9.1)

$h = d + 1\frac{1}{2}$ in. cover $+ \frac{3}{8}$ diameter stirrup + bar radius, say $\frac{1}{2}$ in.

$= d + (2\frac{3}{8}$ to $2\frac{1}{2}$ in.)

$= 31.1 + 2.5 = 33.6$ in.

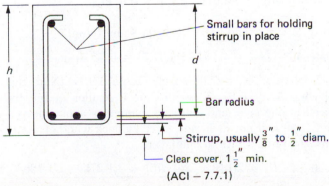

Small bars for holding stirrup in place

d

Bar radius

Stirrup, usually $\frac{3}{8}''$ to $\frac{1}{2}''$ diam.

Clear cover, $1\frac{1}{2}''$ min.

(ACI − 7.7.1)

Figure 3.9.1 Quantities added to effective depth d to get overall depth h for beams with one layer of tension steel.

The overall depth would be in whole inches; so try 34 in. Since the guideline value of R_n = 596 psi for ρ = 0.011 is not a rigorous requirement, the overall depth selected could be somewhat less or somewhat more than the computed requirement in order to obtain a desired dimension. The stirrup is reinforcement to provide shear strength for the beam and should always be allowed for at this stage of the design. The actual size and spacing of stirrups is determined after the cross section and tension bars have been selected, and the subject is treated in Chap. 5. _Try h = 34 in._

(f) Check weight, revise M_u, and select reinforcement.

$$w = \frac{18(34)}{144}(0.15) = 0.638 \text{ kip/ft}$$

$$\text{revised } M_D = \frac{(0.8 + 0.638)}{8}(40)^2 = 288 \text{ ft-kips}$$

$$\text{revised } M_u = 1.4(288) + 442 = 404 + 442 = 846 \text{ ft-kips}$$

$$\text{revised required } M_n = \frac{846}{0.90} = 940 \text{ ft-kips}$$

Compute the value of d from the overall dimension h,

$$\text{actual } d = h - (\approx 2.5 \text{ in.}) \qquad \text{for one layer of bars}$$

$$= 34 - 2.5 = 31.5 \text{ in.}$$

When the overall depth is increased or decreased, the clear cover distance (ACI-7.7.1) is the one that is held constant; thus the effective depth will shift.

$$\text{required } R_n = \frac{\text{required } M_n}{bd^2} = \frac{940(12,000)}{18(31.5)^2} = 632 \text{ psi}$$

The steel requirement may be determined from Eq. (3.8.5), from the curves of Fig. 3.8.1, or approximated by straightline proportion. Using the latter and knowing that

$$R_n = 596 \text{ psi} \qquad \text{for} \quad \rho = 0.011$$

find approximate ρ for R_n = 632 psi,

$$\text{approximate } \rho = 0.011 \left(\frac{632}{596}\right) = 0.0117$$

[using Eq. (3.8.5) would have given the same value]

$$\text{approximate } A_s = \rho bd = 0.0117(18)(31.5) = 6.63 \text{ sq in.}$$

Select 4-#10 and 2-#9 bars, A_s = 7.08 sq in. (Table 3.9.1). (Note that 3-#10 and 3-#9 having A_s = 6.81 sq in. provide a smaller A_s but the bars cannot be placed in one layer and still have symmetry about the vertical beam axis, and 7-#9 will not fit in one layer.) Check whether 4-#10 and 2-#9 will fit into an 18-in. width in one layer.

$$\frac{\text{approx clear spacing}}{\text{between bars}} = \frac{18 - 2(1.5) - 2(0.375) - 4(1.27) - 2(1.128)}{5}$$

$$= 1.38 \text{ in.} > [1.27 \text{ in.} = \text{diameter of #10}] \qquad \text{OK}$$

Subtracted from the overall width are the combined values of the minimum clear cover on both sides (3.0), one stirrup diameter on both sides (0.75), 4-#10 bar diameters (5.08), and 2-#9 bar diameters (2.26). The result is divided by the number of spaces between bars, and this is the *approximate* clearance that must exceed the diameter of the larger bar (ACI-7.6.1). Table 3.9.2 gives the minimum beam width for 6-#10 bars as 18.0 in.

Note that the above clearance computation is approximate because it assumes the #3 stirrup may be bent tightly around the corner longitudinal bar. ACI-7.2.2 requires the inside diameter of bends for stirrups to be not less than four stirrup bar diameters for #5 stirrups and smaller; thus for #3 stirrups the actual curve of the stirrup at the corner has a radius of $\frac{3}{4}$ in., which is larger than the longitudinal bar radius for #11 bars and smaller. Table 3.9.2 is based on the conservative assumption that the diameter of the corner longitudinal bar is located to intersect the horizontal tangent to the stirrup bend (see sketch under Table 3.9.2). Using Table 3.9.2, the minimum required width for 4-#10 and 2-#9 is

$$\text{min } b = 2(7.6) + 4(2.54) = 17.76 \text{ in.}$$

The 7.6 is for 2-#9 and the 2.54 is for each of the additional #10 bars.

(g) Check strength and provide design sketch.

$$C = 0.85f'_c ba = 0.85(4)18a = 61.2a$$

$$T = A_s f_y = 7.08(60) = 425 \text{ kips}$$

$$a = \frac{425}{61.2} = 6.95 \text{ in.}$$

$$M_n = A_s f_y \left(d - \frac{a}{2} \right) = 425(31.5 - 3.47)\tfrac{1}{12} = 994 \text{ ft-kips}$$

$$[\phi M_n = 0.90(994) = 895 \text{ ft-kips}] > [M_u = 846 \text{ ft-kips}] \qquad \text{OK}$$

Use 18 × 34 beam with 4-#10 and 2-#9 bars, as shown in Fig. 3.9.2.

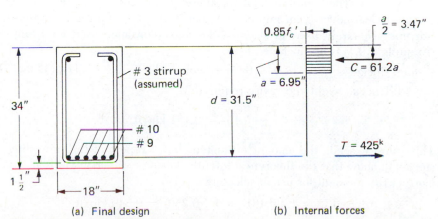

(a) Final design   (b) Internal forces

Figure 3.9.2 Design for Example 3.9.1.

EXAMPLE 3.9.2 Design a floor slab to carry a uniform live load of 350 psf (pounds per square foot) on a simply supported span of 23 ft, given $f'_c = 3000$ psi and $f_y = 60,000$ psi. Use the ACI Code. Keep the reinforcement ratio at about 30% of the amount in the balanced strain condition so that deflection is unlikely to be excessive.

Solution: The design of a slab is usually made by taking a 1-ft wide typical strip for calculation purposes rather than the entire slab width. This is known as a one-way slab that acts as a wide beam. The common situation of the one-way slab continuous across several beams is treated in detail in Chap. 8.

(a) Determine factored moment.

$$U = 1.4D + 1.7L = 1.4(0.15 \text{ kip/ft estimated}) + 1.7(0.350)$$
$$= 0.210 + 0.595 = 0.805 \text{ kip/ft/ft of width}$$

$$M_u = \frac{0.805(23)^2}{8} = 53.2 \text{ ft-kips/ft of width}$$

$$\text{required } M_n = \frac{M_u}{\phi} = \frac{53.2}{0.90} = 59.2 \text{ ft-kips/ft}$$

(b) Since $\rho = 0.3\rho_b$ is desired, using $\rho = 0.0160(0.3/0.75) = 0.0064$ based on Table 3.6.1, determine the corresponding desired R_n,

$$R_n = \rho f_y(1 - \tfrac{1}{2}\rho m)$$

where

$$m = \frac{f_y}{0.85f'_c} = \frac{60}{0.85(3)} = 23.5$$

$$R_n = 0.0064(60,000)[1 - 0.5(0.0064)(23.5)] = 355 \text{ psi}$$

(c) Determine required bd^2 from desired R_n and select trial slab thickness.

$$\text{required } bd^2 = \frac{M_n}{R_n} = \frac{59.2(12,000)}{355} = 2000 \text{ in.}^3$$

Since a slab is designed by using a 1-ft strip, b is 12 in. Then required $d = 12.9$ in. The required total thickness is obtained by adding on the required clear cover ($\frac{3}{4}$-in. minimum as per ACI-7.7.1) and the bar radius. Stirrups are rarely used in slabs so the $\frac{3}{8}$-in. allowance used for beams (Example 3.9.1) is not included here.

total thickness, $h = 12.9 + 0.75 + 0.50 = 14.15$ in.　　　Try $14\frac{1}{2}$ in.

(d) Check weight and revise required M_n.

$$w = \frac{14.5}{12}(0.15) = 0.181 \text{ kip/ft}$$

This value exceeds the amount estimated, but a repeat of the preceding steps will show that the theoretical total thickness is still about 14.5 in. Correct the moment for use in selecting steel.

$$\text{revised } U = 1.4(0.181) + 1.7(0.350) = 0.849 \text{ kip/ft}$$

$$\text{revised required } M_n = 59.2 \left(\frac{0.849}{0.805}\right) = 62.3 \text{ ft-kips/ft}$$

Table 3.9.4 AVERAGE AREA PER FOOT OF WIDTH PROVIDED BY VARIOUS BAR SPACINGS

BAR SIZE NUMBER	NOMINAL DIAMETER (in.)	SPACING OF BARS IN INCHES													
		2	$2\frac{1}{2}$	3	$3\frac{1}{2}$	4	$4\frac{1}{2}$	5	$5\frac{1}{2}$	6	7	8	9	10	12
3	0.375	0.66	0.53	0.44	0.38	0.33	0.29	0.26	0.24	0.22	0.19	0.17	0.15	0.13	0.11
4	0.500	1.18	0.94	0.78	0.67	0.59	0.52	0.47	0.43	0.39	0.34	0.29	0.26	0.24	0.20
5	0.625	1.84	1.47	1.23	1.05	0.92	0.82	0.74	0.67	0.61	0.53	0.46	0.41	0.37	0.31
6	0.750	2.65	2.12	1.77	1.51	1.32	1.18	1.06	0.96	0.88	0.76	0.66	0.59	0.53	0.44
7	0.875	3.61	2.88	2.40	2.06	1.80	1.60	1.44	1.31	1.20	1.03	0.90	0.80	0.72	0.60
8	1.000		3.77	3.14	2.69	2.36	2.09	1.88	1.71	1.57	1.35	1.18	1.05	0.94	0.78
9	1.128		4.80	4.00	3.43	3.00	2.67	2.40	2.18	2.00	1.71	1.50	1.33	1.20	1.00
10	1.270		5.06	4.34	3.80	3.37	3.04	2.76	2.53	2.17	1.89	1.69	1.52	1.27	
11	1.410		6.25	5.36	4.69	4.17	3.75	3.41	3.12	2.68	2.34	2.08	1.87	1.56	

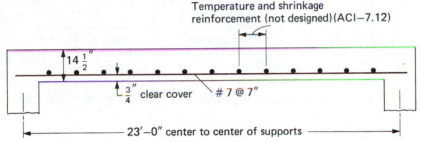

Temperature and shrinkage reinforcement (not designed)(ACI−7.12)

$14\frac{1}{2}''$

$\frac{3}{4}''$ clear cover   #7 @ 7"

23'−0" center to center of supports

Figure 3.9.3 Slab design of Example 3.9.2.

(e) Determine the steel to be used.

$$\text{actual } d = h - 0.75 - 0.5(\text{est}) = 14.5 - 1.25 = 13.25 \text{ in.}$$

$$\text{required } R_n = \frac{\text{required } M_n}{bd^2} = \frac{62.3(12,000)}{12(13.25)^2} = 354 \text{ psi}$$

The actual reinforcement ratio ρ may be determined from Eq. (3.8.5), from the curves of Fig. 3.8.1, or approximated by straightline proportion (see Example 3.9.1). Since the required R_u is nearly exactly that corresponding to $\rho = 0.3\rho_b = 0.0064$, the required ρ value will not change from the 0.0064 value,

$$\text{required } A_s = 0.0064(12)(13.25) = 1.02 \text{ sq in./ft}$$

Try #7 @ 7 in. spacing, $A_s = 1.03$ sq in./ft.

For slabs, bars are not selected by picking a total number; instead they are selected on the basis of an average area provided per foot of width. One #7 gives 0.60 sq in., or 0.60 sq in./ft if a bar is spaced every 12 in. For #7 bars spaced at 7 in., the average area provided is 0.60 (12/7) = 1.03 sq in./ft. Table 3.9.4 gives average areas per foot of width provided by various bar spacings.

(f) Check strength and provide design sketch. Make a check of strength by using statics (not illustrated here; see Example 3.9.1); select steel transverse to the main steel for temperature, shrinkage, and distribution of loading (not illustrated here, but treated in Chap. 8); and draw design sketch (see Fig. 3.9.3). Use $14\frac{1}{2}$-in. thick slab, with #7 @ 7 as main reinforcement.

EXAMPLE 3.9.3 Compute the nominal strength M_n of the beam of Fig. 3.9.4, 500 mm wide and 760 mm deep overall, containing 3-#35M bars in the outer layer and 2-#30M bars in the inner layer. The stirrup is #10M, and minimum clear cover is 40 mm with 25 mm between layers. Use $f'_c = 25$ MPa, $f_y = 400$ MPa, and $E_s = 200,000$ MPa.

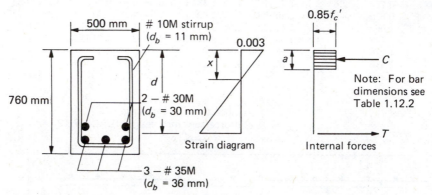

Figure 3.9.4 Computation of beam strength with metric dimensions, Example 3.9.3.

Solution: (a) Determine that the steel is within the permissible limits of the ACI Code. Examine the balanced strain condition (see Sec. 3.5).

$$\epsilon_y = \frac{f_y}{E_s} = \frac{400}{200,000} = 0.0020$$

$$x_b = \frac{0.003}{0.003 + 0.002} d = 0.600d$$

$$d = 760 - 40 - 11 - \frac{3(10.0)18 + 2(7.0)76}{3(10.0) + 2(7.0)} = 673 \text{ mm}$$

$$f'_c = 25 \text{ MPa} < f'_c = 30 \text{ MPa}; \qquad \beta_1 = 0.85 \qquad \text{(see Table 3.6.1)}$$

$$a_b = \beta_1 x_b = 0.85(0.600)673 = 343 \text{ mm}$$

$$C_b = 0.85 f'_c b a_b = 0.85(0.025)(500)(343) = 3640 \text{ kN}$$

$$T_b = A_{sb} f_y$$

$$C_b = T_b$$

$$A_{sb} = \frac{3640(1000)}{400} = 9100 \text{ mm}^2 \ (91.0 \text{ cm}^2)$$

$$\text{max } A_s = 0.75(91.0) = 68.3 \text{ cm}^2$$

Since the actual $A_s = 44.0$ cm^2 is less than the maximum permitted, the beam has an acceptable reinforcement ratio (deflection might be a problem, so it should be checked).

(b) Determine the nominal flexural strength M_n.

$$C = 0.85 f'_c b a = 0.85(0.025)(500)a = 10.6a$$

$$T = A_s f_y = 4400(0.400) = 1760 \text{ kN}$$

$$a = \frac{1760}{10.6} = 166 \text{ mm}$$

$$M_n = T(d - 0.5a) = 1760[673 - 0.5(166)]\tfrac{1}{1000} = 1038 \text{ kN·m}$$

3.10 INVESTIGATION OF RECTANGULAR SECTIONS IN BENDING WITH BOTH TENSION AND COMPRESSION REINFORCEMENT

Rectangular sections with both tension and compression reinforcement are also called "doubly reinforced" sections. Because the compressive strength of concrete is high, the need for compression reinforcement *to obtain adequate strength* is not great. In beams where compression reinforcement might be used in order to reduce the size of the cross section, deflection may be excessive, and there may be difficulty in placing all of the tension reinforcement within the width of the beam, even if two or more layers of bars are used. In addition, shear stress will become high so that a large amount of shear reinforcement might be required.

The use of compression reinforcement for deflection control (to reduce the creep and shrinkage deflection) is the usual reason for its use when the strength method is used.

The investigation of a doubly reinforced section such as that shown in Fig. 3.10.1 involves the determination of the nominal flexural strength M_n with b, d, d', A_s, A'_s, f'_c, and f_y as the given data. The investigation is similar to that for the singly reinforced beam except the compressive force C is composed of two parts, one in the concrete and the other in the steel. The actual stress in the compression steel when the nominal strength M_n is reached may be the yield stress, or it may be something less, depending on the position of the neutral axis. The stress used in the compression steel must be compatible with the strain in the compression steel when nominal strength is reached.

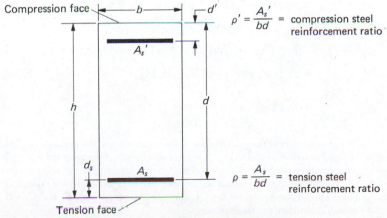

$$\rho' = \frac{A'_s}{bd} = \text{compression steel reinforcement ratio}$$

$$\rho = \frac{A_s}{bd} = \text{tension steel reinforcement ratio}$$

Figure 3.10.1 Doubly reinforced beam cross section.

In the case of the beam with compression steel, the ACI Code states that ρ shall not exceed $0.75\rho_b$ (just as for the singly reinforced beam), but in addition states (ACI-10.3.3) that "the portion of ρ_b equalized by compression reinforcement need not be reduced by the 0.75 factor." The use of the above quoted statement was intended to make for easier computation.

Philosophically the ductility requirement, in terms of how large the strain in the tension steel should be when the extreme concrete compression fiber reaches a strain of 0.003, should be the same for a beam having compression steel as for one without it. Controlling the strain diagram such that the neutral axis distance x cannot exceed $0.75x_b$ is the easiest to understand and the simplest in computation as the way to determine the maximum reinforcement ratio ρ permitted in a beam having compression steel.

The literal application of ACI-10.3.3 containing the above quoted phrase gives the same result (for rectangular sections) as limiting the actual x to a maximum of $0.75x_b$, *when the compression steel yields;* but is more difficult to apply and actually requires less ductility in a beam *when the compression steel does not yield.* Since rarely if ever does a doubly reinforced beam have its maximum limit of tension reinforcement, and since the compression steel most often yields, the subtle distinction between the results of limiting the actual x to a maximum of $0.75x_b$ and a literal use of the quoted clause in ACI-10.3.3 is of little consequence.

EXAMPLE 3.10.1 Determine the nominal moment strength M_n of the rectangular section shown in Fig. 3.10.2(a), given $f'_c = 5000$ psi, $f_y = 60,000$ psi, $b = 14$ in., $d = 26$ in., $d' = 3$ in., $A'_s = 2$-#8 bars and $A_s = 8$-#10 bars.

Solution: (a) Determine A_{sy} in the compression steel yield condition. Refer to Fig. 3.10.2(c) where the strain diagram is defined by points A and B. Point B is obtained by letting the strain in the compression steel exactly equal ϵ_y.

$$x_y = \frac{0.003(3)}{0.003 - 0.00207} = 9.67 \text{ in.}$$

$$a_y = \beta_1 x_y = 0.80(9.67) = 7.74 \text{ in.}$$

$$C_{cy} = 0.85 f'_c b a_y = 4.25(14)(7.74) = 460 \text{ kips}$$

$$C_{sy} = (f_y - 0.85f'_c)A'_s = (60 - 4.25)(1.58) = 88 \text{ kips*}$$

$$T_y = C_{cy} + C_{sy} = 460 + 88 = 548 \text{ kips}$$

$$A_{sy} = \frac{548}{60} = 9.13 \text{ sq in.}$$

---

*It is noted that the components of the compressive force nominally are the portions carried by the steel and by the concrete, respectively. For the convenience of knowing the point of action, however, the compressive force in the concrete is taken larger than the real amount by stressing $0.85f'_c$ over the steel area, whereas the compressive force in the steel is taken less by the same amount. The total compressive force is correct. The components C_c and C_s will hereinafter be computed as shown in this example.

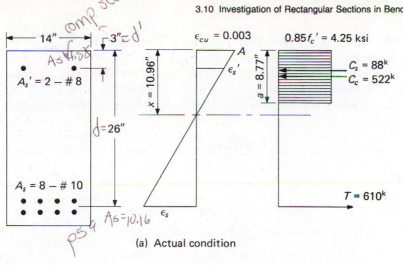

(a) Actual condition

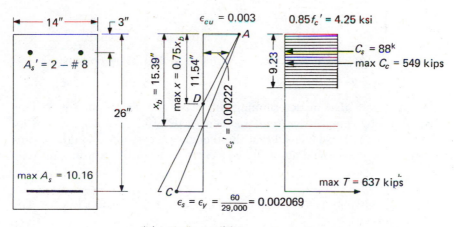

(b) max A_s condition

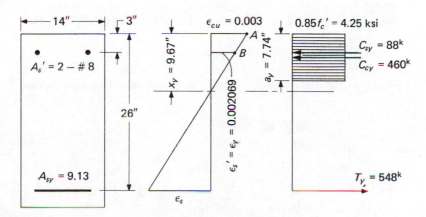

(c) Compression steel yield condition

Figure 3.10.2 Section for Example 3.10.1.

Since the actual A_s is greater than A_{sy}, actual x is larger than x_y, $\epsilon'_s > \epsilon_y$, and compression steel yields. An alternative would be to assume compression steel yields and check it afterwards, as shown in part (c).

(b) Determine the maximum permissible A_s. From Fig. 3.10.2(b), using the balanced strain condition defined by points A and C on the strain diagram,

$$x_b = \frac{0.003(26)}{0.003 + 0.00207} = 15.39 \text{ in.}$$

$$\max x = 0.75x_b = 0.75(15.39) = 11.54 \text{ in.}$$

which defines the strain line through points A and D in Fig. 3.10.2(b).

$$\max a = \beta_1 x = 0.80(11.54) = 9.23 \text{ in.}$$

$$\max C_c = 0.85f'_c ba = 4.25(14)(9.23) = 549 \text{ kips}$$

$$\epsilon'_s = \frac{11.54 - 3.00}{11.54}(0.003)$$

$$= 0.00222 > \epsilon_y \qquad \text{(compression steel yields)}$$

$$C_s = (f_y - 0.85f'_c)A'_s = (60 - 4.25)1.58 = 88 \text{ kips}$$

$$\max T = 549 + 88 = 637 \text{ kips}$$

One may note that in determining the maximum T the portion 88 kips "equalized by compression reinforcement" (ACI-10.3.3) is *not* reduced by the 0.75 factor. It is believed that the thought process illustrated here provides better understanding of the concept than does the wording in ACI-10.3.3.

$$\max \text{ permissible } A_s = \frac{637}{60} = 10.62 \text{ sq in.} > (\text{actual } A_s = 10.16 \text{ sq in.}) \qquad \text{OK}$$

One may note that the tension steel in this case is 78% of the full balanced amount (i.e., if A_{sb} were computed using $x = x_b$), greater than the 75% acceptable for a beam without compression steel. This is reasonable since compression steel provides ductility; so that the greater the proportion of the compression force is carried by compression steel, the closer the maximum tension steel can approach the balanced amount.

(c) Determine M_n in the actual condition. From Fig. 3.10.2(a), using a strain condition defined by point A and the requirement of equilibrium ($C = T$),

$$T = 60(10.16) = 610 \text{ kips}$$

$$C_s = (60 - 4.25)(1.58) = 88 \text{ kips}$$

$$C_c = 0.85f'_c ba = 4.25(14)a$$

$$a = \frac{C_c}{4.25(14)} = \frac{610 - 88}{4.25(14)} = 8.77 \text{ in.}$$

$$x = \frac{8.77}{\beta_1} = \frac{8.77}{0.80} = 10.96 \text{ in.}$$

Instead of doing the work in part (b), this value of $x = 10.96$ in. may be checked against $0.75x_b = 11.54$ in.

$$\epsilon'_s = \frac{10.96 - 3.00}{10.96}(0.003) = 0.00218 > \epsilon_y \qquad\qquad \text{OK}$$

The check on ϵ'_s verifies the conclusion in part (a).

$$M_n = C_c\left(d - \frac{a}{2}\right) + C_s(d - d')$$

$$= 522[26 - 0.5(8.77)]\tfrac{1}{12} + 88(26 - 3)\tfrac{1}{12} = 941 + 169$$

$$= 1110 \text{ ft-kips } (1500 \text{ kN·m})$$

Note that the work shown in parts (a) and (b) is to explain why the checks on (1) if actual $x < 0.75x_b$ and (2) if actual $\epsilon'_s < \epsilon_y$ are needed in part (c). In a usual problem the work shown in part (c) is all that is necessary (if both checks can be verified).

EXAMPLE 3.10.2 Repeat the solution for Example 3.10.1, except that A_s is 4-#11 bars instead of 8-#10 bars. Also determine the percentage increase in the strength of the beam over the same beam without compression steel.

Solution: (a) Determine A_{sy} in the compression steel yield condition. From Example 3.10.1, $A_{sy} = 9.13$ sq in. Since actual A_s (6.24 sq in.) is less than A_{sy}, the compression steel does not yield.

(b) Determine M_n in the actual condition. Let the location of the neutral axis in Fig. 3.10.3 be the unknown. Equating $(C_c + C_s)$ to T,

$$4.25(14)(0.80x) + \left[\frac{x - 3}{x}(0.003)(29,000) - 4.25\right](1.58) = 60(6.24)$$

Solving the quadratic equation for x,

$$x = 6.46 \text{ in.}$$

$$a = \beta_1 x = 0.80(6.46) = 5.17 \text{ in.}$$

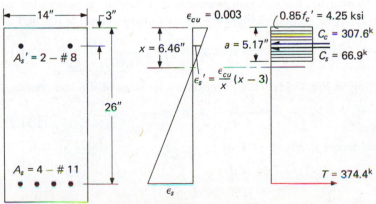

Figure 3.10.3 Section for Example 3.10.2.

Then

$$C_c = 4.25(14)(5.17) = 307.6 \text{ kips}$$

$$\epsilon_s' = \left(\frac{6.46 - 3.00}{6.46}\right) 0.003 = 0.00161 < \epsilon_y \quad \text{(as assumed)}$$

$$C_s = [29(1.61) - 4.25](1.58) = 66.9 \text{ kips}$$

$$C_c + C_s = 307.6 + 66.9 = 374.5 \text{ kips}$$

$$T = A_s f_y = 6.24(60) = 374.4 \text{ kips} \quad \text{(Check)}$$

$$M_n = 307.6[26 - 0.5(5.17)]\tfrac{1}{12} + 66.9(26 - 3)\tfrac{1}{12} = 600 + 128$$

$$= 728 \text{ ft-kips (988 kN·m)}$$

(c) Determine M_n if the compression steel had not existed. From Fig. 3.10.2(b) in Example 3.10.1(a), check whether or not $A_s = 6.24$ sq in. exceeds $0.75 A_{sb}$ without compression steel. Using maximum C_c,

$$\max A_s = \frac{549}{60} = 9.15 \text{ sq in.} > (\text{actual } A_s = 6.24 \text{ sq in.}) \qquad \text{OK}$$

$$C = 0.85 f_c' ba = 4.25(14)a$$

$$T = 6.24(60) = 374.4 \text{ kips}$$

$$a = \frac{374.4}{4.25(14)} = 6.29 \text{ in.}$$

$$M_n = 374.4[26 - 0.5(6.29)]\tfrac{1}{12} = 713 \text{ ft-kips (966 kN·m)}$$

Thus even though in part (b) the 128 ft-kips representing the contribution of compression steel was 17.5% of the total, the addition of the compression steel to the singly reinforced beam actually added only 2.1% [(728 − 713)/713] to the original capacity. This is a common situation; the compression steel was not used because the compression zone without it was inadequate for strength, but instead it was used for deflection control.

3.11 CRITERION FOR THE COMPRESSION STEEL YIELD CONDITION

It may be shown that the criterion to ensure that the compression steel in a doubly reinforced section yields when the nominal strength is reached is

$$\rho - \rho' \left(1 - \frac{0.85 f_c'}{f_y}\right) \geq 0.85 \beta_1 \left(\frac{f_c' d'}{f_y d}\right) \left(\frac{87,000}{87,000 - f_y}\right) \qquad (3.11.1)$$

Referring to Fig. 3.11.1, the criterion for yielding of the compression steel is

$$\epsilon_s' \geq \epsilon_y \qquad (3.11.2)$$

The internal forces in Fig. 3.11.1 are

$$T = \rho b d f_y$$

$$C_c = 0.85 f_c' \beta_1 x b$$

$$C_s = (f_y - 0.85 f_c') \rho' b d$$

Equating T to $C_c + C_s$ and solving for x,

$$x = \frac{f_y d}{0.85\beta_1 f_c'}\left[\rho - \rho'\left(1 - \frac{0.85f_c'}{f_y}\right)\right] \qquad (3.11.3)$$

From Fig. 3.11.1(b)

$$\epsilon_s' = \frac{\epsilon_{cu}}{x}(x - d') \qquad (3.11.4)$$

Substituting Eq. (3.11.4) in Eq. (3.11.2) gives

$$\frac{\epsilon_{cu}}{x}(x - d') \geq \epsilon_y \qquad \text{or} \qquad \frac{87,000}{x}(x - d') \geq f_y$$

or

$$x \geq \frac{87,000 d'}{87,000 - f_y} \text{ KSi} \qquad (3.11.5)$$

Equation (3.11.1) may be obtained by substituting Eq. (3.11.3) in Eq. (3.11.5).

Use of Eq. (3.11.1) is the alternative to use of basic statics as done in Example 3.10.1, part (a), to determine if compression steel does yield when nominal strength M_n is reached. Although solution by basic statics is preferable in longhand computations, a formula like Eq. (3.11.1) may be useful in computer programming.

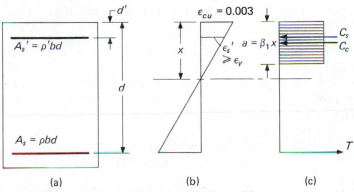

$\epsilon_{cu} = 0.003$

(a)   (b)   (c)

Figure 3.11.1   Compression steel yield condition.

3.12 DESIGN OF RECTANGULAR SECTIONS IN BENDING WITH BOTH TENSION AND COMPRESSION REINFORCEMENT

When a rectangular section of given dimensions is required to have a strength greater than it may have when it is reinforced with the maximum permissible amount of tension reinforcement, compression reinforcement becomes necessary. However, such necessary use of compression steel for strength is rare. The principal reason for using compression reinforcement is to reduce long-time deflection due to creep and shrinkage.

The logical procedure for designing a doubly reinforced section is to determine first whether compression steel is needed for strength. This may be done by comparing the required moment strength with the moment strength of a singly reinforced section in which the maximum permissible amount of tension steel has been used.

Having decided that compression steel is to be used, be it required for strength or desirable for deflection control, the designer now needs to select the appropriate tension steel A_s and compression steel A'_s. For this purpose the two equilibrium equations may be utilized; namely,

$$C_c + C_s = T \tag{3.12.1}$$

and

$$M_n = C_c \left(d - \frac{\beta_1 x}{2} \right) + C_s (d - d') \tag{3.12.2}$$

wherein the symbols are those defined in Fig. 3.10.1. Further, verifications are necessary to see that the tension steel as finally used does not make the neutral axis distance x exceed 75% of the value in the balanced strain condition and that the compression steel does in fact yield [Eq. (3.11.1)] when it has been so assumed in the equilibrium equations [Eqs. (3.12.1) and (3.12.2)]. If the compression steel does not yield, the equilibrium conditions must be reformulated using a stress f'_s in the compression steel proportional to the strain in that steel.

A unique solution is obtainable once the actual x is preset. This actual x has a maximum limit of $0.75x_b$. If deflection control is an important consideration, the actual x may be preset at a much smaller value than $0.75x_b$, perhaps $0.375x_b$ or smaller.

EXAMPLE 3.12.1 Determine the A_s and A'_s required to carry a service live load moment of 390 ft-kips and a service dead load moment of 200 ft-kips, using $b = 14$ in., $d = 26$ in., $d' = 3$ in., $f'_c = 5000$ psi, $f_y = 60,000$ psi, and the ACI Code, as shown in Fig. 3.12.1(a).

Solution: (a) Determine the required strength.

$$M_u = 1.4(200) + 1.7(390) = 280 + 664 = 944 \text{ ft-kips}$$

$$\text{required } M_n = \frac{M_u}{\phi} = \frac{944}{0.90} = 1047 \text{ ft-kips}$$

(b) Determine the maximum strength and reinforcement allowed by ACI Code for a singly reinforced section. The location of the neutral axis at the balanced strain condition [Fig. 3.12.1(b)] may be obtained as

$$x_b = \frac{\epsilon_{cu}}{\epsilon_{cu} + \epsilon_y} d = \frac{3}{3 + 60/29} \, 26 = 15.39 \text{ in.}$$

$$\text{max } x = 0.75(15.39) = 11.54 \text{ in.} \qquad [\text{Fig. } 3.12.1(c)]$$

$$\text{max } C = 0.85f'_c b\beta_1 x = 4.25(14)(0.80)(11.54) = 549 \text{ kips}$$

$$\text{max } A_s \text{ in singly reinforced section} = 549/60 = 9.15 \text{ sq in.}$$

$$\text{max } M_n \text{ of singly reinforced section} = (549)(26 - 4.62)\tfrac{1}{12}$$
$$= 977 \text{ ft-kips} \qquad (1330 \text{ kN·m})$$

The required M_n exceeds the maximum strength obtainable without compression steel; therefore compression steel is needed for strength.

(c) Determine the minimum compression reinforcement required. Maintain max $x = 0.75x_b = 11.54$ in. [Fig. 3.12.1(d)]. Let

$$M_{nc} = 977 \text{ ft-kips} \qquad \text{[from part (b)]}$$

$$M_{ns} = M_n - M_{nc} = 1047 - 977 = 70 \text{ ft-kips}$$

$$\text{required } C_s = \frac{70(12)}{26 - 3} = 36.5 \text{ kips}$$

Will compression steel yield when $x = 11.54$ in.?

$$\epsilon'_s = \frac{11.54 - 3.00}{11.54}(0.003) = 0.0022 > \epsilon_y$$

$$C_s = A'_s(f_y - 0.85f'_c) = 36.5 \text{ kips}$$

$$\text{required } A'_s = \frac{36.5}{60 - 4.25} = 0.65 \text{ sq in.}$$

$$T = \max C_c + C_s = 549 + 36.5 = 585.5 \text{ kips}$$

$$\text{required } A_s = \frac{585.5}{60} = 9.76 \text{ sq in.}$$

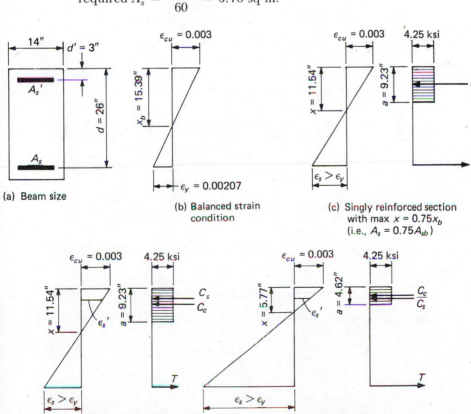

(a) Beam size

(b) Balanced strain condition

(c) Singly reinforced section with max $x = 0.75x_b$ (i.e., $A_s = 0.75A_{sb}$)

(d) Doubly reinforced section with $x = 0.75x_b = 11.54$ in.

(e) Doubly reinforced section with $x = 0.375x_b = 5.77$ in.

Figure 3.12.1 Section for Examples 3.12.1 and 3.12.2.

This amount of tension steel is exactly the maximum permitted by the ACI Code.

In this example the remaining steps of selecting actual bars and making a final check of strength are not shown.

EXAMPLE 3.12.2 Redesign the section of Example 3.12.1 so that the actual neutral axis location is at 0.375 of that in the balanced strain condition. Assume that the purpose is for deflection control.

Solution: (a) Determine M_{nc} corresponding to locating the actual neutral axis at 0.375 of the value in the balanced strain condition. Use of this assumption seems a rational choice consistent with the guideline value for deflection control on a singly reinforced beam, as discussed in Sec. 3.9. Thus, referring to Fig. 3.12.1(e),

$$\text{actual } x = 0.375x_b = 0.375(15.39) = 5.77 \text{ in.}$$
$$\text{actual } a = 0.80x = 0.80(5.77) = 4.62 \text{ in.}$$
$$C_c = 0.85f'_cba = 0.85(5)(14)(4.62) = 275 \text{ kips}$$
$$T_1 = C_c = 275 \text{ kips}$$
$$M_{nc} = 275[26 - 0.5(4.62)]\tfrac{1}{12} = 542 \text{ ft-kips}$$

Let A_{sc} be the part of tension steel to match the concrete in compression; then

$$A_{sc} = \frac{T_1}{f_y} = \frac{275}{60} = 4.58 \text{ sq in.}$$

(b) Determine steel requirements for both faces of the beam.

$$M_{ns} = 1047 - 542 = 505 \text{ ft-kips}$$
$$C_s = T_2 = \frac{505(12)}{26 - 3} = 264 \text{ kips}$$
$$\epsilon'_s = \frac{5.77 - 3.0}{5.77}(0.003) = 0.00144 < \epsilon_y$$

Compression steel does not yield. If $x = 5.77$ in. is still to be maintained, then

$$f'_s = 0.00144(29,000 \text{ ksi}) = 41.8 \text{ ksi}$$

so that the stress used is consistent with the strain on the compression steel.

$$A'_s = \frac{264}{41.8 - 4.25} = 7.02 \text{ sq in.}$$
$$A_{ss} = \frac{264}{60} = 4.40 \text{ sq in.}$$
$$A_s = A_{sc} + A_{ss} = 4.58 + 4.40 = 8.98 \text{ sq in.}$$

Select 9-#9 in two layers for tension steel, and 4-#11 as compression reinforcement. The 4-#11 (A'_s = 6.24 sq in.) is the maximum steel that will fit into one layer.

(c) Check the strength of the section. The above solution of A_s = 8.98 and A'_s = 7.02 is the correct one for x = $0.375x_b$. When A'_s = 6.24 is used instead, both x and A_s will have to change. A trial-and-error procedure is presented below to determine the stress f'_s acting in the compression steel; an alternative to that used in Example 3.10.2 wherein a quadratic equation was solved for the neutral axis location x.

Since a smaller amount of compression steel is used than that required to locate the neutral axis at $0.375\ x_b$, the actual neutral axis will be lower if the section is to carry the same bending moment. Also, the stress f'_s must exceed 41.8 ksi for x = $0.375x_b$. Estimate the compression steel stress to be 45 ksi.

$$C_c = 0.85f'_cba = 0.85(5)(14)a = 59.5a$$

$$C_s = A'_s(f'_s - 0.85f'_c) = 6.24(45 - 4.25) = 254 \text{ kips}$$

$$T = A_sf_y = 9.00(60) = 540 \text{ kips}$$

$$C_c + C_s = T$$

$$a = \frac{540 - 254}{59.5} = 4.81 \text{ in.}; \qquad x = \frac{4.81}{0.80} = 6.00 \text{ in.}$$

$$\epsilon'_s = \frac{6.00 - 3.00}{6.00}(0.003) = 0.0015$$

$$f'_s = 0.0015(29,000) = 43.5 \text{ ksi}$$

Since this does not agree with the 45 ksi assumed, make a new assumption, say try f'_s = 44 ksi:

$$\text{revised } C_s = 6.24(44 - 4.25) = 248 \text{ kips}$$

$$a = \frac{540 - 248}{59.5} = 4.91 \text{ in.}; \qquad x = \frac{4.91}{0.80} = 6.14 \text{ in.}$$

$$\epsilon'_s = \frac{6.14 - 3.00}{6.14}(0.003) = 0.001535$$

$$f'_s = 0.001535(29,000) = 44.5 \text{ ksi}$$

Repeated trials may be made until computed f'_s agrees as closely as desired with the assumed value. In this case assume present agreement is close enough.

$$C_c = 59.5a = 59.5(4.91) = 292 \text{ kips}$$

$$C_s = 248 \text{ kips}$$

$$M_n = 292[26 - 0.5(4.91)]\tfrac{1}{12} + 248(26 - 3)\tfrac{1}{12}$$

$$= 573 + 475 = 1048 \text{ ft-kips}$$

$$\approx 1047 \text{ ft-kips required} \qquad \text{OK}$$

A comparison of three possible designs for this doubly reinforced section as worked out in detail in Examples 3.12.1 and 3.12.2 is tabulated below:

Example 3.12.1, part (c), $x=11.54$ in. $A_s=9.76$ sq in. $A'_s=0.65$ sq in.

Example 3.12.2, part (b), $x=5.77$ in. $A_s=8.98$ sq in. $A'_s=7.02$ sq in.

Example 3.12.2, part (c), $x=6.14$ in. $A_s=9.00$ sq in. $A'_s=6.24$ sq in.

It may be observed that when there is a large amount of compression steel that does not yield due to the linear strain variation across the depth of the section, varying the amount of compression steel has essentially the effect of changing only the proportions in M_{nc} and M_{ns}. It has a negligible effect (say typically 3 to 4%) on total capacity. The choice of 9-#9 (tension steel) and 4-#10 (compression steel) is acceptable. Deflection may still need checking but it will probably be within acceptable limits.

3.13 NON-RECTANGULAR BEAMS

When non-rectangular beams are used, it is the shape of the compression zone that dictates whether the formulas developed in this chapter are applicable. The concept of using the Whitney rectangular compressive stress distribution may be used in accordance with ACI-10.2.7 for any shape of the compression zone in the cross section. For ductility, it is appropriate to limit actual x so that it does not exceed $0.75x_b$, particularly for notched or other shaped beams having less compression concrete area at the extreme fiber than at the neutral axis. The maximum amount of tension reinforcement may be obtained by setting $x = 0.75x_b$ and evaluating the internal forces to be compatible with the strain in the maximum x condition. Then equating C to T the maximum tension steel permitted can be obtained. Alternatively, of course, ACI-10.3.3 may be used by limiting A_s to a maximum of $0.75A_{sb}$. The treatment of compression steel in a non-rectangular compression zone would be identical to that in a rectangular beam, as discussed in Sec. 3.10.

For sections having a non-rectangular compression zone, the ductility (that is, the amount of tension steel strain achieved at maximum A_s condition) obtained by using the literal wording of ACI-10.3.3 may be either *more* or *less* than obtained by a consistent use of maximum $x = 0.75x_b$. When the section has the greater compression area near the extreme fiber, as for a T-section, the tension steel strain achieved will be greater than for the rectangular beam; that is, maximum x will be less than $0.75x_b$. When the section has the greater compression area near the neutral axis, the tension steel strain achieved will be less than for the rectangular beam; that is, maximum x will be greater than $0.75x_b$. Rational treatment would set the same maximum x limit for all beams. ACI-10.3.3 was not specifically worded with the intention of providing varying amounts of tension steel strain achieved when the nominal strength is reached; rather, it was intended to make for easier computation.

Though no examples of non-rectangular sections are provided, there are some problems provided at the end of the chapter for practice.

SELECTED REFERENCES

1. Jack R. Janney, Eivind Hognestad, and Douglas McHenry. "Ultimate Flexural Strength of Prestressed and Conventionally Reinforced Concrete Beams," *ACI Journal, Proceedings,* **52**, February 1956, 601–620.
2. Eivind Hognestad, N. W. Hanson, and Douglas McHenry. "Concrete Stress Distribution in Ultimate Strength Design," *ACI Journal, Proceedings,* **52**, December 1955, 455–479.
3. Hjalmar Granholm. *A General Flexural Theory of Reinforced Concrete.* New York; Wiley, 1965.
4. Eivind Hognestad. "Confirmation of Inelastic Stress Distribution in Concrete," *Journal of Structural Division,* ASCE, **83**, Paper No. 1189, No. ST2, March 1957.
5. C. S. Whitney. "Plastic Theory in Reinforced Concrete Design," *Transactions ASCE,* **107**, 1942, 251–326.
6. C. S. Whitney and Edward Cohen. "Guide for Ultimate Strength Design of Reinforced Concrete," *ACI Journal, Proceedings,* **53**, November 1956, 455–475.
7. ACI-ASCE Joint Committee: "Report on ASCE-ACI Joint Committee on Ultimate-Strength Design," *ASCE, Proceedings,* **81**, Paper No. 809, October 1955. See also *ACI Journal, Proceedings,* **52**, January 1956, 505–524.

PROBLEMS

All problems* are to be done in accordance with the strength design method of the ACI Code (except Probs. 3.1 and 3.2), and all loads given are *service* loads, unless otherwise indicated. Wherever possible basic principles are to be used for solutions, avoiding the direct use of formulas. All design problems require a clear statement of the final choice at the end of the calculations, along with a design sketch drawn to scale.

Review Problems in Mechanics of Materials

3.1 An elastic homogeneous beam of material capable of carrying both tension and compression is simply supported over a span of 20 ft (6 m) center to center of supports. It is carrying a uniformly distributed load of 2 klf (kips per linear ft) (30 kN/m) in addition to a concentrated load of 10 kips (44 kN) located 5 ft (1.5 m) from the right end of the span.

(a) Compute the maximum flexural stress on the section shown in the accompanying figure, using only basic statics and the internal-force method, using $C = T$ and $M = (C$ or $T)$ (moment arm between points of action of C and T).

(b) Check by using the flexure formula, $f = Mc/I$.

---

*Most problems may be solved as problems stated in Inch-Pound units, or as problems in SI units using quantities in parenthesis at the end of the statement. The metric conversions are approximate to avoid implying higher precision for the given information in metric units than that for the Inch-Pound units.

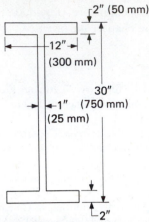

Problem 3.1

3.2 The elastic homogeneous beam of the accompanying figure is simply supported on a span of 25 ft (7.5 m), and carries a uniform loading of 1 kip/ft (15 kN/m) plus a concentrated load of 9 kips (40 kN) at 8 ft (2.4 m) from the left end of the span. Compute maximum tensile and compressive stresses using (a) basic statics of the internal-force method, using $C = T$ to locate neutral axis and $M = (C$ or $T)$ (moment arm between points of action of C and T); and (b) the flexure formula, $f = Mc/I$.

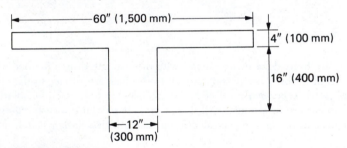

Problem 3.2

Reinforced Concrete Problems

3.3 For each of the beams in the accompanying figure using f'_c = 3500 psi, f_y = 60,000 psi, and the internal-couple method with the Whitney rectangular

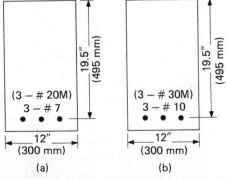

Problem 3.3

stress block, determine the nominal flexural strength M_n. If the loading is 60% live load and the basic overload provision for dead load plus live load controls, what is the service moment capacity? (f'_c = 25 MPa and f_y = 400 MPa.)

3.4 Using basic principles, prove whether or not there is yield of the tension steel in the beam of the accompanying figure according to its behavior when the extreme fiber in compression reaches a strain of 0.003. Does the beam violate the ACI Code? If so, in what respect? Use f'_c = 4000 psi and f_y = 58,000 psi. (f'_o = 30 MPa and f_y = 400 MPa.)

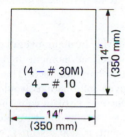

(4 – # 30M)
4 – # 10

14″
(350 mm)

14″
(350 mm)

Problem 3.4

3.5 Determine the required effective size (dimensions b and d) and steel area A_s for a rectangular beam cross section (using a d/b ratio of about 1.75) to carry 1.0 kip/ft live load and 1.3 kips/ft dead load (including beam weight) on a simply supported span of 24 ft. Without making the practical decision of choosing an overall beam size and selecting actual bars, determine (see ACI-10.3.3 and 10.5.1) (a) the largest effective size; and (b) the smallest effective size that would be permitted. Use f'_c = 3000 psi and f_y = 60,000 psi. (Live load = 15 kN/m; dead load = 19 kN/m; span = 7.3 m; f'_c = 20 MPa; f_y = 400 MPa.)

3.6 Repeat Prob. 3.5, except use f'_c = 4000 psi and use a d/b ratio of 2.0. (f'_c = 30 MPa.)

3.7 Repeat Prob. 3.5, except use f'_c = 4000 psi and use a d/b ratio of 1.25. (f'_c = 30 MPa.)

3.8 Design a rectangular beam for a 30-ft simply supported span to carry a live load of 2 kips/ft and a dead load of 1 kip/ft, in addition to the weight of the beam. Use f'_c = 3500 psi, f_y = 40,000 psi. (Span = 9 m; live load = 30 kN/m; dead load = 15 kN/m; f'_c = 25 MPa; f_y = 300 MPa.)
(a) Assuming no deflection limitation, design the smallest singly reinforced size of beam permitted.
(b) Design a singly reinforced beam of such size that deflections would not be expected to be excessive under normal design circumstances. (Hint: Use ρ = approximately one-half of the maximum permitted by the ACI Code.)

3.9 Apply the requirements of Prob. 3.8, except live load is 1.8 kips/ft, dead load (without beam weight) is 0.8 kip/ft, f'_c = 3000 psi, and f_y = 60,000 psi.

3.10 Apply the requirements of Prob. 3.8, except live load is 2.5 kips/ft, dead load (without beam weight) is 1.3 kips/ft, f'_c = 3500 psi, and f_y = 60,000 psi.

3.11 Design a rectangular beam to carry a live load of 1.5 kips/ft and a dead load of 1.5 kips/ft, in addition to the beam weight, for a simple span of 32 ft. Use an approximate steel ratio ρ of about 0.375 of the balanced amount (one-half of the maximum permitted), and also satisfy ACI-Table 9.5a. Use $f'_c = 4000$ psi and $f_y = 60,000$ psi. (Live load = 22 kN/m; dead load = 22 kN/m; span = 9.8 m; $f'_c = 30$ MPa; $f_y = 400$ MPa.)

3.12 Select the economical reinforcement for a beam 12 in. wide by 22 in. deep overall to carry dead load and live load moments of 15 ft-kips and 60 ft-kips, respectively. Use $f'_c = 3000$ psi, $f_y = 40,000$ psi. (Width = 300 mm; depth = 560 mm; dead load moment = 20 kN·m; live load moment = 80 kN·m; $f'_c = 20$ MPa; $f_y = 300$ MPa.)

3.13 Select economical reinforcement for a beam 20 in. wide by 40 in. overall depth to carry a live load moment of 500 ft-kips and a dead load moment (including beam weight) of 300 ft-kips. Use reinforcement in only one face. Though a check of deflection cannot be made with the given information, would you expect such a check to show that deflection is excessive? Explain your answer. Use $f'_c = 4000$ psi and $f_y = 60,000$ psi. (Beam size = 500 mm × 1000 mm; live load moment = 700 kN·m; dead load moment = 400 kN·m; $f'_c = 30$ MPa; $f_y = 400$ MPa.)

3.14 Assuming no deflection limitation, and without using compression reinforcement, select the smallest rectangular section permitted by the ACI Code using the same size for the entire 28-ft beam of the accompanying figure. The live load is 1.5 kips/ft and the dead load is 1 kip/ft in addition to the weight of the beam. Select steel for both maximum positive and negative moment locations. Use $f'_c = 3500$ psi and $f_y = 40,000$ psi. (Note: Live load is always to be applied in the manner to cause the most severe effect; spans may be fully loaded, partially loaded, or unloaded, as is necessary for maximum stress.) (Live load = 20 kN/m; dead load = 15 kN/m; main span = 6 m; cantilever = 2.5 m; $f'_c = 25$ MPa; $f_y = 300$ MPa.)

3.15 Repeat Prob. 3.14, except use 2.0 kips/ft live load, and 1.25 kips/ft dead load, in addition to the beam weight, and increase the main span from 20 to 24 ft. (Live load = 30 kN/m; dead load = 17 kN/m; main span = 7.3 m; cantilever = 2.4 m.)

3.16 Repeat Prob. 3.14, except use 16 ft instead of 20 ft for the main span, and 10 ft instead of 8 ft for the cantilever, and use $f_y = 60,000$ psi.

3.17 Repeat Prob. 3.14, except use 18 ft instead of 20 ft for the main span, use 2.0 kips/ft live load instead of 1.5 kips/ft, and use $f'_c = 4000$ psi and $f_y = 60,000$ psi.

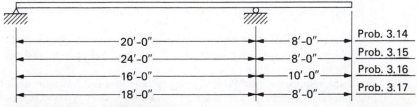

Problems 3.14 through 3.17

3.18 For the double overhanging cantilever beam shown, design the smallest practical rectangular cross section (without compression steel) to be used for entire 50 ft of beam. Select steel for both positive- and negative-moment regions. The live load is 1.5 kips/ft and the dead load is 1.0 kip/ft, in addition to the beam weight. Assume there is no deflection limit. Use $f'_c = 3500$ psi and $f_y = 60,000$ psi. (Refer to note at end of Prob. 3.14.) (Live load = 22 kN/m; dead load = 14 kN/m; main span = 9.2 m; cantilevers = 3 m; $f'_c = 25$ MPa; $f_y = 400$ MPa.)

3.19 Repeat Prob. 3.18, except assume the beam is part of a floor system supporting nonstructural elements likely to be damaged by large deflections. (Hint: Refer to ACI-Table 9.5a and Prob. 3.8.)

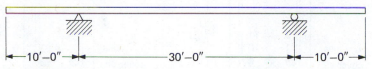

Problems 3.18 and 3.19

3.20 For the beam with compression reinforcement given in the accompanying figure and using $f'_c = 4000$ psi, $f_y = 40,000$ psi, compute the nominal strength M_n using the principles of statics with the internal couple. As a part of the procedure, verify by using basic principles whether or not the compression steel has yielded when nominal strength is reached; if it does not yield, use a compression steel stress proportional to the strain in compression steel. Verify that the tension steel does not exceed the maximum permitted by the ACI Code. ($f'_c = 30$ MPa; $f_y = 300$ MPa.)

3.21 For the beam with compression reinforcement given in the accompanying figure, and using $f'_c = 4000$ psi and $f_y = 60,000$ psi, compute the nominal strength M_n using the principles of statics with the internal couple. Make sure to verify whether or not the compression steel reaches yield when M_n is reached. With the given compression steel, determine the maximum tension steel the ACI Code would permit for this section. ($f'_c = 30$ MPa; $f_y = 400$ MPa.)

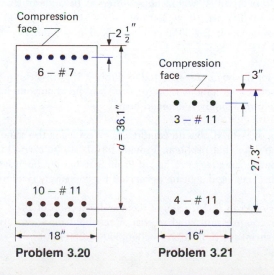

Problem 3.20   **Problem 3.21**

3.22 Apply the requirements of Prob. 3.21 to the beam of the accompanying figure, using $f'_c = 5000$ psi and $f_y = 60{,}000$ psi.

3.23 Apply the requirements of Prob. 3.21 to the beam of the accompanying figure, using $f'_c = 3000$ psi and $f_y = 60{,}000$ psi.

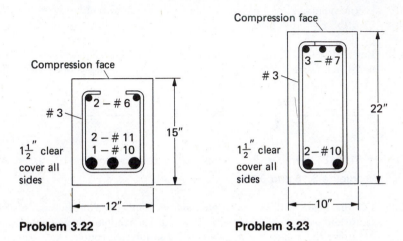

Problem 3.22   **Problem 3.23**

3.24 If a rectangular beam with $b = 12$ in. and $d = 19.5$ in. has 3-#10 for tension reinforcement and 2-#10 (centered $2\frac{1}{2}$ in. from face of beam) for compression reinforcement, what percent increase in the nominal strength M_n is provided by the 2-#10 over that provided without any compression steel? Use $f'_c = 3500$ psi and $f_y = 60{,}000$ psi.

3.25 A rectangular section with $b = 14$ in. and effective depth $d = 21.5$ in. has 4-#10 as tension reinforcement and 2-#10 as compression reinforcement centered 2.5 in. from the face of the beam. Determine the strength M_n for this section. How much can the strength be increased by adding tension steel only? At the point where compression steel is needed to further increase capacity, what ratio of A'_s to A_s would be required to be added? Use $f'_c = 5000$ psi and $f_y = 60{,}000$ psi. ($b = 350$ mm; $d = 546$ mm; tension steel, 4-#30M; compression steel, 2-#30M centered 63.5 mm from face; $f'_c = 35$ MPa; $f_y = 400$ MPa.)

3.26 Redesign the steel for the beam size selected for Prob. 3.15, using compression steel so that the net reinforcement ratio $(\rho - \rho')$ will be about one-half the maximum permitted for a singly reinforced beam.

3.27 Redesign the beam of Prob. 3.16 with compression steel using the same small cross section selected for that problem. Provide for M_{nc} to be carried by the singly reinforced beam using $\rho = 0.375\rho_b$. If Prob. 3.16 has not been worked, use the smallest size rectangular beam permitted for one singly reinforced.

3.28 For the conditions of Prob. 3.18, using a rectangular section of 14×22 in. overall, design for compression steel so that A'_s is about $0.5A_s$, in order to control creep and shrinkage deflection (section 350×560 mm).

3.29 For the beam of the accompanying figure, it is desired to utilize 3-#7 bars as compression reinforcement. The factored moment M_u to be carried is 210 ft-kips; $f_c' = 3000$ psi and $f_y = 60,000$ psi. Determine the tension steel required, select the bars, and check the section.

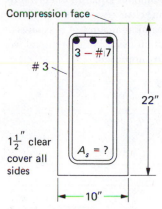

Compression face

3 – #7

3

22"

$1\frac{1}{2}''$ clear cover all sides

$A_s = ?$

10"

Problem 3.29

Non-rectangular Section Problems

3.30 to 3.35 For the beam cross section of the accompanying figure, assuming $f_c' = 3000$ psi, $f_y = 60,000$ psi, and the clear cover from the bottom (tension) face to the bars = 2 in., compute the following:
(a) Compare the given tension reinforcement with the maximum (max $A_s = 0.75A_{sb}$) permitted by the ACI Code, and with the max $x = 0.75x_b$ requirement. Use basic principles starting with the balanced strain condition.
(b) Using basic principles with the Whitney rectangular stress distribution, compute the nominal strength M_n for the cross section.
(c) Neglecting the reinforcement and assuming that the concrete is a homogeneous elastic material, compute the cracking moment M_{cr} when the extreme fiber in tension reaches the modulus of rupture value given by ACI (i.e., $7.5\sqrt{f_c'}$ for normal weight concrete).

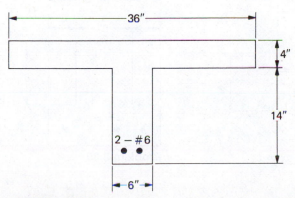

36"

4"

14"

2 – #6

6"

Problem 3.30

(d) Using the result in part (c), determine the reinforcement ratio $\rho = A_s/b_w d$ that makes M_n for a reinforced concrete beam (ignore the given bars for this part) equal to M_{cr}. Give answer in terms of min $\rho = \text{coefficient}/f_y$ and compare with ACI-10.5.1. Use for b_w the narrowest width on the tension side of the neutral axis.

(e) Assuming that compression steel of an amount equal to the given tension steel were to be used and located with 2-in. clear cover to the top (compression) face of the beam, determine the maximum amount of tension steel the ACI Code would permit. Do the computation using both the literal wording of ACI-10.3.3 and the maximum $x = 0.75x_b$ approach.

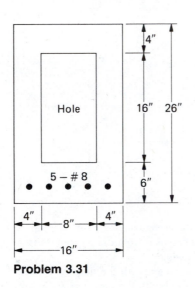

Problem 3.31

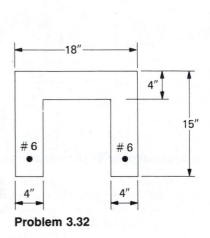

Problem 3.32

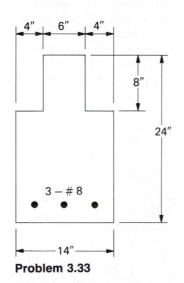

Problem 3.33

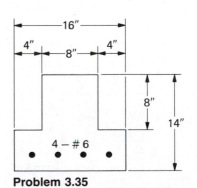

Problem 3.35

4

Rectangular
Sections
in Bending
Under Service
Load Conditions

4.1 GENERAL INTRODUCTION

As discussed in Chap. 2, there have been two generally accepted philosophies of design and investigation—working stress and strength design. In Chap. 3 the strength design method was treated using overload factors, strength reduction factors, and the flexural strength of sections where stress is no longer proportional to strain. This chapter considers the working stress method in which service loads, allowable working stresses, and the linear relationship between stress and strain are used.

The working stress method is referred to by the present ACI Code as the *alternate design method.* With increasing emphasis on clear distinction between the design criteria of *strength* and *serviceability,* the term "strength" has come to mean imminent failure under severe overload, whereas "serviceability" means satisfactory performance under service load (i.e., working stress) conditions. Satisfactory performance may be defined in terms of (1) deflection within acceptable limits so that supported nonstructural elements such as walls, partitions, and ceilings are not damaged; (2) cracking controlled to prevent large crack widths that are either unsightly or may permit water to enter causing corrosion of steel and perhaps deterioration of concrete. Other serviceability requirements such as vibration and noise control may also be important but lie outside the scope of this book. Though its use is not encouraged, the ACI "alternate design method" does provide another way of design which gives about the same results as long as the tension steel used is less than about one-half of the maximum permitted in the strength design method.

City Hall, Toronto, Ontario, Canada. (Courtesy of Portland Cement Association.)

4.2 FUNDAMENTAL ASSUMPTIONS

Four basic assumptions are used in the working stress method for flexural members:

1. Plane sections remain plane after bending; that is, the variation in strain is linear across the depth of the member.
2. Stress is proportional to strain.
3. Concrete does not take tension (concrete cracks under tension).
4. Perfect bond exists between steel and concrete such that no slip occurs.

Of the four assumptions, only the second assumption, that stress is proportional to strain, is unique to the working stress method, because it is reasonably accurate for stresses below about one-half f'_c, the 28-day compressive strength. Regarding the third assumption, it is true that more of the concrete close to the neutral axis on the tension side will not be cracked under service load than under factored load; however, the magnitude of tensile stress transmitted is generally so small that, for reason of simplicity, the steel is assumed to take all of the tensile stress. The subject of bond, or the longitudinal interaction between reinforcing bars and the surrounding concrete, is treated in Chap. 6.

4.3 MODULUS OF ELASTICITY RATIO, *n*

As discussed in Chap. 1, the stress-strain curve for reinforcement steel is linear below the yield stress, but that of concrete is only approximately linear, even at or below the allowable working stress. The modulus of elasticity of steel varies little with its strength, whereas that of concrete varies with its density and strength. In the ACI Code, E_s is taken to be 29,000,000 psi (200,000 MPa), and E_c is expressed as $w_c^{1.5} 33\sqrt{f'_c}$, in which w_c is the density of concrete. Table 1.9.1 gives values for the modulus of elasticity of concrete, E_c.

In the working stress procedure, it will be shown that it is the ratio n of the modulus of elasticity of steel to that of concrete that is needed for computing working stresses rather than the actual values of E_c or E_s.

The modular ratio $n = E_s/E_c$ may be taken (ACI–Appendix B.5.4) as the nearest whole number, but not less than 6. Except in deflection calculations, the value of n for lightweight concrete shall be assumed to be the same as for normal weight concrete of the same strength. It is suggested that the values in Table 4.3.1 be used for normal weight concrete.

4.4 EQUILIBRIUM CONDITIONS

Two equilibrium conditions apply to a section subjected to bending only: (1) the resultant internal compressive force must be equal to the resultant internal tensile force; and (2) the moment of the internal couple, composed

Table 4.3.1 PRACTICAL VALUES FOR MODULAR RATIO, *n*

INCH-POUND		SI	
f'_c (psi)	n	f'_c (MPa)[a]	n
3000	9	20	9
3500	8.5	25	8
4000	8	30	7.5
4500	7.5	35	7
5000	7	40	6.5
6000	6.5		

[a] For practical use, multiply MPa by 10 to obtain value in kgf/cm².

of the resultant compressive and tensile forces, must be equal to the applied bending moment. In fact, these two equilibrium conditions must hold true regardless of whether the service load is acting or failure is imminent under overload, the only difference being that the stress distribution across the depth is linear in one but not in the other.

The resultant compressive force may be entirely from concrete stresses or it may be from a combination of the stresses in concrete and those in the compression reinforcement. The resultant tensile force, of course, comes entirely from the tension reinforcement.

EXAMPLE 4.4.1 Using the equilibrium conditions, determine the working stresses in the steel and on the extreme fiber of concrete in the section of Fig. 4.4.1(a) due to an applied service load moment M_w of 2000 in.-kips. Use $n = 8$ for the ratio of the moduli of elasticity of steel to concrete.

Solution: The assumptions of linear strain and stress proportional to strain are shown in Fig. 4.4.1(b) and (c). The first step in the solution is to locate the neutral axis (NA). The internal compressive force is obtained by integration of the stress times the area on which it acts, equivalent to computing the volume of the stress solid. Thus the internal compressive force in the concrete is the volume of a triangular wedge, as in Fig. 4.4.1(a),

$$C = \tfrac{1}{2}f_c bx = 6.0f_c x$$

The internal tensile force is

$$T = f_s A_s = f_s(4)(1.27) = 5.08f_s$$

Equating C to T gives

$$\frac{f_s}{f_c} = \frac{6.0x}{5.08}$$

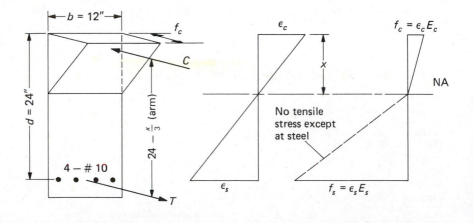

(a) Section   (b) Strains   (c) Stresses

Figure 4.4.1   Section for Example 4.4.1.

The ratio of f_s to f_c may also be obtained using the linear strain relationship and Hooke's law, as shown in Fig. 4.4.1,

$$\frac{\epsilon_s}{\epsilon_c} = \frac{24 - x}{x}$$

$$\frac{f_s}{f_c} = \frac{E_s \epsilon_s}{E_c \epsilon_c} = n\frac{\epsilon_s}{\epsilon_c} = 8\frac{(24 - x)}{x}$$

Note that the actual values of E_s and E_c are not needed. Equating the two expressions for f_s/f_c,

$$\frac{6.0x}{5.08} = \frac{8(24 - x)}{x}$$

$$6x^2 = 40.64(24 - x)$$

$$x^2 + 6.77x = 162.6$$

Solving by completing the square gives

$$(x + 3.38)^2 = 162.6 + (3.38)^2$$

$$x = \sqrt{174.0} - 3.38 = 9.81 \text{ in.}$$

Note that, other than the *ratio* of the moduli of elasticity, only the properties of the section (depth, width, and steel area) affect the position of the neutral axis. The loading does *not* affect the neutral axis location.

Using the second equilibrium condition,

$$M_w = 2000 \text{ in.-kips} = (C \text{ or } T) \times \text{arm}$$

The moment *arm* of the internal couple, or the distance from centroid of compressive solid to centroid of tension steel, equals for the case of the triangular wedge compressive solid,

$$\text{arm} = 24 - \frac{x}{3} = 24 - \frac{9.81}{3} = 20.73 \text{ in.}$$

Then for the given applied bending moment,

$$C = T = \frac{M_w}{\text{arm}} = \frac{2000}{20.73} = 96.5 \text{ kips}$$

The stresses are determined from the expressions for C and T,

$$f_c = \frac{C}{6x} = \frac{96,500}{6(9.81)} = 1640 \text{ psi}$$

$$f_s = \frac{T}{5.08} = \frac{96,500}{5.08} = 19,000 \text{ psi}$$

4.5 METHOD OF TRANSFORMED SECTION

In the method of transformed section, the cross section containing steel and concrete is transformed into a homogeneous section of one material all having the modulus of elasticity of concrete. This requires the replacement of the actual steel area by an equivalent area (i.e., imaginary area) in concrete. In the transformation, two conditions must be satisifed. Let A_s and A_t be the areas of and f_s and f_t be the tensile stresses in the actual steel and the equivalent concrete, respectively. First, the equilibrium condition requires that the total tensile force be the same, or

$$A_s f_s = A_t f_t \tag{4.5.1}$$

Second, the compatibility of deformation condition requires that the unit elongation be the same; or

$$\frac{f_s}{E_s} = \frac{f_t}{E_c} \tag{4.5.2}$$

Utilizing the modular ratio $n = E_s/E_c$ and solving Eqs. (4.5.1) and (4.5.2),

$$A_t = nA_s \tag{4.5.3}$$

$$f_t = \frac{f_s}{n} \tag{4.5.4}$$

Thus the equivalent concrete area A_t is n times the actual steel area, and the equivalent tensile stress f_t (i.e., imaginary stress) is $1/n$ times the actual tensile stress.

Equations (4.5.3) and (4.5.4) are useful in working stress design computations because a reinforced concrete section may then be treated as a section of one material, with the equivalent concrete on the tension side taking tension.

It may be noted that the use of transformed section is particularly convenient in that the common flexure formula Mc/I then applies so long as there is linear variation of stress with strain.

4.6 INVESTIGATION FOR SAFETY OF RECTANGULAR SECTIONS IN BENDING WITH TENSION REINFORCEMENT ONLY

In problems of investigation for safety the cross-sectional dimensions (including reinforcement area and location), the modular ratio, and the allowable working stresses are given. The problem may be (1) to compare the actual working stresses with the allowable working stresses for a given service load bending moment, or (2) to determine the allowable service load bending moment that the section may carry.

Two general approaches to investigation for safety may be used. In the "internal-force" method, as used in Example 4.4.1, the external bending moment is equated to the internal resisting couple. This couple is composed of an internal compressive force C on one side of the neutral axis and an internal tensile force T on the other side of the neutral axis, at a distance apart equal to the moment arm. In the transformed section method, the well-known flexure formula Mc/I is used, in which I is the moment of inertia I_{cr} of the transformed cracked section about its centroidal axis. Note that the neutral axis coincides with the centroidal axis only in a section under pure bending (without axial load).

The detailed procedure for investigating a rectangular section in bending with tension reinforcement only is illustrated in the following example.

EXAMPLE 4.6.1 Given $f'_c = 3000$ psi, allowable $f_s = 20,000$ psi, and the ACI Code, determine the allowable service load bending moment M_w that the rectangular section as shown in Fig. 4.6.1(a) may carry. Find the neutral axis by the transformed section method; then find M_w using both the internal-force and the transformed section methods.

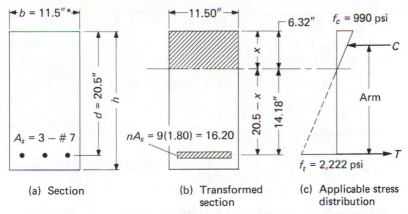

| (a) Section | (b) Transformed section | (c) Applicable stress distribution |

*The $11\frac{1}{2}$-in. beam width is an uncommon dimension, but is used here to provide a number that can be conveniently followed in calculations. For typical beam widths refer to Sec. 3.9.

Figure 4.6.1 Section for Example 4.6.1.

Solution: (a) Locate the neutral axis using transformed area [Fig. 4.6.1(b)]. For $f'_c = 3000$ psi, the modular ratio n may be taken as 9 (see Table 4.3.1). The area $11.50x$ above the neutral axis is in compression, whereas the equivalent tension area $nA_s = 16.20$ sq in. is assumed to be concentrated at a distance $(20.50 - x)$ below the neutral axis.

Equating the first moments of the compression and tension areas about the neutral or centroidal axis,

$$\tfrac{1}{2}(11.50)x^2 = 16.20(20.50 - x)$$

$$x = 6.32 \text{ in.}$$

(b) Determine applicable stress distribution. If the allowable compressive stress (see ACI-Appendix B.3.1) in concrete of $f_c = 0.45f'_c = 1350$ psi is realized at the extreme compressive face, the corresponding actual tensile stress f_t in the equivalent concrete would be

$$\text{actual } f_t = (1350)\frac{14.18}{6.32} = 3029 \text{ psi}$$

which cannot be since it exceeds the allowable f_t of $20,000/9 = 2222$ psi. The applicable stress variation is shown in Fig. 4.6.1(c), in which the actual f_t is made equal to the allowable f_t of 2222 psi. The actual f_c is, by proportion,

$$\text{actual } f_c = 2222(6.32/14.18) = 990 \text{ psi}$$

(c) Use the stress-solid internal-couple method to find allowable M_w.

$$\text{actual } C = \tfrac{1}{2}(0.990)(11.50)(6.32) = 36.0 \text{ kips}$$

$$\text{actual } T = 2.222(16.20) \quad \text{or} \quad 20(1.80) = 36.0 \text{ kips}$$

Note that the two internal forces are equal. This is always a good check on the correctness of x.

$$\text{actual arm} = 20.50 - \frac{6.32}{3} = 18.39 \text{ in.}$$

$$\text{allowable } M_w = 36.0(18.39)\tfrac{1}{12} = 55.2 \text{ ft-kips}$$

(d) Use the transformed section method to find M_w. The transformed cracked section moment of inertia is

$$I_{cr} = \tfrac{1}{3}(11.5)(6.32)^3 + 16.20(14.18)^2$$

$$= 968 + 3260 = 4230 \text{ in.}^4$$

By the flexure formula,

$$\text{allowable } M_w = \frac{0.990(4230)}{6.32(12)} \quad \text{or} \quad \frac{2.222(4230)}{14.18(12)} = 55.2 \text{ ft-kips } (74.9 \text{ kN·m})$$

4.7 UNDERREINFORCED, IDEALLY REINFORCED, AND OVERREINFORCED SECTIONS

The *ideally reinforced* section in the working stress method is one in which the neutral axis is so situated that the allowable stresses for both steel and concrete are reached simultaneously. Accordingly, the *underreinforced section* contains less steel than the ideally reinforced section; thus its neutral (or centroidal) axis is nearer to the compressive face and only the allowable steel stress, not the allowable concrete stress, can be reached under the allowable service loading. In a similar manner the *overreinforced section* contains more steel than if it were ideally reinforced, and only the allowable concrete stress can be reached. Generally, even sections overreinforced from the viewpoint of the working stress method have less tension steel than 75% of the amount in the balanced strain condition as defined in the strength method.

For the purpose of illustration, the three sections in Fig. 4.7.1 having the same size but different amounts of reinforcing steel may be compared.

The first section is the one just investigated in Example 4.6.1; similar work is performed on the other two sections, and the results are compiled in Fig. 4.7.1. The section of Fig. 4.7.1(b) is called "ideally reinforced" because at the allowable service load moment the allowable working stresses of $f_c = 1350$ psi and $f_s = 20,000$ psi are attained simultaneously. The section of Fig. 4.7.1(a) is called "underreinforced" because it contains less steel than the ideally reinforced section. Similarly, the section of Fig. 4.7.1(c) is said to be "overreinforced" because it contains more steel than the ideally reinforced section. Note additionally that the moment arm for the under-reinforced section is slightly greater, and the moment arm for the overreinforced section is slightly less, than on the ideally reinforced section.

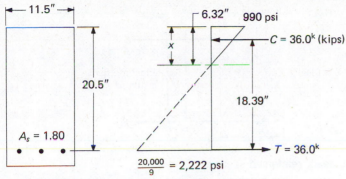

(a) Underreinforced section; $M_w = 55.2$ ft-kips

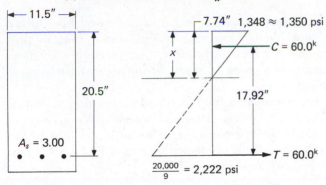

(b) Ideally reinforced section; $M_w = 89.6$ ft-kips

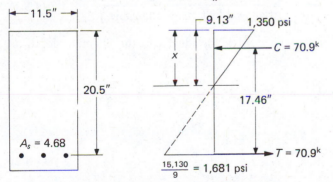

(c) Overreinforced section; $M_w = 103.1$ ft-kips

Figure 4.7.1   Underreinforced, ideally reinforced, and overreinforced sections.

The reinforcement ratio of the overreinforced section in Fig. 4.7.1(c) is 0.0198, while $0.75\rho_b$ for $f'_c = 3000$ psi and $f_y = 40,000$ psi is 0.0278 in strength design.

4.8 DESIGN OF RECTANGULAR SECTIONS IN BENDING WITH TENSION REINFORCEMENT ONLY

In problems of design, the bending moment, the modular ratio, and the allowable working stresses are given. The designer is to select the values of b, d, and A_s. The attainment of an ideally reinforced section can rarely be

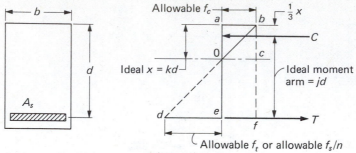

Figure 4.8.1 An ideally reinforced section.

accomplished with exactness, because b and h (d is derived from h) are adjusted to a desirable whole inch (occasionally $\frac{1}{2}$ in.) and A_s must be provided by an actual number of bars. In building construction it is common to use the same b and h for a group of beams having approximately equal loadings and span lengths.

The procedure for determining the theoretical values of b, d, and A_s of an ideally reinforced section will be described, without reference for the time being to the adjustment to convenient values of b and h or to the choice of bars. Consider the ideally reinforced section of Fig. 4.8.1. From similar triangles obc and dbf, $bc/bf = oc/df$ or

$$k = \frac{\text{ideal } x}{d} = \frac{\text{allowable } f_c}{\text{allowable } f_s/n + \text{allowable } f_c} \tag{4.8.1}$$

The term k is defined as the ratio of the ideal x to the effective depth d. Then

$$\text{ideal moment arm } jd = d - \frac{x}{3} = d - \frac{kd}{3}$$

or

$$j = 1 - \frac{k}{3} \tag{4.8.2}$$

The term j is defined as the ratio of the ideal moment arm jd to the effective depth d. Equating the internal forces C and T and calling $A_s = \rho bd$,

$$\tfrac{1}{2}(\text{allowable } f_c)(bkd) = (\text{allowable } f_s)(\rho bd)$$

from which

$$\rho = \frac{k \, (\text{allowable } f_c)}{2 \, (\text{allowable } f_s)} \tag{4.8.3}$$

This ρ is the reinforcement ratio for an ideally reinforced beam. Equating the bending moment M_w to Cjd,

$$M_w = \tfrac{1}{2}(\text{allowable } f_c)(bkd)(jd) = \tfrac{1}{2}(\text{allowable } f_c)jkbd^2 = Rbd^2$$

in which

$$R = \tfrac{1}{2}(\text{allowable } f_c)jk \qquad (4.8.4)$$

Equating the bending moment M_w to Tjd, $M_w = A_s$ (allowable f_s)jd, from which

$$A_s = \frac{M_w}{(\text{allowable } f_s)jd} \qquad (4.8.5)$$

The constants k, j, ρ, and R as expressed in Eqs. (4.8.1) to (4.8.4) will be called the four *design constants* of an ideally reinforced section. Their values depend on the allowable working stresses f_c and f_s and on the value of n. It is to be noted that k, j, and ρ are dimensionless constants, but R is usually in psi.

In general, the ideally reinforced value of ρ is approximately one-half of the maximum permitted by the strength method. Other values for the ideal reinforcement ratio are given in Table 4.8.1.

Table 4.8.1 RATIO ρ FOR IDEALLY REINFORCED RECTANGULAR BEAMS ACCORDING TO WORKING STRESS METHOD[a]

f_y		$f'_c = 3000$ psi (21 MPa) $n = 9$	$f'_c = 3500$ psi (24 MPa) $n = 8.5$	$f'_c = 4000$ psi (28 MPa) $n = 8$	$f'_c = 5000$ psi (35 MPa) $n = 7$
(psi)	(MPa)				
40,000	280	0.0128	0.0158	0.0188	0.0248
50,000	350	0.0128	0.0158	0.0188	0.0248
60,000	410	0.0095	0.0117	0.0141	0.0186

[a]Note: SI equivalents are approximate, given to two significant figures.

The procedure for determining the theoretical values of b, d, and A_s of an ideally reinforced section may be summarized as follows:

1. Find the required value of bd^2 from M_w/R.
2. Assume a value of b and determine d (select b and h and check weight). See Sec. 3.9 for practical selection of beam sizes.
3. Determine A_s from ρbd and check its value from $M_w/[(\text{allowable } f_s)(jd)]$.

It is to be noted that there may be many usable solutions in accordance with the above procedure.

EXAMPLE 4.8.1 Design the cross section for a rectangular beam to carry a uniform live load of 1.9 kips/ft and a uniform dead load of 1.0 kip/ft (not including beam weight) on a simply supported span of 32 ft. Use the ACI Code with $f'_c = 4000$ psi and $f_y = 60,000$ psi (Grade 60 steel).

Solution: (a) Determine design constants for an ideally reinforced section. The allowable stresses (ACI-Appendix B.3) are $f_c = 0.45f'_c = 1800$ psi, $f_s = 24,000$ psi for Grade 60 steel, and $n = 8$. Referring to Fig. 4.8.2(b) and using similar triangles,

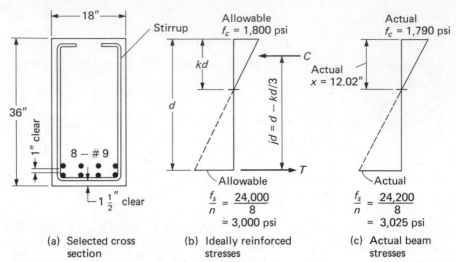

Figure 4.8.2 Section for Example 4.8.1.

$$k = \frac{1800}{1800 + 24{,}000/8} = 0.375$$

$$j = 1 - \frac{k}{3} = 0.875$$

$$R = \tfrac{1}{2} f_c k j = \tfrac{1}{2}(1800)(0.375)(0.875) = 295 \text{ psi}$$

$$\rho = \frac{1}{2} \frac{k f_c}{f_s} = \frac{1}{2}(0.375)\frac{1800}{24{,}000} = 0.01405$$

(b) Determine preliminary size. Estimating beam weight per foot at 0.5 kip/ft,

$$M_w \text{ (live load)} = \tfrac{1}{8}(1.9)(32)^2 = 243 \text{ ft-kips}$$

$$M_w \text{ (dead load)} = \tfrac{1}{8}(1.0 + 0.5)(32)^2 = 192 \text{ ft-kips}$$

$$M_w = 243 + 192 = 435 \text{ ft-kips}$$

Since there is no size limitation, it is desired to make the beam approximately ideally reinforced. Thus

$$\text{required } bd^2 = \frac{M_w}{R} = \frac{435(12{,}000)}{295} = 17{,}700 \text{ in.}^3$$

TRIAL b	REQUIRED d	
12	38.4	
15	34.4	
18	31.4	←Try

Overall depth required (assuming two layers of equal sized bars) = 31.4 + clear cover + stirrup diameter + bar diameter + 0.5 of clear between layers, where clear cover is 1.5 in. (ACI-7.7.1) and stirrup diameter is $\frac{3}{8}$ or $\frac{1}{2}$ in. usually. [Note: Stirrups enclosing the main steel, as shown in Fig. 4.8.2, provide shear capacity. Since nearly all beams contain stirrups, whose size is generally not definitely established until later in the design process (see Chap. 5), provision for them should be made.] Bar diameter is, say, 1 in. One-half clear between layers is 0.5 in. (ACI-7.6.2). Thus for two layers of bars the overall depth is about $3\frac{1}{2}$ to 4 in. greater than the required d.

Overall depth $h = 31.4 + 4.0 = 35.4$ in. Try a beam 18×36.

(c) Check weight, and revise step (b) if necessary.

$$\text{weight} = \frac{18(36)}{144}(0.15) = 0.675 \text{ kip/ft}$$

$$\text{corrected } M_w = 243 + \tfrac{1}{8}(1.675)(32)^2 = 457 \text{ ft-kips}$$

For $b = 18$ in.,

$$\text{revised required } d = 32.1 \text{ in.}$$

$$\text{required } h = 32.1 + \approx 4 = 36.1 \approx 36 \text{ in.} \qquad \text{OK}$$

A beam 18×36 appears acceptable.

(d) Determine steel area required. If #8 bars may possibly be used, 3.5 in. instead of 4.0 might be subtracted from h to obtain d.

$$\text{actual } d \approx 36 - 3.5 = 32.5 \text{ in.}$$

$$\text{required } A_s = \frac{M_w}{f_s jd} = \frac{457(12)}{24(0.875)(32.5)} = 8.03 \text{ sq in.}$$

Try 8-#9 in two layers; $A_s = 8.00$ sq in.

(e) Check the section:

$$d = 36 - 1.5 - 0.5 - 1.128 - 0.5 = 32.37 \text{ in.}$$

$$\underset{\uparrow}{\rule{0pt}{0pt}}\text{(stirrup diameter estimate)}$$

Locate neutral axis:

$$18x\left(\frac{x}{2}\right) = 8.0(8)(32.37 - x)$$

$$x^2 + 7.12x = 230$$

$$x = 12.02 \text{ in.}$$

$$\text{arm} = 32.37 - 12.02/3 = 28.36 \text{ in.}$$

$$C = T = \frac{M_w}{\text{arm}} = \frac{457(12)}{28.36} = 193.5 \text{ kips}$$

$$\text{actual } f_s = \frac{T}{A_s} = \frac{193.5}{8.0} = 24.2 \text{ ksi} \approx 24 \text{ ksi} \qquad \text{OK}$$

$$\text{actual } f_c = \frac{C}{(\frac{1}{2})bx} = \frac{193.5}{0.5(18)12.02} = 1.79 \text{ ksi} < 1.80 \text{ ksi} \qquad \text{OK}$$

Even though the steel stress is slightly high, usual practice would accept a slight overstress.

Selection of beam size and of bars has been done according to the general guidelines of Sec. 3.9. The designer must also be certain that the bars fit into the beam width (see Table 3.9.2) without violating code clearance requirements (ACI-7.6.1). The same cover requirements apply for the side of a beam as for the top or bottom. The selected cross section is shown in Fig. 4.8.2.

If in the preceding example a section 18×34 ($d \approx 30$ in.) instead of 18×36 ($d = 32.37$) is used, it is obvious that the A_s required must be larger than 8.00 in order to resist the same bending moment. The 18×34 section, containing more than its ideal reinforcement, must be an overreinforced section. On the other hand if a section 18×40 ($d \approx 36$ in.) is used, the A_s required will be smaller than 8.00. The 18×40 section, containing less than its ideal reinforcement, must be an underreinforced section. In the following two examples the exact amounts of steel required in the overreinforced and underreinforced sections, respectively, are determined.

Design of an Overreinforced Section

EXAMPLE 4.8.2 For the service load moment $M_w = 457$ ft-kips (final M_w for Example 4.8.1), determine the required A_s if a section 18×34 in. were to be used. Use $f'_c = 4000$ psi, $f_y = 60,000$ psi, and the ACI Code. (The value of M_w could be revised due to reduction in weight of beam, but the purpose here is to illustrate the increase in required A_s for the same M_w.)

Solution: (a) Allowable stresses and design constants. From ACI-Appendix B.3, allowable $f_c = 1800$ psi, allowable $f_s = 24,000$ psi, $n = 8$. Design constants are $k = 0.375$, $j = 0.875$, and $R = 295$ psi (see Example 4.8.1 for computation).

(b) Determine whether design as a singly reinforced beam must be underreinforced, overreinforced, or, by chance, ideally reinforced. (In this case, based on solution to Example 4.8.1, the reduced size section will be overreinforced.) Estimating two layers of steel, $d \approx 30$ in.

$$\text{required } R = \frac{M_w}{bd^2} = \frac{457(12,000)}{18(30)^2} = 339 \text{ psi}$$

Since this exceeds the ideal R, the section would not carry 457 ft-kips if it were ideally reinforced.

As shown in Fig. 4.8.3, the compressive force C must be increased from that of an ideally reinforced section. One way this can be accomplished is by adding extra tension steel for the sole purpose of lowering the neutral axis. Thus the extra steel is there to give the necessary neutral axis location and will be acting under very low actual stress. This is the main drawback of the working stress method; it does not recognize the "strength" of an overreinforced section (as defined in working stress design).

(c) Determine the required stress distribution.

$$M_w = (\text{required } C)(d - x/3)$$
$$457(12) = \tfrac{1}{2}(1.80)(18x)(30 - x/3)$$

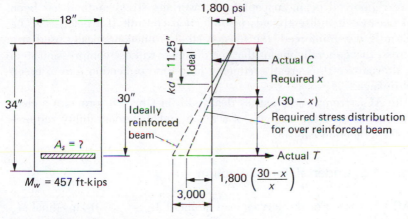

Figure 4.8.3   Section for Example 4.8.2.

Solving the quadratic equation for the required neutral axis distance x,

$$\text{required } x = 13.23 \text{ in.}$$
$$C = \tfrac{1}{2}(1.80)(18)(13.23) = 214 \text{ kips}$$
$$\text{arm} = d - x/3 = 25.59 \text{ in.}$$
$$M_w = C \text{ times (arm)} = 214(25.59)\tfrac{1}{12} = 457 \text{ ft-kips} \quad \text{(Check)}$$

In order for the neutral axis to be at $x = 13.23$ in., the stress used to compute the required steel area must be the value defined by the straight-line stress relationship; a much smaller value than the code-specified allowable f_s of 24 ksi. Thus the reduced allowable value is

$$\text{allowable } f_s = nf_c \left(\frac{d - x}{x} \right)$$

$$= 8(1.80) \left(\frac{30 - 13.23}{13.23} \right) = 18.3 \text{ ksi}$$

(d) Determine required steel area.

$$\text{required } A_s = \frac{T}{\text{allowable } f_s} = \frac{214}{18.3} = 11.7 \text{ sq in.}$$

The selection of bars, recomputation of d, and a final check of stresses are steps omitted here.

As a section becomes more overreinforced, the reduced allowable stress may get so small as to make an overreinforced beam impractical or impossible. The alternative is to use compression steel to increase the compressive force.

Computation by the ACI strength method shows that the use of a maximum reinforcement ratio of $0.75\rho_b$ (a steel area of 11.6 sq in.) would give a nominal strength M_n of 1410 ft-kips, which is 3.1 times the working moment M_w of 457 ft-kips. Even if almost all of the working moment is from live load, the U/ϕ value can only be as large as $1.7/0.90 = 1.89$. However, the overreinforced beam concept in the working stress method has been useful because it indirectly aids in deflection control. If a beam is to be significantly overreinforced, the tension steel is uneconomically used at a low stress; this generally means that the designer will use compression steel as an alternative, thus greatly reducing long-time deflection due to creep and shrinkage.

The ACI strength method, as dealt with in Chap. 3, provides a clear separation between the strength of a beam and its serviceability requirements for deflection control.

Design of an Underreinforced Section

EXAMPLE 4.8.3   For the service load moment $M_w = 457$ ft-kips (final M_w for Example 4.8.1), determine the required A_s if a section 18 × 40 in. were to be used. Use $f'_c = 4000$ psi, $f_y = 60,000$ psi, and the ACI Code. (The value of M_w should be revised due to increase in weight of beam, but the purpose here is to illustrate the decrease in required A_s for the same M_w.)

Solution: (a) Allowable stresses and design constants. From ACI-Appendix B.3, allowable $f_c = 1800$ psi, allowable $f_s = 24,000$ psi, $n = 8$. Design constants are $k = 0.375$, $j = 0.875$, and $R = 295$ psi (see Example 4.8.1 for computation).

(b) Determine whether design as a singly reinforced beam must be underreinforced, overreinforced, or, by chance, ideally reinforced. Estimating two layers of steel, $d \approx 36$ in.,

$$\text{required } R = \frac{M_w}{bd^2} = \frac{457(12,000)}{18(36)^2} = 235 \text{ psi}$$

Since this is less than the ideal R, the section, if ideally reinforced, would carry more bending moment than necessary. Thus less steel will be needed than the ideal amount, and the section should be designed as underreinforced.

Shown in Fig. 4.8.4, the required compressive force C is less than that

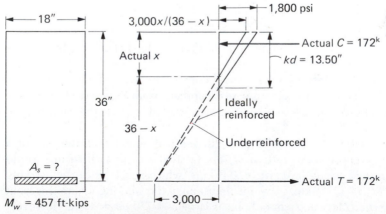

Figure 4.8.4   Section for Example 4.8.3.

of an ideally reinforced section. The neutral axis distance x will be smaller than kd and the actual concrete stress will be less than the allowable.

(c) Determine the required stress distribution.

$$M_w = (\text{actual } C)(d - x/3)$$

$$457(12) = \frac{1}{2}\left(\frac{3.00x}{36 - x}\right)(18x)\left(36 - \frac{x}{3}\right)$$

$$x^3 - 108x^2 - 609x + 22{,}000 = 0$$

Solving the cubic equation for the required neutral axis position,

$$\text{required } x = 12.29 \text{ in.}$$

$$f_c = \frac{3000x}{36 - x} = \frac{3000(12.29)}{36 - 12.29} = 1560 \text{ psi}$$

$$C = \tfrac{1}{2}f_c bx = \tfrac{1}{2}(1.56)(18)(12.29) = 172 \text{ kips}$$

$$\text{available arm} = 36 - 12.29/3 = 31.9 \text{ in.}$$

$$M_w = 172(31.9)\tfrac{1}{12} = 457 \text{ ft-kips} \qquad \text{(Check)}$$

$$\text{required } A_s = \frac{T \text{ or } C}{\text{allowable } f_s} = \frac{172}{24} = 7.16 \text{ sq in.}$$

Since the steel controls for the underreinforced beam, the effort of solving the cubic equation only provides the slightly larger moment arm (31.9 in.) compared with the ideal jd (31.5 in.); thus an approximate value of the A_s required may be determined from the following formula:

$$A_s \left(\begin{array}{c}\text{in } underreinforced\\ \text{sections only}\end{array}\right) = \frac{M_w}{(\text{allowable } f_s)(\text{design constant } j)(\text{actual } d)}$$

$$(4.8.6)$$

For the present problem,

$$A_s = \frac{457(12)}{24(0.875)(36)} = 7.25 \text{ sq in.}$$

This value of A_s is very close to its correct value using the cubic equation, and the error is on the safe side.

The A_s required in an *underreinforced* section with tension reinforcement only is usually determined from Eq. (4.8.6). In practical design the only time the cubic solution value for the moment arm is used is when it is readily available from design aids.

For *overreinforced* sections, however, Eq. (4.8.6) should *never* be used. For instance, had it been used in Example 4.8.2, the A_s required would have been $(457)(12)/[24(0.875)(30)] = 8.7$ sq in., which is far from the correct value of 11.7 sq in.

4.9 DESIGN OF RECTANGULAR SECTIONS IN BENDING WITH BOTH TENSION AND COMPRESSION REINFORCEMENT

When the size of a rectangular section is restricted such that the bending moment is decidedly larger than the resisting moment of the section if ideally reinforced, the working stress method indicates the use of both tension and compression reinforcement. This would, in effect, limit the amount of tension steel to about 0.40 to 0.60 of the maximum permitted in the strength method. The only unknowns in such a case are the values of the tension steel A_s and the compression steel A_s'. In practice there may be complications as to whether all the required A_s can be placed in one layer within the width of the section, because the effective depth d will have to be further reduced from the available h if two layers of steel become necessary. In this discussion it is assumed that the effective depth d is given, and this will be measured from the extreme compressive face to the centroid of the tension steel, regardless of the number of layers used.

The "two-couple" method is commonly used to determine the required values of A_s and A_s' in a doubly reinforced section. In this method the bending moment is considered equal to the sum of two resisting couples, one of which is provided by an ideally reinforced section with tension steel only, and the other by the compression steel and the remainder of the tension steel. Thus the section with compression steel is designed as an ideally reinforced section in that the compression steel and the extra tension steel for the second resisting couple are so proportioned that the neutral axis is maintained in the ideally reinforced position. By referring to Fig. 4.9.1, the procedure for determining A_s and A_s' is as follows:

1. Compute $M_{w1} = Rbd^2$.
2. Compute $A_{s1} = \rho bd$ and check with $A_{s1} = M_{w1}/[(\text{allowable } f_s)(jd)]$.
3. Compute $M_{w2} = M_w - M_{w1}$.
4. Compute $C_2 = T_2 = M_{w2}/(d - d')$.
5. Compute $A_{s2} = T_2/\text{allowable } f_s$.
6. Compute $A_s = A_{s1} + A_{s2}$.
7. Compute $f_{c1} = (\text{allowable } f_c)(kd - d')/kd$.
8. Compare $2nf_{c1}$ with allowable f_s.
9. Compute $A_s' = C_2/[(2n - 1)f_{c1}]$ if $2nf_{c1} < \text{allowable } f_s$; compute $A_s' = C_2/(\text{allowable } f_s - f_{c1})$ if $2nf_{c1} > \text{allowable } f_s$.

Steps 1 to 6 are self-explanatory. Steps 7 to 9 will now be explained.

In a rectangular section with tension reinforcement only, concrete alone takes the compression. In a doubly reinforced section, concrete and steel act together to take the compression. If concrete and steel were both elastic,

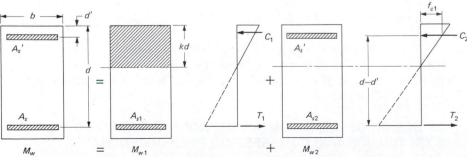

Figure 4.9.1 Two-couple method for doubly reinforced sections.

that is, if all deformation occurred at the instant when load was applied and disappeared when the load was removed, the method of transformed section should still apply; that is, the compression steel would be replaced by an equivalent concrete area equal to n times its actual area. However, as discussed in Sec. 1.10 concrete under stress deforms (creeps) with time and it also is subject to shrinkage over a period of time. These time-dependent effects do not occur in the steel. Hence as concrete deforms, even at or below its allowable working stress, there is a continuous transfer of load from concrete to steel. In order that the straightline variation of stress across the depth of the section may be applied to doubly reinforced sections, one way of approximating the effect of this transfer of load is to increase the equivalent concrete area of compression steel to more than n times the actual steel area.

An effective modular ratio of $2n$ is permitted by ACI-Appendix B.5.5 when transforming the compression reinforcement into equivalent concrete area for *stress computations*. The ACI provision means actually that the stress in the compression reinforcement in an elastic analysis is $2nf_{c1}$ (see Fig. 4.9.1), but it may not exceed the allowable f_s. This accounts approximately for the transfer of load to the compression steel, but also prevents one from assuming a compression capacity of steel that exceeds its tensile capacity. Thus steps 7 and 8 are explained.

The compression steel actually displaces some concrete area which has already been counted on in taking the internal force C_1 in the first resisting couple. Thus the internal force C_2 represents the excess from the compression reinforcement over that required to replenish the little force $f_{c1}A'_s$ already included in the internal force C_1; or $C_2 = [(2nf_{c1}$ or allowable f_s, whichever is smaller) $A'_s - f_{c1}A'_s]$. The expressions in step 9 are obtained by solving the above equation for A'_s.

4.10 INVESTIGATION FOR SAFETY OF RECTANGULAR SECTIONS IN BENDING WITH BOTH TENSION AND COMPRESSION REINFORCEMENT

The investigation for safety of a rectangular section in bending with both tension and compression reinforcement is complicated by the provision that the stress in the latter can be either $2nf_{c1}$ or allowable f_s, whichever is smaller. On the other hand when computing the moment of inertia of the transformed section in deflection calculations, the equivalent area in concrete of the compression steel is definitely to be taken at n times, not $2n$ times, the actual area of the compression steel, because creep and shrinkage (time-dependent effects) are accounted for in a different way (see Chap. 14 on deflections).

Returning to the problem of finding the allowable service load moment for a doubly reinforced section, the important step is to determine the neutral axis distance x. If f_{c1} is the value on the linear stress diagram at the level of the compression steel, then $f_{c1} = f_c(x - d')/x$. At the outset it is known whether the additional compressive force C_s provided by the compression steel is (allowable $f_s - f_{c1}$) times A'_s or $(2nf_{c1} - f_{c1})$ times

A'_s, with the latter prevailing if $2nf_{c1}$ is less than allowable f_s. Thus, it is necessary to select one possibility and make the conformation after x is solved from the equilibrium condition $C_c + C_s = T$; or

$$\frac{1}{2} f_c bx + \left(\begin{matrix} 2nf_{c1} - f_{c1} \\ \text{or} \\ \text{allowable } f_s - f_{c1} \end{matrix} \right) A'_s = f_c \frac{(x - d)}{x} A_s \qquad (4.10.1)$$

Note that the term f_c, which is the actual extreme fiber concrete stress, cancels out in the above equation so that x is the only unknown.

Numerical examples for design and investigation of doubly reinforced sections by the working stress method appeared in earlier editions of this book.

4.11 SERVICEABILITY—DEFLECTIONS

Whether safety of a beam or floor system is established by the working stress method (ACI "alternate design method") or by the strength design method, excessive deflection may make the system unserviceable. When flexural members support or are attached to partitions and other construction likely to be damaged by large deflection, deflection computations *under service load conditions* will usually be necessary. Even in situations in which excessive deflection may not crack or damage anything, large noticeable deflection is psychologically disturbing to humans.

For situations of "members not supporting or attached to partitions or other construction likely to be damaged by large deflections," ACI-Table 9.5(a) provides minimum thickness for beams and one-way slabs unless deflections are computed. When the minimum thickness requirement of ACI-Table 9.5(a) is not satisfied *or* if excessive deflection may cause a problem, deflection must be computed and must satisfy the limits of ACI-Table 9.5(b).

Since the deflection that concerns the designer is a service load phenomenon, the elastic beam properties (such as moment of inertia of the transformed cracked section) are a necessary part of the computations. Properties such as neutral axis location and moment of inertia involve the basic assumptions of the working stress method, and hence are dealt with in this chapter.

Since the applied service load moment usually far exceeds the moment that causes the tension concrete to crack, the so-called properties of the "cracked section" are needed. In earlier sections of this chapter the neutral axis of the cracked section has been obtained from either (a) equating the compressive force C to the tensile force T, or (b) locating the centroid of the transformed section (see Secs. 4.4 and 4.6).

Deflection calculations involve using a formula of the type,

$$\Delta = \beta_a \frac{M_w L^2}{E_c I_e} \qquad (4.11.1)$$

where

$\beta_a =$ constant that depends on the loading and support conditions
$L =$ span length

M_w = maximum service load moment
E_c = concrete modulus of elasticity
I_e = effective moment of inertia

The effective moment of inertia I_e used by the ACI Code (ACI-9.5) involves both the gross section moment of inertia I_g (commonly without steel) and the transformed cracked section moment of inertia I_{cr}. The following example shows the computation of the moment of inertia of the transformed cracked section for a section having both tension and compression steel.

Since deflection computations are based on *elastic* conditions and service loads, it is the *elastic* section properties such as centroid and moment of inertia that are needed. Thus, in obtaining the transformed section all steel, both tension and compression, is transformed into equivalent concrete using the elastic modular ratio n. According to ACI-9.5, the long time deflection under sustained loads due to creep and shrinkage is obtained by multiplying the *elastic* "instantaneous" deflection by a time-dependent multiplier. Thus, even for computation of the time-dependent deflection, the elastic I_{cr} is used.

EXAMPLE 4.11.1 Compute the moment of inertia I_{cr} of the transformed cracked section to be used in deflection calculation for the doubly reinforced section shown in Fig. 4.11.1, using $f'_c = 4000$ psi $(n = 8)$.

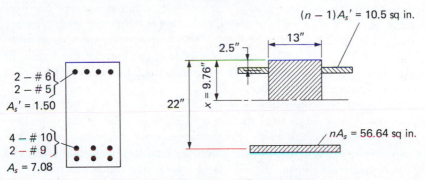

Figure 4.11.1 Doubly reinforced section of Example 4.11.1.

Solution: (a) Locate neutral axis. The steel is transformed into equivalent concrete by using $A_t = nA_s$ for both compression and tension steel. Referring to Fig. 4.11.1,

$$\frac{13x^2}{2} + 10.5(x - 2.5) = 56.64(22 - x)$$

Solving for x,

$$x = 9.76 \text{ in.}$$

(b) Determine transformed cracked section moment of inertia.

$$I_{cr} = \tfrac{1}{3}(13)(9.76)^3 + 10.5(9.76 - 2.5)^2 + 56.64(22 - 9.76)^2$$

$$= 4030 + 550 + 8480 = 13{,}100 \text{ in.}^4$$

For examples of computing deflection according to the ACI Code (ACI-9.5), the reader is referred to Sec. 14.12.

4.12 SERVICEABILITY—FLEXURAL CRACK CONTROL FOR BEAMS AND ONE-WAY SLABS

Cracking in concrete is generally the result of the following actions [1,2]: (1) volumetric change, including that due to drying shrinkage, creep under sustained load, thermal stresses, and chemical incompatibility of concrete components; (2) internal or external direct stress due to continuity, reversible load, long-time deflection, camber in prestressed concrete, or differential movement in structures; and (3) flexural stress due to bending.

Visible cracking is generally initiated by either internal microcracking (volumetric change would usually induce this type) or flexural microcracks. Flexural microcracks are surface cracks that are not visible except by careful close investigation and are generally initiated by flexural stress. Once flexural microcracks have formed, a slight increase in flexural load causes these cracks to open up suddenly to measurable widths. An excellent source for information on control of cracking resulting from flexural and other causes is contained in an ACI Committee 224 report [1].

The use of the strength design method and Grade 60 reinforcement has meant that service load *strains* in the tension steel can be 50% higher than those in existing structures designed by the working stress method. Higher strains mean wider cracks. For example, using an average overload factor of, say, 1.55 divided by strength reduction factor $\phi = 0.90$ for flexure, gives 1.72 as the total factor for safety. This corresponds approximately to $f_y/1.72 = 0.58f_y$, say $0.60f_y$, as the service load steel stress. For Grade 60 steel, $0.60f_y = 36$ ksi, which is 50% more than the 24 ksi allowed for Grade 60 in the working stress method.

Because of the compatibility in strain between concrete and steel, the 50% or more increase in steel strain would seem to indicate the same percentage increase in crack width. Furthermore, even at low steel stress levels, say 9000 psi (63 MPa), flexural microcracking has been found to occur [2]. As larger cracks form, the effects of a corrosive environment may be detrimental to the steel. Such factors as humidity, salt air, alternate wetting and drying, or freezing and thawing may accelerate corrosion and contribute to concrete deterioration in the vicinity of large width cracks. Wide cracks may also be unsightly and contribute to doubt about structural safety. Although cracking cannot be expected to be eliminated, it is generally more desirable to have many fine hair cracks than a few wide cracks. Thus crack control is a matter of controlling the distribution and size of cracks rather than eliminating them.

To control cracking it is better to use several smaller bars at moderate spacing than larger bars of equivalent area. The objective is therefore one of distributing the reinforcement in the concrete tension zone; hence the words in the title of ACI-10.6, "Distribution of flexural reinforcement," which contains the crack control provisions for beams and one-way slabs. Control of cracking is particularly important when reinforcement with a yield stress in excess of 40,000 psi (280 MPa) is used, or when reinforcement percentages exceed the amount of reinforcement traditionally used in the working stress method. For the strength method the comparable percentage is about $0.375\rho_b$ (one-half the maximum permitted amount).

Good bar arrangement in the cross section will usually lead to adequate crack control even when Grade 60 bars are used. Entirely satisfactory structures have been built, particularly in Europe, using design yield stresses exceeding 80,000 psi (560 MPa), which is the current specified limit in ACI-9.4.

Extensive laboratory studies [2–11] have verified the generally accepted belief that crack width is proportional to steel stress. Two other significant variables have been found to be the thickness of concrete cover and the area of concrete surrounding each individual reinforcing bar in the zone of maximum tension.

The ACI Code provisions (ACI-10.6.4) are based on the Gergely-Lutz [4] expression for crack width, as follows:

$$w = C\beta_h f_s \sqrt[3]{d_c A} \qquad (4.12.1)$$

where, referring to Fig. 4.12.1,

w = crack width at the tension face of the beam (in. or mm)

β_h = h_2/h_1, the ratio of the distances to the working stress neutral axis from the extreme tension fiber and from the centroid of the main tension reinforcement

f_s = service load stress in the steel (ksi or MPa)

d_c = thickness of concrete cover measured from the extreme tension fiber to the center of the bar located closest thereto (in. or mm)

A = A_e/m, effective tension area of concrete surrounding the main tension reinforcing bars and having the same centroid as that reinforcement, divided by the number of bars (sq in. or mm²)

m = number of bars; for different sizes use $m = A_s/(A_b$ for largest bar)

C = an experimental constant (76×10^{-6} sq in./kip for Inch–Pound units or 11.0×10^{-6} mm²/N for SI units)

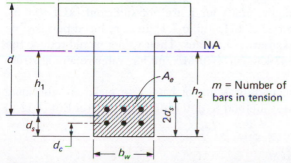

Figure 4.12.1 Dimensional notation for Gergely–Lutz equation (4.12.1).

Equation (4.12.1) represents the most probable maximum crack width on the bottom face of a beam. Because of considerable scatter expected, even under carefully controlled laboratory conditions, crack width calculations should serve only as a guide to good detailing of bars and not as values to use for comparison with measured cracks in a building that is in service. Consequently Eq. (4.12.1) has been converted into a form in which a simple calculation could be made to arrive at reasonable reinforcing details as indicated by experience and laboratory tests, without actually emphasizing crack width. Thus dividing Eq. (4.12.1) by $C\beta_h$, the quantity z is defined,

$$z = f_s \sqrt[3]{d_c A} = \frac{w}{C\beta_h} \tag{4.12.2}$$

For simplification β_h may be taken at an approximate value of 1.2. In order to have numerical values for the quantity $w/(C\beta_h)$ in the ACI Code, crack widths have been limited to 0.016 and 0.013 in. (0.41 and 0.33 mm) for interior and exterior exposure, respectively. Substitution of these values in Eq. (4.12.2) gives

$$z_{\text{limit}} = \frac{16}{0.076(1.2)} = 175 \text{ kips/in. (interior) (30.6 MN/m)}$$

$$z_{\text{limit}} = \frac{13}{0.076(1.2)} = 142.5 \approx 145 \text{ kips/in. (exterior) (25.4 MN/m)}$$

which are limits specified in ACI-10.6.4* when the design yield stress f_y for tension reinforcement exceeds 40,000 psi (280 MPa).

The Gergely-Lutz expression has been found [10,11] to apply also to one-way slabs (wide beams). However, the average value for $\beta_h = h_2/h_1$ (see Fig. 4.12.1) is about 1.35 for floor slabs, rather than 1.2 which applies to beams. Accordingly for the same crack control on one-way slabs as for beams the limits of z should be reduced by the factor 1.20/1.35 [i.e., z should be limited to 155 kips/in. (27.1 MN/m) for interior exposure and about 130 kips/in. (22.9 MN/m) for exterior exposure].

For two-way slabs, the above crack control procedure and limiting z values do not apply; other recommendations are given in the ACI Committee 224 Report [1]. Chapter 16 on two-way slabs also contains brief treatment.

When structures are "subject to very aggressive exposure or designed to be watertight," the requirements for z are not sufficient (ACI-10.6.5). For guidance ACI Committee 224 [1, Table 4.1] gives reduced permissible crack widths for various exposure conditions. Their values would reduce the z limitation to as little as 45 kips/in. (7.9 MN/m) for water-retaining structures.

In recognition of the undesirability of having to make an additional analysis using working stress method to determine f_s for use in Eq. (4.12.2)

*For SI, ACI 318–83M, Sec. 10.6.4, gives

$$z_{\text{limit}} = 30 \text{ MN/m (interior)}$$

$$z_{\text{limit}} = 25 \text{ MN/m (exterior)}$$

when the strength method is otherwise being used, ACI-10.6.4 permits taking f_s as 60% of the specified yield stress f_y in lieu of using the working stress computation. This will be a conservative approach because it will frequently overestimate the stress f_s. When a steel percentage approximately equal to that in an ideally reinforced section in the working stress method is used (corresponds in strength method to $\rho \approx 0.375\rho_b$), the value of f_s for Grade 60 steel would be about 24,000 psi which is only 40% of the yield stress.

EXAMPLE 4.12.1 Check the crack control provisions of the ACI Code against the cross section selected in Example 4.8.1 [Fig. 4.8.2(a)]. The selected beam has $b = 18$ in., $h = 36$ in., 8-#9 bars in two layers, #3 stirrups, 1.5 in. clear cover at bottom, and 1 in. clear between layers; $f'_c = 4000$ psi and Grade 60 steel are used.

Solution: The crack control provision of the ACI Code is given under the heading "Distribution of flexural reinforcement in beams and one-way slabs." Compute z of ACI-10.6.4:

$$z = f_s\sqrt[3]{d_c A}$$

f_s = 24.2 ksi (computed in Example 4.8.1)

d_c = distance to centroid of bottom layer

  = 1.5 (cover) + 0.375 (stirrup) + 0.564 (bar radius)

  = 2.44 in.

$A = A_e/m$

$A_e = 2d_s b$

d_s = distance to centroid of tension bar group

  = 1.5 (cover) + 0.375 (stirrup) + 1.128 (bar diameter) + 0.5

  = 3.50 in.

A_e = 2(3.50)18 = 126 sq in.

m = number of tension bars = 8

A = 126/8 = 15.75 sq in./bar

z = $24.2\sqrt[3]{2.44(15.73)}$ = 81.5 kips/in.

Since this does not exceed the 175 and 145 kips/in. (30.6 and 25.4 MN/m) allowed (ACI-10.6.4) for interior and exterior exposure, respectively, the selected cross section is acceptable.

EXAMPLE 4.12.2 Examine the crack control situation of Example 4.12.1 if 2-#18 bars had been used instead of 8-#9.

Solution: Referring to Fig. 4.12.2 for this cross section

$$d_c = d_s = 1.5 + 0.375 + \frac{2.257}{2} = 3.00 \text{ in.}$$

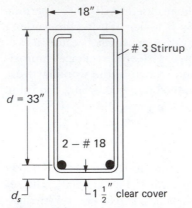

Figure 4.12.2   Cross section for Example 4.12.2 (an unsatisfactory bar selection).

The steel stress f_s should be computed using elastic properties of the cross section, as follows:

$$18x\left(\frac{x}{2}\right) = 8.0(8)(33.0 - x)$$

$$x = 12.17 \text{ in.}$$

$$f_s = \frac{M_w}{A_s(\text{arm})} = \frac{457(12)}{8.0(33 - 12.17/3)} = 23.7 \text{ ksi}$$

If the elastic analysis is not made, f_s is to be taken at $0.60f_y$ which is 36.0 ksi (ACI-10.6.4).

$$A_e = 2(3.0)18 = 108 \text{ sq in.}$$

$$m = 2$$

$$A = \frac{A_e}{m} = \frac{108}{2} = 54 \text{ sq in./bar}$$

$$z = f_s\sqrt[3]{d_cA} = 23.7\sqrt[3]{3.0(54)}$$

$$= 129 \text{ kips/in.} < 145 \text{ kips/in.} \qquad \text{OK}$$

Even this generally undesirable bar arrangement (see guidelines of Sec. 3.9) seems to satisfy the crack control provision. If $0.60f_y$ had been used instead of the computed value for f_s,

$$z = 36.0\sqrt[3]{3.0(54)} = 196 \text{ kips/in.} > 175 \text{ kips/in.} \qquad \text{NG}$$

Because of the low allowable stresses in the working stress method, it does not seem likely that crack control will be a problem when that method is used. When the ACI "strength design method" is used, the computed flexural stress at service load may be significantly higher, because the maximum percentage of reinforcement permitted in that method is nearly twice that for the ideally reinforced section in the working stress method.

SELECTED REFERENCES

1. ACI Committee 224. "Control of Cracking in Concrete Structures," *Concrete International*, **2**, October 1980, 35–76. Disc. **3**, 101–110.
2. Edward G. Nawy. "Crack Control in Reinforced Concrete Structures," *ACI Journal, Proceedings*, **65**, October 1968, 825–836. Disc. **66**, 308–311.
3. Bengt B. Broms and LeRoy A. Lutz. "Effects of Arrangement of Reinforcement on Crack Width and Spacing of Reinforced Concrete Members," *ACI Journal, Proceedings*, **62**, November 1965, 1395–1410. Disc. 1807–1812.
4. Peter Gergely and LeRoy A. Lutz. "Maximum Crack Width in Reinforced Concrete Flexural Members," *Causes, Mechanism, and Control of Cracking in Concrete* (SP-20). Detroit: American Concrete Institute, 1968 (pp. 87–117).
5. Paul H. Kaar. "High Strength Bars as Concrete Reinforcement, Part 8: Similitude in Flexural Cracking of T-Beams Flanges," *Journal PCA Research and Development Laboratories*, 8(2), May 1966, 2–12.
6. G. D. Base, J. B. Reed, A. W. Beeby, and H. P. J. Taylor. *An Investigation of the Crack Control Characteristics of Various Types of Bar in Reinforced Concrete Beams*. Research Report No. 18. London: Cement and Concrete Association, December 1966, (44 pp.).
7. LeRoy A. Lutz, Nand K. Sharma, and Peter Gergely. "Increase in Crack Width in Reinforced Concrete Beams Under Sustained Loading," *ACI Journal, Proceedings*, **64**, September 1967, 538–546.
8. Mete A. Sozen and William L. Gamble. "Strength and Cracking Characteristics of Beams with #14 and #18 Bars Spliced with Mechanical Splices." *ACI Journal, Proceedings*, **66**, December 1969, 949–956.
9. A. W. Beeby. "The Prediction and Control of Flexural Cracking in Reinforced Concrete Members," *Cracking, Deflection, and Ultimate Load of Concrete Slab Systems* (SP-30). Detroit: American Concrete Institute, 1971, (pp. 55–75).
10. John P. Lloyd, Hassen M. Rejali, and Clyde E. Kesler. "Crack Control in One-Way Slabs Reinforced With Deformed Wire Fabric," *ACI Journal, Proceedings*, **66**, May 1969, 366–376.
11. A. W. Beeby. "An Investigation of Cracking in Slabs Spanning One Way," Technical Report No. TRA 433, Cement and Concrete Association, London, April 1970, (32 pp).

PROBLEMS

All problems are to be done in accordance with the ACI Code "alternate design method" (i.e., working stress method) unless otherwise indicated. When selecting bars be certain that they fit into the beam width (see Table 3.9.2) and allow for #3 stirrups (for SI use #10M) around longitudinal bars. For all designs a check of stresses is required. In addition, a design sketch (to scale) is required, showing cross-sectional dimensions, number, size, and location of bars, and amount of protective clear cover.

The SI units in parenthesis provide a problem approximately the same as that in Inch-Pound units; the conversions have been adjusted so that given information indicates comparable precision to that of the original given data. For all metric unit problems, consider the ACI allowable stresses for steel to be $f_s = 150$ MPa for $f_y = 300$ MPa and for $f_y = 350$ MPa; and $f_s = 160$ MPa for $f_y = 400$ MPa.

4.1 A reinforced concrete beam is 12 in. wide and 22 in. in overall depth, with 3-#9 bars centered $2\frac{1}{2}$ in. from the bottom of the beam. The moduli of elas-

ticity are $E_c = 2.9 \times 10^6$ psi and $E_s = 29 \times 10^6$ psi. Determine the maximum stresses in the steel and the concrete (f_s and f_c max) if the applied bending moment is 85 ft-kips. (Beam: width $= 300$ mm, depth $= 560$ mm, 3-#30M bars centered 60 mm from bottom. $E_c = 20,000$ MPa; $E_s = 200,000$ MPa; $M_w = 115$ kN·m.)
(a) Use the stress-solid internal-couple method.
(b) Use the flexure-formula method with transformed section.

4.2 What is the maximum allowable bending moment for the beam of Prob. 4.1 if the maximum allowable stresses are $f_s = 22,000$ psi and $f_c = 1200$ psi? ($f_s = 150$ MPa; $f_c = 8.3$ MPa.)

4.3 What is the maximum allowable bending moment for the beam of Prob. 4.1 if the maximum allowable stresses are $f_s = 19,000$ psi and $f_c = 1350$ psi? ($f_s = 130$ MPa; $f_c = 9.5$ MPa.)

4.4 Consider a beam with a 12-in. width and an effective depth of 19.5 in. which has 3-#7 bars in the tension face. The concrete has $f_c' = 3000$ psi, and the steel has a minimum specified yield stress of 40,000 psi. (Beam: $b = 300$ mm; $d = 500$ mm; 3-#20M bars. $f_c' = 20$ MPa; $f_y = 300$ MPa.)
(a) Determine the allowable bending moment.
(b) Is the beam underreinforced, ideally reinforced, or overreinforced?

4.5 For the beam of Prob. 4.4, use 3-#10 bars instead of 3-#7. (Use 3-#30M bars.)
(a) Determine the allowable bending moment.
(b) Is the beam underreinforced, ideally reinforced, or overreinforced?

4.6 Design a rectangular beam 16 in. wide to resist a service load moment of 400 ft-kips. Use $f_c' = 4000$ psi and $f_y = 60,000$ psi. ($b = 400$ mm; $M_w = 540$ kN·m; $f_c' = 30$ MPa; and $f_y = 400$ MPa.)

4.7 Design a rectangular beam having a 30-in. overall depth to resist a service load moment of 400 ft-kips. Use $f_c' = 4000$ psi and $f_y = 60,000$ psi. ($h = 760$ mm; $M_w = 540$ kN·m; $f_c' = 30$ MPa; and $f_y = 400$ MPa.)

4.8 Design a beam having a 12-in. width to carry service load moments of 70 ft-kips live load and 40 ft-kips dead load. Use $f_c' = 4000$ psi and Grade 50 steel. ($b = 300$ mm; $M_L = 95$ kN·m; $M_D = 54$ kN·m; $f_c' = 30$ MPa; $f_y = 350$ MPa.)

4.9 Design a reinforced concrete beam to carry a live load moment of 225 ft-kips on a simply supported span of 20 ft. Use $f_c' = 4000$ psi and $f_y = 60,000$ psi. Use a beam width of 14 in. ($M_L = 305$ kN·m; span $= 6.1$ m; $f_c' = 30$ MPa; $f_y = 400$ MPa; $b = 350$ mm.)

4.10 In order to reuse forms and eliminate multiple framing details, design a reinforced concrete cross section 15 in. wide × 30 in. deep overall to be used in a location where the maximum live load moment is 100 ft-kips. The simply supported span is 24 ft. Use $f_c' = 3000$ psi and Grade 40 steel. ($b = 380$ mm; $h = 760$ mm; $M_L = 136$ kN·m; span $= 7.3$ m; $f_c' = 20$ MPa; $f_y = 300$ MPa.)
In the process of design, do both of the following:

(a) Determine steel requirement using "exact" method.

(b) Determine steel requirement using "approximate" method.

4.11 Design the cross section of Prob. 4.10 to be used where the live load moment is 200 ft-kips and the simply supported span is 26 ft. The beam is to have tension reinforcement only. (M_L = 270 kN·m; span = 8 m.)

4.12 Design the cross section for a beam to carry a live load of 5 kips/ft on a span of 28 ft. Use f'_c = 3000 psi and Grade 60 steel. (See Sec. 3.9 for practical proportions.) (w_L = 73 kN/m; span = 8.5 m; f'_c = 20 MPa; f_y = 400 MPa.)

4.13 Design a reinforced concrete simply supported one-way floor slab to carry a uniformly distributed live load of 175 psf on a span of 18 ft center to center of supports. Use f'_c = 3000 psi, and Grade 40 steel. (Note: ACI-7.12 and Example 3.9.2.) (w_L = 8.4 kN/m²; span = 5.5 m; f'_c = 20 MPa; f_y = 300 MPa.)

4.14 Design a rectangular beam having compression and tension reinforcement, and having overall dimensions of 20 in. wide and 32 in. overall depth, to carry a bending moment of 475 ft-kips. Assume the usual stirrup will go around the compression steel. Use f'_c = 4000 psi and Grade 60 steel. (b = 500 mm; h = 800 mm; M_w = 645 kN·m; f'_c = 30 MPa; f_y = 400 MPa.)

4.15 Redesign the beam of Prob. 4.11 using compression reinforcement. Assume the usual stirrup will go around the compression steel.

4.16 For the beam of the accompanying figure, determine the allowable moment capacity M_w. Use f'_c = 4000 psi and Grade 60 steel. (f'_c = 30 MPa; f_y = 400 MPa.)

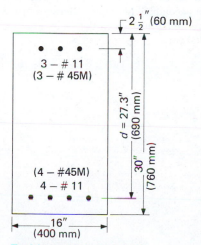

Problems 4.16 and 4.17

4.17 For the beam of Prob. 4.16, compute the transformed cracked section moment of inertia I_{cr} that would be needed for a deflection computation [i.e., used in ACI Formula (9-7)].

4.18   For the given beam of the accompanying figure compute the transformed cracked section moment of inertia I_{cr} that would be needed for a deflection calculation. Use $f'_c = 4000$ psi and $f_y = 60,000$ psi. ($f'_c = 30$ MPa; $f_y = 400$ MPa.)

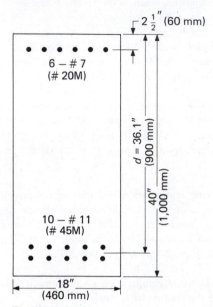

Problem 4.18

4.19   Design an irregular-shaped beam for a building whose floor system is composed of precast slabs. The 3-in. ledges are required for the support of the precast sections. The total moment to be carried is 50 ft-kips, and the cross section to be used is given in the accompanying figure. Because of forming costs, a large number of beams are being made with the same cross section. Use $f'_c = 3000$ psi and $f_y = 40,000$ psi.

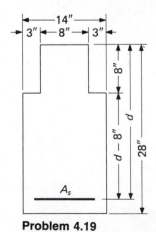

Problem 4.19

4.20   (a) Derive an equation for the neutral axis location in nondimensional form using x/d in terms of ρ and n for investigating rectangular sections with tension reinforcement only. Plot x/d versus ρ for $n = 9$ from $\rho = 0.005$ to 0.035 in increments of 0.005.

(b) Derive an equation for the moment arm of the internal couple, using arm/d as the nondimensional quantity. Plot as in part (a).

4.21 Check the ACI crack control provisions for the tension steel detail of Prob. 4.16.

4.22 Check the ACI crack control provisions for the tension steel detail of Prob. 4.18. Assume 1.5-in. (38-mm) clear cover and that a #3 stirrup is used.

4.23 Check the ACI crack control provisions for the tension steel details of the beams designed for whichever of Probs. 4.8 through 4.13 have already been worked.

5

Shear Strength and Shear Reinforcement

5.1 INTRODUCTION

In this chapter the shear strength of nonprestressed flexural members is treated. Effects of axial compression and tension are included in Sec. 5.13. Consideration is also given to the special requirements for deep beams (Sec. 5.14), the shear-friction concept (Sec. 5.15), and brackets and corbels (Sec. 5.16). Shear strength of prestressed concrete members is treated in Chap. 21. Shear effects arising from torsion and from its combination with bending or shear, or both, are treated in Chap. 19.

For the simple beam shown in Fig. 5.1.1, the bending moment M at section A-A causes compressive stresses in the concrete above the neutral axis, and tensile stresses in the reinforcement and in the concrete below the neutral axis had it not yet been cracked. To satisfy the vertical force equilibrium, the summation of the vertical shear stresses across the section must be equal to the shear force V. Below the neutral axis there is nearly a state of pure shear as shown in Fig. 5.1.2 which gives rise to a tensile stress of equal magnitude on a 45° plane. This diagonal tension constitutes

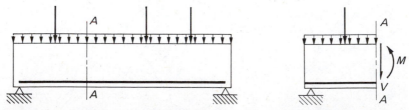

Figure 5.1.1 Shear force and bending moment in a simple beam.

Inclined shear crack; test by D. R. Buettner at the University of Wisconsin, Madison.

the main cause of inclined cracking. Thus the failures in beams commonly referred to as "shear failures" are actually tension failures at the inclined cracks. One of the earliest to recognize this was E. Morsch in Germany in the early 1900s [1].

The factors influencing shear strength and formation of inclined cracks are so numerous and complex that a definitive conclusion regarding the correct mechanism of inclined cracking that results from high shear is difficult to establish. Bresler and MacGregor [2] have introduced an excellent systematic correlation of the basic concepts, and this work has been expanded and updated by ACI–ASCE Committee 426 [3]. In this chapter the

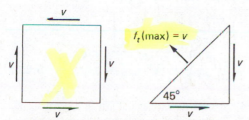

Figure 5.1.2 A state of pure shear (that is, no tensile or compressive stresses on the faces of the element).

concepts of horizontal and vertical shear, with the resulting principal tensile stress, are presented first so that the reader may gain a feeling for the potential direction of inclined cracks. Then follows a presentation of recent knowledge in regard to the variables affecting shear strength, leading up to the ACI Code provisions.

5.2 THE SHEAR STRESS FORMULA BASED ON LINEAR STRESS DISTRIBUTION

A uniformly loaded simple beam is shown in Fig. 5.2.1. Consider the free body of the elemental block *abcd,* as shown in Fig. 5.2.1(b), where kd is the general neutral axis distance* (not necessarily the ideal value used in Chap. 4). Horizontal force equilibrium requires

$$v_y b \, dz = C_2 - C_1 \qquad (5.2.1)$$

in which v_y is the unit horizontal shear stress on a plane at a distance y from the neutral axis. In the working stress range,

$$C_1 = \tfrac{1}{2}(f_{c1} + f_{c1y})b(kd - y)$$

$$f_{c1y} = \frac{y}{kd} f_{c1}$$

$$C_1 = \frac{1}{2} f_{c1} \left(1 - \frac{y}{kd}\right)(kd - y)b = \frac{1}{2} f_{c1}bkd \left[1 - \left(\frac{y}{kd}\right)^2\right] \qquad (5.2.2)$$

Dividing the bending moment M_1 on section 1-1 by the arm of the internal couple gives the full compressive force on section 1-1; or

$$\frac{M_1}{\text{arm}} = \frac{1}{2} f_{c1}b(kd)$$

Substitution of f_{c1} from the above expression into Eq. (5.2.2) gives

$$C_1 = \frac{M_1}{\text{arm}} \left[1 - \frac{y^2}{(kd)^2}\right] \qquad (5.2.3)$$

Similarly,

$$C_2 = \frac{M_2}{\text{arm}} \left[1 - \frac{y^2}{(kd)^2}\right] \qquad (5.2.4)$$

Substituting Eqs. (5.2.3) and (5.2.4) into Eq. (5.2.1) gives

$$v_y = \left(\frac{M_2 - M_1}{dz}\right)\frac{1}{b(\text{arm})}\left[1 - \frac{y^2}{(kd)^2}\right]$$

$$= \frac{V}{b(\text{arm})}\left[1 - \frac{y^2}{(kd)^2}\right] \qquad (5.2.5)$$

*In most places in this book, the symbol x is used for the neutral axis distance from the compression face of a beam in the strength design method as well as in the working stress method. Here y is used to measure the distance from the neutral axis to any point in the compression area.

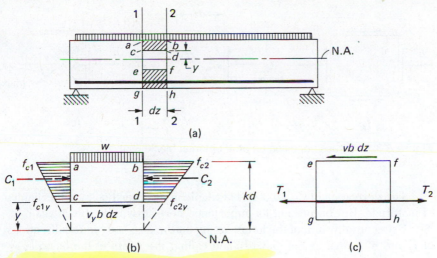

Figure 5.2.1 Horizontal shear stress in a beam.

Equation (5.2.5) is valid from $y = 0$ at the neutral axis to the extreme concrete compression fiber where $y = kd$. Proceeding from the extreme compression fiber to the neutral axis, the differential horizontal force $C_2 - C_1$ increases to a maximum. Thus at the neutral axis $y = 0$, Eq. (5.2.5) gives the maximum shear stress

$$v = \frac{V}{b(\text{arm})} \qquad (5.2.6)$$

The horizontal shear stress v of Eq. (5.2.6) is also the vertical shear stress, because shear stresses on two perpendicular planes must be equal. In regions where bending stress is low or where flexural cracks exist, the stress condition is close to a state of pure shear. In such a case the maximum principal stress (see Fig. 5.1.2), a tensile stress acting at 45°, equals the shear stress.

For strength design, the shear force V is the factored shear force and the distribution of flexural stress is no longer linear in the compression region. The tensile stress v of Eq. (5.2.6) is then *not* the actual tensile stress; however, the relative magnitude obtained from this equation is still a measure of the potential for inclined cracking. In the ACI Code, the denominator in Eq. (5.2.6) is thus replaced by b times the full value of the effective depth d.

5.3 THE COMBINED STRESS FORMULA

If at a certain point below the neutral axis in a homogeneous beam the tensile stress is f_t and the shear stress is v, the principal tensile stress $f_t(\text{max})$ is given by

$$f_t(\text{max}) = \tfrac{1}{2}f_t + \sqrt{(\tfrac{1}{2}f_t)^2 + v^2} \qquad (5.3.1)$$

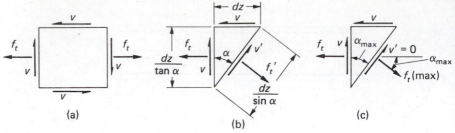

Figure 5.3.1 Stress condition on an elemental block.

The derivation of Eq. (5.3.1) is available in most textbooks on mechanics of materials. But because of its importance here, it will be shown again.

Using equilibrium of the forces acting on the free body in the directions of f'_t and v' shown in Fig. 5.3.1(b) and calling the width of the beam b,

$$f'_t \left(\frac{b\,dz}{\sin\alpha} \right) = f_t \left(\frac{b\,dz}{\tan\alpha} \right) \cos\alpha + v(b\,dz)\cos\alpha + v \left(\frac{b\,dz}{\tan\alpha} \right) \sin\alpha$$

$$v' \left(\frac{b\,dz}{\sin\alpha} \right) = f_t \left(\frac{b\,dz}{\tan\alpha} \right) \sin\alpha + v(b\,dz)\sin\alpha - v \left(\frac{b\,dz}{\tan\alpha} \right) \cos\alpha$$

from which

$$f'_t = \tfrac{1}{2}f_t(1 + \cos 2\alpha) + v \sin 2\alpha$$

$$v' = \tfrac{1}{2}f_t \sin 2\alpha - v \cos 2\alpha$$

The value of α_{max}; that is, α which makes f'_t maximum and at the same time makes v' zero, may be found by differentiating the expression for f'_t with respect to α. Equation (5.3.1) may be obtained by substituting

$$\tan 2\alpha_{max} = \frac{v}{\tfrac{1}{2}f_t} \tag{5.3.2}$$

into the expression for f'_t.

The principal tensile stress $f_t(max)$ in the diagonal direction, which is at an angle α_{max} with the beam axis, is at least as large as either f_t or v. It is nearly equal to the longitudinal tensile stress f_t if the shear stress v is small and its direction is nearly horizontal. It is nearly equal to the shear stress v if the longitudinal tensile stress f_t is small and its direction is nearly at 45° with the beam axis. Since concrete is weak in tension, these principal tensile stresses are undoubtedly correlated to inclined cracking as shown in Fig. 5.3.2.

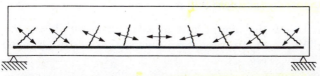

Figure 5.3.2 Directions of potential cracks in a simply supported beam.

5.4 BEHAVIOR OF BEAMS WITHOUT SHEAR REINFORCEMENT

As shown in Sec. 5.3, high shear stress on a beam results in the formation of inclined cracks. This is particularly true for beams having only longitudinal reinforcement; that is, reinforcement designed to carry the flexural tensile and compressive forces arising from bending moment. In order to prevent the formation of inclined cracks, transverse reinforcement (known as "shear reinforcement") in the form of closed or U-shaped stirrups is used in the vertical or inclined directions to enclose the main longitudinal reinforcement along the faces of the beam (see Fig. 5.6.1 in Sec. 5.6).

The following discussion considers the behavior of beams without shear reinforcement and includes the concepts summarized by ACI–ASCE Committee 426 [3]. Further treatment toward a more rational theory has been presented by Collins [19] with his "diagonal compression field theory."

Inclined cracking in the webs of reinforced or prestressed concrete beams may develop either in the absence of flexural cracks in the vicinity or as an extension of a previously developed flexural crack. An inclined crack occurring in a beam that was previously uncracked due to flexure is known as a *web-shear crack* as shown in Fig. 5.4.1(a). An inclined crack originating at the top of and becoming an extension to a previously existing flexural crack is known as a *flexure-shear crack*, as shown in Fig. 5.4.1(b). The critical flexural crack is referred to as the "initiating crack."

Web-shear cracks are relatively rare in nonprestressed beams. These cracks occur in thin-webbed I-shaped beams having relatively large flanges, common only in prestressed concrete construction. This is discussed further in Chap. 21 which is devoted entirely to prestressed concrete. Web-shear cracks may also occur near the inflection points or bar cutoff points on continuous reinforced concrete beams subjected to axial tension [4,5].

Flexure-shear cracks are the usual type found in both reinforced and prestressed concrete. In nonprestressed reinforced concrete beams, flexural cracking is expected under service load. The flexural cracks, usually ex-

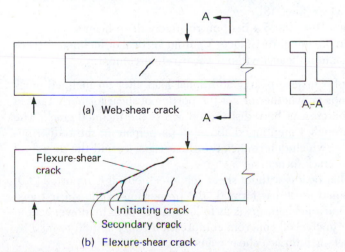

(a) Web-shear crack

A–A

Flexure-shear crack

Initiating crack
Secondary crack

(b) Flexure-shear crack

Figure 5.4.1 Types of inclined cracks (from Ref. 2).

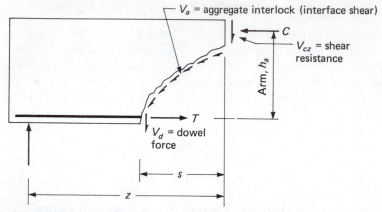

Figure 5.4.2 Redistribution of shear resistance after formation of inclined crack.

tending approximately vertically into the beam, cause no distress to the beam until a critical combination of flexural and shear stresses develops near the interior extremity of one of the cracks. The inclined crack then forms. The rate of transformation of the initiating flexural crack into the flexure-shear crack depends on the rate of growth and height of flexural cracks, as well as the magnitude of shear stresses acting near the tops of flexural cracks.

The transfer of shear in reinforced concrete members occurs by a combination of the following mechanisms [3], as shown in Fig. 5.4.2:

1. Shear resistance of the uncracked concrete, V_{cz}.
2. Aggregate interlock (or interface shear transfer) force V_a, tangentially along a crack [6,7], and similar to a frictional force due to irregular interlocking of the aggregates along the rough concrete surfaces on each side of the crack.
3. Dowel action, V_d, the resistance of the longitudinal reinforcement to a transverse force [8].
4. Arch action [see Fig. 5.4.5(a)] on relatively deep beams.
5. Shear reinforcement resistance, V_s, from vertical or inclined stirrups (not available in beams without shear reinforcement).

The ability of a beam to carry additional load after an inclined crack has formed depends on whether or not the portion of shear formerly carried by uncracked concrete can be redistributed across the inclined crack. The mechanisms 1 through 4 mentioned above all participate in the redistribution, the success of which determines the shear capacity and the degree of seriousness of the crack formation.

For rectangular beams without shear reinforcement, it is reported [3,9] that after an inclined crack has formed, the proportion of the shear transferred by the various mechanisms is as follows: 15 to 25% by dowel action; 20 to 40% by the uncracked concrete compression zone; and 33 to 50% by aggregate interlock or interface shear transfer.

Inclined cracks begin and grow depending on the relative magnitudes of shear stress v and flexural stress f_t, as used in Sec. 5.3. These controlling stresses may be expressed

$$v = k_1 \frac{V}{bd} \tag{5.4.1a}$$

$$f_t = k_2 \frac{M}{bd^2} \tag{5.4.1b}$$

where k_1 and k_2 are proportionality constants. The discussion in Sec. 5.2 may serve to justify the shear stress as proportional to V/bd; and in the presentation of Chaps. 3 and 4 the flexural capacity M has been related to bd^2 through the use of a coefficient of resistance R which has the same units as the flexural stress.

From Sec. 5.3 one may note that the principal tensile stress is a function of the ratio f_t/v. Another expression for f_t/v may be obtained from Eqs. (5.4.1); thus

$$\frac{f_t}{v} = \frac{k_2}{k_1} \frac{M}{Vd} = k_3 \frac{M}{Vd} \tag{5.4.2}$$

For a simple beam symmetrically loaded with two equal concentrated loads (see Fig. 5.4.3), the ratio M/V may be thought of as the distance a over which the shear is constant. This distance a is known as the *shear span*. For the general case where the shear is continually varying, the "shear span" may be expressed as

$$a = \frac{M}{V} \tag{5.4.3}$$

which has a value at every point along a beam.

Using Eq. (5.4.3) in Eq. (5.4.2), the ratio f_t/v becomes

$$\frac{f_t}{v} = k_3 \left(\frac{a}{d} \right) \tag{5.4.4}$$

The shear span to depth ratio a/d has also been shown experimentally to

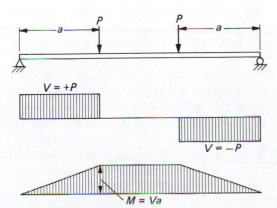

Figure 5.4.3 Basic definition of shear span a.

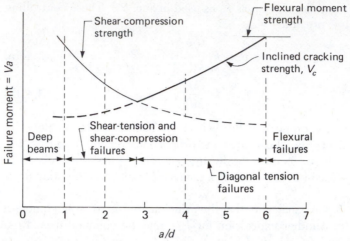

Figure 5.4.4 Variation in shear strength with a/d for rectangular beams (adapted from Ref. 2).

be a highly influential factor in establishing shear strength [1,2,3,10–12]. When factors other than a/d are kept constant, the variation in shear capacity may be illustrated by Fig. 5.4.4 using the results for rectangular beams.

From Fig. 5.4.4 four general categories of failure may be established: (1) deep beams with $a/d < 1$; (2) short beams with a/d ratios from 1 to about $2\frac{1}{2}$, in which the shear strength exceeds the inclined cracking capacity; (3) usual beams of intermediate length having a/d ratios from about $2\frac{1}{2}$ to 6, in which the shear strength equals the inclined cracking strength; and (4) long beams with a/d greater than 6, whose flexural strength is less than their shear strength.

Deep Beams. $a/d \leq 1$. For a deep beam, shear stress has the predominant effect. After inclined cracking occurs, this beam tends to behave like a tied-arch wherein the load is carried by direct compression extending around the shaded area of Fig. 5.4.5(a) and by the tension in the longitudinal steel. Once the shear-related crack develops, the beam transforms quickly into a tied-arch which exhibits considerable reserve capacity. Several modes of failure are possible [13–17] for the tied-arch system, as shown in Fig. 5.4.5(b).

Possible modes of failure are indicated in Fig. 5.4.5(b); they are (1) an anchorage failure; that is, pullout of the tension reinforcement at the support; (2) a crushing failure at the reactions; (3) a "flexural failure" arising from either a crushing of concrete near the top of the arch or a yielding of the tension reinforcement; (4) failure of the arch rib due to an eccentricity of the arch thrust, resulting in either a tension crack over the support at point 4 of Fig. 5.4.5(b) or a crushing of concrete on the underside of the rib at point 5.

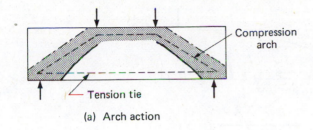

(a) Arch action

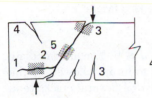

1. Anchorage failure
2. Bearing failure
3. Flexure failure
4. & 5. Arch-rib failure

(b) Types of failure

Figure 5.4.5 Modes of failure in deep beams, $a/d \leq 1.0$ (adapted from Ref. 2).

Short Beams. $1 < a/d \leq 2\frac{1}{2}$. Just as for deep beams, short beams have a shear strength that exceeds the inclined cracking strength. After the flexure-shear crack develops, the crack extends further into the compression zone as the load increases. It also propagates as a secondary crack toward the tension reinforcement and then progresses horizontally along that reinforcement. Failure eventually results, either (1) by an anchorage failure at the tension reinforcement, called a "shear-tension" failure [Fig. 5.4.6(a)]; or (2) by a crushing failure in the concrete near the compression face, called a "shear-compression" failure [Fig. 5.4.6(b)].

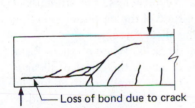

(a) Shear-tension failure

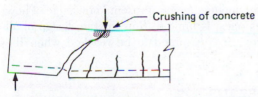

(b) Shear-compression failure

Figure 5.4.6 Typical shear failures in short beams, $a/d = 1$ to $2\frac{1}{2}$ (adapted from Ref. 2).

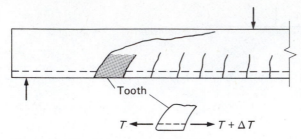

Figure 5.4.7 "Diagonal tension failure" or "tooth cracking failure" on intermediate length beams; a/d about $2\frac{1}{2}$ to 6.

Usual Beams of Intermediate Length. $2\frac{1}{2} < a/d \leq 6$. For intermediate length beams, vertical flexural cracks are the first to form, followed by the inclined flexure-shear cracks. At the beginning several flexural cracks tend to bend over creating beam segments between cracks, the "teeth" shown in Fig. 5.4.7. When the root of the "tooth," as a result of the increasing number of flexural cracks, is so reduced [18] in size that it becomes unable to carry the moment arising from ΔT, it breaks to form the inclined flexure-shear crack. At the sudden occurrence of the inclined crack, the beam is not able to redistribute the load, as in the situation of smaller a/d ratio. In other words the formation of the inclined crack represents the shear strength of beams in this category, for which the term "diagonal tension failure" has been given [2]. This is the usual category for beam design.

Long Beams. $a/d > 6$. The failure of long beams starts with yielding of the tension reinforcement and ends by crushing of the concrete at the section of maximum bending moment. In addition to the nearly vertical flexural cracks at the section of maximum bending moment, prior to failure slightly inclined (from the vertical) cracks may be present between the support and the section of maximum bending moment. Nevertheless, the strength of the beam is entirely dependent on the magnitude of the maximum bending moment and is not affected by the size of the shear force.

In summary, shear tends to cause inclined cracks. When no such cracks form before the nominal flexural strength is reached, the effect of shear is negligible. Beams may fail upon formation of an inclined crack, as for the so-called "diagonal tension" failures when a/d is between about $2\frac{1}{2}$ and 6, or there may be considerable reserve strength. In order for the beams with reserve capacity to maintain a state of equilibrium, stresses must be redistributed after the formation of the inclined crack. Present knowledge of how the redistribution actually takes place is limited; thus for the design of all but deep beams the shear strength is assumed to be reached when the inclined crack forms.

5.5 SHEAR STRENGTH OF BEAMS WITHOUT SHEAR REINFORCEMENT

The strength at which an inclined crack [usually a flexure-shear crack as in Fig. 5.4.1(b)] forms is taken to be the shear strength of a beam without shear reinforcement, according to the intent of the ACI Code. After establishing on a rational basis those variables that are involved, the relationship between them was statistically determined from test results [1].

It is assumed that the strength is reached when the principal tensile stress in Eq. (5.3.1) reaches the tensile strength of concrete, which is proportional to $\sqrt{f_c'}$. Although the exact distributions of the flexural and shear stresses in a cross section are not known, it may be assumed that flexural tensile stress f_t varies as E_c/E_s times the tensile stress in the reinforcement and that v varies as the average shear stress. Assume that E_c is also proportional to $\sqrt{f_c'}$ and let V_n and M_n be the nominal shear force and nominal bending moment at a section when the inclined crack forms.

As already shown by Eq. (5.4.1a), the shear stress may be written

$$v = k_1 \frac{V_n}{bd} \tag{5.5.1}$$

The stress in the steel is proportional to $M_n/A_s d$, and the tensile stress f_t in the concrete then becomes

$$f_t \propto \frac{E_c f_s}{E_s} \propto \frac{E_c M_n}{E_s dA_s} \propto \frac{M_n \sqrt{f_c'}}{E_s dA_s} \propto \frac{M_n}{bd^2} \left(\frac{\sqrt{f_c'}}{\rho E_s} \right)$$

The above expression for f_t may be written as

$$f_t = \frac{k_4}{E_s} \left(\frac{\sqrt{f_c'}}{\rho} \right) \frac{M_n}{bd^2} \tag{5.5.2}$$

in which k_4 is a dimensionless constant and E_s has a definitely known value. Also, the tensile strength of concrete may be represented by

$$f_t(\text{max}) = k_5 \sqrt{f_c'} \tag{5.5.3}$$

Substituting Eqs. (5.5.1) to (5.5.3) into the principal stress equation, Eq. (5.3.1),

$$k_5 \sqrt{f_c'} = \frac{V_n}{bd} \left[\frac{1}{2} \frac{k_4}{E_s} \frac{M_n}{V_n d} \frac{\sqrt{f_c'}}{\rho} + \sqrt{\left(\frac{1}{2} \frac{k_4}{E_s} \frac{M_n}{V_n d} \frac{\sqrt{f_c'}}{\rho} \right)^2 + k_1^2} \right]$$

or

$$\frac{V_n}{bd\sqrt{f_c'}} = k_5 \Bigg/ \left[\frac{1}{2} \frac{k_4}{E_s} \frac{M_n}{V_n d} \frac{\sqrt{f_c'}}{\rho} + \sqrt{\left(\frac{1}{2} \frac{k_4}{E_s} \frac{M_n}{V_n d} \frac{\sqrt{f_c'}}{\rho} \right)^2 + k_1^2} \right] \tag{5.5.4}$$

In Eq. (5.5.4) the variables are observed to be $V_n/(bd\sqrt{f_c'})$ and $M_n\sqrt{f_c'}/(E_s \rho V_n d)$. It may be noted that these two variables are nondimensional quantities, because $\sqrt{f_c'}$ has units of force per unit area. In the statistical study the nominal shear force V_n was defined as that causing the critical inclined crack and M_n as the corresponding moment at the top of the initiating crack (Fig. 5.4.1). On the basis of 440 tests [1], as shown in Fig. 5.5.1, the relationship between these two variables, on using the correct value of E_s, was obtained as follows:

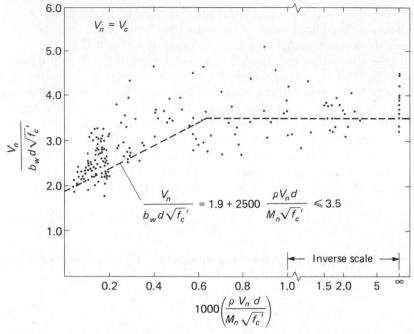

Figure 5.5.1 Derivation of ACI shear strength equation, Eq. (5.5.6), for beams without shear reinforcement, that is, $V_n = V_c$ (adapted from Ref. 1).

$$\frac{V_n}{bd\sqrt{f_c'}} = 1.9 + 2500 \frac{\rho V_n d}{M_n \sqrt{f_c'}} \le 3.5 \qquad (5.5.5)$$

Equation (5.5.5) is generally considered [3,11,18,20–26] to be acceptable for predicting the flexure-shear cracking load, particularly for $M_n/(V_n d)$ (i.e., shear span/depth) ratios of about $2\frac{1}{2}$ to 6, with considerable conservatism for lower $M_n/(V_n d)$ values. Thus the favorable factors for the shear strength of beams without shear reinforcement are a high percentage ρ of longitudinal reinforcement and a high ratio of $V_n d$ to M_n, that is, a low a/d ratio.

Since 1963, the ACI Code has accepted the relationship of Eq. (5.5.5) as the shear (inclined cracking) strength of beams without shear reinforcement. Thus defining V_c as the nominal strength of such beams, Eq. (5.5.5) using the web width* b_w for b becomes

$$V_c = \left[1.9 \sqrt{f_c'} + 2500 \frac{\rho_w V_u d}{M_u} \right] b_w d \le 3.5 \sqrt{f_c'} b_w d \qquad (5.5.6)^\dagger$$

which is ACI Formula (11-6). The use of the factored shear force V_u and the factored moment M_u instead of the values $V_n = V_u/\phi$ and $M_n = M_u/\phi$

---

*The term "web" refers to the width dimension of a beam at its neutral axis. For T-shaped beams the flange width b is distinguished from the stem, or web, width b_w.

†For SI, ACI 318-83M, with f_c' in MPa, gives

$$V_c = \frac{1}{6}\left(\sqrt{f_c'} + 100 \frac{\rho_w V_u d}{M_u} \right) b_w d \le 0.3 \sqrt{f_c'} b_w d \qquad (5.5.6)$$

makes little difference because the ratio V_u/M_u remains approximately equal to the ratio V_n/M_n in spite of some difference in the strength reduction factors ϕ for shear and for moment or combined axial compression with bending. Note also that the reinforcement ratio $\rho_w = A_s/(b_w d)$ is used in the ACI Code formula, where b_w is the web width for a T-section rather than the flange width. The ACI Code defines b_w as "web width." For tapered webs such a definition is unclear. In general, when the web is subject to flexural tension the "average web width" should be used as b_w. However, when the web is subject to flexural compression (as for negative moment regions) the use of average web width may be unsafe [27]. For such negative moment regions the use of a value lower than the average, perhaps the minimum, web width is more appropriate.

The value of $V_u d/M_u$ shall not be taken greater than 1.0 except when axial compression is present, which has the effect of limiting V_c at and near the points of inflection.

Continuous Beams. The application of Eq. (5.5.6) for continuous beams has recently been subject to question [3,28,29]. The compression-strut action on a continuous beam is shown in Fig. 5.5.2. The analogy in Fig. 5.4.3 that M/V equals the shear span a implies that at the point of zero moment there is a support to accommodate a compression-strut [see Fig. 5.4.5(a)]. For a continuous beam as in Fig. 5.5.2, the distances a_1 and a_2 are analogous to a in Fig. 5.4.3; however, there is no support to take a compression-strut reaction at the inflection point. The actual strut would relate to a longer length $a_1 + a_2$. Thus Ferguson [28] recommends using $d/(a_1 + a_2)$ for concentrated loads and 0.25 for uniform load, respectively, as the *effective* depth to shear-span ratio in place of $V_u d/M_u$ in Eq. (5.5.6). Alternatively, he recommends using the simplified constant value of V_c (as discussed in Sec. 5.10) for continuous beams.

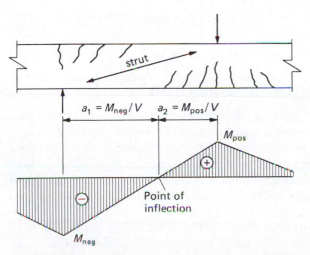

Figure 5.5.2 Crack pattern, compression strut, and bending moment diagram near support on continuous beam.

Lightweight Concrete. It has been shown [30–32] that the same general relationships as those for normal-weight concrete are valid for lightweight concrete, with a slight modification. In lightweight concrete, the tensile strength f_{ct} based on the split cylinder (see Sec. 1.8) provides a better correlation with inclined cracking strength than does the compressive strength f_c'. Since for normal-weight concrete the splitting tensile strength f_{ct} is approximately $6.7 \sqrt{f_c'}$, this value may be substituted into Eq. (5.5.6); thus for lightweight concrete

$$V_c = \left[1.9 \left(\frac{f_{ct}}{6.7} \right) + 2500 \frac{\rho_w V_u d}{M_u} \right] b_w d \leq 3.5 \left(\frac{f_{ct}}{6.7} \right) b_w d \quad (5.5.7)$$

as prescribed in accordance with ACI-11.2.1.1. In order that shear strength for lightweight concrete beams not exceed that for normal-weight concrete, $f_{ct}/6.7$ may not be taken to exceed $\sqrt{f_c'}$.

Tests have shown [30] that inclined cracking strength for lightweight concrete varies from about 60 to 100% of the values for normal-weight concrete of the same nominal compressive strength, depending on the particular aggregates used. More recent studies [32] have shown that using 0.75 to 0.85 of the tensile strength related term $\sqrt{f_c'}$ for normal-weight concrete in shear-strength equations is a reasonable and generally conservative approach.

Thus ACI-11.2.1.2 permits multiplying all values of $\sqrt{f_c'}$ appearing in shear (or torsion) equations by 0.75 for "all-lightweight" concrete and by 0.85 for "sand-lightweight" concrete when the split cylinder strength f_{ct} is not specified. For "all-lightweight" concrete,

$$V_c = \left[0.75(1.9\sqrt{f_c'}) + 2500 \frac{\rho_w V_u d}{M_u} \right] b_w d \leq 0.75(3.5)\sqrt{f_c'} b_w d \quad (5.5.8)$$

For "sand-lightweight" concrete,

$$V_c = \left[0.85(1.9\sqrt{f_c'}) + 2500 \frac{\rho_w V_u d}{M_u} \right] b_w d \leq 0.85(3.5)\sqrt{f_c'} b_w d \quad (5.5.9)$$

Linear interpolation is permitted when partial sand replacement is used.

5.6 FUNCTION OF SHEAR REINFORCEMENT

The common types of shear reinforcement, as shown in Fig. 5.6.1, are (1) stirrups perpendicular to the longitudinal reinforcement; (2) stirrups making an angle of 45° or more with longitudinal reinforcement; (3) longitudinal bars bent so that the axis of the bent portion makes an angle of 30° or more with the axis of the longitudinal portion of the bar; and (4) combinations of (1) or (2) with (3). Spirals, including rectangular helices, are also permitted by the ACI Code.

In a simple steel truss as shown in Fig. 5.6.2(a), the upper and lower chords are in compression and tension, respectively, and the diagonal members, usually called web members, are alternately in compression and tension. The shear strength of a reinforced concrete beam may be increased

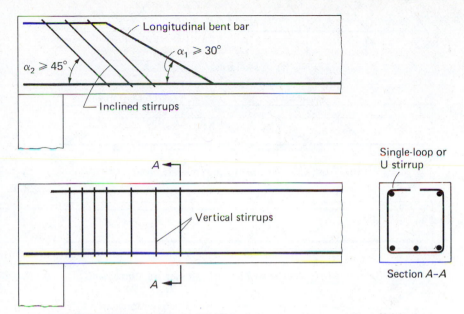

Figure 5.6.1 Types of shear reinforcement.

by the use of shear reinforcement, similar in its action to the tensile web members in a truss; hence, the term "web reinforcement" has been used for shear reinforcement. The shear reinforcement must be anchored in the compression zone of the concrete and is usually hooped around the longitudinal tension reinforcement. The action of inclined and vertical shear reinforcement, as shown in Fig. 5.6.2(c) and (e), may be described by the analogous truss action in Fig. 5.6.2(b) and (d).

Although the truss analogy has formed a simple explanation for stirrup behavior in beams for many years, it does not include several significant components of shear force transmission. Certainly shear reinforcement increases the shear strength of a member, but such reinforcement contributes little to the shear resistance prior to the formation of inclined cracks.

To provide shear strength by allowing a redistribution of internal forces (as described in Sec. 5.4), across any inclined crack that may form, the shear reinforcement has three primary functions [3]; these are (see Fig. 5.6.3) (1) to carry part of the shear, V_s; (2) to restrict the growth of the inclined crack and thus help maintain aggregate interlock (or interface shear transfer) V_a; and (3) to tie the longitudinal bars in place and thereby increase their dowel capacity, V_d. In addition to these primary functions, dowel action on the stirrups may transfer a small force across a crack, and the confining action of the stirrups on the compression concrete may slightly increase its strength.

If the amount of shear reinforcement is too little, it will yield immediately at the formation of the inclined crack, and the beam then fails. If the amount of shear reinforcement is too high, there will be a shear-compression failure before yielding of the web steel. The optimum amount of shear reinforcement should be such that both the shear reinforcement

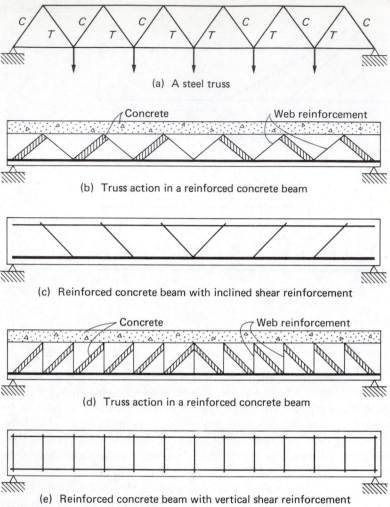

(a) A steel truss

(b) Truss action in a reinforced concrete beam

(c) Reinforced concrete beam with inclined shear reinforcement

(d) Truss action in a reinforced concrete beam

(e) Reinforced concrete beam with vertical shear reinforcement

Figure 5.6.2 Truss analogies.

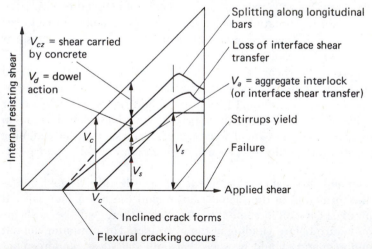

Figure 5.6.3 Distribution of internal shears in beams with shear reinforcement (adapted from Ref. 3).

and the compression zone continue to carry increasing shear after the formation of the inclined crack until yielding of the shear reinforcement, thus ensuring a ductile failure.

5.7 SHEAR STRENGTH OF BEAMS WITH SHEAR REINFORCEMENT

The traditional ACI approach to design for shear strength has been to consider the total nominal shear strength V_n as the sum of two parts,

$$V_n = V_c + V_s \tag{5.7.1}$$

in which V_n is the nominal shear strength; V_c is the shear strength of the beam attributable to the concrete (see Fig. 5.6.3 and Sec. 5.5); and V_s is the shear strength attributable to the shear reinforcement.

An expression for V_s may be developed from the truss analogy. Assume that an inclined crack in the 45° direction extends all the way from the longitudinal reinforcement to the compression surface and that it intersects an average of N shear reinforcing bars, as shown in Fig. 5.7.1. The portion V_s carried across the crack by the shear reinforcement equals the sum of the vertical components of the tensile forces developed in the shear reinforcement. Thus

$$V_s = N A_v f_y \sin \alpha \tag{5.7.2}$$

in which A_v is the area of shear reinforcement within a distance s, and f_y is the tensile yield stress for the shear reinforcement. From the trigonometry,

$$Ns = d(\cot 45° + \cot \alpha) = d(1 + \cot \alpha)$$

Thus

$$V_s = \frac{d(1 + \cot \alpha)}{s} A_v f_y \sin \alpha = \frac{A_v f_y (\sin \alpha + \cos \alpha) d}{s} \tag{5.7.3}$$

or if $\alpha = 90°$,

$$V_s = \frac{A_v f_y d}{s} \tag{5.7.4}$$

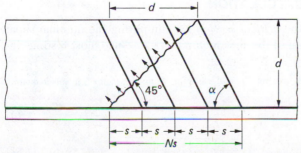

Figure 5.7.1 Shear strength V_s attributable to shear reinforcement.

As reported by the ACI–ASCE Joint Committee, a comparison of calculations based on Eqs. (5.7.1) to (5.7.4) with results of 166 tests shows that the calculated values are conservative [1]. There have been a number of more recent studies [6,26,33–38] attempting to develop a more accurate relationship to account for the behavior of shear reinforcement. The "diagonal compression field theory" of Collins [19] is particularly noteworthy.

The reader may note that bent bars or inclined stirrups are rarely used as shear reinforcement in the United States because of the high fabrication and labor costs involved.

5.8 LOWER AND UPPER LIMITS FOR AMOUNT OF SHEAR REINFORCEMENT

It has been noted in Sec. 5.6 that the amount of shear reinforcement should be neither too low nor too high in order to ensure the yielding of steel when the failure strength in shear is reached. The ACI Code [ACI Formula (11-14)] requires a minimum shear reinforcement area A_v equal to

$$\min A_v = 50 \frac{b_w s}{f_y} \qquad (5.8.1)^*$$

in which b_w is the beam web width.

From Eq. (5.7.4) this minimum amount corresponds to

$$V_s = \frac{A_v f_y d}{s} = \frac{f_y d}{s} \left(50 \frac{b_w s}{f_y} \right) = 50 b_w d \qquad (5.8.2)^*$$

or in terms of nominal unit stress on area $b_w d$,

$$v_s = \frac{V_s}{b_w d} = \frac{50 b_w d}{b_w d} = 50 \text{ psi} \qquad (5.8.3)$$

To ensure that the amount of shear reinforcement is not too high, a maximum value of

$$v_s \le 6\sqrt{f'_c} \text{ to } 8\sqrt{f'_c}$$

is usually specified; ACI-11.5.6.8 gives the upper limit for v_s as $8\sqrt{f'_c}$.

5.9 CRITICAL SECTION FOR NOMINAL SHEAR STRENGTH CALCULATION

In experimental work the *critical section* for computing the nominal shear strength was the location of the first inclined crack. Since most testing was

*For SI, ACI 318-83M, for A_v in mm², b_w and s in mm, f_y in MPa, and V_s in meganewtons (MN), gives

$$\min A_v = \frac{b_w s}{3 f_y} \qquad (5.8.1)$$

$$V_s = \left(\frac{1}{3} \text{ MPa} \right) b_w d \qquad (5.8.2)$$

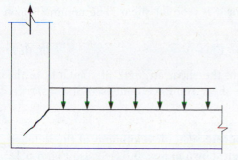

Figure 5.9.1 Inclined crack when the reaction induces tension in the member. Critical crack is at the face of support.

made on simply supported beams under simple loading arrangements, it was difficult to extend such results to generalized loadings on continuous structures.

In order to plot the test points for the development of Eq. (5.5.6) two assumptions based on observations were used: (1) For shear span to depth ratio (a/d) greater than 2, the critical inclined crack is expected at d from the section of maximum moment; and (2) for shear span to depth ratio (a/d) less than 2, an inclined crack is expected at the center of the shear span.

Thus for gradually varying shear force (such as for uniform loading), ACI-11.1.2.1 permits taking the critical section at a distance d from the face of support, in recognition of the fact that the support reaction being in the direction of the applied shear introduces compression into the end region of the member. This compression in the end region would occur when the beam is gravity loaded and supported by columns or walls. Shear reinforcement must be provided, however, between the face of support and the distance d therefrom, using the same requirements as at the critical section.

The critical section must be taken at the face of support when one of the following occurs:

1. Factored shear V_u does gradually decrease from the face of support but the support is itself a beam or girder and therefore does *not* introduce compression into the end region of the member (see MacGregor and Gergely [29] and Fereig and Smith [39]).
2. When a concentrated load occurs between the face of support and the distance d therefrom.
3. When any loading may cause a potential inclined crack to occur *at* the face of support or *extend into* instead of away from the support (see Fig. 5.9.1).

5.10 ACI CODE PROVISIONS FOR SHEAR STRENGTH OF BEAMS

In the ACI strength design method for shear, it is required that

$$\phi V_n \geq V_u \tag{5.10.1}$$

where V_u is the factored shear force and ϕV_n is the provided strength in shear. The strength reduction factor ϕ is 0.85 for shear. The nominal shear strength V_n is

$$V_n = V_c + V_s \tag{5.10.2}$$

where V_c and V_s are the portions of the shear strength attributable to the concrete (see Sec. 5.5) and to the shear reinforcement (see Sec. 5.7), respectively.

Strength V_c Attributable to Concrete. The development of the detailed equation, Eq. (5.5.6), for V_c has been shown in Sec. 5.5 (see also Fig. 5.5.1). Since that equation is not easy to use as a design equation, and because of the wide scatter of test results, ACI-11.3.1 and 11.3.2 permit using either of the following:

1. For the *simplified method*,

$$V_c = 2\sqrt{f_c'}b_w d \tag{5.10.3}*$$

From Fig. 5.5.1 this value appears to be conservative; however, recent studies [3,23,40,41] have shown otherwise when ρ_w is below about 0.012. For values of ρ_w lower than 0.012, the following is suggested [23,29],

$$V_c = (0.8 + 100\rho_w)\sqrt{f_c'}b_w d \tag{5.10.4}\dagger$$

2. For the *more detailed method*,

$$V_c = \left(1.9\sqrt{f_c'} + 2500\rho_w \frac{V_u d}{M_u}\right)b_w d \le 3.5\sqrt{f_c'}b_w d \tag{5.10.5}\ddagger$$

Equation (5.10.5) is identical to Eq. (5.5.6). The value of $V_u d/M_u$ may not exceed 1.0 (ACI-11.3.2.1), and M_u is the factored moment occurring simultaneously with the V_u for which shear strength is being provided. Note the discussion following Eq. (5.5.6) regarding the recommendations for substitute $V_u d/M_u$ values for shear strength on continuous beams.

Strength V_s Attributable to Shear Reinforcement. The contribution of shear reinforcement, as developed in Sec. 5.7, is (ACI-11.5.6)

$$V_s = \frac{A_v f_y d}{s}(\sin \alpha + \cos \alpha) \tag{5.10.6}$$

---

*For SI, ACI 318-83M, with f_c' in MPa, gives

$$V_c = \frac{1}{6}\sqrt{f_c'}\,b_w d \tag{5.10.3}$$

$\dagger$For SI, an approximate conversion with f_c' in MPa is

$$V_c = (0.07 + 8.3\rho_w)\sqrt{f_c'}b_w d \tag{5.10.4}$$

$\ddagger$For SI, ACI 318-83M, with f_c' in MPa, gives

$$V_c = \frac{1}{6}\left(\sqrt{f_c'} + 100\frac{\rho_w V_u d}{M_u}\right)b_w d \le 0.3\sqrt{f_c'}\,b_w d \tag{5.10.5}$$

and when vertical stirrups are used ($\alpha = 90°$),

$$V_s = \frac{A_v f_y d}{s} \tag{5.10.7}$$

Design Categories and Requirements. For design, the shear envelope for V_u is the starting point. From a practical point of view it is better to plot the V_u diagram rather than the "required V_n" diagram (equal to V_u/ϕ). The basic diagrams used should be those of M_u and V_u due to factored load; the ϕ factor is different for moment than for shear and should not be included in the load-related diagrams.

The design for shear may be separated into the following categories:

1.
$$V_u \leq 0.5\phi V_c \tag{5.10.8}$$

In this category, no shear reinforcement is required (ACI-11.5.5.1).

2.
$$0.5\ \phi V_c < V_u \leq \phi V_c \tag{5.10.9}$$

Minimum shear reinforcement is required except for thin slablike flexural members which experience has shown may perform satisfactorily without shear reinforcement. The thin slablike member exceptions include (a) slabs, (b) footings, (c) floor joist construction, and (d) beams where the total depth does not exceed 10 in., $2\frac{1}{2}$ times the flange thickness for T-shaped sections, or one-half of the web width. The requirement of minimum shear reinforcement may be waived if tests are conducted which show that the required flexural and shear strengths can be developed.

For this category, the shear reinforcement must satisfy ACI-11.5.5.3 and 11.5.4.1, as follows:

$$\text{required } \phi V_s = \min\ \phi V_s = \phi(50)b_w d \tag{5.10.10*}$$

and

$$\text{maximum spacing } s \leq \frac{d}{2} \leq 24 \text{ in.} \tag{5.10.11}$$

3.
$$\phi V_c < V_u \leq [\phi V_c + \min\ \phi V_s] \tag{5.10.12}$$

For all flexural members, including those exempted from shear reinforcement in Category 2, shear reinforcement must be provided satisfying Eqs. (5.10.10) and (5.10.11).

4.
$$[\phi V_c + \min\ \phi V_s] < V_u \leq [\phi V_c + \phi(4\ \sqrt{f'_c})b_w d] \tag{5.10.13†}$$

For this category the computed shear reinforcement requirement will exceed the min ϕV_s requirement, and the shear reinforcement must satisfy ACI Formula (11-2), ACI-11.5.6, 11.5.4.1, and 11.5.4.3, as follows:

*For SI, ACI 318-83M, for b_w and d in mm and V_s in MN, gives

$$\min\ \phi V_s = \phi\left(\frac{1}{3}\ \text{MPa}\right)b_w d \tag{5.10.10}$$

†For SI, ACI 318-83M gives in place of $4\sqrt{f'_c}$ psi, $\sqrt{f'_c}/3$ when f'_c is in MPa.

Table 5.10.1 SHEAR STRENGTH OF MEMBERS UNDER BENDING ONLY—ACI CODE

ITEM	STRENGTH DESIGN ($\phi = 0.85$)	CODE
1	$\phi V_n \geq V_u$	Formula (11-1), 11.1.1
	Maximum V_u at a distance d from face of support in usual situations (three exceptions)	11.1.2.1
2	$V_n = V_c + V_s$	Formula (11-2), 11.1.1
3	Simplified method:	
	$V_c = 2\sqrt{f'_c}b_w d$	Formula (11-3), 11.3.1.1
	$V_c = (0.8 + 100\rho_w)\sqrt{f'_c}b_w d$ for $\rho_w < 0.012$	Refs. 23,29
	More detailed method:	
	$V_c = \left(1.9\sqrt{f'_c} + 2500\rho_w \dfrac{V_u d}{M_u}\right)b_w d$ $\leq 3.5\sqrt{f'_c}b_w d$	Formula (11-6), 11.3.2.1
	$V_u d/M_u$ not to exceed unity	
	Allow 10% increase for joists	8.11.8
	Lightweight concrete when f_{ct} is specified: Use smaller of $f_{ct}/6.7$ or $\sqrt{f'_c}$ for $\sqrt{f'_c}$	11.2.1.1
	Lightweight concrete when f_{ct} is not specified: Use $0.75\sqrt{f'_c}$ to $0.85\sqrt{f'_c}$ in cases of all lightweight to sand lightweight concrete	11.2.1.2
4	$V_s = \dfrac{A_v f_y d}{s}(\sin \alpha + \cos \alpha)$	Formula (11-18), 11.5.6.3
	$V_s = A_v f_y \sin \alpha \leq 3\sqrt{f'_c}b_w d$ (single bar)	Formula (11-19), 11.5.6.4
	f_y not to exceed 60,000 psi	11.5.2
	V_s not to exceed $8\sqrt{f'_c}b_w d$	11.5.6.8

$$\text{required } \phi V_s = V_u - \phi V_c \qquad (5.10.14)$$

$$\text{provided } \phi V_s = \frac{\phi A_v f_y d}{s} \quad \text{(for } \alpha = 90°\text{)} \qquad (5.10.15)$$

$$\text{maximum } s = \frac{d}{2} \leq 24 \text{ in.} \qquad (5.10.16)$$

Note that in terms of nominal stress $v_s = V_s/b_w d$, $4\sqrt{f'_c}$ psi is the maximum v_s for which the $d/2$ maximum spacing limit applies.

5. $\quad [\phi V_c + \phi(4\sqrt{f'_c})b_w d] < V_u \leq [\phi V_c + \phi(8\sqrt{f'_c})b_w d] \qquad (5.10.17)^\ddagger$

The difference between categories 4 and 5 is that for all regions of a beam

‡For SI, ACI 318-83M gives in place of $4\sqrt{f'_c}$ and $8\sqrt{f'_c}$ psi, $\sqrt{f'_c}/3$ and $2\sqrt{f'_c}/3$, respectively, when f'_c is in MPa.

Table 5.10.1   *(Continued)*

ITEM	STRENGTH DESIGN ($\phi = 0.85$)	CODE
5	For $0.5\ \phi V_c < V_u \leq \phi V_c$ Use $s = \dfrac{A_v f_y}{50 b_w}$, max $s = \dfrac{d}{2} \leq 24$ in. except for slabs, footings, joists, and small beams shallower than 10 in., $2\frac{1}{2}$ times flange thickness, or $b_w/2$; for these cases no shear reinforcement required unless $V_u > \phi V_c$.	Formula (11-14), 11.5.5.3 11.5.5.1, 11.5.4.1
6	For $\phi V_c < V_u \leq [\phi V_c + \min\ \phi V_s]$ $\min\ \phi V_s = \phi 50 b_w d$ Use $s = \dfrac{A_v f_y}{50 b_w}$, max $s = \dfrac{d}{2} \leq 24$ in.	
7	For $[\phi V_c + \min\ \phi V_s] < V_u$ $\qquad\qquad \leq [\phi V_c + \phi(4\sqrt{f_c'}\,b_w d)]$ $\min\ \phi V_s = \phi 50 b_w d$ Design shear reinforcement, max $s = \dfrac{d}{2} \leq 24$ in.	11.5.4.1
8	For $[\phi V_c + \phi(4\sqrt{f_c'}\,b_w d)] < V_u$ $\qquad\qquad \leq [\phi V_c + \phi(8\sqrt{f_c'}\,b_w d)]$ Design shear reinforcement, max $s = \dfrac{d}{4} \leq 12$ in.	11.5.6.8

where the nominal stress v_s to be taken by shear reinforcement is between $4\sqrt{f_c'}$ and $8\sqrt{f_c'}$ the maximum shear reinforcement spacing s may not exceed $d/4$. The shear reinforcement provided in this category must satisfy ACI Formula (11-2), ACI-11.5.6, 11.5.4.3, and 11.5.6.8, as follows:

$$\text{required } \phi V_s = V_u - \phi V_c \qquad\qquad (5.10.18)$$

$$\text{provided } \phi V_s = \frac{\phi A_v f_y d}{s} \quad (\text{for } \alpha = 90°) \qquad (5.10.19)$$

$$\text{maximum spacing } s \leq \frac{d}{4} \leq 12 \text{ in.} \qquad\qquad (5.10.20)$$

The factored shear V_u may not exceed the upper limit of Eq. (5.10.17) according to ACI-11.5.6.8.

The maximum factored shear that must be provided for on any beam is that occurring at the *critical section*, defined as in Sec. 5.9. The V_u requirement between the face of support and the critical section is to be taken as constant, equal to the value at the critical section.

The ACI Code provisions for shear strength of beams described in this section are summarized in Table 5.10.1.

5.11 WORKING STRESS METHOD—ACI CODE, APPENDIX B

In strength design, the preference is to use shear force rather than shear stress, because stresses are less definable when the loading is above the service load range. It may still be noted, however, that the difference between computing shear force and shear stress (however nominal the stress may be) is only in the multiplier $b_w d$, which is the effective area of the web cross section.

The nominal shear stress in the "alternate design method," as prescribed in Appendix B.7 of the ACI Code is to be based on the service load shears without applying the overload factors U or the strength reduction factors ϕ; thus

$$v = \frac{V_D + V_L}{b_w d} \tag{5.11.1}$$

The allowable concrete stresses and the limiting maximum stresses for shear are taken as approximately 55% for beams, joists, walls, and one-way slabs, and 50% for two-way slabs and footings, respectively, of the stresses used for strength design.

Thus for beams the allowable stresses for concrete, using approximately 55% of Eqs. (5.10.3) and (5.10.5) are (ACI-Appendix B.3.1)

$$v_c = 1.1\sqrt{f_c'} \tag{5.11.2}$$

or (ACI-Appendix B.7.4.4)

$$v_c = \sqrt{f_c'} + 1300\rho_w \frac{Vd}{M} \leq 1.9\sqrt{f_c'} \tag{5.11.3}$$

Note that reducing the allowable stresses to 55% is equivalent to using a U/ϕ value of $1/0.55 = 1.82$. Assuming equal dead and live loads gives an average overload factor of $\frac{1}{2}(1.4 + 1.7) = 1.55$, which with $\phi = 0.85$ results in a U/ϕ value of 1.82.

The procedure for the working stress method is identical to that for the strength method. The minimum area of shear reinforcement is the same as for the strength method, as given by Eq. (5.8.1). The limiting maximum stresses, as discussed in Sec. 5.8 and as related to maximum web reinforcement spacing (ACI-Appendix B.7.5.4), are also reduced to 55 or 50% of their corresponding values in the strength method.

Since the procedures are the same as for the strength method, the use of the alternate method for designing shear reinforcement offers little advantage, except the results could be more or less conservative depending on whether the dead load is larger or smaller than the live load.

5.12 SHEAR STRENGTH OF BEAMS—DESIGN EXAMPLES

Four examples are presented to illustrate the basic procedure of designing vertical stirrups. In the first two examples, the complete design procedure for flexure and shear is shown for a simple beam, showing first the simplified and then the more detailed methods of spacing the stirrups. Some of the more practical aspects are discussed in the third example. Use of metric units is shown for the fourth example. Design of shear reinforcement in the continuous spans of a slab-beam-girder floor system is treated in Chap. 10.

EXAMPLE 5.12.1 For the given beam shown in Fig. 5.12.1(a), first determine the maximum uniform dead and live loads under service condition permitted by the ACI strength method. Then using those maximum service loads design the shear reinforcement using vertical stirrups and the strength method with the simplified procedure using constant V_c. Assume that service live load to dead load ratio is 1.5, $f'_c = 4000$ psi, and $f_y = 60,000$ psi.

Solution: (a) Check whether tension steel exceeds the maximum amount permitted [Fig. 5.12.1(b)].

$$x_b = \frac{0.003(22.5)}{0.003 + 0.00207} = 13.31 \text{ in.}$$

$$\text{max } x = 0.75x_b = 0.75(13.31) = 9.98 \text{ in.}$$

For $x = 9.98$ in.,

$$\epsilon'_s = \frac{9.98 - 2.5}{9.98}(0.003) = 0.00225 > \epsilon_y; \qquad f'_s = f_y$$

$$\text{max } C_c = 0.85f'_c b\beta_1(\text{max } x)$$

$$= 0.85(4)(14)(0.85)(9.98) = 404 \text{ kips}$$

$$C_s = A'_s(f_y - 0.85f'_c) = 2.40(60 - 3.4) = 136 \text{ kips}$$

$$\text{max } T = \text{max } C = 404 + 136 = 540 \text{ kips}$$

$$\text{max } A_s = \frac{\text{max } T}{f_y} = \frac{540}{60} = 9.00 \text{ sq in.} > (8\text{-}\#9 = 8.00) \qquad \text{OK}$$

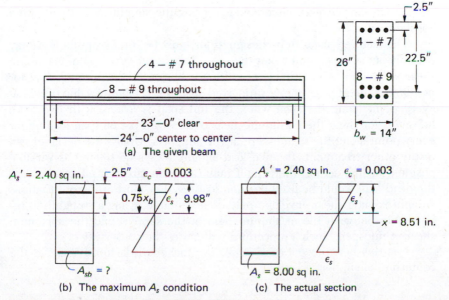

(a) The given beam

4 – # 7 throughout
8 – # 9 throughout

23′–0″ clear
24′–0″ center to center

2.5″
4 – # 7
26″
8 – # 9
22.5″
$b_w = 14$″

$A_s' = 2.40$ sq in.   2.5″ $\epsilon_c = 0.003$

$0.75x_b$ ϵ_s' 9.98″

$A_{sb} = ?$

(b) The maximum A_s condition

$A_s' = 2.40$ sq in. $\epsilon_c = 0.003$

ϵ_s'

$x = 8.51$ in.

$A_s = 8.00$ sq in. ϵ_s

(c) The actual section

Figure 5.12.1 Beam for Example 5.12.1.

(b) Find the nominal flexural strength M_n [Fig. 5.12.1(c)] and allowable service load. Assuming compression steel yields (this is not certain even though compression steel does yield in the max x condition),

$$0.85f'_c b\beta_1 x + A'_s(f_y - 0.85f'_c) = A_s f_y$$

$$0.85(4)(14)(0.85x) + 2.40(60 - 3.4) = 8.00(60)$$

$$x = 8.51 \text{ in.}$$

$$\epsilon'_s = 0.003\,\frac{8.51 - 2.5}{8.51} = 0.00212 > \epsilon_y$$

Assumption is confirmed; compression steel yields ($f'_s = f_y$) in actual condition.

$$C_s = 0.85f'_c b\beta_1 x = 0.85(4)(14)(0.85)(8.51) = 344 \text{ kips}$$

$$C_s = A'_s(f_y - 0.85f'_c) = 2.40(60 - 3.4) = 136 \text{ kips}$$

$$T = A_s f_y = 8.00(60) = 480 \text{ kips}$$

$$d - \frac{a}{2} = 22.5 - \tfrac{1}{2}(0.85)(8.51) = 18.88 \text{ in.}$$

$$M_n = 344(18.88)\tfrac{1}{12} + 136(20)\tfrac{1}{12} = 541 + 227 = 768 \text{ ft-kips}$$

$$M_u = \tfrac{1}{8}(w_u)(24)^2 = \phi M_n = 0.90(768) = 691 \text{ ft-kips}$$

$$w_u = 9.60 \text{ kips/ft}$$

$$w_L = 1.5w_D$$

$$w_u = 1.4w_D + 1.7(1.5w_D)$$

$$\text{service dead load } w_D = \frac{9.60}{1.4 + 2.55} = 2.43 \text{ kips/ft}$$

$$\text{service live load } w_L = 3.64 \text{ kips/ft}$$

(c) Design of shear reinforcement, using the simplified method with a constant value for V_c.

The factored shear to be designed for must be the maximum that may possibly act at each point along the span; that is, an envelope of maximum shear is needed. A knowledge of influence lines tells the designer that for bending moment on a simply supported span the maximum value occurs at every point along the span when the full span is loaded with live load; however, for shear the maximum shear occurs with partial span loading for every point along the span except at the supports. Unless the designer can justify other treatment, the live load should always be treated as variable position loading acting wherever it may cause the greatest effect, whereas the dead load would be fixed position loading. For most ordinary situations an approximate shear envelope may be acceptable, using a straightline relationship between the maximum shear at the support and the maximum shear at midspan. Such a procedure will always be conservative.

For this beam (see Fig. 5.12.2), the maximum factored shear at the centerline of support is

$$V_u = \frac{w_u L}{2} = \frac{[2.43(1.4) + 3.64(1.7)]}{2} = 115.2 \text{ kips}$$

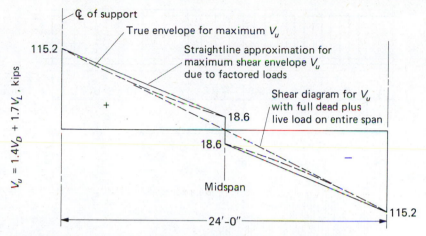

Figure 5.12.2 The factored shear V_u diagram for Example 5.12.1.

The maximum factored shear possible at midspan occurs with live load on half the span,

$$V_u = \frac{w_L L}{8} = \frac{3.64(1.7)(24)}{8} = 18.6 \text{ kips}$$

The dead load shear (with full span loaded) is zero at midspan.

The critical section for determining the closest stirrup spacing may be taken according to ACI-11.1.2.1 at a distance d from *face* of support; in this case $d = 22.5$ in. and the support width is 12 in., making the critical section (22.5 + 6 = 28.5 in.) 2.38 ft from the center of the support. By linear interpolation, V_u at d from the face is

$$V_u = 115.2 - \left(\frac{115.2 - 18.6}{12}\right) 2.38 = 96.0 \text{ kips}$$

The design requirement between the face of support and the critical section is considered to be constant (in this case 96.0 kips).

The factored shear V_u diagram for the left half of this symmetrical structure is shown enlarged in Fig. 5.12.3. The design may be done primarily *on* the factored shear V_u diagram (Fig. 5.12.3).

The design shear strength attributable to the concrete is

$$\phi V_c = \phi(2\sqrt{f_c'}b_w d)$$
$$= 0.85(2\sqrt{4000})(14)(22.5)\tfrac{1}{1000} = 34 \text{ kips}$$

The difference between V_u and ϕV_c must be provided for by shear reinforcement; the portion crosshatched in Fig. 5.12.3. For most beams (ACI-11.5.5.1) shear reinforcement may *not* be terminated when V_u equals ϕV_c, but rather must be continued until $V_u = 0.5\phi V_c$ (Fig. 5.12.3). The distance over which shear reinforcement is required *from the face of support* is in this case the entire span, because $0.5\phi V_c$ never exceeds V_u.

Examination of Fig. 5.12.3 shows that V_u exceeds ϕV_c over most of the span, putting most of the design in categories 3 through 5 of Sec. 5.10. At the critical section,

$$\text{required } \phi V_s = V_u - \phi V_c = 96.0 - 34.0 = 62.0 \text{ kips}$$

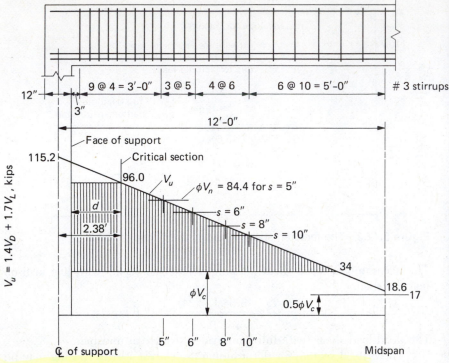

Figure 5.12.3 Design of shear reinforcement in Example 5.12.1 using constant ϕV_c.

Check limits on ϕV_s,

$$\min \phi V_s = 0.85(50)(14)(22.5)\tfrac{1}{1000} = 13.4 \text{ kips}$$

(ACI-11.5.5.3)

This value of ϕV_s corresponds to the location where the factored shear V_u would equal 47.4 kips (that is, $\min \phi V_s + \phi V_c$), in this case 95 in. from face of support (see Table 5.12.1).

$$\max \phi V_s \left(\text{for } s = \frac{d}{2} \right) = 0.85(4\sqrt{4000})(14)(22.5)\tfrac{1}{1000} = 67.7 \text{ kips}$$

(ACI-11.5.4.1 and 11.5.4.3)

$$\max \phi V_s = 0.85(8\sqrt{4000})(14)(22.5)\tfrac{1}{1000} = 135.5 \text{ kips}$$

(ACI-11.5.6.8)

Since the required ϕV_s at the critical section (which is the maximum for the beam) lies between 13.4 and 67.7 kips, design for the portion of the beam between the face of support and the critical section is in category 4 (max $s = d/2$). The strength requirement using ϕV_s computed from V_u will control. Using #3 vertical U stirrups, in which two stirrup bar areas comprise A_v, in Eq. (5.10.15),

$$s \text{ (at critical section)} = \frac{\phi A_v f_y d}{\phi V_s} = \frac{(0.85)(2)(0.11)(60)22.5}{62.0} = 4.1 \text{ in.}$$

Table 5.12.1 SPACING AND STRENGTH RELATIONSHIPS FOR VERTICAL STIRRUPS IN EXAMPLE 5.12.1

s	$\phi V_s = \dfrac{\phi A_v f_y d}{s(1000)} = \dfrac{252}{s}$	$z =$ DISTANCE FROM FACE OF SUPPORT TO INTERSECTION OF ϕV_n (i.e., $\phi V_c + \phi V_s$) WITH V_u $= 22.5 + \dfrac{62.0 - \phi V_s}{96.0 - 18.6}(144 - 28.5)$
4.1 in.	62.0 kips (max)	0 to 22.5 in.
5	50.4	$22.5 + \dfrac{115.5}{77.4}(11.6) = 22.5 + 17.3 = 40$ in.
6	42.0	$(20.0) = 22.5 + 29.8 = 52$
8	31.5	$(30.5) = 22.5 + 45.5 = 68$
10	25.2	$(36.8) = 22.5 + 54.9 = 77$
11.25 max	22.4	$(39.6) = 22.5 + 59.1 = 82$
18.9 (not usable)	13.4 min	$(48.6) = 22.5 + 72.5 = 95$

This means that #3 stirrups may not be spaced farther apart than 4.1 in. for the region from the face of support to the critical section a distance d from the face of support.

The determination of max s at the critical section usually controls what size stirrups are to be used. If the spacing for #3 stirrups were required to be too small, say less than about 3 in., the stirrup bar size would be increased to #4, or occasionally #5. In this case a spacing of 4 in. is acceptable.

The stirrup size is now determined, for which the spacing limitation between a section 81 in. from face of support to center of span is

$$s \text{ (between } z = 81 \text{ in. and } z = 138 \text{ in.)} = \frac{A_v f_y}{50 b_w}$$

$$s = \frac{0.22(60,000)}{50(14)} = 18.9 \text{ in.}$$

which does not control ($d/2$ controls in this case).

As the V_u requirement decreases toward midspan, the stirrup spacing can be increased. A table of potentially acceptable spacings (s vs ϕV_s) should be made, as in the first two columns of Table 5.12.1. Spacings are chosen until the spacing reaches $d/2$ (in this example 11.25 in.) or until ϕV_s gets down to min ϕV_s (in this example 13.4 kips). In the table the $d/2$ spacing limit, not the 18.9-in. limit, happens to control out to $z = 95$ in. In fact, between $z = 95$ in. and midspan, $d/2$ still controls in this example because V_u at midspan is larger than $0.5\ \phi V_c$.

To determine the set of spacings to actually be used, the designer may prefer to draw on the V_u diagram the lines representing the ϕV_n provided by different spacings, as shown on Fig. 5.12.3, and then scale from the diagram the locations z where the various spacings are permissible (marked along the base line of Fig. 5.12.3); or, alternatively, the locations z may be computed as in the third column of Table 5.12.1. Note that in this column, the quantity $(144 - 28.5)/(96.0 - 18.6)$ is the distance in which the value of V_u drops 1 kip. Whether location z is computed or scaled, a table of ϕV_s versus s should be used. The limiting spacings of 4.1 and 11.25 have already been explained. The intermediate spacings of 5, 6, 8, and 10 in. are those chosen by the designer as practical possibilities.

Since spacings of stirrups cannot be varied continuously, they must change by "jumps." One conservative policy is to use a spacing, say of 6 in., only beyond the theoretical point at which a 6-in. spacing may be used (in this case, 52 in. or more from the face of support). In a less conservative manner, the next larger spacing may be used somewhat before the point at which this spacing may be used. With this in mind, the set of spacings to be used is

$$3 \text{ in.} | \overset{10}{9 \text{ @ 4 in.}} | \overset{3}{3 \text{ @ 5 in.}} | \overset{4}{4 \text{ @ 6 in.}} | \overset{6}{6 \text{ @ 10 in.}} | = 138 \text{ in.}$$

$$\text{required}$$

$$3 \qquad\qquad 39 \qquad\qquad 54 \qquad\qquad 78 \qquad\qquad 138 \quad \text{(midspan)}$$

The first stirrup is placed at 3 in. from the support, more than the usual half-space in order to make uniform spacing at midspan. Many designers prefer to place the first stirrup a full space from the face of support for closely spaced stirrups. Note that in the adopted set of spacings, 5-in. spacing is used after $z = 39$ in. ($z = 40$ in. theoretically required); 6-in. spacing is used after $z = 54$ in. ($z = 52$ in. theoretically required); and 10-in. spacing is used after $z = 78$ in. ($z = 77$ in. theoretically required).

EXAMPLE 5.12.2 Repeat the design for shear reinforcement for the beam of Example 5.12.1 using the more detailed procedure for V_c involving $\rho_w V_u d / M_u$ (ACI-11.3.2.1).

Solution: (a) Factored loads and materials. From Example 5.12.1, the factored shear V_u envelope is given by Fig. 5.12.2. Also $f_c' = 4000$ psi and $f_y = 60,000$ psi.

(b) Compute ϕV_c. Since $\rho_w V_u d / M_u$ varies along the span, ϕV_c is not constant. For practical purposes, this method is useful only near the critical section to justify a larger spacing for stirrups than might otherwise be indicated.

At the critical section,

$$V_u = 96.0 \text{ kips} \qquad \text{(Fig. 5.12.4)}$$

$$V_c = \left(1.9\sqrt{f_c'} + 2500\rho_w \frac{V_u d}{M_u} \right) b_w d \leq 3.5\sqrt{f_c'} b_w d$$

$$\rho_w = \frac{A_s}{b_w d} = \frac{8(1.0)}{14(22.5)} = 0.0254$$

The proper loading causing maximum V_u at the critical section is dead load over the entire span plus live load between the critical section and the right end of the beam; however, it is always conservative on a simply supported beam to use the larger value of M_u at the critical section obtained when full dead and live loads are acting over the entire span. Thus using the maximum M_u at the critical section,

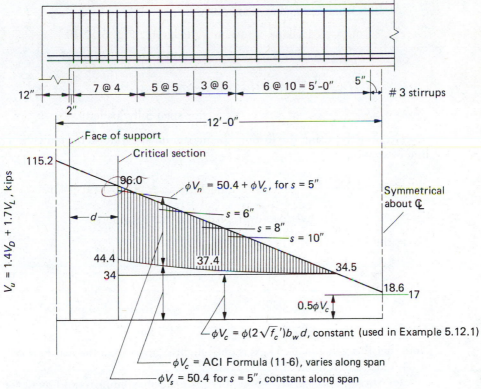

Figure 5.12.4   Design of shear reinforcement in Example 5.12.2 using the more detailed method for ϕV_c involving $\rho_w V_u d/M_u$.

$$M_u = \frac{w_u}{2}(2.38)(24 - 2.38) = \frac{9.60}{2}(2.38)(21.62) = 247 \text{ ft-kips}$$

$$\frac{V_u d}{M_u} = \frac{96.0(22.5)}{247(12)} = 0.729 < 1.0 \text{ max} \qquad\qquad \text{OK}$$

$$V_c = [1.9\sqrt{4000} + 2500(0.0254)(0.729)](14)(22.5)\tfrac{1}{1000}$$

$$= 52.4 \text{ kips} < (3.5\sqrt{f_c'}b_w d = 69.7 \text{ kips}) \qquad \text{OK}$$

$$\phi V_c = 0.85(52.4) = 44.4 \text{ kips}$$

At 5 ft (an arbitrary choice to get a point on the ϕV_c curve) from the centerline of support,

$$V_u = 75.0 \text{ kips} \qquad \text{(from } V_u \text{ diagram, Fig. 5.12.4)}$$

$$\rho_w = 0.0254 \qquad \text{(same as at the critical section)}$$

$$M_u = \frac{9.60}{2}(5.0)(24 - 5.0) = 456 \text{ ft-kips}$$

$$\frac{V_u d}{M_u} = \frac{75.0(22.5)}{456(12)} = 0.308 < 1.0 \text{ max} \qquad\qquad \text{OK}$$

$$V_c = [1.9\sqrt{4000} + 2500(0.0254)(0.308)](14)(22.5)\tfrac{1}{1000} = 44.0 \text{ kips}$$

$$\phi V_c = 0.85(44.0) = 37.4 \text{ kips}$$

At 10 ft from centerline of support,

$$V_u = 34.7 \text{ kips} \qquad \text{(from } V_u \text{ diagram, Fig. 5.12.4)}$$

$$\rho_w = 0.0254$$

$$M_u = \frac{9.60}{2} (10)(24 - 10) = 672 \text{ ft-kips}$$

$$\text{(for span fully loaded)}$$

or more exactly, using the loading for maximum shear at 10 ft from centerline of support,

$$M_u = \frac{2.43(1.4)}{2} (10)(14) + 3.64(1.7)(14) \frac{7}{24} (10) = 491 \text{ ft-kips}$$

$$\text{(for live load over 14 ft from right support)}$$

Using the latter value,

$$\frac{V_u d}{M_u} = \frac{34.7(22.5)}{491(12)} = 0.133 < 1.0 \text{ max} \qquad\qquad \text{OK}$$

$$v_c = 1.9\sqrt{4000} + 2500(0.0254)(0.133) = 129 \text{ psi}$$

$$\phi V_c = 0.85(0.129)(14)(22.5) = 34.5 \text{ kips}$$

Note that this value of 34.5 kips is only slightly larger than the constant ϕV_c of 34.0 kips in Example 5.12.1.

(c) Stirrup spacing for #3 U stirrups. The maximum requirement for ϕV_s is at the critical section,

$$\text{max required } \phi V_s = V_u - \phi V_c = 96.0 - 44.4 = 51.6 \text{ kips}$$

The stirrup spacing based on the strength requirement (ACI-11.5.6.2) at the critical section is

$$\text{max } s \text{ (at critical section)} = \frac{\phi A_v f_y d}{\phi V_s} = \frac{0.85(2)(0.11)(60)22.5}{51.6} = 4.9 \text{ in.}$$

In this case the spacings arrived at by the simplified method could be revised as shown in Fig. 5.12.4 to 2 in., 7 at 4 in., 5 at 5 in., 3 at 6 in., and 6 at 10 in., resulting in a saving of only one stirrup for the entire beam (one less space on each side of midspan but also no stirrup at midspan). Thus the gain from using the more detailed procedure is negligible. Unless the close spacing between the face of support and the critical section can be increased by an inch or more, little saving can be expected from the detailed procedure.

EXAMPLE 5.12.3 Design the locations of #3 vertical U stirrups to be used in the beam of Fig. 5.12.5. The beam is to carry service dead and live loads of 5.2 and 6.0 kips/ft, respectively. Use $f'_c = 3500$ psi and $f_y = 60,000$ psi. Use the alternate approach of plotting v_n (v_n is the nominal shear stress which is the required nominal shear strength V_u/ϕ divided by $b_w d$) diagram and then providing the v_c attributable to the concrete and v_s attributable to shear reinforcement.

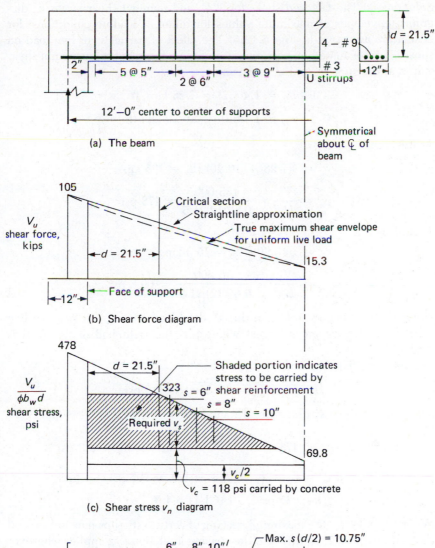

(a) The beam

(b) Shear force diagram

(c) Shear stress v_n diagram

(d) Stirrup requirement diagram

Figure 5.12.5 Stirrup placement for Example 5.12.3.

Solution: (a) The factored shear force V_u and nominal shear stress v_n diagrams (envelopes of maximum values for different loading conditions) for design are as shown in Fig. 5.12.5. The effect of partial span live load on shear is approximated by passing a straight line through the maximum shear values at the support and at midspan.

$$1.4w_D = 1.4(5.2) = 7.28 \text{ kips/ft}$$

$$1.7w_L = 1.7(6.0) = 10.20 \text{ kips/ft}$$

At the support,

$$V_u = \tfrac{1}{2}(7.28 + 10.20)(12) = 105 \text{ kips}$$

$$v_n = \frac{V_u}{\phi bd} = \frac{105,000}{0.85(12)21.5} = 478 \text{ psi}$$

At midspan,

$$V_u = \tfrac{1}{8}(10.20)(12) = 15.3 \text{ kips}$$

$$v_n = \frac{V_u}{\phi bd} = \frac{15,300}{0.85(12)(21.5)} = 69.8 \text{ psi}$$

Note that the shear at midspan due to dead load is zero; but positive live load shear at midspan is largest when only the right half of the span is loaded.

(b) Simplified method.

$$v_c = 2\sqrt{f'_c} = 2\sqrt{3500} = 118 \text{ psi}$$

The nominal shear stress at the distance d from the face of the support is

$$v_n(\text{at } d) = 478 - \frac{(6 + 21.5)}{72}(478 - 69.8) = 323 \text{ psi}$$

$$\text{required } v_s = v_n - v_c = 323 - 118 = 205 \text{ psi}$$

$$4\sqrt{f'_c} = 237 \text{ psi} \qquad (\text{ACI-11.5.4.3})$$

Where $v_s > 4\sqrt{f'_c}$, the maximum spacing of vertical stirrups may not exceed $d/4$, otherwise $d/2$. In this case the spacing limitation $d/2$ applies wherever stirrups are needed. Note that this simple check on whether or not required v_s exceeds $4\sqrt{f'_c}$ is often advantageous over using the criterion shown in item 8 of Table 5.10.1.

Try #3 vertical U stirrups.

$$A_v f_y = 2(0.11)(60,000) = v_s b_w s$$

$$v_s s = \frac{13,200}{12} = 1100$$

which for the v_s of 205 psi at the critical section permits

$$\text{max } s \text{ (at critical section)} = \frac{1100}{205} = 5.4 \text{ in.}$$

The maximum spacing for shear reinforcement to provide for a minimum v_s

of 50 psi is

$$\text{max } s \text{ (for min } A_v) = \frac{1100}{50} = 22 \text{ in.} > \left[\frac{d}{2} = 10.75 \text{ in.}\right]$$

Thus the spacing cannot be more than 10.75 in. even when v_s gets small. If the above computation had given a value less than $d/2$ instead of 22 in., then that lesser value would be the maximum s for the region where stirrups *are* required ($v_n > v_c/2$) but where the loading gives the required v_s (= $v_n - v_c$) to be less than 50 psi.

The placement of stirrups will be done by scaling from the diagram of Fig. 5.12.5(c). To facilitate this, Table 5.12.3 is computed using spacings considered desirable by the designer. The shear stress v_s values are plotted as horizontal lines marked s = 6, 8, and 10 in. on Fig. 5.12.5(c); then their intersections with the maximum shear stress v_n line are projected downward to the base line. Stirrups are then laid out with a scale beginning a distance $s/2$ from the face of support. (Some designers place the first stirrup a full space from the face of support for small spacings.) The same spacing necessary at the critical section is specified by ACI-11.1.2.1 to be used between the face of support and the critical section. Thus one may start at $s/2$ from support with a 5-in. spacing until within $s/2$ of the capacity line for the next desired spacing, which in this case is slightly beyond the vertical line projected from s = 6 in. The ACI Code requires shear reinforcement until the stress $v_n \le v_c/2$. In this case the entire beam must be provided with stirrups, because the smallest v_n (69.8 psi) is larger than $v_c/2$ (59 psi).

Table 5.12.3 SPACING OF VERTICAL STIRRUPS (EXAMPLE 5.12.3)

s	$v_s = \dfrac{1100}{s}$	$v_n = v_c + v_s$
5.4	205	323
6	183	301
8	138	256
10	110	228
max 10.75	102	220
(22)	(50)	(168)

To illustrate clearly what has been done, a diagram showing permitted (actual spacings must be below permitted line) and provided stirrup spacings is given in Fig. 5.12.5(d). In this case some shifting of spacing has been made to avoid leaving a small fragmental space at midspan. One more space at 5 in. was used than necessary and the three spaces adjacent to midspan were reduced to 9 in. from the permitted 10 in. in order to eliminate a small space near midspan.

The final design details are shown in Fig. 5.12.5(a).

EXAMPLE 5.12.4 Determine the vertical stirrup requirement for the beam of Fig. 5.12.6. Use #10M bars (see Table 1.12.2) for U stirrups, $f_c' = 25$ MPa, and $f_y = 300$ MPa. The service live load is 30 kN/m and the service dead load is 43 kN/m (including beam weight).

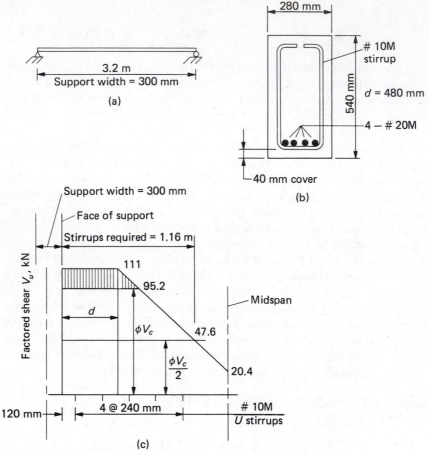

Figure 5.12.6 Beam and stirrup design for Example 5.12.4.

Solution: (a) Determine maximum factored shear V_u envelope.

$$w_u = 1.4(43) + 1.7(30) = 111.2 \text{ kN/m}$$

For maximum shear at d (480 mm) from face of support, place live load on remainder $(3.2 - 0.15 - 0.48 = 2.57 \text{ m})$ of the span.

$$V_u \text{ at critical section} = 1.4(43)(1.6 - 0.15 - 0.48) + 1.7(30)\frac{2.57}{2(3.2)}$$

$$= 58.4 + 52.6 = 111 \text{ kN}$$

$$V_u \text{ at midspan} = \tfrac{1}{8}(51.0)(3.2) = 20.4 \text{ kN}$$

The straightline approximation maximum shear envelope is given as Fig. 5.12.6(c). (Note that in the previous example, the straightline approximation was made between centerline of support and midspan. The present approximation is closer to the "exact" maximum shear curve.)

(b) Determine stirrup spacing. If the factored shear V_u diagram is used,

$$v_c = \sqrt{f_c'}/6 \quad \text{(from ACI 318-83M, with } f_c' \text{ in MPa)}$$
$$= \sqrt{25}/6 = 0.833 \text{ MPa}$$
$$\phi V_c = \phi v_c b_w d = 0.85(0.833)(280)(480)\tfrac{1}{1000} = 95.2 \text{ kN}$$

At the critical section,

$$\text{required } \phi V_s = V_u - \phi V_c = 111 - 95.2 = 15.8 \text{ kN}$$

The limiting $\phi V_s = \phi(\sqrt{f_c'}/12)b_w d$ for $d/2$ stirrup spacing limitation can be conveniently obtained from $2\phi V_c$; thus

$$\text{limiting } \phi V_s = 2\phi V_c = 2(95.2) = 190 \text{ kN} > [\text{required } \phi V_s = 15.8 \text{ kN}]$$

Thus maximum spacing cannot exceed $d/2$.

For #10M U stirrups, $A_v = 2(100) = 200 \text{ mm}^2$. The spacing requirement for strength is

$$s = \frac{\phi A_v f_y d}{\phi V_s} = \frac{(0.85)(200)(0.300)480}{\phi V_s} = \frac{24{,}480}{\phi V_s \text{ (in kN)}}$$

The stirrup spacing requirements can be summarized as follows:

1. Maximum spacing for strength requirement at the critical section,

$$\text{max } s = \frac{24{,}480}{15.8} = 1550 \text{ mm}$$

2. Maximum spacing $d/2$,

$$\text{max } s = \frac{d}{2} = \frac{480}{2} = 240 \text{ mm} \qquad \text{(Controls)}$$

3. Maximum spacing for minimum shear reinforcement (ACI-11.5.5.3),

$$\text{min } \phi V_s = \phi(\tfrac{1}{3} \text{ MPa})b_w d \qquad \text{(ACI 318-83M)}$$
$$= 0.85(0.333)(280)(480)\tfrac{1}{1000} = 38.0 \text{ kN}$$
$$\text{max } s = \frac{24{,}480}{38.0} = 644 \text{ mm} > \frac{d}{2}$$

4. Conclusion. Use 240-mm spacing for the portion where stirrups are required (i.e., from face of support to location where $V_u = 0.5\phi V_c$).

The final stirrup arrangement is shown in Fig. 5.12.6(c).

(c) Examine the stirrup requirement if Eq. (5.10.4) with $v_c = (0.8 + 100\rho_w) \sqrt{f_c'}$ (Inch–Pound units) is used for $\rho_w < 0.012$.

$$v_c = (0.8 + 100\rho)\sqrt{f_c'} \qquad (f_c' \text{ in psi})$$
$$= 0.083(0.8 + 100\rho)\sqrt{f_c'} \qquad (f_c' \text{ in MPa})$$
$$\rho_w = \frac{4(300)}{280(480)} = 0.00893$$
$$v_c = 0.083(0.8 + 0.893)\sqrt{25} = 0.703 \text{ MPa}$$
$$\phi V_c = \phi v_c b_w d = 0.85(0.703)(280)(480)\tfrac{1}{1000} = 80.3 \text{ kN}$$
$$\text{max } \phi V_s = V_u - \phi V_c = 111 - 80.3 = 30.7 \text{ kN}$$

For the strength requirement at the critical section,

$$\max s = \frac{24,480}{\phi V_s} = \frac{24,480}{30.7} = 797 \text{ mm}$$

For this problem, $d/2$ would still control with stirrups spaced at 240 mm maximum. Even though the ACI Code does not require the use of Eq. (5.10.4) at present, the designer should be conservative about the use of stirrups when ρ_w is less than about 0.012.

5.13 SHEAR STRENGTH OF MEMBERS UNDER COMBINED BENDING AND AXIAL LOAD

The presence of an axial compressive load on a reinforced concrete flexural member decreases the longitudinal tensile stress and the resulting tendency for inclined cracking. Conversely the addition of an axial tensile load increases the longitudinal tensile stress and the tendency for inclined cracking. Thus, for the same bending moment, the shear strength of a member is increased by the addition of an axial compressive load and decreased by an axial tensile load. Some experimental work is available on shear strength in the presence of axial load [3,42–44].

Axial Compression. Since inclined cracking is dependent on the combination of tensile or compressive stress due to flexure and shear stress as discussed in Secs. 5.3 through 5.5, the addition of axial compression tends to delay the opening of the shear crack and prevent its extending as far into the beam. Consider the free body of a short length of beam dz as shown in Fig. 5.13.1. Taking moments about point A, the line of action of the internal compressive force C,

$$T\left(d - \frac{a}{2}\right) = M_n - N_n\left(\frac{h}{2} - \frac{a}{2}\right) \tag{5.13.1}$$

As assumed in the development of the basic formula for the flexure-shear

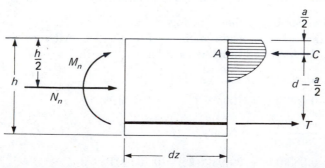

Figure 5.13.1 Member under combined axial compression and bending moment.

inclined cracking capacity,

$$f_t \text{ (tensile stress in concrete)} \propto \frac{f_s \text{ (tensile stress in steel)}}{n}$$

Thus

$$f_t \propto \frac{T}{nA_s} = \frac{M_n - N_n(h/2 - a/2)}{nA_s(d - a/2)} \tag{5.13.2}$$

Letting $n = E_s/E_c$, $A_s = \rho b d$, $(d - a/2)$ be approximated by $7d/8$ (making $a/2 = d/8$), E_c be proportional to $\sqrt{f'_c}$, and k be a proportionality constant, Eq. (5.13.2) may be written

$$f_t = \frac{k}{E_s} \frac{\sqrt{f'_c}}{\rho} \left(\frac{M_n - N_n(h/2 - d/8)}{bd^2} \right) \tag{5.13.3}$$

which is analogous to Eq. (5.5.2) if a moment equivalent to M_n of Eq. (5.5.2) is defined as

$$\text{equivalent } M_n = M_n - N_n \left(\frac{4h - d}{8} \right) \tag{5.13.4}$$

Since Eq. (5.13.4) is to be used in the more detailed expression for V_c, Eq. (5.5.6), where the ratio V_u/M_u (rather than the ratio V_n/M_n) is used M_n and N_n become M_u and N_u. ACI-11.3.2.2 states that M_m shall replace M_u in the expression for V_c, where

$$M_m = M_u - N_u \left(\frac{4h - d}{8} \right) \tag{5.13.5}$$

which is ACI Formula (11-7). Note that N_u is positive for compression and that M_m must never be used as a negative value. Also $V_u d/M_m$ is permitted to have values greater than unity. This equivalent M procedure is to be used in the more detailed method for *combined axial compression and bending only*. Experience [42] has shown the method to be unsafe for axial tension.

When the simplified method is used with axial compression, an alternate equation is given by ACI-11.3.1.2:

$$V_c = 2 \left(1 + \frac{N_u}{2000A_g} \right) \sqrt{f'_c} b_w d \tag{5.13.6}*$$

which is ACI Formula (11-4).

The upper limit for the nominal shear strength V_c of members without shear reinforcement subject to bending only has been prescribed to be $3.5\sqrt{f'_c}b_w d$. This upper limit should be adjusted upward in the presence of

*For SI, ACI 318-83M, for N_u/A_g and f'_c in MPa, gives

$$V_c = \left(1 + \frac{N_u}{14A_g} \right) \left(\frac{\sqrt{f'_c}}{6} \right) b_w d \tag{5.13.6}$$

an axial compression. As explained below, this adjustment factor may be rationally put into the form

$$\text{adjustment factor} = \sqrt{1 + \frac{N_u}{500A_g}} \tag{5.13.7}$$

In Eqs. (5.13.6) and (5.13.7) N_u is the factored axial compressive load, and A_g is the gross area of the concrete section.

The formula for the principal tensile stress f_t (max) in terms of the tensile stress f_t and the shear stress v has been derived in Sec. 5.3 to be

$$f_t \text{ (max)} = \tfrac{1}{2}f_t + \sqrt{(\tfrac{1}{2}f_t)^2 + v^2}$$

Solving this equation for v,

$$v = f_t \text{ (max)} \sqrt{1 - \frac{f_t}{f_t \text{ (max)}}} \tag{5.13.8}$$

It may be seen from Eq. (5.13.8) that the shear strength of a member under bending only becomes a constant if f_t is zero, that is, if the bending moment approaches zero. Empirically this constant is the upper limit $3.5\sqrt{f_c'}$. For a member under an axial compressive load N_u without bending moment, f_t is a constant that is equal to $-N_u/A_g$. Substituting this value of f_t in Eq. (5.13.8) and using an average value of 500 psi for f_t (max), the upper limit of strength V_c in members under combined bending and axial compression becomes

$$V_c \text{ (upper limit)} = vb_wd = 3.5\sqrt{f_c'}\sqrt{1 + \frac{N_u}{500A_g}}\, b_wd \tag{5.13.9}*$$

which is ACI Formula (11-8).

Axial Tension. When a flexural member is subject to axial tension, ACI in the simplified method states in ACI-11.3.1.3 that V_c is to be zero. In the more detailed method, a simple linear reduction for V_c has been specified in ACI-11.3.2.3. Thus

$$V_c = 2\left(1 + \frac{N_u}{500A_g}\right)\sqrt{f_c'}b_wd \tag{5.13.10}†$$

which is ACI Formula (11-9). Note that N_u is negative for tension.

In order to show the relationship between ACI formulas and experimental results, Fig. 5.13.2 is presented. The crosshatched portion on the compression side represents the reasonable range when using the $\rho Vd/M$ procedure.

Parts of the ACI Code are summarized in Table 5.13.1 so that the shear

---

*For SI, ACI 318-83M, for N_u/A_g and f_c' in MPa, gives

$$V_c \text{ (upper limit)} = 0.3\sqrt{f_c'}\sqrt{1 + \frac{0.3N_u}{A_g}}\, b_wd \tag{5.13.9}$$

†For SI, ACI 318-83M, for N_u/A_g and f_c' in MPa, gives

$$V_c = \left(1 + \frac{0.3N_u}{A_g}\right)\left(\frac{\sqrt{f_c'}}{6}\right)b_wd \tag{5.13.10}$$

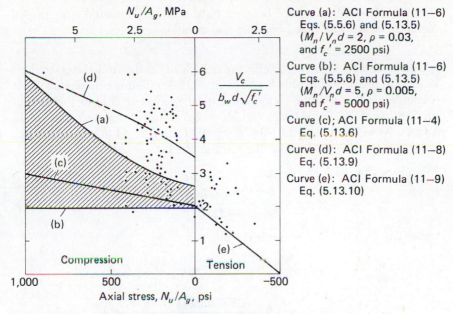

Curve (a): ACI Formula (11−6)
Eqs. (5.5.6) and (5.13.5)
($M_n/V_n d = 2$, $\rho = 0.03$,
and $f_c' = 2500$ psi)

Curve (b): ACI Formula (11−6)
Eqs. (5.5.6) and (5.13.5)
($M_n/V_n d = 5$, $\rho = 0.005$,
and $f_c' = 5000$ psi)

Curve (c); ACI Formula (11−4)
Eq. (5.13.6)

Curve (d): ACI Formula (11−8)
Eq. (5.13.9)

Curve (e): ACI Formula (11−9)
Eq. (5.13.10)

Figure 5.13.2 Effect of axial load on inclined cracking shear stress (dots indicate test results) (adapted from Ref. 42).

Table 5.13.1   EFFECT OF AXIAL LOAD ON THE SHEAR STRENGTH OF MEMBERS WITHOUT SHEAR REINFORCEMENT—ACI CODE

	SIMPLIFIED METHOD	MORE DETAILED METHOD
Bending only	Formula (11-3), 11.3.1.1 $$V_c = 2\sqrt{f_c'}b_w d$$ $$\le 3.5\sqrt{f_c'}b_w d$$	Formula (11-6), 11.3.2.1 $$V_c = \left(1.9\sqrt{f_c'} + 2500\rho_w \frac{V_u d}{M_u}\right) b_w d$$ $$\le 3.5\sqrt{f_c'}b_w d$$ $V_u d/M_u$ not to exceed unity
Bending and axial compression	Formula (11-4), 11.3.1.2 $$V_c = 2\left(1 + \frac{N_u}{2000A_g}\right)\sqrt{f_c'}b_w d$$	Formula (11-7), 11.3.2.2 $$M_m = M_u - N_u\left(\frac{4h - d}{8}\right)$$ Use M_m for M_u in Formula (11-6) $V_u d/M_u$ has no limitation
	Formula (11-8), 11.3.2.2 $$V_c \le 3.5\sqrt{f_c'}b_w d\sqrt{1 + \frac{N_u}{500A_g}}$$ N_u is positive for compression and N_u/A_g is in psi	Formula (11-8), 11.3.2.2 $$V_c \le 3.5\sqrt{f_c'}b_w d\sqrt{1 + \frac{N_u}{500A_g}}$$ N_u is positive for compression and N_u/A_g is in psi
Bending and axial tension	11.3.1.3 $$V_c = 0$$ Design shear reinforcement for total shear	Formula (11-9), 11.3.2.3 $$V_c = 2\left(1 + \frac{N_u}{500A_g}\right)\sqrt{f_c'}b_w d$$ N_u is negative for tension and N_u/A_g is in psi

strength of members under bending only may be compared with that of members under combined bending and axial load.

EXAMPLE 5.13.1 Show the effect of axial load on the ACI shear strength V_c for the beam of Example 5.12.1 (Fig. 5.12.1) when it contains no shear reinforcement. Compute V_c for the critical section at d from the face of support, where $V_u = 96.0$ kips and $M_u = 247$ ft-kips (computed in Example 5.12.2). As in Example 5.12.1 use $f_c' = 4000$ psi, $b = 14$ in., $d = 22.5$ in., $h = 26$ in., and 8-#9 bars for the tension steel.

Solution: (a) Axial compression.

$$\rho_w = \frac{A_s}{bd} = \frac{8(1.00)}{14(22.5)} = 0.0254$$

Using Eq. (5.5.6) [ACI Formula (11-6)],

$$V_c = \left(1.9\sqrt{f_c'} + 2500\rho_w\, \frac{V_u d}{M_u}\right) b_w d$$

$$\frac{V_c}{\sqrt{f_c'}\, b_w d} = 1.9 + \frac{2500(0.0254)(1.88)}{\sqrt{4000}}\left(\frac{V_u}{M_u}\right)$$

$$= 1.9 + 1.89\,\frac{V_u}{M_u}$$

Eq. (5.13.5) [ACI Formula (11-7)] gives M_m to replace M_u in the above equation;

$$M_m = M_u - N_u\left(\frac{4h - d}{8}\right)$$

$$= M_u - N_u\left[\frac{4(26) - 22.5}{8(12)}\right] = M_u - 0.849 N_u$$

where M_u is in ft-kips and N_u is in kips.

At the critical section, $M_u = 247$ ft-kips and $V_u = 96.0$ kips,

$$M_m = 247 - 0.849 N_u = 247 - \frac{N_u}{A_g}(0.849)(14)(26)$$

$$= 247 - 309\,\frac{N_u}{A_g}$$

Values for the $\rho_w V_u d/M_u$ formula as well as the upper limit equation, Eq. (5.13.9) [ACI Formula (11-8)], are tabulated in Table 5.13.2. The results are shown in Fig. 5.13.3.

(b) Axial tension. The simple linear expression, Eq. (5.13.10) [ACI Formula (11-9)], is plotted in Fig. 5.13.3.

5.14 DEEP BEAMS

When the shear span ($a = M/V$) to depth ratio of beams is lower than about 2, the total shear strength significantly exceeds that at inclined cracking, as

Table 5.13.2 SHEAR STRENGTH WITH AXIAL COMPRESSION—
EXAMPLE 5.13.1

N_u/A_g (psi)	M_m (ft-kips)	$1.89V_u/M_m$	$V_c/(\sqrt{f_c'}b_w d)$	$3.5\sqrt{1 + 0.002N_u/A_g}$
50	232	0.78	2.68	3.5(1.05) = 3.67
100	216	0.84	2.74	3.5(1.095) = 3.83
200	185	0.98	2.88	3.5(1.183) = 4.14
400	123	1.48	3.38	3.5(1.342) = 4.70
600	62	2.93	4.83	3.5(1.484) = 5.19
800	0	—	—	3.5(1.612) = 5.64

shown in Fig. 5.4.4. After inclined cracking occurs, the beam can still carry
a considerably larger load primarily as a compression strut in arch action as
shown in Fig. 5.4.5. For this type of beam, the inclined crack usually forms
at an angle with the vertical much less than 45°; often it is nearly vertical.
Because of this, shear reinforcement, when required, must consist of both
horizontal and vertical bars.

If in the design of deep beams the shear strength which exceeds that
at the formation of the inclined crack is to be utilized, shear reinforcement
will have to be relatively more closely spaced than for ordinary beams in
order to control and restrain the crack from opening a wide amount. Finally,
because actual failure (Fig. 5.4.5) frequently relates to an anchorage failure
of the main tension bars, proper embedment is of great importance.

The ACI Code gives shear-related provisions for deep beams in ACI-
11.8. The provisions apply to beams whose ratio L_n/d of clear span L_n to
effective depth d is less than 5 and which are *loaded at the top or compression face* and supported at the bottom. This is referred to as being *directly
loaded*. The higher strength indicated by ACI Formula (11-30) does not

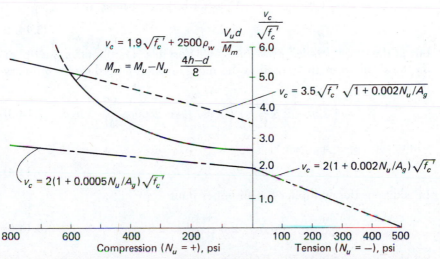

Figure 5.13.3 Shear strength variation with axial load: Example 5.13.1, with
$f_c' = 4000$ psi, $\rho_w = 0.0254$, $M_u = 247$ ft-kips, $V_u = 96.0$ kips, $d = 22.5$ in.,
$h = 26$ in., and $A_g = 364$ sq in. (Note that $v_c = V_c/b_w d$.)

exist when the deep member is primarily loaded by beams framing from the side (said to be *indirectly loaded*). For indirectly loaded beams having a/d less than about 2.5, the vertical compressive stress component between the load and reaction does not exist to increase the shear strength; thus provisions for ordinary beams (ACI-11.3) apply. The deep beam provisions are based on the results of more than 250 tests [3,13,25,45,46]. Additional information with recommendations may be found in the 1979 Report of ACI-ASCE Committee 426 [4,5], and the paper by MacGregor and Hawkins [17].

Nominal Shear Strength. The nominal shear strength is computed as for ordinary beams,

$$V_n = V_c + V_s \qquad (5.14.1)$$

where V_s is the sum of the contributions of horizontal and vertical shear reinforcement.

Since the shear reinforcement required, if any, at the critical section *is to be used throughout the span*, the factored shear V_u is determined at only one location. The *critical section* is defined as located at a distance (say z) from the face of support; thus ACI-11.8.4 states

1. $z = 0.15L_n \leq d$ for uniform loading
2. $z = 0.50a \leq d$ for concentrated loading

where the shear span a is the distance from the face of support to the concentrated load (ACI-ASCE Committee 426 recommends [4,17] limiting a to a maximum of 1.15 times the clear distance from the face of the load to the face of the support).

Strength V_c Attributable to the Concrete. ACI-11.8.5 permits using the *simplified method* with the same V_c as for ordinary beams, or

$$V_c = 2\sqrt{f_c'}b_w d \qquad (5.14.2)$$

but in the *more detailed procedure* ACI Formula (11-30) puts a multiplier on the shear strength V_c used in the more detailed procedure for ordinary beams; or

$$V_c = \left(3.5 - 2.5\frac{M_u}{V_u d}\right)\left(1.9\sqrt{f_c'} + 2500\rho_w \frac{V_u d}{M_u}\right)b_w d \qquad (5.14.3)$$

where the upper limit on V_c is

$$\text{limit } V_c \leq 6\sqrt{f_c'}b_w d \qquad (5.14.4)$$

In addition, the multiplier has an upper limit,

$$\left(3.5 - 2.5\frac{M_u}{V_u d}\right) \leq 2.5 \qquad (5.14.5)$$

which is reached at $M_u/(V_u d) = 0.4$.

The value of M_u is the factored bending moment occurring at the critical section simultaneously with maximum factored shear V_u.

Strength V_s Attributable to Shear Reinforcement. The equation requiring both horizontal and vertical reinforcement is given as ACI Formula (11-31),

$$V_s = \left[\frac{A_v}{s} \left(\frac{1 + L_n/d}{12} \right) + \frac{A_{vh}}{s_2} \left(\frac{11 - L_n/d}{12} \right) \right] f_y d \qquad (5.14.6)$$

where

L_n = clear span
A_v = vertical stirrup area
A_{vh} = longitudinal shear reinforcement area
s = spacing of vertical stirrups
s_2 = vertical spacing of longitudinal shear reinforcement

ACI-11.8.8 and 11.8.9 provide for minimum amounts of shear reinforcement, A_v and A_{vh}.

$$\min A_v = 0.0015 b_w s \qquad (5.14.7)$$

$$s \leq \frac{d}{5} \leq 18 \text{ in.}^*$$

$$\min A_{vh} = 0.0025 b_w s_2 \qquad (5.14.8)$$

$$s_2 \leq \frac{d}{3} \leq 18 \text{ in.}^*$$

where b_w is the web width. The minimum amount of horizontal and vertical shear reinforcement is required throughout the span (ACI-11.8.10) wherever $V_u > 0.5\phi V_c$ (ACI-11.5.5.1).

Limitations on Nominal Shear Strength. The limitations on strength are best itemized in terms of the nominal stress $v_n = V_u/(\phi b_w d)$ since they will then be independent of $b_w d$.
The maximum shear stress v_n allowed by ACI-11.8.3 is

$$\max v_n \leq 8\sqrt{f_c'} \qquad \text{for } \frac{L_n}{d} \leq 2 \qquad (5.14.9)$$

$$\max v_n = \frac{2}{3} \left(10 + \frac{L_n}{d} \right) \sqrt{f_c'} \qquad \text{for } \frac{L_n}{d} > 2 \qquad (5.14.10)$$

This gives a maximum value of v_n varying from $8\sqrt{f_c'}$ to $10\sqrt{f_c'}$. The upper value $10\sqrt{f_c'}$ that is permitted when $L_n/d = 5$ is approximately the same as the limit of $8\sqrt{f_c'}$ on v_s for longer beams (ACI-11.5.6.8).

EXAMPLE 5.14.1 Design the shear reinforcement required for the beam of Fig. 5.14.1, which is to support heavy machinery. The service loading consists of 3 kips/ft dead load and 15.2 kips/ft live load (including impact

*For SI, ACI 318-83M gives the upper limit as 500 mm.

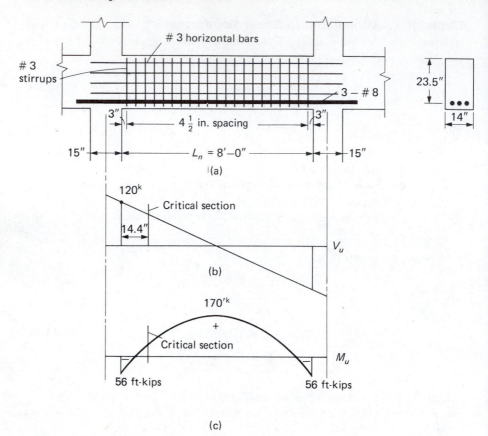

Figure 5.14.1 Design of shear reinforcement in a deep beam, Example 5.14.1.

effect). The factored moment and shear diagrams are given in Fig. 5.14.1. Use $f'_c = 3500$ psi and $f_y = 60,000$ psi.

Solution: Since this is a beam of short span with heavy direct loading at the compression face, it is quite likely a deep beam according to ACI-11.8.

(a) Determine if deep beam provisions apply.

$$\frac{L_n}{d} = \frac{8(12)}{23.5} = 4.08 < 5 \quad \text{(ACI-11.8 applies)}$$

(b) Critical section (ACI-11.8.4). For uniform loading,

$$0.15L_n = 0.15(8)(12) = 14.4 \text{ in.} < [d = 23.5 \text{ in.}]$$

Critical section is taken at 14.4 in. from face of support.

(c) Shear strength V_c at critical section (ACI-11.8.4) without shear reinforcement. Using the *simplified method* (ACI-11.8.5), and in terms of nominal stress,

$$v_c = 2\sqrt{f'_c} = 2\sqrt{3500} = 118 \text{ psi}$$

Using the *more detailed procedure* (ACI-11.8.6),

$$v_c = \left(3.5 - 2.5 \frac{M_u}{V_u d} \right) \left(1.9\sqrt{f_c'} + 2500\rho_w \frac{V_u d}{M_u} \right)$$

$$w_u = 3(1.4) + 15.2(1.7) = 4.2 + 25.8 = 30 \text{ kips/ft}$$

$$V_u = 120 - 30 \left(\frac{14.4}{12} \right) = 84 \text{ kips}$$

$$M_u = -56 + \frac{1}{2}(120 + 84) \left(\frac{14.4}{12} \right) = 66 \text{ ft-kips}$$

$$\frac{M_u}{V_u d} = \frac{66(12)}{84(23.5)} = 0.40 = 0.4 \text{ limit}$$

$$\text{max value for } \left(3.5 - 2.5 \frac{M_u}{V_u d} \right) = 2.5 \text{ just applies}$$

$$v_c = 2.5 \left[1.9\sqrt{f_c'} + 2500\rho_w \frac{1}{0.316} \right]$$

$$= 2.5 \left[1.9\sqrt{3500} + \frac{2500}{0.40} \left(\frac{3(0.79)}{14(23.5)} \right) \right]$$

$$= 2.5[112 + 45] = 342 \text{ psi (controls)}$$

upper limit $v_c = 6\sqrt{f_c'} = 6(59.2) = 355 \text{ psi}$ 　　　　　　　　OK

(d) Nominal shear stress. At critical section,

$$v_n = \frac{V_u}{\phi bd} = \frac{84{,}000}{0.85(14)(23.5)} = 300 \text{ psi}$$

which is less than v_c of 342 psi.

$$v_n \text{ (max allowed)} = \frac{2}{3} \left(10 + \frac{L_n}{d} \right) \sqrt{f_c'}$$

$$= \frac{2}{3} (10 + 4.08)\sqrt{f_c'}$$

$$= 9.4\sqrt{f_c'} = 557 \text{ psi} > 300 \text{ psi} \qquad \text{OK}$$

Since $v_n > \frac{1}{2} v_c$, that is, 300 psi $> 342/2$ psi, the minimum amount of shear reinforcement is required for the entire beam.

(e) Shear reinforcement (ACI-11.8.8, 11.8.9, and 11.8.10). For vertical reinforcement,

$$\text{min } A_v = 0.0015 b_w s$$

which when #3 bars are used becomes

$$\text{max } s = \frac{2(0.11)}{0.0015(14)} = 10.5 \text{ in.}$$

but the spacing s shall not exceed

$$\text{max } s = \frac{d}{5} = \frac{23.5}{5} = 4.7 \text{ in. (controls)} \qquad \text{or 18 in.}$$

For horizontal reinforcement,

$$\min A_{vh} = 0.0025b_w s_2$$

which for #3 bars in pairs on each face of beam gives

$$\max s_2 \doteq \frac{2(0.11)}{0.0025(14)} = 6.3 \text{ in. (controls)}$$

but the spacing shall not exceed

$$\max s_2 = \frac{d}{3} = \frac{23.5}{3} = 7.8 \text{ in.} \quad \text{or} \quad 18 \text{ in.}$$

Use #3 vertical U stirrups spaced at $4\frac{1}{2}$ in. throughout the beam, and #3 bars horizontally in each face spaced at about 5 in. to give proper cover (4-#3 bars horizontally in each face). The results are shown in Fig. 5.14.1(a).

EXAMPLE 5.14.2 Design the shear reinforcement for a simply supported beam that carries two concentrated service live loads of 126 kips each on a clear span of 12 ft, as shown in Fig. 5.14.2. The beam has a width of 14 in. and an effective depth d of 36 in. Use $f'_c = 3500$ psi and $f_y = 60,000$ psi.

Solution: For this beam, $L_n/d = 144/36 = 4$; thus it is a deep beam according to ACI-11.8.

(a) Critical section (ACI-11.8.4). For concentrated loads, using the shear span of $a = 4$ ft,

$$0.50a = 0.50(4.0) = 2.0 \text{ ft} < [d = 3 \text{ ft}]$$

Critical section is taken at 2.0 ft from face of support.

(b) Shear strength of beam without shear reinforcement at critical section (ACI-11.8.4). The factored concentrated load is

$$126(1.7) = 214 \text{ kips}$$

Neglect the uniform dead load which is small compared to the concentrated load.

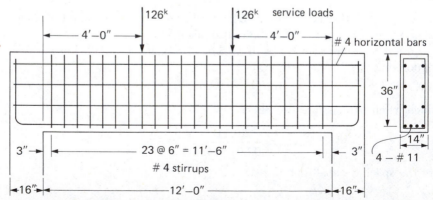

Figure 5.14.2 Beam for Example 5.14.2.

Using the more detailed procedure, at the critical section,

$$\frac{M_u}{V_u d} = \frac{214(2)}{214(3)} = 0.67$$

The multiplier for deep beam action is then

$$3.5 - 2.5 \frac{M_u}{V_u d} = 3.5 - 2.5(0.67) = 1.83 < 2.5 \qquad \text{OK}$$

$$v_c = 1.83 \left[1.9\sqrt{f'_c} + 2500\rho_w \frac{V_u d}{M_u} \right]$$

$$\rho_w = \frac{4(1.56)}{14(36)} = 0.0124$$

$$v_c = 1.83 \left[1.9\sqrt{3500} + \frac{2500(0.0124)}{0.67} \right]$$

$$= 1.83[112 + 46.3] = 290 \text{ psi (controls)}$$

$$\text{upper limit } v_c = 6\sqrt{f'_c} = 6(59.2) = 355 \text{ psi}$$

$$V_c = v_c b_w d = 290(14)(36)\tfrac{1}{1000} = 146 \text{ kips}$$

(c) Nominal shear strength. At critical section,

$$\text{required } V_n = \frac{V_u}{\phi} = \frac{214}{0.85} = 252 \text{ kips}$$

$$v_n \text{ (max allowed)} = \frac{2}{3}\left(10 + \frac{L_n}{d}\right)\sqrt{f'_c}$$

$$= \frac{2}{3}(10 + 4)\sqrt{f'_c} = 9.33\sqrt{f'_c} = 552 \text{ psi}$$

$$\text{max } V_n = (0.552)(14)(36) = 278 \text{ kips} > 252 \text{ kips} \qquad \text{OK}$$

Since required $V_n > V_c$ (that is, $252 > 146$), vertical and horizontal shear reinforcement is required for strength.

(d) Shear reinforcement. According to ACI-11.8.7,

$$\frac{A_v}{s}\left(\frac{1 + L_n/d}{12}\right) + \frac{A_{vh}}{s_2}\left(\frac{11 - L_n/d}{12}\right) = \frac{V_s}{f_y d}$$

or for $L_n/d = 4$, $V_s = 252 - 146 = 106$ kips, $b = 14$ in., and $f_y = 60,000$ psi,

$$\frac{A_v}{s}\left(\frac{5}{12}\right) + \frac{A_{vh}}{s_2}\left(\frac{7}{12}\right) = \frac{106}{60(36)} = 0.049$$

$$\text{min } A_v = 0.0015 b_w s, \qquad \text{max } s = \frac{d}{5} = 7.2 \text{ in.}$$

$$\text{min } A_{vh} = 0.0025 b_w s_2, \qquad \text{max } s_2 = \frac{d}{3} = 12 \text{ in.}$$

Try using #4 bars horizontally in each face at about 11 in.

$$\text{min required } A_{vh} = 0.0025(14)11 = 0.38 \text{ sq in.}$$

$$\text{provided } A_{vh} = (0.20)2 = 0.40 \text{ sq in.} > 0.38 \text{ sq in.} \qquad \text{OK}$$

$$\frac{A_v}{s}\left(\frac{5}{12}\right) + \frac{0.40}{11}\left(\frac{7}{12}\right) = 0.049$$

$$\frac{A_v}{s} = [0.049 - 0.021]\frac{12}{5} = 0.067$$

For the #4 U stirrups, $A_v = 2(0.20) = 0.40$,

$$\text{required } s = \frac{0.40}{0.067} = 6.0 \text{ in.} < \frac{d}{5} \qquad \text{OK}$$

Use #4 vertical stirrups @ 6 in. *throughout span*. Note that the main flexural steel is extended into the supports as far as practical and bent upward to obtain maximum embedment. Anchorage of bars is the subject of Chap. 6.

5.15 SHEAR-FRICTION

Even though uncracked concrete is relatively strong in shear, and shear-related cracks in usual beams are inclined cracks (diagonal tension cracks), such shear-related cracks become more vertical as the member becomes deeper compared to the shear span, as discussed in Sec. 5.4. The ACI Code design procedures for beams as discussed in Secs. 5.5 through 5.13 are intended to prevent *inclined* cracking (diagonal tension cracking).

For situations in which a crack may form and slippage *along* that crack interface might occur if no steel reinforcement crosses the crack, *and* the usual design procedures for shear reinforcement to resist inclined cracking are inappropriate (such as for a/d less than about 1.0), the shear-friction concept of shear transfer should be applied. The shear-friction concept *is* appropriate for providing a shear transfer mechanism in such cases as:

1. At the interface between concretes cast at different times.
2. At the junction of a corbel (bracket) with a column, such as shown in Fig. 5.15.1(a).
3. At the junction of elements in precast concrete construction, such as the bearing detail shown in Fig. 5.15.1(b).
4. At an interface between steel and concrete, such as the steel bracket attachment to a concrete column shown in Fig. 5.15.1(c).

The crack for which shear-friction reinforcement is required *may* not have been caused by shear. It could, for instance, have been caused by shrinkage. However, once the crack has occurred (from whatever source) a shear transfer mechanism must be provided. The design approach is to assume that a crack will occur and then provide reinforcement across the assumed crack to resist relative displacement along the crack.

Consider a block of concrete as in Fig. 5.15.2(a) acted on by collinear

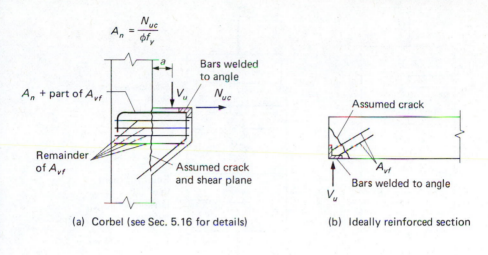

(a) Corbel (see Sec. 5.16 for details)

(b) Ideally reinforced section

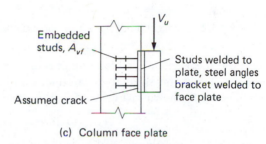

(c) Column face plate

Figure 5.15.1   Uses for the shear-friction concept (adapted from Ref. 45).

shear forces V such that a failure plane would form along plane *a-a*. Since the crack along *a-a* will tend to be rough, the sliding motion will produce a separation, as in Fig. 5.15.2(b). One might imagine slippage along a saw-tooth that forces the crack to open; as it opens the reinforcement is put in tension, with a resulting compression or clamping force on the concrete. A frictional force is then developed equal to the compression in the concrete (or the tension in the bars) times the coefficient of friction. If one may assume that the separation is sufficient to load the steel reinforcement to its yield stress, the shearing resistance then equals the frictional force; thus (ACI-11.7.4) states

$$V_n = A_{vf} f_y \mu \qquad (5.15.1)$$

where A_{vf} is the area of reinforcement extending across the potential crack at 90° to it, and μ is the coefficient of friction between materials along the potential crack. This concept of shear-friction has been verified experimentally [47–55].

ACI-11.7.1 states that shear-friction provisions apply where, "it is appropriate to consider shear transfer across a given plane, such as an existing or potential crack, an interface between dissimilar materials, or an interface between two concretes cast at different times."

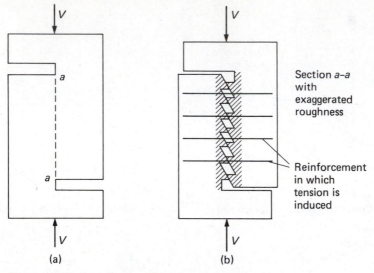

Figure 5.15.2 Idealization of the shear-friction concept.

If the shear-friction reinforcement is inclined at an angle to the assumed crack, such that the shear force produces tension in the shear-friction reinforcement, as shown in Fig. 5.15.3, the shear strength V_n becomes

$$V_n = A_{vf}f_y(\mu \sin \alpha_f + \cos \alpha_f) \qquad (5.15.2)$$

where α_f is the angle between the shear-friction reinforcement and the shear plane.

The logic of Eq. (5.15.2) may be observed from Fig. 5.15.3 where the shear strength provided along the potential crack consists of two parts: the vertical component $T\cos \alpha_f$ of the force in the reinforcement A_{vf} and the frictional force μC, which is the same as $\mu \sin \alpha_f$. Thus,

$$V_n = T \sin \alpha_f + \mu C \qquad (5.15.3)$$

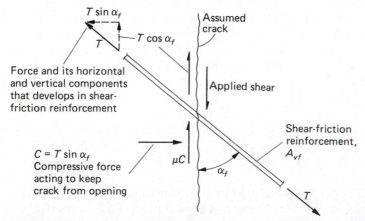

Figure 5.15.3 Action of shear-friction reinforcement when inclined to shear plane.

One should note that the component of the tensile force in the shear-friction reinforcement normal to the potential crack causes a compression at the crack interface, giving rise to the friction force μC.

The required nominal shear-friction strength is $V_n = V_u/\phi$, in which case Eq. (5.15.1) becomes

$$\text{required } A_{vf} = \frac{V_u}{\phi f_y \mu} \qquad (5.15.4)$$

and when α_f is less than $90°$, the required A_{vf} from Eq. (5.15.2) is

$$\text{required } A_{vf} = \frac{V_u}{\phi f_y (\mu \sin \alpha_f + \cos \alpha_f)} \qquad (5.15.5)$$

Note that Eq. (5.15.5) becomes Eq. (5.15.4) when $\alpha_f = 90°$. Just as for regular stirrups, f_y may not be taken greater than 60,000 psi.

The coefficient of friction μ is to be taken (ACI-11.7.4.3) as follows:

1. Concrete cast monolithically. $\qquad \mu = 1.4 \lambda$
2. Concrete placed against hardened concrete with surface intentionally roughened to an "amplitude of approximately $\frac{1}{4}$ in." and the surface must be "free of laitance." $\qquad \mu = 1.0 \lambda$
3. Concrete placed against hardened concrete not intentionally roughened, but "clean and free of laitance." $\qquad \mu = 0.6 \lambda$
4. Concrete anchored to as-rolled structural steel by headed studs or by reinforcing bars, with as-rolled steel "clean and free of paint." $\qquad \mu = 0.7 \lambda$

In the above expressions for μ, the multiplier λ shall be 1.0 for normal-weight concrete, 0.85 for "sand-lightweight" concrete, and 0.75 for "all-lightweight" concrete. Linear interpolation is permitted when partial sand replacement is used.

The maximum nominal shear stress may not exceed $0.2f_c'$ nor 800 psi, which means according to ACI-11.7.5 that

$$\max V_n = v_n A_c = 0.2f_c' A_c \leq 800 A_c \qquad (5.15.6)$$

where A_c is the area of concrete section resisting shear transfer (sq in.).

Since the steel A_{vf} across the potential crack as determined by Eqs. (5.15.4) and (5.15.5) is only that necessary to provide the clamping action that produces friction, any *external* direct tension on the assumed crack must be provided for by additional reinforcement (ACI-11.7.7).

To ensure attainment of a uniform frictional force along the assumed crack, the "shear-friction reinforcement shall be appropriately placed along the shear plane and shall be anchored to develop the specified yield strength on both sides by embedment, hooks, or welding to special devices" (ACI-11.7.8).

The application of the shear-friction provisions to brackets and corbels appears in Sec. 5.16 devoted entirely to that topic.

For guidance in the application of shear-friction to bearings and other special situations commonly encountered in precast concrete construction, the *PCI Design Handbook* [56] should be consulted. The following example may serve to illustrate an application other than for brackets and corbels.

EXAMPLE 5.15.1 Design the reinforcement needed at the bearing region of a precast beam 14 in. wide by 28 in. deep supported on a 4-in. bearing pad. The factored shear V_u is 105 kips. The horizontal force resulting from restraint of volume change movements due to creep, shrinkage, and temperature effects, is 0.3 of the factored shear V_u. Grade 60 steel is to be used for reinforcement.

Solution: (a) Identify the potential crack location. One should assume that a crack will form in the most undesirable manner. According to the *PCI Design Handbook* [56], an appropriate assumption for the crack angle is approximately 20° as shown in Fig. 5.15.4(a). The crack may intersect the bottom of the beam immediately adjacent to the bearing pad, which in this example is taken as 4 in. A rolled structural steel angle is used for confinement across the width of the beam at the bearing.

(b) Determine the shear-friction reinforcement A_{vf} required. Presumably it would be appropriate to resolve the V_u and N_{uc} in Fig. 5.15.4(a) into components parallel and perpendicular to the potential crack. However, it will be simpler and more practical to assume all of V_u will act parallel to the crack, particularly since the 20° angle is an approximation. Thus, using Eq. (5.15.4),

$$\text{required } A_{vf} = \frac{V_u}{\phi f_y \mu} = \frac{105}{0.85(60)1.4} = 1.47 \text{ sq in.}$$

Note that the strength reduction factor ϕ is the 0.85 value used for shear.

(c) Determine the additional reinforcement A_n to provide for the net

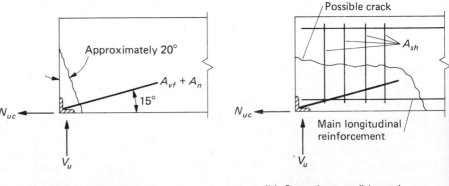

(a) Primary potential crack and reinforcement to provide shear transfer

(b) Secondary possible crack and reinforcement A_{sh} to provide shear transfer

Figure 5.15.4 Shear-friction concept applied to bearing region of a beam.

tension across the potential crack. It will be conservative not to use the sum of components of V_u and N_{uc} perpendicular to the 20° crack, but rather to use N_{uc} as if for a vertical crack. According to ACI-11.7.7,

$$\text{required } A_n = \frac{N_{uc}}{\phi f_y} = \frac{0.3(105)}{0.85(60)} = 0.62 \text{ sq in.}$$

Note that N_{uc} was given as 0.3 of V_u, presumably as the result of an analysis for volume change effects. It is recommended [56,57] that unless all tensile N_{uc} can be eliminated by appropriate design the value of N_{uc} should not be taken less than $0.2V_u$. The ϕ factor of 0.85 for shear is considered appropriate for the above calculation even though 0.90 for axial tension is indicated by ACI-9.3.2. For brackets and corbels, in the similar situation of reinforcement for the tensile N_{uc}, ACI-11.9.3.1 specifies taking ϕ as 0.85.

(d) Total reinforcement to restrain primary crack [Fig. 5.15.4(a)]. The total reinforcement required is

$$A_s = A_{vf} + A_n = 1.47 + 0.62 = 2.09 \text{ sq in.}$$

Use 5-#6, $A_s = 2.20$ sq in. Distribute as shown in Fig. 5.15.5, place at the recommended [56] 15° with the horizontal, weld to the steel angle on one end, and embed the other end into the beam to develop the tensile strength of the #6 bars beyond the potential crack. (Development length requirements are treated in Chap. 6).

(e) Reinforcement for the potential secondary horizontal crack that may form as shown in Fig. 5.15.4(b). If a vertical crack begins near the corner region where the main shear-friction reinforcement exists, then either with or without the tensile force N_{uc} acting, there would be a potential horizontal crack due to the tensile force developed in the *main* shear-friction reinforcement. The maximum shear that could act along such a failure plane would be the tensile capacity of the main shear-friction reinforcement, which makes this vertical stirrup shear-friction reinforcement A_{sh} [Fig. 5.15.4(b)],

$$max\ V_u = (A_{vf} + A_n)f_y = 2.09(60) = 125.4 \text{ kips}$$

$$\text{required } A_{sh} = \frac{V_u}{\mu f_y} = \frac{125.4}{1.4(60)} = 1.49 \text{ sq in.}$$

Use 4-#4 U stirrups, $A_{sh} = 4(0.4) = 1.60$ sq in.

(f) Additional confinement reinforcement. A conservative approach, recommended by Mast [48], is to provide reinforcement to prevent splitting in the vertical plane of the beam equal to 25% of the support reaction. This confinement reinforcement is divided equally into horizontal A_{ch} and vertical A_{cv} portions. Thus,

$$A_{ch} = A_{cv} = \frac{V_u}{8f_y} = \frac{105}{8(60)} = 0.22 \text{ sq in.}$$

Use 2-#4 vertical and 2-#3 U stirrups horizontal. The final design is shown in Fig. 5.15.5.

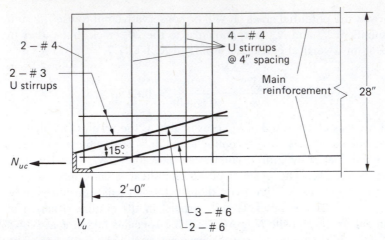

Figure 5.15.5   Final design for Example 5.15.1.

5.16 BRACKETS AND CORBELS

Brackets and corbels projecting from the faces of columns are widely used in precast concrete construction to support beams and girders, as shown in Fig. 5.16.1. It is inappropriate to design brackets and corbels as cantilever beams using the usual beam provisions for shear as described in Secs. 5.5 through 5.13. As discussed in Sec. 5.4 and shown in Fig. 5.4.4, when a/d is less than about 1.0, deep beam theory, rather than simple flexural theory, should apply. Brackets and corbels, furthermore, differ from the deep beams discussed in Sec. 5.14 in that design calculations for horizontal forces must also be made. Because the beams are attached to the bracket, the restraint on the beams due to creep, shrinkage, and temperature deformations give rise to horizontal forces (N_{uc} in Fig. 5.16.1).

Typically, reinforcement for brackets or corbels has in the past consisted of several bars across the width of the bracket bent as shown in Fig. 5.16.2(a). When minimum bend radii are considered, the actual arrange-

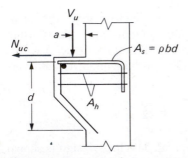

Figure 5.16.1   Bracket or corbel.

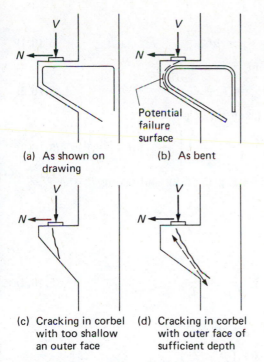

(a) As shown on drawing   (b) As bent

Potential failure surface

(c) Cracking in corbel with too shallow an outer face   (d) Cracking in corbel with outer face of sufficient depth

Figure 5.16.2 Corbel details and possible failure modes (from Kriz and Raths [57]).

ment is as in Fig. 5.16.2(b), in which case a potential failure surface is indicated by the dashed line. When the outer face is too shallow, the critical inclined crack will form in the location shown in Fig. 5.16.2(c). When the bracket is deep enough, the crack will tend to extend back into the column [Fig. 5.16.2(d)] with the portion between the crack and the sloping face acting as a compression element. If the compression strut can be developed, the bracket will have reserve capacity after the crack forms; if the strut cannot develop [as in Fig. 5.16.2(c)], failure will be instantaneous upon formation of the crack.

ACI Code Provisions. Research by Mattock et al. [54,58] has shown that the shear-friction concept can be applied to bracket (corbel) design for a/d as high as 1.0, thus eliminating the necessity for using the formulas of the 1977 ACI Code that were derived from the work of Kriz and Raths [57]. The design recommendations of Mattock [59], the suggestions of ACI-ASCE Committee 426 [4], and the further discussion of MacGregor and Hawkins [17] are the basis for the present ACI-11.9. Additional discussion of the subject is provided by Shaikh [60].

Referring to Fig. 5.16.3, the strengths in shear V_n and in tension N_{uc} are related to the internal forces such that statics is satisfied. From vertical force equilibrium,

$$V_n = \mu C \qquad (5.16.1)$$

From horizontal force equilibrium,

$$N_{nc} = T - C \qquad (5.16.2)$$

and from moment equilibrium, taken about point A,

$$V_n a + N_{nc}\left[h - d + \left(d - \frac{a_1}{2} \right) \right] = T\left(d - \frac{a_1}{2} \right) \qquad (5.16.3)$$

Substituting Eq. (5.16.1) for C into Eq. (5.16.2), and taking $T = A_s f_y$,

$$N_{nc} = A_s f_y - \frac{V_n}{\mu}$$

or

$$A_s = \frac{V_n}{f_y \mu} + \frac{N_{nc}}{f_y} \qquad (5.16.4)$$

Substitution of $A_s f_y$ for T in Eq. (5.16.3) gives

$$V_n a + N_{nc}(h - d) + N_{nc}\left(d - \frac{a_1}{2} \right) = A_s f_y\left(d - \frac{a_1}{2} \right)$$

and solving for A_s gives

$$A_s = \frac{V_n a + N_{nc}(h - d)}{f_y(d - a_1/2)} + \frac{N_{nc}}{f_y} \qquad (5.16.5)$$

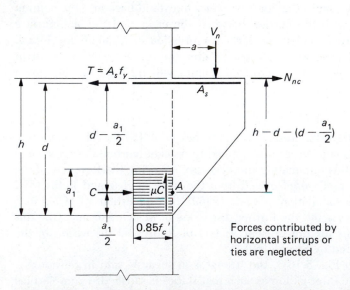

Figure 5.16.3   Equilibrium of forces acting on a bracket or corbel.

Equations (5.16.4) and (5.16.5) give the formulas for A_s to provide the required strengths V_n and N_{nc}.

For design, the factored loads V_u and N_{uc} divided by ϕ are the required strengths V_n and N_{nc}, respectively. Thus, Eq. (5.16.4) becomes

$$\text{required } A_s = \frac{V_u}{\phi f_y \mu} + \frac{N_{uc}}{\phi f_y} \qquad (5.16.6)$$

and Eq. (5.16.5) becomes

$$\text{required } A_s = \frac{V_u a + N_{uc}(h - d)}{\phi f_y (d - a_1/2)} + \frac{N_{uc}}{\phi f_y} \qquad (5.16.7)$$

Note that $N_{uc}/\phi f_y$ is the reinforcement A_n required for axial tension (using the symbol A_n used in Example 5.15.1) and $V_u/(\phi f_y \mu)$ is the shear-friction reinforcement A_{vf} given by Eq. (5.15.4). Further, observe that if the numerator of the first term in Eq. (5.16.7) is treated as an "equivalent" moment, it would represent the reinforcement A_f required for a beam, corresponding to A_s of Eq. (3.3.4).

To summarize the steel area requirements for brackets and corbels, the following may be stated:

$$\text{required } A_s = A_{vf} + A_n \qquad (5.16.8)$$

or

$$\text{required } A_s = A_f + A_n \qquad (5.16.9)$$

in which

$$A_{vf} = \frac{V_u}{\phi f_y \mu} \qquad (5.16.10)$$

$$A_n = \frac{N_{uc}}{\phi f_y} \qquad (5.16.11)$$

$$A_f = \frac{\text{equivalent } M_u}{\phi f_y (d - a_1/2)} \qquad (5.16.12)$$

and

$$\text{equivalent } M_u = V_u a + N_{uc}(h - d) \qquad (5.16.13)$$

In addition to the steel A_s in Fig. 5.16.3, stirrups in the horizontal plane are needed across the vertical potential crack in order to prevent premature diagonal tension failure. These stirrups or ties must be closed hoops, having a total area A_h. It was conservative to neglect this steel A_h in the development of Eqs. (5.16.8) and (5.16.9) since it could have been deducted from the right side of those equations. Thus Eq. (5.16.8) could become

$$\text{required } A_s = A_{vf} + A_n - A_h \qquad (5.16.14)$$

Tests [58] on brackets (corbels) indicate that minimum A_h for the horizontal stirrups must be

$$\min A_h \geq \frac{1}{2}A_f \tag{5.16.15}$$

and

$$\min A_h \geq \frac{1}{3}A_{vf} \tag{5.16.16}$$

ACI-11.9.3.5 requires the area of primary tension reinforcement A_s to be the greater of the following:

$$\text{required } A_s = \frac{2}{3}A_{vf} + A_n \tag{5.16.17}$$

$$\text{required } A_s = A_f + A_n \tag{5.16.18}$$

and, according to ACI-11.9.4, closed stirrups or ties parallel to A_s must be used, having a total area A_h not less than the following:

$$\text{required } A_h \geq 0.5(A_s - A_n) \tag{5.16.19}$$

Note that the three ACI requirements, expressed by Eqs. (5.16.17), (5.16.18), and (5.16.19) automatically satisfy Eqs. (5.16.14), (5.16.9), (5.16.15), and (5.16.16). That Eq. (5.16.16) is always satisfied can be proved by substituting Eq. (5.16.17) into Eq. (5.16.19); or

$$\text{required } A_h \geq 0.5\left(\frac{2}{3}A_{vf} + A_n - A_n\right)$$

$$\text{required } A_h \geq \frac{1}{3}A_{vf}$$

Equation (5.16.17) satisfies Eq. (5.16.14) in recognition that A_{vf} is at least $\frac{1}{3}A_{vf}$.

Other limitations and requirements in the design of brackets and corbels are:

1. Shear span to depth ratio a/d may not exceed 1.0 (ACI-11.9.1).
2. Factored tensile force N_{uc} may not exceed factored shear V_u (ACI-11.9.1).
3. Factored tensile force N_{uc} may not be taken less than $0.2V_u$ unless special precautions are taken to avoid tensile forces (ACI-11.9.3.4).
4. Critical section is *at* face of support, where the effective depth d is to be measured (ACI-11.9.1), as shown in Fig. 5.16.4. The effective depth d may not be less than twice the depth d_1 at the outer edge of the bearing area (ACI-11.9.2).
5. The strength reduction factor ϕ is to be taken as 0.85 for all calculations relating to the design of brackets and corbels (ACI-11.9.3.1).
6. The maximum strength V_n for which brackets and corbels may be designed using normal weight concrete is

$$\max V_n \leq 0.2f'_c b_w d \leq (800 \text{ psi})b_w d \tag{5.16.20}$$

according to ACI-11.9.3.2.1. For "all-lightweight" or "sand-light-weight" concrete, the maximum (ACI-11.9.3.2.2) is

$$\max V_n \leq \left(0.2 - 0.07\frac{a}{d}\right)f'_c b_w d \leq \left(800 \text{ psi} - \frac{(280 \text{ psi})a}{d}\right)b_w d \tag{5.16.21}$$

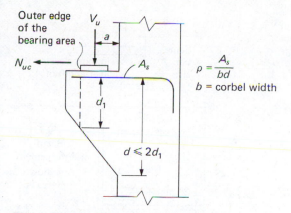

Figure 5.16.4   Effective depth of bracket or corbel.

7. The flexure calculations for A_f are to be in accordance with ACI-10.2 and 10.3 (that is, using the basic assumptions of Chap. 3 of this book).

8. The shear-friction calculations for A_{vf} are to be in accordance with ACI-11.7 (that is, using the procedures discussed in Sec. 5.15 of this book).

9. The minimum reinforcement ratio ρ for the main tension steel A_s is

$$\min \rho = 0.04 \frac{f_c'}{f_y} \qquad (5.16.22)$$

Recommendation for Good Practice in Detailing. Kriz and Raths [57] provide several recommendations for good detailing, some of which are also in the ACI Code and Commentary [45] as follows:

1. The primary tension reinforcement A_s must be anchored as close to the front face of the corbel as cover requirements permit (ACI-11.9.6). A recommended way of accomplishing this is by welding a crossbar to the ends of the tension reinforcing bars, with the size of such crossbar at least equal to the maximum size of the main tension bars (see Fig. 5.16.5, Detail A). Alternatively, a confinement angle, as in Fig. 5.16.6 can be used, to which the main tension bars are welded at the underside. The use of the confinement angle is one of the recommendations of the ACI Commentary [45] and the *PCI Design Handbook* [56]. Another alternative is to bend the primary tension bars to form a loop in the horizontal plane.

2. The total depth of a corbel under the outer edge of a bearing plate resting on the corbel should be not less than half the total depth of the corbel at the face of the column (ACI-11.9.2).

3. The outer edge of a bearing plate resting on a corbel should be placed not closer than 2 in. from the outer edge of the corbel [57]. The *PCI Design Handbook* [56] recommends a 1-in. minimum setback when *no* confinement angle is used but does not prescribe a setback when a confinement angle is used.

4. When corbels are designed to resist horizontal forces, steel bearing

Note:
   Distance x should be great
   enough to prevent contact
   between outer corbel edge
   and beam due to possible
   rotations

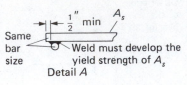

Detail A

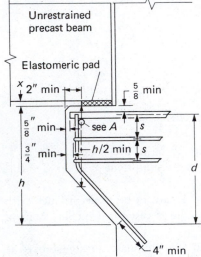

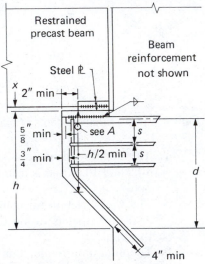

(a)   Corbels subject to vertical load only

(b)   Corbels subject to vertical load and
      restrained creep and shrinkage force.
      Steel ℄'s welded or not welded.

Figure 5.16.5   Recommended corbel details (from Kriz and Raths [57]).

plates welded to the tension reinforcement should be used to transfer the horizontal forces directly to the tension reinforcement.

Details and design aids for brackets and corbels are given in the *PCI Design
Handbook* [56].

EXAMPLE 5.16.1   Design a typical interior bracket that projects from a 14-
in. square tied column. It must support a dead load reaction of 30 kips and

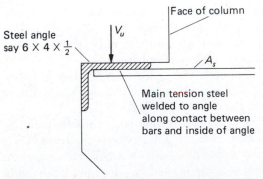

Figure 5.16.6   Anchorage of main steel provided by welding to a confinement
angle.

a live load reaction of 50 kips, resulting from gravity loads. Assume that suitable bearings are provided for the supported prestressed concrete girder so that horizontal restraint forces are eliminated. The tolerance gap between the beam end and column face is 1 in. Use $f'_c = 5000$ psi, $f_y = 60,000$ psi, and the ACI Code.

Solution: (a) Factored loads:

$$V_u = 1.4V_D + 1.7V_L = 1.4(30) + 1.7(50) = 127 \text{ kips}$$

(b) Preliminary bracket size. The shear span a is dependent on the bearing length required to support the reaction on the concrete. ACI-10.15 gives nominal bearing strength as $0.85f'_c A_1$, so that using $\phi = 0.70$ (ACI-9.3.2),

$$V_u = \phi(0.85f'_c)A_1$$

$$\text{bearing plate width} = \frac{127,000}{0.70(0.85)(5000)14} = 3.1 \text{ in.}$$

Use $3\frac{1}{2}$ in. for bearing plate width. Allowing the tolerance gap of 1-in. clear between face of column and beam for possible overrun in beam length and in case the beam might also be 1 in. too short, the shear span is

$$a = 2 + \tfrac{1}{2}(\text{bearing plate width}) = 2 + 1.75 = 3.75 \text{ in.}$$

(c) Determine depth of bracket. Based on the maximum strength V_n (Eq. 5.16.20) permitted by ACI-11.9.3.2.1,

$$\max V_n = 0.2f'_c b_w d \le (800 \text{ psi})b_w d \qquad \text{[5.16.20]}$$

Since $0.2f'_c = 1000$ psi, max $v_n = 800$ psi; then

$$\min d = \frac{V_u}{\phi b(\max v_n)} = \frac{127,000}{0.85(14)800} = 13.3 \text{ in.}$$

If overall $h = 15$ in., $d \approx 13.5$ in. (allowing 1-in. cover), check

$$\frac{a}{d} = \frac{3.75}{13.5} = 0.28 < 1.0 \qquad (\text{ACI-11.9.1})$$

(d) Determine the shear-friction reinforcement A_{vf}. Using Eq. (5.15.4) according to ACI-11.7.4.1,

$$A_{vf} = \frac{V_u}{\phi f_y \mu} = \frac{127}{0.85(60)1.4} = 1.78 \text{ sq in.}$$

using $\mu = 1.4$ for monolithically cast concrete.

(e) Determine main tension reinforcement A_s. Calculate first the requirement A_f for flexure.

$$M_u = V_u a + N_{uc}(h - d)$$

$$= V_u a = 127(3.75)\tfrac{1}{12} = 39.7 \text{ ft-kips}$$

$$\text{required } R_n = \frac{M_u}{\phi bd^2} = \frac{39.7(12,000)}{0.85(14)(13.5)^2} = 220 \text{ psi}$$

Note that $\phi = 0.85$ according to ACI-11.9.3.1. The use of $\phi = 0.90$ if it had been already included in design aids would make little difference. From Eq. (3.8.5) or from Fig. 3.8.1,

$$\text{required } \rho = 0.0035$$

From ACI-11.9.5,

$$\min \rho = 0.04 \frac{f'_c}{f_y} = 0.04 \left(\frac{5}{60}\right) = 0.0033$$

Then,

$$\text{required } A_f = 0.0035(14)(13.5) = 0.66 \text{ sq in.}$$

From ACI-11.9.3.5, the main steel A_s requirement is the larger of Eqs. (5.16.17) and (5.16.18), as follows:

$$A_s = \frac{2}{3}A_{vf} + A_n \text{ (zero in this example)} = \frac{2}{3}(1.78) = 1.19 \text{ sq in.}$$

or

$$A_s = A_f + A_n \text{ (zero in this example)} = 0.66 \text{ sq in.}$$

Use 4-#5 $(A_s = 1.24 \text{ sq in.})$.

(f) Design closed stirrups or ties. In accordance with ACI-11.9.4, Eq. (5.16.19) requires

$$\text{required } A_h = 0.5(A_s - A_n) = 0.5(1.19) = 0.60 \text{ sq in.}$$

Note that in this design, with small $a/d = 0.28$, the flexure requirement A_f does not affect the design; the shear-friction requirement A_{vf} for V_u controls with $A_{vf} = A_s + A_h$.

Use 3-#3 closed hoops $[A_h = 2(3)0.11 = 0.66 \text{ sq in.}]$. The spacing of the hoops must be within the upper two-thirds of the effective depth (ACI-11.9.4).

(g) Final design. Referring to Fig. 5.16.7 the overall depth at the face of column is

$$h = 13.5 + 1(\text{cover}) + 0.3125(\text{bar radius}) = 14.8 \text{ in.}$$

Use $h = 15$ *in.*

The 1-in. cover is used here but it could be as little as the $\frac{5}{8}$-in. bar diameter for precast columns (ACI-7.7.2) but would presumably have to be $1\frac{1}{2}$ in. for cast-in-place members.

At the outer edge of the bearing area, the effective depth d_1 must be at least half of that used at the face of column. In this case, making the outer face 8 in. will just about provide the required d_1 of 13.7/2 (see Fig. 5.16.4).

Another important aspect of the bracket design is the provision of adequate anchorage into the column so that the tensile force $A_s f_y$ is available at the face of the column. Development of reinforcement is treated in Chap. 6.

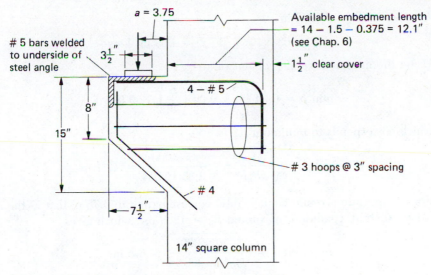

Figure 5.16.7   Corbel designed for Example 5.16.1.

EXAMPLE 5.16.2 Design a bracket that is to support gravity dead and live loads of 15 and 25 kips, respectively. The vertical reaction is 10 in. from the face of a 14-in. square column. Provide a horizontal reaction of 10 kips due to creep and shrinkage of a restrained beam. Use $f'_c = 5000$ psi, $f_y = 40,000$ psi, and the ACI Code.

Solution: (a) Factored loads:

$$V_u = 1.4(15) + 1.7(25) = 21 + 42.5 = 63.5 \text{ kips}$$

$$N_{uc} = 1.7(10) = 17 \text{ kips}$$

ACI-11.9.3.4 states that N_{uc} is to be regarded as live load when it results from creep, shrinkage, or temperature change.

$$\frac{N_{uc}}{V_u} = \frac{17}{63.5} = 0.268 > 0.20 \text{ min} \qquad\qquad \text{OK}$$

(b) Depth of bracket for shear.

$$\max V_n = 0.2 f'_c b_w d \le (800 \text{ psi}) b_w d$$

Since $0.2 f'_c = 1000$ psi is larger than 800 psi, use 800 psi; then

$$\min d = \frac{V_u}{\phi b_w (800)} = \frac{63,500}{0.85(14)(800)} = 6.7 \text{ in.}$$

Since this is very small, perhaps the flexure requirement will require a larger d (this is a good possibility because the load on the bracket is 10 in. from the face of column).

(c) Depth of bracket for flexure.

$$M_u = V_u a + N_{uc}(h - d)$$

$$= 63.5(10) + 17(h - d)$$

Estimating $(h - d)$ at 2 in.,

$$M_u = 63.5(10) + 17(2) = 669 \text{ in.-kips}$$

Using the minimum reinforcement ratio

$$\text{min } \rho = 0.04 \frac{f'_c}{f_y} = 0.04 \left(\frac{5}{40}\right) = 0.005$$

which corresponds to minimum $R_n = 193$ psi,

$$\text{required } d = \sqrt{\frac{M_u}{\phi R_n b}} = \sqrt{\frac{669,000}{0.85(193)14}} = 17.1 \text{ in.}$$

For the maximum reinforcement ratio corresponding to 0.75 of that in the balanced strain condition, maximum $R_n = 1209$ psi, which gives

$$\text{required } d = \sqrt{\frac{669,000}{0.85(1209)14}} = 6.8 \text{ in.}$$

(d) Select bracket depth. Since the ACI rules for bracket and corbel design only apply when a/d does not exceed 1.0,

$$\text{min } d = a = 10 \text{ in.}$$

Try a bracket somewhat deeper, say 15 in. overall. This would make $d \approx 13.5$ in.

(e) Determine the shear-friction reinforcement A_{vf}. Using Eqs. (5.15.4) or (5.16.10) according to ACI-11.7.4.1,

$$\text{required } A_{vf} = \frac{V_u}{\phi f_y \mu} = \frac{63.5}{0.85(40)1.4} = 1.33 \text{ sq in.}$$

where $\mu = 1.4$ for monolithic concrete.

(f) Determine the flexure reinforcement A_f. Following ACI-11.9.3.3,

$$\text{required } R_n = \frac{M_u}{\phi b d^2}$$

where

$$M_u = V_u a + N_{uc}(h - d) = 63.5(10) + 17(1.5) = 661 \text{ in.-kips}$$

$$\text{required } R_n = \frac{661,000}{0.85(14)(13.5)^2} = 305 \text{ psi}$$

$$\text{required } \rho = 0.0081 \quad \text{[from Eq. (3.8.5) or Fig. 3.8.1]}$$

$$\text{required } A_f = 0.0081(14)13.5 = 1.53 \text{ sq in.}$$

(g) Determine additional reinforcement A_n for axial tension. In accordance with ACI-11.9.3.4, using Eq. (5.16.11),

$$\text{required } A_n = \frac{N_{uc}}{\phi f_y} = \frac{10}{0.85(40)} = 0.29 \text{ sq in.}$$

(h) Total main tension reinforcement A_s. According to ACI-11.9.3.5, the required A_s is the larger of the values from Eqs. (5.16.17) and (5.16.18), as follows:

$$\text{required } A_s = A_f + A_n = 1.53 + 0.29 = 1.82 \text{ sq in.}$$

or

$$\text{required } A_s = \frac{2}{3}A_{vf} + A_n = \frac{2}{3}(1.33) + 0.29 = 1.18 \text{ sq in.}$$

The required $A_s = 1.82$ sq. in.

Use 3-#7 for main tension steel, $A_s = 1.80$ sq in.

(i) Determine stirrup (or tie) requirements. According to ACI-11.9.4,

$$\text{required } A_h = 0.5(A_s - A_n) = 0.5(1.82 - 0.29) = 0.77 \text{ sq in.}$$

Use 3-#4 closed stirrups, $A_h = 0.40(3) = 1.20$ sq in., the spacing of which should be $(\frac{2}{3})13.5/3 = 3.0$ in. (ACI-11.9.4). Use 3-in. spacing.

(j) Overall bracket dimensions. Assuming that a 1-in. thick bearing plate is to be welded to the main tension reinforcement, the overall depth is

$$h = \text{bearing plate} + \text{bar radius} + \text{effective depth, } d$$
$$= 1 + 0.44 + 13.5 = 14.94 \text{ in., say 15 in.}$$

$$\text{bearing plate length} = \frac{V_u}{\phi 0.85 f_c' \text{ (column width)}}$$

$$= \frac{63,500}{0.70(0.85)5000(14)} = 1.53 \text{ in.}$$

Use a 3-in. plate length as the practical minimum.

$$\text{length of bracket projection} = 2 \text{ in.} + \tfrac{1}{2} \text{ bearing plate} + \text{shear span, } a$$
$$= 2 + 1.5 + 10 = 13.5 \text{ in.}$$

$$\text{depth of outer face of bracket} = \tfrac{1}{2} \text{ overall depth} = 7\tfrac{1}{2} \text{ in.}$$

Final design is shown in Fig. 5.16.8.

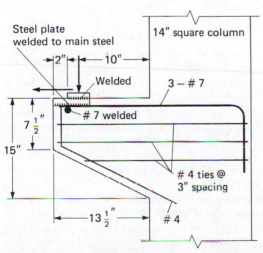

Figure 5.16.8 Final design for Example 5.16.2.

SELECTED REFERENCES

1. ACI-ASCE Committee 326. "Shear and Diagonal Tension," *ACI Journal, Proceedings,* **59,** January, February, and March 1962, 1–30, 277–344, and 352–396.
2. Boris Bresler and James G. MacGregor. "Review of Concrete Beams Failing in Shear," *Journal of Structural Division,* ASCE, **93,** February 1967 (ST1), 343–372.
3. ACI-ASCE Committee 426. "The Shear Strength of Reinforced Concrete Members—Chapters 1 to 4," *Journal of Structural Division,* ASCE, **99,** June 1973, 1091–1187.
4. ACI–ASCE Committee 426. *Suggested Revisions to Shear Provisions for Building Codes.* Detroit, Michigan: American Concrete Institute, 1979. 82 pp.
5. ACI–ASCE Committee 426. "Suggested Revisions to Shear Provisions for Building Codes," (Abstract), *ACI Journal, Proceedings,* **74,** September 1977, 458–469.
6. R. C. Fenwick and Thomas Paulay. "Mechanisms of Shear Resistance of Concrete Beams," *Journal of the Structural Division,* ASCE, **94,** (ST10), October 1968, 2325–2350.
7. T. Paulay and P. J. Loeber. "Shear Transfer by Aggregate Interlock," *Shear in Reinforced Concrete,* Vol. 1 (SP–42). Detroit: American Concrete Institute, 1974 (pp. 503–537).
8. David W. Johnston and Paul Zia. "Analysis of Dowel Action," *Journal of the Structural Division,* ASCE, **97,** May 1971 (ST5), 1611–1630.
9. H. P. J. Taylor. "The Fundamental Behavior of Reinforced Concrete Beams in Bending and Shear," *Shear in Reinforced Concrete,* Vol. 1 (SP–42). Detroit: American Concrete Institute, 1974 (pp. 43–77).
10. JoDean Morrow and I. M. Viest. "Shear Strength of Reinforced Concrete Frame Members Without Web Reinforcement," *ACI Journal, Proceedings,* **53,** March 1957, 833–869.
11. G. N. J. Kani. "Basic Facts Concerning Shear Failure," *ACI Journal, Proceedings,* **63,** June 1966, 675–692. Disc. pp. 1511–1528.
12. R. Diaz de Cossio and S. Loera. Discussion of "Basic Facts Concerning Shear Failure," by G. N. J. Kani, *ACI Journal, Proceedings,* **63,** December 1966, 1511–1514.
13. H. A. Rawdon de Paiva and Chester P. Siess. "Strength and Behavior of Deep Beams in Shear," *Journal of the Structural Division,* ASCE, **91,** October 1965 (ST5), 19–42.
14. V. Ramakrishnan and Y. Ananthanarayana. "Ultimate Strength of Deep Beams in Shear," *ACI Journal, Proceedings,* **65,** February 1968, 87–98.
15. Howard P. J. Taylor. "Shear Strength of Large Beams," *Journal of the Structural Division,* ASCE, **98,** November 1972 (ST11), 2473–2490.
16. R. F. Manual. "Failure of Deep Beams," *Shear in Reinforced Concrete,* Vol. 2 (SP–42). Detroit: American Concrete Institute, 1974 (pp. 425–440).
17. J. G. MacGregor and N. M. Hawkins. "Suggested Revisions to ACI Building Code Clauses Dealing with Shear Friction and Shear in Deep Beams and Corbels," *ACI Journal, Proceedings,* **74,** November 1977, 537–545. Disc., **75,** May 1978, 221–224.
18. G. N. J. Kani. "The Riddle of Shear Failure and Its Solution," *ACI Journal, Proceedings,* **61,** April 1964, 441–467.
19. Michael P. Collins. "Towards a Rational Theory for RC Members in Shear," *Journal of the Structural Division,* ASCE, **104,** April 1978 (ST4), 649–666.
20. G. N. J. Kani. "How Safe Are Our Large Reinforced Concrete Beams," *ACI Journal, Proceedings,* **64,** March 1967, 128–141.

21. James G. MacGregor. Discussion of "How Safe Are Our Large Reinforced Concrete Beams," G. N. J. Kani, *ACI Journal, Proceedings*, **64**, September 1967, 603–604.
22. William J. Krefeld and Charles W. Thurston. "Studies of the Shear and Diagonal Tension Strength of Simply Supported Reinforced Concrete Beams," *ACI Journal, Proceedings*, **63**, April 1966, 451–476. Disc. 1469–1476.
23. K. S. Rajagopalan and P. M. Ferguson. "Exploratory Shear Tests Emphasizing Percentage of Longitudinal Steel," *ACI Journal, Proceedings*, **65**, August 1968, 634–638. Disc. **66**, 150–154.
24. Theodore C. Zsutty. "Beam Shear Strength Prediction by Analysis of Existing Data," *ACI Journal, Proceedings*, **65**, November 1968, 943–951.
25. Theodore C. Zsutty. "Shear Strength Prediction for Separate Categories of Simple Beam Tests," *ACI Journal, Proceedings*, **68**, February 1971, 138–143.
26. Gerhard T. Suter and Robert F. Manuel. "Diagonal Crack Width Control in Short Beams," *ACI Journal, Proceedings*, **68**, June 1971, 451–455.
27. John M. Hanson. *Square Openings in Webs of Continuous Joists* (RD001.01D). Skokie, Illinois: Portland Cement Association, 1969.
28. Phil M. Ferguson. *Reinforced Concrete Fundamentals* (4th ed.). New York: Wiley, 1979 (pp. 138–139).
29. J. G. MacGregor and P. Gergely. "Suggested Revisions to ACI Building Code Clauses Dealing with Shear in Beams," *ACI Journal, Proceedings*, **74**, October 1977, 493–500.
30. J. A. Hanson. "Tensile Strength and Diagonal Tension Resistance of Structural Lightweight Concrete," *ACI Journal, Proceedings*, **58**, July 1961, 1–40.
31. E. Hognestad, R. C. Elstner, and J. A. Hanson. "Shear Strength of Reinforced Structural Lightweight Aggregate Concrete Slabs," *ACI Journal, Proceedings*, **61**, June 1964, 643–656.
32. Don L. Ivey and Eugene Buth. "Shear Capacity of Lightweight Concrete Beams," *ACI Journal, Proceedings*, **64**, October 1967, 634–643.
33. G. N. J. Kani. "A Rational Theory for the Function of Web Reinforcement," *ACI Journal, Proceedings*, **66**, March 1969, 185–197. Disc. 769–774.
34. Fung-Keu Kong, Peter J. Robins, and David F. Cole. "Web Reinforcement Effects in Deep Beams," *ACI Journal, Proceedings*, **67**, December 1970, 1010–1017.
35. Fung-Keu Kong and Peter J. Robins. "Web Reinforcement in Lightweight Concrete Deep Beams," *ACI Journal, Proceedings*, **68**, July 1971, 514–520.
36. P. E. Regan and M. H. Khan. "Bent-Up Bars as Shear Reinforcement," *Shear in Reinforced Concrete*, Vol. 1 (SP–42). Detroit: American Concrete Institute, 1974 (pp. 249–266).
37. H. C. Sorensen. "Efficiency of Bent-Up Bars as Shear Reinforcement," *Shear in Reinforced Concrete*, Vol. 1 (SP–42). Detroit: American Concrete Institute, 1974 (pp. 267–283).
38. Kitcha Leksukhum and R. B. L. Smith. "Comparative Study of Bent-Up Bars with Other Forms of Secondary Reinforcement in Beams," *ACI Journal, Proceedings*, **68**, January 1971, 32–35.
39. S. M. Fereig and K. N. Smith. "Indirect Loading on Beams with Short Shear Span," *ACI Journal, Proceedings*, **74**, May 1977, 220–222.
40. Barrington deV. Batchelor and Mankit Kwun. "Shear in RC Beams Without Web Reinforcement," *Journal of the Structural Division*, ASCE, **107**, May 1981 (ST5), 907–921.
41. Michael N. Palaskas, Emmanuel K. Attiogbe, and David Darwin. "Shear Strength

of Lightly Reinforced T-Beams," *ACI Journal, Proceedings*, **78**, November–December 1981, 447–455.

42. James G. MacGregor and John M. Hanson. "Proposed Changes in Shear Provisions for Reinforced and Prestressed Concrete Beams," *ACI Journal, Proceedings*, **66**, April 1969, 276–288. Disc. 849–851.

43. Alan H. Mattock. "Diagonal Tension Cracking in Concrete Beams With Axial Forces," *Journal of the Structural Division*. ASCE, **95**, September 1969 (ST9), 1887–1900.

44. Munther J. Haddadin, Sheu-Tien Hong, and Alan M. Mattock. "Stirrup Effectiveness in Reinforced Concrete Beams With Axial Force," *Journal of the Structural Division*, ASCE, **97**, September 1971 (ST9), 2277–2297.

45. ACI Committee 318. *Commentary on Building Code Requirements for Reinforced Concrete (ACI 318–83)*. Detroit: American Concrete Institute, 1983.

46. K. N. Smith and S. M. Fereig. "Effect of Loading and Supporting Conditions on the Shear Strength of Deep Beams," *Shear in Reinforced Concrete*, Vol. 2 (SP–42). Detroit: American Concrete Institute, 1974 (pp. 441–460).

47. Philip W. Birkeland and Halvard W. Birkeland. "Connections in Precast Concrete Construction," *ACI Journal, Proceedings*, **63**, March 1966, 345–368.

48. R. F. Mast. "Auxiliary Reinforcement in Precast Concrete Connections," *Journal of the Structural Division*, ASCE, **94**, June 1968 (ST6), 1485–1504.

49. J. A. Hofbeck, I. O. Ibrahim, and Alan H. Mattock. "Shear Transfer in Reinforced Concrete," *ACI Journal, Proceedings*, **66**, February 1969, 119–128. Disc. 678–680.

50. A. H. Mattock and N. M. Hawkins. "Research on Shear Transfer in Reinforced Concrete," *PCI Journal*, **17**, March–April 1972, 55–75.

51. Bjorn R. Hermansen and John Cowan. "Modified Shear-Friction Theory for Bracket Design," *ACI Journal, Proceedings*, **71**, February 1974, 55–60.

52. A. H. Mattock. Disc. of "Modified Shear-Friction Theory for Bracket Design," by B. R. Hermansen and J. Cowan, *ACI Journal, Proceedings*, **71**, August 1974, 421–423.

53. A. H. Mattock. "Shear Transfer in Concrete Having Reinforcement at an Angle to the Shear Plane," *Shear in Reinforced Concrete*, Vol. 1 (SP–42). Detroit: American Concrete Institute, 1974 (pp. 17–42).

54. Alan H. Mattock, L. Johal, and H. C. Chow. "Shear Transfer in Reinforced Concrete with Moment or Tension Acting Across the Shear Plane," *PCI Journal*, **20**, July–August 1975, 76–93.

55. Alan H. Mattock, W. K. Li, and T. C. Wang. "Shear Transfer in Lightweight Reinforced Concrete," *PCI Journal*, **21**, January–February 1976, 20–39.

56. *PCI Design Handbook* (2nd ed.). Chicago: Prestressed Concrete Institute, 1978.

57. L. B. Kriz and C. H. Raths. "Connections in Precast Concrete Structures—Strength of Corbels," *PCI Journal*, **10** (1), February 1965, 16–47.

58. Alan H. Mattock, K. C. Chen, and K. Soongswang. "The Behavior of Reinforced Concrete Corbels," *PCI Journal*, **21**, March–April 1976, 52–77.

59. Alan H. Mattock. "Design Proposals for Reinforced Concrete Corbels," *PCI Journal*, **21**, May–June 1976, 18–42. Disc., 22, March/April 1977, 90–109.

60. A. Fattah Shaikh. "Proposed Revisions to Shear-Friction Provisions," *PCI Journal*, **23**, March/April 1978, 12–21.

PROBLEMS

All problems* are to be worked in accordance with the strength method of the ACI Code unless otherwise indicated, and all stated loads are service loads. All shear diagrams are to be drawn to scale in terms of V_u directly below the diagram of the beam, also drawn to scale. Use the V_u diagram for design by scaling values from it; also scale from it the locations where stirrup spacings are permitted. Computation of V_u values and locations of stirrup spacings may be made to confirm scaled information.

5.1 The simply supported beam of 16-ft span is to carry a uniform dead load of 1.5 kips/ft (including beam weight) and a uniform live load of 2.4 kips/ft. Use $f'_c = 3500$ psi and $f_y = 40,000$ psi.
(a) Determine the adequacy of the #3 U stirrups that are spaced at 8 in. Use the simplified method with constant V_c.
(b) Draw superimposed on the factored shear V_u diagram the diagram of strength provided ϕV_n. From the comparison of ϕV_n with V_u, determine if the stirrups are adequate for the entire beam. If not, make recommendations to obtain a beam having adequate shear strength.

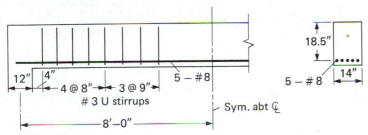

Problem 5.1

5.2 The beam of the accompanying figure carries a uniform live load of 3.0 kips/ft in addition to its own weight. Assume a support width of 12 in., and use $f'_c = 3000$ psi and $f_y = 40,000$ psi,
(a) Draw the maximum factored shear V_u envelope. Calculate the value at the critical section, at the $\frac{1}{4}$ point, and at midspan; then use a French curve to draw the diagram for design use.

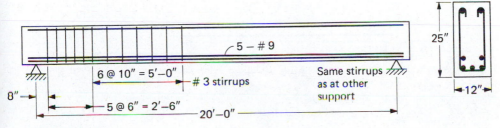

Problem 5.2

*Most problems may be solved as problems stated in Inch-Pound units, or as problems in SI units using quantities in parenthesis at the end of the statement. The metric conversions are approximate to avoid implying higher precision for the given information in SI units than that given for the Inch-Pound units.

(b) Draw the curve of required stirrup spacing using the simplified procedure permitted by ACI-11.3.1.1; show on the same diagram the spacings provided.
(c) Evaluate whether or not the spacings used satisfy the ACI Code using the simplified method of constant V_c.
(d) Repeat (b) using the more detailed procedure of ACI-11.3.2.1; show also the spacings provided.
(e) Evaluate whether or not the spacings used are satisfactory according to the analysis of (d).

5.3   The beam of the accompanying figure is to carry 1.6 kips/ft live load and 0.90 kips/ft dead load (including beam weight). Using $f_c' = 3000$ psi and $f_y = 40,000$ psi, investigate the beam for stirrup adequacy according to the simplified method using constant V_c. If design is not adequate, indicate what revision is necessary.

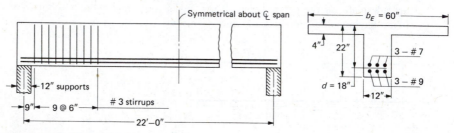

Problems 5.3 and 5.11

5.4   For the portion of the continuous beam shown, with the given portion of the factored shear V_u envelope, determine the spacings to be used for #3 U stirrups. Dimension and show the stirrups on the given portion of beam. Use the simplified method of constant V_c, with $f_c' = 3500$ psi and $f_y = 60,000$ psi. (Beam: $b = 300$ mm; $d = 530$ mm; support width $= 300$ mm; half-span $= 2.7$ m; V_u at support $= 260$ kN; V_u at midspan $= 80$ kN; use #10M stirrups; $f_c' = 25$ MPa; $f_y = 400$ MPa.)

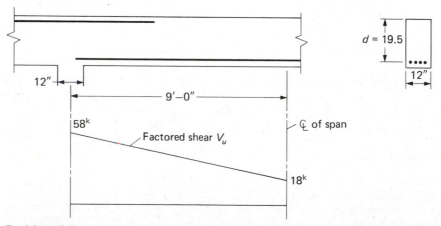

Problem 5.4

5.5   The given 20-ft simply supported beam must carry a dead load of 6 kips/ft (including beam weight) and a live load of 10 kips/ft. Use $f'_c = 4000$ psi and $f_y = 40,000$ psi.
(a) Using the simplified method of ACI-11.3.1.1, design the stirrups, dimension their locations, and show them on the beam. Use a bar size for the U stirrups such that the spacing will not be closer than 3 in. Explicitly state the length over which stirrups are required.
(b) Repeat (a) but use the more detailed procedure of ACI-11.3.2.1.

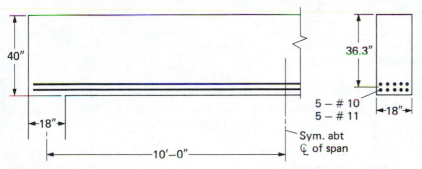

Problem 5.5

5.6   For the simply supported beam of the accompanying figure, having a span of 32 ft, with uniform dead load of 2.3 kips/ft (including beam weight) and live load of 3.7 kips/ft, design #4 vertical U stirrups using 1-in. multiples. Show the stirrups in a side view of the beam and dimension their locations. Use $f'_c = 3750$ psi and $f_y = 50,000$ psi, and the simplified method with constant V_c. The support width is 12 in.

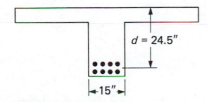

Problem 5.6

5.7   A reinforced concrete simply supported beam ($b = 250$ mm, $d = 410$ mm) must carry on a span of 5.5 m the single concentrated moving load of 45 kN plus a uniform dead load of 30 kN/m (including beam weight). Design and detail the stirrup spacing for #10M vertical U stirrups; support width is 300 mm, $f'_c = 25$ MPa; $f_y = 400$ MPa. Apply the simplified method using a constant value for V_c. See SI footnote to Eq. (5.10.3).

5.8   The beam in the accompanying figure is to carry dead load of 1.4 kips/ft (including beam weight) and live load of 1.6 kips/ft. Use $f'_c = 3500$ psi, $f_y = 40,000$ psi, and neglect any compression steel effect. Use a "correct" factored shear V_u envelope. *(Continued next page.)*

(a) Using the simplified procedure, design and detail the #3 single U-stirrup spacing for the beam.

(b) Repeat (a) using the more detailed procedure involving $\rho V d/M$. How many stirrups could be eliminated as compared to the simplified method?

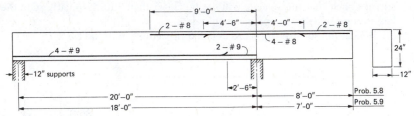

Problems 5.8 and 5.9

5.9 Repeat Prob. 5.8, except use an 18-ft main span and a 7-ft cantilever, with dead load of 1.5 kips/ft (including beam weight) and live load of 1.7 kips/ft. All other dimensions and reinforcing bars are the same as in Prob. 5.8.

5.10 The beam in the accompanying figure is to carry dead load of 55 kN/m (including beam weight) and live load of 72 kN/m. Use $f_c' = 25$ MPa, $f_y = 400$ MPa.

(a) Using the simplified procedure, design and detail on the beam the single U-stirrup spacing for the beam. Use stirrups of a size such that spacing will not be closer than 75 mm.

(b) Repeat (a) using the more detailed procedure involving $\rho V d/M$.

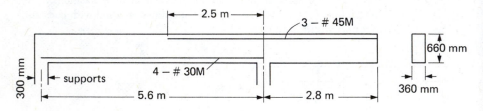

Problem 5.10

5.11 Completely design and detail the stirrups for the beam of Prob. 5.3 (ignore the spacings given), using the more detailed $\rho V d/M$ method of ACI-11.3.2.1.

5.12 For a simply supported beam of 16-ft span, having support widths of 12 in., design and detail on the beam U stirrups (use no spacing less than 3 in.). The dead load is 1.6 kips/ft (including weight of beam) and the live load is 3.0 kips/ft. Use $f_c' = 3000$ psi and $f_y = 60,000$ psi. Use the ACI Code simplified procedure for V_c. Assume $b_w = 14$ in. and $d = 21.5$ in.

5.13 The beam of the accompanying figure carries a live load of 2.7 kips/ft in addition to the weight of the slab and beam. Design and detail on the beam the vertical U-stirrup spacing for #3 stirrups. Use $f_c' = 3000$ psi and $f_y = 40,000$ psi.

(a) Use simplified procedure of ACI-11.3.1.1.

(b) Use more detailed procedure of ACI-11.3.2.1.

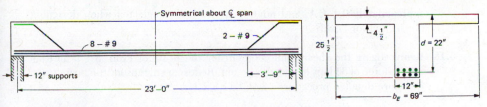

Problem 5.13

5.14 A reinforced concrete simply supported beam of span 5.5 m carries a concentrated dead load of 23 kN at 1.8 m from the left support, and a uniform dead load of 90 kN/m. The width of support is 300 mm. The rectangular beam has a 300-mm width and a 650 mm effective depth d. Use $f'_c = 30$ MPa and $f_y = 400$ MPa. Design and specify by dimensioning, the spacings to be used for #10M U stirrups. Use the simplified procedure of ACI-11.3.1.1.

5.15 For a rectangular beam of 14 in. width and effective depth 22.5 in. with $f'_c = 4000$ psi and $f_y = 60,000$ psi, determine the maximum factored shear V_u for this beam for the following conditions:
(a) When no stirrups are to be used.
(b) When minimum percentage of shear reinforcement (#3 U stirrups) is used according to ACI-11.5.5.3; specify the spacing to be used.
(c) When maximum percentage of stirrups is used (#4 U stirrups); specify the spacing to be used.

5.16 The beam of the accompanying figure is to carry a uniform live load of 0.6 kips/ft in addition to the concentrated load shown and the beam weight. Use $f'_c = 4000$ psi and $f_y = 50,000$ psi.
(a) If the bent-up longitudinal bars are counted on to resist shear, is the beam adequate as shown?
(b) If the beam is not adequate, prescribe the number and location of #3 U stirrups required to act along with the bent-up bars.

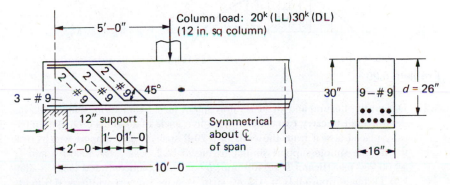

Problem 5.16

5.17 If a factored axial compression N_{uc} of 140 kips is acting additionally on the beam of Example 5.12.1, determine the number of stirrups that may be eliminated by taking the compression into account when computing V_c by the more detailed procedure involving $\rho V d/M$. (Note: This compressive force is approximately $0.1 f'_c A_g$ and might reasonably be neglected when designing the section for flexure.)

5.18 Reinvestigate the shear reinforcement for the beam of Prob. 5.1 if an axial tensile force of 35 kips live load is acting. Redesign stirrups for the beam using the more detailed procedure of ACI-11.3.2.1.

5.19 Redesign the stirrups for the beam of Prob. 5.5 if an axial compressive force of 70 kips dead load and 120 kips live load is acting.
(a) Use simplified method using ACI Formula (11-4).
(b) Use more detailed $\rho Vd/M$ method.

5.20 Design the shear reinforcement for the beam shown. The rectangular beam is to support heavy machinery, and the service loading is 12 kips/ft dead load and 40 kips/ft live load (including impact). The factored moment M_u diagram is given in the accompanying figure. Use $f'_c = 4000$ psi and $f_y = 60,000$ psi.

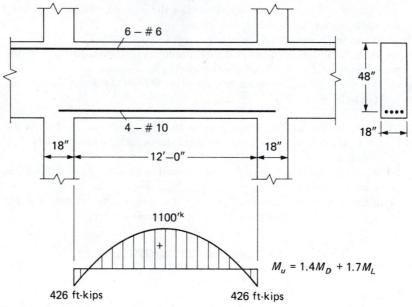

Problem 5.20

5.21 Design the flexural and shear reinforcement for a beam 16 in. wide × 52 in. deep overall to carry two concentrated live loads of 200 kips each symmetrically placed at 4 ft from the ends of a 20-ft span having support widths of 2 ft. Assume simple support similar to Fig. 5.14.2. Use $f'_c = 4000$ psi and $f_y = 60,000$ psi. (Beam: $b = 400$ mm; $h = 1.3$ m; concentrated loads $= 890$ kN; Distance from ends $= 1.2$ m; span $= 6$ m; support width $= 0.6$ m; $f'_c = 30$ MPa; $f_y = 400$ MPa.)

5.22 Design the details of the bearing shoe on a prestressed girder of 12 in. width. Assume an angle will be used across the width for bearing as in Fig. 5.16.1(b). The reaction is 35 kips dead load and 40 kips live load. The girder concrete has $f'_c = 6000$ psi. Assume no horizontal restraint is developed.

5.23 Design a bracket (corbel) that projects from one side of a 16 × 16 column to support a vertical load of 35 kips dead load and 65 kips live load. Assume that suitable bearings are provided so that horizontal restraint is eliminated. The reaction is located 5 in. from the column face. Use $f'_c = 5000$ psi and $f_y = 60,000$ psi. (Column size = 400 × 400 mm; dead load = 160 kN; live load = 290 kN; reaction 130 mm from column face; $f'_c = 35$ MPa; $f_y = 400$ MPa.)

5.24 Redesign the bracket (corbel) of Prob. 5.23 if the reaction is 3.5 in. (90 mm) from the column face.

5.25 Design for the conditions of Prob. 5.23 except take the reaction location 9 in. (230 mm) from the column face.

5.26 Repeat Prob. 5.23 if the reaction is from a restrained beam that induces a horizontal tension equal to 50% of the total gravity reaction.

5.27 Redesign the bracket (corbel) of Prob. 5.26 if the reaction is 3.5 in. (90 mm) from the column face.

5.28 Repeat Prob. 5.25 if the reaction is from a restrained beam that induces a horizontal tension equal to 40% of the total gravity reaction.

5.29 Redesign the bracket (corbel) of Example 5.16.1 considering that the supported prestressed girder is welded to the bracket. Creep, shrinkage, and temperature effects on the restrained girder induce a horizontal force of 50 kips (unfactored) on the bracket.

6

Development of Reinforcement

6.1 GENERAL

A basic requirement in reinforced concrete construction is that there is *bond* between reinforcement and the surrounding concrete, which means that under service load there is no slip of the steel bar relative to its surrounding concrete. From the strength point of view, slippage of bars relative to surrounding concrete may or may not result in overall failure of the beam. Even though there may be complete separation of bars and concrete over much of the length, a beam may continue to carry load as long as the bars cannot pull out at the ends. Mechanical end anchorages may be used to accomplish integrity of the system, or wherever possible, bars should be anchored by embedment beyond the point where the loading causes maximum tension a distance adequate to develop the full tensile capacity of the bar.

The concepts for the transfer of force between reinforcement and the surrounding concrete, either by development of reinforcement or mechanical anchorage, are presented in the next several sections. An excellent summary also appears in ACI Committee 408 reports [1,2], and more recently the mechanics of bond and slip has been explained by Lutz and Gergely [3], Orangun, Jirsa, and Breen [4], Losberg and Olsson [5], and Jirsa, Lutz, and Gergely [6].

6.2 ANCHORAGE BOND

Design of longitudinal and shear reinforcement to accommodate the moment and shear at sections along a beam has been treated in Chaps. 3, 4, and 5. For resisting the bending moment, an area of longitudinal steel is provided to carry a tensile force. It must also be realized that no matter

Brunswick Building, Chicago. (Courtesy of Portland Cement Association.)

what area is provided, if the bars are not anchored in the concrete sufficiently so that the applied tensile force can be developed by bond between steel and concrete, the bars will pull out. The moment capacity of a beam is, therefore, a three-dimensional relationship involving not only the cross-sectional properties at a location along the span, but also the embedment lengths of the steel bars in both directions therefrom.

Consider a uniformly loaded cantilever beam as shown in Fig. 6.2.1(a), which has been properly proportioned for flexural strength. To illustrate the principle, assume that the tension reinforcement consists of one single bar of diameter d_b. When the bar segment AB is considered as a free body as shown in Fig. 6.2.1(b), the tensile force at B, which is $f_s(\pi d_b^2/4)$, must

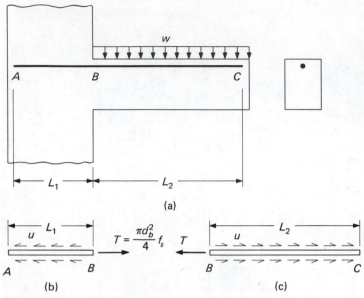

Figure 6.2.1 Anchorage bond in tension bars.

be transmitted to the concrete by the interaction between the bar and the surrounding concrete in the embedment length $L_1 = AB$. If u is the average unit bond stress over the nominal surface area $\pi d_b L_1$, then

$$u \pi d_b L_1 = f_s \pi \frac{d_b^2}{4}$$

or

$$u = \frac{f_s d_b}{4 L_1} \tag{6.2.1}$$

Thus if the average allowable bond stress u under service conditions can be assumed, the development length required for a given f_s becomes

$$\text{development length } L_1 = \frac{f_s}{4u} d_b \tag{6.2.2}$$

The same situation exists in free body BC, as shown in Fig. 6.2.1(c). Thus the maximum tensile force at B has to be developed by embedment through the distances BA or BC. Where space limitations prevent anchoring the bars the proper amount by straight embedment, such bars may be terminated in standard hooks. A standard hook may be considered as contributing to an equivalent development length by mechanical action that is proportional to the tensile strength of concrete (ACI-12.5), thus reducing the total embedment dimension required.

In the strength method of design, the objective is to develop the yield stress f_y in the steel; therefore f_s of Eq. (6.2.2) becomes f_y. Also the bond stress u is the nominal unit stress when a pullout failure is imminent; that is, the ultimate bond stress capacity u_u. Thus the *development length* L_d required for the anchoring of bars acting at yield stress is

$$\text{development length } L_d = \frac{f_y d_b}{4u_u} \tag{6.2.3}$$

Adequate development length must be provided for a reinforcing bar in compression as well as in tension.

6.3 FLEXURAL BOND

As moment varies along a span, the tensile force in the steel also varies; this induces a longitudinal interaction between the bars and the surrounding concrete, known as *flexural bond*. High flexural bond requirement exists at locations along the span where the rate of change of tensile force in the bars is high, such as at points of inflection within continuous spans and at simply supported ends of beams, even though the tensile force to be developed at such locations is zero.

Consider a segment DD' of the reinforcing bar in the same cantilever beam used in Sec. 6.2. As shown by the free body of DD' in Fig. 6.3.1(b), T_D is slightly greater than $T_{D'}$. The bending moment M_D equals the internal forces C and T times the moment arm between them; thus

$$T_D = \frac{M_D}{\text{arm}} \quad \text{and} \quad T_{D'} = \frac{M_{D'}}{\text{arm}} \tag{6.3.1}$$

Also, from horizontal force equilibrium,

$$u\pi d_b \, dz = T_D - T_{D'} \tag{6.3.2}$$

in which u is the average flexural bond stress requirement over the nominal contact area between the steel bar and the concrete, and d_b is the diameter

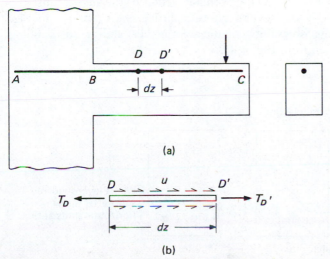

Figure 6.3.1 Flexural bond in a tension bar.

of the single bar. Substituting Eq. (6.3.1) into Eq. (6.3.2),

$$u = \frac{M_D - M_{D'}}{dz}\left(\frac{1}{\pi d_b(\mathrm{arm})}\right) = \frac{dM}{dz}\left(\frac{1}{\pi d_b(\mathrm{arm})}\right) = \frac{V}{\pi d_b(\mathrm{arm})} \quad \textbf{(6.3.3)}$$

Although the flexural bond stress formula, Eq. (6.3.3), does serve to describe to a certain degree the elastic behavior at service load, the present ACI Code procedures are adequate in providing the proper embedment length at the support of simple spans and at the points of inflection of continuous beams to prevent bars from pulling out when under factored loading. Further discussion on the reasons for not using flexural bond stress in strength design is to be found in Sec. 6.5, after the nature of bond failure is described in Sec. 6.4.

6.4 THE NATURE OF BOND FAILURE

Formerly when plain bars (relatively smooth bars without lug deformations) where used for reinforcement, bond was thought of as an adhesion between concrete paste and the surface of the bar. Even with low tensile stress in the reinforcement, there would be sufficient slippage to break the adhesion immediately adjacent to a crack in the concrete, leaving only friction to resist bar movement relative to the surrounding concrete over the slip length. Shrinkage can also cause frictional drag against the bars. Typically, a hot rolled *plain* bar may either pull loose by longitudinal splitting if the adhesion and friction resistances are high enough, or just pull out leaving a round hole when adhesion and friction resistances are low.

Deformed bars were designed to change the behavior pattern so that there would be less reliance on friction and adhesion (though they still exist) and more reliance on the bearing of the lugs against the concrete. A so-called "bond failure" with deformed bars in normal weight concrete is nearly always a splitting failure [1,5]. In a splitting failure the concrete splits into two or three segments due to the wedging action of the lugs against the concrete, as shown by the several typical splitting crack patterns in Fig. 6.4.1. The interacting forces between the deformed bar and the surrounding concrete may be seen from Fig. 6.4.2.

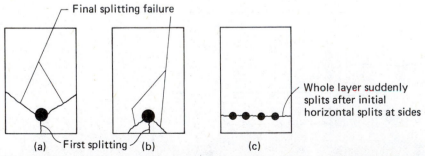

Figure 6.4.1   Splitting cracks and ultimate splitting failure modes (from Ref. 1).

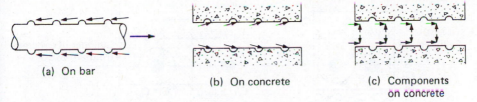

(a) On bar

(b) On concrete

(c) Components
on concrete

Figure 6.4.2   Forces between bar and concrete (from Ref. 15).

When small size bars are used with large cover, or when there may tend to be air pockets at the underside of bars cast in the top of beams, the lugs bearing against the concrete may crush it and result in a pullout failure without splitting the concrete. This nonsplitting failure has also been reported for larger bars on lightweight concrete [1]. Although splitting is the usual bond failure mode, it is noted that an initial splitting crack on one face of a beam does *not* constitute failure. Progressive splitting is the first sign of bond distress and may be considered the usual cause of bond collapse. Confinement of tension steel by stirrups, ties, or spirals may significantly delay bond collapse until several splitting cracks have formed.

Studies of bond strength (combining splitting and friction) originally were made on pull-out tests [7,8] of plain bars. Results from pull-out tests [9–14] of deformed bars including the related load-slip data have provided the basis for the current ACI design procedures for development of reinforcement. As the concrete is confined in pull-out tests, the results may provide data concerning anchorage but probably little about the factors that relate to splitting. The focus on splitting as the important bond failure mode was given impetus by the work of Ferguson, Turpin, and Thompson [15]. Recent studies of the splitting failure mode have been made by Orangun, Jirsa, and Breen [4], Untrauer and Warren [35], Losberg and Olsson [5], Kemp and Wilhelm [16], Morita and Kaku [17], and Jimenez, White, and Gergely [18].

Because of the complex interrelationship between bond, shear, and moment, present design practice makes use of a large number of experimental studies. In general, bond strength is directly proportional to $\sqrt{f'_c}$ (i.e., proportional to the tensile strength of the concrete) and inversely proportional to the bar diameter.

Recent studies [4,34,35] have hypothesized that the action of splitting arises from a stress condition analogous to a concrete cylinder surrounding a reinforcing bar and acted upon by the outward radial components [Fig. 6.4.2(c)] of the bearing forces from the bar. The cylinder would have an inner diameter equal to the bar diameter d_b and a thickness C equal to the smaller of C_b, the clear bottom cover, or C_s, half of the clear spacing to the next adjacent bar (see Fig. 6.4.3). The tensile strength of this concrete cylinder determines the strength against splitting. If $C_s < C_b$, a side-split type of failure occurs [Fig. 6.4.1(c)]. When $C_s > C_b$, longitudinal cracks through the bottom cover form first [first splitting cracks in Fig. 6.4.1(a),(b)]. If C_s is only nominally greater than C_b, the secondary splitting will be side

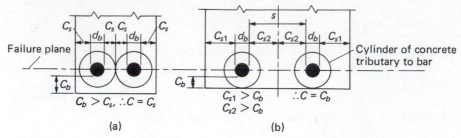

Figure 6.4.3 Concrete cylinder hypothesis for splitting failure (from Ref. 4).

splitting along the plane of the bars. If C_s is significantly greater than C_b, the secondary splitting will also be through the bottom cover to create a V-notch failure [Fig. 6.4.1(b)].

ACI Committee 408 has proposed [6,19] changes in the ACI Code to reflect the cylinder hypothesis for splitting failure. Parts of this proposal relating to hooks and splices (see Secs. 6.11 and 6.17–6.20) have been adopted for the 1983 ACI Code, but adoption of the proposed changes in the basic development length procedure has been postponed.

6.5 REASONS FOR NOT USING FLEXURAL BOND STRESS IN STRENGTH DESIGN

When evaluating the strength of a beam, flexural bond stress does not provide an adequate measure of the margin of safety against a splitting (or bond) failure. The reasons may be divided into two general categories: (1) inability of the flexural bond stress equation, Eq. (6.3.3), to measure accurately bond stress along tension reinforcement in a beam; and (2) lack of correlation between *localized* slip from high flexural bond stress and the strength of a beam as represented by splitting and subsequent loss of anchorage of the tension bars.

Several situations may be identified where flexural bond stress as computed by Eq. (6.3.3) is inaccurate:

1. The tension concrete is uncracked in a region of low bending moment; thus the flexural bond stress is overestimated as concrete still carries part of the tensile force.
2. At a point where high bending moment exists, according to Eq. (6.3.3) the low shear would indicate low flexural bond stress; however, *at* a flexural crack (see Fig. 6.5.1) in such a region the steel carries all of the tensile force; but adjacent to such a crack, bond stress is likely to be high since the concrete shares in carrying the tension [8,11].
3. In the vicinity of a shear-related inclined crack, such as that of Fig. 5.4.2, not only does high bond stress exist adjacent to the crack but in addition the tensile force T computed at z from the support ac-

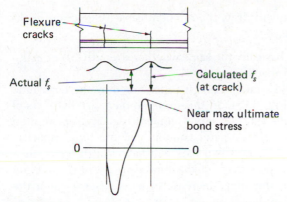

Figure 6.5.1   Probable bond stress between cracks when beam shear is zero (from Ref. 8).

      tually acts at a much closer distance from the support; that is, at the shear crack intersection with the longitudinal steel.

4. At locations where some bars are terminated in a tension zone, there is an abrupt change in the distribution of the total tensile force among the bars, causing stress concentrations [20,21] and bond stresses greatly in excess of formula predictions.

    Thus the computed flexural bond stress represents only a nominal stress and does not correctly reflect true behavior. If the bars remain adequately anchored so as to continue to carry their required tensile force, no reduction in strength should result from a so-called "local failure." The design approach of the ACI Code is to utilize the strength concept of requiring adequate anchorage, that is, development length, and eliminate computation of nominal flexural bond stress which had been a major design consideration since the early 1900s.

6.6 MOMENT CAPACITY DIAGRAM—BAR BENDS AND CUTOFFS

As stated in Sec. 6.2, the moment capacity of a beam at any section along its length is a function of its cross section and the embedment length of its reinforcement. The concept of a diagram showing this three-dimensional relationship can be a valuable aid in determining cutoff or bend points of longitudinal reinforcement. It may be recalled from Chap. 3 that in terms of the cross section, the moment capacity (i.e., strength) for a singly reinforced rectangular beam may be expressed

$$M_n = A_s f_y (d - a/2) \qquad [3.8.1]$$

    Equation (3.8.1) assumes that the steel reinforcement comprising A_s is adequately embedded *in each direction* from the section where M_n is computed such that the stress f_y can develop.

EXAMPLE 6.6.1 Compute and draw the moment capacity diagram qualitatively for the beam of Fig. 6.6.1.

Solution: The procedure is basically the same whether strength (M_n or ϕM_n)
or working stress moment capacity is desired.

The maximum capacity in each region is represented by the horizontal
portions of the diagram in Fig. 6.6.1. In this example there are five bars of
one size in section *C-C*; thus the maximum moment capacity represented
by each bar is in this case approximately one-fifth of the total capacity.
Actually, the sections with four and two bars will have a little more than
four-fifths and two-fifths, respectively, of the total capacity of the section
containing five bars, due to the slight increase in moment arm when the
number of bars in the section decreases.

At point *a*, the location where the fifth bar terminates, this bar has zero

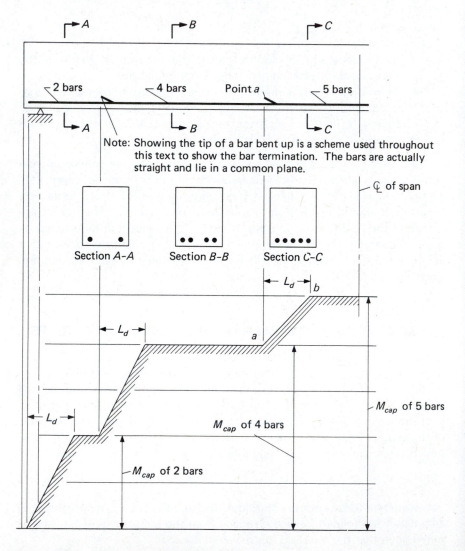

Figure 6.6.1 Moment capacity diagram.

embedment length to the left and thus has zero capacity. Proceeding to the right from point a, the bar may be counted on to carry a tensile force proportional to its embedment from point a up to the development length L_d where, according to Eq. (6.2.3),

$$L_d = \frac{f_y d_b}{4u_u}$$

For instance, if the yield stress $f_y = 40,000$ psi is to be developed and #8 bars are used with an ultimate bond stress capacity of 520 psi, the development length necessary would be

$$L_d = \frac{40,000(1)}{4(520)} = 19.2 \text{ in.}$$

In Fig. 6.6.1, point b represents the point where the fifth bar is anchored a distance L_d and can therefore carry its full tensile capacity. The other cutoff points are treated in the same way.

EXAMPLE 6.6.2 Demonstrate qualitatively the practical use of the moment capacity ϕM_n diagram for verification of the locations of cutoff or bend points in a design. Assume that the main cross section with five equal-sized bars provides exactly the required strength at midspan for this simply supported beam with uniform load, as shown in Fig. 6.6.2.

Solution: The factored moment M_u diagram and the moment capacity ϕM_n diagram for the chosen arrangement are shown in Fig. 6.6.2.

(a) Compute the actual ϕM_n for each potential bar grouping that may be used; in the present case, for five bars, four bars, and two bars.

(b) Decide what bars must extend entirely across the span and into the support; ACI-12.11.1 states that "At least one-third the positive moment reinforcement in simple members . . . shall extend along the same face of member into the support." In beams the reinforcement must extend into the support at least 6 in. In this case two bars would have to extend into the support.

(c) Decide on the order of cutting or bending the remaining bars. The least amount of longitudinal reinforcement will be obtained when the resulting moment capacity ϕM_n diagram is closest to the factored moment M_u diagram. With that thought in mind, and proceeding from maximum moment region to the support, cut off one bar as soon as permissible.

(d) Cutoff restrictions. Point A of Fig. 6.6.2 is the theoretical location to the left of which the capacity represented by the remaining four bars is adequate. To provide for a safety factor against shifting of the moment M_u diagram (especially in continuous spans) and to provide partially for the difficulty arising from a potential diagonal crack, the ACI Code provides that there must be an extension beyond the point where a bar theoretically may be terminated, or it may be bent into the compression face. In ACI-12.10.3 is the statement, "Reinforcement shall extend beyond the point at which it is no longer required to resist flexure for a distance equal to the effective depth of the member or 12 bar diameters, whichever is greater, except at supports of simple spans and at free end of cantilevers."

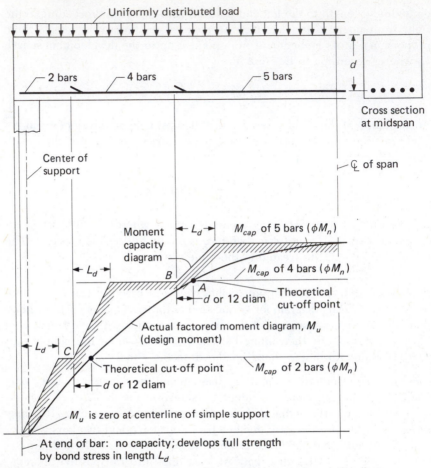

Figure 6.6.2 Verification of bar cutoffs with the moment capacity diagram.

(e) Once cutoff or bend points are located, a check is made by drawing the moment capacity ϕM_n diagram to ensure no encroachment on the factored moment M_u diagram.

(f) Other restrictions. Since points B and C of Fig. 6.6.2 are bar terminations in a tension zone, the stress concentrations described in Sec. 6.5 are present, effectively reducing the shear strength of the beam [20,21]. Thus *one* of the three special conditions of ACI-12.10.5 must be satisfied for cutoffs to be acceptable. However, if these bars were bent up and anchored in the compression zone, no further investigation would be necessary.

6.7 BASIC DEVELOPMENT LENGTH
FOR TENSION REINFORCEMENT

The term "development length" has been defined in Sec. 6.2 as the length of embedment needed to develop the yield stress in the reinforcement. Equation (6.2.3) gives an expression for the development length L_d in terms of the yield stress f_y, the bar diameter d_b, and the ultimate bond stress capacity u_u as follows:

$$L_d = \frac{f_y d_b}{4u_u} \qquad\qquad [6.2.3]$$

The four formulas for the basic development length L_{db} of tension reinforcement as stated in ACI-12.2.2 and shown below may be derived by using Eq. (6.2.3) and a set of values of ultimate bond stress capacity u_u for different bar sizes; thus, when L_d is the basic value L_{db},

1. For #11 or smaller bars, the larger of

$$L_{db} = 0.04 \frac{A_b f_y}{\sqrt{f_c'}} \quad \text{or} \quad 0.0004 d_b f_y \qquad (6.7.1)^*$$

2. For #14 bars,

$$L_{db} = 0.085 \frac{f_y}{\sqrt{f_c'}} \qquad (6.7.2)^*$$

3. For #18 bars,

$$L_{db} = 0.11 \frac{f_y}{\sqrt{f_c'}} \qquad (6.7.3)^*$$

4. For deformed wire (but not deformed wire fabric),

$$L_{db} = 0.03 \frac{f_y d_b}{\sqrt{f_c'}} \qquad (6.7.4)^*$$

In Eqs. (6.7.1) to (6.7.4), the dimensional units of L_{db} and d_b are inches; of A_b, square inches; and of f_y and $\sqrt{f_c'}$, pounds per square inch. Basic development lengths for tension reinforcement in some of the common situations are given in Table 6.7.1. Modifications of these basic formulas for special situations are discussed in Sec. 6.8. Except for lap splices (Sec. 6.17) and anchorage of stirrups (Sec. 6.16), the minimum L_d to be used after all modifications are applied is 12 in.

---

*For SI, ACI 318-83M, with L_{db} and d_b in mm, A_b in mm², and f_y and f_c' in MPa, gives for #35M bars or smaller,

$$L_{db} = 0.019 \frac{A_b f_y}{\sqrt{f_c'}} \quad \text{or } 0.058\, d_b f_y \qquad (6.7.1)$$

for #45M,

$$L_{db} = \frac{26 f_y}{\sqrt{f_c'}} \qquad (6.7.2)$$

for #55M,

$$L_{db} = \frac{34 f_y}{\sqrt{f_c'}} \qquad (6.7.3)$$

for deformed wire,

$$L_{db} = \frac{0.36 d_b f_y}{\sqrt{f_c'}} \qquad (6.7.4)$$

Table 6.7.1 BASIC DEVELOPMENT LENGTH L_{db} FOR TENSION REINFORCEMENT

INCH-POUND BARS WITH L_{db} IN INCHES

BAR	$f_y = 40,000$ psi			$f_y = 60,000$ psi		
	f'_c (psi)			f'_c (psi)		
	3000	4000	5000	3000	4000	5000
#3	12^a	12^a	12^a	9^b	9^b	9^b
#4	12^a	12^a	12^a	12	12	12
#5	10^b	10^b	10^b	15	15	15
#6	12.8	12	12	19.2	18	18
#7	17.5	15.2	14	26.2	22.8	21
#8	23.1	20.0	17.9	34.6	30.0	26.8
#9	29.2	25.3	22.6	43.8	38.0	33.9
#10	37.1	32.2	28.8	55.6	48.2	43.1
#11	45.5	39.5	35.4	68.4	59.2	53.0
#14	62.1	53.8	48.1	93.1	80.6	72.1
#18	80.3	69.6	62.2	121	104	93.3

METRIC BARS^c WITH L_{db} IN CENTIMETERS

BAR	$f_y = 300$ MPa			$f_y = 400$ MPa		
	f'_c (MPa)			f'_c (MPa)		
	25	30	35	25	30	35
#10M	30^a	30^a	30^a	26.2^b	26.2^b	26.2^b
#15M	27.8^b	27.8^b	27.8^b	37.1	37.1	37.1
#20M	34.2	33.9	28.9^b	45.6	45.2	45.2
#25M	57.0	52.0	48.2	76.0	69.4	64.2
#30M	79.8	72.8	67.4	106	97.1	89.9
#35M	114	104	96.3	152	139	129
#45M	156	142	132	208	190	176
#55M	204	186	172	272	248	230

^{a}Development length L_d is the 12 in. (30 cm) minimum even with the 1.4 multiplier for top bar modification (see Sec. 6.8).
^{b}If no modification factors greater than 1.0 are to be used, the development length L_d is 12 in. (30 cm) minimum.
^{c}Metric bars are from Table 1.12.2.

The ACI expressions for L_{db} involve reducing the average ultimate unit bond stress capacities u_u given in the 1963 ACI Code (see Table 6.7.2) by a factor 5/6 so as to decrease the possibility of a splitting failure until the resistance developed between the bar and its surrounding concrete can become nearly uniform along the length L_{db}. Using, then, 5/6 of the bond stress capacities itemized in Table 6.7.2 for u_u in the development length formula, Eq. (6.7.1) for #11 and smaller bars may be substantiated as follows:

$$L_{db} = \frac{f_y d_b}{4u_u} = \frac{f_y d_b}{4[(5/6)(9.5\sqrt{f'_c}/d_b)]} = 0.0402\frac{A_b f_y}{\sqrt{f'_c}}$$

$$L_{db} \geq \frac{f_y d_b}{4u_u} = \frac{f_y d_b}{4[(5/6)800]} = 0.000375 d_b f_y$$

Table 6.7.2   AVERAGE ULTIMATE BOND STRESS CAPACITIES AS GIVEN
BY 1963 ACI CODE

DEFORMED BARS	BASIC BOND STRESS, u_u	TOP BARS, u_u
Tension: #11 and smaller	$\dfrac{9.5\sqrt{f'_c}}{d_b} \leq 800$ psi	$\dfrac{6.7\sqrt{f'_c}}{d_b} \leq 560$ psi
Tension: #14 and #18	$6\sqrt{f'_c}$	$4.2\sqrt{f'_c}$
Compression: All sizes	$13\sqrt{f'_c} \leq 800$ psi	$13\sqrt{f'_c} \leq 800$ psi

Likewise, for #14 bars (Eq. 6.7.2),

$$L_{db} = \frac{f_y d_b}{4u_u} = \frac{f_y(1.693)}{4[(5/6)6\sqrt{f'_c}]} = 0.0847 \frac{f_y}{\sqrt{f'_c}}$$

and for #18 bars (Eq. 6.7.3),

$$L_{db} = \frac{f_y d_b}{4u_u} = \frac{f_y(2.257)}{4[(5/6)6\sqrt{f'_c}]} = 0.113 \frac{f_y}{\sqrt{f'_c}}$$

For *deformed* wire, using 5/6 of a basic bond stress capacity of $10\sqrt{f'_c}$ (see
Ref. 25),

$$L_{db} = \frac{f_y d_b}{4u_u} = \frac{f_y d_b}{4[(5/6)10\sqrt{f'_c}]} = 0.03 \frac{d_b f_y}{\sqrt{f'_c}}$$

Some of the numerical constants as obtained above have been rounded off
to obtain those adopted in the ACI Code.

6.8 FACTORS TO MODIFY BASIC DEVELOPMENT LENGTH

Modifications to the basic development lengths for tension reinforcement
are prescribed by ACI-12.2.3 and 12.2.4 to account for conditions that may
either decrease or increase the tendency for splitting.

Unfavorable conditions indicating a weaker situation so as to require
an increased development length over the basic value are (1) when horizontal bars are placed so that more than 12 in. of concrete is cast in the
member below the bars, known as *top bars;* (2) when the reinforcement
yield stress f_y exceeds 60,000 psi; and (3) when lightweight aggregate concrete is used.

Favorable conditions that permit a reduction in the required development length from the basic value are (1) when reinforcement is laterally
spaced at least 6 in. on center with at least 3 in. from the edge bar to the
face of the member measured in the direction of the spacing (i.e., typically
slabs); (2) when more reinforcement than required for strength has been
used in the flexural member; and (3) when bars are confined by enclosure
within a spiral (such as in a beam-column member; see Chap. 13).

A summary of the modification factors given by ACI-12.2.3 and 12.2.4
appears in Table 6.8.1. These multipliers are to be applied cumulatively,
as illustrated in Example 6.8.1.

Table 6.8.1 FACTORS TO MODIFY BASIC DEVELOPMENT LENGTH
FOR STRAIGHT EMBEDMENT (ACI-12.2.3 AND 12.2.4)

CONDITION	MULTIPLIER
1. Top bars; horizontal reinforcement with more than 12 in. of concrete cast beneath the bars	1.4
2. Reinforcement with f_y greater than 60,000 psi	$2 - \dfrac{60,000}{f_y}$
3. Lightweight concrete: a. "All-lightweight" concrete "Sand-lightweight" concrete (Linear interpolation may be used when partial sand replacement is used) or b. When average splitting tensile strength f_{ct} is specified, and concrete is proportioned according to ACI-4.2	 1.33 1.18 $\dfrac{6.7\sqrt{f'_c}}{f_{ct}} \geq 1.0$
4. Wide lateral spacing of bars; at least 6 in. on center and at least 3 in. from the edge bar to the face of member	0.8
5. Excess reinforcement is used for a flexural member	$\dfrac{\text{required } A_s}{\text{provided } A_s} \leq 1.0$
6. Bars enclosed within a spiral not less than $\frac{1}{4}$ in. diameter and not more than 4 in. pitch	0.75

The requirement of extra development length for top bars has long been a provision of the ACI Code in recognition of the fact that top cast bars exhibit reduced strength apparently due to the settling away of concrete from the bar on its underside [1,10]. The exact variation of bond strength with depth of concrete below the bars has not been established. Limited tests with concrete depths of 12 to 18 in. and with high-strength steel bars have indicated the ultimate splitting resistance to be lowered about 10 to 20% [1].

For lightweight concrete, the increased development lengths are necessary because bars may pull out without splitting the concrete [2].

The reductions in development length permitted for wide lateral spacing and for confinement in a spiral are in recognition of the favorable influence of these factors in restricting the propagation of splitting; however, the specific multipliers are somewhat arbitrary. Bars within stirrups and ties also show improved resistance to splitting but no quantitative relationship has been established.

EXAMPLE 6.8.1 Determine the development length L_d required for the #9 bars A on the top of a 15-in. slab, as shown in Fig. 6.8.1. Use $f_y = 70,000$ psi, and $f'_c = 4000$ psi.

Solution: The basic development length for a #9 bar is

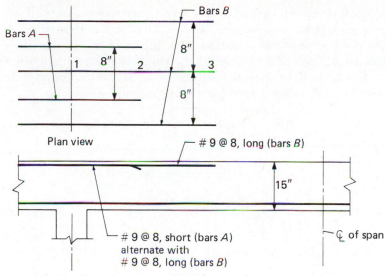

Figure 6.8.1 Top bars for Example 6.8.1.

$$L_{db} = 0.04 \frac{A_b f_y}{\sqrt{f_c'}} = 0.04 \frac{1.0(70,000)}{\sqrt{4000}} = 44.3 \text{ in.} \qquad \text{(Controls)}$$

or

$$L_{db} = 0.0004 d_b f_y = 0.0004(1.128)(70,000) = 31.6 \text{ in.}$$

Since the concrete thickness below the top bars exceeds 12 in., the 1.4 multiplier must be used.

The correction factor for high yield stress is

$$2 - \frac{60,000}{70,000} = 2 - 0.84 = 1.16$$

Referring to Fig. 6.8.1 the bars A are developed over the distance 1–2, while the bars B are developed over the distance 2–3. The spacing to be used for the comparison of the wide spacing requirement in Case 4, Table 6.8.1, is the spacing of the *closest bars that terminate at the same point*. In other words, the spacing for both bars A and B is 8 in. Thus the 0.8 multiplier may be used.

The development length for this situation is

$$L_d = 44.3(1.4)(1.16)(0.8) = 57.5 \text{ in.}$$

6.9 DEVELOPMENT LENGTH
FOR COMPRESSION REINFORCEMENT

Relatively little is known about bond on compression bars, except that the weakening effect of flexural tension cracks is not present and there is the beneficial effect of the end bearing of the bars on the concrete. The ultimate bond stress capacity of compression bars in all sizes has been taken as $13\sqrt{f_c'}$ but not to exceed 800 psi in the 1963 ACI Code, as listed in Table 6.7.2. Without the necessity of reducing the bond stress capacity to 5/6 of its value, the basic development length expression for compression reinforcement may be obtained from Eq. (6.2.3) (calling $L_d = L_{db}$) as

$$L_{db} = \frac{f_y d_b}{4u_u} = \frac{f_y d_b}{4(13)\sqrt{f_c'}} = 0.0192 \left(\frac{f_y d_b}{\sqrt{f_c'}} \right)$$

which ACI-12.3 gives as the basic development length L_{db},

$$L_{db} = 0.02 \left(\frac{f_y d_b}{\sqrt{f_c'}} \right) \tag{6.9.1}*$$

Using the upper limit of 800 psi for u_u, Eq. (6.2.3) gives

$$L_{db} = \frac{f_y d_b}{4u_u} = \frac{f_y d_b}{4(800)} = 0.000312 f_y d_b$$

which ACI-12.3 gives as

$$L_{db} = 0.0003 f_y d_b \tag{6.9.2}*$$

The basic development length is to be taken as the larger of Eqs. (6.9.1) and (6.9.2).

When excess bar area is provided so that the A_s provided exceeds the A_s required, Eqs. (6.9.1) or (6.9.2), whichever controls, may be reduced by applying the multiplier (required A_s/provided A_s).

When reinforcement is enclosed by spirals (typically in columns; see Chap. 13) which are not less than $\frac{1}{4}$-in. in diameter and not more than 4-in. pitch, Eqs. (6.9.1) or (6.9.2) may be reduced by 25%. Confinement by spirals has been shown to increase the strength of all types of concrete members; hence the somewhat arbitrary 25% reduction in the required development length. The development length L_d after any modification factors are applied may not be less than 8 in. (200 mm). Thus, in general, the development length for compression reinforcement is

$$L_d = \begin{bmatrix} \text{Eqs. (6.9.1)} \\ \text{or (6.9.2)} \end{bmatrix} \begin{bmatrix} \text{required } A_s \\ \text{provided } A_s \end{bmatrix} \begin{bmatrix} 0.75 \text{ for enclosure} \\ \text{by spirals} \end{bmatrix} \geq 8 \text{ in.} \tag{6.9.3}$$

---

*For SI, ACI 318-83M, for L_{db} and d_b in mm, and f_c' and f_y in MPa, gives

$$L_{db} = 0.24 \left(\frac{f_y d_b}{\sqrt{f_c'}} \right) \tag{6.9.1}$$

$$L_{db} = 0.044 f_y d_b \tag{6.9.2}$$

Figure 6.10.1 Bundled bar arrangements (bars are in contact with each other).

6.10 DEVELOPMENT LENGTH FOR BUNDLED BARS

When space for proper clearance is restricted and large steel areas are required, groups of parallel bars are sometimes bundled. Not more than four bars bundled in contact (ACI-7.6.6), enclosed by stirrups or ties, may be arranged with no more than two bars in the same plane into typical bundle shapes, such as triangular, square, or L-shaped for three- and four-bar bundles (see Fig. 6.10.1). Bars larger than #11 shall not be bundled in beams or girders, primarily to ensure proper control of cracking (see ACI Commentary-7.6.6). In flexural members, termination of individual bars within a bundle at points along the span must be at different points offset by at least 40 bar diameters. Where spacing requirements and minimum clear cover are based on bar size, one bundle of bars shall be treated as a single bar of an equivalent diameter derived from the total area of the bars in the bundle. In applying the crack control provisions of ACI-10.6.4 (see Chap. 4, Sec. 4.12) to bundled bars, this assumption of treating a bundle of bars as a single large bar may be overly conservative. An alternative is suggested by Lutz [22].

When considering bond strength for a bundle of bars, the three-bar triangular pattern and the four-bar arrangement will have $16\frac{2}{3}$ and 25% reduction, respectively, in the total surface contact of bars with surrounding concrete. In order to allow further for the difficulty of getting good bond at the reentrant corner where the bars in the bundle touch each other, additional reduction appears logical.

The requirements of ACI-12.4 specify that the development length of the bundle shall be based on that for the individual bar in the bundle, increased by 20% for a three-bar bundle and 33% for a four-bar bundle.

Experimental results for beams and columns with bundled reinforcement are reported in Ref. 23, and some practical applications are presented in Ref. 24.

6.11 DEVELOPMENT LENGTH FOR A TENSION BAR TERMINATING IN A STANDARD HOOK

When straight embedment is inadequate to provide for the necessary development length of a tension bar or when it is desired to have the full capacity of a bar available in the shortest distance of embedment, a standard hook as defined in ACI-7.1 and 7.2 and shown in Fig. 6.11.1 may be used. Hooks are not considered effective in adding to the compressive resistance of reinforcement.

In general the anchorage capacity of a hook *in mass concrete* is about the same as that of a straight bar with the same total length of embedment

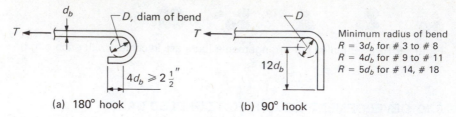

Figure 6.11.1 Standard hooks for development of main tension reinforcement.

[25,26]. Quite commonly hooks in *structural members* are located close to a free surface where splitting forces proportional to the tensile capacity of the bar will govern the hook capacity. Because of the tendency for splitting failures, hooks may have less capacity than provided by an equal length of straight embedment.

In the past it was customary to consider the actual length around the bend of a hook to the end of the bar as an alternative to the mechanical anchorage provided by the hook. With 90° hooks particularly, designers have often assumed that satisfactory anchorage is obtained by adding length to the end of the bar in excess of the 12 bar diameters [Fig. 6.11.1(b)] required as part of the hook. This is an unsatisfactory practice.

ACI Code Procedure. Based on the works of Hribar and Vasko [25], Orangun, Jirsa, and Breen [4], Minor and Jirsa [26], Marques and Jirsa [27], and Pinc, Watkins, and Jirsa [28], ACI Committee 408 recommended [19] changes from the longstanding ACI procedure for tension development provided by a hook. Instead of prescribing required development length L_d as defined in Secs. 6.7 and 6.8 using the sum of available straight embedment

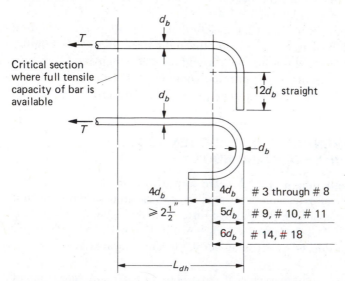

Figure 6.11.2 Development length L_{dh} for hooked bar.

Table 6.11.1 FACTORS TO MODIFY BASIC DEVELOPMENT LENGTH L_{hb} FOR A HOOKED BAR (ACI-12.5.3)

CONDITION	MULTIPLIER
1. *Yield stress:* For f_y other than 60,000 psi	$\dfrac{f_y}{60,000}$

2. *Concrete cover:*
     180° hook, #11 bars and smaller having side cover
         (normal to plane of hook) $\geq 2\frac{1}{2}$ in.
     90° hook, same as for 180° hook *plus* cover on bar
         extension beyond hook ≥ 2 in.
	0.7

3. *Ties or stirrups:*
     For bars #11 and smaller, hook enclosed vertically
     or horizontally within ties or stirrup-ties spaced
     along the full development length L_{dh} not greater
     than $3d_b$ (d_b = diameter of hooked bar). [Note:
     Multiplier cannot be used if the hook occurs at the
     discontinuous end of a member when clear cover
     to hooked bar is less than $2\frac{1}{2}$ in. (ACI-12.5.4).]
	0.8

4. *Excess reinforcement:*
     Where anchorage or development for f_y is not
     specifically required, or where there is
     reinforcement in excess of that required
	$\dfrac{\text{required } A_s}{\text{provided } A_s} \leq 1.0$

5. *Lightweight aggregate concrete*	1.3

plus an equivalent straight length attributable to the mechanical anchorage of the hook, the 1983 ACI Code prescribes in a direct manner the required embedment (development length L_{dh} as in Fig. 6.11.2) to develop f_y in a hooked bar, uncoupled from the embedment required (development length L_d) to develop f_y in a straight bar. A summary of the rationale for the new procedure is given by Jirsa, Lutz, and Gergely [6].

The procedure as prescribed by ACI-12.5 is to compute a *basic development length* L_{hb} required to develop the yield strength f_y of a hooked bar, measured from the location where the yield strength is needed to the extreme outside face of the hook, as follows:

$$L_{hb} = \frac{1200d_b}{\sqrt{f_c'}} \qquad (6.11.1)^*$$

for f_y = 60,000 psi.

Modification factors are then applied in a manner similar to what is done for straight bar embedment (see Sec. 6.8). The modification factors are given in Table 6.11.1.

*For SI, ACI 318-83M, for L_{hb} and d_b in mm, and f_c' in MPa, gives

$$L_{hb} = \frac{100d_b}{\sqrt{f_c'}} \qquad (6.11.1)$$

for f_y = 400 MPa.

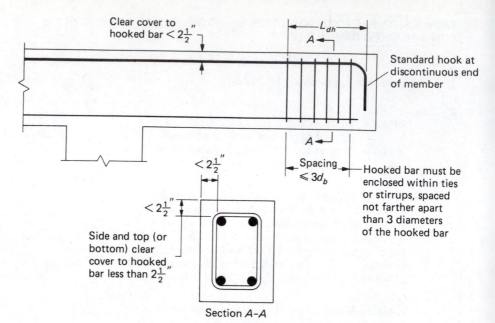

Figure 6.11.3   Special requirements for standard hooks developed at the discontinuous end of a member (ACI-12.5.4).

Table 6.11.2   DEVELOPMENT LENGTH L_{dh} (IN.) FOR HOOKED BARS[a]

| HOOKED BAR | $f_y = 40,000$ psi | | | $f_y = 60,000$ psi | | |
| | f'_c (psi) | | | f'_c (psi) | | |
	3000	4000	5000	3000	4000	5000
#3	5.5[b]	4.7[b]	4.3[b]	8.2	7.1	6.4
#4	7.3	6.3	5.7[b]	11.0	9.5	8.5
#5	9.1	7.9	7.1	13.7	11.9	10.6
#6	10.9	9.5	8.5	16.4	14.2	12.7
#7	12.8	11.1	9.9	19.2	16.6	14.8
#8	14.6	12.7	11.3	21.9	19.0	17.0
#9	16.5	14.3	12.7	24.7	21.4	19.1
#10	18.5	16.1	14.4	27.8	24.1	21.6
#11	20.6	17.9	15.9	30.9	26.8	23.9
#14	24.7	21.4	19.1	37.1	32.1	28.7
#18	32.9	28.5	25.5	49.4	42.8	38.3

[a]Assumes normal-weight concrete and side cover to hooked bar less than $2\frac{1}{2}$ in., computed in accordance with Eq. (6.11.1). For $f_y = 40,000$ psi, modification 1 from Table 6.11.1 has been applied to Eq. (6.11.1).
[b]If no modification greater than 1.0 is used, value is 6 in.

Once the basic value L_{hb} has been multiplied by the appropriate modification factors, the resulting required development length L_{dh} for the hooked bar is the largest of (1) the modified L_{hb}, (2) 8 hooked bar diameters, and (3) 6 in.

Most hooked bars are embedded in joints where other concrete members frame in at the sides and therefore provide some lateral confinement, in addition to the vertical confinement provided by the force in the column. When these other members are not present, such as at the discontinuous end of a cantilever as shown in Fig. 6.11.3, ACI-12.5.4 requires either $2\frac{1}{2}$ in. or more of cover to the hooked bar or that the hooked bar be enclosed in stirrups or ties over the development length L_{dh}.

Values of the development length L_{dh}, assuming that L_{dh} equals the basic development length L_{hb}, according to Eq. (6.11.1), for $f_y = 60{,}000$ psi, and L_{hb} modified by $(f_y/60{,}000)$ for $f_y = 40{,}000$ psi.

6.12 BAR CUTOFFS IN NEGATIVE MOMENT REGION OF CONTINUOUS BEAMS

The general concept of drawing the moment capacity diagram to investigate the adequacy of a given beam has been presented in Sec. 6.6. In this and the next two sections, several situations are discussed in which the ACI Code provisions are applied to establish cutoff points, aided by the drawing of the moment capacity diagram. Three factors are involved: (1) adequate horizontal offset from theoretical cutoff points to provide safety against the possible shifting of the moment diagram due to unusual loading arrangements; (2) adequate embedment lengths so that full bar capacity is available where needed; and (3) sufficient relief from stress concentrations when bars are terminated in a tension zone.

Cutoffs in the negative moment region of continuous beams can best be explained by means of a specific case. (For an example of a complete factored moment M_u envelope for a continuous beam, the reader is referred to Fig. 10.6.1.) Referring to the partial envelope for M_u in Fig. 6.12.1, assume that it is desired to cut two out of four bars as close as possible (point C) to the support, and then terminate the remaining two as soon as feasible (point E). In accordance with ACI-12.12.3, the area provided by bars $R2$ must exceed one-third the total reinforcement area provided for negative moment. In a general situation the horizontal distance between points C and E is greater than the development length L_d for bars $R2$; the special case wherein this is not so will be discussed later.

Point A represents the maximum factored moment M_u at the face of support, the critical location for bending moment. The higher factored moment (shown dashed in Fig. 6.12.1) within the support is resisted by a larger cross section than the one at the face of support; thus, the factored moment M_u at point A is considered the largest value for which strength must be provided. The full moment capacity ϕM_n provided by bars $R1$ and $R2$ is

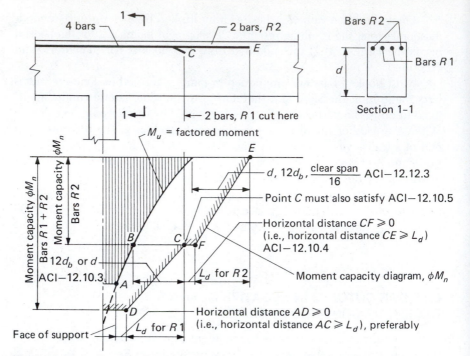

Figure 6.12.1   Bar cutoff in negative moment region of continuous beams.

somewhat greater than required. The theoretical cutoff for the two $R1$ bars is at point B. Point C is then located (ACI-12.10.3) horizontally to the right from point B a distance equal to the effective depth of the member or 12 diameters of bars $R1$. Point D is located horizontally to the left of point C a distance equal to the development length L_d for bars $R1$. Since points A and D will generally be approximately horizontal with one another, point D must lie at or to the right of point A in order to provide adequate capacity at the face of support. When extra capacity is provided, the safety may be considered adequate when the line CD intersects the face of support line at or below point A (satisfying ACI-12.12.2).

Point E is next established by extension, according to ACI-12.12.3, beyond the point of inflection (point of zero moment) "not less than effective depth of the member, $12d_b$, or one-sixteenth the clear span, whichever is greater." Point F, where the bars $R2$ will become capable of carrying their full capacity, is located L_d horizontally to the left of point E. In order that the moment capacity diagram does not encroach closer than $12d_b$ or d to point B, point F must lie to the right of point C.

The special case wherein the horizontal distance between points C and E is smaller than the development length L_d for bars $R2$ is shown in Fig. 6.12.2. The moment capacity ϕM_n diagram is indicated by $EC'F'D$, because the variation between the moment capacities at C' and F' has to be linear since development of capacity is assumed to be proportional to the amount

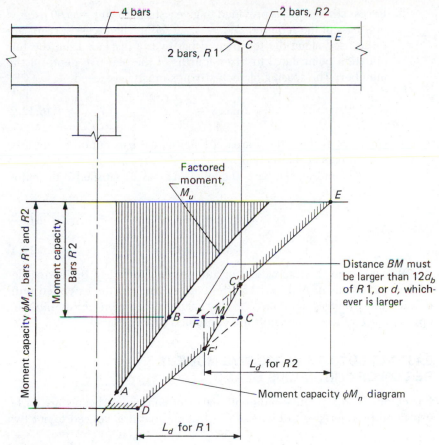

Figure 6.12.2 Special case of Fig. 6.12.1 wherein the distance *CE* is less than L_d for bars *R*2.

of embedment up to a maximum length of L_d. The requirement to allow for the shifting of the factored moment M_u curve is to be satisfied by making sure that the distance *BM* is larger than the effective depth *d* or 12 diameters of bars *R*1. If the distance *BC* were *equal* to the distance *d* or $12d_b$, as it was indicated to be in Fig. 6.12.1, then of course the distance *BM* would not satisfy ACI-12.10.3.

Cutting Bars in the Tension Zone. Since point *C* in Fig. 6.12.1 is still in the tension zone (though only slightly), the provisions of ACI-12.10.5 to minimize stress concentrations must be checked. *One* of the following three conditions must be satisifed *at point C*:

1. The factored shear V_u at the cutoff point does not exceed two-thirds of the shear strength ϕV_n, or

$$V_u \leq [\tfrac{2}{3}\phi V_n = \tfrac{2}{3}\phi(V_c + V_s)] \tag{6.12.1}$$

2. Excess stirrup area is provided to give a strength $\phi V_s = \phi(60 \text{ psi})b_w d$ in excess of the required $\phi V_s = V_u - \phi V_c$. The excess stirrups are to be used along the terminated bar over a distance from the termination point equal to three-fourths of the effective depth of the member. The spacing of such stirrups must not exceed

$$\max s = \frac{d}{8\beta_b} \tag{6.12.2}$$

where β_b is the ratio of area of longitudinal bars cut off to the total area of bars at the section.

3. For #11 and smaller bars, the following may be satisfied at the cutoff point:

$$V_u \leq [\tfrac{3}{4}\phi V_n = \tfrac{3}{4}\phi(V_c + V_s)] \tag{6.12.3}$$

and

$$\phi M_n \geq 2M_u \tag{6.12.4}$$

When it may be impractical or undesirable to satisfy the above described provisions of ACI-12.10.5, the cutoff point may be extended until it is in the compression zone, or the bars may be bent across into the opposite face of the beam and then continued or terminated.

6.13 BAR CUTOFFS IN POSITIVE MOMENT REGION OF CONTINUOUS BEAMS

Bar cutoffs in the positive moment region of continuous beams are to be explained by reference to Fig. 6.13.1. Assume that it is desired to cut two

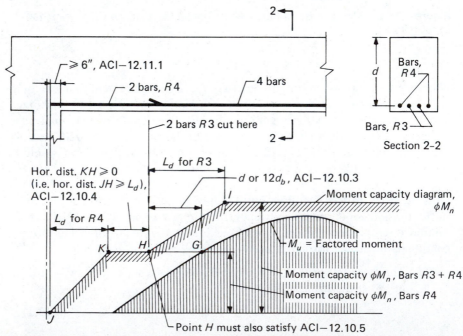

Figure 6.13.1   Bar cutoff in positive moment region of continuous beams.

of the four bars that are used for resisting the maximum positive factored moment M_u. The area provided by bars $R4$ must exceed one-fourth (one-third for simple spans) of the total reinforcement area provided for positive moment, in accordance with ACI-12.11.1. In a general situation the horizontal distance between points J and H is greater than the development length L_d for bars $R4$; the special case wherein this is not so will be discussed later in the section.

Point G is first located by computing the moment capacity ϕM_n of the continuing bars $R4$. The cutoff point H must lie to the left of point G at least 12 diameters of bars $R3$ or the effective depth d, whichever is larger. Point I is then located horizontally to the right from point H a distance equal to the development length L_d for the bars $R3$. Point J is located at the end of the bars $R4$, and point K is located horizontally to the right of point J, a distance equal to the development length L_d for the bars $R4$. The cutoff at point H must satisfy ACI-12.10.3 by giving a moment capacity ϕM_n diagram that is offset horizontally from the factored moment diagram at every point (except at the support) by 12 bar diameters or the effective depth d, whichever is greater.

It is to be noted that whereas ACI-12.11.1 requires $R4$ bars to extend into the support a distance of 6 in. minimum, ACI-12.11.2 requires full development of the tensile yield strength of such bars at the face of support when the flexural member is part of the primary lateral load resisting sys-

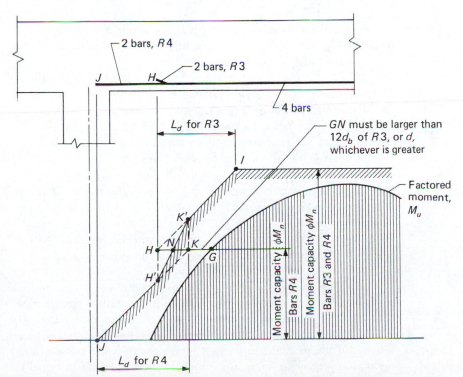

Figure 6.13.2 Special case of Fig. 6.13.1 wherein the distance JH is less than L_d for bars $R4$.

tem. Such would be the case if significant moments are developed in the member as a result of wind or earthquake loading.

The special case wherein the horizontal distance between points J and H is smaller than the development length L_d for bars $R4$ is shown in Fig. 6.13.2. The moment capacity ϕM_n diagram is indicated by $JH'K'I$. The requirement to allow for the shifting of the factored moment curve is satisfied by making sure that the distance GN is larger than the effective depth d or 12 diameters of bars $R3$.

In addition, since the cutoff at point H lies in the tension zone, one of the conditions of ACI-12.10.5 must be satisfied [see Eqs. (6.12.1) through (6.12.4)].

6.14 BAR CUTOFFS IN UNIFORMLY LOADED CANTILEVER BEAMS

Bar cutoffs in uniformly loaded cantilever beams can best be discussed by examining Fig. 6.14.1. Assume that of the six bars provided for the maximum factored moment M_u, it is desired to cut two bars $R5$ as soon as possible, then cut two more bars $R6$, and run the remaining two bars $R7$ out to the end of the cantilever.

The following steps illustrate the cutoff determination and check:

1. Locate the theoretical cutoff point A where the moment capacity ϕM_n of four bars (bars $R6$ and $R7$) is adequate; extend 12 diameters of bars $R5$ or the effective depth of the member, whichever is greater, to arrive at point B.
2. Determine the development length L_d for the bars $R5$ that are intended to be cut. Full capacity from the $R5$ bars is available at the distance L_d to the left of the cutoff point.
3. Since the horizontal distance CB is less than L_d, less than full capacity of the $R5$ bars will be available at the support had they been cut at point B; therefore extend cutoff location to point D so that the horizontal distance CD equals L_d. (ACI-12.10.5 must also be satisfied since the proposed cut location lies in a tension zone.)
4. Locate point E, the theoretical location where only the two $R7$ bars are required for moment; extend 12 diameters of bars $R6$ or the effective depth of the member, whichever is greater, to arrive at point F. (ACI-12.10.5 must also be satisfied for point F since it lies in a tension zone.)
5. Determine the development length L_d required for the $R6$ bars being cut; these two bars will have their full capacity available at point G, a distance L_d to the left of point F.
6. In the region from G to D the moment capacity consists of partial contributions from the bars $R5$ and $R6$ plus the full contribution of bars $R7$. The combined moment capacity (represented by line IH) is the sum of the linear contributions. The intent of ACI-12.10.3 is satisfied if the moment capacity diagram encloses the factored moment diagram by an adequate amount of horizontal offset equal to the effective depth of the member or 12 bar diameters (of the cutoff bars), whichever is greater. At a support or the end of a cantilever,

this horizontal offset need not be provided because the factored moment curve cannot shift left or right at such locations. In this case, the line *IH* passes by chance through point *B*, satisfying the offset requirement. Point *H* should also have the necessary horizontal offset.

7. When any part of the line *IH* encroaches on the factored moment M_u diagram closer than $12d_b$ of the *R*5 bars or *d*, whichever is greater, then either or both of the proposed cut locations (points *D* and *F*) must be extended toward the end of the cantilever. A convenient and practical procedure is to extend point *F* to the right until point *G* coincides with point *D*.

EXAMPLE 6.14.1 For the cantilever beam shown in Fig. 6.14.2 determine the distance L_1 from the support to the point where 2-#8 bars may be cut off. Assume the #4 stirrups shown (solid; not the dashed ones) have been preliminarily designed. Draw the resulting moment capacity ϕM_n diagram for the entire beam. Use $f'_c = 3000$ psi and $f_y = 60,000$ psi.

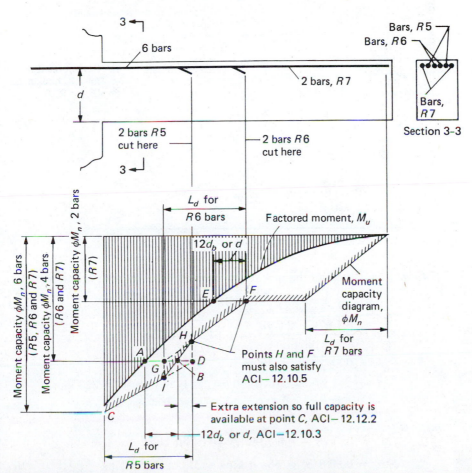

Figure 6.14.1 Bar cutoff in a uniformly loaded cantilever beam.

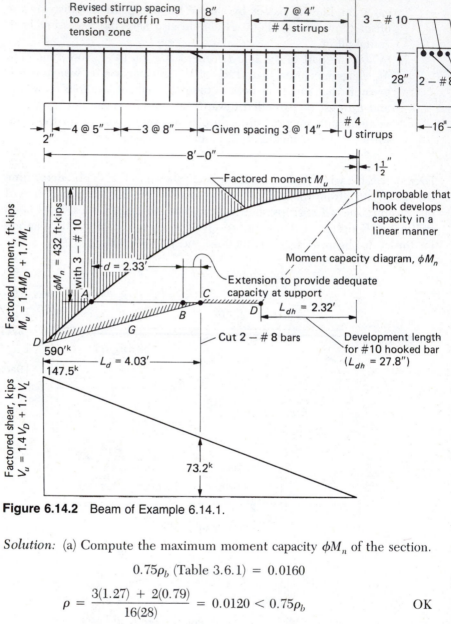

Figure 6.14.2 Beam of Example 6.14.1.

Solution: (a) Compute the maximum moment capacity ϕM_n of the section.

$$0.75\rho_b \text{ (Table 3.6.1)} = 0.0160$$

$$\rho = \frac{3(1.27) + 2(0.79)}{16(28)} = 0.0120 < 0.75\rho_b \qquad \text{OK}$$

$$C = 0.85(3)16a = 40.8a$$

$$T = [3(1.27) + 2(0.79)]60 = (3.81 + 1.58)60 = 323 \text{ kips}$$

$$a = \frac{323}{40.8} = 7.92 \text{ in.}$$

$$M_n = 323[28 - 0.5(7.92)]\tfrac{1}{12} = 648 \text{ ft-kips}$$

$$\phi M_n = 0.90(648) = 584 \text{ ft-kips} \approx M_u = 590 \text{ ft-kips} \qquad \text{OK}$$

(b) Determine the theoretical cutoff point for 2-#8 bars. The moment capacity ϕM_n remaining with 3-#10 bars is

$$C = 40.8a$$
$$T = 3.81(60) = 229 \text{ kips}$$
$$a = \frac{229}{40.8} = 5.60 \text{ in.}$$
$$\phi M_n = 0.90(229)[28 - 0.5(5.60)]\tfrac{1}{12} = 432 \text{ ft-kips}$$

Plot on the factored moment M_u diagram and locate the theoretical cutoff point A. Extend to the right 12 bar diameters (of the #8 bars that are to be cut) or the effective depth of the member, whichever is greater, to arrive at point B.

$$d = 28 \text{ in. } (2.33 \text{ ft}) > [12d_b = 12(1.0) = 12 \text{ in.}]$$

(c) Determine development length for #8 bars. The basic development length is obtained by using Eqs. (6.7.1) or (6.7.2); thus

$$L_{db} = 0.04 \frac{f_y A_b}{\sqrt{f_c'}} = 0.04 \left(\frac{60,000(0.79)}{\sqrt{3000}} \right) = 34.6 \text{ in.}$$

or

$$L_{db} = 0.0004 f_y d_b = 0.0004(60,000)1.0 = 24 \text{ in.}$$

The larger value governs and agrees with the value given by Table 6.7.1. For top bars the development length must be increased by 40% (as per Table 6.8.1).

$$L_d \text{ (for #8)} = 1.4 L_{db} = 1.4(34.6) = 48.4 \text{ in. } (4.03 \text{ ft})$$

Since point B, the proposed cutoff point, lies only about 3.5 ft from the support, the #8 bars would not have full capacity at the support. Thus extend the proposed cutoff to point C which is located at L_d (for #8) from the support (satisfying ACI-12.1 and 12.12.2).

(d) Check ACI-12.10.5 for cutting bars at point C in the tension zone. The shear strength, including contribution of stirrups, is first computed. Using the simplified method of constant V_c,

$$V_c = 2\sqrt{f_c'}b_w d = 2\sqrt{3000}(16)(28)\tfrac{1}{1000} = 49.1 \text{ kips}$$

For the 14-in. spaced #4 stirrups in the vicinity of the potential cut point C,

$$V_s = \frac{A_v f_y d}{s} = \frac{2(0.20)(60)28}{14} = 48.0 \text{ kips}$$

The shear strength ϕV_n at point C is

$$\phi V_n = \phi(V_c + V_s) = 0.85(49.1 + 48.0) = 82.5 \text{ kips}$$

$$\text{percent stressed in shear} = \frac{V_u}{\phi V_n} = \frac{73.2}{82.5} = 89\% > 75\% \qquad \text{NG}$$

Even when only 50% of the moment strength ϕM_n is used by M_u, the percent stressed in shear cannot exceed 75% (see Condition 3, Eqs. 6.12.3 and 6.12.4). Try using one more 8-in. stirrup spacing to cover the potential cut at point C, and see whether or not Condition 1, Eq. (6.12.1), is satisfied.

$$V_s = 48.0 \left(\frac{14}{8}\right) = 84.0 \text{ kips}$$

$$\text{percent stressed in shear} = \frac{73.2}{0.85(49.1 + 84)} = 65\% \qquad \text{OK}$$

This is acceptable since the two-thirds limit of ACI-12.10.5.1 (Eq. 6.12.1) is now satisfied.

(e) Check whether the continuing #10 bars have adequate embedment to the right of point C.

$$L_d \text{ (for \#10)} = L_d \text{ (for \#8)} \frac{A_b \text{ (\#10)}}{A_b \text{ (\#8)}}$$

$$= 48.4 \left(\frac{1.27}{0.79}\right) = 77.7 \text{ in. (6.47 ft)}$$

Straight embedment will give a moment capacity ϕM_n diagram encroaching closer to the factored moment M_u diagram than 12 bar diameters or the effective depth of the member. The #10 bars would satisfy literally the statement of ACI-12.10.4, which requires "Continuing reinforcement shall have an embedment length not less than the development length L_d beyond the point where bent or terminated tension reinforcement is no longer required to resist flexure." In other words, the distance from point A to the end of the cantilever must be at least L_d (for #10). The authors believe in a somewhat more conservative approach, requiring the moment capacity ϕM_n diagram to have an offset from the factored moment M_u diagram, except at or near a simple support or the free end of a cantilever, equal to 12 bar diameters or the effective depth d, whichever is greater.

In this case, try standard 90° hooks (see Fig. 6.11.1) on the ends of the #10 bars. Since the beam has the usual 1.5-in. clear cover and #4 stirrups, the cover to the hooked bars is 2 in., which is less than the $2\frac{1}{2}$ in. required by ACI-12.5.4; thus, the special provisions of that Code section must be satisfied.

The development length L_{dh} for the #10 hooked bar is the basic value L_{hb} (i.e., no modification to L_{hb} applies) given by Eq. (6.11.1) and Table 6.11.2. Thus, for #10 hooked bar,

$$L_{dh} = L_{hb} = \frac{1200 d_b}{\sqrt{f_c'}} = \frac{1200(1.27)}{\sqrt{3000}} = 27.8 \text{ in.}$$

which exceeds the minimum $8d_b$ or 6 in., whichever is greater. The L_{dh} of 27.8 in. is dimensioned from the outside face of the tail of the hook, as shown in Fig. 6.14.2. The 27.8-in. embedment required to develop the strength of the #10 bar is much less than what was required by the 1977 ACI Code. Of course, here stirrups spaced at not more than $3d_b$ (4.23 in.)

must be provided along the 27.8 in. of development distance, in accordance with ACI-12.5.4.

(f) Moment capacity ϕM_n diagram. The full strength ϕM_n for the beam with 3-#10 hooked bars will be available at 27.8 in. from the outside of the hook on the end of the beam. Assuming 1.5 in. cover, full capacity is available at 29.3 in. (2.44 ft) from end of beam (point D). The dashed line in Fig. 6.14.2 has been drawn from zero strength at the end of the hook to full strength $\phi M_n = 423$ ft-kips at 2.44 ft from end of beam; however, it is *not* intended to imply that the hooked bar develops its strength linearly since that is highly improbable.

(g) Final decision. Cut 2-#8 bars at 4 ft–0 in. from the support; use 90° standard hooks on the 3-#10 bars; use 8-#4 U stirrups as confinement over the L_{dh} distance, as shown in Fig. 6.14.2 by the dashed stirrups.

The use of #10 bars in this cantilever beam is not a practical design but did serve to illustrate the need for extending the cut location from B to C and the need for and treatment of hooked bars.

6.15 DEVELOPMENT OF REINFORCEMENT AT SIMPLE SUPPORTS AND AT POINTS OF INFLECTION

The concept of requiring the development of reinforcement on both sides of a section where the bars are to be fully stressed may also be applied to the continuation of positive moment tension reinforcement beyond either the centerline of a simple support or a point of inflection.

Simple Supports. Referring to Fig. 6.15.1, consider the point A on the factored moment M_u curve near a simple support, where the factored moment M_u equals the moment capacity ϕM_n of the bars continuing into the support. The distance from point A to the end of the bars must be at least equal to the required development length L_d as computed from ACI-12.2 [Eqs. (6.7.1) to (6.7.4) with modifications of Sec. 6.8]. This requirement given in ACI-12.11.3 is that the available embedment length must equal or exceed L_d, or

$$\text{available embedment length} = L_a + 1.3\frac{M_n}{V_u} \geq L_d \qquad (6.15.1)$$

where

M_n = nominal flexural strength of the remaining bars

$\quad = A_s f_y \left(d - \dfrac{a}{2} \right)$, or

$\quad$ = moment capacity in the working stress method (ACI-Appendix B.4 and B.5),

V_u = factored shear at the support, or

$\quad$ = the service load shear in the working stress method (ACI-Appendix B.4),

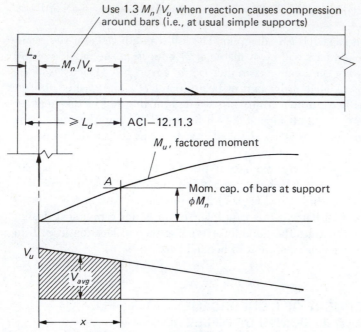

Use 1.3 M_n/V_u when reaction causes compression around bars (i.e., at usual simple supports)

L_a

M_n/V_u

$\geqslant L_d$ ACI–12.11.3

M_u, factored moment

A

Mom. cap. of bars at support ϕM_n

V_u

V_{avg}

x

Figure 6.15.1 Development of reinforcement at a simple support.

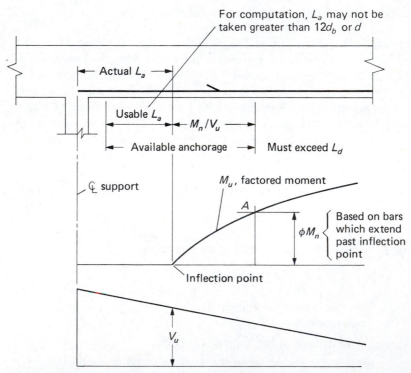

For computation, L_a may not be taken greater than $12d_b$ or d

Actual L_a

Usable L_a

M_n/V_u

Available anchorage

Must exceed L_d

$\mathcal{C}$ support

M_u, factored moment

A

ϕM_n { Based on bars which extend past inflection point

Inflection point

V_u

Figure 6.15.2 Development of reinforcement at an inflection point.

L_a = the straight embedment length beyond the centerline of support to the end of the bars [when the bars are hooked, Eq. (6.15.1) is considered automatically satisfied].

The distance x between the point A and the centerline of support in Fig. 6.15.1 is approximately equal to

$$x = \frac{M_n}{V_u} \qquad (6.15.2)$$

because the area of the shaded portion of the shear diagram equals the *change* in moment between the center of support and point A; thus the ordinate on the factored moment diagram at A is

$$V_{avg} \text{ (say, } 0.9\, V_u)x = \phi M_n$$

Comparing Eq. (6.15.1) and Eq. (6.15.2), it is seen that Eq. (6.15.1) is identical to

$$L_a + 1.30x \geq L_d$$

wherein the 1.30 factor is in recognition of the fact that the bars extending into a simple support have less tendency to cause splitting when confined by a compressive reaction [29].

The 1983 ACI Code exempts the requirement of ACI-12.11.3 "for reinforcement terminating beyond centerline of simple supports by a standard hook, or a mechanical anchorage at least equivalent to a standard hook."

Inflection Points. Since an inflection point is a point of zero moment located away from a support (refer to Fig. 6.15.2), bars in that region are not confined by a compressive reaction; therefore the 1.30 factor is interpreted as not to apply. In this case the embedment length that must exceed the required development length L_d (ACI-12.11.3) may be stated as

$$\text{available embedment length} = \begin{bmatrix} \text{actual } L_a, \text{ but} \\ \text{not exceeding the} \\ \text{larger of } 12d_b \text{ or } d \end{bmatrix} + \frac{M_n}{V_u} \geq L_d \quad (6.15.3)$$

wherein M_n and V_u refer to the nominal flexural strength and the factored shear at the point of inflection. The limitation of the usable L_a to 12 bar diameters or the effective depth has been applied because there is no experimental evidence to show that long anchorage length will be fully effective in developing a bar in a short length between the point of inflection and a point of maximum stress [30].

Additional development of reinforcement at the face of support is required by ACI-12.11.2 when the flexural member is part of the primary lateral load resisting system.

6.16 DEVELOPMENT OF SHEAR REINFORCEMENT

Reinforcement in the web of a beam, whether it be for shear or for torsion (see Chap. 19), must be properly anchored so that its full tensile capacity is available at or near the middepth of a beam. For proper function, the

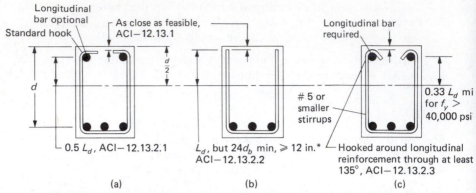

*The 12-in. minimum does not apply to welded deformed wire fabric satisfying ACI-12.7.2

Figure 6.16.1 Development of deformed bar or deformed wire stirrups.

shear reinforcement must be "carried as close to compression and tension surfaces of member as cover requirements and proximity of other reinforcement will permit" (ACI-12.13.1). It is especially important to extend the stirrups as close to the compression face as possible because the flexural tension cracks may extend deeply into the compression zone when the nominal strength of the member is approached.

The ends of single leg, simple U, or multiple U stirrups of deformed bars or deformed wire fabric must be anchored by one of the means shown in Fig. 6.16.1. Such stirrups may be inclined but in accordance with ACI-11.5.1.2 the angle between the stirrups and the longitudinal bars must be at least 45°.

The standard hooks for stirrups and ties have somewhat relaxed requirements as compared with the standard hooks used for main bars as shown in Fig. 6.11.1. Also hooked stirrups and hooked ties are not permitted for bars larger than #8. As noted in Chap. 5, stirrups and ties are usually #3 or #4 bars, occasionally #5 or #6, and rarely, if ever, larger than #8, so the 1983 ACI Code limit on size of hooked stirrups should rarely apply. Note also that the 180° hook is not used for stirrups and ties. The requirements for standard hooks at ends of ties and stirrups are shown in Fig. 6.16.2.

When *closed* stirrups are desired, one practical procedure is to use a pair of U stirrups without hooks (Fig. 6.16.1b) placed to form a closed unit. If this is done, ACI-12.13.5 requires laps of $1.7L_d$ for proper splicing. When members are at least 18 in. deep and the tensile capacity $A_b f_y$ of the stirrup does not exceed 9 kips, splices are adequate if the legs extend the full available depth of the member.

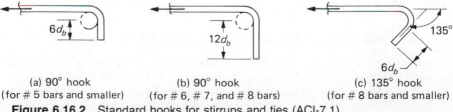

(a) 90° hook (for # 5 bars and smaller)   (b) 90° hook (for # 6, # 7, and # 8 bars)   (c) 135° hook (for # 8 bars and smaller)

Figure 6.16.2 Standard hooks for stirrups and ties (ACI-7.1).

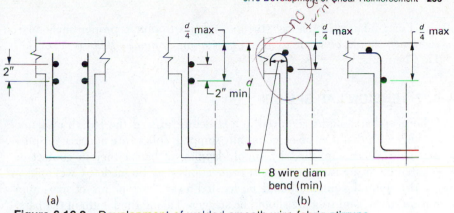

no always turn inwards

$\frac{d}{4}$ max $\qquad$ $\frac{d}{4}$ max $\qquad$ $\frac{d}{4}$ max

2"

d

2" min

8 wire diam
bend (min)

(a) (b)

Figure 6.16.3 Development of welded smooth wire fabric stirrups
(ACI-12.13.2.4).

For U stirrups of welded smooth wire fabric, anchorage may be accomplished (ACI-12.13.2.4) using either "(a) Two longitudinal wires spaced at a 2-in. spacing along the member at the top of the U," or "(b) One longitudinal wire located not more than $d/4$ from the compression face and a second wire closer to the compression face and spaced not less than 2 in. from the first wire. The second wire may be located on the stirrup leg beyond a bend, or on a bend with an inside diameter of bend not less than $8d_b$." These provisions are illustrated in Fig. 6.16.3.

For *single-leg* stirrups of welded smooth or deformed wire fabric, the 1983 ACI Code has included (ACI-12.13.2.5) the recommendations [31] of the PCI/WRI Ad Hoc Committee on Welded Wire Fabric for Shear Reinforcement. Single leg stirrups are practical and common in precast prestressed concrete T-sections. The provisions of ACI-12.13.2.5 are detailed in Fig. 6.16.4.

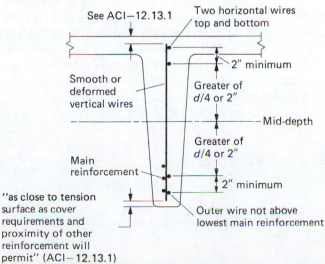

See ACI—12.13.1

Two horizontal wires
top and bottom

2" minimum

Smooth or
deformed
vertical wires

Greater of
$d/4$ or 2"

Mid-depth

Greater of
$d/4$ or 2"

Main
reinforcement

2" minimum

"as close to tension
surface as cover
requirements and
proximity of other
reinforcement will
permit" (ACI—12.13.1)

Outer wire not above
lowest main reinforcement

Figure 6.16.4 Development of single-leg smooth or deformed welded wire fabric shear reinforcement (ACI-12.13.2.5) (adapted from Ref. 31).

For guidance in placing stirrups and ties to obtain proper anchorage, reference should be made to the *ACI Detailing Manual—1980* [p. 33 of Ref. 17, Chap. 2].

6.17 TENSION LAP SPLICES

Wherever bar lengths required in a structure exceed the length available or the length that may be economically shipped, splices are necessary. Splicing may be accomplished by simple lapping of bars either in contact or separated. As an alternative, butt connections may be made by welding.

In general, splices should be located away from points of maximum tensile stress, and splicing should be staggered along the length of the bars. In other words all of the bars should not be spliced at one location.

The beam with splices should be as ductile as one without splices. The ACI Code provisions are intended to assure that no splice failure will occur when the full nominal strength in flexure is reached at the spliced location. Requirements for minimum clear spacing of *contact* splices (ACI-7.6.4) are to ensure adequate amount of concrete for the development of full anchorage capacity; but in *noncontact* lap splices the individual bars should *not* be spaced transversely too far apart (ACI-12.14.2.3).

The overlap distance required in tension lap splices should be equal to or larger than the development length L_d of the bar because stress concentrations near the splice ends tend to produce splitting at early stages of loading unless special precautions are taken [4,32,33]. Classes A, B, and C tension lap splices are defined by ACI-12.15.1 to have overlap distances of $1.0L_d$, $1.3L_d$, and $1.7L_d$, respectively, but a total lap of not less than 12 in. The class of tension lap splice to be used depends on (1) the percentage of bars being spliced of the total number within the required lap distance, and (2) the stress level in the unspliced bars at the splice location.

A summary of the requirements for the three classes of tension lap splices appears in Table 6.17.1. The provisions apply equally to deformed bar or deformed wire splices. Tension lap splices may not be used for bars larger than #11 (ACI-12.14.2.1). As stated previously in Sec. 6.10, the lap lengths prescribed in Table 6.17.1 shall be increased 20% for a three-bar bundle and 33% for a four-bar bundle (ACI-12.14.2.2). Bars spliced by noncontact lap splices in flexural members shall not be spaced transversely farther apart than one-fifth of the required lap length nor 6 in. (ACI-12.14.2.3).

The required overlap distances shown in Table 6.17.1 are determined

Table 6.17.1 TENSION LAP SPLICES (ACI-12.15)

$\left(\dfrac{A_s \text{ REQUIRED}}{A_s \text{ PROVIDED}}\right)$ AT THE SPLICE	PERCENT OF A_s SPLICED	SPLICE CLASS	REQUIRED LAP	NOTES
≤0.5	≤75	A	L_d	Desirable
	>75	B	$1.3L_d$	OK
>0.5	≤50	B	$1.3L_d$	OK
	>50	C	$1.7L_d$	Avoid if possible

by the basic development length requirements of ACI-12.2.2 multiplied by the modification factors (ACI-12.2.3 and 12.2.4) shown previously in Table 6.8.1, but the resulting lap requirement may not be less than 12 in.

The ratio $(A_s$ required)/$(A_s$ provided) column in Table 6.17.1 refers to the percentage of available capacity that is utilized. The ratio may also be considered as the percent of f_y to which the bars are stressed. When the factored moment M_u is only 50% of the moment capacity ϕM_n, the ratio would be considered 0.5. In general, temperature, shrinkage, and load distribution reinforcement should be considered as fully stressed for the purpose of designing splices.

Though not yet adopted for the ACI Code, recent studies [4,32] on splices have shown that, in general, splice lengths may be reduced from those required by ACI-12.15. The recommendations [4,6] of ACI Committee 408 are under study by ACI Committee 318.

Members Under Compression and Bending. For compression members there are two categories of special splice provisions (ACI-12.17):

1. Design in the upper part of the "compression-controls" region such that (Fig. 13.6.2) the reinforcement at the face of the member *opposite* the face subject to maximum compression has (when strength is reached) a tensile stress of $0.5f_y$ or less (it may be compressive): (a) Any type of splice may be used appropriate to the stress acting when strength is reached. (b) A minimum tensile strength from the splices in combination with any continuing unspliced bars must be provided in each face of the member; this minimum strength must be twice the calculated tension in the face of the member but not less than one-quarter of the maximum tensile strength $A_s f_y$ of all bars in that face of the member.

2. Design in the lower part of "compression controls" region or in "tension controls" region (Fig. 13.6.2) where the reinforcement at the face of the member *opposite* the face subject to maximum compression has (when strength is reached) a tensile stress greater than $0.5f_y$: (a) Lap splices (Class B or C) may be used and must develop the full yield stress f_y. (b) Full welded splices or full positive connections may be used (see Sec. 6.18).

The analysis of compression members is treated in Chap. 13. For use with compression member analysis the ratio $(A_s$ required)/$(A_s$ provided) may also be considered as a percentage of f_y acting on the bars when strength of the section is reached. The stress in the tension bars is that obtained from a strength analysis, satisfying compatibility of stress and strain at each bar or layer of bars.

6.18 WELDED TENSION SPLICES AND MECHANICAL CONNECTIONS

A welded tension butt splice or mechanical connection is used in situations where large tensile forces are to be transmitted across the splice or large bars need to be spliced and the lap splice may be impractical or prohibited. Bars larger than #11 may not be lap spliced to carry tension (ACI-12.14.2.1). Tension tie members also may not be lap spliced (ACI-12.15.5). A tension tie is a member (a) carrying a tensile force large enough to cause tension over the entire section; (b) having a stress level in the reinforcement high enough to require every bar to be fully effective; and (c) having limited concrete cover on its sides [30].

These tension splices, referred to as *full welded splices* or *full mechanical connections* (ACI-12.14.3), are required to develop in tension at least 125% of the specified yield strength of the bar when used in regions of high stress.

The full welded splice is intended primarily for relatively large bars (#6 and larger) in main members (ACI Commentary-12.14.3.3). The tensile capacity required is intended to ensure sound full penetration welds—that is, to produce splices capable of developing the strength $A_s f_y$ of the bars spliced. According to the ACI Commentary-12.14.3.3 [30], "the 25 percent increase above the specified yield strength was selected as both an adequate minimum for safety and a practical maximum for economy."

In regions of low stress (i.e., where less than 50% of available capacity is being used) welded splices or mechanical connections of less capacity than 125% are permitted (ACI-12.14.3.5 and 12.15.4). In these situations welded lap joints of reinforcing bars, either with or without backup material, or other welded bar arrangements, may be allowed. However, such splices must be "staggered at least 24 in. and in such manner as to develop at every section at least twice the calculated tensile force at that section but not less than 20,000 psi for total area of reinforcement provided."

In computing the capacity, spliced bars may be rated at the specified splice strength (assuming it is less than the strength of the bars), whereas any unspliced bars are to be rated in proportion to their length of bar development to the splice point (but not to exceed the maximum bar capacity at embedment L_d). For example, a welded splice may be specified to provide 90% of the full capacity of the bars being spliced, in which case that bar area is taken as 90% of its actual area for computing capacity; for unspliced bars that have been embedded, say only $0.5L_d$, the bar area used for strength calculation should be taken as one-half the actual area.

6.19 COMPRESSION LAP SPLICES

Whereas bars larger than #11 acting in tension are not permitted to be lap spliced; however, in *compression* #14 and #18 bars may be lap spliced to #11 and smaller bars (ACI-12.16.2).

The minimum overlap in compression lap splices when f'_c is not less than 3000 psi must be at least equal to the following* (ACI-12.16.1):

*For SI, ACI 318-83M, for lap, L_d, and d_b in mm, f'_c and f_y in MPa, gives

For $f_y \leq 400$ MPa,

$$\text{lap} = 0.073 \, f_y d_b \qquad \text{or } L_d \qquad \text{or } 300 \text{ mm}$$

For $f_y > 400$ MPa,

$$\text{lap} = (0.13 f_y - 24) d_b \qquad \text{or } L_d \qquad \text{or } 300 \text{ mm}$$

where

$$L_d = \frac{0.24 f_y d_b}{\sqrt{f'_c}} \leq 0.044 f_y d_b \leq 200 \text{ mm}$$

Table 6.19.1 BAR DIAMETERS REQUIRED
FOR COMPRESSION LAP SPLICES FOR
$f_c' \geq 3000$ PSI (ACI-12.16.1)

YIELD STRESS f_y (ksi)	BAR DIAMETERS[a]		
	SPIRAL COLUMN	TIED COLUMN	OTHERS
40	15	16.6	20
50	18.75	20.75	25
60	22.5	24.9	30
75	32.6	36.2	43.5
80	36.0	39.9	48.0

[a]When computing splice length, the minimum to be used is 12 in.

For $f_y \leq 60,000$ psi,

$$\text{lap} = 0.0005 f_y d_b \quad \text{or } L_d \quad \text{or 12 in.} \quad \text{(whichever is largest)}$$

For $f_y > 60,000$ psi,

$$\text{lap} = (0.0009 f_y - 24) d_b \quad \text{or } L_d \quad \text{or 12 in.} \quad \text{(whichever is largest)}$$

where from ACI-12.3,

$$L_d = \frac{0.02 f_y d_b}{\sqrt{f_c'}} \leq 0.0003 \, f_y d_b \leq 8 \text{ in.}$$

less 25% for spiral enclosure satisfying ACI-12.3.3.2. When f_c' is less than 3000 psi, the lap length is to be increased by one-third. When lapping bars of two different sizes, the lap splice length is to be the larger of (1) the compression splice length of the smaller bar, or (2) the compression development length L_d of the larger bar.

For compression members whose main steel is surrounded by closed ties throughout the lap length, the required lap may be taken at 0.83 of that otherwise required, but not less than 12 in. A minimum percentage of column tie area is also required by ACI-12.16.2. Column ties are discussed in Chap. 13 (Sec. 13.8).

For members whose main steel is surrounded by a closely wound spiral, the required lap may be taken at 0.75 of that otherwise required, but not less than 12 in. Spiral reinforcement is discussed in Chap. 13 (Sec. 13.9).

The number of bar diameters required for the overlap in compression lap splices is summarized in Table 6.19.1.

6.20 COMPRESSION END BEARING CONNECTIONS, WELDED SPLICES, AND MECHANICAL CONNECTIONS

End bearing connections are allowed for compression only, wherein the load in the bars is transmitted by bearing of square cut ends held in concentric contact by a suitable device. According to ACI-12.16.5.2 bar ends must terminate in flat surfaces within $1\frac{1}{2}°$ of right angles to the axis of the bars and be fitted within 3° of full bearing after assembly. End bearing splices are only permitted when the member contains closed ties, closed stirrups, or spirals.

When welded splices or mechanical connections are used in compression, the requirements are the same as for tension splices—that is, the development of 125% of the yield strength of the bars, except where less than 50% of the full unspliced bar capacity is required by the design load (ACI-12.14.3 and 12.15.4).

6.21 DESIGN EXAMPLES

Two complete examples in the design of reinforced concrete flexural members are presented here for the purpose of showing the design for flexure, shear, and development of reinforcement, all in the same beam.

EXAMPLE 6.21.1 Design the simply supported beam shown in Fig. 6.21.1(a). The dead load is 0.9 kip/ft, not including the weight of the beam. The live load consists of a concentrated load of 13 kips at midspan. Use $f'_c = 3000$ psi, $f_y = 40,000$ psi, and the ACI strength method.

Solution: (a) Design for flexure. Assume that a rectangular section with tension reinforcement only will be used at a reinforcement ratio somewhat lower than the maximum permissible value. Using basic principles as illustrated in Sec. 3.6, or the value from Table 3.6.1,

$$\text{max } \rho = 0.75\rho_b = 0.0278$$

Arbitrarily selecting an approximate $\rho = 0.025$ and using Eq. (3.8.4) or the direct statics as illustrated in Sec. 3.8,

$$m = \frac{f_y}{0.85f'_c} = 15.7$$

$$R_n = \rho f_y(1 - \tfrac{1}{2}\rho m) = 804 \text{ psi}$$

Assume weight of beam is 0.2 kip/ft,

$$w_u = 1.4(0.9 + 0.2) = 1.54 \text{ kips/ft (dead load)}$$

$$W_u = 1.7(13) = 22.1 \text{ kips (live load)}$$

$$M_u = \tfrac{1}{8}(1.54)(20)^2 + \tfrac{1}{4}(22.1)(20) = 77 + 110.5 = 188 \text{ ft-kips}$$

$$\text{required } M_n = \frac{M_u}{\phi} = \frac{188}{0.90} = 209 \text{ ft-kips}$$

$$\text{required } bd^2 = \frac{M_n}{R_n} = \frac{209(12,000)}{804} = 3120 \text{ in.}^3$$

If $b = 12$ in.

$$d = \sqrt{\frac{3120}{12}} = 16.1 \text{ in.}$$

$$\text{required } h = d + \text{approx } 2\tfrac{1}{2} \text{ in. for one layer of bars}$$

$$= 16.1 + 2.5 = 18.6 \text{ in.}$$

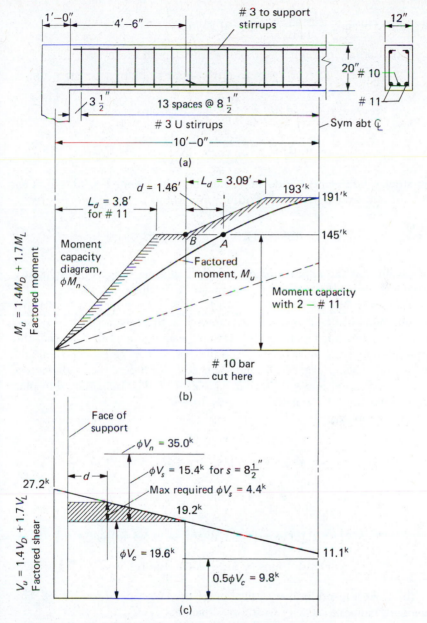

Figure 6.21.1 Simply supported beam of Example 6.21.1.

The minimum thickness for deflection control is (ACI-Table 9.5a),

$$\min h = \frac{L}{16} = \frac{20(12)}{16} = 15 \text{ in.}$$

if the member is *not* supporting or attached to partitions or other construction likely to be damaged by large deflections. Use $h = 20$ in., thus $d =$ approx 17.5 in.

$$\text{weight of beam} = \frac{12(20)}{144}(0.15) = 0.25 \text{ kip/ft}$$

$$\text{revised } w_u = 1.4(0.9 + 0.25) = 1.61 \text{ kips/ft (dead load)}$$

$$\text{revised } M_u = \tfrac{1}{8}(1.61)(20)^2 + 110.5 = 80.5 + 110.5 = 191 \text{ ft-kips}$$

$$\text{revised required } M_n = \frac{191}{0.90} = 212 \text{ ft-kips}$$

$$\text{required } R_n = \frac{M_n}{bd^2} = \frac{212(12,000)}{12(17.5)^2} = 692 \text{ psi}$$

The steel reinforcement ratio may then be found from Eq. (3.8.5), Fig. 3.8.1, or approximately by straightline proportion,

$$A_s \approx 0.025(12)(17.5)\left(\frac{692}{804}\right) = 4.52 \text{ sq in.}$$

Note again that the approximation using straightline proportion is slightly on the safe side when revised R_n is smaller, as in this case.

Try 2-#11 and 1-#10 bars ($A_s = 4.39$ sq in.). Assuming #3 U stirrups, the minimum width of beam to accommodate these bars is 10.8 in. (see Table 3.9.2) which is less than the width of beam being used. Note that the entry of Table 3.9.2 must be with the two corner bars (#11) and then add for the one smaller bar (#10), and then add the difference in diameters between the larger and smaller bars, giving 10.64 in. from the table plus 0.14 in., making a total of 10.78 in. (say 10.8 in.).

Check capacity.

$$C = 0.85f'_cba = 0.85(3)12a = 30.6a$$

$$T = A_s f_y = 4.39(40) = 176 \text{ kips}$$

$$a = \frac{176}{30.6} = 5.75 \text{ in.}$$

$$M_n = T\left(d - \frac{a}{2}\right) = 176[17.5 - 0.5(5.75)]\tfrac{1}{12}$$

$$= 214 \text{ ft-kips} > 212 \text{ ft-kips required} \qquad \text{OK}$$

(b) Make the preliminary selection of the cutoff point for 1-#10. The remaining moment capacity with 2-#11 bars is

$$C = 30.6a$$

$$T = 2(1.56)40 = 125 \text{ kips}$$

$$a = \frac{125}{30.6} = 4.08 \text{ in.}$$

$$\phi M_n = 0.90(125)[17.5 - 0.5(4.08)]\tfrac{1}{12} = 145 \text{ ft-kips}$$

The value of 145 ft-kips is plotted on the factored M_u diagram to locate the theoretical cutoff point A in Fig. 6.21.1. The actual potential cutoff location (point B) is found by extending from point A toward the support a distance of 12 bar diameters or the effective depth of the member, whichever is greater (ACI-12.10.3).

$$12d_b = 12 \left(\frac{1.27}{12}\right) = 1.27 \text{ ft}$$

$$d = \frac{17.5}{12} = 1.46 \text{ ft} \qquad \text{(Controls)}$$

The cutoff at point B will be acceptable only if the shear does not exceed two-thirds of the shear strength at point B (ACI-12.10.5.1). An alternative is to provide extra stirrups in accordance with ACI-12.10.5.2.

(c) Determine development lengths. Using Eq. (6.7.1) or ACI-12.2 gives

$$L_{db} (\#10) = 0.04 \frac{A_b f_y}{\sqrt{f_c'}} = 0.04(1.27) \frac{40,000}{\sqrt{3000}} = 37.1 \text{ in.} \qquad \text{(Controls)}$$

but not less than

$$L_{db} (\#10) = 0.0004 d_b f_y = 0.0004(1.27)(40,000) = 20.3 \text{ in.}$$

Thus,

$$L_d (\#10) = L_{db} = 37.1 \text{ in.} \ (3.09 \text{ ft})$$

$$L_d (\#11) = 37.1 \left(\frac{1.56}{1.27}\right) = 45.5 \text{ in.} \ (3.80 \text{ ft})$$

(d) Design of shear reinforcement; simplified method with constant V_c.

$$V_u \text{ (at centerline of support)} = 1.61(10) + 11.1 = 27.2 \text{ kips}$$

$$V_u \text{ (at } d \text{ from face of support)} = 27.2 - 1.96(1.61) = 24.0 \text{ kips}$$

$$\phi V_c = \phi(2\sqrt{f_c'}b_w d) = 0.85(2\sqrt{3000})(12)(17.5)\tfrac{1}{1000} = 19.6 \text{ kips}$$

$$\text{required } \phi V_s = V_u - \phi V_c = 24.0 - 19.6 = 4.4 \text{ kips}$$

$$\text{min } \phi V_s = \phi(50 b_w d) = 0.85(50)(12)(17.5)\tfrac{1}{1000} = 8.9 \text{ kips}$$

Since required $\phi V_s = 4.4$ kips is less than min $\phi V_s = 8.9$ kips to satisfy the requirement of ACI-11.5.5.3 based on $v_s = 50$ psi, use min $\phi V_s = 8.9$ kips to determine the stirrup spacing at the critical section. For #3 U stirrups, using Eq. (5.10.7),

$$\phi V_s = \frac{\phi A_v f_y d}{s}$$

$$\text{max } s = \frac{\phi A_v f_y d}{\phi V_s} = \frac{0.85(0.22)(40)(17.5)}{8.9} = 14.7 \text{ in.}$$

However, the stirrup spacing may not exceed $d/2 = 8.75$ in.

Try #3 stirrups @ $8\tfrac{1}{2}$ in. spacing. Stirrups at this spacing must be used until $V_u \leq 0.5\phi V_c$ which for this beam [see Fig. 6.21.1(c)] means the entire span.

(e) Check cutoff point for satisfying the shear requirement of ACI-12.10.5 for cutting bars in the tension zone. The shear strength provided by #3 stirrups at $8\frac{1}{2}$ in. spacing is

$$\phi V_n = \phi V_c + \phi V_s = 19.6 + \frac{0.85(0.22)(40)(17.5)}{8.5} = 35.0 \text{ kips}$$

$$V_u \text{ at proposed cutoff point } B = 19.2 \text{ kips}$$

$$\text{percent stressed in shear} = \frac{V_u}{\phi V_n} = \frac{19.2}{35.0} = 55\% < 66\frac{2}{3}\% \qquad \text{OK}$$

Use 1-#10, 10′–0″ long placed symmetrically about midspan.

(f) Check development length requirement at the support. According to ACI-12.11.3, it is required that

$$1.30 \frac{M_n}{V_u} + L_a \geq L_d$$

$$M_n \text{ for 2-#11 bars} = \frac{145}{0.9} = 161 \text{ ft-kips}$$

V_u = factored shear at centerline of support = 27.2 kips

L_a = embedment length beyond the center of support; assume zero here

$$1.30 \frac{161(12)}{27.2} = 92.5 \text{ in.} > L_d = 45.5 \text{ in.} \qquad \text{OK}$$

Actually the moment capacity ϕM_n diagram provides this same check but more conservatively (i.e., without the 1.30 factor) since the horizontal distance from the center of support to point A exceeds L_d.

(g) Design sketch. The final conclusions are presented in Fig. 6.21.1(a). Since $f_y = 40,000$ psi, the crack control provisions of ACI-10.6.4 need not be checked. If deflection control is important to prevent damage to partitions or other construction, the deflection must be checked according to ACI-9.5. Computation for deflections is treated in Chap. 14.

EXAMPLE 6.21.2 Design the overhanging beam shown in Fig. 6.21.2. The superimposed service uniform dead load is 4 kips/ft. Use $f'_c = 4000$ psi, $f_y = 60,000$ psi, and the ACI strength method. (Note that live load is not used in this example because the purpose here is to show how flexure, shear, and development of reinforcement are to be considered together in one simple case.)

Solution: (a) Design for flexure. Since tension reinforcement will be required in the top of the overhang, it may be desirable to run some bars straight across the top of the entire beam. This would also help reduce creep and shrinkage deflection under sustained load.

For deflection control, a guideline value of ρ equal to one-half the maximum permissible value may serve to establish the beam size.

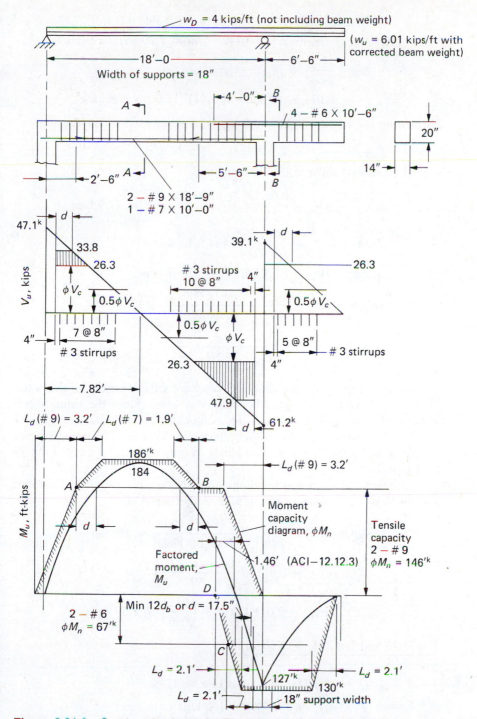

w_D = 4 kips/ft (not including beam weight)

(w_u = 6.01 kips/ft with corrected beam weight)

18'-0
Width of supports = 18"

6'-6"

4'-0"

A

4 – # 6 × 10'-6"

20"

14"

2'-6"   A

5'-6"

B

2 – # 9 × 18'-9"
1 – # 7 × 10'-0"

47.1^k

d

33.8
26.3

ϕV_c
0.5ϕV_c

39.1^k

d

26.3

3 stirrups
10 @ 8"   4"

0.5ϕV_c

0.5ϕV_c

V_u, kips

4"

7 @ 8"

3 stirrups

0.5ϕV_c
ϕV_c

5 @ 8"
4"

3 stirrups

7.82'

26.3

47.9

61.2^k

d

L_d (# 9) = 3.2' L_d (# 7) = 1.9'

186'^{k}
184

A

B

L_d (# 9) = 3.2'

M_u, ft-kips

d

d

Moment
capacity
diagram, ϕM_n

Factored
moment,
M_u

D

1.46' (ACI–12.12.3)

Tensile
capacity
2 – # 9
ϕM_n = 146'^{k}

2 – # 6
ϕM_n = 67'^{k}

Min 12d_b or d = 17.5"

C

L_d = 2.1'

L_d = 2.1'

127'^{k}

130'^{k}

L_d = 2.1'

18" support width

Figure 6.21.2 Overhanging beam for Example 6.21.2.

$$\text{max } \rho = 0.75 \, \rho_b = 0.0214 \qquad \text{(Table 3.6.1)}$$

Arbitrarily choose $\rho = 0.011$. $\qquad m = f_y/0.85f_c' = 17.6$.

$$R_n = \frac{M_u}{\phi b d^2} = \rho f_y(1 - \tfrac{1}{2}\rho m)$$

$$= 0.011(60,000)[1 - \tfrac{1}{2}(0.011)(17.6)] = 596 \text{ psi}$$

Estimating the beam weight at 0.3 kip/ft, the factored uniform loading is

$$w_u = 1.4(4.3) = 6.02 \text{ kips/ft}$$

The factored shear at the left support is

$$V_u = \frac{6.02(18)}{2} - \frac{0.5(6.02)(6.5)^2}{18} = 47.1 \text{ kips}$$

$$\text{max } (+)M_u = \frac{(47.1)^2}{2(6.02)} = 184 \text{ ft-kips}$$

$$\text{max } (-)M_u = \tfrac{1}{2}(6.02)(6.5)^2 = 127 \text{ ft-kips}$$

$$\text{required } bd^2 = \frac{M_u}{\phi R_n} = \frac{184(12,000)}{0.90(596)} = 4120 \text{ in.}^3$$

b	d	h
12	18.5	21
14	17.2	20

In selecting a size, the designer need not adhere rigidly to the value of $R_n = 596$ psi but may go either higher or lower, since the value of ρ is about in the middle of the permissible range. Use a rectangular section with $b = 14$ in. and $h = 20$ in., which gives $d \approx 17.5$ in. The revised beam weight is $14(20)(0.15)/144 = 0.29$ kip/ft, giving $w_u = 1.4(0.29 + 4.00) = 6.01$ kips/ft.

The factored shear and bending moment diagrams (i.e., using factored service loads) are shown in Fig. 6.21.2. At section A-A,

$$\text{required } R_n = \frac{M_u}{\phi b d^2} = \frac{184(12,000)}{0.90(14)(17.5)^2} = 572 \text{ psi}$$

$$A_s \approx 0.011bd \left(\frac{\text{required } R_n}{596} \right)$$

$$A_s = 0.011(14)(17.5) \left(\frac{572}{596} \right) = 2.59 \text{ sq in.}$$

Use 2-#9 and 1-#7 ($A_s = 2.60$ sq in.)

Check:

$$C = 0.85f_c'ba = 0.85(4)(14)a = 47.6a$$

$$T = A_s f_y = 2.60(60) = 156 \text{ kips}$$

$$a = 156/47.6 = 3.28 \text{ in.}$$

$$M_n = T(d - a/2) = 156(17.5 - 1.64)\tfrac{1}{12} = 206 \text{ ft-kips}$$

$$\phi M_n = 0.90(206) = 186 \text{ ft-kips} > M_u = 184 \text{ ft-kips} \qquad \text{OK}$$

At section B-B, obtain by proportion

$$\text{required } R_n = \frac{M_u}{\phi b d^2} = 572 \left(\frac{127}{184}\right) = 395 \text{ psi}$$

$$A_s \approx 2.59 \left(\frac{395}{572}\right) = 1.79 \text{ sq in.}$$

Use 4-#6 bars (A_s = 1.76 sq in.)
Check:

$$C = 47.6a$$
$$T = 1.76(60) = 106 \text{ kips}$$
$$a = 106/47.6 = 2.22 \text{ in.}$$
$$\phi M_n = 0.90(106)(17.5 - 1.11)\tfrac{1}{12} = 130 \text{ ft-kips} > 127 \text{ ft-kips} \quad \text{OK}$$

The reader is reminded that for a given R_n, the required ρ expression is a quadratic function; however, for practical use it may be approximated to be linear (see Fig. 3.8.1). A check of strength provided is then made for verification.

For simplicity in bar arrangement, no bending of bars is proposed. The arrangement of main reinforcement finally selected is shown in Fig. 6.21.2.

Note that in the design of the flexural reinforcement for the maximum negative moment, the center of support value of 127 ft-kips was used instead of the lesser value at the face of support, as could have been used in accordance with the discussion of the negative moment on _continuous_ beams (Sec. 6.12). On this statically determinate structure, the authors prefer the more conservative approach of using the center of support value. For beams on wider supports, corrections to the face of support would be appropriate.

(b) Determine whether or not the crack control criterion of ACI-10.6.4 is satisfied at maximum positive moment region. Using ACI Formula (10-4) (see also Sec. 4.12),

$$z = f_s \sqrt[3]{d_c A}$$

where A is equal to the effective tension area of concrete surrounding the main tension reinforcing bars divided by the number of bars. Using 1.5 in. of clear cover,

$$d_c = 1.5 \text{ (cover)} + 0.375 \text{ (stirrup)} + 0.52 \text{ (avg radius)} = 2.40 \text{ in.)}$$

$$\text{number of bars} = \frac{A_s}{\text{area of largest bar}} = \frac{2.60}{1.0} = 2.6$$

$$A = \frac{2(2.40)(14)}{2.6} = 25.8 \text{ sq in./bar}$$

$$f_s = 0.60 f_y = 0.60(60) = 36 \text{ ksi}$$
$$z = 36\sqrt[3]{2.40(25.8)} = 36(3.96) = 143 \text{ kips/in.}$$

which is less than the limit of 145 allowed by ACI-10.6.4 for exterior exposure. If 2-in. cover is used (ACI-7.7.1), computed z is 162 kips/in., which would be unacceptable for exterior exposure. The actual service load stress is usually less than $0.60f_y$ and the lesser value could be computed if needed to satisfy the crack control limitation (although little can be gained in this case because of the use of a load factor equal to 1.4 for all the loads).

(c) Development of reinforcement and bar cutoff. According to Eq. (6.7.1), the basic development lengths (ACI-12.2) are

$$L_{db} = 0.04 \frac{A_b f_y}{\sqrt{f_c'}} = 0.04 A_b \left(\frac{60,000}{\sqrt{4000}}\right) = 38 A_b$$

but not less than

$$L_{db} = 0.0004 d_b f_y = 24 d_b$$

$$L_d \, (\#6) = L_{db} \, (\#6) = 38(0.44) = 16.7 \text{ in.} < 24 d_b = \underline{18 \text{ in.}}$$

$$L_d \, (\#7) = L_{db} \, (\#7) = 38(0.60) = \underline{22.8 \text{ in.}} \ (1.9 \text{ ft})$$

$$L_d \, (\#9) = L_{db} \, (\#9) = 38(1.0) \ \ = \underline{38.0 \text{ in.}} \ (3.2 \text{ ft})$$

For top bars, according to ACI-12.2.3.1,

$$L_d \, (\#6) = 1.4 L_{db} = 1.4(18.0) = 25.2 \text{ in.} \ (2.1 \text{ ft})$$

Examine the feasibility of cutting the 1-#7 in the positive moment zone. The remaining 2-#9 would provide the following capacity,

$$C = 47.6a; \qquad T = 2(1.0)60 = 120 \text{ kips}$$

$$a = 2.52 \text{ in.}$$

$$\phi M_n = 0.90(120)(17.5 - 1.26)\tfrac{1}{12} = 146 \text{ ft-kips}$$

Based on moment capacity alone, the #7 bar could be cut at points A and B of Fig. 6.21.2; however, ACI-12.10.5 must also be satisfed to cut bars in the tension zone. This is examined in combination with designing stirrups [see item (e) below].

In trying to cut the 2-#6 bars from the negative moment region near section B-B, the bars are extended into the span *to the left of the support* farther than 12 bar diameters or the effective depth d in order to have full development of the provided steel at the face of support. Since this potential cut at point C is not in a tension zone, no further investigation of this cutoff location is required. On determining that the remaining 2-#6 bars can be cut off only a short distance farther into the span at point D, the decision is made to cut all 4-#6 bars at point D. If for some reason, such as the desire to have compression steel for deflection control, two of the #6 bars were extended across the 18-ft span instead of being cut at D, then the other 2-#6 bars would be cut at point C. On the *right* side of the support in the negative moment region, any potential cut location would be so near the end of the cantilever that it was decided to run all 4-#6 bars to the end (say 1 in. clear).

At the simple support at the left end of the 18-ft span, and at the point of inflection near the right end, ACI-12.11.3 must be checked for positive moment reinforcement. For the inflection point it is required that

$$\frac{M_n}{V_u} + L_a \geq L_d$$

Assuming that 1-#7 will be terminated (a conservative assumption for this computation), only the 2-#9 bars contribute to M_n. The actual L_a from the inflection point to the end of the bars exceeds d (17.5 in.); thus use $L_a = 17.5$ in.

$$\frac{146(12)}{0.90(47.0)} + 17.5 = 58.9 \text{ in.} > L_d = 38 \text{ in.} \qquad\qquad \text{OK}$$

For the simple support, ACI-12.11.3 requires

$$\frac{1.30 M_n}{V_u} + L_a \geq L_d$$

In this case, $L_a = 7.5$ in. since the bars extend this amount beyond the center of the support. Assuming the #7 bar does not extend into the support,

$$\frac{1.30(146)12}{0.90(47.0)} + 7.5 = 61.3 \text{ in.} > L_d = 38 \text{ in.} \qquad\qquad \text{OK}$$

(d) **Shear reinforcement.** At d from the face of right support on the 18-ft span,

$$V_u = 61.2 - 6.01 \left(\frac{17.5 + 9}{12} \right) = 47.9 \text{ kips}$$

$$\phi V_c = \phi(2\sqrt{f_c'})b_w d$$

$$= 0.85\,(2\sqrt{4000})(14)(17.5)\tfrac{1}{1000} = 26.3 \text{ kips}$$

$$\text{max required } \phi V_s = V_u - \phi V_c = 47.9 - 26.3 = 21.6 \text{ kips}$$

$$\text{max permitted } \phi V_s \left(\text{for } s = \frac{d}{2} \right) = 2\phi V_c = 52.6 \text{ kips}$$

Note that maximum ϕV_s (ACI-11.5.4.3) is based on the nominal stress $v_s = 4\sqrt{f_c'}$, which gives maximum $\phi V_s = 2\phi V_c$ when the simplified V_c expression is used. For minimum percentage of stirrups (ACI-11.5.5.3),

$$\text{min } \phi V_s = \phi 50 b_w d = 0.85(50)(14)(17.5)\tfrac{1}{1000} = 10.4 \text{ kips}$$

For strength, using #3 U stirrups,

$$\phi V_s = \frac{\phi A_v f_y d}{s} = \frac{0.85(0.22)(60)(17.5)}{s} = \frac{196}{s}$$

s	ϕV_s
9.1 in.	21.6 kips max

Since the strength requirement permits stirrups at 9.1 in. and is less restrictive than the $d/2$ limit, the maximum spacing at $d/2 = 8.75$ in. controls wherever stirrups are required.

Use #3 U stirrups at 8 in. spacing, as dimensioned on the shear diagram in Fig. 6.21.2. (Ordinarily the stirrups should be dimensioned on the design sketch; that is, the side view of the beam in Fig. 6.21.2.)

(e) Check ACI-12.10.5 for cutting #7 bar at points A and B in the tension zone. Since ϕM_n for the continuing #9 bars *is not* twice the factored moment M_u at the cutoff point, ACI-12.10.5.3 cannot be satisfied.

Check ACI-12.10.5.1, for #3 U stirrups at 8 in. as provided,

$$\phi V_s = \frac{196}{s} = \frac{196}{8} = 24.5 \text{ kips}$$

V_u at points A and B = 30 kips (scaled)

$$\text{percent stressed} = \frac{V_u}{\phi V_n} = \frac{V_u}{\phi (V_c + V_s)}$$

$$= \frac{30}{26.3 + 24.5} = 59\% < 67\% \qquad \text{OK}$$

Cut 1-#7 bar at points A and B (length = 10 ft).

(f) Moment capacity diagram. Figure 6.21.2 shows the comparison between the factored moment M_u diagram and the moment capacity ϕM_n diagram provided by the selected reinforcing bars. In computing the moment capacity, any effect of steel in the compression side of the beam has been neglected because it would have negligible effect since it was not required for strength. At the left end of the beam, the 2-#9 have been extended as far as feasible into the support in order to have the bars develop their full capacity at point A. The available embedment to the left of point A was nearly exactly the 3.2 ft required.

(g) Deflection. Even though the reinforcement ratio ρ being used was expected to control the deflection, nevertheless the deflection must be investigated if excessive deflection may cause damage to partitions or other construction.

(h) Design sketch. The final arrangement of longitudinal steel and stirrups is shown in Fig. 6.21.2. The stirrup locations are dimensioned on the V_u diagram. Omitted from the elevation view of the beam are the nominal sized (say #3 or #4) longitudinal bars arbitrarily added for the stirrups to wrap around wherever the stirrups would otherwise have no support to hold them in vertical position. These pairs of bars would be located in this beam at both top and bottom faces of the beam where no longitudinal bars are shown in the figure. Cross-section views showing bars have been omitted here but are always part of the design sketch.

SELECTED REFERENCES

1. ACI Committee 408. "Bond Stress—The State of the Art," *ACI Journal, Proceedings*, **63**, November 1966, 1161–1190. Disc. 1569–1570.
2. ACI Committee 408. "Opportunities in Bond Research," *ACI Journal, Proceedings*, **67**, November 1970, 857–869 (contains 67 references).
3. LeRoy A. Lutz and Peter Gergely. "Mechanics of Bond and Slip of Deformed Bars in Concrete," *ACI Journal, Proceedings*, **64**, November 1967, 711–721. Disc., **65**, 412–414.
4. C. O. Orangun, J. O. Jirsa, and J. E. Breen. "A Reevaluation of Test Data on Development Length and Splices," *ACI Journal, Proceedings*, **74**, March 1977, 114–122. Disc., **74**, September 1977, 470–475.

5. Anders Losberg and Per-Ake Olsson. "Bond Failure of Deformed Reinforcing Bars Based on Longitudinal Splitting Effect of the Bars," *ACI Journal, Proceedings*, **76**, January 1979, 5–18.

6. James O. Jirsa, LeRoy A. Lutz, and Peter Gergely. "Rationale for Suggested Development, Splice, and Standard Hook Provisions for Deformed Bars in Tension," *Concrete International*, **1**, July 1979, 47–61.

7. Herbert J. Gilkey, Stephen J. Chamberlain, and Robert W. Beal. "Bond Between Concrete and Steel," *Iowa Engineering Experiment Station Bulletin*, No. 147, Iowa State College, 1940.

8. T. D. Mylrea. "Bond and Anchorage," *ACI Journal, Proceedings*, **44**, March 1948, 521–552.

9. Phil M. Ferguson and J. Neils Thompson. "Development Length for Large High Strength Reinforcing Bars," *ACI Journal, Proceedings*, **62**, January 1965, 71–93. Disc. 1153–1156.

10. Phil M. Ferguson, John E. Breen, and J. Neils Thompson. "Pullout Tests on High Strength Reinforcing Bars," *ACI Journal, Proceedings*, **62**, August 1965, 933–950.

11. Ervin S. Perry and J. Neils Thompson. "Bond Stress Distribution on Reinforcing Steel in Beams and Pullout Specimens," *ACI Journal, Proceedings*, **63**, August 1966, 865–875.

12. E. L. Kemp, F. S. Brezny, and J. A. Unterspan. "Effect of Rust and Scale on the Bond Characteristics of Deformed Reinforcing Bars," *ACI Journal, Proceedings*, **65**, September 1968, 743–756. Disc., **66**, 224–226.

13. Saeed M. Mirza and Jules Houde. "Study of Bond Stress-Slip Relationships in Reinforced Concrete," *ACI Journal, Proceedings*, **76**, January 1979, 19–46.

14. S. Soretz and H. Holzenbein. "Influence of Rib Dimensions of Reinforcing Bars on Bond and Bendability," *ACI Journal, Proceedings*, **76**, January 1979, 111–125.

15. Phil M. Ferguson, Robert D. Turpin, and J. Neils Thompson. "Minimum Bar Spacing as a Function of Bond and Shear Strength," *ACI Journal, Proceedings*, **50**, June 1954, 869–887.

16. E. L. Kemp and W. J. Wilhelm. "Investigation of the Parameters Influencing Bond Cracking," *ACI Journal, Proceedings*, **76**, January 1979, 47–72.

17. Shiro Morita and Tetsuzo Kaku. "Splitting Bond Failures of Large Deformed Reinforcing Bars," *ACI Journal, Proceedings*, **76**, January 1979, 93–110.

18. R. Jimenez, R. N. White, and P. Gergely. "Bond and Dowel Capacities of Reinforced Concrete," *ACI Journal, Proceedings*, **76**, January 1979, 73–92.

19. ACI Committee 408. "Suggested Development, Splice, and Standard Hook Provisions for Deformed Bars in Tension," *Concrete International*, **1**, July 1979, 44–46.

20. Phil M. Ferguson and Farid N. Matloob. "Effect of Bar Cutoff on Bond and Shear Strength of Reinforced Concrete Beams," *ACI Journal, Proceedings*, **56**, July 1959, 5–23.

21. Anthony M. Kao and Raymond E. Untrauer. "Shear Strength of Reinforced Concrete Beams with Bars Terminated in Tension Zones," *ACI Journal, Proceedings*, **72**, December 1975, 720–722.

22. LeRoy A. Lutz. "Crack Control Factor for Bundled Bars and for Bars of Different Sizes," *ACI Journal, Proceedings*, **71**, January 1974, 9–10.

23. N. W. Hanson and Hans Reiffenstuhl. "Concrete Beams and Columns with Bundled Reinforcement," *Journal of the Structural Division*, ASCE, **84**, October 1958, (ST6), 1–23.

24. Frank D. Steiner. "Suggested Applications for Bundled Bars," *ACI Journal, Proceedings*, **64**, April 1967, 213–214.

25. John A. Hribar and Raymond C. Vasko. "End Anchorage of High Strength Steel Reinforcing Bars," *ACI Journal, Proceedings,* **66,** November 1969, 875–883. Disc, **67,** 423–424.

26. John Minor and James O. Jirsa. "Behavior of Bent Bar Anchorages," *ACI Journal, Proceedings,* **72,** April 1975, 141–149.

27. Jose L. G. Marques and James O. Jirsa. "A Study of Hooked Bar Anchorages in Beam-Column Joints," *ACI Journal, Proceedings,* **72,** May 1975, 198–209.

28. Robert L. Pinc, Michael D. Watkins, and James O. Jirsa. "Strength of Hooked Bar Anchorages in Beam-Column Joints," CESRL Report No. 77-3, Department of Civil Engineering, University of Texas, Austin, Texas, November 1977.

29. Raymond E. Untrauer and Robert L. Henry. "Influence of Normal Pressure on Bond Strength," *ACI Journal, Proceedings,* **62,** May 1965, 577–586.

30. ACI Committee 318. *Commentary on Building Code Requirements for Reinforced Concrete* (ACI 318–83). Detroit, Michigan: American Concrete Institute, 1983.

31. Joint PCI/WRI Ad Hoc Committee on Welded Wire Fabric for Shear Reinforcement, PCI Technical Activities Committee (Leslie D. Martin, Chairman). "Welded Wire Fabric for Shear Reinforcement," *PCI Journal,* **25,** July/August 1980, 32–36.

32. Phil M. Ferguson and John E. Breen. "Lapped Splices for High Strength Reinforcing Bars," *ACI Journal, Proceedings,* **62,** September 1965, 1063–1078.

33. John P. Lloyd and Clyde E. Kessler. "Splices and Anchorages in One-Way Slabs Reinforced with Deformed Wire Fabric," *ACI Journal, Proceedings,* **67,** August 1970, 636–642.

34. M. A. Thompson, J. O. Jirsa, J. E. Breen, and D. F. Meinheit, "Behavior of Multiple Lap Splices in Wide Sections," *ACI Journal, Proceedings,* **76,** February 1979, 227–248.

35. Raymond E. Untrauer and George E. Warren. "Stress Development of Tension Steel in Beams," *ACI Journal, Proceedings,* **74,** August 1977, 368–372.

36. Phil M. Ferguson. "Small Bar Spacing or Cover—A Bond Problem for the Designer," *ACI Journal, Proceedings,* **74,** September 1977, 435–439.

PROBLEMS

All problems* are to be done in accordance with the strength method of the ACI Code, and all loads given are *service* loads, unless otherwise indicated. All moment capacity ϕM_n diagrams used in these problems must be drawn to scale directly below a side view of the beam drawn to the same longitudinal scale. Moment capacity ϕM_n diagrams must have critical numerical values stated on the diagram, and horizontal distances to the critical points dimensioned.

6.1   Draw the free-body diagram for the 3-in. slice shown crosshatched on the beam given in the accompanying figure. Using working stress assumptions, compute and show on the diagram values for the internal forces (tension, compression, and shear) on each side of the slice.

(a) Compute the average flexural bond stress on the bars over the 3-in. slice.

---

*Most problems may be solved as problems stated in Inch-Pound units, or as problems in SI units using quantities in parenthesis at end of the statement. The SI conversions are approximate to avoid implying higher precision for the given information in SI units than that given for the Inch-Pound units.

(b) Compute the average flexural bond stress on the bars over the distance from section A to the left end of the beam.

(c) What is the average anchorage bond stress resisting the tensile force at section A?

(d) Explain what happens if the flexural bond stress in part (a) is so high that slippage occurs over the 3-in. slice. What determines the adequacy of the beam? Assume loading is within the usual service range.

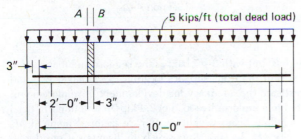 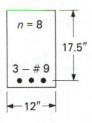

Problem 6.1

6.2 For the simply supported beam shown, draw the moment capacity (ϕM_n) diagram. Assume that the cutoff location 3 ft from the support satisfies the requirements of ACI-12.10.5, and that ACI-12.11.3 is satisfied. What is the maximum uniformly distributed service load that the beam may be permitted to carry (assume 50% live load and 50% dead load)? Use $f'_c = 3500$ psi and $f_y = 40,000$ psi.

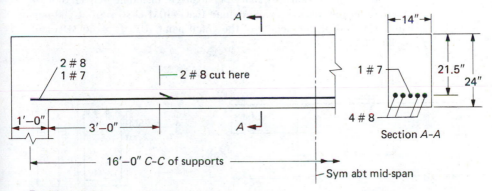

Problem 6.2

6.3 For the cantilever beam of the accompanying figure having $f'_c = 3000$ psi and $f_y = 40,000$ psi:

(a) Draw the moment capacity ϕM_n diagram. Be sure to include the 2-ft embedment into the support, and assume that the full cross section capacity in that region is based on that at the support.

(b) Investigate the adequacy for development of reinforcement if the beam is subjected to uniform dead and live loads of 1.8 kips/ft (including weight of beam) and 2.1 kips/ft, respectively. Neglect any concern about cutting bars in the tension zone (ACI-12.10.5).

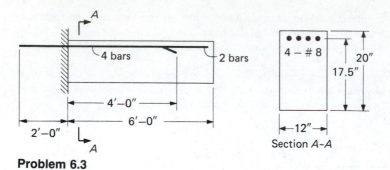

Problem 6.3

6.4 For the beam of Prob. 6.3, but with 3-#7 bars as the reinforcement at section A-A, 1-#7 is cut at 4 ft-0 in. from support location, f'_c = 4000 psi, and f_y = 60,000 psi; investigate the adequacy for development of reinforcement if the beam is subjected to a uniform live load of 2.4 kips/ft and uniform dead load of 2.1 kips/ft (including beam weight). Compare factored moment M_u with the moment capacity ϕM_n by superimposing both diagrams. Consider all factors involved, including ACI-12.10.5 for cutting bars in the tension zone. Assume #3 U stirrups at 8-in. spacing are used in the vicinity of the cutoff point.

6.5 For the beam of the accompanying figure, investigate the adequacy for development of reinforcement if the beam must carry uniform dead and live loads of 30 kN/m (including weight of beam) and 34 kN/m, respectively. Compare factored moment M_u with the moment capacity ϕM_n by superimposing both diagrams. Consider all factors involved, including ACI-12.10.5 for cutting bars in the tension zone. Assume that #10M U stirrups at 200-mm spacings are used in the vicinity of the cutoff point. Use f'_c = 30 MPa and f_y = 400 MPa.

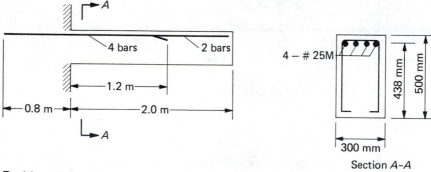

Problem 6.5

6.6 For the cantilever beam of the accompanying figure, determine the safe uniformly distributed load w (dead load, DL, plus live load, LL) that the beam may be permitted to carry, if the dead load to live load ratio is 0.8. Consider all factors involved, including ACI-12.10.5 for cutting bars in the tension zone. Use f'_c = 4000 psi and f_y = 60,000 psi. Show the comparison of factored moment M_u with the moment capacity ϕM_n diagram. (For SI problem use bars: 3-#25M; #10M stirrups; f'_c = 30 MPa; f_y = 300 MPa.)

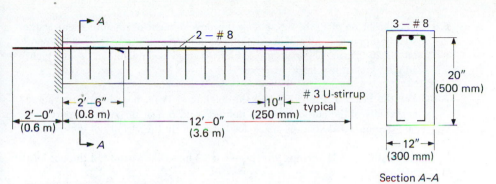

Problem 6.6

6.7   If the beam of the accompanying figure is to carry uniformly distributed loads of 4.0 kips/ft live load and 1.9 kips/ft dead load (including beam weight), determine the adequacy of the bar cutoffs and the development of reinforcement. Stirrups are #3 at 6 in. spacing where bars A are cut, and #3 at 8 in. spacing where bars B are cut. As part of the solution, compare the factored moment M_u with the moment capacity ϕM_n by superimposing both diagrams. Use $f'_c = 5000$ psi and $f_y = 60,000$ psi.

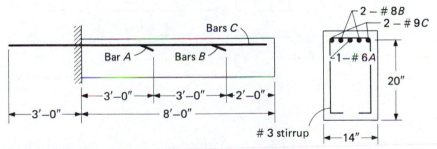

Problem 6.7

6.8   Neglecting any compression reinforcement effect, plot the moment capacity ϕM_n diagram (positive moment over the 20-ft span and negative moment for the steel over the support) for the beam of the accompanying figure. Use $f'_c = 3000$ psi and $f_y = 40,000$ psi.

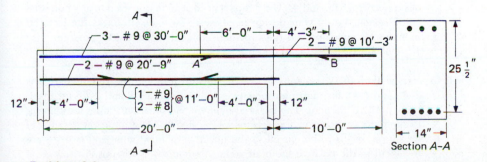

Problem 6.8

6.9 For the beam of Prob. 6.8, check the development of reinforcement at the simply supported end and at the point of inflection closest to the right support on the 20-ft span. The loads are 1.4 kips/ft dead load (including beam weight) and 2.8 kips/ft live load.

6.10 For the beam of Prob. 6.8 and the loading of Prob. 6.9, determine the acceptability of the cut locations at points A and B near the right support.
(a) Assume #3 U stirrups are spaced at 10 in. in the vicinity of the cut locations.
(b) Assume #3 U stirrups are spaced at 12 in. in the vicinity of the cut locations.

6.11 Investigate the adequacy of the beam shown in the accompanying figure for (a) bar cutoffs and (b) stirrups. The uniform dead and live loading is 9.3 kN/m and 42 kN/m, respectively. Use $f_c' = 30$ MPa and $f_y = 400$ MPa.

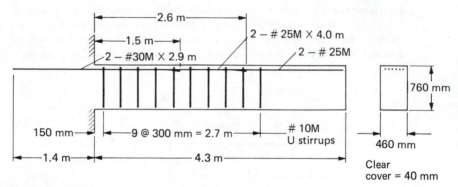

Problem 6.11

6.12 A 14-in. wide by 24-in. deep beam (effective depth = 21.5 in.) is used as the section for a 20-ft simply supported span having an 8-ft cantilever at one end. The positive moment reinforcement is 4-#9 bars and the negative moment reinforcement is 4-#8 bars. The loading to be carried is a live load of 1.9 kips/ft and a dead load of 1.6 kips/ft (including beam weight). Determine the lengths of bars (3-in. increments) if two of the four bars in both the positive and the negative moment regions are to be terminated as soon as practicable. The remaining bars are to be extended as required by the ACI Code. Width of supports is 15 in. and #3 U stirrups spaced at 10 in. are used in any potential cutoff region. Verify your design by showing for one set of axes the factored moment M_u envelope and the provided moment capacity ϕM_n diagram. Use $f_c' = 3500$ psi and $f_y = 40,000$ psi.

6.13 Repeat Prob. 6.12, except use a 12-in. beam width and the reinforcement is 2-#8 and 1-#9 for maximum negative moment and 2-#9 and 2-#8 for maximum positive moment. The factored moment and shear envelopes for uniform loading are as shown in the accompanying figure. Determine the lengths of bars if one #9 is to be cut off in the negative moment zone and two #8 are to be cut in the positive moment zone. Assume #3 U stirrups are spaced at 10 in. in any potential cutoff region. Use a moment capacity ϕM_n diagram to verify your answer.

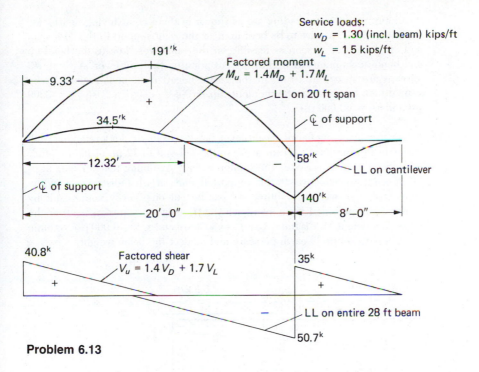

Problem 6.13

6.14 A 12-in. wide by 24-in. (overall size) beam is used as the section for a 20-ft simply supported span having an 8-ft cantilever at one end. Support widths are 12 in. The positive moment reinforcement is 2-#8 and 2-#7 and the negative moment reinforcement is 4-#7. The factored moment and shear

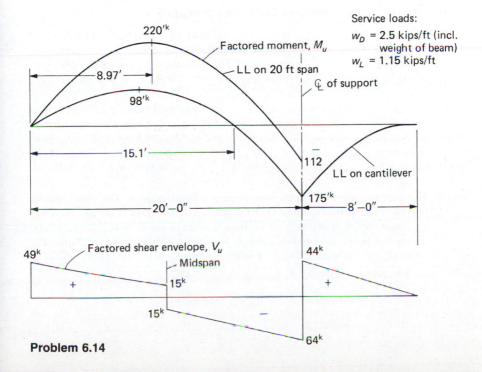

Problem 6.14

envelopes for uniform loading are as shown in the accompanying figure. As-sume that 2-#7 bars are to be bent up near the right support in the 20-ft span and are to be cut as soon as feasible on the cantilever. Locate the bend-up and bend-down points as well as the cut point on the cantilever. Assume #3 stirrups are spaced at 10 in. in the entire negative moment region. Verify your design by showing the moment capacity ϕM_n diagram. Use $f'_c = 3500$ psi and $f_y = 60,000$ psi.

6.15 In the accompanying figure, a cantilever slab (for example, for a retaining wall) varies in thickness from 24 in. at its supported end to 12 in. at its free end, with $2\frac{1}{2}$ in. clear cover over the reinforcement. Assuming that #7 bars at 6-in. spacing are effective at the supported end, at what distance from the supported end may every other #7 bar be cut off? Verify your result by showing the provided moment capacity ϕM_n diagram superimposed on the factored moment M_u diagram. Use $f'_c = 3000$ psi and $f_y = 40,000$ psi. Assume the loading is entirely earth pressure and neglect the beam weight.

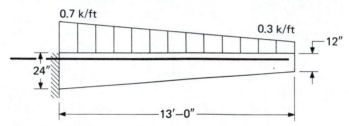

Problem 6.15

Problems Utilizing Concepts of Chapters 1 through 6

6.16 Design, including design sketch, a rectangular cantilever beam 14 ft long to carry a live load of 2.5 kips/ft. The beam size may not exceed 15 in. wide and 24 in. deep. Use $f'_c = 4000$ psi and $f_y = 60,000$ psi. (For SI problem use length = 4.3 m; live load = 36 kN/m; $f'_c = 30$ MPa; $f_y = 400$ MPa.)

6.17 Design, including design sketch, a cantilever beam 14 ft long to carry a live load of 3.0 kips/ft. Without actually satisfying a deflection limit, design a rectangular beam of such size that no serious problem with excessive deflection is to be expected. Use $f'_c = 4000$ psi and $f_y = 60,000$ psi. (For SI problem use live load = 43 kN/m; length = 4.3 m; $f'_c = 30$ MPa; $f_y = 400$ MPa.)

6.18 The floor system shown in the accompanying figure is given. The floor consists of 6-in. precast slab sections of 10 ft span. The live load is 50 psf, and the maximum depth available is 30 in. from the top of the floor slab. Completely design the beam indicated as a simply supported one. Use 1-in. multiples for beam depth and width. Use $f'_c = 4000$ psi and $f_y = 60,000$ psi.

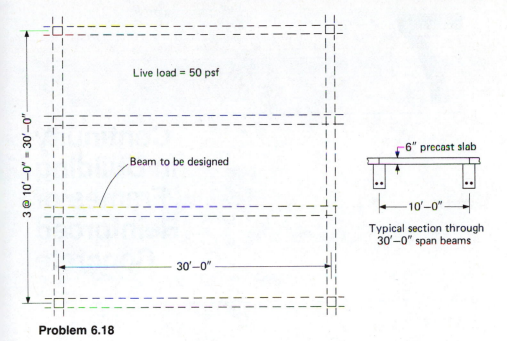

Problem 6.18

6.19 Design, including design sketch, a simply supported rectangular beam to carry a dead load of 1.6 kips/ft (not including beam weight) and a live load of 3.4 kips/ft on a span of 22 ft. Use a reinforcement ratio roughly 0.3 of the value at the balanced strain condition. Assume support widths are 18 in. Use $f'_c = 4000$ psi and $f_y = 60,000$ psi. (For SI problem, use dead load = 23.3 kN/m; live load = 50 kN/m; span = 6.7 m; $f'_c = 30$ MPa; $f_y = 400$ MPa; support width = 460 mm.)

6.20 Repeat Prob. 6.19, except use dead load 1.7 kips/ft instead of 1.6, and live load 3.8 kips/ft instead of 3.4, and a span of 16 ft. (For SI, dead load = 25 kN/m; live load = 55 kN/m; span = 5 m.)

6.21 Design, including design sketch, a simply supported rectangular beam on a span of 18 ft, and having a cantilever at the left end of span 6 ft. The dead load (not including beam weight) is 2.2 kips/ft and the live load is 3.5 kips/ft. Assume the support widths are 20 in. Use a reinforcement ratio for positive moment roughly 0.3 of the value at the balanced strain condition. Use $f'_c = 3000$ psi and $f_y = 60,000$ psi. (For SI, use dead load = 32 kN/m; live load = 52 kN/m; support widths = 500 mm; span = 5.5 m; cantilever = 1.8 m; $f'_c = 20$ MPa; $f_y = 400$ MPa.)

6.22 Repeat Prob. 6.21, except use a 16-ft main span and an 8-ft cantilever. (For SI, use span = 5 m; cantilever = 2.4 m.)

7

Continuity in Building Frames of Reinforced Concrete

7.1 COMMON BUILDING FRAMES

Reinforced concrete building construction commonly has floor slabs, beams, girders, and columns continuously placed to form a monolithic system. Consider the plan of typical slab–beam–girder floor construction shown in Fig. 7.1.1. Section A-A through the slab shows that the slab is supported on 10 beams. Intermediate beams such as B1, B2, and B3 are supported on the girders, whereas beams on the column lines such as B4, B5, and B6 are supported directly by the columns. Girders such as G1, G2, and G3 also go directly into the columns.

Beams such as B4, B5, and B6 are not only continuous beams, they are also built integrally with the upper and lower columns. For correct analysis, then, the complete frame in this plane, which may consist of, say, 10 or 15 stories or more should be analyzed as a rigid frame. In the analysis for gravity load on regular building frames, according to ACI-8.9.1 the beams and the adjacent columns may be isolated and treated as a subassembly (Fig. 7.1.2), with the far ends of the columns assumed as fixed. This assumption should not be used for wind load, however. For wind analysis, except for very tall structures, a simplified approximate method may be used [1]. Although titled "Equivalent Frame Analysis for Two-Way Floor Systems in Unbraced Frames," Chap. 17 gives the general treatment of building frame analysis for both vertical (gravity) and lateral loads by the matrix displacement method.

Rigid frame bridge piers, I-75 near Detroit. (Photo by C. G. Salmon.)

Girders $G1$, $G2$, and $G3$ may be treated in the same manner as the beams. Relative stiffnesses for columns, beams, and girders must first be assumed or established by preliminary design and later reviewed, as would be done in the analysis and design of any statically indeterminate rigid frame.

The continuous slab is supported at the beams, and the beams $B1$, $B2$, and $B3$ are supported at the girders. The supporting beams or girders possess torsional rigidity which may be approximated by using equivalent supporting columns with fixed far ends. The stiffness factors to be used for such equivalent columns are discussed in Chap. 10 (Sec. 10.3).

7.2 POSITIONS OF LIVE LOAD FOR MOMENT ENVELOPE

The live load positions that cause the largest bending moments in slabs, beams, and girders are discussed in this section. The reader should review the subject of influence lines [2] in order to be able to establish the loading conditions appropriate for obtaining the bending moment and shear envelopes. Bending moments to be used in the design of columns are treated separately later.

Consider the continuous beam $ABCDEFGH$ with its adjacent columns as shown in Fig. 7.2.1(a). The influence lines for bending moment at any point in the central portion of span CD and at a section an infinitesimal distance to the left of support D are shown in Figs. 7.2.1(b) and (c), re-

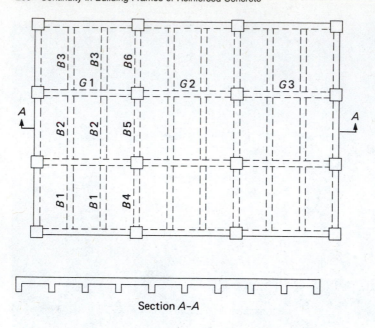

Section *A–A*

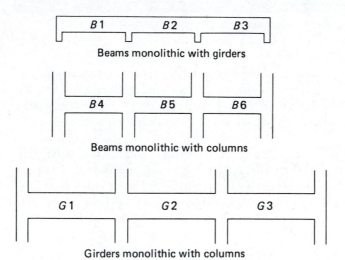

B 1   B 2   B 3

Beams monolithic with girders

B 4   B 5   B 6

Beams monolithic with columns

G 1   G 2   G 3

Girders monolithic with columns

Figure 7.1.1   Slab—beam—girder floor.

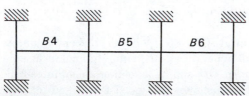

B 4   B 5   B 6

Figure 7.1.2   Beams and adjacent columns.

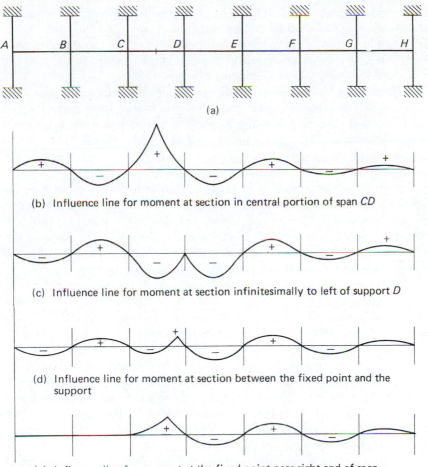

(a)

(b) Influence line for moment at section in central portion of span *CD*

(c) Influence line for moment at section infinitesimally to left of support *D*

(d) Influence line for moment at section between the fixed point and the
      support

(e) Influence line for moment at the fixed point near right end of span

Figure 7.2.1   Influence lines for continuous spans.

spectively. From these influence lines the following cases for uniform live
load are indicated:

1. For maximum positive moment within a span, load that span and
   all other alternate spans.
2. For maximum negative moment within a span, load the two spans
   adjacent to that span and all other alternate spans (all the spans not
   loaded in 1).
3. For maximum negative moment at a support, load the two spans
   adjacent to that support and all other alternate spans.
4. For maximum positive moment at a support, load the two spans
   beyond each of the two spans adjacent to that support and all other
   alternate spans (all the spans not loaded in 3).

It may be noted that loading cases 1 and 2 in the preceding paragraph
are complementary; that is, their combination results in *all* spans being
loaded. Loading cases 3 and 4 are also complementary. It may be noted

further that loading cases 1 and 3 are primary; that is, they result in moments of the same sign as dead load. Loading cases 2 and 4, on the other hand, are secondary; they result in moments opposite in sign to those due to dead load. When the secondary live load moment is numerically larger than the dead load moment, there is moment reversal.

The loading cases thus far mentioned involve loading entire spans, that is, no partially loading of a span. These full span loading cases correctly give maximum and minimum bending moment in the midspan region (roughly the middle 50 to 60%) of a span as well as *at* the support. The correct maximum (or minimum) values are not obtained, however, for approximately 20 to 25% of the span nearest the supports.

A qualitative examination of Figs. 7.2.1(b) and (c) shows that if an influence line were drawn for a series of specific points along span *CD* between the midspan and the support, the peak ordinate in the positive portion of such influence line would get smaller and smaller. Also, the *slope* of the influence line at point *C* goes upward to the right for the midspan influence line but it is downward to the right for the support *D* influence line.

As successive influence lines are drawn for the points along the span from midspan to the support, there must be some location for which the slope of the influence line at *C* is horizontal. Such point is called a *fixed point*, giving an influence line illustrated by Fig. 7.2.1(e). Any loading on spans to the left of the span under study will cause no moment at this fixed point. The fixed point is the closest location to the support for which full span loadings give the correct maximum or minimum bending moments. The influence line for a section between a fixed point and the support is as shown by Fig. 7.2.1(d), indicating partial loading for the span in question to obtain maximum or minimum bending moment. See Ref. 2 for extended treatment of fixed points.

From a practical point of view, partial span loading for maximum or minimum bending moments is rarely actually done in the design of building frames because the effect of doing so on the design is small enough as to not justify the effort. For the design of long spans, such as highway bridges, the full consideration of the proper loading (loading partial span as indicated by influence lines) for locations between the fixed point and the support would usually be made.

7.3 METHOD OF ANALYSIS

When using the strength method of design, the analysis of the continuous concrete structure is made using *factored loads*; that is, the service loads multiplied by appropriate overload factors in accordance with ACI-9.2. Thus the structural analysis is to be made assuming an elastic system even though the factored load may cause inelastic effects. After obtaining the moments and shears assuming an elastic structure under the factored load condition, each section is proportioned to provide adequate strength. Although the procedure may seem somewhat inconsistent, it has been found to provide safe and adequate designs. The other possible approach would be to utilize the true ultimate or collapse condition, using the so-called limit design (discussed in Sec. 10.12) for continuous beams or frames, and yield-line theory (Chap. 18) for slabs. These two methods are not incorporated in the ACI Code, although the Code does contain provisions for adjustments to the results of elastic analysis to reflect present knowledge of collapse behavior.

There are numerous methods of statically indeterminate structural analysis that may be used. When the computer is used, matrix methods will usually be preferred [2,3]. The reader is referred to Chap. 17 for the treatment of the global stiffness matrix, the load matrix, the displacement matrix, and the end-moment matrix. Using this method, either the subassembly or the entire frame may be analyzed. The subassembly analysis is useful in preliminary design, while the entire frame may be preferred for the final review of the design.

For computation using nonprogrammable or small programmable calculators the moment distribution method is probably the most convenient method for analyzing the rigid frame involving several continuous spans with the far ends of upper and lower columns fixed. The moment distribution is described in most textbooks [2] on structural analysis, so it will not be developed here. The application to a six-span continuous beam with upper and lower columns is illustrated in the following example.

EXAMPLE 7.3.1 Determine the maximum and minimum moments at the middle and the ends of each span in a rigid frame with six equal spans as shown in Fig. 7.3.1. Assume that the uniform live load w_L is twice the uniform dead load w_D, and that the stiffness factor K_c (representing $4EI/L$) of the column is twice the stiffness factor K_b of the beam span. Express all moments in terms of wL^2 in which $w = w_D + w_L$ and L is the span length.

Solution: All moments will be expressed numerically in terms of $wL^2(10^{-4})$. The FEM (fixed-end moment) due to w_D and $w_L = \pm\frac{1}{12}wL^2 = \pm 833wL^2(10^{-4})$, whereas the FEM due to w_D only $= \pm\frac{1}{12}w_D L^2 = \pm\frac{1}{36}wL^2 = \pm 278wL^2(10^{-4})$.

Seven loading conditions as shown in Fig. 7.3.2 need to be investigated. Three cycles of moment distribution should provide sufficiently accurate results for design use. The moment distribution and the relative slope check are shown completely only for loading condition No. 1 in Table 7.3.1. For the other six loading conditions only the FEM and the final end moments are shown in Tables 7.3.2 to 7.3.7.

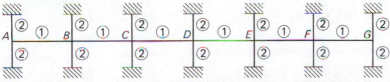

Figure 7.3.1 A six-span continuous frame showing relative stiffnesses.

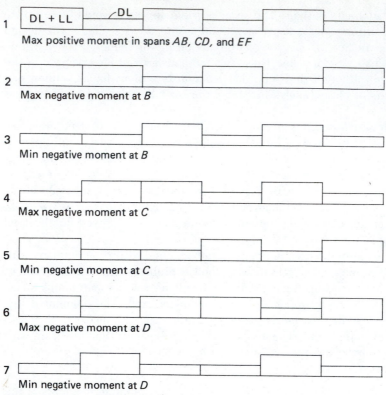

Figure 7.3.2 Loading conditions.

For prismatic members the final moment at end A of a member AB expressed in slope deflection terminology is

$$M_{ab} = \text{FEM}_{ab} + \left(\frac{4EI}{L}\right)\theta_a + \left(\frac{2EI}{L}\right)\theta_b$$

where FEM_{ab} is the original fixed-end moment at end A, and θ_a and θ_b are the slopes of the elastic curve at ends A and B, respectively. The quantity $4EI/L$ is the stiffness K at one end of a prismatic member with its far end fixed. The change from fixed-end moment to final moment, ΔM_{ab}, is

$$\Delta M_{ab} = M_{ab} - \text{FEM}_{ab} = \left(\frac{4EI}{L}\right)\theta_a + \left(\frac{2EI}{L}\right)\theta_b$$

Also

$$\Delta M_{ba} = M_{ba} - \text{FEM}_{ba} = \left(\frac{4EI}{L}\right)\theta_b + \left(\frac{2EI}{L}\right)\theta_a$$

Solving for θ_a and θ_b gives

$$\theta_a = \frac{\Delta M_{ab} - \frac{1}{2}\Delta M_{ba}}{3EI/L} \quad \text{and} \quad \theta_b = \frac{\Delta M_{ba} - \frac{1}{2}\Delta M_{ab}}{3EI/L}$$

Table 7.3.1 LOADING CONDITION NO. 1 (FIG. 7.3.2)

JOINT	A	B		C		D		E		F		G
MEMBER	AB	BA	BC	CB	CD	DC	DE	ED	EF	FE	FG	GF
Distribution factor	$\frac{1}{5}$	$\frac{1}{6}$	$\frac{1}{6}$	$\frac{1}{6}$	$\frac{1}{6}$	$\frac{1}{6}$	$\frac{1}{6}$	$\frac{1}{6}$	$\frac{1}{6}$	$\frac{1}{6}$	$\frac{1}{6}$	$\frac{1}{5}$
FEM	−833	+833	−278	+278	−833	+833	−278	+278	−833	+833	−278	+278
Balance	+167	−92	−92	+92	+92	−92	−92	+92	+92	−92	−92	−56
Carryover	−46	+83	+46	−46	−46	+46	+46	−46	−46	+46	−28	−46
Balance	+9	−22	−22	+15	+15	−15	−15	+15	+15	−3	−3	+9
Carryover	−11	+4	+8	−11	−8	+8	+8	−8	−1	+8	+4	−1
Balance	+2	−2	−2	+3	+3	−3	−3	+1	+1	−2	−2	0
Final M	−712	+804	−340	+331	−777	+777	−334	+332	−772	+790	−399	+184
Change	+121	−29	−62	+53	+56	−56	−56	+54	+61	−43	−121	−94
− = change	+14	−60	−26	+31	+28	−28	−27	+28	+22	−30	+47	+60
sum	+135	−89	−88	+84	+84	−84	−83	+82	+83	−73	−74	−34
θ_{rel} = sum/K	+135	−89	−88	+84	+84	−84	−83	+82	+83	−73	−74	−34

Table 7.3.2 LOADING CONDITION NO. 2 (FIG. 7.3.2)

MEMBER	AB	BA	BC	CB	CD	DC	DE	ED	EF	FE	FG	GF
FEM	−833	+833	−833	+833	−278	+278	−833	+833	−278	+278	−833	+833
Final M	−668	+911	−888	+729	−325	+337	−778	+777	−331	+340	−804	+712

Table 7.3.3 LOADING CONDITION NO. 3 (FIG. 7.3.2)

MEMBER	AB	BA	BC	CB	CD	DC	DE	ED	EF	FE	FG	GF
FEM	−278	+278	−278	+278	−833	+833	−278	+278	−833	+833	−278	+278
Final M	−227	+293	−241	+374	−785	+774	−333	+332	−772	+790	−399	+184

Table 7.3.4 LOADING CONDITION NO. 4 (FIG. 7.3.2)

MEMBER	AB	BA	BC	CB	CD	DC	DE	ED	EF	FE	FG	GF
FEM	−278	+278	−833	+833	−833	+833	−278	+278	−833	+833	−278	+278
Final M	−187	+390	−745	+878	−882	+732	−325	+336	−772	+790	−399	+184

Table 7.3.5 LOADING CONDITION NO. 5 (FIG. 7.3.2)

MEMBER	AB	BA	BC	CB	CD	DC	DE	ED	EF	FE	FG	GF
FEM	−833	+833	−278	+278	−278	+278	−833	+833	−278	+278	−833	+833
Final M	−709	+813	−385	+225	−228	+379	−786	+774	−330	+340	−804	+712

Table 7.3.6 LOADING CONDITION NO. 6 (FIG. 7.3.2)

MEMBER	AB	BA	BC	CB	CD	DC	DE	ED	EF	FE	FG	GF
FEM	−833	+833	−278	+278	−833	+833	−833	+833	−278	+278	−833	+833
Final M	−712	+805	−343	+323	−731	+883	−883	+731	−323	+343	−805	+712

Table 7.3.7 LOADING CONDITION NO. 7 (FIG. 7.3.2)

MEMBER	AB	BA	BC	CB	CD	DC	DE	ED	EF	FE	FG	GF
FEM	−278	+278	−833	+833	−278	+278	−278	+278	−833	+833	−278	+278
Final M	−184	+398	−787	+780	−378	+228	−228	+378	−780	+787	−398	+184

Thus

$$\theta_a \text{ (relative)} = \left(\frac{\Delta M_{ab} - \frac{1}{2}\Delta M_{ba}}{K} \right) \tag{7.3.1}$$

$$\theta_b \text{ (relative)} = \left(\frac{\Delta M_{ba} - \frac{1}{2}\Delta M_{ab}}{K} \right) \tag{7.3.2}$$

Equations (7.3.1) and (7.3.2) are used to compute the relative slopes in the last four lines of Table 7.3.1.

The controlling values of final moments at the left and right ends of each span, M_L and M_R, for the various critical conditions are taken from Tables 7.3.1 through 7.3.7 and entered in Table 7.3.8. Note that the designer's sign convention for bending moment, in which a positive moment causes compression on the top side of the beam, is used in Table 7.3.8.

The moment at the midspan M_s may be determined by superposition of the effect of end moments with that of the simple beam moment due to transverse loading,

$$M_s = M_0 - \tfrac{1}{2}(M_L + M_R)$$

in which M_0 is the moment at the midspan for a simple beam. When the end moments are not equal, the maximum moment in the span does not occur at midspan, but its value is close to that at midspan.

Table 7.3.8 SUMMARY OF RESULTS OF MOMENT DISTRIBUTION ANALYSIS

LINE NUMBER		SPAN	M_L	M_R	VALUES OF M_L AND M_R FROM TABLE NO.	$M_s = M_0 - \tfrac{1}{2}(M_L + M_R)$
1	For maximum positive	AB	-712	-804	7.3.1	$+492$
2	moment at	BC	-790	-772	7.3.1	$+469$ ($M_0 = +1250$)
3	midspan	CD	-777	-777	7.3.1	$+473$
4	For minimum positive or maximum	AB	-184	-399	7.3.1	$+125$
5	negative	BC	-340	-331	7.3.1	$+81$ ($M_0 = +417$)
6	moment at midspan	CD	-332	-334	7.3.1	$+84$
7		A of AB	-712	(-804)	7.3.1	
8		B of AB	(-668)	-911	7.3.2	
9	For maximum negative	B of BC	-888	(-729)	7.3.2	
10	moment at	C of BC	(-745)	-878	7.3.4	
11		C of CD	-882	(-732)	7.3.4	
12		D of CD	(-731)	-883	7.3.6	
13		A of AB	-184		7.3.1	
14	For minimum	B of AB		-293	7.3.3	
15	negative or maximum	B of BC	-241		7.3.3	
16	positive	C of BC		-225	7.3.5	
17	moment at	C of CD	-228		7.3.5	
18		D of CD		-228	7.3.7	

NOTE: Values of moments in $wL^2(10^{-4})$. Numbers in parenthesis are to be used in shear envelope computations (see Sec. 7.6).

For the purpose of illustration, the moment diagram for the maximum and minimum positive moments at midspan, using results of the first loading condition in Fig. 7.3.2, is shown in Fig. 7.3.3. First the simple beam moment diagrams for the total load and for dead load only are drawn to scale in Fig. 7.3.3(a). Next the end moments for each span are taken from Table 7.3.8 and plotted in Fig. 7.3.3(b). Note that because of symmetry, only the moment diagrams for the first three spans are shown. The final moment diagrams in Fig. 7.3.3(b) are drawn by rotating the base lines for zero end moments in Fig. 7.3.3(a) to those connecting the end moments in Fig. 7.3.3(b).

In this example the dead load has been applied on all the spans in each of the seven loading conditions of Fig. 7.3.2. This has been done because an important purpose of the example is to justify the use of approximate moment coefficients as discussed in the next section. An alternative approach would be to use eight loading conditions; one condition for dead load only plus the same seven live load conditions but without the dead load applied simultaneously. Another alternative for the live load cases would be to load one span at a time and carry out the moment distribution. Because of symmetry only three distributions would be required; live load in spans 1, 2, and 3 separately. Moments and shears due to load on spans 4, 5, or 6 can be deduced from load on spans 3, 2, or 1, respectively. The procedure of loading one span at a time with live load will require less moment distribution operations but more effort in combining the appropriate cases to obtain the maximum and minimum moments at the various locations. Sepa-

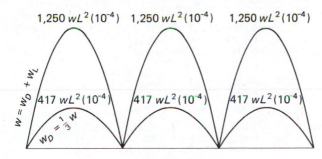

(a) Simple beam moment diagrams

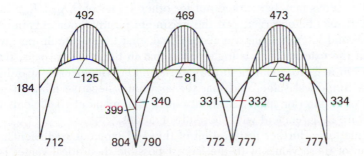

(b) Moment diagrams for maximum and minimum positive midspan moments

Figure 7.3.3 Moment diagrams for Example 7.3.1.

ration of dead and live load reactions to columns is frequently desirable, because if a loading combination that includes wind or earthquake must be considered, different live load factors U must be used according to ACI-9.2.

7.4 ACI MOMENT COEFFICIENTS

ACI-8.3 specifies that (1) the theory of elastic analysis is to be used in analyzing frames or continuous construction; (2) except for prestressed concrete, approximate methods of frame analysis may be used for buildings of usual types of construction, spans, and story heights; and (3) except for prestressed concrete, design for the moments and shears as listed in ACI-8.3.3 is satisfactory in the case of two or more approximately equal spans (the larger of two adjacent spans not exceeding the shorter by more than 20%) with loads uniformly distributed, where the unit live load does not exceed three times the unit dead load. One approximate method would be the method of isolating one floor at a time (with its upper and lower columns) as permitted by ACI-8.9.1 and discussed in Sec. 7.3. The ACI Code leaves open to the designer the precise definition of the term "approximate methods." The reader is referred to the subassembly model defined in Fig. 7.1.2.

It seems desirable to examine the moment coefficients in ACI-8.3.3. Observation of Table 7.3.8 in the preceding section shows that for a six-span frame in which the ratio of ΣK_{col} to K_{bm} is 4 and the ratio of w_L to w_D is 2, critical values of moments may vary within the following limits:

Exterior span:	
Exterior end	$-0.0184wL^2$ and $-0.0712wL^2$
Midspan	$+0.0125wL^2$ and $+0.0492wL^2$
Interior end	$-0.0293wL^2$ and $-0.0911wL^2$
First interior span:	
Exterior end	$-0.0241wL^2$ and $-0.0888wL^2$
Midspan	$+0.0081wL^2$ and $+0.0469wL^2$
Interior end	$-0.0225wL^2$ and $-0.0878wL^2$
Second interior span:	
Exterior end	$-0.0228wL^2$ and $-0.0882wL^2$
Midspan	$+0.0084wL^2$ and $+0.0473wL^2$
Interior end	$-0.0228wL^2$ and $-0.0883wL^2$

Similar values may be worked out for other values of $(\Sigma K_{col})/K_{bm}$ and of w_L/w_D. It may be observed that the maximum positive moments in the first and second interior spans are about equal, that the maximum positive moment in the exterior span is higher than that in the interior spans, that the maximum negative moment at the interior end of the exterior span has the largest numerical value, and that the maximum negative moments at both ends of all interior spans are about equal. The moment coefficients in ACI-8.3.3 are in agreement with these observations.

As a matter of further justification of the ACI moment coefficients, a comparison of these values with the largest possible theoretical values [4] is shown in Table 7.4.1. Certainly the largest possible theoretical values will be for the case of $w_L/w_D = 3$, which is the limit set forth in ACI-8.3.3.

Table 7.4.1 COMPARISON OF ACI MOMENT COEFFICIENTS WITH THEORETICAL VALUES [4]

| LOCATION OF SECTION | ACI | THEORETICAL COEFFICIENTS | | | |
		VALUE	NUMBER OF SPANS	$(\Sigma K_{col})/K_{bm}$	w_L/w_D
Positive moment					
End spans					
If discontinuous end is unrestrained	$+\frac{1}{11}$	$+0.094$	3	0	3
If discontinous end is integral with the support	$+\frac{1}{14}$	$+0.073$	3	0.5	3
Interior spans	$+\frac{1}{16}$	$+0.063$	4 or more	0.5	3
Negative moment at exterior face of first interior support					
Two spans	$-\frac{1}{9}$	-0.111	2	0.5	3
More than two spans	$-\frac{1}{10}$	-0.107	4 or more	0.5	3
Negative moment at other faces of interior supports	$-\frac{1}{11}$	-0.092	4 or more	2	3
Negative moment at face of all supports for (a) slabs with spans not exceeding 10 ft and (b) beams and girders where $(\Sigma K_{col})/K_{bm}$ exceeds 8 at each end of the span	$-\frac{1}{12}$	-0.083	any number	∞	any ratio
Negative moment at interior faces of exterior supports for members built integrally with their supports					
Where the support is a spandrel beam or girder	$-\frac{1}{24}$	-0.036	4 or more	0.5	3
		-0.050	4 or more	1	3
Where the support is a column	$-\frac{1}{16}$	-0.064	4 or more	2	3

In this instance, secondary live load moments with signs opposite to that of dead load occur infrequently; or, if they do occur, their values are small. Thus as long as the ratio of live load to dead load is well within 3 and span lengths do not differ considerably, there will be no moment reversal so that the ACI moment coefficients are reasonably close and, in general, on the safe side. It may be noted, however, that the ACI moment coefficients are in terms of wL_n^2, in which L_n is the clear span for positive moment and the average of the two adjacent clear spans for negative moment, negative moments being those at the face of supports and not at the centerline of support. On the other hand, the theoretical coefficients are in terms of wL^2, in which L is the distance between centerlines of supports, and coefficients for negative moments refer to those at the centerlines of support. Although span lengths between centerlines of supports are always used in elastic analysis, ACI-8.7.3 states that for beams built integrally with supports, moments at faces of supports may be used for design.

Logically, any approximate analysis that provides adequate strength and no excessive deflection at service load should be acceptable. Furlong [5] has proposed an alternative to the use of ACI-8.3. The alternate is based on limit analysis as discussed in Sec. 10.12, with requirements that every component be ductile and strong enough, and that no reinforcement yields at service load.

7.5 ACI MOMENT DIAGRAMS

In designing any span in a multispan continuous rigid frame subjected to live load where moment coefficients are used, two primary sets of shear and moment diagrams are inherently being assumed. In the general case, one will result from the loading position that causes maximum positive moment

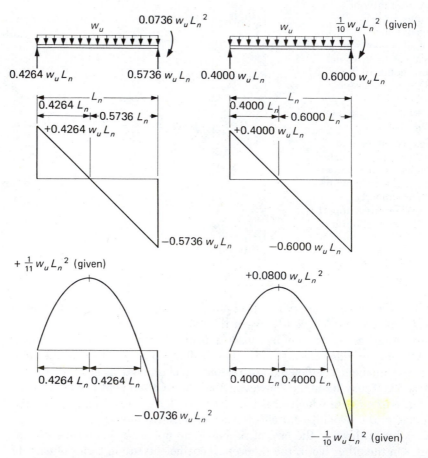

(a) Maximum in the positive zone (b) Maximum in the negative zone

Figure 7.5.1 Exterior span with discontinuous end unrestrained.

within the span, and the other will result from assuming that the maximum negative moments occur simultaneously at both ends. Actually the loading position that causes maximum moment at one end is different from that which causes maximum negative moment at the other end; however, by assuming that both maximum negative end moments occur simultaneously, a critical curve having greater magnitude than either of the two actual curves is obtained.

The ACI moment coefficients (ACI-8.3.3) as shown in Table 7.4.1 are the common values from the two primary conditions as described in the preceding paragraph. No secondary moment coefficients are suggested by the Code, the reason being that so long as the design live to dead load ratio is limited to 3, no moment reversal would occur; that is, there can be only positive moment in the midspan region and only negative moment in the support region.

In Figs. 7.5.1 to 7.5.4, inclusive, are shown the two primary sets of

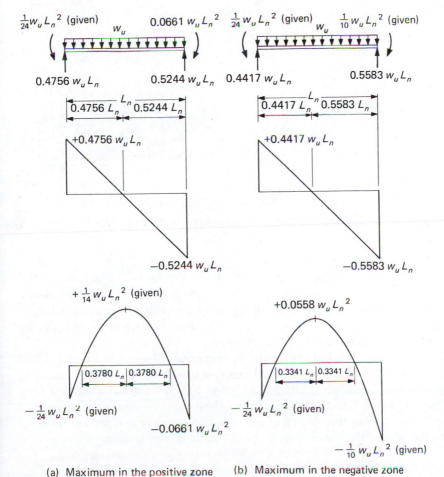

(a) Maximum in the positive zone   (b) Maximum in the negative zone

Figure 7.5.2 Exterior span with exterior support built integrally with spandrel beam or girder.

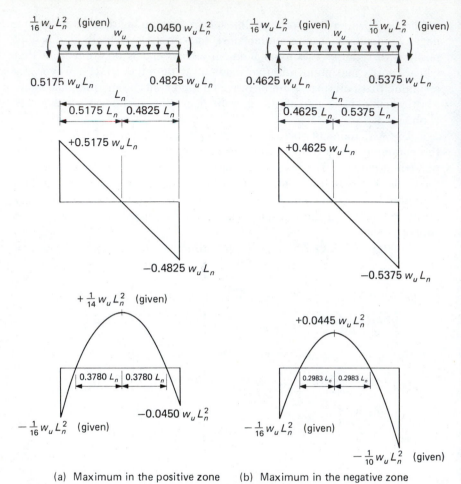

(a) Maximum in the positive zone   (b) Maximum in the negative zone

Figure 7.5.3 Exterior span with exterior support built integrally with column.

shear and moment diagrams, for the various conditions, to be used in the design of continuous spans in accordance with the ACI moment coefficients. Note that these diagrams are applicable to the actual *clear* span, which is also used to compute the positive moment. For negative moment, L_n is the average of adjacent clear spans (ACI-8.0).

The reader should utilize the fundamentals of shear and moment diagrams to verify the numerical ordinates on these diagrams. For instance, in the case of maximum positive moment in Fig. 7.5.2(a) the distance x from the left support to the point of zero shear may be determined from the relationship that the change of moment between any two sections is equal to the area of the shear diagram between these two sections. Thus

$$\frac{w_u x^2}{2} = \left(\frac{1}{24} + \frac{1}{14}\right) w_u L_n^2$$

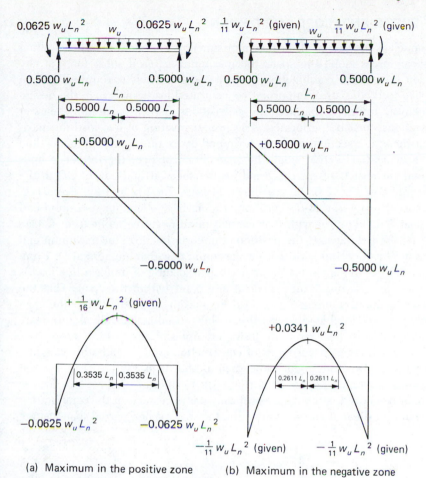

(a) Maximum in the positive zone   (b) Maximum in the negative zone

Figure 7.5.4 Interior span.

from which

$$x = 0.4756L_n$$

Also, the distance x between the section of maximum positive moment to the point of zero moment is

$$\frac{w_u x^2}{2} = \frac{1}{14} w_u L_n^2$$

from which

$$x = 0.3780L_n$$

Any time a designer uses moment coefficients for determining the factored moments, as permitted by the approximate method of ACI-8.3.3, the moment diagrams that correspond to such coefficients should be used when establishing bar bend or cutoff locations. The use of a given moment coefficient implies a statically compatible moment diagram.

7.6 SHEAR ENVELOPE FOR DESIGN

Inasmuch as the design of shear reinforcement is dependent on the variation of shear forces along the span, the maximum factored shear force at any section due to the combination of dead and live loads is required. When the live load consists of important concentrated loads such as in the design of highway bridge spans, accurate calculations of such maximum shears at all sections must be performed. The proper position of live load for maximum shear at a section can be determined by examining the influence line for shear at that section along the span. For instance, the influence lines for end shear at C of span CD and for the shear at midspan of CD in the rigid frame of Fig. 7.6.1(a) are shown in Figs. 7.6.1(b) and (c). From Fig. 7.6.1(b), it is noted that the position of uniform live load for maximum end shear at C is identical with that for maximum negative moment at C [see Fig. 7.2.1(c)]. Likewise, the position of uniform live load for maximum end shear at D is identical with that for maximum negative moment at D. From Fig. 7.6.1(c), it is apparent that partial span loading of uniform live load is indicated to give maximum shear at any point within the span. That the correct position of uniform live load for maximum shear within the span involves partial span loading has already been used in statically determinate situations in the shear strength design examples of Sec. 5.12.

In buildings of usual types of construction, spans, and story heights, wherein the idealized rigid frame such as shown in Fig. 7.6.1(a) is taken into consideration, the use of partial span loading of uniform live load is commonly ignored, although theoretically it is necessary for the computation of maximum shear at any section within the span. When partial span loading

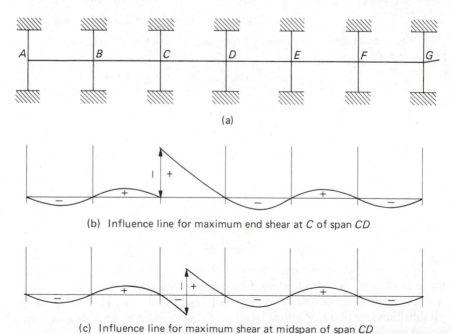

(b) Influence line for maximum end shear at C of span CD

(c) Influence line for maximum shear at midspan of span CD

Figure 7.6.1   Influence lines for shear in rigid frames.

is not considered necessary, the maximum shears at only the ends can be used to establish an approximate shear envelope. Note that the loading condition for maximum shear at one end is different from that for maximum shear at the other end. These two critical shear diagrams for each span, when a continuity analysis is performed, may be easily obtained by using the values of the negative end moments (including those in parenthesis) such as those contained in lines 7 through 12 of Table 7.3.8.

When the ACI moment coefficients are used, it is generally assumed that the shear diagrams accompanying the critical moment diagrams as shown in Fig. 7.5.1 to 7.5.4 may be used in the design.

SELECTED REFERENCES

1. *Continuity in Concrete Building Frames* (4th ed.). Chicago: Portland Cement Association, 1959.
2. Chu-Kia Wang and Charles G. Salmon. *Introductory Structural Analysis*. Englewood Cliffs, N.J.: Prentice-Hall, Inc., 1984 (Chaps. 6 and 8).
3. Chu-Kia Wang. *Matrix Methods of Structural Analysis* (2nd ed.). Madison, Wisconsin: American Publishing Company, 1970.
4. A. J. Boase and J. T. Howell. "Design Coefficients for Building Frames," *ACI Journal, Proceedings,* **36**, September 1939, 21–36.
5. Richard W. Furlong and Carlos Rezende. "Alternate to ACI Analysis Coefficients," *Journal of the Structural Division,* ASCE, **105**, November 1979 (ST 11), 2203–2220.

PROBLEMS

7.1 Compute and draw to scale the bending moment and shear diagrams for the loading conditions 1 through 7 of Fig. 7.3.2; that is, verify the results given in Tables 7.3.2 through 7.3.7, by using any structural analysis method assigned by the instructor.

7.2 Compute and draw to scale the envelope of bending moments (diagram showing range over which bending moment may vary) due to factored loads for the beams of the frame of the accompanying figure, using a continuity analysis method such as moment distribution. The uniform factored dead load is 1 kip/ft and the uniform factored live load is 2 kips/ft.

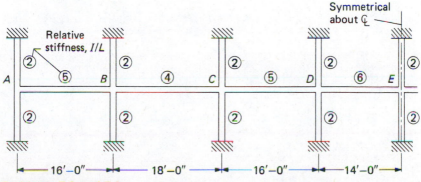

Problems 7.2 and 7.3

7.3 For the beams of the frame in the accompanying figure, compute and draw the bending moment envelope using the coefficients of ACI-8.3.3. (If Prob. 7.2 has also been solved, compare the moments by giving the percentage difference in the maximum values obtained by coefficients as compared with the more exact values of a continuity analysis.)

7.4 Consider an equal-span, uniform-section continuous beam over many supports. Compute and show diagrams for dead load coefficients of wL^2 for moments at critical locations in the exterior and first interior spans. Could the coefficients for the first interior span be applied appropriately to the other interior spans? Recommend dead load coefficients for equal spans.

7.5 Repeat Prob. 7.4 for the case where alternate spans are 20% longer than the others ($1.2L$), taking the exterior span as a short one (L).

7.6 For an equal-span, uniform-section continuous beam over many supports, compute and show diagrams for live load coefficients in terms of wL^2 for maximum negative moments at (**a**) the first interior support; (**b**) the second interior support; and (**c**) the typical interior support. Recommend live load coefficients.

7.7 For an equal-span, uniform-section continuous beam over many supports, compute and show diagrams for live load coefficients in terms of wL^2 for the maximum positive moment in (**a**) the exterior span; (**b**) the first interior span; and (**c**) the typical interior span. Recommend live load coefficients.

8

Design of One-Way Slabs

8.1 DEFINITION

One of the most common types of floor construction is the slab–beam–girder system, as has been briefly described in Sec. 7.1. The slab panel, bounded on its two long sides by the beams and on its two short sides by the girders, is usually at least twice as long as it is wide. In such a condition the dead and live load acting on the slab area may be considered as being entirely supported in the short direction by the beams, hence the term "one-way slab." Two-way floor systems with or without beams on column lines are treated in Chap. 16; and ribbed-joist floor construction is described in Secs. 10.10 and 10.11.

The determination of an optimum floor framing plan—that is, the spacing of columns, beams, and girders—depends on both the functional and the structural requirements. In most cases preliminary calculations are necessary for several different layouts, and after comparison the most suitable and economical plan is chosen.

8.2 DESIGN METHODS

Since the loading on a one-way slab is nearly all transferred in the short direction, such a slab continuous over several supports may be treated as a beam. Because sufficiently accurate results are obtained, ACI-8.3.3 permits the use of moment and shear coefficients in the case of two or more approximately equal spans (the larger of two adjacent spans not exceeding the shorter by more than 20%) with loads uniformly distributed, where the unit

Lennox Square Buildings, Atlanta. (Courtesy of Portland Cement Association.)

live load does not exceed three times the unit dead load. These coefficients are in terms of clear span L_n and the values given are for critical locations, that is, faces of support for shears and negative moments and midspan regions for positive moments.

When the conditions of ACI-8.3.3 are not satisfied, an elastic analysis is required. Approximate methods are permitted in "buildings of usual types of construction, spans, and story heights."

One such approximate procedure applicable to equal spans, where dead load and live load are to be treated separately (say, when $w_L/w_D > 3$), is to use the coefficients given in Appendix 3 of the 1940 Joint Committee Report [1]. Other approximate methods are available for determining critical moments and shears, short of a formal elastic analysis according to a procedure such as that described in Chap. 7, which does not seem justified on a slab. The Portland Cement Association publication, *Continuity in Concrete Building Frames* [2], provides moment coefficients that are nearly as exact as those which would be obtained by actually carrying out an elastic analysis.

Generally, span lengths between centers of supports are used in elastic analysis, notably in the computation for fixed-end moments and stiffness factors. Similarly, the results of an elastic analysis show only the negative bending moments at centers of supports. It will be shown that the negative bending moment at the face of support, the value of which should be used in the design of the member itself, may be obtained approximately by reducing from the maximum value at the center of support by a quantity equal to $Vb/3$, where V is the shear force at either the center or at the face of support, and b is the width of support.

Consider an ideal situation where a member AB is fixed at points A and B and subjected to a uniform load, as shown in Fig. 8.2.1. The elastic curve is shown as $AA'B'B$ in Fig. 8.2.1(a), where there are rotation and deflection at A' and B'. The bending moment at A' (Fig. 8.2.1) is

$$M = -\frac{wL^2}{12} + \frac{wL}{2}\left(\frac{b}{2}\right) - \frac{w(b/2)^2}{2} = -\left(\frac{wL^2}{12} - \frac{wLb}{4} + \frac{wb^2}{8}\right) \quad (8.2.1)$$

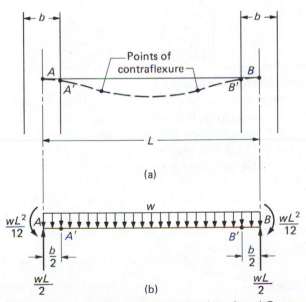

Figure 8.2.1 Single span fixed at points A and B.

Now if the same member is considered to be fixed at A' and B' as shown in Fig. 8.2.2, the bending moment at A' (Fig. 8.2.2) is

$$M = -\frac{w(L-b)^2}{12} = -\left(\frac{wL^2}{12} - \frac{wLb}{6} + \frac{wb^2}{12}\right) \quad (8.2.2)$$

The quantity in Eq. (8.2.2) is numerically larger than in Eq. (8.2.1) by an amount $wLb/12$, provided the terms involving b^2 are neglected. However, the condition in Fig. 8.2.2 is believed to be more correct because at a relatively stiff support, such as a column or girder, there can be very little change in rotation or deflection between the points A and A'. Thus the moment at the face of support is less than that at the center of support by a quantity equal to $Vb/3$ [Eq. (8.2.2)] rather than $Vb/2$ [Eq. (8.2.1)] if V is taken as $wL/2$ and, again, if the term involving b^2 is neglected (see also Ref. 2).

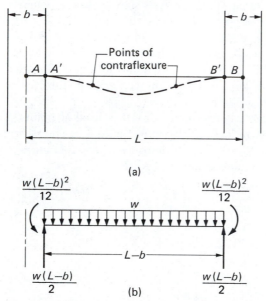

(a)

(b)

Figure 8.2.2 Single span fixed at points A' and B'.

On the other hand, the moment at the center of span in Fig. 8.2.2 is less than that in Fig. 8.2.1. Inasmuch as the elastic analysis results are associated with the concept of Fig. 8.2.1, midspan moments so obtained are on the safe side, and, in general, no attempt is necessary to correct them to the concept of Fig. 8.2.2.

As the span L becomes large compared to the support width b, the common practice of using the moment envelope value at the face of support (scaled, for example) is reasonable. That practice is equivalent to subtracting $Vb/2$ from the center of support value.

8.3 THICKNESS OF SLAB

In designing a one-way slab, a typical imaginary strip 12 in. wide is usually considered. The continuous slab may then be designed as a continuous beam having a known width of 12 in.; the slab thickness is the only unknown.

The thickness of the slab depends on the deflection, bending, and shear requirements. Deflection requirements are imposed to prevent excessive deformations that might adversely affect the serviceability of the structure. According to ACI-Table 9.5a, one-way slabs must have at least a minimum slab thickness (for Grade 60 steel) of $L/20$, $L/24$, $L/28$, or $L/10$ depending on whether L is the length of a simply supported, a one end continuous, a both ends continuous, or a cantilever span. If the slab supports or is attached to construction likely to be damaged by large deflections, deflections must be computed and shown to satisfy the limits of ACI-Table 9.5b. An extensive treatment of deflections will be found in Chap. 14.

The flexural strength M_n required to resist the factored moment M_u is provided in accordance with the principles of Chap. 3. The desired coefficient of resistance R_n can be computed for the desired reinforcement ratio ρ. Then the required $M_n = M_u/\phi = R_n bd^2$. In equal continuous spans the negative moment at the exterior face of the first interior support is the largest; therefore this negative moment should be used to establish the slab thickness.

The shear requirement does not usually control, but it must be checked. Because of practical space limitations, shear reinforcement is not used in a slab; thus the governing factored shear V_u, which in equal continuous spans occurs at the exterior face (the distance d therefrom) of the first interior support, must be kept below ϕV_c of ACI-11.2 for lightweight concrete and ACI-11.3 for nonprestressed normal-weight concrete.

According to ACI-7.7.1c, the concrete protective covering for reinforcement (#11 and smaller) in slabs shall be not less than $\frac{3}{4}$ in. at surfaces not exposed directly to the ground or the weather. Also, when the top of a monolithic slab is the wearing surface and when unusual wear is expected as in buildings of the warehouse or industrial class, it has been customary to use an additional depth of $\frac{1}{2}$ in. of concrete protective covering over that required by the design of the member. The ACI Code no longer specifies such extra slab thickness for wearing surface and in ACI-8.12 *permits*, on the discretion of the designer, a monolithic floor finish to be considered as part of the structural member. For nonstructural purposes, any concrete floor finish may be considered as part of the required cover.

EXAMPLE 8.3.1 Establish the thickness of the floor slab shown in the second-floor framing plan of Fig. 8.3.1 (see Fig. 8.4.2 for a section through slabs 2S1-2S2-2S3) for a service live load of 100 psf. Use $f'_c = 3000$ psi, $f_y = 40,000$ psi, and the strength method of the ACI Code. Assume an exterior staircase so that no openings are to be made in the slabs.

Solution: Since live load does not appear to exceed three times the dead load, the design will be done in accordance with ACI moment coefficients and their corresponding shear and moment diagrams (see Sec. 7.5).

(a) Minimum thickness. For spans with one end and both ends continuous, the respective minimum thicknesses, h, from ACI-Table 9.5a are $L/24$ and $L/28$. Since the table values are for $f_y = 60,000$ psi, they must be corrected for $f_y = 40,000$ psi by multiplying by 0.8.

$$\min h = \frac{L}{24}(0.8) = \frac{L}{30} = \frac{13(12)}{30} = 5.2 \text{ in. (for 2S1)}$$

$$\min h = \frac{L}{28}(0.8) = \frac{L}{35} = \frac{13(12)}{35} = 4.5 \text{ in. (for 2S2 and 2S3)}$$

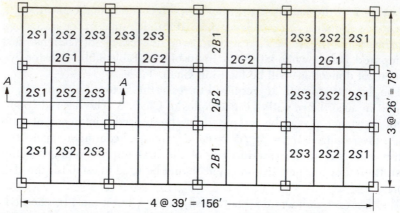

Figure 8.3.1 Second-floor framing plan (for section *A-A*, see Fig. 8.4.2).

Assume a $4\frac{1}{2}$-in. slab. The weight of the slab is $(4.5/12)(0.15) = 0.056$ kips/sq-ft. If deflection is of concern in the end span, that span could be made 5 in. thick or compression reinforcement could be used (see Example 14.12.2 for deflection calculation for a similar end span).

(b) Bending moment requirement. Assume width of supporting beams to be 13 in. Note that 13 in., an uncommon dimension for width, is used rather than 12 in. to permit easier following of the numerical calculations.

$$w_D = 0.056(1.4) = 0.079 \text{ kip/ft/ft of width}$$

$$w_L = 0.100(1.7) = 0.170 \text{ kip/ft/ft of width}$$

$$\text{clear span} = 13 - \tfrac{13}{12} = 11.92 \text{ ft}$$

$$M_u = \tfrac{1}{10}(0.079 + 0.170)(11.92)^2 = 3.54 \text{ ft-kips/ft of width}$$

Choose a reinforcement percentage ρ equal to about $0.375\rho_b$, or one-half the maximum permitted by the ACI Code, in order to have reasonable deflection control. From Table 3.6.1,

$$0.375\rho_b = 0.5(0.0278) = 0.0139$$

Then using Eq. (3.8.4) or Fig. 3.8.1, find corresponding R_n,

$$m = \frac{f_y}{0.85f'_c} = \frac{40,000}{0.85(3000)} = 15.7$$

$$R_n = \rho f_y(1 - \tfrac{1}{2}\rho m)$$

$$= 0.0139(40,000)[1 - 0.5(0.0139)(15.7)] = 495 \text{ psi}$$

$$\text{required } d = \sqrt{\frac{M_u}{\phi R_n b}} = \sqrt{\frac{3.54(12,000)}{0.90(495)12}} = 2.82 \text{ in.}$$

Assume #5 bars ($d_b = 0.625$) and $\frac{3}{4}$-in. cover,

$$\text{required } h = 2.82 + 0.31 + 0.75 = 3.88 \text{ in.}$$

Use $h = 4\frac{1}{2}$ in.

$$\text{provided } d = 4.50 - 0.31 - 0.75 = 3.44 \text{ in.}$$

(c) Shear requirement.

$$\max V_u = 1.15\frac{w_uL_n}{2} = 1.15\frac{0.249(11.92)}{2} = 1.71 \text{ kips/ft of width}$$

The shear strength, ϕV_c, for a member without shear reinforcement is

$$\phi V_c = \phi[2\sqrt{f_c'}bd]$$
$$= 0.85(2\sqrt{3000})(12)(3.44)\tfrac{1}{1000} = 3.84 \text{ kips/ft} > 1.71 \text{ kips/ft}$$

The slab is acceptable for a member without stirrups. Note that to be strictly correct the shear force at a distance d from the face of support should have been used, in which case the factored shear would have been even less than 1.71 kips/ft.

8.4 CHOICE OF REINFORCEMENT

The choice of reinforcement depends primarily on the steel area and secondarily on development length requirements. The steel areas required at the principal sections, namely those at the middle and at the ends of each span, are first computed. Then a *tentative* choice of reinforcement may be made. A common arrangement of reinforcement in floor slabs is as shown in Fig. 8.4.1, in which the straight and bent (trussed) bars are placed alternately. The straight bars across the bottom are ordinarily one size smaller than the bent bars. Frequently, in thin slabs of under 5 in. of thickness, straight bars are used in both the top and bottom in all spans. Some designers prefer straight bars in all cases.

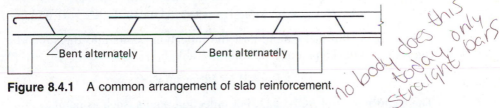

Bent alternately   Bent alternately

Figure 8.4.1   A common arrangement of slab reinforcement.

no body does this today. only straight bars

Development length requirements should next be examined. The positive moment bars that are bent up into the negative moment region have ample embedment. The bars that continue along the bottom of the slab and extend into the support at least 6 in. (ACI-12.11.1) must be checked for adequate development at points of inflection (ACI-12.11.3). In addition, the bars extending straight across the top beyond the bend-down points must be checked for satisfactory embedment (ACI-12.12.3).

Some other limitations affecting the choice of reinforcement in slabs are (1) that in structural slabs of uniform thickness (ACI-10.5.3) the minimum amount of reinforcement in the direction of the span shall not be less than that required for shrinkage and temperature reinforcement (ACI-7.12), and (2) that the principal reinforcement shall be centered not farther apart than three times the slab thickness nor more than 18 in. (ACI-7.6.5).

EXAMPLE 8.4.1 Choose the arrangement of reinforcement in the $4\frac{1}{2}$-in. floor slab 2S1, 2S2, and 2S3, as designed in Example 8.3.1.

Solution: (a) Area requirements. The ACI moment coefficient, the bending moment, the steel area required, and the tentative choice of reinforcement at the critical section are shown in lines 1 to 6 of Table 8.4.1. The arrangement of reinforcement is shown in Fig. 8.4.2. Choice of this steel should begin at the typical interior support. In this case, #4 bars at 6 in. are chosen. At the middle of the first and typical interior spans, the required area would be furnished by #3 straight (st) bars at 12 in. in addition to the #4 bent (bt) bars at 12 in. which were already there. As a matter of practical judgment, however, some designers would not use bars smaller than #4 for main reinforcement. Since the decision has now been made on the 6-in. and 12-in. spacings, the bent-bar size in the exterior span follows automatically. These spacings satisfy the limitations of ACI-7.6.5.

(b) Development length requirements at inflection points. In order to confirm the choice of longitudinal reinforcement made on the basis of the required areas only, it is necessary to review the development length requirements. The requirements of ACI-12.11.3 must be checked at the exterior supported end (point 1 of Fig. 8.4.2) and at the inflection points 2 to 6. In addition, embedment equal to the required development length must be provided in both directions from the maximum moment points at the faces of supports (points 7 through 12). Finally, embedment of the straight portion of the bars at the top of the slab beyond extreme points of inflection (points 13 through 17) must satisfy ACI-12.12.3.

From the ACI shear and moment diagrams in Figs. 7.5.2 and 7.5.4, the shears at inflection points on the typical 1-ft width of slab are

$$V_1 = V_2 = 0.3780 w_u L_n = 0.3780(0.249)(11.92) = 1.13 \text{ kips}$$

$$V_3 = V_4 = V_5 = V_6 = 0.3535 w_u L_n = 0.3535(0.249)(11.92) = 1.05 \text{ kips}$$

For the bars (#3) that extend past the inflection points into the supports, the required development length (from Table 6.7.1) is

$$L_d \ (\#3) = 12 \text{ in. min}$$

The requirement of ACI-12.11.3 at inflection points (such as point 2) is

$$\frac{M_n}{V_u} + L_a \geq L_d$$

For #3 @ 12 in.,

$$C = 0.85(3)12a = 30.6a$$

$$T = 0.11(40) = 4.4 \text{ kips}$$

$$a = \frac{4.4}{30.6} = 0.14 \text{ in.}$$

$$M_n = 4.4(3.44 - 0.07)\tfrac{1}{12} = 1.24 \text{ ft-kips/ft}$$

$$V_u = 1.13 \text{ kips (computed above)}$$

$$L_a = 12 d_b \text{ max} = 4.5 \text{ in.}$$

$$\frac{1.24(12)}{1.13} + 4.5 = 13.1 + 4.5 = 17.6 \text{ in.} > L_d = 12 \text{ in.} \qquad \text{OK}$$

This calculation applies identically to other inflection points (3 through 6).

Table 8.4.1 REINFORCEMENT FOR ONE-WAY SLAB OF EXAMPLE 8.4.1, USING ACI MOMENT COEFFICIENTS

LINE NUMBER	2S1			2S2			2S3		
	SUPPORT	MIDDLE	SUPPORT	SUPPORT	MIDDLE	SUPPORT	SUPPORT	MIDDLE	SUPPORT
1. ACI moment coefficient	$-\dfrac{1}{24}$	$+\dfrac{1}{14}$	$-\dfrac{1}{10}$	$-\dfrac{1}{11}$	$+\dfrac{1}{16}$	$-\dfrac{1}{11}$	$-\dfrac{1}{11}$	$+\dfrac{1}{16}$	$-\dfrac{1}{11}$
2. M_u (ft-kips) $=$ line (1) $\times\,0.249\,(11.92)^2$ $\quad=$ line (1) $\times\,35.4$	-1.48	$+2.53$	-3.54	-3.22	$+2.21$	-3.22	-3.22	$+2.21$	-3.22
3. Required $R_n = \dfrac{\text{line (2)} \times 12{,}000}{0.9(12)(3.44)^2}$ (psi) $\quad=$ line (2) $\times\,94.0$	139	238	333	303	208	303	303	208	303
4. Required $\rho \approx \dfrac{\text{line (3)} \times 0.0139}{495}$ $\quad= \dfrac{\text{line (3)}}{35{,}600}$	0.0039 (0.005 min.)	0.0067	0.0094	0.0085	0.0059	0.0085	0.0085	0.0059	0.0085
5. Required $A_s =$ line (4) $\times$ 12(3.44) (sq in./ft) $\quad=$ line (4) $\times\,41.3$	0.21	0.28	0.39	0.35	0.25	0.35	0.35	0.25	0.35
6. Provided A_s (sq in./ft)	#4 @ 12 st (0.20)	#4 @ 12 bt #3 @ 12 st (0.31)	#4 @ 12 bt #4 @ 12 st (0.39)	#4 @ 12 bt #4 @ 12 bt (0.39)	#4 @ 12 bt #3 @ 12 st (0.31)	#4 @ 12 bt #4 @ 12 bt (0.39)	#4 @ 12 bt #4 @ 12 bt (0.39)	#4 @ 12 bt #3 @ 12 st (0.31)	

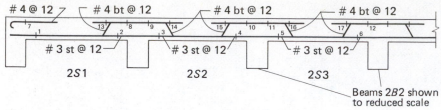

Figure 8.4.2 Longitudinal reinforcement in Example 8.4.1.

At the exterior end, which might be considered a simple support, the requirement of ACI-12.11.3

$$1.30 \frac{M_n}{V_u} + L_a \geq L_d$$

is satisfied by inspection.

(c) Cutoff points for negative moment reinforcement. The distance from the face of support to the cutoff location (beyond point 13, for instance) must be (1) greater than L_d, and (2) adequate to satisfy ACI-12.12.3. For #4 bars, from Table 6.7.1

$$L_d = 12 \text{ in.}$$

To satisfy ACI-12.12.3, the moment diagram of Fig. 7.5.2(b) corresponding to the ACI coefficients may be used to locate the point of inflection. The required distance to the cut location is

$$(0.5583 - 0.3341)(11.92) + \frac{11.92}{16} = 3.42 \text{ ft}$$

which compares favorably with the commonly used value of 0.3 of the clear span, as suggested by the *ACI Detailing Manual—1980* [3] (see pp. 21–23).

(d) Bend-up or bend-down location. Since the bend slope is usually about 45°, the bend can be defined by locating either the bend-up or the bend-down location. The moment diagram of Fig. 7.5.2(a) can be used to determine the bend-up location near the right end of slab 2S1.

The percent of moment capacity remaining after the bend-up is approximately proportional to the percent of total positive moment steel area remaining in the bottom of the slab.

Moment corresponding to #3 @ 12 in.

$$= \frac{0.11}{0.31} \left(\frac{w_u L_n^2}{14} \right) = 0.0253 w_u L_n^2$$

The theoretical location, measured from the face of support, where the moment is $0.0253 w_u L_n^2$ is

$$0.5244 L_n - 0.5 L_n \sqrt{(\tfrac{1}{14} - 0.0253)8} = 0.22 L_n$$

This compares reasonably close to the suggested value in the *ACI Detailing Manual—1980* [3] of 0.25 of the clear span from the face of support to the bend-up location. The theoretical bend-down location may be similarly determined by referring to the right end portion of the moment diagram shown in Fig. 7.5.2(b).

8.5 CONTINUITY ANALYSIS

When the necessary conditions for using the ACI-8.3.3 moment and shear coefficients are not satisfied, an elastic analysis is required. As discussed in Sec. 8.2, several approaches to the elastic analysis requirement may be used. The adjustment of negative moments permitted (ACI-8.4) under certain conditions should not be applied with any of the moment coefficient methods used to design one-way slabs. The subject of redistribution of bending moment to conform partially with true ultimate load behavior is treated in Chap. 10.

In the interest of obtaining a comparison of answers, an elastic analysis will be made to determine the required steel area at all critical sections of the slab used in Examples 8.3.1 and 8.4.1. It is noted that the choice of reinforcement based on either the ACI coefficients or an elastic analysis will be practically identical, as should be expected for the conditions of this design.

EXAMPLE 8.5.1 Using the results of the slightly approximate elastic method as given in *Continuity in Concrete Building Frames* [2], determine the required steel areas at all critical sections of the continuous slab of Examples 8.3.1 and 8.4.1.

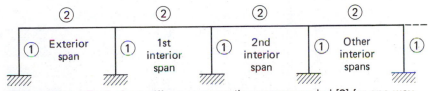

Figure 8.5.1 Equivalent stiffness assumption recommended [2] for one-way continuous slab on beam supports.

Solution: The 12-span continuous slab is supported on 13 beams. The restraining action of the beams on the slab may be accounted for by approximating the torsional stiffness of the supporting beam as a column having half the flexural stiffness as the slab, as shown in Fig. 8.5.1. Using this equivalent stiffness assumption and a ratio of w_L to w_D equal to 2, the theoretical moment coefficients could be determined by a formal elastic analysis as demonstrated in Chap. 7. The results of such an analysis are available [2], as presented in Table 8.5.1. Note that these coefficients are in terms of wL^2, in which L is the distance between centers of supporting beams. Although the moments within the span may be thus computed, the moments at the face of support may be determined by deducting $Vb/3$ (see Fig. 8.5.2) as discussed in Sec. 8.2 from the negative moment at the center of support. The required computations are shown in Table 8.5.1.

The results generally show higher moments than those obtained by the ACI coefficients; the choice of reinforcement may be revised by changing the 6- and 12-in. spacings used in Example 8.4.1 to $5\frac{1}{2}$ and 11 in., respectively.

Table 8.5.1 REINFORCEMENT AREAS REQUIRED USING CONTINUITY ANALYSIS

ITEM	2S1			2S2			2S3		
	SUPPORT	MIDDLE	SUPPORT	SUPPORT	MIDDLE	SUPPORT	SUPPORT	MIDDLE	SUPPORT
Theoretical coefficient from Table 8, *Continuity in Concrete Building Frames* [2]	−0.035	+0.070	−0.106	−0.103	+0.057 −0.007	−0.095	−0.096	+0.061 −0.004	−0.098
M_u (ft-kips) = coefficient × 0.249(13)² = 42.1	−1.47	+2.95	−4.47	−4.34	+2.40 −0.30	−3.58	−4.04	+2.57	−4.21
M_u (ft-kips) at face of support (see Fig. 8.5.2)	−1.07		−3.90	−3.83		−3.12	−3.54		−3.70
Required $R_n = M_u × 94.0$ (see Table 8.4.1)	101	277	367	360	226	294	333	242	348
Required A_s (sq in./ft)	0.12	0.32	0.43	0.42	0.26	0.34	0.39	0.28	0.41

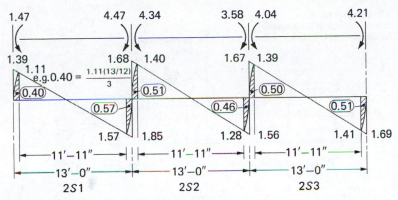

Figure 8.5.2   Shear diagram for Example 8.5.1.

The moment reversal of 0.30 ft-kips may cause a tensile stress at the extreme top of the midspan in 2S2 of

$$\text{tensile stress} = \frac{0.30(12{,}000)}{(12)(4.5)^2/6} = 89 \text{ psi}$$

which is well below $5\phi\sqrt{f_c'} = 5(0.65)\sqrt{3000} = 178$ psi for plain concrete.*

8.6 SHRINKAGE AND TEMPERATURE REINFORCEMENT

Reinforcement for shrinkage and temperature stresses normal to the principal reinforcement is required in structural floor and roof slabs, where the principal reinforcement extends in one direction only (ACI-7.12.1). Further, such reinforcement shall provide (ACI-7.12.2) for the following minimum ratios of reinforcement area to gross concrete area, but in no case shall such reinforcing bars be placed farther apart than five times the slab thickness nor more than 18 in. (ACI-7.12.3):

1. Slabs where Grades 40 or 50 deformed bars are used 0.0020
2. Slabs where Grade 60 deformed bars or welded wire fabric (smooth or deformed) are used 0.0018
3. Slabs where reinforcement with yield strength exceeding 60,000 psi measured at a yield strain of 0.35% is used ... $\dfrac{0.0018(60{,}000)}{f_y}$

EXAMPLE 8.6.1 Design the shrinkage and temperature reinforcement in the floor slab of Examples 8.3.1 and 8.4.1.

*See *Building Code Requirements for Structural Plain Concrete* (ACI 318.1-83). Detroit: American Concrete Institute, 1983 (see Sec. 6.2.1).

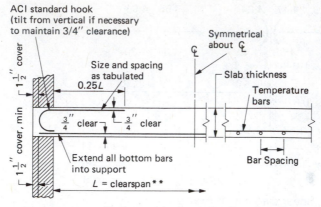

ACI standard hook
(tilt from vertical if necessary
to maintain 3/4" clearance)

Symmetrical
about ₵

Size and spacing
as tabulated
0.25L

Slab thickness

Temperature
bars

$\frac{3}{4}''$ clear — $\frac{3}{4}''$ clear

Extend all bottom bars
into support

Bar Spacing

L = clearspan**

Single span, simply supported

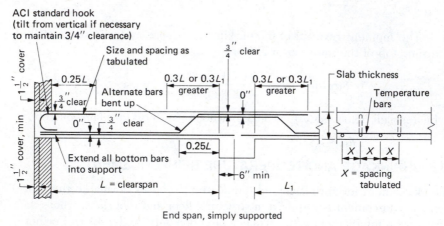

ACI standard hook
(tilt from vertical if necessary
to maintain 3/4" clearance)

Size and spacing as
tabulated

$\frac{3}{4}''$ clear

0.25L

0.3L or 0.3L₁
greater

0''

0.3L or 0.3L₁
greater

Slab thickness

Temperature
bars

Alternate bars
bent up

0'' — $\frac{3}{4}''$ clear

0.25L

X X X

Extend all bottom bars
into support

6'' min

L = clearspan

L₁

X = spacing
tabulated

End span, simply supported

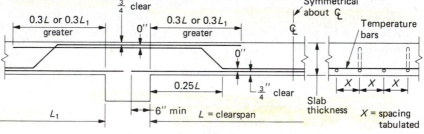

0.3L or 0.3L₁
greater

$\frac{3}{4}''$ clear

0''

0.3L or 0.3L₁
greater

Symmetrical
about ₵

Temperature
bars

0''

0.25L

$\frac{3}{4}''$ clear

X X X

L₁

6'' min

L = clearspan

Slab
thickness

X = spacing
tabulated

Interior span, continuous

Figure 8.7.1 Typical details for one-way solid slabs (from *ACI Detailing Manual*—1980 [3]).

Solution: Area of shrinkage and temperature reinforcement is

$$A_s = \rho bh = (0.0020)(12)(4.5) = 0.11 \text{ sq in.}$$

Use #3 bars at 12-in. spacing.

8.7 BAR DETAILS

Consistent with the shear and moment diagrams given in Chap. 7 for the ACI Code coefficients, acceptable standard bar bend distances, extensions, and anchorage lengths have been developed. For instance, typical bar details in end and interior spans of one-way slabs as presented in the *ACI Detailing Manual-1980* [3] are reproduced in Fig. 8.7.1.

It must be noted that although typical bar details are of importance in office practice, they must be used with discretion and care so that they conform with the prevailing shear and moment diagrams.

SELECTED REFERENCES

1. Joint Committee. *Recommended Practice and Standard Specifications for Concrete and Reinforced Concrete,* submitted to constituent organizations, June 1940, American Concrete Institute, among others.
2. *Continuity in Concrete Building Frames* (4th ed.). Chicago: Portland Cement Association, 1959.
3. ACI Committee 315. *ACI Detailing Manual—1980* (SP-66). Detroit: American Concrete Institute, 1980.

PROBLEMS

All problems are to be worked in accordance with the ACI Code and all stated loads are service loads, unless otherwise indicated.

8.1 Design for a warehouse a continuous one-way slab supported on beams 12 ft on centers as shown in the accompanying figure. Assume that beam stems are 12 in. wide. The dead load is 25 psf in addition to the slab weight, and the live load is 200 psf. Use $f'_c = 3000$ psi, $f_y = 40,000$ psi, and the strength method. Use ACI coefficients if permissible, and use only straight reinforcing bars. (For SI problem, use 3.7-m spans, 300-mm beam widths; superimposed dead load = 1.2 kN/m; live load = 9.6 kN/m; $f'_c = 25$ MPa; $f_y = 300$ MPa.)

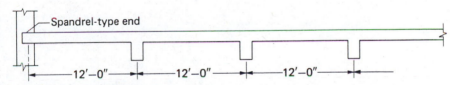

Problems 8.1 through 8.4

8.2 Repeat Prob. 8.1 using a live load of 250 psf with $f'_c = 4000$ psi and $f_y = 60,000$ psi. (For SI: live load = 12.0 kN/m; $f'_c = 30$ MPa; $f_y = 400$ MPa.)

8.3 Repeat Prob. 8.1 using a live load of 300 psf with $f'_c = 4000$ psi and $f_y = 60,000$ psi.

8.4 Repeat Prob. 8.1 using a live load of 350 psf with $f'_c = 4000$ psi and $f_y = 60,000$ psi.

8.5 Design a one-way slab for the conditions shown in the accompanying figure. Assume that beam stems are 12 in. wide. The live load is 175 psf and the slab will not be the final wearing surface. Use $f_c' = 4000$ psi, $f_y = 60,000$ psi, and the strength method. Use alternate bent and straight bar reinforcement if it seems practical.

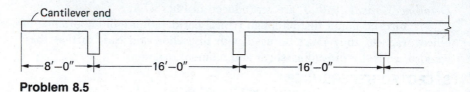

Problem 8.5

9

T-Sections in Bending

9.1 GENERAL

The T-section treatment in this chapter provides the link between the basic concepts of flexural and shear strength in Chaps. 3 through 6 and the monolithic statically indeterminate slab–beam–girder system in Chap. 10.

Chapter 7 contains a brief review of the necessary concepts of statically indeterminate analysis and positioning of live load for obtaining moment and shear envelopes, as well as the ACI moment coefficients for common situations. The design of the one-way slab in the slab–beam–girder system is treated in Chap. 8. Finally, the design of the beams (such as 2B1-2B2-2B1 in Fig. 8.3.1) is treated in this chapter. These beams are built monolithically with the slab, hence the name *T-beams*.

The monolithic multiple T-section in a slab–beam–girder system shown in Fig. 8.4.2 (section A-A of Fig. 8.3.1) has several stems and includes as its flange the entire one-way slab that spans transversely between the stems. For the purpose of design, the multiple T-section is divided into individual T-sections that have a portion of the slab projecting from each side of the stem as flanges. For negative bending moment the flange is on the tension side of the neutral axis, thus the T-section is in effect a rectangular section. For positive bending moment the flange does provide considerably more compression area than in the negative bending moment portion of the span. Just how much of the slab projecting from the stem may be considered as part of the individual T-section, as well as the methods of design and analysis for the flexural strength, appear in subsequent sections.

Exterior rigid frames; Water Tower Place, Chicago. (Courtesy of Portland Cement Association.)

9.2 COMPARISON OF RECTANGULAR AND T-SECTIONS

A comparison of the two sections shown in Fig. 9.2.1 indicates that the flexural strength of a rectangular section is identical with that of the T-section as long as they possess the same compression area above the neutral axis (NA) and the same steel area at the same effective depth. Thus, as far as bending is concerned, any T-section with a rectangular compression area, such as shown in Fig. 9.2.1(b) may be regarded as a rectangular section.

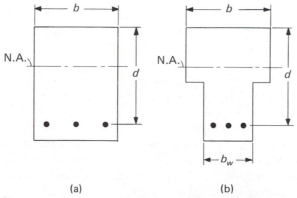

      (a)   (b)

Figure 9.2.1 Two equivalent sections in bending.

When the neutral axis of a T-section is located below the flange, as shown in Fig. 9.2.2, computation of its flexural strength requires different treatment than for a rectangular section. Beams with T-sections may be individually built as such; however, they usually occur in the positive moment regions of floor beams and girders that are built monolithically with a slab.

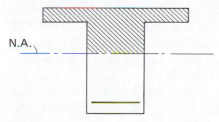

N.A.

Figure 9.2.2 A T-section in bending.

9.3 EFFECTIVE FLANGE WIDTH

Very wide beams do not conform in behavior to the assumption of the elementary theory of bending. In ordinary theory, bending stresses are assumed not to vary across the beam width. Simple theory, therefore, would dictate a constant stress at, say, the extreme fiber over the entire flange width of a T-section, no matter how great the overhang from the stem. More precisely, the stress based on the plate theory decreases the more distant a point is from the stem of the beam. Thus, for a flange of infinite width the compressive stress in the flange varies as shown in Fig. 9.3.1.

Theoretical investigations for an infinitely long continuous beam on equidistant supports, with an infinitely large flange width and a small thickness compared to beam depth, have determined an effective flange width

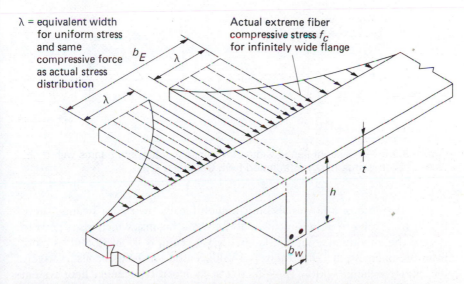

λ = equivalent width for uniform stress and same compressive force as actual stress distribution

b_E λ

Actual extreme fiber compressive stress f_c for infinitely wide flange

t

h

b_W

Figure 9.3.1 Actual and equivalent stress distribution over flange width.

b_E over which the compressive stress may be considered constant. The total compressive force carried by the equivalent system is the same as that carried by the actual system. For such assumptions the equivalent width of overhang λ depends only on the type of loading and the span length of the beam. The theory, first developed by T. von Kármán, along with certain results is summarized by Timoshenko and Goodier [1] and Girkmann [2].

Whereas the aforementioned theory gives the equivalent width b_E for an infinite flange width as a function of the span length L, in practical situations other variables are important as well. These variables are the spacing of the beams, the width of the stem of the beam, and the relative thickness of the slab with respect to the total beam depth t/h. To illustrate the effect of loading, several cases (from Girkmann [2]) showing the variation of effective flange projection for a flange of zero stiffness between beams $(t/h = 0)$ are given in Fig. 9.3.2.

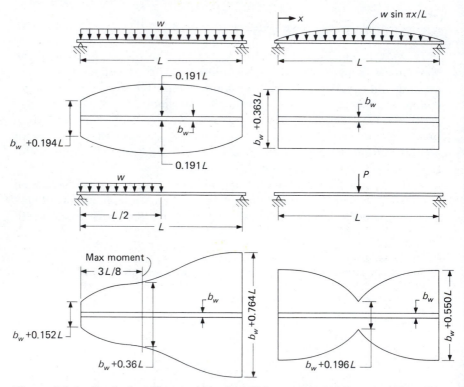

Figure 9.3.2 Equivalent flange widths for infinite actual width beams with a rib cross-sectional area of $0.1tL$ (adapted from Girkmann [2]).

In the case of a floor slab built monolithically over floor beams, there is still transverse bending in the slab between beams, which also tends to reduce the effectiveness of the slab in carrying compression at points remote from the beam stem. Thus there is a valid reason for using a conservatively low effective flange width. The effective section of a T-beam in a floor system is shown crosshatched in Fig. 9.3.3.

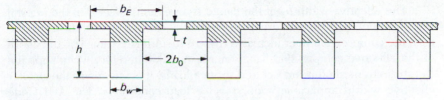

Figure 9.3.3   Effective section in compression of a T-beam in a floor system.

The ACI Code (ACI-8.10.2) prescribes a limit on the effective flange width b_E of *interior* T-sections to the *smallest* of the following:

1. $b_E = \dfrac{L}{4}$ **(9.3.1a)**

2. $b_E = b_w + 16t$ **(9.3.1b)**

3. b_E = center-to-center spacing of beams   **(9.3.1c)**

where L is the span length of the beam, and b_w and t are as shown in Fig. 9.3.3.

For *exterior* T-sections, ACI-8.10.3 prescribes for effective width b_E the *smallest* of the following:

1. $b_E = b_w + \dfrac{L}{12}$ **(9.3.2a)**

2. $b_E = b_w + 6t$ **(9.3.2b)**

3. $b_E = b_w + \dfrac{1}{2}$ (clear distance to next beam)   **(9.3.2c)**

For *isolated* T-sections, ACI-8.10.4 gives

$$b_E \leq 4b_w \qquad\qquad\qquad\qquad\qquad \textbf{(9.3.3a)}$$

and also requires that

$$t \geq \tfrac{1}{2}b_w \qquad\qquad\qquad\qquad\qquad \textbf{(9.3.3b)}$$

Since the true effective width is very much dependent on the type of loading and on the relationships (Fig. 9.3.3) t/h, L/b_w, and L/b_0, the ACI criteria are very much simplified. They seem properly adequate for certain loading cases and are unduly conservative for others. The European Concrete Committee [3,4] has presented detail procedures for determining the values of effective flange width.

9.4 INVESTIGATION OF T-SECTIONS IN BENDING— STRENGTH METHOD

In computing the strength of a T-section in bending, the neutral axis location determines whether the compression zone is T-shaped or rectangular. Since the option (ACI-10.2.6 and 10.2.7) of using Whitney's rectangular stress distribution (Sec. 3.3) is also available for non-rectangular sections, use of the rectangular stress distribution would mean that as long as its depth a does not exceed the flange thickness t, the beam strength would be the same as for a rectangular section with $b = b_E$.

The effective width b_E of the flange is an essential factor in the neutral axis location. Available information indicates that as the strain ϵ_c at the extreme compression fiber increases toward its ultimate value ϵ_{cu}, the width of the effective compression flange increases [4]. Therefore when applying the strength method using factored service loads, it is safe to use the smaller effective widths applicable under service load conditions. The ACI Code therefore uses the same effective width for the strength method as it has for many years used for the working stress method.

The computation for the nominal moment strength M_n of a T-section occurs in two categories: (1) the depth a of the rectangular stress distribution is equal to or less than t; and (2) the depth a is greater than t.

Case 1: $a \leq t$ [Fig. 9.4.1(a)]. For this case to occur, the tension steel area A_s must not exceed

$$A_s \leq \frac{0.85 f_c' b_E t}{f_y} \quad \text{for} \quad x \leq \frac{t}{\beta_1} \tag{9.4.1a}$$

obtained from equating C to T, where

$$C = 0.85 f_c' b_E a \tag{9.4.1b}$$

$$T = A_s f_y \tag{9.4.1c}$$

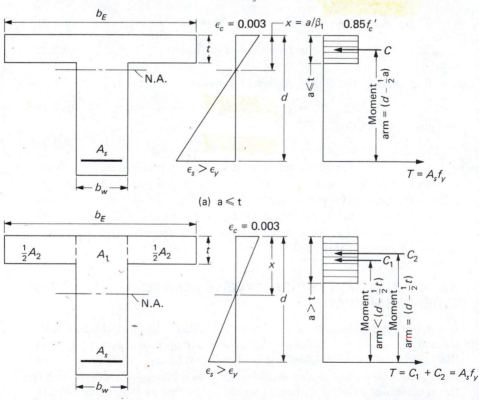

(a) $a \leq t$

(b) $a > t$

Figure 9.4.1 Strength of T-sections in bending.

The analysis is exactly as presented for rectangular sections in Chap. 3, using b_E for b.

Case 2: $a > t$ [Fig. 9.4.1(b)]. For this case, where the neutral axis distance x is larger than t/β_1, the area on which the uniform stress $0.85f'_c$ acts is T-shaped. Thus it is desirable to separate the total compressive force into forces C_1 and C_2; C_1 resulting from the stress on area A_1 and C_2 on area A_2. The moment arm for C_2 is equal to $d - t/2$ but that of C_1 is less than $d - t/2$. Thus

$$M_n = C_1 \left(d - \frac{a}{2} \right) + C_2 \left(d - \frac{t}{2} \right) \qquad (9.4.2a)$$

in which

$$C_1 = 0.85f'_c b_w a \qquad (9.4.2b)$$

$$C_2 = 0.85f'_c (b_E - b_w) t \qquad (9.4.2c)$$

and

$$a = \frac{T - C_2}{0.85f'_c b_w} \qquad (9.4.2d)$$

The tensile force T and the tension steel area A_s may also be separated into T_1 and T_2 and A_{s1} and A_{s2}, respectively.

Because the compressive strength of concrete is considered useful to a much greater extent in strength design than was the case for the working stress method, Case 2 seldom occurs in reinforced concrete building floor beams. This is due to the large equivalent width of the compression area that is available.

EXAMPLE 9.4.1 Determine the nominal moment strength M_n within the span of a floor beam (Fig. 9.4.2) whose projection below a $4\frac{1}{2}$-in. slab is 13 × 24 in. (effective depth is 25 in. for two layers of steel). Tension reinforcement is 8-#8 bars. The span length of the beam is 26 ft and the beams are centered 13 ft apart. Use $f'_c = 3000$ psi and $f_y = 50,000$ psi.

Solution: Following ACI-8.10.2, the effective flange width b_E is the smallest

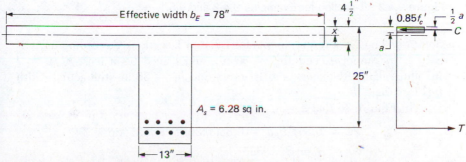

Figure 9.4.2   T-section for Example 9.4.1.

of $(26)(12)/4 = 78$ in.; $13 + 16(4.5) = 85$ in.; or $(13)(12) = 156$ in. Thus $b_E = 78$ in.

(a) Find A_s so that $a = 4\frac{1}{2}$ in.

$$C = 0.85f'_c b_E a = 0.85(3)(78)(4.5) = 895 \text{ kips}$$

$$A_s = \frac{T \text{ or } C}{f_y} = \frac{895}{50} = 17.9 \text{ sq in.}$$

Thus the depth a of the rectangular stress distribution is less than t for $A_s = 6.28$ sq in.

(b) Treat as a rectangular section.

$$T = f_y A_s = 50(6.28) = 314 \text{ kips}$$

$$a = \frac{T \text{ or } C}{0.85f'_c b_E} = \frac{314}{0.85(3)(78)} = 1.58 \text{ in.}$$

$$\text{moment arm} = d - \frac{a}{2} = 25 - 0.79 = 24.21 \text{ in.}$$

$$M_n = 314(24.21)\tfrac{1}{12} = 633 \text{ ft-kips}$$

EXAMPLE 9.4.2 Determine the nominal moment strength M_n of the isolated T-section shown in Fig. 9.4.3 when $A_s = 12.48$ sq in. (8-#11). Use $f'_c = 3000$ psi and $f_y = 50,000$ psi.

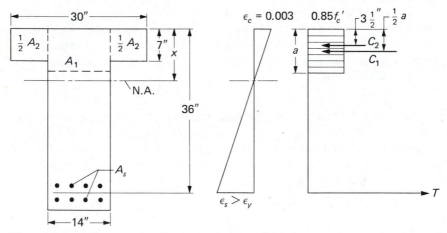

Figure 9.4.3 T-section for Examples 9.4.2 and 9.5.1.

Solution: (a) Check requirements for isolated T-sections. According to ACI-8.10.4, b_E cannot exceed $4b_w = 56$ in., and t must be at least $b_w/2 = 7$ in. Thus, flange thickness is satisfactory and $b_E = 30$ in. (the actual width b is less than $4b_w$).

(b) Find A_s so that $a = t$.

$$C = 0.85f'_c b_E a = 0.85(3)(30)(7) = 535 \text{ kips}$$

$$A_s = \frac{T \text{ or } C}{f_y} = \frac{535}{50} = 10.70 \text{ sq in.}$$

Thus, actual a exceeds t when $A_s = 12.48$ sq in.

(b) Treat by the two-couple method (Fig. 9.4.3).

$$T = A_s f_y = 12.48(50) = 624 \text{ kips}$$

$$C = C_1 + C_2 = 0.85 f_c' A_1 + 0.85 f_c' A_2$$

$$624 = 2.55(14a) + 2.55(16)(7)$$

$$a = 9.48 \text{ in.}$$

$$C_1 = 2.55(14)(9.48) = 338 \text{ kips}$$

$$C_2 = 2.55(16)(7) = 286 \text{ kips}$$

$$M_n = C_1[36 - 0.5(9.48)]\tfrac{1}{12} + C_2(36 - 3.50)\tfrac{1}{12}$$

$$= 882 + 774 = 1656 \text{ ft-kips}$$

9.5 MAXIMUM TENSION REINFORCEMENT PERMITTED IN T-SECTIONS

As discussed in Sec. 3.13, the literal wording of ACI-10.3.3 will not be identical to a maximum $x = 0.75x_b$ for non-rectangular sections. The authors are convinced that the precise code wording was used for practical purposes and that different ductility requirements for different shapes of cross section were not intended. For practical purposes, the maximum tension reinforcement permitted for T-sections would rarely if ever be used, so that the argument as to what limit is specified or what was intended is purely academic.

Since the limit based on maximum $x = 0.75x_b$ is not conservative for T-sections, the literal use of maximum $\rho = 0.75\rho_b$ is appropriate. The following example illustrates the procedure.

EXAMPLE 9.5.1 For the T-section of Fig. 9.4.3 determine the maximum reinforcement ratio ρ and area A_s permitted by the strength method of the ACI Code. Use $f_c' = 3000$ psi and $f_y = 50,000$ psi.

Solution: (a) Determine the neutral axis distance x_b and the depth a_b of the rectangular stress distribution in the balanced strain condition. Referring to the strain diagram of Fig. 3.5.1,

$$x_b = \left(\frac{0.003}{0.003 + f_y/E_s}\right)d = \left(\frac{0.003}{0.003 + 0.00172}\right)d$$

$$= 0.635d = 0.635(36) = 22.9 \text{ in.}$$

$$a_b = \beta_1 x_b = 0.85(22.9) = 19.4 \text{ in.}$$

(b) Determine the amount of steel A_{sb} in the balanced strain condition. Since $a_b > t$, the two-couple approach may be used, using C_1 and C_2 acting on areas A_1 and A_2, respectively, similar to what is shown in Fig. 9.4.3, except that $x = x_b$ and $a = a_b$.

$$C_1 = C_{1b} = 0.85f'_c b_w a_b = 0.85(3)(14)(19.4) = 693 \text{ kips}$$

$$A_{s1b} = \frac{693}{50} = 13.9 \text{ sq in.}$$

$$C_2 = 0.85f'_c(b_E - b_w)t = 0.85(3)(16)(7) = 286 \text{ kips}$$

$$A_{s2} = \frac{286}{50} = 5.72 \text{ sq in.}$$

Note that $C_2 = 286$ kips even if $x < x_b$, as long as $a \geq t$.
The balanced amount of steel is

$$A_{sb} = A_{s1b} + A_{s2} = 13.9 + 5.72 = 19.6 \text{ sq in.}$$

(c) Determine the maximum reinforcement permitted by ACI-10.3.3.

$$\max A_s = 0.75(19.6) = 14.7 \text{ sq in.}$$

$$\max \rho = \frac{A_s}{b_E d} = \frac{14.7}{30(36)} = 0.0136$$

(d) Compare the ductility requirement for a T-section with that for a singly reinforced beam as discussed in Sec. 3.5. Locate the neutral axis when the beam contains the maximum steel permitted by the ACI Code. Assume $a > t$,

$$C_1 = 0.85f'_c b_w a = 0.85(3)(14)a = 35.7a$$

$$C_2 = 0.85f'_c(b_E - b_w)t = 0.85(3)(16)7 = 286 \text{ kips}$$

$$T = A_s f_y = 14.7(50) = 735 \text{ kips}$$

$$C = T$$

$$a = \frac{735 - 286}{35.7} = 12.6 \text{ in.} > t, \qquad \text{as assumed}$$

$$x = \frac{a}{\beta_1} = \frac{12.6}{0.85} = 14.8 \text{ in.}$$

$$\frac{x}{x_b} = \frac{14.8}{22.9} = 0.65$$

The neutral axis for this T-section actually is restricted to 0.65 of x_b, in effect requiring more ductility for the T-section than for singly or doubly reinforced rectangular beams.

Again, the reader is reminded that the beam of Fig. 9.4.2 is far more typical of T-sections than the beam of Fig. 9.4.3. Even for the beam of Fig. 9.4.3 it would be unlikely that more than the 8-#11 shown would be used. The maximum $A_s = 14.7$ sq in. would require three layers of steel. For the section of Fig. 9.4.2, the maximum $A_s = 0.75A_{sb}$ is 18.35 sq in., a very unlikely amount to be contemplated for use in that section.

9.6 DESIGN OF T-SECTIONS IN BENDING—STRENGTH METHOD

The design of T-shaped isolated beams involves the dimensions of the flange and web and the area of tension steel, or a total of five unknowns—one more if compression steel is used. Thus there are many possible solutions to the problem. Because of the large compression area in the flange, these T-sections, when used, are usually deep and wide enough so that the tension steel may be placed within the width of the web in not more than two or three layers.

The more common T-sections are those in the region of positive bending moment in continuous monolithic slab–beam–girder systems. The size of the available flange is therefore definitely known after the slab has been designed, and only the web size needs to be designed. The selection of web size for such beams is treated in Chap. 10.

Once the overall dimensions have been established, the design problem is to determine the amount of positive moment reinforcement. Thus it is important first to ascertain the location of the neutral axis associated with the positive moment. This may be done by first computing the moment strength at which the depth a of the rectangular compressive stress block is equal to the flange thickness t.

EXAMPLE 9.6.1 Determine the amount of tension steel required in the T-section of Fig. 9.4.3 to take a dead load moment of 370 ft-kips and a live load moment of 520 ft-kips, using $f'_c = 3000$ psi, $f_y = 50,000$ psi, and the strength method of the ACI Code.

Solution: (a) Determine the factored moment M_u.

$$M_u = 1.4(370) + 1.7(520) = 1402 \text{ ft-kips}$$

$$\text{required } M_n = \frac{M_u}{\phi} = \frac{1402}{0.90} = 1560 \text{ ft-kips}$$

(b) Determine if the depth a of the rectangular stress distribution will be greater than $t = 7$ in. For $a = t$,

$$C = 0.85f'_c b_E t = 0.85(3)(30)(7) = 536 \text{ kips}$$

$$M_n = C\left(d - \frac{t}{2}\right) = 536(36 - 3.50)\tfrac{1}{12} = 1450 \text{ ft-kips}$$

Since the required M_n exceeds 1450 ft-kips the actual a must exceed t.

(c) Use the two-couple method (Fig. 9.4.3) to obtain A_s.

$$M_n = 0.85f'_c A_1\left(d - \frac{a}{2}\right) + 0.85f'_c A_2\left(d - \frac{t}{2}\right)$$

$$1560(12) = 2.55(14a)\left(36 - \frac{a}{2}\right) + 2.55(16)(7)(36 - 3.50)$$

$$a^2 - 72a = -527$$

$$a = 8.3 \text{ in.}$$

$$C_1 = 0.85f'_c b_w a = 0.85(3)(14)(8.3) = 296 \text{ kips}$$

$$A_{s1} = \frac{T_1}{f_y} = \frac{296}{50} = 5.92 \text{ sq in.}$$

$$C_2 = 0.85f'_c(b_E - b_w)t = 0.85(3)(16)(7) = 286 \text{ kips}$$

$$A_{s2} = \frac{T_2}{f_y} = \frac{286}{50} = 5.72 \text{ sq in.}$$

$$A_s = 5.92 + 5.72 = 11.64 \text{ sq in.}$$

Since the 11.64 sq in. required is less than the maximum permitted amount $(0.75A_{sb} = 14.7$ sq in. from Example 9.5.1), the design would be acceptable. The only practical selection of steel is the 8-#11 originally used in the analysis for Example 9.4.2.

EXAMPLE 9.6.2 Design the steel reinforcement for the section of Fig. 9.4.2 to carry a factored moment M_u of 735 ft-kips. Use $f'_c = 3000$ psi and $f_y = 50,000$ psi.

Solution: (a) Determine whether or not the depth a of the rectangular stress distribution will be greater than t. For $a = t$,

$$C = 0.85f'_c b_E t = 0.85(3)(78)(4.5) = 895 \text{ kips}$$

$$M_n = C\left(d - \frac{t}{2}\right) = 895\left(25 - \frac{4.5}{2}\right)\frac{1}{12} = 1597 \text{ ft-kips}$$

$$\text{required } M_n = \frac{M_u}{\phi} = \frac{735}{0.90} = 817 \text{ ft-kips}$$

Since the required M_n is less than the amount necessary to cause $a = t$, the section can be designed as a rectangular section.

(b) Design as a rectangular section. The computation of the coefficient of resistance $R_n = M_u/(\phi b_E d^2)$ and the solving of the quadratic equation (or its equivalent as in item 4 of Sec. 3.8) for ρ may be made exactly as described in Chap. 3. However, because of the wide flange ($b_E = 78$ in.) and the likelihood of a relatively small value for a, a trial procedure may be preferred. First use $0.9d$ as a conservative trial value for the moment arm $(d - a/2)$; then

$$\text{required } A_s = \frac{\text{required } M_n}{(d - a/2)f_y} = \frac{817(12)}{0.9(25)50} = 8.71 \text{ sq in.}$$

Try 4-#9 and 4-#10, $A_s = 9.08$ sq in.
  Check:

$$C = 0.85f'_c b_E a = 0.85(3)(78)a = 198.9a$$

$$T = A_s f_y = 9.08(50) = 454 \text{ kips}$$

$$a = \frac{454}{198.9} = 2.28 \text{ in.}$$

$$\text{arm} = 25 - 0.5(2.28) = 23.86 \text{ in.} \qquad (\text{i.e., } 0.95d)$$

A second trial gives

$$\text{required } A_s = \frac{817(12)}{23.86(50)} = 8.22 \text{ sq in.}$$

Revise to 4-#8 and 4-#10, $A_s = 8.24$ sq in. (minimum width = 12.9 in. from Table 3.9.2).

Check:

$$C = 198.9a \qquad \text{(as before)}$$

$$T = 8.24(50) = 412 \text{ kips}$$

$$a = \frac{412}{198.9} = 2.07 \text{ in.}$$

$$\text{arm} = 25 - 0.5(2.07) = 23.96 \text{ in.} \approx 23.86 \text{ in. used}$$

No further saving can be made.

$$M_n = 412(23.96)\tfrac{1}{12} = 823 \text{ ft-kips}$$

$$[\phi M_n = 0.90(823) = 740 \text{ ft-kips}] > [M_u = 735 \text{ ft-kips}] \qquad \text{OK}$$

Use 4-#8 and 4-#10.

9.7 INVESTIGATION OF T-SECTIONS IN BENDING— WORKING STRESS METHOD

T-sections in bending must be investigated by the working stress method when service load deflection calculations are needed. It is first necessary to determine whether or not the neutral or effective centroidal axis is below the flange. This may be done by comparing the first moment of the flange area about the base of the flange with that of the equivalent concrete area of steel. In the event that the neutral axis is within the flange, the T-section is to be treated as a rectangular section.

When it is found that the neutral axis is definitely below the flange, the exact transformed section should include the small compression area between the base of flange and the neutral axis. It will be shown in the following example, however, that this area may be ignored in ordinary cases without appreciably affecting the results.

EXAMPLE 9.7.1 Determine the transformed cracked section moment of inertia I_{cr} under service load conditions for the floor beam in Example 9.4.1. Use $f_c' = 3000$ psi, $f_y = 50,000$ psi, and the ACI Code. Solve the problem by first considering and then ignoring the compression area in the web.

Solution: (a) Determine whether the neutral axis is above or below the bottom of the slab.

The effective flange width is 78 in. as shown in Fig. 9.4.2. The transformed area of steel is $(9)(6.28) = 56.52$ sq in. Comparing the first moment of the flange area about its base with that of the transformed area of steel (Fig. 9.7.1),

$$[(78)(4.5)(2.25) = 790] < [(56.52)(20.5) = 1159]$$

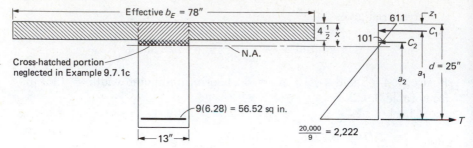

Figure 9.7.1   T-section for Example 9.7.1, including web contribution.

Therefore the neutral axis is below the flange.

(b) Compute I_{cr} considering the compression area in the web. Taking moments about the unknown actual neutral axis (Fig. 9.7.1),

$$\frac{13x^2}{2} + (65)(4.5)(x - 2.25) = 56.52(25 - x)$$

$$x^2 + 53.70x = 318.63$$

$$x = 5.39 \text{ in.}$$

The transformed cracked section moment of inertia may then be computed about the neutral axis,

$$I_{cr} = \frac{78(4.5)^3}{12} + 78(4.5)(3.14)^2 + \frac{13(0.89)^3}{3} + 56.52(19.61)^2$$

$$= 25,800 \text{ in.}^4$$

The serviceability check of deflection will utilize I_{cr} as is illustrated in Chap. 14 on deflections.

(c) Compute I_{cr} ignoring the compression area in the web. Taking moments about the unknown actual neutral axis (Fig. 9.7.1),

$$78(4.5)(x - 2.25) = 56.52(25 - x)$$

$$408x = 2203$$

$$x = 5.40 \text{ in.}$$

It is apparent that the neutral axis location remains essentially the same as when the compression area in the web was included. Computing the moment of inertia I_{cr}, also neglecting the compression area in the web, gives

$$I_{cr} = \frac{78(4.5)^3}{12} + 78(4.5)(3.15)^2 + 56.52(19.60)^2 = 25,800 \text{ in.}^4$$

For deflection computation purposes, $I_{cr} = 25,800 \text{ in.}^4$ is all that can be justified (three significant figures at most).

The service load moment capacity M_w according to the working stress method may be obtained by using either (a) the internal-force method or (b) the transformed section method with the flexure formula, the general principles for which have been treated in Chap. 4.

The applicable stress distribution is determined when the allowable stress is reached at one of the extreme fibers. Usually in T-sections the steel controls (i.e., underreinforced according to the working stress method) because of the abundance of concrete compression area.

SELECTED REFERENCES

1. S. Timoshenko and J. N. Goodier. *Theory of Elasticity* (2d ed.). New York: McGraw-Hill, 1951 (pp. 171–177).
2. Karl Girkmann. *Flachentragwerke* (3d ed.). Vienna: Springer-Verlag, 1954 (pp. 116–123).
3. Franco Leve. "Work of the European Concrete Committee," *ACI Journal, Proceedings,* **57,** March 1961, 1049–1054.
4. Gottfried Brendel. "Strength of the Compression Slab of T-Beams Subject to Simple Bending," *ACI Journal, Proceedings,* **61,** January 1964, 57–76.

PROBLEMS

All problems* are to be done according to the ACI Code and all loads given are service loads unless otherwise indicated.

9.1 (a) Determine the nominal moment strength M_n for the beam cross section shown. (b) Determine the maximum tension reinforcement A_s permitted for this beam by ACI-10.3.3. What is the maximum x permitted, given in terms of a proportion of x_b? The beam span is 30 ft. Use $f'_c = 4000$ psi and $f_y = 60,000$ psi. (Span $= 9.1$ m; $f'_c = 30$ MPa; $f_y = 400$ MPa; slab $t = 125$ mm; beam $h = 900$ mm; $b_w = 380$ mm.)

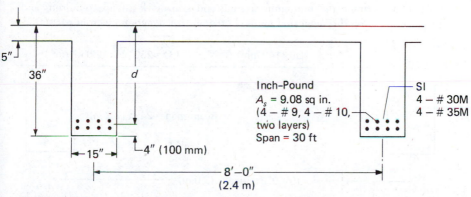

Problem 9.1

9.2 (a) Design the reinforcement for the beam shown, according to the strength method, if the dead load moment is 65 ft-kips and the live load moment is 100 ft-kips. (b) Determine the maximum tension reinforcement A_s permitted for this beam by ACI-10.3.3. What is the maximum x permitted, given

*Problems may be solved as problems stated in Inch-Pound units, or as problems in SI units using the quantities in parenthesis at the end of the statement. The SI values are approximate conversions to avoid implying higher precision for given information in metric units than for Inch-Pound units.

in terms of a proportion of x_b? Use $f_c' = 3000$ psi and $f_y = 40,000$ psi. $(M_D = 88$ kN·m; $M_L = 135$ kN·m; $f_c' = 20$ MPa; $f_y = 300$ MPa.)

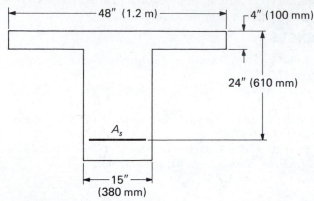

Problem 9.2

9.3 Make a partial design of a simply supported T-beam floor system to meet the conditions shown in the accompanying figure. Use $f_c' = 3000$ psi, $f_y = 40,000$ psi, and the strength method. $(f_c' = 20$ MPa; $f_y = 300$ MPa.)
  (a) Select the size of stem. Assume that a stem area consistent with a nominal shear stress $V_u/(\phi b_w d)$ equal to approximately $6\sqrt{f_c'}$ psi $(\sqrt{f_c'}/2$ with f_c' in MPa) will result in an economical section.
  (b) Determine the longitudinal reinforcement, and determine bar lengths and cutoff or bends, if any.
  (c) Determine the minimum spacing and size of U stirrups required, and comment on the suggested method of stem size selection given in part (a).

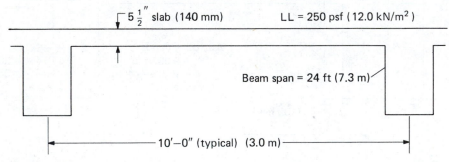

Problem 9.3

9.4 For the beam of Prob. 9.1, use the transformed section method to determine the transformed cracked section moment of inertia I_{cr} **(a)** neglecting any compression in the web below the slab; and **(b)** including any compression in the web below the slab.

9.5 For the beam of Prob. 9.1, use the transformed section method to compute the allowable service load moment M_w, using the principles of Chap. 4. Compare the result for M_w by using the internal-force method, **(a)** neglecting any compression in the web below the slab, and **(b)** considering any compression in the web below the slab.

10

Continuous Slab–Beam–Girder and Concrete Joist Floor Systems

10.1 INTRODUCTION

The design of rectangular and T-sections has been treated in Chaps. 3, 4, and 9; shear strength and stirrup design in Chap. 5; development of reinforcement in Chap. 6; and continuity analysis in Chap. 7. The slab–beam–girder type of floor construction has been described generally and the design of one-way slabs illustrated in Chap. 8. In this chapter, complete designs of a typical floor beam and girder in a monolithic slab–beam–girder system are shown. Primarily, what has been developed in the preceding chapters is applied. Thus the reader may consider the material of this chapter as an integrated review of the subjects in the aforementioned chapters.

Also included in this chapter is the design of one-way concrete joist floors. On continuous spans, the concrete joists should be regarded as having rectangular sections near the supports and T-sections in the positive moment region within the span.

It should be noted that in recent years a marked increase has occurred in the use of precast slabs, either conventionally reinforced or prestressed, placed on a monolithic beam and girder framing system. In such systems the slab rests on, but does not act with, the beams and girders; thus T-sections are not involved. The procedures discussed in this chapter are also generally applicable to this simpler system of rectangular beams.

Reinforced concrete building design may be a complicated venture, involving irregular floor plans and intricate structural framing. The simple floor framing plan used in the examples does not necessarily reflect a typical

Monolithic slab–beam–girder system; Liggett and Myers Tobacco Company, Richmond, VA. (Courtesy of Portland Cement Association.)

practical situation, but it does serve to illustrate the basic essentials of design.

10.2 SIZE OF BEAM WEB

The size of the beam web for continuous T-shaped sections is usually controlled by the flexural and shear strength requirements at the exterior face of the first interior support. For typical conditions of equal spans and uniform dead and live loads, a negative moment of $\frac{1}{10}wL_n^2$ and a shear of $0.60wL_n$ (see Fig. 7.5.1) may be used in estimating the size of the beam. In the following discussion are given the detailed considerations involved in selecting the beam cross section based on the critical bending moment and shear.

Negative Moment Requirement. The section of the beam resisting the negative bending moment at the face of the support is a rectangular section, even in T-section construction, because compression is at the lower part of the section. The tension reinforcement is provided (a) usually by straight bars extending across the top of the beam as required for negative moment; (b) sometimes by bent bars from both adjacent spans; or (c) sometimes by a combination of bent bars and additional straight bars across the top. The development of positive moment reinforcement (ACI-12.11.1) requires some straight bars to continue along the bottom of the beam into the support; thus some portion of them could be utilized as compression reinforcement. Although compression reinforcement will rarely be required for strength, it is frequently desired for added ductility and for deflection control.

The guideline percentage ρ to be used for deflection control of singly reinforced rectangular beams has been suggested in Chap. 3 to be about

$0.375\rho_b$ (one-half of the maximum permissible value). For the negative moment requirement on continuous T-sections, a higher value for ρ should be acceptable, because the gross moment of inertia of a typically proportioned T-section is roughly twice that of its rectangular portion. Thus it may be reasonable to design the negative moment region of a T-shaped section using twice the percentage of reinforcement that would be used for reasonable deflection control on a completely rectangular beam, that is, the full $0.75 \, \rho_b$ permitted by the ACI Code (or more if compression steel is utilized).

Positive Moment Requirement. In the positive moment region the flange of the T-section is in compression. Since the effective flange width is large, it is rare that the depth a of the Whitney rectangular stress distribution will extend below the bottom of the flange. On the tension side, the amount of steel required will be inversely proportional to the depth of the section.

For sizing the beam web, the only consideration utilizing the positive moment is to establish the width required to maintain adequate clearances for a given number of bars. This should be examined because a somewhat deeper or shallower beam might still permit the required steel to fit into one or two layers, as the case may be.

Shear Strength Requirement. The designer may wish to establish the beam web size to achieve a certain maximum nominal shear stress. This may be desirable for economical stirrup size and spacings. The ranges of five categories of design for shear reinforcement are given in Sec. 5.10. The upper limit of category 4 ($V_s = 4\sqrt{f_c'}b_w d$) serves as a practical guideline for ordinary design, typically permitting stirrup spacing from 3 in. to a maximum of $d/2$.

When V_c is taken as $2\sqrt{f_c'}b_w d$ according to the simplified procedure, the total V_n at the upper limit in category 4 is

$$V_n = V_c + V_s$$
$$= 2\sqrt{f_c'}b_w d + 4\sqrt{f_c'}b_w d = 6\sqrt{f_c'}b_w d \qquad \textbf{(10.2.1)}$$

or, for a given factored shear V_u, the $b_w d$ required to satisfy this guideline shear strength is

$$\text{required } b_w d = \frac{\text{required } V_n}{6\sqrt{f_c'}} = \frac{V_u}{\phi(6\sqrt{f_c'})} \qquad \textbf{(10.2.2)}$$

Except for unusual conditions involving short spans, heavy concentrated loads, heavily doubly reinforced sections, or combinations of these, the maximum nominal shear stress $V_u/(\phi b_w d)$ should be about $6\sqrt{f_c'}$ for reasonable stirrup size and spacing.

EXAMPLE 10.2.1 Using the ACI strength method of design, establish the preliminary size for the floor beams 2B1-2B2-2B1 supported by girders as shown in the floor framing plan of Fig. 8.3.1. Use information previously described in Chap. 8.

Solution: For $f'_c = 3000$ psi and $f_y = 40,000$ psi, one may determine the maximum reinforcement ratio $(0.75\rho_b)$ allowed by the ACI Code using basic principles, or from Table 3.6.1,

$$\max \rho = 0.75\rho_b = 0.0278$$

(a) Negative moment requirement. Estimating the weight of the stem (portion of the web below the slab) at 0.3 kip/ft (2 sq ft of area), and applying the overload factors,

$$w_D = 1.4[0.056(13) + 0.3] = 1.44 \text{ kips/ft}$$

$$w_L = 1.7(0.100)(13) = 2.21 \text{ kips/ft}$$

Using basic principles, or Eq. (3.8.4) as follows, the R_n corresponding to $\rho = 0.0278$ is

$$m = \frac{f_y}{0.85f'_c} = 15.7$$

$$R_n = \rho f_y(1 - \tfrac{1}{2}\rho m) = 870 \text{ psi}$$

Assume width of supporting girders to be 18 in., in which case the clear span L_n is

$$L_n = 26 - 1.5 = 24.5 \text{ ft}$$

Using the moment coefficient from ACI-8.3,

$$\max M_u = \tfrac{1}{10}(1.44 + 2.21)(24.5)^2 = 219 \text{ ft-kips}$$

$$\text{required } bd^2 = \frac{M_u}{\phi R_n} = \frac{219(12,000)}{0.90(870)} = 3360 \text{ in.}^3$$

If $b = 13$ in.,

$$\text{required } d = \sqrt{\frac{3360}{13}} = 16.1 \text{ in.}$$

The minimum effective size permitted is 13×16.1, for which the steel area required would be

$$A_s = 0.0278(13)(16.1) = 5.82 \text{ sq in.}$$

In the negative moment region the flange is available so that the steel does not have to fit within the web width.

Before making a decision, the shear stress and the steel requirement for positive moment should be examined.

(b) Shear requirement. Using the shear coefficient from ACI-8.3,

$$\max V_u = 1.15\left(\frac{wL_n}{2}\right) = 1.15\left(\frac{3.65}{2}\right)(24.5) = 51.4 \text{ kips}$$

If it is desired that the nominal shear stress $v_n = V_u/(\phi b_w d)$ does not exceed $6\sqrt{f'_c} = 329$ psi,

$$\text{required } b_w d = \frac{51,400}{0.85(329)} = 184 \text{ sq in.}$$

For $b = 13$ in., $d = 14.1$ in. (Actually the maximum shear may be taken at a distance d from the face of the support.)

(c) Positive moment requirement. The effective flange width b_E is the smallest of the following:

1. $b_w + 16t = b_w + 16(4.5) = b_w + 72$
2. $\dfrac{L}{4} = 26 \left(\dfrac{12}{4} \right) = 78$ in.   (Controls)
3. center-to-center spacing $= 13(12) = 156$ in.

Try $d = 20$ in.; estimate moment arm at 19 in.

$$M_u = \frac{1}{14} \, wL_n^2 = \frac{3.65}{14} \, (24.5)^2 = 156 \text{ ft-kips}$$

$$\text{required } A_s = \frac{M_u}{\phi f_y \, (\text{arm})} \approx \frac{156(12)}{0.90(40)(\approx 19)} = 2.74 \text{ sq in.}$$

This amount of steel should easily fit into a 13-in. wide beam in one layer.

(d) Minimum depth. According to ACI-Table 9.5a the minimum depth cannot be less than

$$\min h = \frac{L}{18.5} \, (0.8) = \frac{26(12)(0.8)}{18.5} = 13.5 \text{ in.}$$

The negative moment region gives the most severe requirement. The designer must also keep in mind that if excessive deflection may cause damage, the deflection must be computed and satisfy ACI-Table 9.5b. The above check is only a minimum requirement where deflection is *not* likely to cause damage to nonstructural construction.

Frequently the size is chosen larger than the requirements would dictate. Sometimes this occurs because of the designer's desire that a large number of beams have the same external dimensions for economy of forming or perhaps because of ductwork or pipes that are to pass through the beams, necessitating larger sizes.

Use $b_w = 13$ in., $h = 22.5$ in. (which gives $d \approx 20$ in.). It is common to make the stem portion below the flange a whole inch increment, such as 18 in. for this case. The arbitrary size selected is larger than necessary; any size at least equal to that indicated by steps (a) through (d) would serve to illustrate the design procedure.

10.3 CONTINUOUS FRAME ANALYSIS FOR BEAMS

The shear and moment diagrams to be used in the design of the floor beams in the preceding example could have followed the ACI moment coefficients. However, since these coefficients are more suitable for use in frames involving, say, more than four continuous spans, and as a matter of illustration, an elastic analysis will be used to obtain the bending moment and shear envelopes for beams 2B1-2B2-2B1.

Some designers would assume the girders to provide only vertical support to those floor beams supported on girders; thus a pure continuous beam analysis is made. The girders, however, have torsional stiffness that acts to restrain the rotation of the beam over these girder supports. Prior to 1971, neglect of torsional stiffness was permitted by the ACI Code where such stiffness did not exceed 20% of the flexural stiffness at the joint. Since 1971 the ACI Code requires that torsion be considered whenever the nominal torsion stress exceeds $1.5\sqrt{f'_c}$ psi ($0.12\sqrt{f'_c}$ for f'_c in MPa). In the case with relatively large and stiff girders, the stiffness of the beam may reasonably be taken as twice the torsional stiffness provided by the girder, or if one thinks in terms of an equivalent beam and column frame system, it is as shown in Fig. 10.3.1(b). More details with regard to torsion are to be found in Chap. 19.

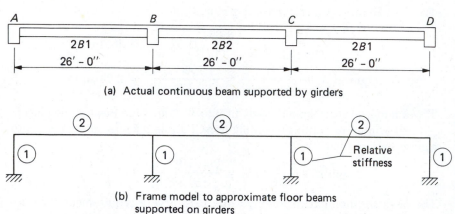

(a) Actual continuous beam supported by girders

(b) Frame model to approximate floor beams supported on girders

Figure 10.3.1 Actual continuous beam and analytical model.

For the beams that frame into columns, however, the combination of the bending stiffness of the columns plus the torsional stiffness of the girders means that the relative end restraint is greater than in the aforementioned case. To design such beams properly, the sizes of the members must be estimated and the $(\Sigma K_{col}/\Sigma K_{bm})$ ratio computed therefrom.

With regard to a T-section, there is considerable difference of opinion as to how its stiffness in a continuous beam should be computed. A T-section certainly provides more stiffness in the positive moment region where the flange is in compression than it does as a rectangular section in the negative moment region. The ACI Code prescribes no specific method but only that "Any reasonable assumptions may be adopted for computing relative flexural and torsional stiffnesses of columns, walls, floors, and roof systems" (ACI-8.6.1). It has been common practice to use the gross moment of inertia, neglecting reinforcement, in computing the flexural stiffness of such elements. For T-sections usually the gross section of the effective flange width is included.

One of the commonly accepted methods [1] is to use a T-section moment of inertia equal to $2(\frac{1}{12}b_w h^3)$, which is equivalent to using an effective flange width of about 6 times the web width. Since the true stiffness is that of a span with variable moment of inertia along its length with the T-section stiffness over, say, the middle one-half of the span and the rectangular section stiffness over the end quarters, the equivalent system is approximately obtained by using the T-section with a flange width only twice the web width over the entire span. Examination of analyses using various stiffness ratios will show that a fairly wide variation in stiffness ratio may be accommodated with relatively small changes in bending moments.

In the following example, only the analysis of the floor beams supported on girders is shown, using the analytical frame model in Fig. 10.3.1(b). An overestimate of the beam stiffness will give a conservative design for the beam; but such assumption should be reevaluated when designing the columns, if any.

EXAMPLE 10.3.1 Using an elastic statically indeterminate analysis, determine the shear and moment diagrams to be used in compiling the moment and shear envelopes for the design of $2B1$ and $2B2$ in Example 10.2.1.

Solution:

$$\text{stem weight} = \frac{13(18)}{144}(0.15) = 0.244 \text{ kip/ft}$$

$$w_D = 1.4[0.056(13) + 0.244] = 1.36 \text{ kips/ft}$$

$$w_L = 1.7(0.100)(13) = 2.21 \text{ kips/ft}$$

$$\text{FEM due to } w_D = \tfrac{1}{12}(1.36)(26)^2 = 76.6 \text{ ft-kips}$$

$$\text{FEM due to } w_L = \tfrac{1}{12}(2.21)(26)^2 = 124.5 \text{ ft-kips}$$

Loading conditions 1 through 5 as shown in Fig. 10.3.2 are established by visualizing the influence lines for positive moment in the midspan region of each span and for negative moment at each support.

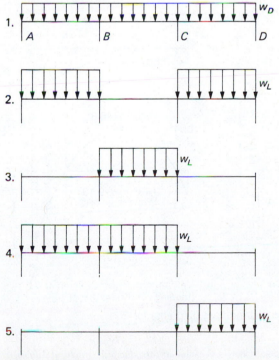

Figure 10.3.2 Loading conditions.

The structural analysis may be performed by any method of statically indeterminate analysis, such as the matrix displacement method* if the digital computer is used, or by moment distribution* if the computations are made using simple arithmetic. Since the specific structural analysis technique is outside the scope of this text, only the end moments resulting from the analysis for each of the five loading cases are given in Table 10.3.1.

Table 10.3.1 FINAL END MOMENTS[a] (FT-KIPS) FOR THE FIVE LOADING CONDITIONS OF FIG. 10.3.2

JOINT	A	B		C		D	
MEMBER	AB	BA	BC	CB	CD	DC	
DF[b]	0.667	0.400	0.400	0.400	0.400	0.667	
Case 1, M =	−27.2	+90.0	−83.7	+83.7	−90.0	+27.2	Dead load only
Case 2, M =	−54.1	+89.4	−46.8	+46.8	−89.4	+54.1	Live load spans 1 and 3
Case 3, M =	+10.0	+56.8	−89.3	+89.3	−56.8	−10.0	Live load span 2
Case 4, M =	−42.5	+159.3	−156.1	+69.4	−43.5	−8.3	Live load spans 1 and 2
Case 5, M =	−1.7	−13.3	+19.9	+66.7	−102.6	+52.5	Live load span 3

[a]Clockwise moments acting on member ends are taken positive.
[b]DF = distribution factor.

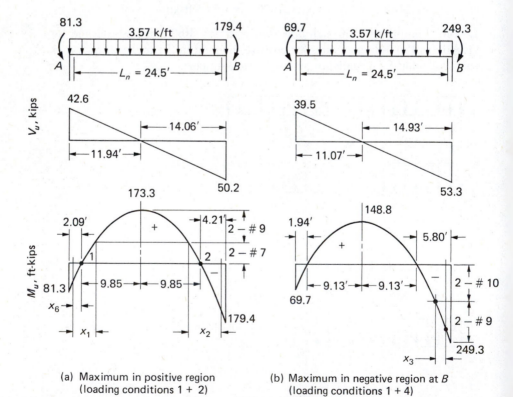

(a) Maximum in positive region
   (loading conditions 1 + 2)

(b) Maximum in negative region at B
   (loading conditions 1 + 4)

Figure 10.3.3 Primary shear and moment diagrams for 2B1.

---

*See, for example, Chu-Kia Wang and Charles G. Salmon. *Introductory Structural Analysis.* Englewood Cliffs, N.J.: Prentice-Hall, 1984.

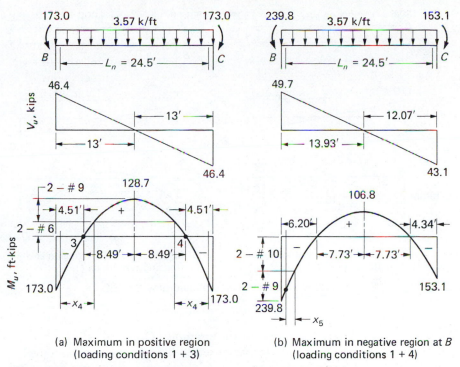

(a) Maximum in positive region
    (loading conditions 1 + 3)

(b) Maximum in negative region at B
    (loading conditions 1 + 4)

Figure 10.3.4 Primary shear and moment diagrams for 2B2.

The primary shear and moment diagrams for maximum positive and
negative moments for beams 2B1 and 2B2 are given individually in Figs.
10.3.3 and 10.3.4. The secondary shear and moment diagrams are shown
in Figs. 10.3.5 and 10.3.6. The primary and secondary moment diagrams

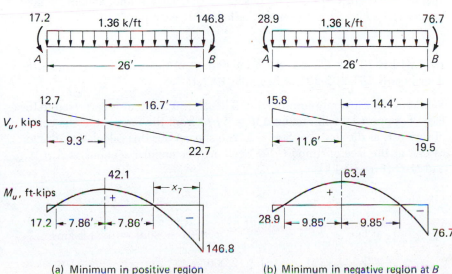

(a) Minimum in positive region
    (loading conditions 1 + 3)

(b) Minimum in negative region at B
    (loading conditions 1 + 5)

Figure 10.3.5 Secondary shear and moment diagrams for 2B1.

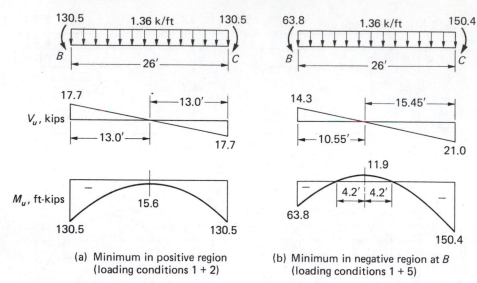

(a) Minimum in positive region
(loading conditions 1 + 2)

(b) Minimum in negative region at B
(loading conditions 1 + 5)

Figure 10.3.6 Secondary shear and moment diagrams for 2B2.

are drawn separately for clarity in the example. Ordinarily they are not drawn separately but instead superimposed to become the moment envelope as shown in Fig. 10.6.1.

10.4 CHOICE OF LONGITUDINAL REINFORCEMENT IN BEAMS

It will not be overemphasis to repeat here that the choice of longitudinal reinforcement depends on both the steel area and development length (or anchorage) requirements. The design of the main reinforcement in floor beams 2B1-2B2-2B1 is shown in the following example.

EXAMPLE 10.4.1 Choose the arrangement of main reinforcement in the floor beams 2B1-2B2-2B1 of Example 10.2.1.

Solution: (a) Flexural requirements. The critical moment is corrected to the face of support by subtracting $\Delta M = Vb/3$ from the moment at the center of support, as discussed in Sec. 8.2. The shear V for this calculation is taken as that at the face of support. For sections (rectangular sections) A-A, C-C, and D-D (Fig. 10.4.1),

Section A-A:

$$M_u \text{ at face} = 81.3 - \frac{40.0(1.5)}{3} = 81.3 - 20.0 = 61.3 \text{ ft-kips}$$

$$\text{required } R_n = \frac{M_u}{\phi bd^2} = \frac{61.3(12,000)}{0.90(13)(20)^2} = 157 \text{ psi}$$

Using Eq. (3.8.5), or Fig. 3.8.1,

$$\text{required } \rho = \frac{1}{m}\left(1 - \sqrt{1 - \frac{2mR_n}{f_y}}\right)$$

$$m = \frac{f_y}{0.85f'_c} = \frac{40}{0.85(3)} = 15.7$$

$$\text{required } \rho = \frac{1}{15.7}\left(1 - \sqrt{1 - \frac{2(15.7)(157)}{40,000}}\right) = 0.004$$

$$\text{min } \rho = \frac{200}{f_y} = \frac{200}{40,000} = 0.005 < \left[\frac{4}{3}\text{ (required } \rho) = 0.0053\right]$$

Use $\rho = 0.005$.

$$\text{required } A_s = 0.005(13)(20) = 1.30 \text{ sq in.}$$

Section C-C:

$$M_u \text{ at face} = 249.3 - \frac{50.7(1.5)}{3} = 249.3 - 25.3 = 224 \text{ ft-kips}$$

$$\text{required } R_n = \frac{224(12,000)}{0.90(13)(20)^2} = 574 \text{ psi}$$

From Fig. 3.8.1, or Eq. (3.8.5),

$$\text{required } \rho = 0.0165 > \text{min } \rho \qquad\qquad \text{OK}$$
$$\text{required } A_s = 0.0165(13)(20) = 4.30 \text{ sq in.}$$

Section D-D:

$$M_u \text{ at face} = 239.8 - \frac{47.1(1.5)}{3} = 239.8 - 23.5 = 216 \text{ ft-kips}$$

$$\text{required } R_n \doteq \frac{216(12,000)}{0.90(13)(20)^2} = 554 \text{ psi}$$

$$\text{required } A_s \approx 4.30\left(\frac{554}{574}\right) = 4.15 \text{ sq in.}$$

For sections (T-sections) B-B and E-E (Fig. 10.4.1),

$$\text{effective flange width } b_E = \frac{26(12)}{4} \qquad \text{or } 13 + 16(4.5) \qquad \text{or } 13(12)$$

$$b_E = 78 \text{ in.}$$

Section B-B:

$$\text{estimate moment arm} = 0.9d = 18 \text{ in.}$$

$$\text{required } A_s = \frac{M_u}{\phi f_y(\text{arm})} = \frac{174.2(12)}{0.90(40)(\approx 18)} = 3.23 \text{ sq in.}$$

Check:

$$C = 0.85 f'_c b_E a = 0.85(3)(78)a = 199a$$

$$T = 3.23(40) = 129 \text{ kips}$$

$$a = \frac{129}{199} = 0.65 \text{ in.}$$

$$\text{arm} = 20 - \frac{0.65}{2} = 19.7 \text{ in.}$$

$$\text{revised required } A_s = \frac{173.3(12)}{0.90(40)19.7} = 2.93 \text{ sq in.}$$

Section *E-E*:

$$\text{required } A_s = \frac{128.7(12)}{0.90(40)(\approx 19.7)} = 2.18 \text{ sq in.}$$

Check:

$$C = 199a$$

$$T = 2.20(40) = 88 \text{ kips}$$

$$a = \frac{88}{199} = 0.44 \text{ in.}$$

$$\text{arm} = 20 - \frac{0.44}{2} = 19.8 \text{ in.} \approx 19.7 \text{ in.} \qquad \text{OK}$$

On the basis of the areas required at sections *A-A*, *B-B*, *C-C*, *D-D*, and *E-E*, the arrangement of main reinforcement as shown in Fig. 10.4.1 is tentatively chosen. Note that the 2-#7 straight bars in 2B1 and the 2-#6 straight bars in 2B2 furnish the minimum one-fourth of the positive moment reinforcement that must be extended into the support in each span.

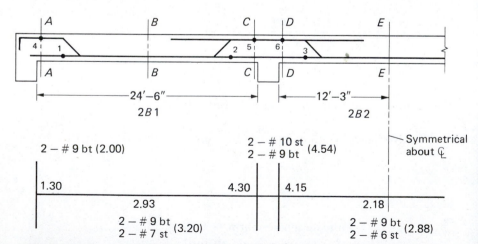

Figure 10.4.1 Longitudinal reinforcement areas required (sq in.) and bars selected for beams 2B1 and 2B2.

In order to confirm the choice of main reinforcement, the development length requirements (ACI-12.11.3) must be checked, and, if f_y were greater than 60,000 psi, the crack control provisions of ACI-10.6. Also the bars at the face of the exterior support (location 4 in Fig. 10.4.1) should be fully developed.

(b) Development length requirements. (For checking ACI-12.11.3, the required development lengths are (using ACI-12.2 or from Table 6.7.1)

$$L_d \text{ (for \#6)} = L_{db} = 12.8 \text{ in.}$$

$$L_d \text{ (for \#7)} = L_{db} = 17.5 \text{ in.}$$

$$L_d \text{ (for \#9)} = 1.4L_{db} = 1.4(29.2) = 41 \text{ in.}$$

where the 1.4 factor is for top bars (see Table 6.8.1).

At location 1 (Fig. 10.4.1): For 2-#7, extending 6 in. beyond face of support (actual length from inflection point = 2.09 − 0.25 = 1.84 ft),

$$C = 0.85f'_c ba = 0.85(3)(78)a = 199a$$

$$T = 2(0.60)40 = 48 \text{ kips}$$

$$a = \frac{48}{199} = 0.24 \text{ in.}$$

$$M_n = 48[20 - 0.5(0.24)]\tfrac{1}{12} = 79.5 \text{ ft-kips}$$

$$V_u = 42.6 - 3.57(2.09) = 35.1 \text{ kips}$$

$$L_a = d = 1.67 \text{ ft} < \text{actual length } 1.84 \text{ ft}$$

$$\frac{M_n}{V_u} + L_a = \frac{79.5}{35.1} + 1.67 = 3.93 \text{ ft} > L_d \text{ (\#7)} \qquad \text{OK}$$

At location 2: For 2-#7, since M_n and V_u are identical to their values at location 1, this check is made by inspection.

At location 3: For 2-#6,

$$C = 199a; \qquad T = 2(0.44)40 = 35.2 \text{ kips}; \qquad a = 0.18 \text{ in.}$$

$$M_n = 35.2[20 - 0.5(0.18)]\tfrac{1}{12} = 58.5 \text{ ft-kips}$$

$$V_u = 46.4 - 3.57(4.51) = 30.3 \text{ kips}$$

$$\frac{M_n}{V_u} + L_a = \frac{58.5}{30.3} + 1.67 = 3.60 \text{ ft} > L_d \text{ (\#6)} \qquad \text{OK}$$

In order to develop properly 2-#9 at location 4, either a straight embedment of 41 in. is required or a standard hook may be used.

For a #9 standard hook, the basic development length L_{hb}, according to Eq. (6.11.1) (ACI-12.5), is

$$L_{hb} = \frac{1200d_b}{\sqrt{f'_c}} = \frac{1200(1.128)}{\sqrt{3000}} = 24.7 \text{ in.}$$

which is to be modified by $f_y/60,000$. Assume concrete cover of at least $2\frac{1}{2}$ in. normal to the plane of the hook, *and* that on the bar extension beyond the 90° hook of at least 2 in. Thus, using modifications 1 and 2 from

Table 6.11.1, the development length L_{dh} becomes

$$L_{dh} = 24.7 \left(\frac{40,000}{60,000}\right) 0.7 = 11.5 \text{ in.}$$

The available L_{dh} (see Fig. 6.11.2) is

available L_{dh} = 18(support width) − 2(cover on tail) = 16 in.

Thus, the 2-#9 bars are assumed to be fully effective at point 4 of Fig. 10.4.1 when 90° hooks are used. Since a girder frames in at each side of beam 2B1 there will automatically be at least $2\frac{1}{2}$-in. side cover on the hooks.

(c) Crack control requirements. The most critical location to check is where the largest bars are put into the smallest area, that is, at section B-B (Fig. 10.4.1). Using Eq. (4.12.2) or ACI-10.6, and referring to Fig. 10.4.2,

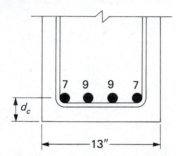

Figure 10.4.2 Cross section for crack control investigation.

$$z = f_s \sqrt[3]{d_c A} \leq 175 \text{ kips/in.} \qquad \text{(interior exposure)}$$

d_c = 1.5 (cover) + 0.375 (stirrup) + 0.4375 (bar radius) = 2.31 in.

$$A = \frac{2d_c b_w}{\Sigma A_s/(A_b \text{ for } \#9)} = \frac{2(2.31)13}{(3.20/1.0)} = 14.4 \text{ sq in./bar}$$

$$z = 0.6(40)\sqrt[3]{2.31(14.4)} = 139 \text{ kips/in.} < 175 \qquad\qquad \text{OK}$$

Actually, because in this example f_y does not exceed 40,000 psi, the crack control investigation is not required. However, the calculation is presented to illustrate the procedure.

10.5 SHEAR REINFORCEMENT IN BEAMS

Although the portions of the bent bars in the body of the beam would provide some shear strength in their vicinity, it will generally be a more clean-cut procedure to depend on the vertical stirrups alone to provide the entire shear strength requirement and welcome the bent bars as giving additional assistance. The design of shear reinforcement in 2B1-2B2-2B1 is shown in the following example.

EXAMPLE 10.5.1 Design the shear reinforcement in the floor beams 2B1-2B2-2B1 of Example 10.2.1.

Solution: The factored shear V_u envelope for which shear reinforcement will be provided is taken from Figs. 10.3.3 and 10.3.4 and shown in Fig. 10.5.1. The maximum factored shear should be taken at a distance equal to the effective depth from the face of support. Let V_1, V_2, and V_3 be the maximum factored shear in regions 1, 2, and 3 as shown in Fig. 10.5.1; then using $w_u = 3.57$ kips/ft,

$$\phi V_c = \phi(2\sqrt{f_c'})\, b_w d = 0.85(2\sqrt{3000})(13)(20)\tfrac{1}{1000} = 24.2 \text{ kips}$$

$$V_1 = 3.57(9.52) = 34.0 \text{ kips}; \qquad x_1 = 114.2 \left(\frac{34.0 - 24.2}{34.0}\right) = 32.9 \text{ in.}$$

$$V_2 = 3.57(12.51) = 44.7 \text{ kips}; \qquad x_2 = 150.1 \left(\frac{44.7 - 24.2}{44.7}\right) = 68.8 \text{ in.}$$

$$V_3 = 3.57(11.51) = 41.1 \text{ kips}; \qquad x_3 = 138.1 \left(\frac{41.1 - 24.2}{41.1}\right) = 56.8 \text{ in.}$$

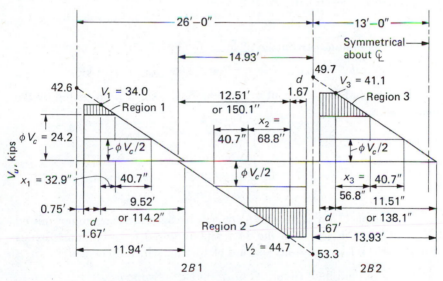

Figure 10.5.1   Factored shear V_u envelope used for 2B1 and 2B2.

The shaded shear areas in Fig. 10.5.1 plus an additional distance of 40.7 in., equal to the distance to the location where $V_u = \phi V_c/2$, represent the portions of the beam where shear reinforcement is required. These portions are also designated as regions 1, 2, and 3 in Fig. 10.5.2.

Assume #3 vertical U stirrups. From Fig. 10.5.2 the maximum required ϕV_s anywhere on the beam is 20.5 kips. Since the largest required ϕV_s does not exceed that based on a nominal stress v_s of $4\sqrt{f_c'}$, the maximum permissible stirrup spacing may not exceed $d/2$.

$$\text{limit } \phi V_s = \phi(4\sqrt{f_c'}b_w d) = 2(\phi V_c) = 2(24.2) = 48.4 \text{ kips}$$

$$\text{max required } \phi V_s = 20.5 \text{ kips} < \text{limit } \phi V_s$$

Thus, the upper limit on stirrup spacing is $d/2 = 10$ in.

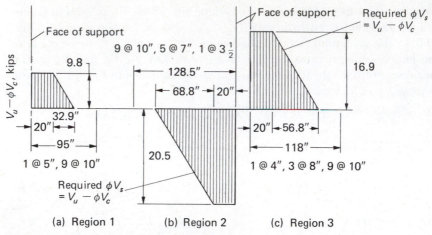

(a) Region 1   (b) Region 2   (c) Region 3

Figure 10.5.2 Portions of factored shear V_u in excess of ϕV_c (that is, required ϕV_s diagram) for beams 2B1 and 2B2.

For a minimum percentage of shear reinforcement,

$$\text{min } \phi V_s = \phi 50 b_w d = 0.85(50)(13)(20)\tfrac{1}{1000} = 11.1 \text{ kips}$$

The strength requirement to provide for the shear represented by the shaded areas of Fig. 10.5.2 is

$$V_s = \frac{A_v f_y d}{s}$$

$$s = \frac{\phi A_v f_y d}{\phi V_s} = \frac{0.85(0.22)(40)20}{\phi V_s} = \frac{149.6 \text{ kips in.}}{\phi V_s}$$

For region 1, a spacing of $d/2$ governs because the maximum required $\phi V_s = 9.8$ kips is less than the 11.1 kips required for minimum stirrups and $\phi V_s = 11.1$ kips permits spacing of 13.5 in. which exceeds $d/2$.

Use 1 @ 5 in. and 9 @ 10 in. (95 in. from face of support).

For region 2 the strength requirement is shown in Table 10.5.1.

Table 10.5.1 DISTANCE x FROM FACE OF SUPPORT TO LOCATION WHERE ϕV_s EQUALS REQUIRED ϕV_s

s	ϕV_s	MAX REQUIRED ϕV_s $- \phi V_s$	x (DISTANCE FROM FACE OF SUPPORT) (in.)
7.3	20.5	$20.5 - 20.5 = 0$	
8	18.7	$20.5 - 18.7 = 1.8$	$20 + \dfrac{68.8}{20.5}(1.8) = 20 + 6 = 26$
9	16.6	$20.5 - 16.6 = 3.9$	$20 + \dfrac{68.8}{20.5}(3.9) = 20 + 13 = 33$
10	15.0	$20.5 - 15.0 = 5.5$	$20 + \dfrac{68.8}{20.5}(5.5) = 20 + 18 = 38$

Use 1 @ $3\frac{1}{2}$ in.; 5 @ 7 in.; and 9 @ 10 in. (128.5 in. from face of support).

For region 3, to satisfy the strength requirement at the critical section,

$$\text{max } s = \frac{149.6}{16.8} = 8.9 \text{ in.}$$

A spacing of 10 in. can be used at $20 + (16.9 - 15.0)(56.8)/16.9 = 26$ in. from the face of support.

Use 1 @ 4 in.; 3 @ 8 in.; and 9 @ 10 in. (118 in. from face of support).

As a practical matter, many designers would place stirrups by scaling from the factored shear envelope (in terms of force or unit stress), rather than accurately compute distances as has been illustrated here. Further, the simplified procedure of ACI-11.3.1.1 for the value of V_c has been used in this example. The more detailed procedure involving $\rho V d/M$ does not seem appropriate unless an accurate shear envelope has been determined for the entire span. Such an envelope would include use of partial span loading for the maximum shear in the central portions of the beam spans, but it is not considered necessary for most ordinary beams in building frames.

10.6 DETAILS OF BARS IN BEAMS

For typical conditions of equal spans and uniform load where moment and shear coefficients of ACI-8.3 are used, standard bar details such as provided in the *ACI Detailing Manual—1980* [2] may be used. When the moment and shear envelopes are available, bar bend or cutoff locations should be determined therefrom, as illustrated for this example.

EXAMPLE 10.6.1 Determine the bar dimensions of the main reinforcing bars in 2B1 and 2B2 of Example 10.2.1 (see Fig. 10.6.1). $f_c' = 3000$ psi and $f_y = 40,000$ psi.

Solution: (a) Maximum moment capacities. In determining bar cutoff or bend locations it is necessary first to locate the theoretical points where the bars are no longer required. Lines representing full strength ϕM_n of the various bar combinations are computed. The effect of any compression steel will be small and is neglected. At section A-A, 2-#9

$$C = 0.85(3)(13)a = 33.1a; \qquad T = 2(1.0)40 = 80.0 \text{ kips}$$

$$\phi M_n = 0.90(80.0)(20 - 1.20)\tfrac{1}{12} = 113 \text{ ft-kips}$$

Note that it is ϕM_n (not M_n) that must be compared with the factored moment M_u on Fig. 10.6.1.
At section B-B, 2-#7

$$C = 0.85(3)(78)a = 199a; \qquad T = 2(0.60)40 = 48 \text{ kips}$$

$$\phi M_n = 0.90(48)(20 - 0.12)\tfrac{1}{12} = 71.5 \text{ ft-kips}$$

At section B-B, 2-#7 and 2-#9

$$C = 199a; \qquad T = 3.20(40) = 128 \text{ kips}$$

$$\phi M_n = 0.90(128)(20 - 0.32)\tfrac{1}{12} = 189 \text{ ft-kips}$$

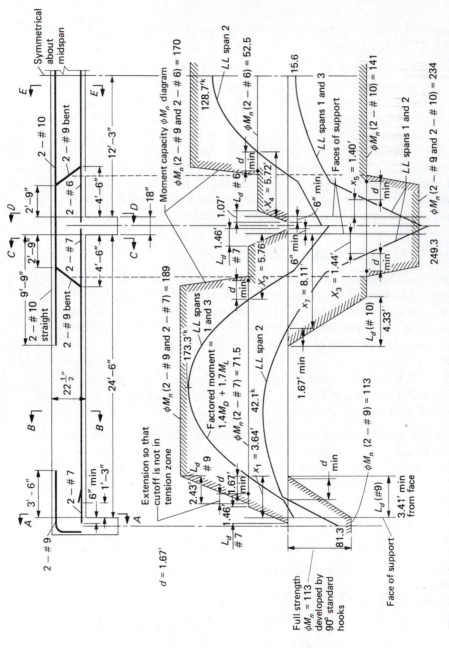

Figure 10.6.1 Moment envelope, moment capacity diagram, and bar arrangement for beams 2B1 and 2B2 (all moments M_u and ϕM_n are given in ft-kips).

At sections C-C and D-D, 2-#10

$$C = 33.1a; \qquad T = 2.54(40) = 101.6 \text{ kips}$$
$$\phi M_n = 0.90(101.6)(20 - 1.53)\tfrac{1}{12} = 141 \text{ ft-kips}$$

At sections C-C and D-D, 2-#9 and 2-#10

$$C = 33.1a; \qquad T = 4.54(40) = 181.6 \text{ kips}$$
$$\phi M_n = 0.90(181.6)(20 - 2.74)\tfrac{1}{12} = 235 \text{ ft-kips}$$

At section E-E, 2-#6

$$C = 199a; \qquad T = 2(0.44)40 = 35.2 \text{ kips}$$
$$\phi M_n = 0.90(35.2)(20 - 0.09)\tfrac{1}{12} = 52.5 \text{ ft-kips}$$

At section E-E, 2-#6 and 2-#9

$$C = 199a; \qquad T = 2.88(40) = 115.2 \text{ kips}$$
$$\phi M_n = 0.90(115.2)(20 - 0.29)\tfrac{1}{12} = 170 \text{ ft-kips}$$

(b) Extension of positive moment reinforcement into supports. In order to satisfy ACI-12.11.1, the #7 straight bars in 2B1 must extend at least 6 in. into both supports. Since there is no tensile requirement for these bars at the first interior support, the 6-in. minimum embedment is sufficient unless the bars are to be utilized as compression reinforcement, in which case the compression capacity of such bars must be developed at the face of support. In this case no compression capacity is required for strength or desired to increase ductility.

At the exterior support, the 6-in. embedment may not be adequate if the support is a column and the flexural member is part of the primary lateral load resisting system; ACI-12.11.2 requires development of the full yield stress in tension at the face of support for such cases. In this example, the supporting member is a girder and is not a part of the primary lateral load resisting system so that 6-in. embedment is sufficient. Similarly, the 2-#6 in beam 2B2 are extended into the supports 6 in. The total length of the #6 and #7 straight bars [24 ft 6 in. + 2(6 in.)] = 25 ft 6 in. (Bar lengths are usually specified in 3-in. increments.)

(c) Theoretical bend or cutoff locations. In detailing the bend-up and bend-down points for the #9 bars, and for establishing the cutoff for the #9 bars near section A-A, the distances x_1 through x_5 as marked in Fig. 10.6.1 [also shown as x_1, x_2 in Fig. 10.3.3(a), x_3 in Fig. 10.3.3(b), x_4 in Fig. 10.3.4(a), and x_5 in Fig. 10.3.4(b)] are in practice determined more frequently by graphical means, although they are computed in this example. Thus referring to Fig. 10.3.3(a),

$$\frac{3.57(11.19 - x_1)^2}{2} = \frac{3.57(13.31 - x_2)^2}{2} = 173.3 - 71.5 = 101.8 \text{ ft-kips}$$

which gives

$$x_1 = 3.64 \text{ ft}; \qquad x_2 = 5.76 \text{ ft}$$

In accordance with ACI-12.10.4, a distance of 12 bar diameters or the effective depth of the member, whichever is larger, must be subtracted from these distances. Thus the bend or cutoff locations could be

$$\text{bend-up or cutoff based on } x_1 = 3.64 - 1.67 = 1.97 \text{ ft}$$

$$\text{bend-up or cutoff based on } x_2 = 5.76 - 1.67 = 4.09 \text{ ft}$$

From Fig. 10.3.3(b),

$$\frac{3.57(14.18 - x_3)^2}{2} = 148.8 + 141 = 289.8 \text{ ft-kips}$$

$$x_3 = 1.44 \text{ ft}$$

$$\text{bend-down or cutoff based on } x_3 = 1.44 + 1.67 = 3.11 \text{ ft}$$

From Fig. 10.3.4(a),

$$\frac{3.57(12.25 - x_4)^2}{2} = 128.7 - 52.5 = 76.2 \text{ ft-kips}$$

$$x_4 = 5.72 \text{ ft}$$

$$\text{bend-up or cutoff based on } x_4 = 5.72 - 1.67 = 4.05 \text{ ft}$$

From Fig. 10.3.4(b),

$$\frac{3.57(13.18 - x_5)^2}{2} = 106.8 + 141 = 247.8 \text{ ft-kips}$$

$$x_5 = 1.40 \text{ ft}$$

$$\text{bend-down or cutoff based on } x_5 = 1.40 + 1.67 = 3.07 \text{ ft}$$

The distances x_1 through x_5 with the effective depth $d = 1.67$ ft either added or subtracted locate correctly the theoretical *cutoff* points. For bending bars, the objective is to have the resulting moment capacity diagram maintain an offset from the factored moment diagram equal to, in this case, 1.67 ft (effective depth). Thus the actual bend-up or bend-down points will be offset from the theoretical cutoff points an amount equal to one-half the horizontal projection of the sloping portion of the bar. If the bend is at 45°, the offset would be $(d - d')/2$. Typically, the bend is at an approximately 45° angle but the locations of the bend-up and bend-down points are made 3-in. increment dimensions from the face of support.

For example, if the bend is at 45°, the horizontal projection will be

$$d - d' = 20 - 2.5 = 17.5 \text{ in. } (1.46 \text{ ft})$$

The bend-down position in beam 2B1 could be no closer than $x_3 + d - 1.46/2$, or 2.38 ft, from the face of support. The selected distance of 2.75 ft exceeds this and is acceptable. The bend-up position in beam 2B1 must be no farther from the face of support than $x_2 - d + 1.46/2$, or 4.82 ft. The selected distance of 4.5 ft is closer than 4.82 ft and is therefore acceptable. The situation in beam 2B2 is similar.

Use the bends shown in Fig. 10.6.1. The potential cutoff relating to x_1 requires further checking to satisfy ACI-12.10.5 for cutting bars in a tension zone.

(d) Inflection point extensions. ACI-12.12.3 requires an extension beyond the extreme point of inflection a distance equal to the largest of one-sixteenth of the clear span, the effective depth of the member, or 12 bar diameters. In this case the effective depth of the member, 1.67 ft, controls. The #6 and #10 bars in member 2B1 are terminated after the 1.67-ft extension. The cut bars develop their full capacity after an embedment equal to the development length L_d required for top bars (see Secs. 6.7 and 6.8).

(e) Check cutoff for 2-#9 bars in the tension zone in the bottom of beam 2B1. ACI-12.10.5 must be satisfied for the cutoff at $x_1 - d$ to be acceptable. Since the continuing bars (2-#7) provide more than double the area required for flexure at the potential cutoff point, it is necessary only to check whether or not the factored shear V_u exceeds three-fourths of the shear strength ϕV_n (ACI-12.10.5.3).

$$V_u = w_u (11.94 - 0.75 - x_1 + d)$$

$$V_u = 3.57(11.94 - 0.75 - 3.64 + 1.67) = 32.9 \text{ kips}$$

It is required that

$$V_u \leq [0.75\phi V_n = 0.75\phi(V_c + V_s)]$$

The strength provided, including stirrups, is

$$\phi V_c + \phi V_s = 24.2 + \frac{149.6}{10} = 39.2 \text{ kips}$$

$$\text{percent strength utilized} = \frac{32.9}{39.2} = 84\% > 75\% \text{ permitted} \qquad \text{NG}$$

If the cutoff is to be made at the indicated location, the stirrup spacing must be reduced so that V_s increases. Alternatively, the 2-#9 bars may be extended to the inflection point where they will no longer be in the tension zone. Thus the #9 bars are cut off at 1 ft–3 in. from the face of support. In this case, there is little material saved by cutting so close to the support, but it serves to illustrate the procedure.

The summary of bar arrangement, lengths, dimensions, and moment capacity provided is shown in Fig. 10.6.1. The cross sections for the final choice at each designated section of Fig. 10.6.1 are shown in Fig. 10.6.2.

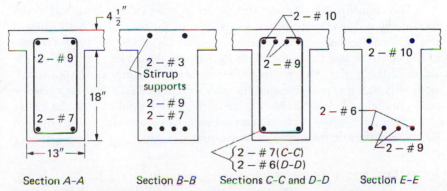

Section A–A   Section B–B   Sections C–C and D–D   Section E–E

Figure 10.6.2   Typical sections for 2B1 and 2B2 (refer to Fig. 10.6.1).

(f) Stirrup development. For stirrup supports, 2-#3 bars are provided in the central part of beam 2B1. The dimensions of the stirrups are shown in Fig. 10.6.3. The embedment required (ACI-12.13) beyond the middepth for development of the stirrups is

$$\text{required } L_d \ (\#3) \ = \ 12 \text{ in.} \qquad \text{(Table 6.7.1)}$$

for which a length of only $\frac{1}{2}(19.5)$ = 9.75 in. is available. ACI-12.13.2.1 requires a straight embedment of $0.5L_d$, or 6 in. in this case, if a standard hook is used. Since 6 in. of straight embedment *is* available, a 90° hook is used. If $0.5L_d$ is not available in straight embedment, a 135° hook around the longitudinal reinforcement would have to be used (ACI-12.13.2.3) plus, if f_y exceeds 40,000 psi, a straight embedment of $0.33L_d$ measured from middepth to the point of tangency at the start of the hook.

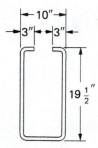

Figure 10.6.3 Stirrup details in beams 2B1 and 2B2.

10.7 SIZE OF GIRDER WEB

For the purpose of determining the *size* of the girder web, the maximum negative moment at the exterior face of the first interior support may be used, taken as 0.8 of the maximum positive moment in a simple span with span length equal to the clear span of the girder and with loadings identical to those on the girder. The maximum shear at the same location may be taken as 1.20 times the reaction to the simple span described above. Note that the coefficients of ACI-8.3.3 cannot be used in lieu of an elastic analysis, since concentrated loads are involved.

The concentrated loads on the girder may be taken as one-half of the total dead and live loads on the clear span of the beams on both sides of the girder. The dead and live uniform load on the girder will be the weight of the concrete and the live load on the floor, respectively, both within the width of the girder. This loading transfer is a sufficiently good approximation, although it is probable that the slab weight and floor load on a narrow strip parallel and close to either edge of the girder would act on the girder as uniform load instead of being carried by the one-way slab to the adjacent beams and then to the girder as concentrated loads.

EXAMPLE 10.7.1 Design the floor girders 2G1-2G2-2G2-2G1 as shown in the floor framing plan of Fig. 8.2.1. Use information previously described in Chap. 8 and in the present chapter. Work through the choice of the size of the girder web in this example according to the strength method of the ACI Code.

Solution: The concentrated reactions from the beams using the 24.5 ft clear span are

$$dead\ load = 1.4(24.50)[(0.056)(13) + 0.244] = 33.4\ kips$$
$$live\ load = 1.7(24.50)(0.100)(13) = 54.2\ kips$$
$$total\ load = 33.4 + 54.2 = 87.6\ kips$$

Assuming 18×36 in. web, the uniform loads on the girder are

$$uniform\ dead\ load = 1.4 \left[\frac{18(40.5)(0.150)}{144} \right] = 1.06\ kips/ft$$
$$uniform\ live\ load = 1.7(0.100)(1.5) = 0.255\ kip/ft$$
$$uniform\ total\ load = 1.06 + 0.255 = 1.315\ kips/ft$$

Maximum positive moment on a simple span (see Fig. 10.7.1) is

$$M_u = 87.6(12.25) + \tfrac{1}{8}(1.315)(37.5)^2 = 1306\ ft\text{-}kips$$

The estimated maximum negative moment in the girder at the exterior face of the first interior support is $0.8(1306) = 1045$ ft-kips, say 1050 ft-kips. From Fig. 10.7.1, estimate

$$d_{neg} = 40.50 - 4.75 = 35.75\ in.$$
$$d_{pos} = 40.50 - 3.50 = 37\ in.$$

The negative moment requirement is

$$required\ R_n = \frac{M_u}{\phi b d^2} = \frac{1050(12,000)}{0.90(18)(35.75)^2} = 609\ psi$$

which is less than the maximum $R_n = 870$ psi for $\rho = 0.75\rho_b$, and is an acceptable value. The girder system should be relatively stiff and will be so if this smaller R_n is used.

$$required\ A_s\ for\ (-M) \approx 0.75\rho_b \left(\frac{609}{870}\right)bd$$
$$= 0.0278 \left(\frac{609}{870}\right)(18)(35.75) = 12.5\ sq\ in.$$

For the positive moment requirement, using the coefficients of ACI-8.3.3 to obtain the approximate proportion between positive and negative moments,

$$estimated\ (+M) \approx \frac{10}{14}(-M) = \frac{10}{14}(1050) = 750\ ft\text{-}kips$$

$$required\ A_s \approx \frac{M_u}{\phi f_y\ (arm)} = \frac{750(12)}{0.90(40)(\approx 33)} = 7.58\ sq\ in.$$

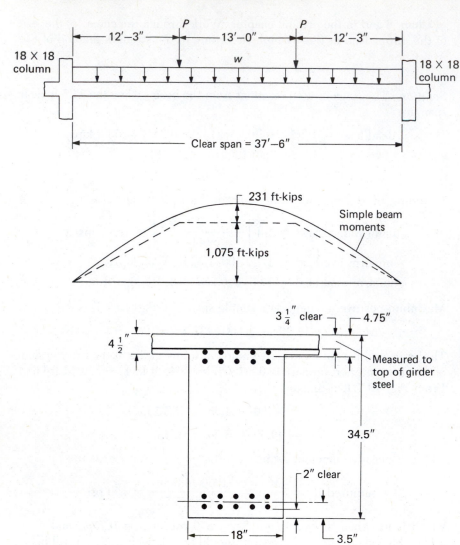

Figure 10.7.1 Loading information for floor girders 2G1 and 2G2.

This may well fit into one layer (6-#10 requires a beam width of 18 in. according to Table 3.9.2).

For the shear requirement,

$$\max V_u = 1.20[87.6 + 1.315(18.75)] = 135 \text{ kips}$$

$$v_n = \frac{135,000}{0.85(18)(35.75)} = 247 \text{ psi} = 4.5\sqrt{f_c'}$$

The stem could be made smaller; with this large depth it is unlikely that deflection would be excessive. The 18×36 stem appears to be somewhat large though it could certainly be used. In this case because of the large depth, reduce the stem size to 18×30. Estimated effective depths

become

$$d_{\text{neg}} = 29.75 \text{ in.}$$

$$d_{\text{pos}} = 31 \text{ in.}$$

$$\text{revised girder weight} = 1.4\left[\frac{18(34.5)}{144}(0.15)\right] = 0.91 \text{ kip/ft}$$

$$w_u = w_D + w_L = 0.91 + 0.255 = 1.165 \text{ kips/ft}$$

$$\text{revised } M_u \text{ (simple beam)} = 1075 + \tfrac{1}{8}(1.165)(37.5)^2 = 1280 \text{ ft-kips}$$

$$\text{required } R_n = \frac{0.8(1280)(12,000)}{0.90(18)(29.75)^2} = 860 \text{ psi} < 870 \text{ psi max} \qquad \text{OK}$$

$$v_n \text{ at support} = \frac{131,500}{0.85(18)(29.75)} = 289 \text{ psi} = 5.3\sqrt{f_c'}$$

Use the 18 × 30 stem section.

10.8 CONTINUOUS FRAME ANALYSIS FOR GIRDERS

Although floor girders subjected to large concentrated loads are structural members of common occurrence, the ACI Code makes no mention of moment coefficients for these cases. In the following example, elastic statically indeterminate structural analysis is used for 2G1-2G2-2G2-2G1.

EXAMPLE 10.8.1 By the use of an elastic statically indeterminate structural analysis, determine the shear and moment diagrams to be used in the design of 2G1 and 2G2 in Example 10.7.1.

Solution: As discussed in Sec. 10.3, there are various ideas regarding what constitutes the correct stiffness for continuous T-sections. In accordance with the thought that the true stiffness is that of a span with a variable cross section, the effective flange width for an equivalent uniform moment of inertia section may be assumed to be twice the web width (Fig. 10.8.1). The principal effect of changing the stiffness of the girder (K_{gr}) occurs at the exterior support where the relative stiffness of the columns (K_{col}) compared

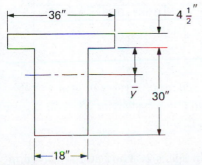

Figure 10.8.1   T-section for stiffness computation.

to that of the girder is greatest. The centroid of the gross area of the T-section is at

$$\bar{y} = \frac{18(30)15 - 36(4.5)(2.25)}{18(30) + 36(4.5)} = \frac{7736}{702} = 11.0 \text{ in.}$$

The gross moment of inertia I_g is

$$I_g = \frac{36(4.5)^3}{3} + \frac{18(30)^3}{3} - 702(11.0)^2 = 78{,}100 \text{ in.}^4$$

$$K_{gr} = \frac{78{,}100}{39} = 2000 \text{ in.}^4/\text{ft}$$

If the size of the upper and lower columns were 18×18 in. and the column height were 15 ft,

$$K_{col} = \frac{18(18)^3/12}{15} = 583 \text{ in.}^4/\text{ft}$$

$$\frac{K_{gr}}{K_{col}} = \frac{2000}{583} = 3.43$$

In the rigid frame of Fig. 10.8.2 are shown the distribution factors

$$\frac{3.43}{3.43 + 1 + 1} = 0.632 \quad \text{and} \quad \frac{3.43}{3.43 + 3.43 + 1 + 1} = 0.387$$

The fixed-end moments are

$$\text{FEM due to dead load} = \frac{33.4(13)(26)}{39} + \frac{0.91(39)^2}{12} = 405 \text{ ft-kips}$$

$$\text{FEM due to live load} = \frac{54.2(13)(26)}{39} + \frac{0.255(39)^2}{12} = 502 \text{ ft-kips}$$

The results of moment distribution for the various loading conditions are given in Table 10.8.1. Note that identical results may be obtained by applying the matrix displacement method, or any other statically indeterminate structural analysis method.

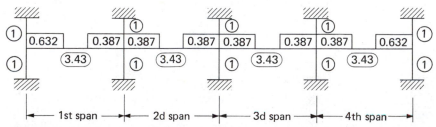

Figure 10.8.2 Elastic analysis model of rigid frame system for 2G1-2G2-2G2-2G1.

The controlling moment envelopes for spans 2G1 and 2G2 are given in Figs. 10.8.3 and 10.8.4. In practice this composite diagram should be made instead of the individual moment and shear diagrams for each loading case.

Table 10.8.1 SUMMARY OF MOMENT DISTRIBUTION RESULTS USING FOUR CYCLES

JOINT	A	B		C		D		E
MEMBER	AB	BA	BC	CB	CD	DC	DE	ED
DF	0.632	0.387	0.387	0.387	0.387	0.387	0.387	0.632
DEAD LOAD ONLY								
FEM	−405	+405	−405	+405	−405	+405	−405	+405
Total	−159	+487	−459	+380	−380	+459	−487	+159
LIVE LOAD ONLY SPANS 1 AND 3								
FEM	−502	+502	0	0	−502	+502	0	0
Total	−244	+376	−182	+146	−326	+384	−229	−48
LIVE LOAD ONLY SPANS 1, 2, AND 4								
FEM	−502	+502	−502	+502	0	0	−502	+502
Total	−188	+648	−641	+242	−96	+197	−385	+242
LIVE LOAD ONLY SPAN 3								
FEM	0	0	0	0	−502	+502	0	0
Total	−8	−44	+73	+230	−375	+369	−220	−46
LIVE LOAD ONLY SPANS 2 AND 3								
FEM	0	0	−502	+502	−502	+502	0	0
Total	+38	+176	−296	+605	−605	+296	−176	−38
LOAD LIVE ONLY SPANS 1 AND 4								
FEM	−502	+502	0	0	0	0	−502	+502
Total	−234	+429	−270	−133	+133	+270	−429	+234

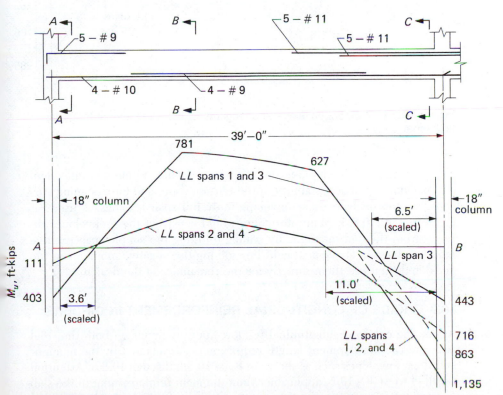

Figure 10.8.3 Moment envelope and longitudinal bar arrangement for girder 2G1 (span 1).

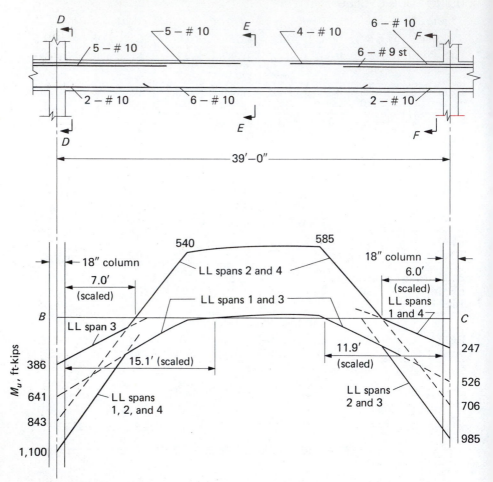

Figure 10.8.4 Moment envelope and longitudinal bar arrangement for girder 2G2 (span 2).

For investigating development length requirements at points of inflection, for checking cutoff acceptability in the tension zone, and for designing stirrups, an approximate shear envelope (only full span loadings) is used, as given in Fig. 10.8.5. Many designers would construct these envelopes directly from the end moments by scaling (plotting the simple span positive moment diagram above a straight line joining the negative moments at the ends) and then use them directly for the remainder of the design.

10.9 CHOICE OF LONGITUDINAL REINFORCEMENT IN GIRDERS

Again, the choice of longitudinal reinforcement depends on both the steel area and the development length requirements as dictated by the moment and shear envelopes such as shown in Figs. 10.8.3 through 10.8.5. Attention is called to ACI-8.10.5, applicable when the main reinforcement in the slab is parallel to the girder. For this situation transverse steel must be provided in the top of the slab to carry the load on the portion of the slab acting

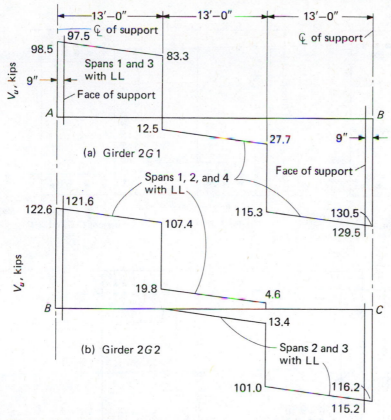

Figure 10.8.5 Shear envelopes for girders 2G1 and 2G2 (full-span loadings only).

effectively as the flange of the girder. The overhanging flange is assumed to act as a cantilever. The spacing of the transverse bars may not exceed five times the thickness of the flange or, in any case, 18 in.

EXAMPLE 10.9.1 Choose the arrangement of the main reinforcement in the floor girders 2G1-2G2-2G2-2G1 of Example 10.7.1.

Solution: (a) Sections (rectangular sections) A-A, C-C, D-D, and F-F (Figs. 10.8.3, 10.8.4, and 10.9.1). Section A-A:

$$M_u \text{ at face} = 403 - \frac{97.5(1.5)}{3} = 403 - 49 = 354 \text{ ft-kips}$$

$$\text{required } R_n = \frac{M_u}{\phi b d^2} = \frac{354(12,000)}{0.90(18)(30.7)^2} = 278 \text{ psi}$$

Either from Eq. (3.8.5), or from Fig. 3.8.1,

$$\text{required } \rho = 0.0075$$

$$\text{required } A_s = 0.0075(18)30.7 = 4.14 \text{ sq in.}$$

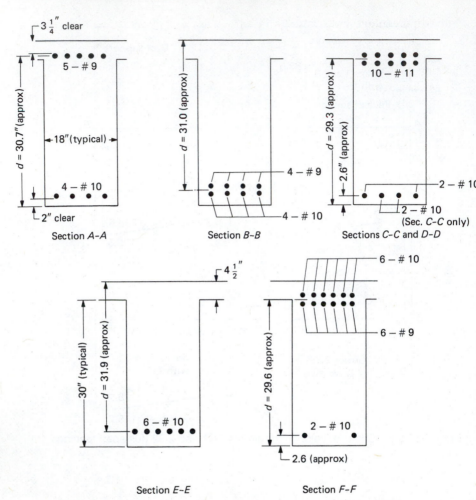

Figure 10.9.1 Typical sections for 2G1 and 2G2 (refer to Figs. 10.8.3 and 10.8.4.)

Sections C-C and D-D: The larger moment is at section *C-C*,

$$M_u \text{ at face} = 1135 - \frac{129.5(1.5)}{3} = 1070 \text{ ft-kips}$$

$$\text{required } R_n = \frac{1070(12,000)}{0.90(18)(29.3)^2} = 923 \text{ psi}$$

This value exceeds the maximum value of 870 psi allowed without using compression steel. In this case, only a small amount of compression steel is required; thus the positive moment steel extended into the support at the bottom of the beam will be developed by proper embedment in order to utilize it as compression steel. Since the use of compression steel influences the tension steel requirement hardly at all, the tension requirement is computed just as if no compression steel were to be used. Estimating the required percentage from Fig. 3.8.1,

$$\text{required } A_s \approx 0.030(18)(29.3) = 15.8 \text{ sq in.}$$

Section *F-F*:

$$M_u \text{ at face} = 985 - \frac{115.2(1.5)}{3} = 985 - 58 = 927 \text{ ft-kips}$$

$$\text{required } R_n = \frac{927(12,000)}{0.90(18)(29.6)^2} = 785 \text{ psi}$$

Using straightline approximation,

$$\text{required } A_s \approx 0.0278 \left(\frac{785}{870}\right)(18)29.6 = 13.4 \text{ sq in.}$$

(b) Sections (T-sections) *B-B* and *E-E* (Figs. 10.8.3, 10.8.4, and 10.9.1): Available flange width b_E,

$$\frac{39(12)}{4} = 117; \qquad \text{or } \underline{18 + 16(4.5) = 90}; \qquad \text{or } (26)(12) = 312$$

It is likely that the depth *a* of the rectangular stress distribution will be less than the flange thickness *t*. Estimate $a = 2$ in. (about one-half flange thickness).

Section *B-B*: estimated $d = 31.0$ in.

$$\text{required } A_s = \frac{M_u}{\phi f_y (\text{arm})} = \frac{781(12)}{0.90(40)(31.0 - 1)} = 8.7 \text{ sq in.}$$

Check:

$$C = 0.85(3)(90)a = 229.5a$$
$$T = 8.7(40) = 348 \text{ kips}; \qquad a = 1.52 \text{ in.}$$
$$\text{revised required } A_s = \frac{781(12)}{0.90(40)(31.0 - 0.76)} = 8.6 \text{ sq in.}$$

Section *E-E*:

$$\text{required } A_s = \frac{585(12)}{0.90(40)(31.9 - 1)} = 6.3 \text{ sq in.}$$

Check:

$$C = 229.5a; \qquad T = 6.3(40) = 252 \text{ kips}$$
$$a = 1.10 \text{ in.}$$
$$\text{revised required } A_s = \frac{585(12)}{0.90(40)(31.9 - 0.55)} = 6.2 \text{ sq in.}$$

(c) Confirmation of the tentative arrangement of main reinforcement as summarized in Fig. 10.9.2 awaits the check of development of reinforcement at the positive moment inflection points (ACI-12.11.3), crack control when

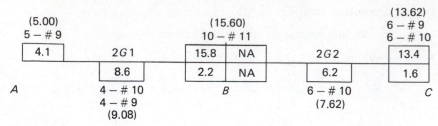

Figure 10.9.2 Longitudinal reinforcement areas (sq in.) required and bars selected for girders 2G1 and 2G2.

f_y exceeds 40,000 psi (does not apply for this problem), and deflection if excessive deflection will cause cracking of attached nonstructural elements.

Note in Fig. 10.9.2 that the requirements indicated in the compression zone at B and C are merely the minimums of one-fourth of the positive moment steel required to satisfy ACI-12.11.1. It is likely that at B, which is the only place compression steel is required for strength, the minimum satisfying ACI-12.11.1 is sufficient.

The details of determining bar lengths are not shown; only the general arrangement is shown in Figs. 10.8.3 and 10.8.4.

(d) Check maximum percentage reinforcement in negative moment region at B (sections C-C and D-D), including effect of compression reinforcement. At balanced condition,

$$x_b = \left(\frac{0.003}{0.003 + f_y/E_s}\right)d = \left(\frac{0.003}{0.003 + 0.00138}\right)d$$

$$= 0.685d = 0.685(29.3) = 20.1 \text{ in.}$$

$$\text{max } x = 0.75x_b = 0.75(20.1) = 15.1 \text{ in.}$$

$$a = 0.85x = 0.85(15.1) = 12.8 \text{ in.}$$

At max A_s condition,

$$\epsilon'_s = \left(\frac{15.1 - 2.6}{15.1}\right)(0.003) = 0.0025 > \epsilon_y$$

Thus compression steel yields at max A_s condition.

$$\text{max } C_c = 0.85f'_c ba = 0.85(3)(18)(12.8) = 586 \text{ kips}$$

Assume 2-#10 that extend through the support region at C-C and D-D are the only bars developed for compression reinforcement.

$$C_s = A'_s(f_y - 0.85f'_c) = 2.54(40 - 2.55) = 95 \text{ kips}$$

$$\text{max } T = \text{max } C_c + C_s = 586 + 95 = 681 \text{ kips}$$

$$\text{max } A_s = \frac{\text{max } T}{f_y} = \frac{681}{40} = 17.0 \text{ sq in.}$$

Since the actual steel used (15.60) is less than max A_s, the design is acceptable.

Check the development length requirement at the positive moment inflection points. From Fig. 10.8.5 (shear envelope), for span 1,

$$V_u \text{ (near left end)} = 97.5 - 3.6(1.165) = 93 \text{ kips}$$
$$V_u \text{ (near right end)} = 129.5 - 6.5(1.165) = 122 \text{ kips}$$

For span 2,

$$V_u \text{ (near left end)} = 121.6 - 7.0(1.165) = 113 \text{ kips}$$
$$V_u \text{ (near right end)} = 115.2 - 6.0(1.165) = 108 \text{ kips}$$

At the span 1 inflection points, 4-#10 bars continue along the bottom of the beam. The critical location is at the right inflection point where maximum V_u occurs.

$$V_u = 122 \text{ kips}$$

For 4-#10,

$$C = 0.85(3)(90)a = 229.5a$$
$$T = 5.08(40) = 203 \text{ kips}$$
$$a = 0.89 \text{ in.}$$
$$M_n = 203[31.9 - 0.5(0.89)]\tfrac{1}{12} = 532 \text{ ft-kips}$$

$$L_d \text{ for #10 bars} = L_{db} = 37.1 \text{ in. } (3.09 \text{ ft}) \qquad \text{(Table 6.7.1)}$$

$$\left[\frac{M_n}{V_u} + L_a = \frac{532(12)}{122} + \frac{31.9}{12} = 7.02 \text{ ft} \right] > 3.09 \text{ ft} \qquad \text{OK}$$

Note that L_a = effective depth, d, for this situation.

In span 2, only 2-#10 continue past the inflection point into the support. For 2-#10,

$$C = 229.5a; \qquad T = 2(1.27)(40) = 102 \text{ kips}$$
$$a = 0.44 \text{ in.}$$
$$M_n = 102[31.9 - 0.5(0.44)]\tfrac{1}{12} = 269 \text{ ft-kips}$$

$$\left[\frac{M_n}{V_u} + L_a = \frac{269}{113} + \frac{31.9}{12} = 5.02 \text{ ft} \right] > 3.09 \text{ ft} \qquad \text{OK}$$

(f) The transverse steel required in the top of the flange of the girder (ACI-8.10.5) may be computed as follows (Fig. 10.9.3):

$$M_u = \tfrac{1}{2}[0.100(1.7) + 0.056(1.4)](3)^2 = 1.12 \text{ ft-kips/ft}$$

$$\text{required } R_n = \frac{1.12(12,000)}{0.90(12)(2.94)^2} = 144 \text{ psi}$$

Since from observing Fig. 3.8.1 this is less than the value for min ρ, the minimum value controls.

$$A_s = \frac{200}{f_y} bd = \frac{200}{40,000} (12)(2.94) = 0.18 \text{ sq in.}$$

Use #5 @ at 18 in. spacing. Generally the amount required for temperature and shrinkage in the slab according to ACI-7.12 is adequate for this purpose.

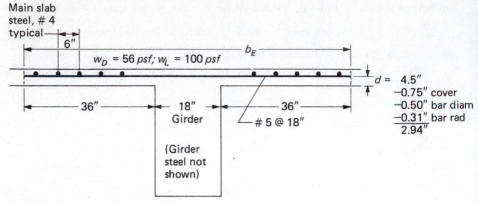

Figure 10.9.3 Transverse steel across the top of girder.

(g) The bar arrangement is shown in Figs. 10.8.3 and 10.8.4 and cross sections are shown in Fig. 10.9.1. The shear reinforcement and dimensions of longitudinal reinforcement are not shown, because little new will be involved in their presentation.

10.10 ONE-WAY JOIST FLOOR CONSTRUCTION

One-way concrete joist floor construction (Fig. 10.10.1), sometimes called "ribbed-slab construction," consists of regularly spaced ribs monolithically built with a top floor slab and arranged to span in one direction. Such a system may also be designed as a two-way system (waffle slab) according to the procedures for two-way floor systems treated in Chaps. 16 and 17. The dimensions of the one-way joist system are usually such that only temperature and shrinkage reinforcement is required in the slab. The slab is usually in the range of 2 to 4 in. (50 to 100 mm) thick but may occasionally be as much as 6 in. (150 mm). The ribs (joists) of at least 4 in. (100 mm) width are usually tapered [Fig. 10.10.1(b)] and are spaced so that the clear spacing

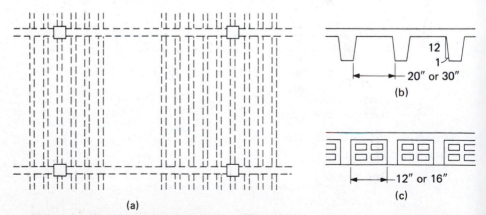

Figure 10.10.1 Concrete joist floor construction.

between adjacent ribs does not exceed 30 in. (760 mm). During construction removable and reusable form fillers are used in spaces between the joists. Such fillers may be standard-sized steel "pans" in 20 or 30 in. widths and 6, 8, 10, 12, 14, 16, and 20 in. depths. Sometimes form fillers are made from hardboard, fiberboard, glass-reinforced plastic, or corrugated cardboard. Occasionally, permanent fillers are used consisting of lightweight or normal-weight concrete blocks or clay tile blocks, as shown in Fig. 10.10.1(c).

Concrete joist floor construction is referred to in ACI-8.11. Some of the requirements are as follows:

1. The joists shall not be farther apart than 30 in. face to face. The ribs shall be not less than 4 in. wide and of a depth not more than $3\frac{1}{2}$ times the width.
2. The vertical shells of permanent fillers in contact with the joists may be included in strength calculations involving shear or negative bending moment provided the filler material has a unit compressive strength at least equal to that of the concrete in the joists. In this case the minimum slab thickness is $1\frac{1}{2}$ in. or $\frac{1}{12}$ of the clear distance between joists, whichever is smaller.
3. When removable forms or fillers having less compressive strength than required under (2) are used, the thickness of the concrete slab shall not be less than $\frac{1}{12}$ of the clear distance between joists, nor less than 2 in.

For more information on concrete joist construction see Chap. 10 of *Manual of Standard Practice* [3], Chap. 4 of Rice and Hoffman [4], and Chap. 8 of *CRSI Handbook* [5].

10.11 DESIGN OF CONCRETE JOIST FLOORS

The design of concrete joist floors involves (1) the slab, (2) the joists, and (3) the girders.

Generally, shrinkage reinforcement is placed at right angles to the joists, and the concrete slab is treated as if it were of plain concrete. The short clear span between joists may be considered as being fixed at both ends.

The joist itself may be designed as a floor beam having a rectangular section in the region of negative bending and a T-section in the region of positive bending. The critical design moment curves for each span may either follow the ACI moment coefficients or be determined by a continuity analysis. Largely because of the interaction of slab with the closely spaced joists, ACI-8.11.8 permits the shear strength V_c provided by the concrete to be 10% higher than for regular beams.

The girder is designed as a floor girder, but the load from the joists may be considered as being uniformly distributed along the span.

In the case of joist floors over removable steel pans, tapered end forms are available that increase the effective joist width 2 in. on each side for 20-in. wide forms and $2\frac{1}{2}$ in. on each side for 30-in. wide forms in a distance of 3 ft from the end. This increased width may be necessary to take the large shear or negative bending moment near the end of the span.

In order to limit deflection on floor joist construction the minimum depth requirements of ACI-Table 9.5a for ribbed one-way slabs must be applied. Whenever excessive deflection may cause cracking or other adverse effects, deflections must be computed even if ACI-Table 9.5a has been satisfied.

EXAMPLE 10.11.1 Design the typical interior span of a concrete joist floor, using 30-in. wide forms, for a center-to-center span of 26 ft and clear span of 24 ft–6 in. The live load is 80 psf and the additional superimposed dead load is 20 psf. Use $f'_c = 3000$ psi, $f_y = 60,000$ psi, and the strength method of the ACI Code. Use the moment and shear coefficients of ACI-8.3.

Solution: (a) Slab design. Assume a 3-in. slab for weight estimation purposes. The factored load is

$$w_u = 1.4 \left[\frac{3.0(0.150)}{12} \right] + 1.4(0.020) + 1.7(0.080) = 0.22 \text{ ksf}$$

Assuming that the slab is fixed at its junction with the joist,

$$M_u \approx \frac{1}{12}(0.22) \left(\frac{30}{12} \right)^2 = 0.115 \text{ ft-kips/ft}$$

The design strength ϕM_n of plain concrete, computed using $f_r = 5\sqrt{f'_c}$ as the modulus of rupture* is

$$\phi M_n = \phi f_r \left(\frac{1}{6} bh^2 \right) = 0.65(5\sqrt{f'_c}) \left(\frac{1}{6} bh^2 \right)$$

Setting $\phi M_n = M_u$ and $b = 12$ in. for a typical 1-ft strip, gives for the required slab thickness,

$$h = \sqrt{ \frac{0.115(12,000)}{\frac{1}{6}(0.65)(5\sqrt{3000})12} } = 2.0 \text{ in.}$$

The 3-in. slab will be thick enough; perhaps $2\frac{1}{2}$ in. would be sufficient. *Use h = 3 in.*

Shrinkage and temperature reinforcement (ACI-7.12) must be provided, equal to

$$A_s = 0.0018(12)(3.0) = 0.065 \text{ sq in./ft}$$

Use welded wire fabric, 4 × 12–W2.5 × W1.4 ($A_s = 0.075$ sq in.).

The selection is made from Table 10.11.1; A_s in direction perpendicular to joists is 0.075 sq in./ft, and A_s in direction parallel to joists is 0.014 sq in./ft. The "4 × 12" means 4-in. spacing of the wires running perpendicular to the joist and 12-in. spacing of the wires running parallel to the joist.

---

*Note that for a strength calculation the modulus of rupture is taken conservatively lower than the value $7.5\sqrt{f'_c}$ used in deflection calculations (ACI-9.5). The value $5\sqrt{f'_c}$ is given by Sec. 6.2.2, *Building Code Requirements for Structural Plain Concrete* (ACI 318.1–83) *and Commentary*. Detroit: American Concrete Institute, 1983.

(b) Joist design. The overall depth of the joist floor must satisfy the minimum requirement of ACI-Table 9.5a unless deflections are computed; thus,

$$\min h = \frac{L}{21} = \frac{26(12)}{21} = 14.9 \text{ in.}$$

If excessive deflection may cause damage to nonstructural elements, deflection must be computed in all cases. Assuming that deflection must later be computed it will be prudent to make joists deeper than indicated by Table 9.5a. Assume use of joists 12 in. deep below the bottom of the slab with a width of 5 in. at the bottom (i.e., average 6 in.). The weight of the joist will then be

$$w_D = [(3.0 + 12)6 + 30(3.0)]\frac{0.150}{144} = 0.19 \text{ kip/ft}$$

The factored load is

$$w_u = 1.4(0.19) + 1.4(0.020)\left(\frac{35}{12}\right) + 1.7(0.080)\left(\frac{35}{12}\right) = 0.744 \text{ kip/ft}$$

The maximum negative factored moment is

$$M_u = \tfrac{1}{11}(0.744)(24.5)^2 = 40.6 \text{ ft-kips}$$

Allowing the minimum cover of $\frac{3}{4}$-in. (ACI-7.7.1c) and assuming #5 bars for main steel,

$$d = 15 - 0.75 - 0.31 = 13.94 \text{ in.}$$

The required R_n then becomes

$$\text{required } R_n = \frac{M_u}{\phi b d^2} = \frac{40.6(12,000)}{0.90(5)(13.94)^2} = 557 \text{ psi}$$

Using Eq. (3.8.5) or Fig. 3.8.1,

$$\text{required } \rho = 0.011$$

This exceeds the minimum $200/f_y = 0.0033$ and does not exceed the maximum of $0.75\rho_b = 0.0160$; therefore it is acceptable. Thus,

$$\text{required } A_s \text{ (over support)} = 0.011(5)(13.94) = 0.77 \text{ sq in.}$$

With a reinforcement ratio ρ so low, deflection should not be a problem even if excessive deflection may cause cracking of nonstructural elements. The shear at a distance d from the face of support is

$$V_u = 0.744(12.25 - 1.16) = 8.25 \text{ kips}$$

The shear strength of a joist without shear reinforcement (ACI-8.11.8) is

$$\phi V_c = \phi(1.10)(2\sqrt{f_c'})b_w d$$
$$= 0.85(1.10)(2\sqrt{3000})(5)(13.94)\tfrac{1}{1000} = 7.14 \text{ kips} < 8.25 \text{ kips} \quad \text{NG}$$

Use a taper on the joist near the support, as shown in Fig. 10.11.1. When

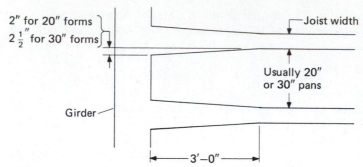

Figure 10.11.1 Plan showing ends of tapered pan joists.

a taper is required to satisfy the shear strength requirement, the 30 in. wide form tapers to 25 in. in the last 3 ft near the support. At d from the face, the beam width then becomes

$$b_w = 5 + \left(\frac{3.0 - 1.16}{3.0}\right) 5 = 8.1 \text{ in.}$$

$$\phi V_c = 7.14 \left(\frac{8.1}{5.0}\right) = 11.6 \text{ kips} > [V_u = 8.25 \text{ kips}] \qquad \text{OK}$$

Note that the *minimum* width (instead of the average) of web should probably be used when the web is in compression (not an ACI requirement).

The maximum factored bending moment near midspan is

$$M_u = \tfrac{1}{16}(0.744)(24.5)^2 = 27.9 \text{ ft-kips}$$

For the T-section, assume that the depth of the rectangular stress distribution falls within the flange; estimate $a \approx 1$ in. Then, by trial,

$$\text{required } A_s \approx \frac{M_u}{\phi f_y (\text{arm})} = \frac{27.9(12)}{0.90(60)(13.94 - 0.5)} = 0.46 \text{ sq in.}$$

Determine moment arm more accurately using $b_E = 8(3) + 5 = 29$ in. (ACI-8.10.2),

$$C = 0.85 f_c' b a = 0.85(3)(29)a = 74.0a$$

$$T = A_s f_y = 0.46(60) = 27.6 \text{ kips}$$

$$a = 27.6/74.0 = 0.37 \text{ in.}$$

$$\text{revised required } A_s = \frac{27.9(12)}{0.90(60)(13.94 - 0.19)} = 0.45 \text{ sq in.}$$

Since a is less than $t/2$, the effective section of the compression zone is rectangular. Use 1-#5 bottom bar, 1-#4 truss bar, and 1-#5 top bar (see Fig. 10.11.2).

$$\text{provided } A_s \text{ over support} = 0.71 \text{ sq in. (2-\#4 + 1-\#5)}$$

$$\text{provided } A_s \text{ at midspan} = 0.51 \text{ sq in. (1-\#4 + 1-\#5)}$$

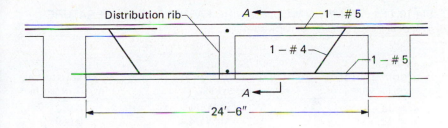

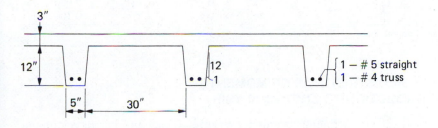

Section *A-A*

Figure 10.11.2 Concrete joist floor using removable pans, Example 10.11.1.

Consistent with a loading for the maximum positive moment in an interior span, the inflection point is at $0.354L$ from centerline of span [see Fig. 7.5.4(a)]. Thus the development length requirement of ACI-12.11.3 must be checked. Only the #5 straight bar extends beyond the inflection point and into the support at least 6 in.

$$C = 0.85f'_c b_E a = 0.85(3)(29)a = 74.0a$$

$$T = 0.31(60) = 18.6 \text{ kips}$$

$$a = 0.25 \text{ in.}$$

$$M_n = 18.6[13.94 - 0.5(0.25)]\tfrac{1}{12} = 21.4 \text{ ft-kips}$$

$$V_u = 0.741(0.354)(24.5) = 6.4 \text{ kips}$$

$$L_a = 12d_b \text{ (controls)} = 7.5 \text{ in.}$$

$$L_d \text{ (#5)} = 12 \text{ in.}$$

The equivalent embedment length provided is

$$\frac{M_n}{V_u} + L_a = \frac{21.4(12)}{6.4} + 7.5 = 40.1 + 7.5 = 47.6 \text{ in.} > L_d \qquad \text{OK}$$

Frequently a transverse distribution rib (Fig. 10.11.2) is used having a 4 in. minimum width and containing at least one #4 bar both top and bottom. Such a rib would be located at the third points of the span for spans greater than 30 ft. For spans between 20 and 30 ft one rib should be used near midspan, and for spans less than 20 ft the rib may be omitted [5].

Table 10.11.1 COMMON WELDED WIRE FABRIC FOR TEMPERATURE AND SHRINKAGE REINFORCEMENT

	DESIGNATION	
SPACING (in.) OF LONGITUDINAL[a] AND TRANSVERSE WIRES	WIRE SIZE DESIGNATION LONGITUDINAL × TRANSVERSE	A_s, LONGITUDINAL DIRECTION (sq in.)
4 × 12 —	W1.4 × W1.4	0.042
4 × 12 —	W2 × W1.4	0.060
4 × 12 —	W2.5 × W1.4	0.075
4 × 12 —	W3 × W2	0.090
4 × 12 —	W3.5 × W2	0.105

[a]Longitudinal refers to the slab span, which is perpendicular to the joist span.

10.12 REDISTRIBUTION OF MOMENTS— INTRODUCTION TO LIMIT ANALYSIS

Methods of proportioning beams for flexure, shear, and bar development requirements according to the strength method have been discussed in Chaps. 3, 5, and 6, and further illustrated in Secs. 10.1 through 10.11. When the inelastic behavior of concrete at a particular location has been accounted for in the design of that cross section based on its strength, it may seem somewhat illogical to have used an elastic analysis to determine the design moments and shears. However, because the evaluation of the true ultimate strength, or limit strength, of an entire structure requires difficult and elaborate analysis, the present safe conservative procedure is used.

The concept of redistribution of moments first introduced into the ACI Code in 1963 was the result of considerable research [6–12] into the limit behavior of the entire structure (primarily continuous beams, at present) beyond the elastic range to the point where the collapse load is reached. Concrete design, therefore, has moved toward limit design, or, as it is called in steel structures, "plastic" design.

The use of a plastic theory requires that the material involved actually behave plastically. Figure 10.12.1(a) shows the ideal relationship of moment M to curvature θ, where there is a perfectly elastic portion and an ideal plastic portion. Reinforced concrete exhibits an M/θ curve as shown in Fig. 10.12.1(b) which, although it differs markedly from the ideal, may be approximated by such an ideal system. It has been found that the lower the net reinforcement ratio $(\rho - \rho')$, the closer is the actual moment-rotation behavior to the ideal.

Consider the simply supported beam of Fig. 10.12.2. The limit load on such a system is reached when the rotation angle under the load reaches the value θ_u which corresponds to the extreme fiber of concrete in compression reaching the crushing strain ϵ_{cu} (taken by ACI as 0.003). In limit analysis, the moment strength achieved when ϵ_{cu} is reached at the extreme concrete fiber is referred to as M_n (nominal moment strength), which corresponds to M_p, the plastic moment in the ideal system. For such simply supported beams the achieving of M_p or M_n at one location along the span

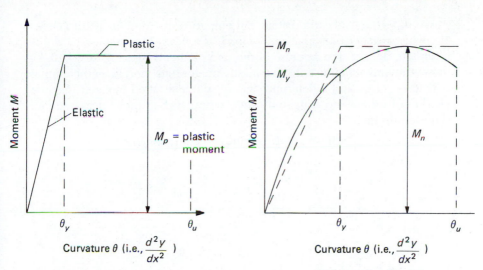

(a) Ideal elastic-plastic   (b) Reinforced concrete

Figure 10.12.1 Moment-curvature characteristics.

represents the limit of the system, and the beam becomes a mechanism, deforming further without inducing further resistance. It will be shown that in a continuous structure (statically indeterminate system), the material at a section where M_p is first reached must undergo additional strain before that structure achieves its limit condition. This additional strain, with regard to bending of a section, is referred to as "rotation capacity." Reinforced concrete has a reasonable amount of rotation capacity when the net reinforcement percentage ρ or $\rho - \rho'$ is 50% or less of the balanced amount ρ_b for a singly reinforced beam.

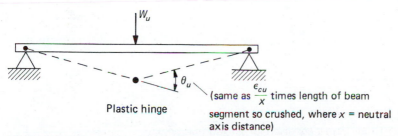

(same as $\dfrac{\epsilon_{cu}}{x}$ times length of beam segment so crushed, where x = neutral axis distance)

Figure 10.12.2 Limit condition for simply supported beam.

Next, consider the statically indeterminate fixed-end beam of Fig. 10.12.3(a), showing its moment diagram in the elastic range. As the load is increased, the moments at the fixed ends achieve M_p, while at all other points moments are still in the elastic range. Thus, at the point of reaching M_p at the ends,

$$\frac{w_y L^2}{12} = M_p \quad \text{and} \quad w_y = \frac{12M_p}{L^2}$$

where w_y is the load carried at an end curvature θ_y (see Fig. 10.12.1) when M_p is just reached. In carrying this load, the beam is as stable as a simple beam and deflects less because of the end moments M_p. Figure 10.12.3(b) shows the limit condition of three plastic hinges (one more than the number of degrees of statical indeterminacy) and the associated moment diagram. The limit load may be computed using statics, such as the midspan equilibrium requirement

$$+M_p = -M_p + \text{simple beam moment}$$

$$2M_p = \frac{w_u L^2}{8}$$

$$w_u = \frac{16M_p}{L^2}$$

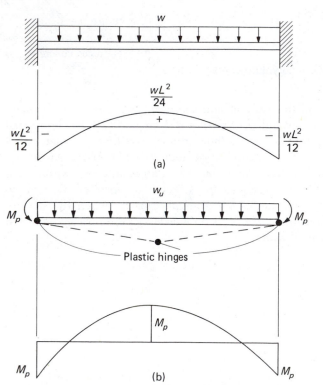

Figure 10.12.3 Limit condition for a fixed-end beam.

Thus the capacity may be increased 33% after the plastic moment has been reached at the fixed ends, provided that sufficient additional deformation (rotation) can be accommodated at the fixed ends to permit development of the plastic moment at midspan. In other words, rotation capacity permitting, the positive and negative moments under any particular loading condition tend to equalize (assuming, of course, the strength of the section at the two regions is the same). For a thorough treatment of limit analysis the reader is referred to the work of Baker [11].

The aforementioned limit behavior has been thoroughly verified for structural steel, and its use under the term "plastic design" has become widespread. Steel being a very ductile material, rotation capacity in beams is available to a high degree. On the other hand, concrete, being a relatively brittle material, has traditionally been considered to have no appreciable plastic deformability. Reinforced concrete beams, when the net percentage of reinforcement $\rho - \rho'$ is low, will have their moment strength controlled by yielding of the steel while the concrete strain is still of low magnitude (see Fig. 10.12.4). Reserve rotation $\theta_u - \theta_y$ is then available for a redistribution of moments to occur before the ultimate concrete strain of about 0.003 is reached.

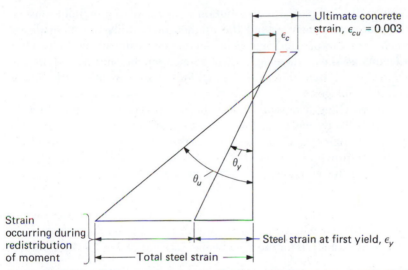

Figure 10.12.4   Strain diagram for reinforced concrete beams wherein steel reaches ϵ_y before concrete $\epsilon_c = \epsilon_{cu}$.

Research has shown that such a redistribution does occur in reinforced concrete beams when certain rotation capacity related conditions are met; therefore the ACI Code (ACI-8.4) allows the negative moments at the supports of continuous flexural members, calculated by elastic theory, to be increased or decreased by not more than

$$20\left(1 - \frac{\rho - \rho'}{\rho_b}\right)\%$$

where ρ_b is the balanced percentage of reinforcement for a beam containing tension reinforcement only, given by Eq. (3.5.4) or ACI Formula (8-1),

$$\rho_b = \frac{0.85\beta_1 f'_c}{f_y}\left(\frac{87,000}{87,000 + f_y}\right)$$

The limits of applicability of this provision may be summarized as follows:

1. Application is limited to continuous flexural members.
2. No more than 20% of the negative moments for any given loading arrangement may be adjusted.
3. Bending moments used in such an adjustment must be obtained by an elastic analysis; moments from use of coefficients or other approximate methods may not be adjusted.
4. The *net* reinforcement ratio $\rho - \rho'$ at the cross section where the moment is *reduced* must not exceed one-half the balanced percentage ρ_b, as defined by ACI Formula (8-1).
5. Adjustment, when permitted, is made for each given loading condition. The envelope of the adjusted diagrams from all loading conditions is then used to proportion the members.

This method of partial redistribution of moments is generally conservative. Future codes may extend the method or go fully to a limit design approach once the plastic hinge behavior is more extensively understood. Such factors as shear, development of reinforcement, and deflection may still control the design; thus, limit design for flexure may offer little, if any, economic advantage.

In approaching a design, it may be advantageous to apply ACI-8.4 directly. It is likely, however, that more frequent use will occur when the design is in progress and the designer realizes that the conditions of the redistribution provision are met and savings appear possible. The following example illustrates an application of this provision.

EXAMPLE 10.12.1 Show the effects of redistribution on the moments obtained by elastic analysis for the floor beams 2B1 and 2B2 in Example 10.3.1.

Solution: The beams along with the critical moment diagrams are shown in Fig. 10.12.5. The elastic end moments M_u are those computed under factored loads and tabulated in Table 10.3.1. The elastic moment diagrams are shown in Figs. 10.3.3 through 10.3.6.

(a) Investigate negative moment region at B to determine the maximum percent moment adjustment. Referring to Fig. 10.4.1, and noting that $0.75\rho_b = 0.0278$ as used for design in Example 10.2.1,

$$\text{actual } \rho = \frac{4.54}{13(20)} = 0.0174$$

$$0.5\rho_b = 0.5 \left(\frac{0.0278}{0.75} \right) = 0.0185 > 0.0174 \qquad\qquad \text{OK}$$

Moment redistribution is permissible at B.

$$\text{percent adjustment permitted} = 20 \left(1 - \frac{\rho}{\rho_b} \right)$$

$$= 20 \left(1 - \frac{0.0174}{2(0.0185)} \right) = 10.6\%$$

Since the steel in the compression zone was not fully developed at the faces of support (see Fig. 10.6.1), it may not be counted as compression steel. It might well be economical to develop the capacity of the steel in the compression zone at the support. This would increase the ductility in that region and allow a higher percentage of moment redistribution.

(b) Make adjustments to elastic moments. Examine first the loading for maximum positive moment in span *AB*, Fig. 10.12.5(a). Increasing the negative moments by 10.6% reduces the maximum positive moment for this

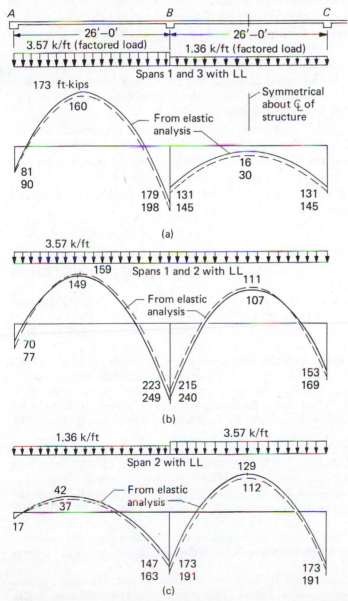

Figure 10.12.5 Redistribution of elastically computed moments according to ACI-8-4.

loading from 173 to 160 ft-kips. The increased negative moment for case (a) is still less than the negative moment occurring under other loadings. The 10.6% adjustment for case (b) is made by reducing the negative moment at B and increasing those at A and C, thus minimizing the effect on the positive moments. The increased positive moment in span 1 is still slightly less than the reduced controlling positive moment of 160 ft-kips from case (a). In case (c) the negative moments are increased 10.6%, thus reducing the positive moment in span 2 from 130 to 112 ft-kips.

The adjustment of the negative moments may be either an increase or decrease so long as the positive moments are also adjusted to satisfy static equilibrium. The envelope of adjusted moments would then be used to design the sections by the strength method. The net effect on the envelope is a reduction for both positive and negative moments. This is not actually a reduction in the safety factor below that used for a simply supported beam. It is a reduction in the excess strength that the system has by virtue of its continuity, one span with another. Since the redistribution does occur, partial utilization of it seems reasonable.

A simple application of limit design concepts to practical design has been presented by Furlong [13]. His method satisfies the ductility and strength requirements of limit design without the usual complexities that have impeded the acceptance of limit design into the ACI Code. Further discussion of Furlong's proposal has been presented [14] as a proposed alternate to the use of ACI-8.3.3.

SELECTED REFERENCES

1. *Continuity in Concrete Building Frames* (4th ed.). Chicago: Portland Cement Association, 1959.
2. ACI Committee 315. *ACI Detailing Manual—1980* (SP-66). Detroit, Michigan: American Concrete Institute, 1980.
3. *Manual of Standard Practice* (23rd ed.). Chicago: Concrete Reinforcing Steel Institute, 1980.
4. Paul F. Rice and Edward S. Hoffman. *Structural Design Guide to the ACI Building Code* (2nd ed.). New York: Van Nostrand Reinhold, 1979.
5. *CRSI Handbook* (3rd ed.). Chicago: Concrete Reinforcing Steel Institute, 1978.
6. A. H. Mattock. "Limit Design for Structural Concrete," *Journal of the Research and Development Laboratories*. Portland Cement Association, **1**, May 1959, 14–24.
7. W. G. Corley. "Rotational Capacity of Reinforced Concrete Beams," *Journal of Structural Division*, ASCE, **92**, October 1966 (ST5), 121–146. (Also PCA Development Department Bulletin D108.)
8. ACI-ASCE Committee 428. "Progress Report on Code Clauses for 'Limit Design,'" *ACI Journal, Proceedings*, **65**, September 1968, 713–720. Disc, **66**, 221–223.
9. M. Z. Cohn. "Limit Design for Reinforced Concrete Structures: An Annotated Bibliography," *ACI Bibliography No. 8*. Detroit, Michigan: American Concrete Institute, 1970.
10. Harold W. Conner, Paul H. Kaar, and W. Gene Corley. "Moment Redistribution in Precast Concrete Frame," *Journal of the Structural Division*, ASCE, **96**, March 1970 (ST3), 637–661.

11. A. L. L. Baker. *Limit State Design of Reinforced Concrete.* London: Cement and Concrete Association, 1971.
12. M. Z. Cohn. "Inelasticity of Reinforced Concrete and Structural Standards," *Journal of the Structural Division,* ASCE, **105**, November 1979 (ST11), 2221–2241.
13. Richard W. Furlong. "Design of Concrete Frames by Assigned Limit Moments," *ACI Journal, Proceedings,* **67**, April 1970, 341–353.
14. Richard W. Furlong and Carlos Rezende. "Alternate to ACI Analysis Coefficients," *Journal of the Structural Division,* ASCE, **105**, November 1979 (ST11), 2203–2220.

PROBLEMS

All problems* are to be worked in accordance with the strength method of the ACI Code, and all stated loads are service loads, unless otherwise indicated. A design sketch on $8\frac{1}{2} \times 11$ paper showing all design decisions is required for each problem. For verifying bar lengths draw a moment capacity ϕM_n diagram directly below a side view of the beam showing the bars, as in Fig. 10.6.1.

Continuous Beam Problems

10.1 Design a rectangular beam continuous over three spans as shown in the accompanying figure. The live load is 2.75 kips/ft, and the dead load is 1.0 kip/ft in addition to the beam weight. The floor is to be a prefabricated system. Assume the supports to be 15 in. wide. Use $f'_c = 4000$ psi and $f_y = 60,000$ psi, and do not apply ACI-8.4 for moment redistribution. (Live load = 40 kN/m; dead load = 15 kN/m plus beam; supports 380 mm wide; $f'_c = 30$ MPa; $f_y = 400$ MPa.)
(a) Determine the moment envelope using factored loads.
(b) Determine the bar bends or cutoff locations, or both, directly from the moment envelope.
(c) Use only full-span loadings for computing the shear envelope and use U stirrups of #3 size if possible.

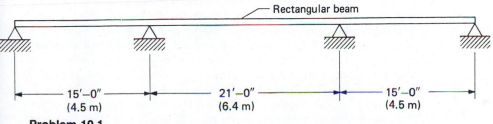

Problem 10.1

10.2 Redesign the beam of Prob. 10.1 as a monolithic T-section floor system. The beams are spaced 8 ft on centers and the slab is 6 in. thick. The 1.0 kip/ft dead load includes the slab but not the beam stem. (Beam spacing = 2.4 m; slab thickness = 150 mm.)

---

*Many problems may be solved as problems stated in Inch-Pound units or as problems in SI units using quantities in parenthesis at the end of the statement. The SI conversions are approximate to avoid implying higher precision for the given information in metric units than that for the Inch-Pound units.

10.3 Design the beam *ABC* of the frame shown in the accompanying figure in which the relative stiffnesses EI/L are given. The beams are T-sections having a 6-in. slab. The dead load is 0.40 kip/ft (not including beam stem or slab) and the live load is 3.75 kips/ft. Assume the supports to be 15 in. wide. Use $f'_c = 4000$ psi and $f_y = 60,000$ psi. (150 mm slab; dead load = 6 kN/m; live load = 55 kN/m; supports 380 mm wide; $f'_c = 30$ MPa; $f_y = 400$ MPa.)

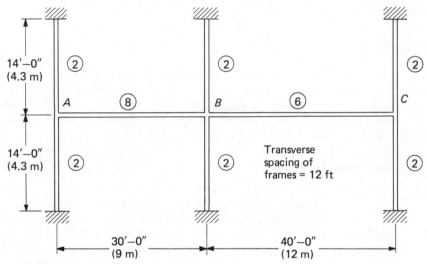

Problem 10.3

10.4 Design the transverse beam indicated for the floor plan given in the accompanying figure. Assume that a warehouse live load of 375 psf is to be used. Assume a 5-in. slab placed monolithically with beams and girders, a width of support at longitudinal girders of 18 in., and that only a nominal minimum of moment restraint is provided by the exterior wall support (i.e., assume hinge for elastic analysis). Use $f'_c = 3500$ psi and $f_y = 60,000$ psi. For a comparison of the effect of considering torsional stiffness of longitudinal girders, divide the class into three parts, each using one of the following assumptions: **(a)** zero torsional stiffness of the two longitudinal girders; **(b)** torsional stiffness equal to 25% of the bending stiffness of the 21-ft span beam; and **(c)** torsional stiffness equal to 50% of the bending stiffness of the 21-ft span beam. (Live load = 18 kN/m²; 130 mm slab thickness; support width = 460 mm; $f'_c = 25$ MPa; $f_y = 400$ MPa.)

10.5 Redesign the transverse beams of Prob. 10.4, except use spans 22-20-22 instead of the original spans 21-18-21, and use $f'_c = 4000$ psi and $f_y = 60,000$ psi. All other details are the same as in Prob. 10.4. ($f'_c = 30$ MPa; $f_y = 400$ MPa.)

10.6 Redesign the transverse beams of Prob. 10.4, except use spans 24-21-24 instead of original spans 21-18-21, and use $f'_c = 4000$ psi and $f_y = 60,000$ psi. All other details are the same as in Prob. 10.4. ($f'_c = 30$ MPa; $f_y = 400$ MPa.)

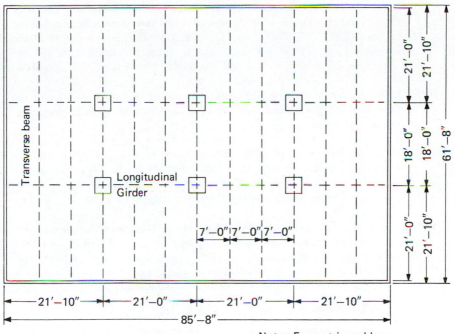

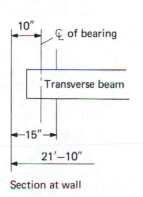

Note: For metric problems
use all lengths in meters
= 0.305 times lengths in
feet rounded to two significant
figures.

Section at wall

Problem 10.4

10.7 Using the moment and shear envelopes of Figs. 10.8.3 and 10.8.5, design
the actual lengths of reinforcement from the faces of support to the cutoff
points for girder 2*G*1. Verify by drawing the moment capacity ϕM_n diagram
superimposed on the factored moment M_u envelope.

10.8 Same as Prob. 10.7 except use Fig. 10.8.4 for 2*G*2 instead of Fig. 10.8.3 for
2*G*1.

10.9 Design the four-span longitudinal girder indicated for the floor system of
Prob. 10.4. In lieu of using the more accurate loadings from the results of
Prob. 10.4, use concentrated dead and live loads of 15 and 60 kips, respec-
tively, coming to the girder from *each* side, plus the weight of the girder.
Assume that the columns are 18 in. square and 15 ft high and that the beam

receives equivalent restraint from monolithic attachment to the 15-in. reinforced concrete exterior wall. Assume also that the columns are fixed at the far ends. Use f'_c = 3500 psi and f_y = 60,000 psi. (Concentrated dead and live loads from each side, 67 and 268 kN, plus girder; columns = 460 mm square and 4.5 m high; wall = 380 mm thick; f'_c = 25 MPa; f_y = 400 MPa.)

10.10 Redesign the floor beams 2B1-2B2-2B1 used in Example 10.2.1 and the following examples, taking the beam stem as 14 in. wide by 14 in. deep (i.e., use h = 18.5 in.).

10.11 Redesign the floor beams 2B1-2B2-2B1 used in Example 10.2.1 and the following examples, taking f'_c = 4000 psi and f_y = 60,000 psi. Make the stem size exactly as indicated by the text design guidelines (that is, do not make it arbitrarily larger as was done in the chapter example).

10.12 Design the continuous 26-ft span floor beams along a column line in Fig. 8.3.1. Assume the columns are 15 in. square. Use f'_c = 4000 psi, f_y = 60,000 psi, and use the 100 psf live load. Make a preliminary design without structural analysis; then using the preliminary size make the structural analysis to obtain the factored moment and factored shear envelopes for the final design.

Concrete Joist Problems

10.13 Design a concrete joist, using 30-in. wide removable pans, for a typical interior span of 28 ft center to center of supporting girders. Assume a support width of 18 in. Use a live load of 100 psf, f'_c = 4000 psi, and f_y = 40,000 psi. (800-mm wide pans; span = 8.5 m; support width = 460 mm; f'_c = 30 MPa; f_y = 300 MPa.)

10.14 Determine the service live load capacity for a single-span joist of 20 ft clear span, using a 2½-in. slab, 20 in. wide and 8-in. deep forms, and 4-in. wide joists with 2-#5 bars. No taper is used. Use f'_c = 3000 psi and f_y = 40,000 psi.

10.15 Determine the service live load capacity for an interior span joist of 32-ft span (clear span), using a 2-in. slab, 20-in. wide and 14-in. deep form, and 5-in. wide joists with #7 bars for the bottom, truss, and top steel. The bottom steel is properly embedded in the support to develop its compression capacity at the face of support. The joist has a standard taper. Use f'_c = 3000 psi and f_y = 40,000 psi.

10.16 Design an end-span joist for a continuous system to carry a live load of 225 psf, using 20-in. wide removable pans, for a clear span of 18 ft. Allow an extra ½ in. of thickness for dead load purposes only, since the concrete slab is to serve as the final wearing surface. Use f'_c = 4500 psi and f_y = 60,000 psi.

Moment Redistribution Problem

10.17 Redesign the beam of Prob. 10.1 taking into account permissible moment redistribution. Compare with Prob. 10.1.

11

Monolithic
Beam-to-Column Joints

11.1 MONOLITHIC JOINTS

Considerable emphasis has been made concerning design of flexural members for bending, shear, and development of reinforcement in Chaps. 3 through 10. Some attention has been given to development of reinforcement at exterior supports, including the use of hooks (Secs. 6.11 and 6.14). The design of compression members is treated in Chap. 13. Often in design not enough attention is given to the details of connections: how the forces in beams and the forces in columns interact and get transmitted through the joint. The ACI Code provides little guidance specifically directed to joint details, except in Appendix A–Special Provisions for Seismic Design, Sec. A.6.

The state-of-the-art regarding the design of beam-to-column joints has been summarized by ACI-ASCE Committee 352, Joints and Connections in Monolithic Reinforced Concrete Structures [1]. In that report there are detailed provisions for the design of two classes of beam-to-column joints:

Type 1 joints, primarily for static loading, where strength is the primary criterion and no significant inelastic deformations are expected; and

Type 2 joints, usually for earthquake or blast loading, where there is need for sustained strength through stress reversals into the inelastic range.

In general, Type 1 joints require only nominal ductility, whereas Type 2 joints require significant ductility, such as would be required in seismic design. The Committee 352 report [1] has utilized the relatively limited

Construction of tapered rigid frame knee for University of Wisconsin Stadium. (Photo by C. G. Salmon.)

research on joints to establish its recommendations. Some of the more readily available references [2–10] are included at the end of this chapter.

In the following sections of this chapter the general concepts are outlined and two examples are presented using the Committee 352 recommendations. No attempt will be made to provide a complete theoretical treatment or to make a complete statement of the Committee 352 recommendations. The reader should make use of the Committee 352 report for design guidance until such time as rules are incorporated into the ACI Code.

11.2 FORCES ACTING ON A JOINT

Just like the members themselves, the joints need to be designed for all types of forces that may act on them: axial load, bending moment, torsion, shear, as well as effects of creep, shrinkage, temperature, or settlement of supports. Assuming that the *members* have themselves been properly designed, the critical factor in joint design is the transmission of the forces that are present at the ends of the members into and through the joint. Figure 11.2.1 shows an interior joint with beams framing into it from all sides of a column.

Referring to Fig. 11.2.2, the forces T_1 and C_1 represent negative bending in a beam framing to a joint from the right side; the forces C_2 and T_2 represent positive bending in a beam framing in from the left side; the

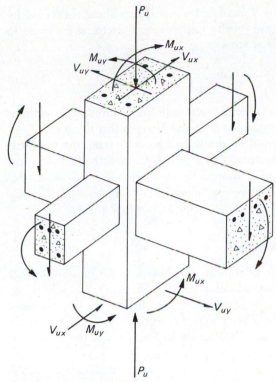

Figure 11.2.1 Forces on members at a joint.

forces V_u (col.) represent the shears in the column just outside the joint. The shear within the joint that potentially may cause the shear crack shown may be expressed

$$V_u = T_1 + T_2 - V_u \text{ (col.)} \tag{11.2.1}$$

or

$$V_u = f_y A_{st} + f_y A_{sb} - V_u \text{ (col.)} \tag{11.2.2}$$

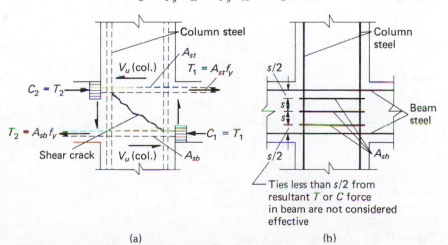

(a)   (b)

Figure 11.2.2 Shear in a beam-to-column joint.

After obtaining the shear V_u within the joint, that shear is divided by the effective shear area A_{cv} and by the undercapacity factor ϕ for shear to give the nominal ultimate shear stress v_n,

$$v_n = \frac{V_u}{\phi A_{cv}} \tag{11.2.3}$$

Since the forces in Eq. (11.2.1) are those due to the factored loads without reference to the undercapacity factor ϕ, it would seem that the terms $f_y A_{st}$ and $f_y A_{sb}$ in Eq. (11.2.2) should be multiplied by $\phi = 0.90$, the undercapacity factor for bending. However, ACI–ASCE Committee 352 [1] indicates that Eq. (11.2.2), as such, should be used. In this situation as described, of course, it is conservative not to use the undercapacity factor (see Sec. 3.1.1, Ref. 1) in front of the positive quantities in Eq. (11.2.2).

11.3 CONFINEMENT AT A JOINT

In the design or investigation of a joint, the forces in the two orthogonal directions oriented to the longitudinal axes of the horizontal members are examined separately. In that process, joint "confinement" by members transverse to the plane of the forces under consideration is an important factor in the strength of the joint. Referring to Fig. 11.3.1, the *joint core*

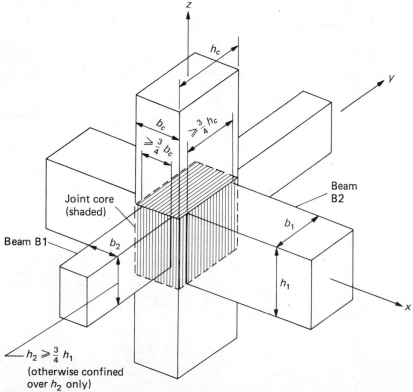

Figure 11.3.1 Confinement at a joint.

(shown shaded) is a volume the three dimensions of which are the cross-sectional dimensions of the column and the depth of the deepest beam framing into the column. When the forces in the xz-plane are transferred through or into the joint (i.e., the forces on beam $B2$ are acting on the joint), the spandrel beams $B1$ on each side of the column provide transverse restraint on the joint core. However, the lateral restraint (confinement) provided by the spandrel beam will depend on the proportion of the face of the joint core that is covered by the spandrel beam.

Confinement by members framing to the face of the joint core (perpendicular to the plane in which the forces are being considered) is considered sufficient if the confining member covers at least three-fourths of the width and three-fourths of the depth of the joint face (see Fig. 11.3.1).

When P_u exceeds $0.4P_b$ on a *Type 1 joint* and where it is *not* confined by a pair of members on opposite faces of the column, special transverse reinforcement should be provided as follows:

1. Where a spiral is used, the requirement is identical with ACI-10.9.3 (see Sec. 13.10),

$$\rho_s = 0.45 \left(\frac{A_g}{A_c} - 1\right)\frac{f'_c}{f_y} \tag{11.3.1}$$

2. Where rectangular hoop and cross-tie reinforcement is used, the required area is (see Fig. 11.3.2)

$$A_{sh} \geq 0.3L_h s_h \left(\frac{A_g}{A_{ch}} - 1\right)\frac{f'_c}{f_y} \tag{11.3.2}$$

where

A_{sh} = total area of all hoop and extra cross-tie legs crossing middepth of the section, within the spacing s_h, in the direction considered

L_h = width, measured to outside of tie reinforcement, in the direction perpendicular to that of shear force being considered

s_h = spacing of tie reinforcement measured along column bars

A_g = gross area of column

A_{ch} = area of rectangular core measured to the outside of the hoop or tie

For a *Type 2 joint*, the minimum transverse reinforcement requirements of Eqs. (11.3.1) and (11.3.2) apply whether or not adequate confinement by members is provided. In addition,

$$\min \rho_s \geq 0.12 \frac{f'_c}{f_y} \tag{11.3.3}$$

or

$$\min \frac{A_{sh}}{L_h s_h} \geq 0.12 \frac{f'_c}{f_y} \tag{11.3.4}$$

For seismic design, ACI Appendix-A.6.2.1 and A.6.2.3 require transverse hoop reinforcement (including additional cross-ties) to satisfy Eqs. (11.3.1) through (11.3.4) only when a joint *is not* confined. When the joint

Direction of shear being considered

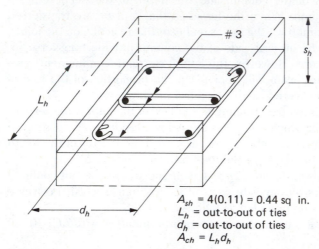

A_{sh} = 4(0.11) = 0.44 sq in.
L_h = out-to-out of ties
d_h = out-to-out of ties
$A_{ch} = L_h d_h$

Figure 11.3.2 Rectangular hoop and cross-tie reinforcement in height s_h of column core.

is confined in terms of beam width (see Fig. 11.3.1), the minimum transverse reinforcement requirement within the depth of the shallowest framing member is reduced to one-half the requirements of Eqs. (11.3.1) and (11.3.3) (ACI Appendix-A.6.2.2). This 1983 ACI Code provision is somewhat liberalized from the Committee 352 Report [1]. Note also that ACI Appendix-A.4.4 uses h_c measured to the *center* of confining reinforcement instead of L_h measured to the *outside* of confining reinforcement, a liberalized difference of little consequence considering the degree of certainty with which the joint behavior is predictable.

11.4 DESIGN EXAMPLES

Instead of completely listing all of the requirements indicated in the Committee 352 report, two examples will be presented to illustrate the recommendations as they apply to Type 1 joints used in ordinary building construction. The requirements are similar but more stringent for Type 2 joints where ductility and dissipation of energy into the inelastic range are required.

EXAMPLE 11.4.1 Design the exterior beam–column joint shown in Fig. 11.4.1. The joint is to be a Type 1 joint where strength is the primary criterion. Use $f'_c = 4000$ psi and $f_y = 60,000$ psi.

Solution: (a) Examine the embedment situation for the 4-#10 bars in the main beam. From Table 6.7.1, the L_{db} required is 48.2 in. which would be increased 40% to give $L_d = 67.5$ in. for top bars. In the 20-in. column, however, where a hook must be used, ACI-12.5 makes no reference to "top" hooks. Referring to Fig. 6.11.1 for the definition of L_{dh}, the required development length L_{dh} equals a basic value L_{hb} multiplied by modification factors. For a #10 hooked bar, according to Eq. (6.11.1),

$$L_{hb} = \frac{1200d_b}{\sqrt{f_c'}} = \frac{1200(1.27)}{\sqrt{4000}} = 24.1 \text{ in.}$$

When minimum side cover of $2\frac{1}{2}$ in. and minimum cover of 2 in. on the tail of a 90° hook are both available, the modification factor 0.7 may be used. For this joint (Fig. 11.4.1), since the 15-in. wide beam is much narrower than the 20-in. column, the side cover will be satisfied, and the 2-in. cover on the tail (Fig. 11.4.2) also will be satisfied. Thus, the requirement is

$$L_{dh} = 24.1(0.7) = 16.9 \text{ in.}$$

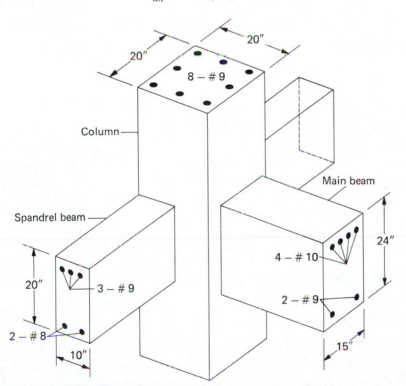

Figure 11.4.1 Design Example 11.4.1.

If ties (vertical in this case) were to enclose the hooks in accordance with ACI-12.5.3.3, an additional modification factor of 0.8 could be used, reducing required L_{dh} to $16.9(0.8) = 13.5$ in. Vertical ties would overcrowd this joint and should not be used. Allowing 2 in. for cover on the tail and assuming the critical section to be at the exterior face of the column steel (see Fig. 11.4.2) would leave available $L_{dh} = 20 - 4 = 16$ in., slightly less than the 16.9 in. required.

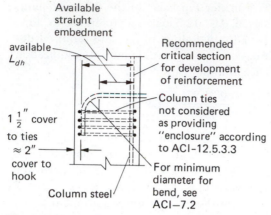

Figure 11.4.2 Dimensions for embedment into column.

Try 5-#9 bars,

$$L_{hb} = \frac{1200(1.0)}{\sqrt{4000}} = 19.0 \text{ in.}$$

$$L_{dh} = L_{hb} (0.7) = 13.3 \text{ in.} < [\text{available } L_{dh} = 16 \text{ in.}] \qquad \text{OK}$$

(b) Examine the shear on the column to be transmitted through the joint. Referring to Fig. 11.4.3, the moment on the columns may be assumed to be zero at midheight (or, preferably, the actual moment diagram would be used). In this case, with 12-ft and 10-ft column lengths, the factored column shear times 11 ft equals the factored moment at the end of the beam.

$$V_u \text{ (for column)} \left(\frac{h_1 + h_2}{2} \right) = M_u = 0.90 \, M_n \text{ (for beam)}$$

The strength of the beam (5-#9, $d = 21.6$ in.) is

$$C = 0.85(4)(15)a = 51.0a; \qquad T = 5(1.0)60 = 300 \text{ kips}$$
$$a = 5.88 \text{ in.}$$
$$M_n = 300(21.6 - 2.94)\tfrac{1}{12} = 466 \text{ ft-kips}$$
$$M_u = 0.90(466) = 420 \text{ ft-kips}$$

$$V_u \text{ (for column)} = \frac{420}{11} = 38.2 \text{ kips}$$

Note that in cases where there is lateral loading on the building frames, V_u for column would have been obtained independently from structural analysis; and the V_u value above the joint may be different from the V_u value below the joint. The shear on the column through the joint is, in this case,

$$\text{joint } V_u = 0.90(300) - 38.2 = 232 \text{ kips}$$

The Committee 352 recommendation is

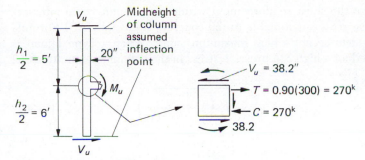

Figure 11.4.3   Column shear at joint.

$$\text{joint } V_u = A_{st}f_y - V_u \text{ (for column)}$$
$$= (5.00)(60) - 38.2 = 262 \text{ kips}$$

The nominal shear stress is computed as

$$v_n = \frac{\text{joint } V_u}{\phi A_{cv}} \tag{11.4.1}$$

where A_{cv} is equal to the effective cross-sectional area $b_E d$ resisting the shear (Fig. 11.4.4). For confined sections, the effective width b_E is taken to the outside of the column. For all others, the effective width b_E is measured to the outside of the ties (or the column bars if no ties are used). The effective depth d is measured to the centroid of the column steel. For this example, the confinement requirement is not satisfied because the 10-in. width of the spandrel beam is less than three-fourths of the 20-in. column width; thus

$$b_E = 20 - 2(1.5) = 17 \text{ in.}; \qquad d = 20 - 2.5 = 17.5 \text{ in.}$$
$$v_n = \frac{\text{joint } V_u}{\phi A_{cv}} = \frac{262,000}{0.85(17)(17.5)} = 1036 \text{ psi}$$

The Committee 352 report [1] (Sec. 4.2.3.5) indicates the maximum nominal shear stress permitted (even with well-designed shear reinforcement) to be

$$\max v_n \le 20\sqrt{f'_c} = 1265 \text{ psi} > 1036 \text{ psi} \qquad \text{OK}$$

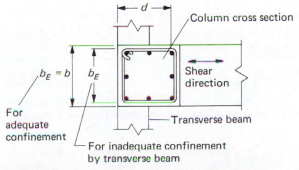

Figure 11.4.4   Effective shear area in joint core.

(c) Design the shear reinforcement within the column–beam intersection. The shear stress attributable to the concrete is given by ACI Formula (11-8) modified to accept a high maximum value. When a Type 1 joint is being designed for a high compressive stress in the column, Committee 352 (Sec. 4.2.3.2) gives

$$v_c = 3.5\beta\gamma \sqrt{f'_c\left(1 + \frac{N_u}{500A_g}\right)} \qquad (11.4.2)$$

where

$\beta = 1.4$ for a Type 1 joint and 1.0 for a Type 2 joint
$\gamma = 1.4$ if joint is adequately confined laterally by members framing in
$\quad = 1.0$ if joint is not adequately confined

For this example, N_u (includes overload factors) is the column load P_u and is given as 530 kips; thus

$$\frac{N_u}{A_g} = \frac{530,000}{20(20)} = 1325 \text{ psi}$$

The joint is not adequately confined by the spandrel beams; thus

$$v_c = 3.5(1.4)(1.0)\sqrt{4000[1 + 0.002(1325)]} = 592 \text{ psi}$$

Since $v_n > v_c$, transverse reinforcement is required.

When shear reinforcement is required, it can be designed by following ACI-11.5.6.2, except $b_w = b_E$.

$$A_v = \frac{V_s s}{f_y d} = \frac{(v_n - v_c)b_E s}{f_y}$$

$$\frac{A_v}{s} = \frac{(1036 - 592)(17)}{60,000} = 0.126$$

For #4 bars, $A_v = A_s$ times number of legs $= 0.20N$. Try four legs $(N = 4)$,

$$s = \frac{A_v}{0.126} = \frac{4(0.20)}{0.126} = 6.3 \text{ in.}$$

Try $s = 6$ in.

(d) Check minimum transverse steel according to Committee 352 (Sec. 4.2.2.2) using Eq. (11.3.2). This provision applies when $P_u > 0.4P_b$. Assume that is the case for this example.

$$A_{sh} \geq 0.3L_h s_h \left(\frac{A_g}{A_{ch}} - 1\right)\frac{f'_c}{f_y} \qquad [11.3.2]$$

For this example, $L_h = 20 - 3 = 17$ in., $A_g = 400$ sq in., $A_{ch} = 17(17) = 289$ sq in.

$$\frac{A_{sh}}{s_h} \geq 0.3(17)\left(\frac{400}{289} - 1\right)\frac{4}{60} = 0.13$$

This slightly exceeds the amount required for computed shear in part (c) and therefore the 0.13 controls. For four #4 legs,

$$s_h = \frac{4(0.20)}{0.13} = 6.2 \text{ in.}$$

Try 4-#4 legs @ 6-in. spacing.

(e) Final choice for spacing of horizontal ties. The placement of these ties must be between the tensile force (300 kips represented by 5-#9 bars) and the compressive force (represented by concrete and the 2-#9 in the compression zone). The distance between the centroids of the bars in the two faces of the 15 × 24 beam is

$$d - d' \approx 21.6 - 2.5 = 19.1 \text{ in.}$$

Using a 6-in. spacing for the shear reinforcement in the core would indicate three spaces. However, shear reinforcement is not considered effective when closer to the main tension or compression reinforcement than $s/2$ [see Fig. 11.2.2(b)]. In this case, the number of spaces is

$$N = \frac{19.1}{6} = 3.2$$

The final detail is shown in Fig. 11.4.5 with a 5-in. spacing.

It is assumed there is no computed shear in the spandrel beam direction because there are no large unbalanced moments. However, since the spandrels are not adequate to provide confinement, the 4-#4 legs need to be provided in the spandrel direction also, as shown in Fig. 11.4.5.

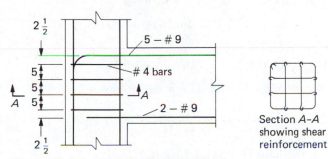

Figure 11.4.5 Joint for Example 11.4.1.

EXAMPLE 11.4.2 For the interior joint shown schematically in Fig. 11.4.6, determine the shear reinforcement required if the joint is Type 1. Use $f'_c = 4000$ psi and $f_y = 60,000$ psi. Assume P_u of 250 kips exceeds $0.4P_b$ for the column.

Solution: (a) Development of reinforcement. All longitudinal steel is to be extended through the joint.

(b) Shear in direction of 16 × 24 beams. Due to lateral loading (from left to right) on the structure, the moments [Fig. 11.4.7(a)] M_{ut} and M_{ub}

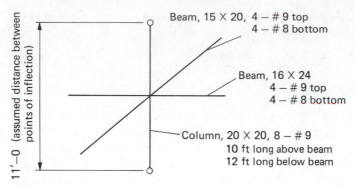

Figure 11.4.6 Joint for Example 11.4.2.

have the same rotational direction and give the highest shear through the joint.

Again, it is assumed that the factored moment M_u acting equals the usable strength ϕM_n. The strength M_{nt} based on tension in the 4-#9 bars is

$$C = 0.85 f'_c ba = 0.85(4)(16)a = 54.4a$$

$$T = 4(1.0)60 = 240 \text{ kips}$$

$$a = 4.41 \text{ in.}$$

$$M_{nt} = 240(21.5 - 2.20)\tfrac{1}{12} = 386 \text{ ft-kips}$$

$$M_{ut} = \phi M_{nt} = 0.90(386) = 347 \text{ ft-kips}$$

The strength M_{nb} based on tension in the 4-#8 bars is

$$C = 54.4a; \qquad T = 4(0.79)60 = 190 \text{ kips}$$

$$a = 3.48 \text{ in.}$$

$$M_{nb} = 190(21.5 - 1.74)\tfrac{1}{12} = 313 \text{ ft-kips}$$

$$M_{ub} = \phi M_{nb} = 0.90(313) = 282 \text{ ft-kips}$$

$$V_u \text{ (on col.)} = \frac{M_{ut} + M_{ub}}{11} = \frac{347 + 282}{11} = 57 \text{ kips}$$

The net shear through the joint [Fig. 11.4.7(b)] is

$$V_u = 190 + 240 - 57 = 373 \text{ kips}$$

For this joint, the 15 × 20 beams do provide confinement for shear transmitted in the direction of the 16 × 24 beams. The 15-in. width = 0.75 (20-in. column dimension) and the 20-in. depth exceeds 0.75 (24-in. beam depth). Thus confinement is provided; use $b_E = b$.

$$A_{cv} = 20(17.5) = 350 \text{ sq in.}$$

$$v_n = \frac{V_u}{\phi A_{cv}} = \frac{373,000}{0.85(350)} = 1250 \text{ psi}$$

$$\max v_n = 20\sqrt{f'_c} = 1260 \text{ psi} > 1250 \text{ psi} \qquad \text{OK}$$

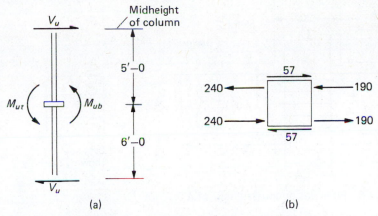

Figure 11.4.7 Forces in direction of 16 × 24 beams, Example 11.4.2.

The shear capacity attributable to the concrete, Eq. (11.4.2)

$$v_c = 3.5\beta\gamma \sqrt{f_c'(1 + 0.002N_u/A_g)}$$

$\beta = 1.4$ for a Type 1 joint, and $\gamma = 1.4$ since lateral confinement is available. Note also that $N_u = P_u = 250$ kips

$$v_c = 3.5(1.4)(1.4)\sqrt{4000[1 + 0.002(250,000/400)]} = 651 \text{ psi}$$

$$v_n - v_c = 1250 - 651 = 599 \text{ psi} < 15\sqrt{f_c'} \quad \text{OK (Committee 352, 4.2.3.5)}$$

The transverse steel required in the direction of the 16 × 24 beam is

$$\frac{A_v}{s} = \frac{(v_n - v_c)b_E}{f_y} = \frac{599(20)}{60,000} = 0.20$$

Try 4 legs with #4 bars,

$$s = \frac{4(0.20)}{0.20} = 4.0 \text{ in.}$$

Try $3\frac{1}{2}$-in. spacing.

(c) Minimum transverse steel requirements. Since $P_u > 0.4P_b$, use Eq. (11.3.2),

$$\frac{A_{sh}}{s_h} \geq 0.3L_h \left(\frac{A_g}{A_{ch}} - 1\right) \frac{f_c'}{f_y}$$

$$\frac{4(0.20)}{3.5} \geq 0.3(17) \left(\frac{400}{289} - 1\right) \frac{4000}{60,000}$$

$$0.23 > 0.13 \qquad\qquad\qquad\qquad\qquad \text{OK}$$

The ties are shown in Fig. 11.4.8.

(d) Shear in the direction of the 15 × 20 beam. If the lateral loading on the structure requires unbalanced moments in this direction, similar to the force system shown in Fig. 11.4.7, then the process illustrated in steps (b) and (c) would be repeated for this direction.

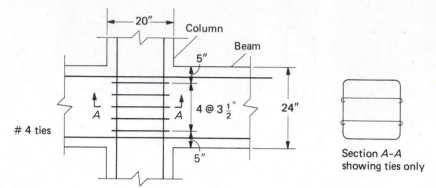

Figure 11.4.8 Final design for Example 11.4.2.

The important factors involved in the design of joints intended to resist primarily static loads (Type 1 joints) have been discussed and illustrated in two examples. For the design of Type 2 joints requiring greater ductility, the reader is referred to the Committee 352 report [1]. Other details regarding bar size and spacing limitations are also to be found in the report [1].

Beam–column connections resisting seismic loads have been given considerable attention by researchers, with the recent state of knowledge well presented by Meinheit and Jirsa [11]. The behavior of beam–column connections in precast concrete has been studied by Pillai and Kirk [12].

SELECTED REFERENCES

1. ACI–ASCE Committee 352. "Recommendations for Design of Beam-Column Joints in Monolithic Reinforced Concrete Structures," *ACI Journal, Proceedings,* **73**, July 1976, 375–393.
2. Norman W. Hanson and Harold W. Connor. "Seismic Resistance of Reinforced Concrete Beams-Columns Joints," *Journal of the Structural Division,* ASCE, **93**, October 1967 (ST5), 533–560.
3. John A. Hribar and Raymond C. Vasko. "End Anchorage of High Strength Steel Reinforcing Bars," *ACI Journal, Proceedings,* **66**, November 1969, 875–883.
4. Brian Mayfield, Fung-Kew Kong, Alan Bennison, and Julian C. D. Twiston-Davies. "Corner Joint Details in Structural Lightweight Concrete," *ACI Journal, Proceedings,* **68**, May 1971, 366–372.
5. Eric F. P. Burnett and Rajendra P. Jajoo. "Reinforced Concrete Beam-Column Connections," *Journal of the Structural Division,* ASCE, **97**, September 1971 (ST9), 2315–2335.
6. Brian Mayfield, Fung-Kew Kong, and Alan Bennison. "Strength and Stiffness of Lightweight Concrete Corners," *ACI Journal, Proceedings,* **69**, July 1972, 420–427.
7. John Minor and James O. Jirsa. "Behavior of Bent Bar Anchorages," *ACI Journal, Proceedings,* **72**, April 1975, 141–149.
8. James K. Wight and Mete A. Sozen. "Strength Decay of Reinforced Concrete Columns under Shear Reversals," *Journal of the Structural Division,* ASCE, **101**, May 1975 (ST5), 1053–1065.

9. Jose G. L. Marques and James O. Jirsa. "A Study of Hooked Bar Anchorages in Beam-Column Joints," *ACI Journal, Proceedings*, **72**, May 1975, 198–209.

10. Ingvar H. E. Nilsson and Anders Losberg. "Reinforced Concrete Corners and Joints Subjected to Bending Moment," *Journal of the Structural Division*, ASCE, **102**, June 1976 (ST6), 1229–1254.

11. Donald F. Meinheit and James O. Jirsa. "Shear Strength of R/C Beam-Column Connections," *Journal of the Structural Division*, ASCE, **107**, November 1981 (ST11), 2227–2244.

12. S. U. Pillai and D. W. Kirk. "Ductile Beam-Column Connection in Precast Concrete," *ACI Journal, Proceedings*, **78**, November–December 1981, 480–487.

PROBLEMS

All problems are to be worked using the ACI Code and the recommendations of ACI Committee 352 [1].

11.1 Check the anchorage of bars and determine the shear reinforcement for the exterior beam–column joint of the accompanying figure. Assume lateral loading provides the critical condition at the joint; assume columns bent in double curvature with an inflection point at midheight. Consider as a Type 1 joint and use $f'_c = 4000$ psi and $f_y = 60,000$ psi. Show detail of shear reinforcement in joint.

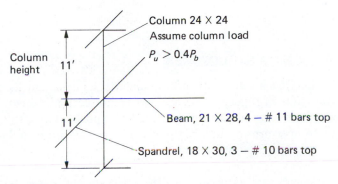

Problem 11.1

11.2 Repeat Prob. 11.1 except consider the column to be 18 in. square, the spandrel 14 × 24 with 3-#9, and the beam 18 × 28 with 4-#10 bars.

11.3 Repeat Prob. 11.1 except consider the joint to be an interior one with the 21 × 28 beam on both sides of the column. The spandrel then becomes an interior beam. Assume the 21 × 28 beam has 4-#9 in the bottom in addition to the bars in the top and that all bars are continuous through the column. The 18 × 30 beam has 3-#9 in the bottom in addition to the top bars, all bars being continuous through the column. The critical loading condition for the joint is with clockwise moments from the 21 × 28 beam acting on the joint at both sides of the column.

12

Cantilever Retaining Walls

12.1 TYPES OF RETAINING STRUCTURES

Retaining structures hold back soil or other loose material and prevent its assuming the natural angle of repose at locations where an abrupt change in ground elevation occurs. The retained material exerts a push on a structure and thus tends to overturn or slide it or both. There may be several types of retaining structures (Fig. 12.1.1) as follows.

1. Gravity Wall. A gravity wall is usually of plain concrete and depends entirely on its weight for stability. It is used for walls up to about 10 ft high.

2. Cantilever Retaining Wall. The cantilever is the most common type of retaining structure and is used for walls in the range of 10 to 25 ft in height. The stem, heel, and toe of such a wall each acts as a cantilever beam.

3. Counterfort Wall. In the counterfort wall the stem and slab are tied together by counterforts, which are transverse walls spaced at intervals and act as tension ties to support the stem wall. Counterfort walls are often economical for heights over about 25 ft.

4. Buttress Wall. A buttress wall is similar to a counterfort wall except that the transverse support walls are located on the side of the stem opposite to the retained material and act as compression struts. Buttresses, as compression elements, are more efficient than the tension counterforts and are economical in the same height range. A counterfort is more widely used

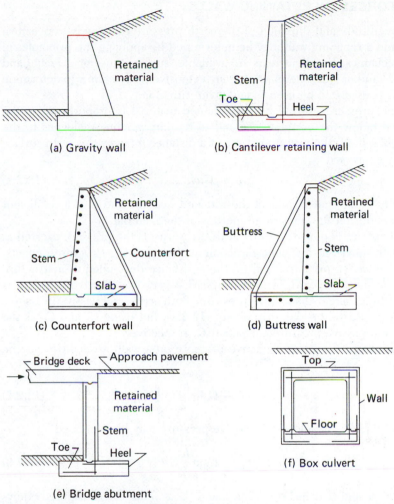

Figure 12.1.1 Types of retaining structures.

than a buttress because the counterfort is hidden beneath the retained material whereas the buttress occupies what may otherwise be usable space in front of the wall.

5. Bridge Abutment. A wall-type bridge abutment acts similarly to a cantilever retaining wall except that the bridge deck provides an additional horizontal restraint at the top of the stem. Thus this abutment is designed as a beam fixed at the bottom and simply supported or partially restrained at the top.

6. Box Culvert. The box culvert, with either single or multiple cells, acts as a closed rigid frame that must not only resist lateral earth pressure but also vertical load from either the soil that it supports or from both the soil and the highway vehicle loads.

12.2 FORCES ON RETAINING WALLS

The magnitude and direction of the earth pressure that tends to overturn and slide a retaining wall may be determined by applying the principles of soil mechanics. Excellent texts are available, such as those of Terzaghi and Peck [1] and of Huntington [2], for any extensive study of how to determine the soil pressure to be used in any given situation.

The pressure exerted by the retained material is proportional to the distance below the earth surface and to its unit weight. Analogous to the action of a fluid, the unit pressure p at a distance h below the earth surface may be expressed

$$p = Cwh \tag{12.2.1}$$

where w is the unit weight of the retained material and C is a coefficient that depends on the physical properties of the material.

There are two categories of earth pressure: (1) the pressure exerted as the earth moves in the same direction as the retaining structure deflects, known as *active pressure*, and (2) the resistance developed as a structure moves against the earth, known as *passive pressure*. Passive pressure is several times larger than active pressure. Both active and passive pressures may be expressed in the form of Eq. (12.2.1), but using C_a and C_p as the active and passive pressure coefficient C, respectively.

The force P_a caused by active pressure on a wall of height h may be expressed

$$P_a = C_a w \frac{h^2}{2} \tag{12.2.2}$$

and the force P_p developed in passive pressure may be expressed

$$P_p = C_p w \frac{h^2}{2} \tag{12.2.3}$$

where $C_a w$ and $C_p w$ in Eqs. (12.2.2) and (12.2.3) may be considered equivalent fluid pressures. Typical values for C_a and C_p are 0.3 and 3.3, respectively, for granular material such as sand. Roughly, $C_p = 1/C_a$.

The proper evaluation of earth pressures on retaining structures is outside the scope of this book; the preceding brief comments are offered to provide some logic for the earth pressures that are used in the design example of this chapter.

The factors that may affect the pressure (active pressure) on a wall are as follows:

1. Type of backfill used.
2. Seasonal condition of the backfill material, such as wet, dry, or frozen.
3. Drainage of backfill material.
4. Possibility of backfill overload, such as trucks and equipment near the wall.
5. Degree of care exercised in backfilling.
6. Degree of rotational restraint between various components of the retaining structure.

7. Possibility of vibration in the vicinity of the wall (especially in the case of granular soil).
8. Type of material beneath the footing of the retaining structure.
9. Level of the water table.

Probably the most important single factor is that water must be prevented from accumulating in the backfill material. Walls are rarely designed to retain saturated material, which means that proper drainage must be provided.

When vehicles may travel near and exert their loads or when buildings are constructed near the top of a retaining wall, the lateral pressure against such a wall is increased. In the case of a fixed static load such as a building, the weight of the building can be converted into an additional height (surcharge) of backfill soil material. The effect of a highway or railroad passing over the retained material near the wall causes a dynamic reaction that cannot accurately be converted into a static effect. However, the AASHTO highway bridge specification (Sec. 1.2.19) [3] and the AREA railroad retaining wall specification [4] prescribe an equivalent static surcharge corresponding to a number of additional feet of backfill material.

12.3 STABILITY REQUIREMENTS

The first step in retaining wall design is to establish the proportions such that the stability of the structure (see Fig. 12.3.1) under active pressure is assured. Three requirements must be satisfied: (a) the overturning moment $P_{ah}(h'/3)$ must be more than balanced by the resisting moment $Wx_1 + P_{av}L$, so that an adequate factor of safety against overturning is provided, usually about 2.0; (b) sufficient frictional resistance F in combination with any reliable passive resistance P_p against the toe must provide an adequate factor of safety (usually 1.5) against sliding caused by P_{ah}; and (c) the base width L must be adequate to distribute the load R to the foundation soil without causing excessive settlement or rotation.

Typically, referring to the pressure distribution of Fig. 12.3.1, the overturning factor of safety (FS) would be computed

$$\text{FS} = \frac{\text{resisting moment}}{\text{overturning moment}} = \frac{Wx_1 + P_{av}L}{P_{ah}(h'/3)} \qquad (12.3.1)$$

or, perhaps more frequently, neglecting the vertical component of P_a,

$$\text{FS} = \frac{Wx_1}{P_{ah}(h'/3)} \qquad (12.3.2)$$

where W represents the weight of the concrete wall and footing and of the soil resting on the footing.

The factor of safety against sliding may be computed, using the notation of Fig. 12.3.1,

$$\text{FS} = \frac{\mu R + P_p}{P_{ah}} \qquad (12.3.3)$$

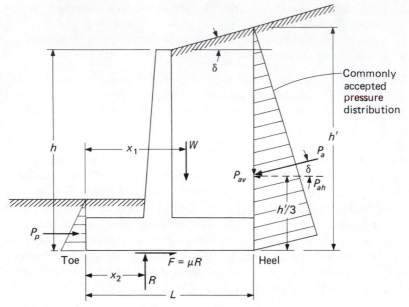

Figure 12.3.1 Forces on retaining wall.

where μ is the coefficient of friction between the soil and the footing. Table 12.3.1 gives the coefficients of friction as recommended by the 1958 AREA specification [4] and these may serve as a guide to typical values in lieu of more accurate ones.

Table 12.3.1 VALUES OF
COEFFICIENT OF FRICTION
BETWEEN SOIL AND CONCRETE,
1958 AREA SPECIFICATIONS

SOIL	μ
Coarse-grained soils (without silt)	0.55
Coarse-grained soils (with silt)	0.45
Silt	0.35
Sound rock (with rough surface)	0.60

The inclusion of some passive resistance P_p on the toe of the footing may or may not be justified. Certainly, to actually develop passive pressure in the soil in front of the wall, the concrete must have been placed without using forms for the toe and without disturbing the soil against which the concrete is placed.

Referring to Fig. 12.3.2, the ordinary passive resistance against the toe is

$$P_{p1} = \tfrac{1}{2}C_p w h_1^2 \tag{12.3.4}$$

but frequently is neglected, and in nearly all cases at least the value of h_1 is reduced under the assumption that during construction, or after, the earth surface cannot be expected to remain undisturbed at its final designated elevation. If when all reliable resistances have been included, the factor of safety remains inadequate, a base key (Fig. 12.3.2) may be used. Essentially, the base key, when placed in an unformed excavation against undisturbed material, may be expected to develop an additional passive force P_{p2} and shift the possible failure plane from line 1 to line 2.

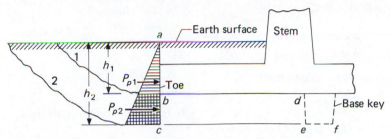

Figure 12.3.2   Passive resistance and effect of base key.

The base key develops the additional resistance

$$P_{p2} = \tfrac{1}{2}C_p w(h_2^2 - h_1^2) \tag{12.3.5}$$

and also, an inert region, *bced* of Fig. 12.3.2, is created which moves the friction plane from *bd* to *ce*. Thus the frictional force developed along *ce* is based on the angle α, the angle of internal friction of the soil, rather than on the friction angle between soil and concrete. Normally, $\tan \alpha > \mu$ for granular material, so that, by making the base key sufficiently deep to create an inert block, additional frictional resistance is developed.

Finally, the magnitude and distribution of the soil pressure requires investigation. Usual practice is to require the resultant vertical force R to be inside the middle third of the footing for sand and gravel subbases and within the middle half for rock subbase. In addition, the maximum pressure may not exceed the allowable value. Comments regarding allowable bearing capacity as well as some typical safe bearing values are to be found in Chap. 20.

Referring to Fig. 12.3.3(a), when the entire footing is under compression, the basic equation for combined bending and axial compression acting on a 1-ft strip along the wall is

$$p = \frac{R}{L} \pm \frac{Re(L/2)}{L^3/12} = \frac{R}{L}\left(1 \pm \frac{6e}{L}\right) \tag{12.3.6}$$

For the limiting condition of zero stress at the heel, $e = L/6$; thus Eq. (12.3.6) is valid for all positions of R within the middle third.

When the resultant R is outside the middle third [Fig. 12.3.3(b)], vertical force equilibrium requires

$$R = \tfrac{1}{2}p_t(3x_2)$$

$$p_t = \frac{2R}{3x_2} \tag{12.3.7}$$

for $0 < 3x_2 < L$.

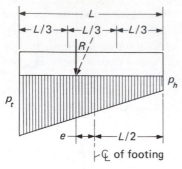

(a) Resultant within middle third

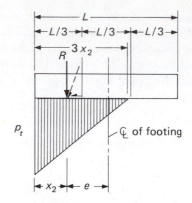

(b) Resultant outside middle third

Figure 12.3.3   Soil pressure distribution.

12.4 PRELIMINARY PROPORTIONING OF CANTILEVER WALLS

To begin the design it is necessary to apply certain "rules" along with an approximate use of statics and estimate such dimensions as the length and thickness of the base footing and the relative position of the wall with respect to the footing.

Height of Wall. Since the bottom of the footing must be below the frost penetration depth, say 3 to 4 ft in the northern United States, the overall height equals the desired difference in elevation plus the frost penetration depth.

Position of Stem on the Base Footing. The following demonstration using approximate statics will show that the front face of the wall should coincide with the desired position of the resultant soil pressure beneath the base. Take the most typical situation of a vertical wall with level backfill, as shown in Fig. 12.4.1. Assume an average unit weight w for all material (concrete and earth) enclosed within $abcd$ and neglect entirely the concrete in the toe. Thus

$$R = W = wh\gamma L$$

Satisfying rotational equilibrium about the heel,

$$P_a \frac{h}{3} + W \frac{\gamma L}{2} = R\xi L$$

$$C_a \frac{wh^2}{2} \left(\frac{h}{3}\right) + wh\gamma L \frac{\gamma L}{2} = wh\gamma L(\xi L)$$

Solving for L/h gives

$$\frac{L}{h} = \sqrt{\frac{C_a}{3\gamma(2\xi - \gamma)}} \qquad (12.4.1)$$

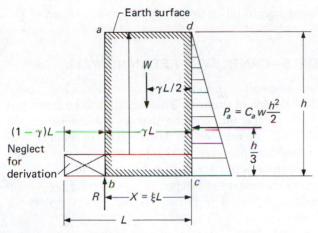

Figure 12.4.1 Data for economical proportioning of wall.

The variable ξ must be selected by the designer, based on the type of soil and desired pressure distribution. For good granular soil and acceptance of a triangular pressure distribution with the resultant at the outer edge of the middle third, $\xi = \frac{2}{3}$. For clay where a uniform distribution might be desired, ξ would be $\frac{1}{2}$.

Minimizing the L/h in Eq. (12.4.1) such that the base width is a minimum,

$$\gamma = \xi \qquad\qquad (12.4.2)$$

Thus the front face of the wall should line up with the desired position of the soil pressure resultant.

Length of Base. A preliminary length of base may be obtained by the application of static moments with respect to point b in Fig. 12.4.1, employing the same assumptions that underlie Eq. (12.4.1). By this process a more exact result may be obtained since one can easily estimate the stem thickness and footing thickness as fractions of L or of h.

Thickness of Footing. The base thickness is usually 7 to 10% of the total height h, with a minimum of about 1 ft. It should be about equal to the base thickness of the stem.

Thickness of Stem. The thickness at the top of the wall is arbitrary; however, cover requirements and construction constraints will dictate how thin it may be. Generally 10 or 12 in. minimum is preferred, though no minimum is prescribed by the ACI Code.

The base thickness of the stem is determined as required for bending moment and shear, though it may be estimated as 12 to 16% of the base width or 10 to 12% of the wall height. The stem thickness should not be too skimpy, since a thin wall deflects considerably at the top and savings in reinforcement due to larger effective depth will tend to offset the cost of any extra concrete used. It is recommended that a batter of $\frac{1}{4}$ in./ft of height be provided on the front face to offset deflection or forward tilting of the structure.

12.5 DESIGN EXAMPLE—CANTILEVER RETAINING WALL

Design a cantilever retaining wall to support a bank of earth 16 ft high above the final level of earth at the toe of the wall. The backfill is to be level, but a building is to be built on the fill. Assume that an 8-ft surcharge will approximate the lateral earth pressure effect. Data: weight of retained material = 120 pcf; equivalent fluid pressure = 32 pcf; angle of internal friction = 35°; coefficient of friction between masonry and soil = 0.40; for passive pressure use equivalent fluid pressure of 400 pcf; f'_c = 3000 psi; f_y = 40,000 psi; maximum soil pressure = 5 ksf (kips per square foot). Use the strength method of the ACI Code.

(a) Basic design data. For adequate deflection control on the cantilever wall, choose to use a reinforcement percentage ρ about one-half the maximum permitted by the ACI Code. Choose

$$\rho \approx 0.375\rho_b = 0.0139 \qquad \text{(Table 3.6.1)}$$

This corresponds to R_n = 495 psi.

(b) Height of wall. Allowing 4 ft for frost penetration to the bottom of the footing in front of the wall, the total height becomes

$$h = 16 \text{ ft} + 4 \text{ ft} = 20 \text{ ft}$$

(c) Thickness of footing. The thickness may be estimated at this stage of design to be 7 to 10% of the overall height h. Assume a uniform footing thickness, t = 2 ft (about 10% of h).

(d) Base length. One could determine the base length by considering the equilibrium of factored loads (using the factor 1.7 for the horizontal earth pressure and using 0.9 for the earth and concrete weight); this practice will give a conservative result. However, it seems within the intent of the ACI Code to establish base dimensions for foundation structures using actual unfactored service loads (see ACI-15.2.2).

Using a 1-ft length of wall and letting the unit weight of material bounded by points a, b, c, and d in Fig. 12.5.1 equal 120 pcf,

$$P_1 = 0.256(20)(1) = 5.12 \text{ kips}$$

$$P_2 = \tfrac{1}{2}(0.640)(20)(1) = 6.40 \text{ kips}$$

$$W = 0.120(20 + 8)x = 3.36x \text{ kips}$$

Summation of moments about point b gives

$$W\left(\frac{x}{2}\right) = P_1(10.0) + P_2(6.67)$$

$$\frac{3.36x^2}{2} = 5.12(10.0) + 6.40(6.67)$$

$$1.68x^2 = 93.9$$

$$x = 7.47 \text{ ft}$$

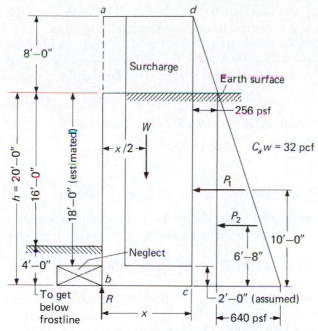

Figure 12.5.1 Preliminary proportioning for cantilever wall in design example.

Since for this granular material the resultant soil pressure is desired to be at the outer edge of the middle third of the footing,

$$\text{base length} = 1.5x = 1.5(7.47) = 11.2 \text{ ft}$$

Try 11 ft 3 in.

(e) Stem thickness. Prior to computing soil pressures and stability factors of safety, a more accurate knowledge of the concrete dimensions is necessary. The thickness of the base of the stem is selected with due regard for the bending moment and shear requirements.

Bending moment will normally provide the governing criterion, for which the general expression (y measured from the top of the wall) is

$$M_y = \frac{0.256y^2}{2} + \frac{0.032y^3}{6} = 0.128y^2 + 0.00533y^3 \quad (12.5.1)$$

At the bottom of the assumed 18-ft high stem wall, using a 1.7 overload factor,

$$M_u = 1.7[0.128(18)^2 + 0.00533(18)^3] = 123 \text{ ft-kips}$$

For a desired $R_n = 495$ psi,

$$\text{required } d = \sqrt{\frac{M_u}{\phi R_n b}} = \sqrt{\frac{123(12,000)}{0.90(495)12}} = 16.6 \text{ in.}$$

Total thickness = 16.6 + 2 (cover) + 0.5 (estimated bar radius) = 19.1 in. Try 21 in. for the stem base thickness and select 12 in. as the practical minimum for the top of the wall. Note that the selected R_n of 495 psi is a *chosen* guideline value and need not be rigidly adhered to. In this case the 3-in. multiple of 21 in. is preferred to the computed value of 20 in.

The critical section for shear strength is taken at a distance d from the bottom of the stem (ACI-11.1.2.1). Assume the critical section at $y = 16.5$ ft ($d =$ approximately 18 in.). The general expression for shear is

$$V_y = 0.256y + \tfrac{1}{2}(0.032)y^2 = 0.256y + 0.016y^2 \qquad (12.5.2)$$

At the approximate critical section, using a 1.7 overload factor

$$V_u = 1.7[0.256(16.5) + 0.016(16.5)^2] = 14.6 \text{ kips}$$

$$\phi V_c = \phi(2\sqrt{f_c'})bd$$

$$= 0.85(2\sqrt{3000})(12)(\approx 18)\tfrac{1}{1000} = 20.1 \text{ kips} > 14.6 \text{ kips} \qquad \text{OK}$$

Since $V_u < \phi V_c$, in accordance with ACI-11.5.5.1 shear reinforcement is not required for slabs. The cantilever is treated as a slab in applying any ACI Code limitations.

This thickness will be arbitrarily made 21 in. in order to minimize deflection.

Where appearance on the front face is important, a batter of that face should be provided so as to counteract the effect of deflection. The usual batter is about $\tfrac{1}{4}$ in./ft of wall height. Thus the minimum batter here is

$$\tfrac{1}{4}(18) = 4\tfrac{1}{2} \text{ in.}$$

In this case the thickness increases by 9 in. from the top to the bottom of the wall, so batter the front face 5 in. and the rear face 4 in. When the wall is in place, it will deflect forward and become nearly vertical; therefore the

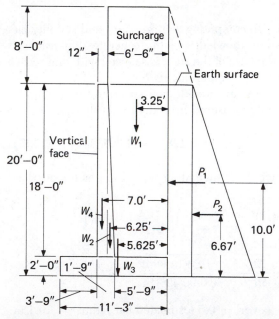

Figure 12.5.2 Dimensions for computing resultant soil pressure in design example.

analysis from this point will consider the front face to be vertical. Economics may dictate having only one sloping face, or in some cases no sloping face.

(f) Factor of safety against overturning. Using the dimensions given in Fig. 12.5.2, locate the resultant of vertical forces (without overload factors) with respect to the heel, as in Table 12.5.1:

Table 12.5.1

FORCE			ARM	MOMENT
$W_1 = 0.120(18 + 8)(6.5)$	$= 20.3$		3.25	66
$W_2 = 0.030(\frac{1}{2})(18)(0.75)$	$= 0.2$		6.25	1
$W_3 = 0.150(11.25)(2.0)$	$= 3.4$		5.63	19
$W_4 = 0.150(1.0)(18)$	$= 2.7$		7.00	19
Totals	26.6			105

$$\text{resultant, from heel} = \frac{105}{26.6} = 3.95 \text{ ft}$$

$$\text{resisting moment} = 26.6(11.25 - 3.95) = 194 \text{ ft-kips}$$

$$\text{overturning moment} = P_1(10) + P_2(6.67)$$

$$= 5.12(10) + 6.40(6.67)$$

$$= 93.9 \text{ ft-kips}$$

$$\text{FS against overturning} = \frac{194}{93.9} = 2.07 > 2.0 \qquad \text{OK}$$

Alternatively, if the stability check is to be made using factored loads, according to ACI-9.2.4,

$$U = 0.9D + 1.7H$$

where H is the horizontal earth pressure. In such a case, it would be required that

$$\text{resisting moment} \geq \text{overturning moment}$$

$$0.9(194) \geq 1.7(93.9)$$

$$175 \geq 160 \qquad \text{OK}$$

In effect the required factor of safety against overturning under service load conditions is $1.7/0.9 = 1.89$, which is not much less than the traditional value of 2.0.

(g) Location of resultant and footing soil pressures. Referring to Fig. 12.5.3, and using service loads because the maximum soil pressure limitation is given for that condition,

$$R = 26.6 \text{ kips}$$

$$\bar{x} = \frac{105 + 93.9}{26.6} = 7.48 \text{ ft}$$

$$e = 7.48 - \frac{11.25}{2} = 1.85 \text{ ft}$$

$$\frac{6e}{\text{base length}} = \frac{6(1.85)}{11.25} = 0.99 < 1.0$$

The resultant lies just outside of the middle third.

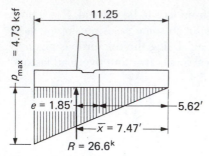

Figure 12.5.3 Soil pressure and location of resultant under service load in design example.

The service load soil pressure diagram is essentially a triangle; thus

$$R = \tfrac{1}{2}(p_{max})(\text{base length})$$

$$26.6 = \tfrac{1}{2}(p_{max})(11.25)$$

$$p_{max} = 4.73 \text{ ksf} < 5 \text{ ksf} \qquad \text{OK}$$

(h) Factor of safety against sliding. Neglecting passive pressure against the toe of the footing and using service loads,

$$\text{force causing sliding} = P_1 + P_2 = 5.12 + 6.40 = 11.52 \text{ kips}$$

$$\text{frictional force} = \mu R = 0.40(26.6) = 10.64 \text{ kips}$$

$$\text{factor of safety} = \frac{10.64}{11.52} = 0.92 < 1.5 \qquad \text{NG}$$

Sliding resistance may also be checked using factored loads, $U = 0.9D + 1.7H$, in which case it is required that

$$\text{resisting force} \geq \text{sliding force}$$

$$0.9(10.64) \geq 1.7(11.52)$$

$$9.58 \text{ kips} < 19.6 \text{ kips} \qquad \text{NG}$$

Thus ACI-9.2.4 would require the same factor of safety against sliding as against overturning, that is, $1.7/0.9 = 1.89$. This exceeds the traditional factor to resist sliding of 1.5.

Since the resistance provided does not give an adequate safety factor, a key (Figs. 12.5.4 and 12.5.5) against sliding is required. Such a key is intended to develop passive pressure in the region in front of and below the bottom of the footing. The procedure for determining the required size of a key is one that is debated by designers. Generally it seems desirable to place the front face of the key about 5 in. in front of the back face of the stem. This will permit anchoring the stem reinforcement in the key.

With full realization of the tendency of practitioners to try to oversimplify this problem, various procedures may reasonably be used for granular cohesionless materials. The maximum effect of a key would be to develop the passive resistance over the depth BC of Fig. 12.5.4, with any resulting failure occurring along a curved path such as $C'C$. Fisher and Mains [5]

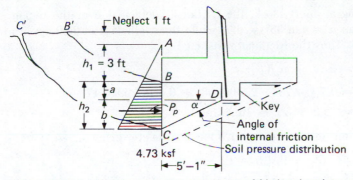

Figure 12.5.4 Passive resistance concept of frictional resistance.

have advocated the inert-block concept of frictional resistance. In this method, any key used must be extended deep enough below the footing to develop an inert block of soil, $BCDE$ of Fig. 12.5.5; it will then have a failure plane approximately as $C'CDGFH$, so that the passive resistance is developed over only the depth BC. In determining the passive resistance, the top 1 ft of overburden is usually neglected in the height h_1, of Fig. 12.5.4 or Fig. 12.5.5.

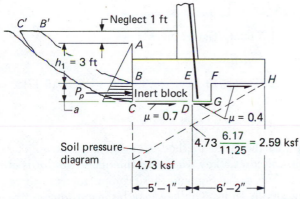

Figure 12.5.5 Inert block concept of frictional resistance.

For this design, a factor of safety of 1.5 against sliding under service loads has been considered proper. When passive resistance against the toe is also included, a higher factor of safety, say 2.0, should be used.

Using the inert-block concept, the passive force P_p developed over the distance BC of Fig. 12.5.5 is

$$\text{equivalent fluid pressure} = 400 \text{ pcf} \qquad \text{(given)}$$

$$P_p = \frac{0.400(h_1 + a)^2}{2} - \frac{0.400(h_1)^2}{2}$$

which for $h_1 = 3$ ft gives

$$P_p = 0.200(6a + a^2)$$

By inducing an inert block, the frictional coefficient over the region CD becomes $\tan \alpha = \tan 35° = 0.7$, while the coefficient of 0.40 applies on DG and FH. Thus the frictional resistance

$$F = \mu_1 R_1 + \mu_2 R_2$$

$$= 0.7(\tfrac{1}{2})(4.73 + 2.59)(5.08) + 0.4(\tfrac{1}{2})(2.59)(6.17)$$

$$= 13.0 + 3.2 = 16.2 \text{ kips}$$

Force equilibrium, incorporating a 1.5 factor of safety, gives

$$(P_1 + P_2)1.5 = P_p + F$$

$$11.52(1.5) = 0.200(6a + a^2) + 16.2$$

$$a^2 + 6a - 5.25 = 0$$

$$a = 0.8 \text{ ft}$$

Neglecting the frictional force in front of the key and considering the passive resistance developed below the toe as shown in Fig. 12.5.4, the depth of key required may be computed as

$$b = 5.08 \tan \alpha = 5.08(0.7) = 3.55 \text{ ft}$$

$$P_p = 0.400 \frac{(h_1 + a + b)^2}{2} - \frac{0.400(h_1)^2}{2}$$

$$= 0.200(a^2 + 13.1a + 33.9)$$

Force equilibrium, using a 1.5 safety factor, gives

$$(P_1 + P_2)1.5 = P_p + \mu_2 R_2$$

$$11.52(1.5) = 0.200(a^2 + 13.1a + 33.9) + 0.4(\tfrac{1}{2})(2.59)(6.17)$$

$$a^2 + 13.1a - 33.9 = 0$$

$$a = 2.4 \text{ ft}$$

This would indicate that the latter method is more conservative. The *CRSI Handbook* [6] arbitrarily uses a key whose depth equals two-thirds of the footing thickness. With this discussion and these computations to serve as a guide, make the key depth 18 in. The key may be made square; use 1 ft–6 in. × 1 ft–6 in. Reinforcement will rarely be required, but it is common practice to extend some of the stem steel down into the key.

(i) Design of heel cantilever. The loads considered are included in Fig. 12.5.6, where the effect of the base key is neglected.

For bending moment, the critical section is taken at the center of the stem steel. It seems that any possible plane of weakness will occur at the stem steel rather than at the face of support. Thus the downward uniform loading due to earth overburden, surcharge, and footing concrete using an overload factor of 1.4 for the earth and footing concrete and 1.7 for the surcharge is, from Fig. 12.5.6,

$$w_u = 1.4(2.16 + 0.30) + 1.7(0.96) = 5.08 \text{ kips/ft}$$

$$M_u \text{ (downward)} = \tfrac{1}{2}(5.08)(5.96)^2 = 90.2 \text{ ft-kips}$$

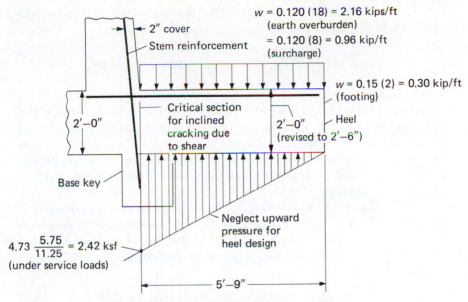

Figure 12.5.6 Design of heel cantilever in design example.

The upward soil pressure would reduce this value. However, the upward soil pressure may not actually act in the linear fashion assumed; in fact it might not be there at all. Applying overload conditions under ACI-9.2.4, the most critical situation results when $0.9D$ (i.e., gravity dead load) and $1.7L$ (i.e., horizontal earth pressure) are considered. This would eliminate pressure under the heel entirely.

For shear, ACI-11.1.2 allows the critical section to be taken at the distance d from the face only when ". . . support reaction, in direction of applied shear, introduces compression into the end regions of member, ..." In this case there is tension induced in the concrete where the heel joins the stem and the inclined crack could extend into the region ahead of the back face of the stem. The factored shear at the face without including upward soil pressure is

$$V_u = 5.08(5.75) = 29.2 \text{ kips}$$

The design shear strength, unless shear reinforcement is used, is

$$\phi V_c = \phi(2\sqrt{f_c'})bd$$
$$= 0.85(2\sqrt{3000})(12)(21.5)\tfrac{1}{1000} = 24.0 \text{ kips} < 29.2 \text{ kips} \qquad \text{NG}$$

Shear appears to control. The effective depth must be increased in the ratio $29.2/24.0$ unless shear reinforcement is to be used.

$$\text{required } d = \frac{21.5(29.2)}{24.0} = 26.2 \text{ in.}$$

Use heel thickness of 30 in. ($d \approx 27.5$ in.).

For flexural strength, using $M_u = 92.0$ ft-kips (corrected for the weight of the 30-in. footing),

$$\text{required } R_n = \frac{M_u}{\phi b d^2} = \frac{92.0(12,000)}{0.90(12)(27.5)^2} = 135 \text{ psi}$$

This is less than the R_n corresponding to the minimum percentage of reinforcement. Though ACI-10.5 exempts slabs of uniform thickness from the minimum requirement, the retaining wall is a major beamlike structure and the minimum is recommended to apply. Increasing the heel thickness from the preliminary 24 in. to 30 in. reduces the stem height and would permit reducing its thickness. The extra heel thickness will also improve stability against overturning.

From Fig. 3.8.1 or Eq. (3.8.5), the required reinforcement ratio is determined for $R_n = 135$ psi to be 0.00347, which is less than the minimum $200/f_y$ prescribed by ACI-10.5.1. Less than $200/f_y$ may be used provided the amount is one-third more than required for strength; that is, 1.33(0.00347) $= 0.0046$.

$$\text{required } A_s = 0.0046(12)(27.5) = 1.52 \text{ sq in./ft}$$

Use #8 @ 6 ($A_s = 1.57$ sq in./ft).

The development length required for #8 top bars is 23.1(1.4) $= 32.4$ in. (see Tables 6.7.1 and 6.8.1 based on ACI-12.2). Thus these bars should be embedded at least 32.4 in. into the toe of the footing measured from the stem reinforcement. Use an embedment length of 3 ft from back face of wall.

(j) Design of toe cantilever. The toe of the footing is also treated as a cantilever beam, with the critical section for moment at the front face of the wall and the critical section for shear (inclined cracking) at a distance d (approximately 20.5 in.) from the front face of the wall (one-way action according to ACI-11.11.1.1 will govern). Shear will usually control the required toe thickness. The thickness need not be the same as the heel, though many engineers would make them the same. In this example the heel is unusually thick due to the heavy surcharge. Try a toe thickness somewhat less than the heel, say 2 ft.

Referring to Fig. 12.5.7 and neglecting the earth on the toe, the factored shear and bending moment are

$$V_u = 1.7 \left(\frac{4.73 + 3.87}{2} - 0.300 \right) (2.04) = 13.9 \text{ kips}$$

$$M_u = 1.7[\tfrac{1}{2}(4.73)(3.75)^2(\tfrac{2}{3}) + \tfrac{1}{2}(3.15)(3.75)^2(\tfrac{1}{3}) - \tfrac{1}{2}(0.300)(3.75)^2]$$

$$= 1.7\left[\frac{(3.75)^2}{6}(9.46 + 3.15 - 0.90) \right] = 46.7 \text{ ft-kips}$$

According to ACI-9.2.4, the 1.7 is to be used for the horizontal earth pressure and for live load, while 1.4 may be used for the weights of the concrete and the earth. Here it has been considered that the toe soil pressure is

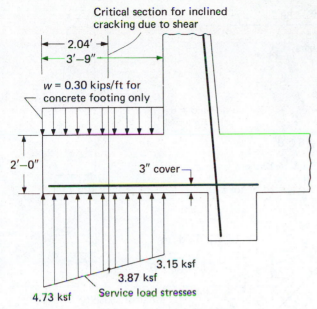

Figure 12.5.7   Design of toe cantilever in design example.

primarily the result of horizontal earth pressure; hence the conservative procedure of using the 1.7 factor.

$$\text{required } R_n = \frac{46.7(12,000)}{0.90(12)(20.5)^2} = 123 \text{ psi}$$

From Fig. 3.8.1, the required $\rho = 0.00318$ is less than $200/f_y$. Use $1.33(0.00318) = 0.0042$.

$$\text{required } A_s = 0.0042(12)(20.5) = 1.04 \text{ sq in./ft}$$

Use #7 @ 7 ($A_s = 1.03$ sq in./ft).
The shear strength is

$$\phi V_c = \phi(2\sqrt{f_c'})bd$$

$$= 0.85(2\sqrt{3000})(12)(20.5)\tfrac{1}{1000} = 22.9 \text{ kips} > 13.9 \text{ kips} \qquad \text{OK}$$

The required development length for the #7 bottom bars is (Tables 6.7.1 and 6.8.1; ACI-12.2)

$$L_d = 17.5(0.8) = 14 \text{ in.}$$

Use embedment of 1 ft 6 in. from front face of wall.

(k) Reinforcement for wall. The wall height (17.5 ft) is now 6 in. less than used in the preliminary design calculations. Retain the 21-in. thickness at the base of the stem. The factored moment diagram is shown in Fig. 12.5.8, using Eq. (12.5.1) with maximum $y = 17.5$ ft along with an overload factor of 1.7. The bending moment M_u diagram and the moment capacity

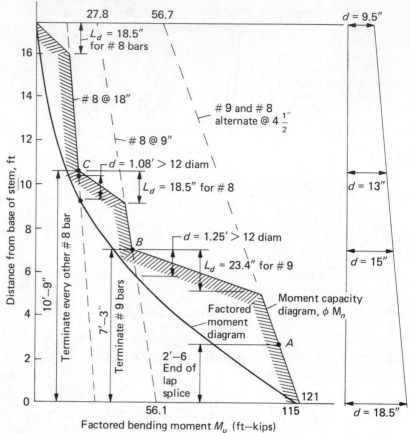

Figure 12.5.8 Determination of stem reinforcement for design example.

ϕM_n diagram of the selected wall steel are shown in Fig. 12.5.8. The steel area required at the base of stem is

$$\text{required } R_n = \frac{115(12,000)}{0.90(12)(18.5)^2} = 373 \text{ psi}$$

$$\text{required } A_s = 0.0103(12)(18.5) = 2.29 \text{ sq in./ft}$$

<u>*Use #8 and #9 bars alternated @ $4\frac{1}{2}$ (A_s = 2.38 sq in./ft).*</u> The shear has previously been checked and found satisfactory.

The embedment of the #9 bars into the footing must be the development length L_d of 29.2 in. (according to Table 6.7.1 or computed from equations in ACI-12.2). Use an embedment of 2 ft 6 in. into the footing for both #8 and #9 bars. The base key is available for embedment of stem bars if it might be necessary.

In an endeavor to economize, the quantity of steel per foot of wall should be reduced in the upper parts of the wall so that the capacity provided approximately equals that required. Bar spacing and cutoff are done in accordance with the ACI Code, and the resulting design is analyzed by drawing the moment capacity ϕM_n diagram as described in Chap. 6.

In this design, proceeding from the stem base the #9 bars embedded in the footing should not be extended more than about 8 to 10 ft out of the footing, which is placed and cured first. When bars extend out too far, they are often bent or broken off during construction. Thus splice the #8 bars at the base of the stem.

For #8 and #9 alternated at $4\frac{1}{2}$ in., the moment capacity is

$$C = 0.85(3)(12)a = 30.6a$$
$$T = 2.38(40) = 95.2 \text{ kips}$$
$$a = 3.11 \text{ in.}$$

At top of wall,

$$\phi M_n = 0.90(95.2)[9.5 - 0.5(3.11)]\tfrac{1}{12} = 56.7 \text{ ft-kips}$$

At bottom of wall,

$$\phi M_n = 0.90(95.2)[18.5 - 0.5(3.11)]\tfrac{1}{12} = 121 \text{ ft-kips}$$

The two values of moment capacity computed above are used to locate the outer dashed line in Fig. 12.5.8.

Since it is proposed to lap the #8 bars at the base of the wall, ACI-12.15 must be applied to determine the lap distance required. When no more than one-half of the bar area is to be lap spliced within the lap length, the tension splice must meet the requirements for a Class B splice; in this case less than one-half of the total bar area is to be spliced. Referring to Table 6.17.1 (which summarizes ACI-12.15), the lap required is

$$\text{required lap} = 1.3L_d$$

where L_d is the development length required for unspliced bars. In this case for #8 bars,

$$L_d = 23.1 \text{ in.} \qquad \text{(ACI-12.2 or Table 6.7.1)}$$
$$\text{required lap} = 1.3(23.1) = 30 \text{ in.}$$

Use a splice lap of 2 ft 6 in. terminating at point A.

Next, using the moment diagram (Fig. 12.5.8) locate the point where the #9 bars may be terminated, leaving the remaining moment capacity based on #8 @ 9. This moment capacity represented by #8 @ 9 is shown in Fig. 12.5.8 by an inclined dashed line that is plotted by using the following moment capacities:

$$C = 30.6a, \qquad T = 42.0 \text{ kips}, \qquad a = 1.37 \text{ in.}$$
$$\phi M_n \text{ (at top)} = 0.90(42.0)[9.5 - 0.5(1.37)]\tfrac{1}{12} = 27.8 \text{ ft-kips}$$
$$\phi M_n \text{ (at bottom)} = 0.90(42.0)[18.5 - 0.5(1.37)]\tfrac{1}{12} = 56.1 \text{ ft-kips}$$

The termination point B is found by extending beyond the intersection of the remaining capacity line (#8 @ 9) with the factored moment diagram a distance of either the effective depth d or 12 bar diameters, whichever is greater. In this case the theoretical termination point and the practical location in 3-in. increments coincide.

Whenever tension bars are terminated in the tension zone, stress concentrations occur; therefore a check of ACI-12.10.5 must also be made. Since stirrups are not used in retaining walls, either condition 1 or condition 3 of that section of the code must be satisfied. The conditions, one of which must be satisfied in this case are (1) that the factored shear V_u at the cutoff point must not exceed two-thirds of the shear strength ϕV_n; and (2) that the continuing bars must provide at least twice the area required for bending moment at the cutoff point, *and* the factored shear V_u at the cutoff point must not exceed three-fourths of the shear strength ϕV_n.

From an inspection of Fig. 12.5.8, it is obvious that item (2) above is not satisfied. Check item (1). Using Eq. (12.5.2), the shear at $y = 10.25$ ft from the top is

$$V_u = 1.7[0.256(10.25) + 0.016(10.25)^2] = 7.32 \text{ kips}$$

$$\phi V_c = \phi(2\sqrt{f_c'})bd$$

$$= 0.85(2\sqrt{3000})(12)(15)\tfrac{1}{1000} = 16.8 \text{ kips}$$

$$\tfrac{2}{3}(16.8) > 7.32 \qquad\qquad \text{OK}$$

Thus terminate #9 bars at 7 ft–3 in. from the top of heel.

Additional economy may be achieved by cutting every other #8 bar, leaving #8 @ 18 to extend to the top of the wall. It should be noted that ACI-7.6.5 states that the principal reinforcement shall be spaced not farther apart than three times the wall or slab thickness nor more than 18 in. Figure 12.5.8 shows the actual cutoff point C for every other #8. In this case the extension d gives a point slightly below point C. A tension zone cutoff check (ACI-12.10.5) at point $C(y = 6.75$ ft from top) shows

$$V_u = 1.7[0.256(6.75) + 0.016(6.75)^2] = 4.18 \text{ kips}$$

$$\phi V_c = \phi(2\sqrt{f_c'})bd$$

$$= 0.85(2\sqrt{3000})(12)(13)\tfrac{1}{1000} = 14.5 \text{ kips}$$

$$\tfrac{2}{3}(14.5) > 4.18 \qquad\qquad \text{OK}$$

Terminate every other #8 at 10 ft–9 in. from the top of heel. Actual cut points are located to make bar lengths in the usual 3-in. increments. The complete bar arrangement for the stem is shown in Fig. 12.5.9.

The reader may note that the moment decreases rapidly from the base of the stem toward the top of the wall. At about 5 ft from the base, the factored moment has decreased about 50%. Somewhere in this vicinity the reinforcement ratio ρ that is required for strength equals the minimum of $200/f_y$. For the upper portion of the wall, ACI-10.5 would require the actual ρ to be either 0.005 (i.e., $200/f_y$) or four-thirds of the required ρ based on the factored moment. As an example, the moment capacity ϕM_n at point C is indeed approximately equal to four-thirds of the factored moment at C. Thus, in general, wherever actual ρ is less than $200/f_y$, the requirement of ACI-10.5 may be reviewed by seeing that the moment capacity diagram is offset at one-third or more from the factored moment curve.

It is further noted that the stem has been designed for bending moment and shear only. However, the weight of the stem causes compression in it, so that strictly speaking it might be treated as a compression member (under

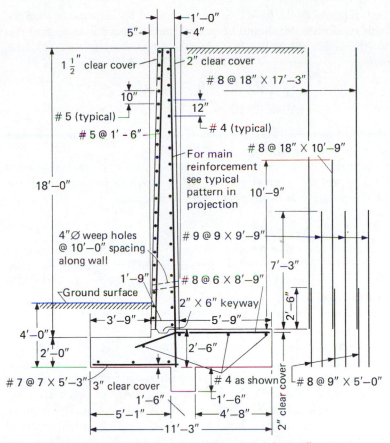

Figure 12.5.9   Design sketch for cantilever retaining wall.

large moment) in accordance with the concepts of Chap. 13. In this design problem the compressive force P_u in the wall under factored loads would be

$$P_u = 1.4 \left(\frac{12 + 21}{12}\right)(0.5)(17.5)(0.15) = 5.1 \text{ kips}$$

The combination of this P_u with $M_u = 115$ ft-kips will give a design near the bottom of the "tension controls" region according to Chap. 13 (see Fig. 13.6.2). Although the action of the compressive force increases the nominal bending strength M_n, the ϕ factor will be slightly decreased from 0.90 toward 0.70. In general, treatment of such walls as compression members is rarely justified but, if done, would permit some slight reduction in reinforcement or thickness of the section.

(l) Temperature and shrinkage reinforcement. Horizontal bars along the length of the wall must be provided, in accordance with ACI-14.1.2 and 14.3.3. Accordingly, the total amount required is

$$A_s = 0.0025bh = 0.0025(12)(16.5) = 0.50 \text{ sq in./ft}$$

where the average wall thickness is used for h.

Since it is primarily the front face that is exposed to temperature changes, more of this reinforcement should be placed there. Thus it is suggested that about two-thirds be put in the front face and one-third in the rear face. Accordingly,

$$\tfrac{2}{3}A_s = \tfrac{2}{3}(0.50) = 0.33 \text{ sq in./ft}$$

$$\tfrac{1}{3}A_s = \tfrac{1}{3}(0.50) = 0.17 \text{ sq in./ft}$$

Use on the front face #5 @ 10 ($A_s = 0.37$ sq in./ft). *Use on the rear face #4 @ 12 ($A_s = 0.20$ sq in./ft).*

For vertical reinforcement on the front face, use any nominal amount (ACI-14.3.5 is probably a good guide) that is adequate for supporting the horizontal temperature and shrinkage steel in that face.

Use #5 @ 1 ft–6 in. spacing.

(m) Drainage and other details. Adequate drainage of backfill must be provided since the pressures used are for drained material. A common minimum provision is for weep holes (say 4-in. diameter tile) every 10 to 15 ft along the wall.

Construction of a retaining wall is accomplished in at least two stages; the footing is placed first and then the wall. A shear key is necessary for positive shear transfer between wall and footing (see Fig. 12.5.9). Such a key is made by embedding a beveled 2 × 4 or 2 × 6 timber in the top of the footing. This key may be designed using the shear-friction provisions of ACI-11.7 (discussed in Sec. 5.15). This is a situation in which it is not appropriate to consider shear as a measure of diagonal tension. The nominal ultimate shear strength of the shear key is based on a nominal stress of $0.2f'_c$ but not to exceed 800 psi, as given by ACI-11.7.5.

In this design no bent bars are used. Frequently, the bar arrangement in the toe can be conveniently bent up into the stem and meshed with additional stem reinforcement to give an economical system.

(n) Design sketch. The final details of this design are presented in a design sketch (Fig. 12.5.9) which must accompany any set of design computations.

SELECTED REFERENCES

1. Karl Terzaghi and Ralph B. Peck. *Soil Mechanics in Engineering Practice* (2nd ed.). New York: Wiley, 1968.
2. Whitney Clark Huntington. *Earth Pressures and Retaining Walls*. New York: Wiley, 1957.
3. *Standard Specification of Highway Bridges* (12th ed.). Washington, D.C.: American Association of State Highway and Transportation Officials, 1977. (Also *Interim Specifications*, 1978–83.)
4. "Retaining Walls and Abutments," *American Railway Engineering Association Manual*, Vol. 1. Chicago: American Railway Engineering Association, 1958 (Chap. 8).
5. G. P. Fisher and R. M. Mains. "Sliding Stability of Retaining Walls," *Civil Engineering*, July 1952, 490.
6. *CRSI Handbook* (4th ed.). Chicago: Concrete Reinforcing Steel Institute, 1983 (Chap. 14).

PROBLEMS

Note: For all problems assume that frost penetration depth is 4 ft.

12.1 Given the cantilever retaining wall of the accompanying figure with the active pressure coefficient $C_a = 0.27$, the weight of retained soil $= 100$ pcf, and the coefficient of friction between concrete and earth $= 0.40$. Using service load conditions:

(a) Determine the factor of safety provided against overturning. Would you consider this to be adequate?

(b) Determine the factor of safety provided against sliding. Neglect any passive pressure resistance on the toe. Would you consider this factor of safety adequate?

(c) Determine the location of the resultant bearing pressure under footing. Is it within the middle third of the base? Would the position you determined probably be permissible if the foundation material is rock?

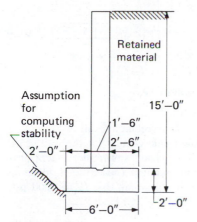

Problem 12.1

12.2 Determine the adequacy of the retaining wall of the accompanying figure with regard to stability (overturning, sliding, and soil pressure magnitude and distribution). Assume that the wall is on good granular soil with a maximum safe soil pressure of 5000 psf under service load conditions.

12.3 Determine the adequacy of the retaining wall of the accompanying figure with regard to stability if the backfill is level. Assume an allowable soil pressure of 4000 psf under service load conditions. Draw a conclusion regarding what would happen if the backfill became saturated to an equivalent fluid active pressure of 65 pcf. See the accompanying figure.

12.4 Reconsider the retaining wall conditions of Sec. 12.5 if the surcharge is changed to 10 ft to approximate the effect of a railroad located parallel to and near the top of the wall. Sometimes under such conditions the lateral effect of a surcharge is included but the beneficial stabilizing effect of the vertical surcharge weight is omitted. Compare the wall proportions for both conditions using a minimum length of base and keeping the soil pressure resultant under service load conditions within the middle third of the base. Verify the adequacy of the dimensions, but do not design the reinforcement. Use $f'_c = 4000$ psi, $f_y = 60,000$ psi, and the ACI Code.

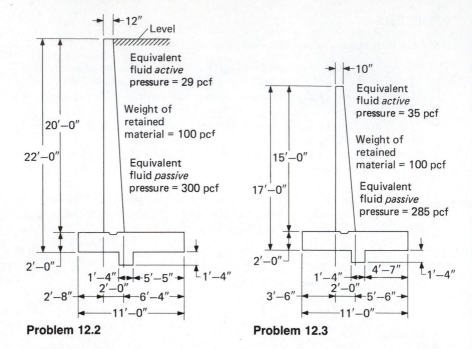

Problem 12.2 **Problem 12.3**

12.5 Assuming that the overall proportioning of the retaining wall of Prob. 12.2 is adequate for earth stability, design the reinforcement for the wall cantilever. Besides dowels at the base, use three changes of reinforcement over the 20-ft wall height. Draw the resulting moment capacity ϕM_n diagram superimposed on the factored moment M_u diagram. Use $f'_c = 4000$ psi, $f_y = 60,000$ psi, and the strength method of the ACI Code.

12.6 Design the heel cantilever for the wall of Prob. 12.2 if the backfill is level. Use $f'_c = 4000$ psi, $f_y = 60,000$ psi, and the strength method of the ACI Code. Revise the thickness from that shown if it seems desirable.

12.7 Design the heel cantilever for the wall of Prob. 12.3 if the backfill slopes 30° to the horizontal. Use $f'_c = 4000$ psi, $f_y = 60,000$ psi, and the strength method of the ACI Code. Revise the thickness if it seems desirable.

12.8 Design the toe cantilever for the wall of Prob. 12.2 if the backfill is level. Use $f'_c = 4000$ psi, $f_y = 60,000$ psi, and the strength method of the ACI Code. Revise the thickness if it seems desirable.

12.9 Design the toe cantilever for the wall of Prob. 12.3 if the backfill slopes at 30° to the horizontal. Use $f'_c = 4000$ psi, $f_y = 60,000$ psi, and the strength method of the ACI Code. Revise the thickness if it seems desirable.

12.10 Design a cantilever retaining wall to support a bank of earth 17 ft high above the final level of earth at the toe of the wall. The retained material is level and has a weight of 100 pcf with *active* and *passive* pressure coefficients of 0.28 and 3.6, respectively. The allowable soil pressure under service load is 5 ksf and the coefficient of friction between concrete and soil is 0.60. Use $f'_c = 3500$ psi, $f_y = 60,000$ psi, and the strength method of the ACI Code.

12.11 Design a cantilever retaining wall to support a bank of earth 22 ft high above the final level at the toe of the wall. The retained material is level and has a weight of 110 pcf and exerts an equivalent fluid active pressure of 30 pcf and an equivalent fluid passive pressure of 350 pcf. The coefficient of friction is 0.45 between concrete and soil. The maximum soil pressure under service load shall not exceed 3500 psf. Use $f'_c = 3000$ psi, $f_y = 60,000$ psi, and the strength method of the ACI Code.

12.12 Redesign the cantilever retaining wall of Prob. 12.11 if the differential elevation is 24 ft instead of 22 ft.

13

Members in Compression and Bending

13.1 INTRODUCTION

The compression member subjected to pure axial load rarely, if ever, exists. All columns are subjected to some moment, which may be due to end restraint arising from the monolithic placement of floor beams and columns or due to accidental eccentricity from imperfect alignment and variable materials.

This chapter considers first the column having minimal bending moment, commonly called "axially" loaded, and later considers the effect of medium and large amounts of bending. Only the basic strength of short compression members is considered, however. The reduction of the strength of compression members due to length effects is treated in Chap. 15.

13.2 TYPES OF COLUMNS

A column has been defined as a member used primarily to support axial compressive load with a ratio of height to least lateral dimension of 3 or greater (ACI-2.1). Shorter concrete compression members may be unreinforced and treated as pedestal footings. Reinforced concrete columns are principally of two types, classified according to the manner in which the longitudinal reinforcing bars are laterally supported. A *tied* column is one, usually of square, rectangular, or circular shape, in which the longitudinal reinforcing bars are held in position by separate lateral ties typically spaced about 12 to 24 in. (300 to 600 mm) apart, as shown in Fig. 13.2.1(a). A *spirally reinforced* column is one, usually of square or circular shape, in

Reinforced concrete tied columns under construction; Engineering Library, University of Wisconsin, Madison, Wis. (Photos by C. G. Salmon.)

which the longitudinal reinforcing bars are arranged in a circle and wrapped
by a continuous closely spaced spiral typically at a pitch of about 2 to 3 in.
(50 to 75 mm), as shown in Fig. 13.2.1(b). A *composite* column is one in
which a structural steel shape, pipe, or tubing is used, with or without
additional longitudinal bars. One common composite column arrangement
may contain a structural steel shape completely encased in concrete, which
is further reinforced with both longitudinal and lateral reinforcement (spiral
or ties) as shown in Fig. 13.2.1(c). In a second kind of composite column
the steel may encase a concrete core, which may or may not contain longitudinal reinforcing bars, as shown in Fig. 13.2.1(d).

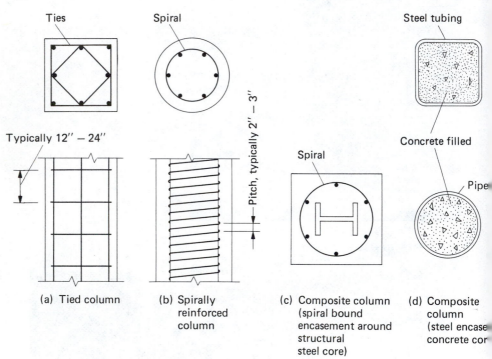

Figure 13.2.1 Types of columns.

13.3 BEHAVIOR OF AXIALLY LOADED COLUMNS

Many tests were made in the early 1900s on reinforced concrete columns
under axial loads [1–5], but loading was generally of short duration. As early
as 1911, however, Withey [5] at the University of Wisconsin observed that
as load increased beyond the service load range, a transfer of load from
concrete to steel took place. In the early 1930s ACI Committee 105 reported
[6] on the results of 564 column tests, primarily at Lehigh University and
the University of Illinois, where attention was given to column size, quality
of concrete, quality and amount of longitudinal and lateral reinforcement,
rate of application of load, and shrinkage and creep under sustained loads.

By 1940, design procedures for axially loaded columns became based on the ultimate strength results of the aforementioned extensive investigation. Joint ACI-ASCE Committee 441 has provided an excellent annotated bibliography on reinforced concrete column studies [7].

When concrete and steel act together in compression, the proportion of loading carried by each changes continuously during the time the load is acting. Initially the stress in the steel is E_s/E_c times the stress in the concrete, according to the elastic theory. As the time-dependent effects of creep and shrinkage occur, the steel gradually takes over relatively more load than its elastic share. Creep and shrinkage deformations have been introduced in Sec. 1.10, discussed in regard to beams in Secs. 4.9 and 4.10, and treated in detail with regard to deflections in Chap. 14.

Members that are subjected to axial compression, either alone or in combination with bending, frequently have a substantial portion of the total load sustained. Consequently the transfer of load to the steel from the concrete due to the time-dependent deformation is more pronounced in these members than in beams. However, even though actual stresses under service loads cannot be meaningfully computed, the strength can be determined. The experimenters verified that the nominal strength P_n for an axially loaded column may be properly expressed [6] by

$$P_n = k_c f'_c A_c + f_y A_{st} + k_s f_{sy} A_{sp} \qquad (13.3.1)$$

where

P_n = nominal strength for a tied column (with third term omitted, i.e., no spiral)

or

P_n = yield strength for a spirally reinforced column

k_c = coefficient (0.85) to account for the difference between concrete in the column and that in a test cylinder

f'_c = standard 28-day cylinder strength

A_c = net area of concrete, based on gross area for tied column and core area for spirally reinforced column

A_{st} = area of longitudinal reinforcement

f_y = yield stress for longitudinal reinforcement

k_s = constant that varies from 1.5 to 2.5 with an average of 1.95

f_{sy} = yield stress of spiral steel

A_{sp} = volume of spiral steel per unit length of column

Equation (13.3.1) represents the yield load for the spirally reinforced column that exhibits a marked yielding followed by considerable deformation before complete failure, as shown in Figs. 13.3.1 and 13.3.2. On reaching the yield point, the shell spalls off and the spiral begins to act to confine the crushed concrete in the core. The spiral therefore adds little to the strength prior to reaching yield, but it provides ductility. From Eq. (13.3.1) it may be observed that spiral reinforcement is from 1.5 to 2.5 times as effective in adding strength as is the main longitudinal steel, but the spiral does not work at all until the deformation is excessive. Recent studies of

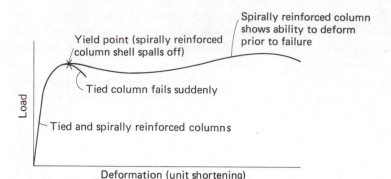

Figure 13.3.1 Typical load-deformation curves for tied and spirally reinforced columns.

spirally reinforced column ductility under cyclic loading have been made by Priestley, Park, and Potangaroa [8].

The tied column behaves exactly as a spirally reinforced column up to the yield point of the spirally reinforced column, at which point failure occurs suddenly in a manner similar to the failure in a compression cylinder test. At that instant the longitudinal bars buckle between points that are restrained by lateral ties. The tied column behavior is illustrated in Figs. 13.3.1 and 13.3.3. Recent studies by Sheikh and Uzumeri [9] and Scott, Park, and Priestley [55] show how the ductility of tied columns subject to large compressive strain varies with tie and longitudinal bar arrangements.

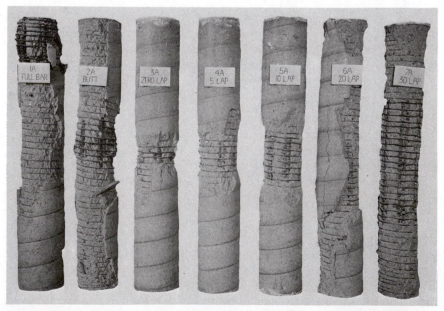

Figure 13.3.2 Spirally reinforced column behavior. (Courtesy of Portland Cement Association.)

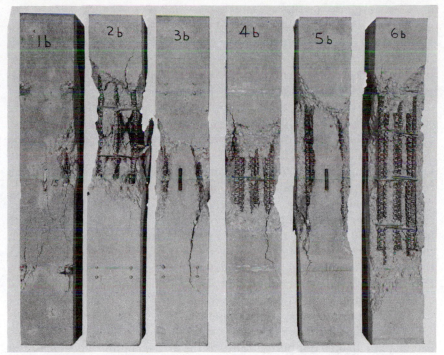

Figure 13.3.3   Tied column behavior. (Courtesy of Portland Cement Association.)

As will be shown later, though both types of column have the same strength, a higher factor of safety should be provided for the tied column than for the spirally reinforced column because of the sudden failure and lack of toughness (energy absorption).

13.4 SAFETY PROVISIONS

The safety provisions for overload in the design of compression members as prescribed by the ACI Code are the same as for any other types of members; these are (ACI-9.2):

For gravity loads,

$$U = 1.4D + 1.7L \tag{13.4.1}$$

For loadings including wind,

$$U = 0.75(1.4D + 1.7L + 1.7W) \tag{13.4.2}$$

or

$$U = 0.9D + 1.3W \tag{13.4.3}$$

The most severe situation among Eqs. (13.4.1) through (13.4.3) will control the design.

Traditionally, under most building codes, members subject to stresses produced by wind or earthquake forces combined with other loads have been proportioned in the working stress method for unit stresses $33\frac{1}{3}\%$ greater than those specified, provided that the section thus required is not less than that required for the combination of dead and live load. In strength design, the infrequent occurrence of maximum, wind load with maximum live load is taken into account by the smaller overload factors when wind effects are included. This reduced safety provision for the temporary effect of wind or earthquake, or both, appears in ACI-Appendix B for the working stress method, and in ACI-9.2.2 and 9.2.3 for the strength method.

Additionally, the possibility of undercapacity is accommodated by the ϕ requirement of ACI-9.3. For spirally reinforced columns $\phi = 0.75$, and for tied columns $\phi = 0.70$, the difference being related partially to the reserve capacity to deform before failure which is exhibited by the spirally reinforced column.

The lower strength reduction factors ϕ for compression members arise from statistical variations in observed strength and are a rough indication that the strength variability to be expected in tied columns is slightly greater than in spiral columns and that the variability expected in columns is greater than in beams. The difference in the behavior of tied and spirally reinforced *axially* loaded columns is further accounted for by the use of a different maximum compression capacity as discussed in Sec. 13.11.

For combined compression and bending, the ϕ value may be variable and increase to 0.90 (for pure flexure) as the axial compression decreases to zero. This variation of ϕ is treated later in Sec. 13.17.

13.5 CONCENTRICALLY LOADED SHORT COLUMNS

In accordance with Eq. (13.3.1) without the term representing the contribution of the spiral, the maximum nominal strength $P_n = P_0$ for a concentrically loaded short column (Fig. 13.5.1) is

$$P_0 = 0.85f'_c(A_g - A_{st}) + f_y A_{st} \tag{13.5.1}$$

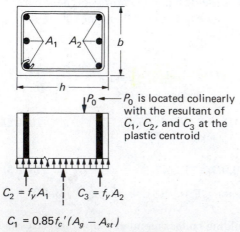

P_0 is located colinearly with the resultant of C_1, C_2, and C_3 at the plastic centroid

$C_2 = f_y A_1$ $C_3 = f_y A_2$

$C_1 = 0.85 f'_c (A_g - A_{st})$

Figure 13.5.1 A concentrically loaded column.

where A_g is the gross area bh and A_{st} is the total longitudinal reinforcement $(A_1 + A_2)$. This equation is also in agreement with the rectangular stress block assumptions of Sec. 3.4 where the entire cross section is subject to a failure compressive strain of 0.003. P_0 may also be expressed as

$$P_0 = A_g[0.85f_c'(1 - \rho_g) + f_y\rho_g] \qquad (13.5.2)$$

where $\rho_g = A_{st}/A_g$.

When the terms including ρ_g are combined, Eq. (13.5.2) becomes

$$P_0 = A_g[0.85f_c' + \rho_g(f_y - 0.85f_c')] \qquad (13.5.3)$$

13.6 GENERAL DISCUSSION ON COMBINED BENDING AND AXIAL LOAD

A structural member may be subjected to combined bending and axial load in many ways. It is common in reinforced concrete buildings that bending moments act on all columns. These moments are generally due to unbalanced floor loads on both exterior and interior columns, to eccentric loads such as crane loads in industrial buildings, and to lateral loading such as from wind or earthquake. In determining bending moments in columns due to unbalanced floor loads, ACI-8.8 provides the following simplifying assumptions:

1. The far ends of columns that are monolithic with the structure may be considered fixed in continuity analysis for gravity loading.
2. Maximum bending in a column is due to factored loads on a single adjacent span of the floor under consideration. This bending is in addition to axial forces from factored loads on all floors.
3. The loading condition causing maximum ratio of bending moment to axial load shall also be considered.

Concrete construction is usually monolithic; thus reinforced concrete rigid frames and arches (Fig. 13.6.1) are common and advantageous. All sections in the two structures shown in Fig. 13.6.1 are subjected to combined bending and axial load. The vertical members in Fig. 13.6.1(a) and sections near the supports in Fig. 13.6.1(b) are subjected to a high ratio of axial force to bending moment, while the horizontal member in Fig. 13.6.1(a) and sections near the crown in Fig. 13.6.1(b) may be subjected to a low ratio of axial force to bending moment.

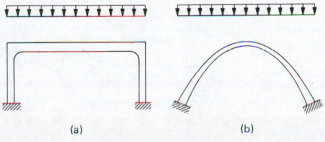

Figure 13.6.1   Reinforced concrete rigid frame and arch.

When combined axial compression and bending moment act on a member having a low slenderness ratio (unbraced length L_u to radius of gyration r) where column buckling is not a possible mode of failure, the strength of the member is governed by the material strength (corresponding to yielding in a homogeneous elastic material) of the cross section. For this so-called short column the strength is achieved (according to ACI-10.2.3) when the extreme concrete compression fiber reaches the strain 0.003. Depending on the ratio of M_n to P_n (see Fig. 13.6.2), the strain diagram will exhibit two distinct categories. There may be (1) compression over most or all of the section such that the compressive strain in the concrete reaches 0.003 before the tension steel yields, known as the "compression controls" region, or (2) tension in a large portion of the section such that the strain in the tension steel is greater than the yield strain ($\epsilon_y = f_y/E_s$) when the compressive strain in the concrete reaches 0.003, known as the "tension controls" region. Separating the "compression controls" region and the "tension controls" region is the balanced strain condition wherein the concrete strain of 0.003 and the steel yield strain $\epsilon_y = f_y/E_s$ are reached simultaneously.

It is important to note that for the same cross section there is an infinite number of strength combinations at which P_n and M_n act together. These strength combinations lie on a curve as shown in Fig. 13.6.2, which is called the *strength interaction diagram*. The balanced strain condition in combined bending and axial load is represented by the point $P_n = P_b$ and $M_n = M_b$ on this diagram.

This maximum strength interaction relationship for short members has been verified by research [10–14]. The major emphasis for this chapter is the analysis and design of sections whose nominal strength (P_n and M_n) lies at various points on the interaction diagram (Fig. 13.6.2).

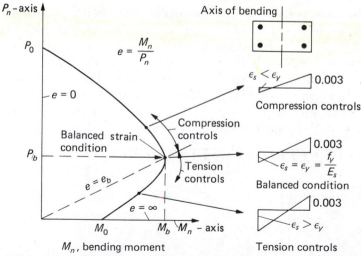

Figure 13.6.2 Typical strength interaction diagram for axial compression and bending moment about one axis.

In studying Fig. 13.6.2, the reader may note that radial lines from the origin ($P_n = 0$, $M_n = 0$) represent constant ratios of M_n to P_n; that is, they represent eccentricities e of the load P_n from the *plastic centroid*, which is defined in Fig. 13.5.1. The fact that e is equal to M/P may be observed from Fig. 13.6.3, where it is shown that a column subjected to an eccentric load is statically equivalent to a member under the combined action of an axial load and a bending moment. Thus in Fig. 13.6.2 the vertical axis represents $e = 0$ and the horizontal axis represents $e = \infty$. This concept of replacing axial load and bending moment by a single eccentric load provides the basis for the practical approach to analysis and design computations in reinforced concrete.

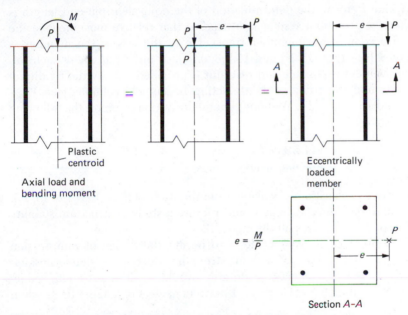

Figure 13.6.3   Eccentrically loaded column statically equivalent to member subject to axial load and bending moment.

13.7 LENGTH EFFECTS

When the height of an upright compression member does not exceed three times its least lateral dimension, it is considered a pedestal and should be designed as such.* For longer members, the effect of slenderness ratio (ratio of unbraced length L_u to radius of gyration r) must be considered. It is well known that buckling may control the strength of any compression member (see Fig. 15.2.3 for curve typical of reinforced concrete sections). The strength design method, together with better understanding of concrete compression

---

*According to *Building Code Requirements for Structural Plain Concrete* (ACI-318.1–83) *and Commentary.* Detroit: American Concrete Institute, 1983 (Sec. 7.3). (Formerly in ACI-15.11.)

member behavior, has produced slimmer members; thus the stability problem has become increasingly important. In addition to the stability problem, smaller members will deflect more under any primary bending moment and thus have a larger secondary moment, which is the product of the axial compression and deflection (as shown in Fig. 15.1.1).

The magnitude of the slenderness ratio determines whether the strength reduction is sufficiently important that it cannot be neglected. The reference condition for slenderness ratio is that of a column with hinged ends (i.e., no resistance to rotation at either end), as shown in Fig. 15.3.1(a). Equivalent pin-end lengths of columns with end restraints can be expressed by kL_u, where k is the effective length factor and L_u is the actual unsupported length.

A vital factor in the determination of the equivalent pin-end length is whether the structural system is *braced* so that relative movement of the ends of a compression member transverse to the axis of the member is prevented [see Figs. 15.3.1 and 15.3.3(a),(c)], or *unbraced* where such relative movement is possible and restraint is provided only by the rigidity of the joints and the stiffness of interacting beams and columns [see Figs. 15.3.2 and 15.3.3(b),(d)]. Without general derivation or proof, the following may be stated:

For *braced* systems, $\qquad k \leq 1.0$
For *unbraced* systems, $\qquad k \geq 1.0$

A qualitative explanation is available from the study of Figs. 15.3.1 to 15.3.3, but a theoretical development would require a study of structural stability (see Chap. 15 references 30 through 34).

The intent of the ACI Code is to permit the design of compression members as short columns, without strength reduction for slenderness effect, when the length effect consideration would result in a strength reduction not exceeding 5%. ACI-10.11.4 permits neglect of length effects when

$$\frac{kL_u}{r} < 34 - 12 \frac{M_{1b}}{M_{2b}} \qquad \text{(for } braced \text{ systems)} \qquad (13.7.1)$$

where M_{1b} and M_{2b} are numerically the smaller and larger bending moments, respectively, at the ends of the member and the ratio M_{1b}/M_{2b} is positive for single curvature and negative for double curvature; and when

$$\frac{kL_u}{r} < 22 \qquad \text{(for } unbraced \text{ systems)} \qquad (13.7.2)$$

Studies of existing structures in 1970 indicated [15] that over 90% of the columns in braced frames and over 40% of the columns in unbraced frames would fall within the limits of ACI-10.11.4 and thus strength reduction due to length effects could be ignored. Since 1970, columns exceeding the ratios indicated by Eqs. (13.7.1) and (13.7.2) have become relatively more common, so that a higher percentage of designs today (1984) will likely require slenderness consideration.

The remainder of this chapter treats only the basic strength of short compression members. The detailed consideration of the strength reduction due to length effects in both unbraced and braced systems appears separately in Chap. 15.

13.8 LATERAL TIES

The lateral ties are used to hold the vertical bars in position, providing lateral support so that individual bars could have the tendency to buckle only *between* the tie supports. Ties do not contribute to the strength, as indicated by the column studies in the early 1930s [6]. Studies [9,16–21] have indicated that present tie requirements are conservative for ordinary columns with Grade 40 reinforcement, but may not be conservative for columns with high-strength reinforcement, with large or bundled bars, or of unusual dimensions.

The effect of ties on the behavior of columns is complex [16]. As a tied column is loaded to failure, the first occurrence is the spalling off of the exterior cover, which in turn causes transfer of load to the concrete core and the longitudinal steel. The loss of stiffness of the longitudinal steel, which begins to yield or to buckle outward, causes additional stress on the concrete core. Once the core achieves its crushing strength, the column suddenly fails. The above sequence usually takes place rapidly, giving the so-called "sudden" failure. Ties placed at a sufficiently close spacing provide confinement and increase the strain at which concrete crushes to values well above the maximum of 0.003 used by the ACI Code [55,56].

The following provisions have been prescribed for lateral ties in columns by the ACI Code (ACI-7.10.5):

1. All nonprestressed bars for tied columns shall be enclosed by lateral ties, at least #3 in size for longitudinal bars #10 or smaller, and at least #4 in size for #11, #14, #18, and bundled longitudinal bars.
2. The spacing of the ties shall not exceed 16 longitudinal bar diameters, 48 tie bar diameters, or the least dimension of the column.
3. The ties shall be so arranged that every corner and alternate longitudinal bar shall have lateral support provided by the corner of a tie having an included angle of not more than 135° and no bar shall be farther than 6 in. clear on either side from such a laterally supported bar.
4. Where the bars are located around the periphery of a circle, a complete circular tie may be used.

The ACI *Detailing Manual—1980* [22] suggests tie arrangements for various numbers of bars, some common ones being shown in Fig. 13.8.1.

The provisions of ACI-7.10.5.3 permitting the included angle where a tie "corners" the longitudinal reinforcement (that is, a corner bar or every other bar) to be as large as 135° became part of the ACI Code as a result of the Bresler and Gilbert tests [17], a liberalization from earlier codes in which a 90° corner was required for every bar. However, since spliced bars and

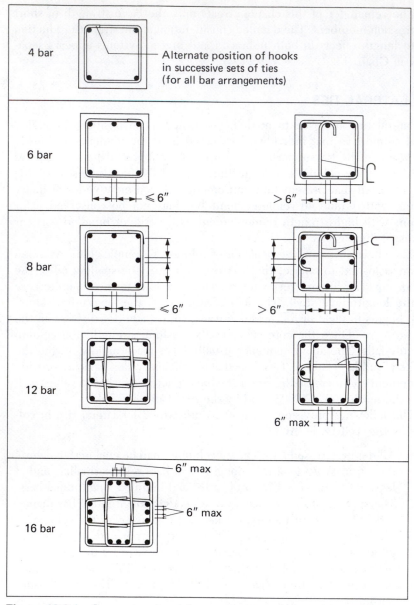

Figure 13.8.1 Common column tie arrangements (adapted from Ref. 22).

bundled bars have not been included in tests, ACI Commentary-7.10.5 recommends [23] that "it would be prudent to provide at least a set of ties at each end of lap spliced bars, above and below end-bearing splices, and at minimum spacings immediately below sloping regions of offset bent bars."

It may also be noted that ACI-7.10.3 waives the lateral reinforcement requirements "where tests and structural analysis show adequate strength and feasibility of construction."

13.9 SPIRAL REINFORCEMENT AND LONGITUDINAL BAR PLACEMENT

The spiral provides the column with the ability to absorb considerable deformation prior to failure [8]. This toughness is the principal gain that is achieved by the use of spirally reinforced columns. The knowledge of spiral behavior is based on the column research of the early 1930s [6]. Although the spiral does actually contribute strength to the column (as early as 1903 Considere [1, 2] indicated that the spiral was 2.4 times as effective in providing column capacity as was longitudinal reinforcement), the conservative policy of ACI specifications since about 1940 has been to provide spiral reinforcement sufficient to increase the capacity of the core by an amount equal to the capacity of the shell, thus maintaining the column yield capacity when the shell spalls off.

Using the third term of Eq. (13.3.1), with an average k_s of 2, the strength represented by the spiral reinforcement is

$$P_n = 2.0 f_{sy} A_{sp} \tag{13.9.1}$$

An alternative approach to the acceptance of an experimental value for k_s has been presented by Huang [24], who considers the triaxial loading condition that exists when the spiral is acting in tension.

Let ρ_s be the ratio of the volume of spiral reinforcement to the total volume of the core (out-to-out of spirals), or $\rho_s = A_{sp}/A_c$. Equation (13.9.1) now becomes

$$P_n = 2.0 f_{sy} \rho_s A_c \tag{13.9.2}$$

Equating Eq. (13.9.2) to the strength of the shell and taking the concrete strength of the shell as about 90% of that inside the core, or $0.75 f'_c$,

$$2.0 f_{sy} \rho_s A_c = 0.75 f'_c (A_g - A_c)$$

from which

$$\rho_s = 0.375 \left(\frac{A_g}{A_c} - 1 \right) \frac{f'_c}{f_{sy}} \tag{13.9.3}$$

By providing an additional factor of safety of 1.20 to assure that the spiral strength exceeds the shell strength, Eq. (13.9.3) becomes

$$\rho_s = 0.45 \left(\frac{A_g}{A_c} - 1 \right) \frac{f'_c}{f_{sy}} \tag{13.9.4}$$

which is specified in ACI-10.9.3. It is noted that the yield strength f_{sy} of the spiral may not exceed 60,000 psi.

The design relationship may be obtained by referring to the definition of ρ_s following Eq. (13.9.1).

$$\rho_s = \frac{A_{sp}}{A_c} = \frac{\text{volume of spiral in one loop}}{\text{volume of core for a length } s}$$

$$= \frac{a_s \pi (D_c - d_b)}{(\pi D_c^2 / 4) s} \tag{13.9.5}$$

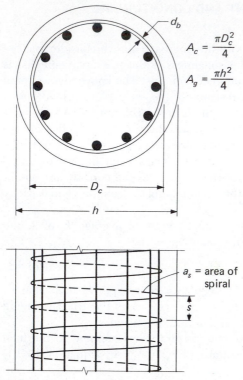

$$A_c = \frac{\pi D_c^2}{4}$$

$$A_g = \frac{\pi h^2}{4}$$

a_s = area of spiral

Figure 13.9.1 Spirally reinforced column.

in which D_c is the diameter of the core, a_s is the area of the spiral, and d_b is the diameter of the spiral wire (Fig. 13.9.1).

According to ACI-7.10.4 the clear spacing between spirals must be at least 1 in. but shall not exceed 3 in. The spiral in cast-in-place construction shall have a diameter not less than $\frac{3}{8}$ in. Anchorage of spiral reinforcement shall be provided by $1\frac{1}{2}$ extra turns of spiral bar or wire at each end of the spiral unit. Splices, when necessary, shall be made in the spiral bar or wire by welding or by a lap of 48 diameters minimum but not less than 12 in. The spiral reinforcement is to be protected everywhere by a $1\frac{1}{2}$ in. covering of concrete cast monolithically with the core.

In order to hold the spiral in place and maintain the desired pitch, vertical spacer bars with small hooks at the desired pitch are used. These spacer bars support the spiral reinforcement in the forms before the concrete is placed; the spiral is not supported by the main longitudinal bars. ACI-7.10.4 requires spacers in accordance with Table 13.9.1.

Since spirally reinforced columns are usually more heavily reinforced than tied columns, it sometimes becomes necessary at splice points to lap bars (see ACI-12.14.2.1, 12.15, and 12.16 on splices) inside the main circle of bars. This is done in order to maintain a desirable minimum center-to-center spacing of longitudinal bars at $2\frac{1}{2}$ times the bar diameter and a minimum clear distance (ACI-7.6.3) between individual longitudinal bars of $1\frac{1}{2}$ times the nominal bar diameter, but not less than $1\frac{1}{2}$ in. The preferred and

Table 13.9.1   NUMBER OF SPACERS REQUIRED PER SPIRAL
(FROM ACI-7.10.4)

SPIRAL CORE DIAMETER	MINIMUM NUMBER OF SPACERS
Bar or wire smaller than $\frac{5}{8}$-in. (16 mm) diameter:	
less than 20 in.	2
20 to 30 in.	3
more than 30 in.	4
Bar or wire $\frac{5}{8}$-in. (16 mm) diameter and larger:	
24 in. or less	3
more than 24 in.	4

the alternate bar arrangements are shown in Fig. 13.9.2. Bars may also be spliced by butt welding (ACI-12.14.3), which permits more utilization of available space. In a large column an inner core of bars wrapped with a spiral or by ties may also be used.

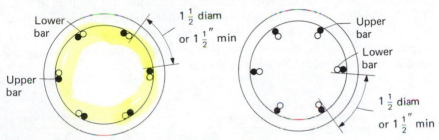

Figure 13.9.2   Bar arrangement in spiral columns.

13.10 LIMITS ON PERCENTAGE OF REINFORCEMENT

The percentage of longitudinal reinforcement in terms of the gross cross-sectional area must be between 1 and 8% (ACI-10.9.1). However, ACI-10.8.4 permits basing the percentage on a reduced concrete area A_g in cases where the gross concrete area is in excess of that needed for load considerations, but in no case may the reinforcement ratio ρ_g be less than 0.005 based on the gross area provided. The primary purpose of these provisions for minimum steel is to prevent the failure mode from becoming that of a plain concrete column, which might be more disastrous than the sudden failure of tied columns previously described. The upper limit on the amount of longitudinal reinforcement is a practical one in that if proper clearances are maintained between bars, little more than $\rho_g = 0.08$ can be put into the section. Thus the maximum ρ_g is, in a way, a double check on the minimum spacing restrictions of ACI-7.6.3.

There are no restrictions on bar size or on dimensions in order to allow for "wider utilization of reinforced concrete compression members in smaller size and lightly loaded structures, such as low rise residential and light office buildings" [23]. It is recommended [23] that "the engineer should recognize the need for careful workmanship, as well as the increased significance of shrinkage stresses with small sections."

13.11 MAXIMUM STRENGTH IN AXIAL COMPRESSION—ACI CODE

Since a truly concentrically loaded column is rare, if not nonexistent, some minimal eccentricity should be provided for. This accidental eccentricity may occur due to end conditions, inaccuracy of manufacture, or variation in materials even when the load is theoretically concentric.

Prior to 1977, the ACI Code provided that no matter how small the computed e is from the actual loading, compression members had to be designed for an eccentricity not less than $0.05h$ for spirally reinforced or composite steel encased columns, or $0.10h$ for tied columns; but at least 1 in. in any case. The minimum eccentricity was measured with respect to either principal axis, with h defined as the overall dimension of the column.

This minimum eccentricity requirement meant that every compression member, even though carrying a small computed bending moment, must be designed to have a strength defined, for example, by point A of Fig. 13.11.1. In other words, not only could the axial strength not exceed that defined by point A but a bending moment M_B must be considered as acting simultaneously; that is, the horizontal portion of the strength curve through point A was not actually available in design.

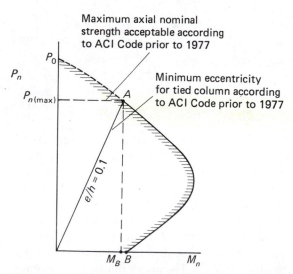

Figure 13.11.1 Maximum axial strength—ACI Code prior to 1977.

The minimum eccentricity procedure was reasonable as long as the cross-sectional dimensions were small and the slenderness ratios were relatively large; a moment corresponding to 1 in. or so eccentricity on a 12-in. column was not unreasonable. However, an $e/h = 0.1$ minimum becomes unreasonable as accidental eccentricity if a column is a power plant machinery pedestal 6 ft square and 20 ft high. A 6-in. accidental eccentricity is unlikely, although a maximum axial nominal strength less than P_0 may be reasonable.

Since 1977, the ACI Code prescribes (ACI-10.3.5) that for members where the effects of slenderness may be neglected, the maximum axial load nominal strength $P_{n(max)}$ may not exceed $0.80P_0$ for tied columns and $0.85P_0$ for spirally reinforced columns, with P_0 given by Eq. (13.5.1) or (13.5.2). This procedure makes available for design use the entire horizontal portion of the strength interaction diagram defined by $P_{n(max)}$. In other words, point C [Fig. 13.11.2(a)] becomes an acceptable point to use in design, which gives a maximum value of P_n comparable to that corresponding to a minimum $e/h = 0.1$ for tied columns or $e/h = 0.05$ for spirally reinforced columns; but yet the design is not explicitly made for a bending moment of P_n times $0.1h$ or P_n times $0.05h$.

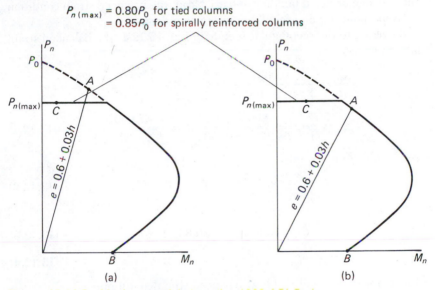

Figure 13.11.2   Maximum axial strength—1983 ACI Code.

The current procedure is also simpler for those cases that are clearly low bending moment cases (moments less than what corresponds to $e/h = 0.1$ on tied columns, for example). Taking $P_{n(max)}$ as a proportion of P_0 is a simple calculation; whereas previously the only way $P_{n(max)}$ could be determined was to solve for point A in Fig. 13.11.1, the intersection of the minimum e/h with the strength interaction diagram.

When the slenderness ratio is high enough to require consideration of the length effects, a minimum eccentricity of $(0.6 + 0.03h)$ is to be used in the evaluation of the magnified factored moment, but in no case may the resulting available nominal strength P_n be taken to exceed $P_{n(max)}$ (ACI-10.11.5.4).

In an unlikely situation as shown by Fig. 13.11.2(b), where $P_{n(max)}$ is vertically higher than point A, the value of P_n at point A on the interaction diagram must be used for design when the effects of slenderness must be considered; but where slenderness may be neglected, the horizontal portion through point C may be used in design for axial compression.

13.12 BALANCED STRAIN CONDITION—RECTANGULAR SECTIONS

The balanced strain condition represents the dividing point between the compression controls and tension controls regions of the strength interaction diagram (Fig. 13.6.2). Defined in the same manner as in Chap. 3, it is the simultaneous occurrence of a strain of 0.003 in the extreme fiber of concrete and the strain $\epsilon_y = f_y/E_s$ on the tension steel.

It may be noted that in the case of bending moment without axial load, the balanced strain condition is not permitted by the ACI Code. However, in the case of combined bending and axial load, the balanced strain condition is only one point on an acceptable interaction diagram.

Referring to the rectangular section in Fig. 13.12.1, the balanced strain condition gives

$$\frac{x_b}{d} = \frac{0.003}{f_y/E_s + 0.003}$$

$$x_b = \frac{0.003}{f_y/[29(10^6)] + 0.003} d = \frac{87,000d}{f_y + 87,000} \qquad (13.12.1)$$

Force equilibrium requires

$$P_b = C_c + C_s - T \qquad (13.12.2)$$

where

$$C_c = 0.85f'_c ab = 0.85f'_c \beta_1 x_b b \qquad (13.12.3)$$

$$T = A_s f_y \qquad (13.12.4)$$

and if compression steel yields at balanced strain condition,

$$C_s = A'_s(f_y - 0.85f'_c) \qquad (13.12.5)$$

Thus Eq. (13.12.2) becomes

$$P_b = 0.85f'_c \beta_1 x_b b + A'_s(f_y - 0.85f'_c) - A_s f_y \qquad (13.12.6)$$

The eccentricity e_b is measured from the plastic centroid, which has been defined in Fig. 13.5.1. For symmetrical sections the plastic centroid is at the mid-depth of the section.

Rotational equilibrium of the forces in Fig. 13.12.1 is satisfied by taking moments about any point such as the plastic centroid,

$$P_b e_b = C_c\left(d - \frac{a}{2} - d''\right) + C_s(d - d' - d'') + Td'' \qquad (13.12.7)$$

Equations (13.12.6) and (13.12.7) may be solved simultaneously to obtain P_b and e_b.

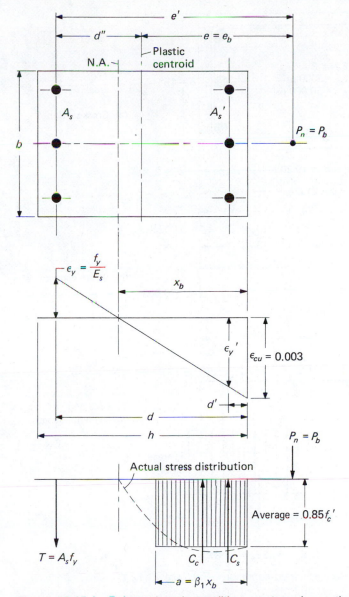

Figure 13.12.1 Balanced strain condition—rectangular section.

EXAMPLE 13.12.1 Determine the eccentric compressive strength $P_n = P_b$ and the eccentricity e_b for a balanced strain condition on the section of Fig. 13.12.2. Use $f_c' = 3000$ psi, $f_y = 50,000$ psi, and the ACI Code.

Solution: (a) Locate the neutral axis for the balanced strain condition.

$$x_b = \frac{0.003(21.6)}{0.00172 + 0.003} = 13.72 \text{ in.}$$

$$a_b = \beta_1 x_b = 0.85(13.72) = 11.67 \text{ in.}$$

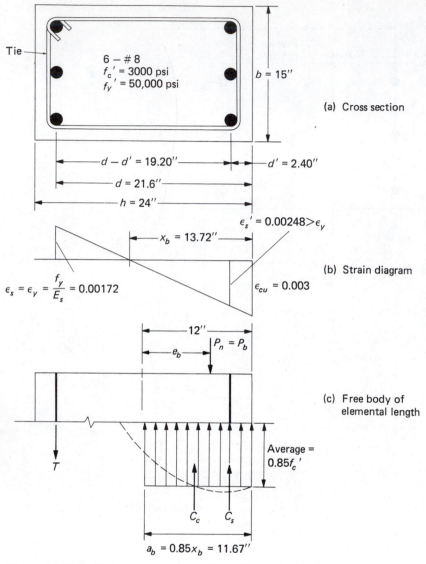

(a) Cross section

(b) Strain diagram

(c) Free body of elemental length

Figure 13.12.2   Balanced strain condition—Example 13.12.1.

The value β_1 is to be taken at 0.85 for $f'_c \leq 4000$ psi (ACI-10.2.7).

   (b) Compute the forces C_c, C_s, and T.

$$C_c = 0.85(3.0)(11.67)(15) = 446 \text{ kips}$$

$$T = 50.0(2.37) = 118 \text{ kips}$$

$$\epsilon'_s = 0.003\left(\frac{13.72 - 2.4}{13.72}\right) = 0.00248 > \frac{f_y}{E_s}, \qquad \text{compression steel yields}$$

$$C_s = 50.0(2.37) - 0.85(3.0)(2.37) = 118 - 6 = 112 \text{ kips}$$

(c) Compute P_b and e_b.

$$P_b = C_c + C_s - T = 446 + 112 - 118 = 440 \text{ kips}$$

For rotational equilibrium about the plastic centroid,

$$P_b e_b = [446(12 - 11.67/2) + 112(12 - 2.4) + 118(12 - 2.4)]\tfrac{1}{12}$$

$$= 229 + 90 + 95 = 414 \text{ ft-kips}$$

$$e_b = \frac{414(12)}{440} = 11.3 \text{ in.}$$

On the given section, if $P_n > 440$ kips (or $e < 11.3$ in.), the member is more a column than a beam and is referred to as a case where compression controls; if $P_n < 440$ kips (or $e > 11.3$ in.), the member is more a beam than a column and is referred to as a case where tension controls.

13.13 INVESTIGATION OF STRENGTH IN COMPRESSION CONTROLS REGION—RECTANGULAR SECTIONS

When the compression strength P_n exceeds the balanced strength P_b or when the eccentricity e is less than the balanced value e_b, the member acts more as a column than as a beam; this case is known as "compression controls." The tensile force T (see Fig. 13.13.1) will then be based on a strain less than ϵ_y (see Fig. 13.6.2) and may actually be a compressive force if the eccentricity is very small.

The nominal strength P_n for a given eccentricity $e < e_b$ may be obtained by considering the actual strain variation as the unknown and using the principles of statics. This is the most rational approach. As an alternative to the direct solution for the neutral axis as shown in the following examples, Reed [57] has presented an iterative procedure for analysis.

EXAMPLE 13.13.1 Determine the nominal compressive strength P_n for the section of Fig. 13.13.1 for an eccentricity $e = 8$ in. by direct application of statics. Use $f'_c = 3000$ psi, $f_y = 50,000$ psi, and the ACI Code.

Solution: (a) Estimate whether the eccentricity will be in the compression controls or the tension controls region. Since the balanced strain condition was computed in Example 13.12.1 as

$$P_b = 440 \text{ kips}, \qquad e_b = 11.3 \text{ in.}$$

it is known that compression controls for $e < 11.3$ in. For $e = 8$ in., the position of the neutral axis x is not known.

(b) Determine the location of the neutral axis. Since the actual x for $e = 8$ in. should exceed the value of $x_b = 13.72$ in. (see Fig. 13.12.2) and the value of ϵ'_s exceeds ϵ_y at the balanced strain condition, it is certain that $\epsilon'_s > \epsilon_y$. Thus, referring to Fig. 13.12.2 for the forces,

$$C_s = A'_s(f_y - 0.85f'_c) = 2.37(50.0 - 2.55) = 112.5 \text{ kips}$$

$$C_c = 0.85f'_c b(0.85x) = 2.55(15)(0.85x) = 32.5x$$

$$T = A_s f_s = 2.37\frac{(29,000)(0.003)(21.60 - x)}{x} = \frac{4450 - 206x}{x}$$

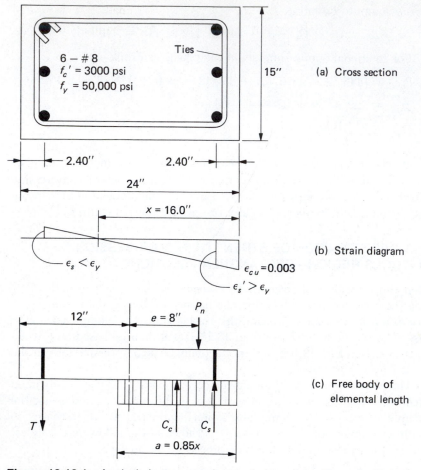

(a) Cross section

6 − #8
$f_c' = 3000$ psi
$f_y = 50,000$ psi

Ties

15″

(b) Strain diagram

2.40″   2.40″

24″

$x = 16.0''$

$\epsilon_s < \epsilon_y$

$\epsilon_{cu} = 0.003$

$\epsilon_s' > \epsilon_y$

(c) Free body of
elemental length

12″ $e = 8''$ P_n

T C_c C_s

$a = 0.85x$

Figure 13.13.1 Analysis in compression controls region—Example 13.13.1.

Taking moments about P_n,

$$0 = 112.5(4.0 - 2.40) - 32.5x\left(\frac{0.85x}{2} - 4\right) + \frac{4450 - 206x}{x}(21.6 - 4.0)$$

$$0 = x^3 - 9.41x^2 + 249.6x - 5673$$

$$x = 16.0 \text{ in.}$$

(c) Compute internal forces and strength P_n.

$$C_s = 112.5 \text{ kips}$$

$$C_c = 32.5(16) = 520 \text{ kips}$$

$$T = \frac{4450 - 206(16)}{16.0} = 72.2 \text{ kips}$$

$$P_n = 520 + 112.5 - 72.4 = 560 \text{ kips}$$

(d) Check by a moment equation about the plastic centroid.

$$560(8) \stackrel{?}{=} 112.5(9.6) + 72.2(9.6) + 520(12 - 6.80)$$
$$4480 \approx 4477$$

OK

EXAMPLE 13.13.2 Repeat the calculation of the nominal strength P_n for the section of Fig. 13.13.1, except consider that 1-#8 is located in each of the 24-in. faces of the member, at the plastic centroid 12 in. from the compression face of the member (that is, at mid-depth) (see Fig. 13.13.2). The eccentricity of P_n is 8 in., $f'_c = 3000$ psi, and $f_y = 50,000$ psi.

Solution: (a) Estimate whether or not the eccentricity will be in the compression or tension controls region of the strength interaction diagram. Since the extra 2-#8 bars are at the center of the section, they are unlikely to have an important effect on e_b. Assume that the analysis will be in the

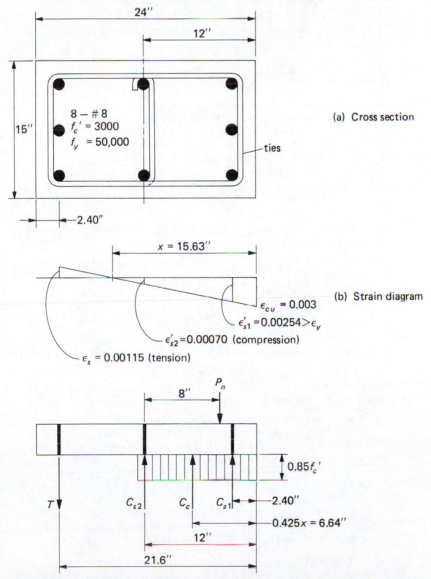

(a) Cross section

8 – # 8
$f'_c = 3000$
$f_y = 50,000$

ties

2.40"

$x = 15.63''$

(b) Strain diagram

$\epsilon_{cu} = 0.003$
$\epsilon'_{s1} = 0.00254 > \epsilon_y$
$\epsilon'_{s2} = 0.00070$ (compression)
$\epsilon_s = 0.00115$ (tension)

P_n

8"

$0.85 f'_c$

T C_{s2} C_c C_{s1} 2.40"

$0.425x = 6.64''$

12"

21.6"

Figure 13.13.2 Analysis of section containing intermediate layer of steel—Example 13.13.2.

compression controls region since e_b for Example 13.13.1 exceeds the 8-in. eccentricity. Note that even if a wrong assumption is made, the neutral axis location thus obtained will reveal $\epsilon'_s < \epsilon_y$ or $\epsilon_s > \epsilon_y$, in which case the revised expressions for C_s or T will have to be used in a new solution.

(b) Estimate for each layer of reinforcement whether it is in compression or tension, and whether the strain at the layer will be greater or less than $\epsilon_y = 0.00172$. The assumptions are as follows:

1. 3-#8 at 2.4 in. from compression face; bars in compression and *assumed* to yield

$$C_{s1} = A'_{s1}(f_y - 0.85f'_c) = 2.37(50 - 2.55) = 112.5 \text{ kips}$$

2. 2-#8 at 12 in. from compression face; bars *assumed* in compression and *assumed* not to yield

$$\epsilon'_{s2} = \frac{0.003}{x}(x - 12.0)$$

$$C_{s2} = A'_{s2}\left[\frac{0.003}{x}(x - 12.0)29{,}000 - 0.85f'_c\right]$$

$$= 1.58\left[\frac{87}{x}(x - 12.0) - 2.55\right]$$

$$= 137.6 - \frac{1649.5}{x} - 4.03$$

3. 3-#8 at 21.6 in. from compression face; bars *assumed* in tension and *assumed* not to yield

$$T = A_s\left[\frac{0.003}{x}(21.6 - x)29{,}000\right]$$

$$= 2.37\left[\frac{87}{x}(21.60 - x)\right] = \frac{4453}{x} - 206.20$$

(c) Determine the compressive force C_c. From Example 13.13.1,

$$C_c = 32.5x$$

(d) Compute the neutral axis distance x. Taking moments about P_n,

$$0 = 112.5(4.0 - 2.40) - 32.5x\left(\frac{0.85x}{2} - 4\right)$$

$$- \frac{133.57x - 1649.5}{x}(12.0 - 4.0)$$

$$+ \frac{4453 - 206.2x}{x}(21.60 - 4.0)$$

$$0 = x^3 - 9.41x^2 + 327.1x - 6629$$

$$x = 15.63 \text{ in.}$$

(e) Verify assumptions (see Fig. 13.13.2).

1. 3-#8 at 2.4 in. from compression face,

$$\epsilon'_{s1} = \frac{0.003}{15.63}(15.63 - 2.40) = 0.00254 > \epsilon_y \qquad \text{(as assumed)}$$

2. 2-#8 at 12 in. from compression face,

$$\epsilon'_{s2} = \frac{0.003}{15.63}(15.63 - 12.0) = 0.00070 < \epsilon_y \qquad \text{(as assumed)}$$

$$f'_{s2} = \epsilon'_{s2}E_s = 0.00070(29{,}000) = 20.3 \text{ ksi}$$

These bars are in compression as assumed; thus, the correction for displaced concrete (i.e., the minus $0.85f'_c$ in the C_{s2} force) was appropriate.

3. 3-#8 at 21.60 in. from compression face,

$$\epsilon_s = \frac{0.003}{15.63}(21.60 - 15.63) = 0.00115 < \epsilon_y \qquad \text{(as assumed)}$$

These bars are in tension as assumed.

(f) Compute internal forces and strength P_n.

$$C_{s1} = 112.5 \text{ kips}$$

$$C_{s2} = A'_{s2}(f'_{s2} - 0.85f'_c) = 1.58(20.3 - 2.55) = 28.0 \text{ kips}$$

$$T = A_s\epsilon_s E_s = 2.37(0.00115)29{,}000 = 79.0 \text{ kips}$$

$$C_c = 32.5x = 32.5(15.63) = 508.0 \text{ kips}$$

$$P_n = 112.5 + 28.0 - 79.0 + 508.0 = 569.5 \text{ kips}$$

(g) Check by a moment equation about the plastic centroid.

$$569.5(8) \stackrel{?}{=} 112.5(9.6) + 28.0(0) + 79.0(9.6) + 508.0(12 - 6.64)$$
$$4556 \approx 4561 \qquad\qquad\qquad\qquad \text{OK}$$

Note that the additional two bars placed at the plastic centroidal axis increase P_n from 560 kips (Example 13.13.1) to 569.5 kips (this example) for the same eccentricity of 8 in.

Whitney Formula—Compression Controls Case. Approximate procedures may sometimes be desirable, particularly as an aid in selecting section sizes, as will be discussed in Secs. 13.15 and 13.16.

One approximate procedure that may be applied to the case when the reinforcement is symmetrically placed in single layers parallel to the axis of bending is the one proposed by Whitney [25]. Taking moments of the forces in Fig. 13.13.3 about the tension steel gives

$$P_n\left(e + \frac{d - d'}{2}\right) = C_c\left(d - \frac{a}{2}\right) + C_s(d - d') \qquad \textbf{(13.13.1)}$$

In estimating the compressive force C_c in the concrete, Whitney used for the depth of the rectangular stress distribution an average value based on

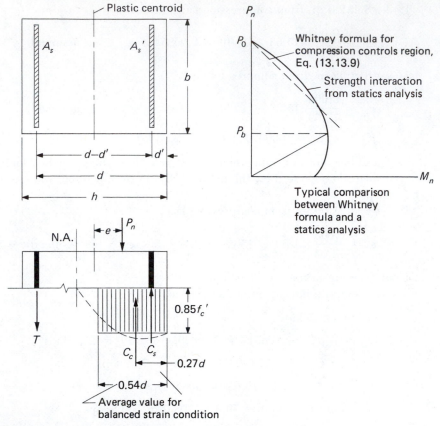

Figure 13.13.3 Whitney formula—compression controls case.

the balanced strain condition, $a = 0.54d$; thus,

$$C_c = 0.85f'_c ba = 0.85f'_c b(0.54d) = 0.459bdf'_c$$

and

$$C_c\left(d - \frac{a}{2}\right) = 0.459bdf'_c\left(d - \frac{0.54d}{2}\right) = \frac{1}{3}f'_c bd^2 \quad \textbf{(13.13.2)}$$

When compression controls, compression steel usually yields when $\epsilon_{cu} = 0.003$ at the extreme fiber in compression. Neglecting displaced concrete,

$$C_s = A'_s f_y \quad \textbf{(13.13.3)}$$

Substituting Eqs. (13.13.2) and (13.13.3) in Eq. (13.13.1) gives

$$P_n = \frac{\frac{1}{3}f'_c bd^2}{e + \frac{1}{2}(d - d')} + \frac{A'_s f_y(d - d')}{e + \frac{1}{2}(d - d')}$$

from which

$$P_n = \frac{f'_c bh}{\dfrac{3he}{d^2} + \dfrac{3(d - d')h}{2d^2}} + \frac{A'_s f_y}{\dfrac{e}{d - d'} + \dfrac{1}{2}} \quad \textbf{(13.13.4)}$$

One of the boundary conditions of this relationship is that it must satisfy the condition

$$P_n = P_0 \qquad \text{at } e = 0 \qquad (13.13.5)$$

in which

$$P_0 = 0.85f'_c bh + 2f_y A'_s \qquad (13.13.6)$$

if the correction for the displaced concrete is not made. Substituting the boundary condition as expressed by Eqs. (13.13.5) and (13.13.6) in Eq. (13.13.4) requires

$$\frac{3(d - d')h}{2d^2} = \frac{1}{0.85} = 1.18 \qquad (13.13.7)$$

Making this necessary substitution, Eq. (13.13.4) becomes

$$P_n = \frac{bhf'_c}{\dfrac{3he}{d^2} + 1.18} + \frac{A'_s f_y}{\dfrac{e}{d - d'} + 0.5} \qquad (13.13.8)$$

which is the Whitney formula for symmetrical steel with no correction for concrete displaced by compression steel.

In using an approximate formula such as Eq. (13.13.8), it is desirable to be on the conservative side. This would be the case for small eccentricities, because the actual depth of the compressive stress block would then be larger than the assumed value of $0.54d$ at the balanced strain condition.

A more useful expression for Eq. (13.13.8) in terms of dimensionless ratios may be obtained by letting $A_g = bh$, $\xi h = d$, $A_s = A'_s$ (for symmetrical reinforcement), $\rho_g = 2A'_s/A_g$, and $\gamma h = d - d'$; thus

$$P_n = A_g \left[\frac{f'_c}{\left(\dfrac{3}{\xi^2}\right)\left(\dfrac{e}{h}\right) + 1.18} + \frac{\rho_g f_y}{\left(\dfrac{2}{\gamma}\right)\left(\dfrac{e}{h}\right) + 1} \right] \qquad (13.13.9)$$

EXAMPLE 13.13.3 Determine the nominal compressive strength P_n for the section of Fig. 13.12.1 for an eccentricity of 8 in. by using the Whitney formula, Eq. (13.13.8). Use $f'_c = 3000$ psi, $f_y = 50,000$ psi, and the ACI Code.

Solution: Since this loading is in the compression controls region, Eq. (13.13.8) may be used to obtain an approximate strength.

$$P_n = \frac{bhf'_c}{3he/d^2 + 1.18} + \frac{A'_s f_y}{e/(d - d') + 0.5}$$

$$= \frac{360(3.0)}{[3(24)(8)]/(21.6)^2 + 1.18} + \frac{2.37(50)}{8/(19.20) + 0.50}$$

$$= \frac{1080}{2.415} + \frac{118.5}{0.916} = 448 + 129 = 577 \text{ kips}$$

It may be noted that the Whitney formula value (577 kips) is not conservative in this case compared to the more exact statics solution (560 kips). In general, Eq. (13.13.8) is conservative for small eccentricity loading and unconservative as the eccentricity approaches the balanced value, as schematically shown in Fig. 13.13.3.

13.14 INVESTIGATION OF STRENGTH IN TENSION CONTROLS REGION—RECTANGULAR SECTIONS

When the compressive strength P_n is less than the balanced value P_b, or the eccentricity e is greater than the balanced value e_b, the member acts more as a beam than as a column; this case is known as "tension controls." The strain in the steel most distant from the neutral axis (near the tension face) will then be greater than the strain $\epsilon_y = f_y/E_s$ (see Fig. 13.6.2).

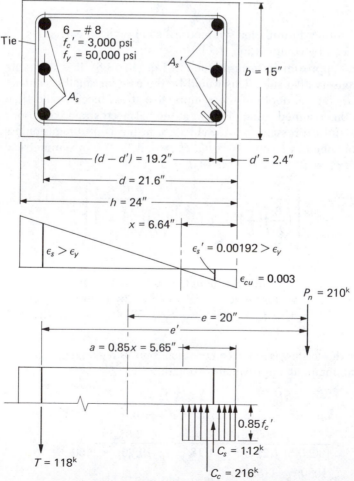

Figure 13.14.1 Analysis in tension controls region—Example 13.14.1.

The most rational approach is to consider the location of the actual neutral axis distance x as the unknown and use the principles of statics.

EXAMPLE 13.14.1 Determine the nominal compressive strength P_n for the member shown in Fig. 13.14.1 for an eccentricity $e = 20$ in., using the method of statics for $f'_c = 3000$ psi, $f_y = 50,000$ psi, and the ACI Code.

Solution: From the result of Example 13.12.1 it is known that $e = 20$ in. exceeds the $e_b = 11.3$ in. for the balanced strain condition; therefore tension controls and the strain ϵ_s on the tension steel exceeds ϵ_y. It is assumed (initially) that the strain ϵ'_s on the compression steel is at least equal to the yield strain ϵ_y although the validity of the assumption must be verified before the solution is accepted. Referring to Fig. 13.14.1, the forces T, C_c, and C_s are

$$T = A_s f_y = 3(0.79)(50.0) = 118 \text{ kips}$$

$$C_c = 0.85 f'_c ab = 0.85(3.0)(0.85)(x)(15) = 32.5x$$

$$C_s = A'_s(f_y - 0.85 f'_c) = 3(0.79)(50 - 2.55) = 112 \text{ kips}$$

Force equilibrium requires

$$P_n = C_c + C_s - T = 32.5x + 112 - 118 = 32.5x - 6.0$$

Taking moments arbitrarily about the tension steel, rotational equilibrium gives

$$P_n \left(e + \frac{d - d'}{2} \right) = C_c \left(d - \frac{a}{2} \right) + C_s (d - d')$$

$$(32.5x - 6.0)(20 + 9.60) = 32.5x (21.6 - 0.425x) + 112(19.20)$$

$$x^2 + 18.82x - 169.2 = 0$$

$$x = 6.64 \text{ in.}$$

Therefore

$$C_c = 32.5(6.64) = 216 \text{ kips}$$

$$P_n = 216 - 6.0 = 210 \text{ kips}$$

Verifying the correctness of the strain condition on the compression steel,

$$\epsilon'_s = \epsilon_c \frac{x - d'}{x} = 0.003 \left(\frac{6.64 - 2.40}{6.64} \right) = 0.00192$$

$$\epsilon_y = \frac{50}{29,000} = 0.00172 < 0.00192 \qquad \text{OK}$$

Therefore compression steel yields as assumed. When compression steel does yield, the solution using statics is the same as would be obtained from Eq. (13.14.7), which is derived below.

The complete strength interaction diagram for the section is given in Fig. 13.14.2.

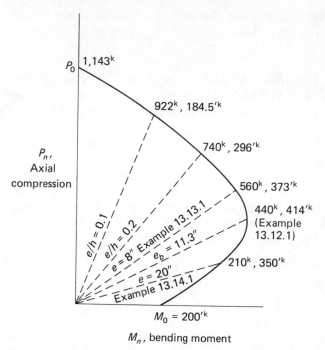

P_0 1,143^k

922^k, 184.5$^{'k}$

P_n,
Axial
compression

740^k, 296$^{'k}$

560^k, 373$^{'k}$

440^k, 414$^{'k}$
(Example
13.12.1)

210^k, 350$^{'k}$

$e/h = 0.1$

$e/h = 0.2$

$e = 8''$ Example 13.13.1

$e_b = 11.3''$

$e = 20''$

Example 13.14.1

$M_0 = 200^{'k}$

M_n, bending moment

Figure 13.14.2   Strength interaction diagram for section of Figure 13.14.1.

Approximate Formulas—Tension Controls Case for Rectangular Sections. Referring to Fig. 13.14.1 and assuming that the strain in the compression steel is larger than the yield strain, one finds that the forces T, C_s, and C_c are

$$T = A_s f_y$$
$$C_s = A_s'(f_y - 0.85f_c')$$
$$C_c = 0.85f_c'(\beta_1 x)b$$

From force equilibrium

$$P_n = 0.85f_c'\beta_1 xb + A_s'(f_y - 0.85f_c') - A_s f_y \qquad (13.14.1)$$

Let $m = f_y/(0.85f_c')$, $\rho = A_s/bd$, and $\rho' = A_s'/bd$; thus

$$P_n = 0.85f_c' [\beta_1 xb + \rho'(m - 1)bd - \rho mbd]$$

$$= 0.85f_c'bd \left[\frac{\beta_1 x}{d} + \rho'(m - 1) - \rho m \right] \qquad (13.14.2)$$

From moment equilibrium with respect to the tension steel,

$$P_n e' = 0.85f_c'(\beta_1 x)b \left(d - \frac{\beta_1 x}{2} \right) + A_s'(f_y - 0.85f_c')(d - d') \qquad (13.14.3)$$

Making the same substitutions as for Eq. (13.14.1) gives

$$P_n e' = 0.85 f'_c bd \left[\beta_1 x - \frac{(\beta_1 x)^2}{2d} + \rho'(m - 1)(d - d') \right] \quad \textbf{(13.14.4)}$$

Substituting Eq. (13.14.2) in Eq. (13.14.4) gives

$$e' \left[\frac{\beta_1 x}{d} + \rho'(m - 1) - \rho m \right] = \left[\beta_1 x - \frac{(\beta_1 x)^2}{2d} + \rho'(m - 1)(d - d') \right]$$

or

$$x^2 + \left(\frac{2\beta_1 e'}{\beta_1^2} - \frac{2\beta_1 d}{\beta_1^2} \right) x$$

$$+ \frac{e'm(\rho' - \rho) - e'\rho' - \rho'(m - 1)(d - d')}{\beta_1^2} 2d = 0$$

Solving the preceding quadratic equation for x gives

$$x = \frac{d - e'}{\beta_1}$$

$$+ \sqrt{ \left(\frac{d - e'}{\beta_1} \right)^2 + \frac{2d[\rho'(m - 1)(d - d') + e'\rho' + e'm(\rho - \rho')]}{\beta_1^2} }$$

or

$$\frac{x}{d} = \frac{1 - e'/d}{\beta_1}$$

$$+ \sqrt{ \left(\frac{1 - e'/d}{\beta_1} \right)^2 + \frac{2[\rho'(m - 1)(1 - d'/d) + \rho'(e'/d) + (e'/d)m(\rho - \rho')]}{\beta_1^2} }$$

$$\textbf{(13.14.5)}$$

Substituting Eq. (13.14.5) in Eq. (13.14.2) gives

$$P_n = 0.85 f'_c bd \left\{ \rho'(m - 1) - \rho m + \left(1 - \frac{e'}{d} \right) \right.$$

$$\left. + \sqrt{ \left(1 - \frac{e'}{d} \right)^2 + 2 \left[\left(\frac{e'}{d} \right)(\rho m - \rho'm + \rho') + \rho'(m - 1) \left(1 - \frac{d'}{d} \right) \right] } \right\}$$

$$\textbf{(13.14.6)}$$

For cases where the tension and compression faces are reinforced the same $\rho' = \rho$, Eq. (13.14.6) reduces to

$$P_n = 0.85 f'_c bd \left\{ -\rho + 1 - \frac{e'}{d} \right.$$

$$\left. + \sqrt{ \left(1 - \frac{e'}{d} \right)^2 + 2\rho \left[(m - 1) \left(1 - \frac{d'}{d} \right) + \frac{e'}{d} \right] } \right\}$$

$$\textbf{(13.14.7)}$$

When no compression reinforcement is present, Eq. (13.14.6) may be simplified by making $\rho' = 0$; thus

$$P_n = 0.85f'_c bd \left[-\rho m + 1 - \frac{e'}{d} + \sqrt{\left(1 - \frac{e'}{d}\right)^2 + \frac{2e'\rho m}{d}} \right] \quad (13.14.8)$$

13.15 DESIGN FOR STRENGTH—REGION I, MINIMUM ECCENTRICITY

The approach to design of compression members in bending in accordance with the strength design method of the ACI Code may be divided into three categories (see Fig. 13.15.1): (a) design in Region I for a member having small or negligible bending moment (i.e., maximum permitted axial strength $P_{n(max)}$ governs); (b) design for Region II in which compression controls but the strength P_n is less than the $0.80P_0$ (tied) or $0.85P_0$ (spirally reinforced) maximum; and (c) design for Region III in which tension controls.

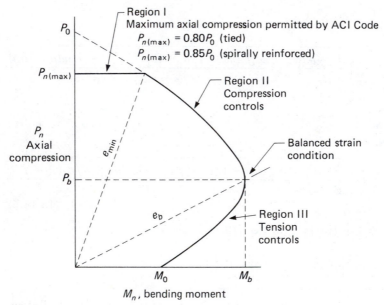

Figure 13.15.1 Design categories for strength of section under combined axial compression and bending moment.

Design in Region I occurs under the following conditions:

1. For members having low slenderness ratio kL_u/r such that the *effects of slenderness may be neglected* according to ACI-10.11.4 (see also Sec. 13.7):
   (a) The member is subject to axial compression where the bending moment is considered negligible and is not computed.
   (b) The bending moment on the member is computed, but the corresponding eccentricity $e = M_u/P_u$ is less than e_{min} (Fig. 13.15.1) corresponding to the maximum axial capacity.

2. For members where the *effects of slenderness must be considered,* computation of bending moment is required, with the result magnified by the factor δ in accordance with ACI-10.11.5 (see Chap. 15).

   (a) When the eccentricity $e = M_u/P_u$ at any end of a member is computed and found to be less than the code-specified $(0.6 + 0.03h)$ in., the value $(0.6 + 0.03h)$ is to be used as the basic value to be magnified by the factor δ. When $\delta(0.6 + 0.03h)$ is less than e_{min} (Fig. 13.15.1), design is in Region I.

   (b) When computed end eccentricity $e = M_u/P_u$ exceeds $(0.6 + 0.03h)$ in., the computed e is magnified to give δe. When δe is less than e_{min} (Fig. 13.15.1), design is in Region I.

The reader may note particularly that the member designed in Region I has its axial compressive force P_u limited by the maximum given by ACI-10.3.5, but need not be explicitly designed to carry the moment based on e_{min} as was required prior to 1977. When slenderness effects can be neglected, there is no requirement for bending strength if factored bending moment was not computed; but if the factored bending moment was computed, it must be provided for in design.

When design is in Region I and slenderness must be considered, the required bending strength for the section is based on the magnified actual eccentricity δe or the magnified minimum $\delta(0.6 + 0.03h)$, whichever is greater. Ordinarily any reasonable arrangement of steel will provide sufficient bending strength when design in Region I is based on just using the maximum axial strength prescribed by ACI-10.3.5.

EXAMPLE 13.15.1 Design an axially loaded spirally reinforced circular column for a gravity dead load of 320 kips and a live load of 360 kips using approximately $3\frac{1}{2}\%$ reinforcement. The column is of average height, and it will be assumed that there is no reduction in strength due to the effects of slenderness. Use $f'_c = 4000$ psi, $f_y = 60,000$ psi, and the ACI Code.

Solution: (a) Determine the required nominal strength P_n.

$$P_u = 1.4(320) + 1.7(360) = 448 + 612 = 1060 \text{ kips}$$

$$\text{required } P_n = \frac{P_u}{\phi} = \frac{1060}{0.75} = 1413 \text{ kips, say } 1410 \text{ kips}$$

(b) Determine the column size. ACI-10.3.5 gives the maximum nominal strength in axial compression as

$$P_{n(max)} = 0.85P_0$$

where P_0 is given by Eqs. (13.5.1) or (13.5.3). Thus

$$P_{n(max)} = 0.85A_g [0.85f'_c + \rho_g(f_y - 0.85f'_c)]$$
$$1410 = 0.85A_g [0.85(4) + 0.035(60 - 3.40)]$$

$$\text{required } A_g = \frac{1410}{4.57} = 308 \text{ sq in.,} \qquad h \text{ (diameter)} = 19.8 \text{ in.}$$

Try $h = 20$ in. with $A_g = 314$ sq in.

(c) Determine reinforcement. Solve the $P_{n(max)}$ equation for ρ_g.

$$1410 = 0.85(314)[3.40 + \rho_g(60 - 3.40)]$$

required $\rho_g = 0.0332$

required $A_{st} = \rho_g A_g = 0.0332(314) = 10.45$ sq in.

Use 20-in. diameter column with 7-#11 bars ($A_s = 10.92$ sq in.).

No further check is required since no moment has been computed and the maximum nominal axial strength given by ACI-10.3.5 governs in this case.

(d) Design the spiral reinforcement. Using Eq. (13.9.4) which is ACI Formula (10-5),

$$\rho_s = 0.45 \left(\frac{A_g}{A_c} - 1\right) \frac{f'_c}{f_{sy}}$$

$$A_g = 314 \quad \text{and} \quad A_c = \frac{(20 - 3)^2 \pi}{4} = 227$$

$$\rho_s = 0.45 \left(\frac{314}{227} - 1\right) \frac{4.0}{60.0} = 0.0115$$

Applying Eq. (13.9.5) gives

$$s_{max} = \frac{a_s \pi (D_c - d_b)}{\rho_s A_c} = \frac{a_s \pi (17 - d_b)}{0.0115(227)}$$

which gives the data in Table 13.15.1.

Limitations (ACI-7.10.4.3):

1. Clear spacing ≤ 3 in.
2. Clear spacing ≥ 1 in.

Use #3 spiral at 2 in. spacing.

Table 13.15.1

BAR	a_s (sq in.)	s_{max} (in.)	MAXIMUM CLEAR SPACING (in.)
#3	0.11	2.20	1.57
#4	0.20	3.97	3.02

13.16 DESIGN FOR STRENGTH—REGION II, COMPRESSION CONTROLS ($e_{min} < e < e_b$)

EXAMPLE 13.16.1 Design a square tied column with about 3% reinforcement for a dead load axial load of 214 kips and bending moment of 47 ft-kips, and a live load axial load of 132 kips and bending moment of 23 ft-kips. Use $f'_c = 3000$ psi, $f_y = 40,000$ psi, and the ACI Code.

Solution: (a) Compute the required nominal strength.

$$\text{required } P_n = \frac{P_u}{\phi} = \frac{1.4(214) + 1.7(132)}{0.70} = 750 \text{ kips}$$

$$\text{required } M_n = \frac{M_u}{\phi} = \frac{1.4(47) + 1.7(23)}{0.70} = 150 \text{ ft-kips}$$

(b) Compute the eccentricity for $P_u/\phi = 750$ kips.

$$e = \frac{M_u}{P_u} = \frac{\text{required } M_n}{\text{required } P_n} = \frac{150(12)}{750} = 2.4 \text{ in.}$$

Although it cannot be certain that $e = 2.4$ in. exceeds e_{min}, it is expected that e_{min} corresponding to $0.80P_0$ (Fig. 13.15.1) will be approximately $0.1h$. Thus for a section smaller than 24 in. the design will be in Region II.

(c) Determine approximate size if required $P_n = 750$ kips equals P_b.

$$x_b = \left(\frac{\epsilon_{cu}}{\epsilon_{cu} + \epsilon_y}\right) d = \left(\frac{0.003}{0.003 + 40/29,000}\right) d = 0.685d$$

$$P_b = C_c + C_s - T$$

$$= 0.85f'_c b\beta_1 x_b + A'_s f_y(\text{approx}) - A_s f_y$$

If the column is made symmetrical, then $A'_s = A_s$; therefore

$$P_b \approx 0.85f'_c b\beta_1 x_b = 0.85(3.0)b(0.85)(0.685d) = 1.483bd$$

For $P_b = \text{required } P_n = 750$ kips,

$$\text{balanced } bd \approx \frac{750}{1.483} = 506 \text{ sq in.}$$

Assuming $d \approx 0.9h$,

$$A_g \text{ (balanced)} = \frac{506}{0.9} = 562 \text{ sq in. (23.7 in. square)}$$

which means (referring to Fig. 13.16.1) that if an area larger than 562 sq in. is used, tension will probably control; and if an area less than 562 sq in. is provided, compression will probably control.

Since it is preferred to use a column smaller than the approximate balanced size, this design will probably be within Region II, in which compression controls. For the purpose of estimating dimensionless ratios, assume a section about 22 in. square. As an alternative to the design approach using the balanced strain condition in combination with the approximate Whitney formula, the reader may prefer the method of Young [58], wherein the moment is approximately converted into equivalent axial compression.

(d) Select size for about 3% reinforcement. Assuming compression con-

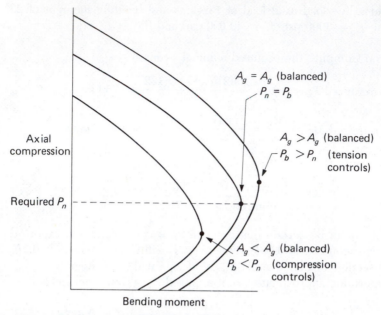

Figure 13.16.1 Variation of strength with gross concrete area, assuming constant percentage of steel reinforcement.

trols for this small eccentricity and using Eq. (13.13.9),

$$P_n = A_g \left[\frac{f_c'}{(3/\xi^2)(e/h) + 1.18} + \frac{\rho_g f_y}{(2/\gamma)(e/h) + 1} \right]$$

$$\left(\frac{e}{h} \right)_{approx} = \frac{2.4}{22} = 0.109, \qquad \text{estimated } \frac{e}{h} \approx 0.12$$

$$(d - d')_{approx} = 19.5 - 2.5 = 17.0, \qquad \gamma \approx \frac{17}{22} \approx 0.77$$

$$\xi = \frac{d}{h} \approx \frac{19.5}{22} = 0.89$$

$$P_n = A_g \left[\frac{3.0}{(3/0.792)(0.12) + 1.18} + \frac{0.03(40)}{(2/0.77)(0.12) + 1} \right]$$

$$750 = A_g(1.845 + 0.915) = 2.76 A_g$$

$$A_g = \frac{750}{2.76} = 272 \text{ sq in.}$$

Try 17 in. square column.

(e) Estimate reinforcement.

$$A_{st} = 0.03(272) = 8.2 \text{ sq in.}$$

Try 10-#9 bars.

(f) Check design. An analysis may be made any one of the three ways: statics, approximate Eq. (13.13.8), or from nondimensional strength interaction diagrams, such as in the *ACI Strength Design Handbook*, Volume 2 [26].

If an approximate formula is to be used as a check, one must be certain that the formula correctly applies. To do this, the balanced condition (P_b and e_b) must be determined for the selected section. In this example, following the procedure of Sec. 13.12, the balanced strain condition gives

$$P_b = 354.7 \text{ kips}, \qquad e_b = 11.03 \text{ in.}$$

Since $P_n > 354.7$ and $e < 11.03$, it is verified that compression controls. The approximate equation (13.13.8) gives for the nominal strength,

$$\begin{aligned}
P_n &= \frac{bhf'_c}{3he/d^2 + 1.18} + \frac{A'_s f_y}{e/(d - d') + 0.5} \\
&= \frac{289(3)}{3(17)(2.4)/(14.56)^2 + 1.18} + \frac{5(40)}{2.4/12.12 + 0.5} \\
&= 493 + 286 = 779 \text{ kips} > 750 \text{ kips required} \qquad \text{OK}
\end{aligned}$$

An analysis using basic statics on the section of Fig. 13.16.2, in the same manner as illustrated in Example 13.13.1, gives $x = 16.10$ in., and $P_n = 809$ kips for $e = 2.4$ in. Another analysis by statics for the section using 8-#9 bars shows $x = 15.90$ in. and $P_n = 755$ kips; thus 8-#9 could have been used. Note that when x is larger than 14.56 in., both layers of steel are in compression, requiring the consideration of displaced concrete in both C_s forces.

Use 17-in. square column and 10-#9 bars, with 5 bars in each face (Fig. 13.16.2).

Note that a better arrangement of bars, especially when the eccentricity is small as in this case, is to distribute them on all four faces, preferably the same arrangement on all four faces to minimize construction errors.

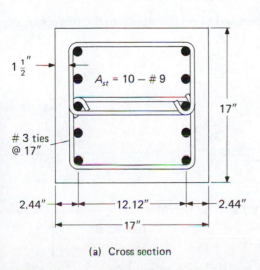

(a) Cross section

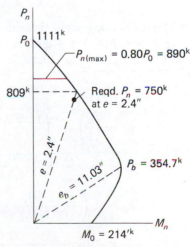

(b) Strength interaction diagram

Figure 13.16.2   Section for Example 13.16.1.

(g) Select lateral ties. Applying the provisions of ACI-7.10.5.2, try #3 ties. Spacing limitations:

1. Least lateral dimension = 17 in.
2. 16 bar diameters = 16(1.128) = 18 in.
3. 48 tie diameters = 48($\frac{3}{8}$) = 18 in.

Use #3 ties, 2 ties per set, at 17 in. spacing. The final cross section and the strength interaction diagram are shown in Fig. 13.16.2.

EXAMPLE 13.16.2 Redesign the column of Example 13.16.1 to be the smallest square tied column permitted by the ACI Code.

Solution: Use Eq. (13.13.9) to estimate size. Since $A_g < A_g$ (balanced), compression controls (see Fig. 13.16.1). In this equation assume that $\xi^2 = (d/h)^2 \approx 0.65$ and $\gamma = (d - d')/h \approx 0.70$, which are reasonable values for $h < 16$ in. For the smallest column, the maximum ρ_g of 0.08 has been assumed.

$$\text{required } P_n = 750 \text{ kips}, \qquad e = 2.4 \text{ in.}$$

$$750 = A_g\left[\frac{3.0}{(3/0.65)(e/h) + 1.18} + \frac{0.08(40)}{(2/0.70)(e/h) + 1}\right]$$

$$750 = A_g\left[\frac{3.0}{4.6(e/h) + 1.18} + \frac{3.2}{2.9(e/h) + 1}\right]$$

Solving by trial:

$$\text{for } \frac{e}{h} = \frac{2.4}{14} = 0.17, \qquad A_g = \frac{750}{3.665} = 204, \qquad h = 14.3 \text{ in.}$$

Try $14\frac{1}{2}$-in. square column, $A_g = 210$ sq in. (Note: This is not a common size for columns and is used here only to illustrate the procedure.)

$$\text{required } A_{st} = 0.08(204) = 16.3 \text{ sq in.}$$

The only arrangement that can be used to get the steel into the two opposite faces is 4-#18. Try 4-#18, $A_s = 16.00$ sq in.

Check (see Fig. 13.16.3):

$$\frac{e}{h} = \frac{2.4}{14.5} = 0.165$$

$$\xi^2 = \left(\frac{d}{h}\right)^2 = \left(\frac{11.50}{14.5}\right)^2 = 0.628$$

$$\gamma = \frac{d - d'}{h} = \frac{11.50 - 3.00}{14.5} = 0.586$$

$$P_n = A_g\left(\frac{f'_c}{(3/\xi^2)(e/h) + 1.18} + \frac{\rho_g A_g}{(2/\gamma)(e/h) + 1}\right)$$

$$= 210\left[\frac{3.0}{(3/0.628)(0.165) + 1.18} + \frac{0.0763(40)}{(2/0.586)(0.165) + 1}\right]$$

$$= 210(1.525 + 1.95) = 210(3.475) = 730 \text{ kips}$$

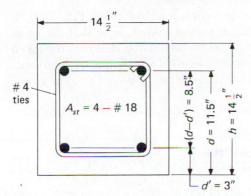

Figure 13.16.3 Trial section for Example 13.16.2.

This is about 2.7% under the required capacity. If this is not acceptable, a more exact statics check could be made in an attempt to verify the adequacy. The practical solution is to increase the column size to 15 in. For ties, #4 size is minimum with #18 bars (ACI-7.10.5),

$$16 \text{ bar diameters} = 16(2.25) = 36 \text{ in.}$$

$$\text{least column dimension} = 15 \text{ in.}$$

$$48 \text{ tie diameters} = 48(\tfrac{1}{2}) = 24 \text{ in.}$$

Use a 15 × 15 in. square column, with 4-#18 bars, symmetrically placed. Provide #4 ties spaced at 15 in.

Practical Design Approach. In design practice, compression member design is rarely done in the detailed manner illustrated throughout this chapter. Instead, use is usually made of design aids giving the strength interaction diagram in nondimensional format. Various design aids are available such as the *ACI Strength Design Handbook*, Volume 2 (containing nondimensional interaction diagrams) [26], the *CRSI Handbook* [27], and for SI, the *Canadian Metric Design Handbook* [28]. One typical interaction chart is given as Fig. 13.16.4. For L-shaped columns, Marin [29] has provided design aids.

EXAMPLE 13.16.3 Design a square tied column containing about 2% reinforcement to carry a dead load axial compression of 710 kN and bending moment of 63 kN·m, and a live load axial compression of 510 kN and bending moment of 33 kN·m. Use $f'_c = 30$ MPa, $f_y = 400$ MPa, and the ACI Code.

Solution: (a) Apply the overload factors and compute the eccentricity.

$$P_u = 1.4(710) + 1.7(510) = 1861 \text{ kN}$$

$$M_u = 1.4(63) + 1.7(33) = 144 \text{ kN·m}$$

$$e = \frac{144(1000)}{1861} = 77.4 \text{ mm}$$

(b) Estimate e/h and use the strength interaction chart, Fig. 13.16.4. The chart has $P_u = \phi P_n = 0.70 P_n$ already included. Note that the chart values of $f'_c = 4000$ psi and $f_y = 60,000$ psi correspond closely to the given metric data, or the *Canadian Metric Design Handbook* [28] can be used. This chart is for $\gamma = 0.75$. Try $e/h = 0.2$ with $\rho_g = 0.02$. From the chart, obtain

$$K = \frac{\phi P_n}{f'_c A_g} = 0.52$$

$$\text{required } A_g = \frac{1861(1000)}{30(0.52)} = 119,300 \text{ mm}^2$$

Try a section 360 mm square ($A_g = 129,600$ mm²).

$$\frac{e}{h} = \frac{77.4}{360} = 0.215$$

$$\gamma = \frac{360 - \approx 2(65)}{360} = 0.64 < 0.75$$

The result so far obtained is slightly on the nonconservative side since it is based on a chart for which the bars at the faces are placed at a relatively farther distance apart than they actually are in this design.

$$\text{actual } K = \frac{P_u}{f'_c bh} = \frac{1861}{30(129.6)} = 0.48$$

$$\text{actual } K\frac{e}{h} = 0.48(0.215) = 0.103$$

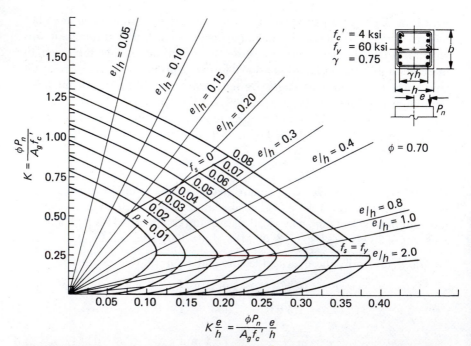

Figure 13.16.4 Typical nondimensional strength interaction diagram (adapted from Ref. 26).

Enter Fig. 13.16.4 with these two coordinates; find $\rho_g = 0.02$.

$$\text{required } A_{st} = 0.02 \, (129{,}600) = 2592 \text{ mm}^2$$

Select 6-#25M bars from Table 1.12.2, $A_{st} = 3000 \text{ mm}^2$.

(c) Check by statics. Based on the e/h ratio and the curves of Fig. 13.16.4, the solution is expected to be in the compression controls region. Referring to Fig. 13.16.5,

$$C_c = 0.85 f_c' b a = 0.85(30)(360)a \tfrac{1}{1000} = 9.18a$$

$$a = \beta_1 x = 0.85x \text{ (see footnote, Sec. 3.3)}$$

$$C_c = 7.80x$$

Assume compression steel yields,

$$C_s = 3(500) \, [400 - 0.85(30)] \, \tfrac{1}{1000} = 562 \text{ kN}$$

Taking $E_s = 200{,}000$ MPa,

$$T = 3(500) \left(\frac{296 - x}{x} \right)(0.003) \, (200{,}000) \tfrac{1}{1000}$$

$$T = \frac{266{,}400 - 900x}{x}$$

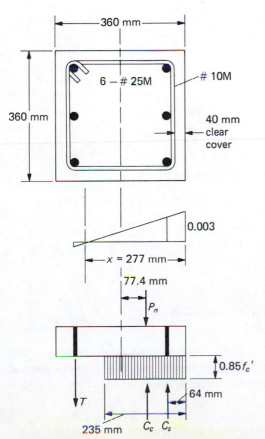

Figure 13.16.5   Section for Example 13.16.3.

Taking moments about P_n gives

$$T(296 - 102.6) + C_s(102.6 - 64) + C_c (102.6 - 0.425x) = 0$$

$$\frac{266,400 - 900x}{x} (193.4) + 562(38.6) + 7.80x(102.6 - 0.425x) = 0$$

$$x^3 - 241x^2 + 45,960x - 15,542,000 = 0$$

$$x = 277 \text{ mm}$$

Check strains:

$$\epsilon_y = \frac{f_y}{E_s} = \frac{400}{200,000} = 0.00200$$

At C_s,

$$\epsilon'_s = 0.003 \left(\frac{277 - 64}{277}\right) = 0.00231 > \epsilon_y \text{ (compression steel yields)}$$

At T,

$$\epsilon_s = 0.003 \left(\frac{296 - 277}{277}\right) = 0.00021 < \epsilon_y$$

This steel is in tension, but does not yield; the strength of the section is in the compression controls region.

Compute forces:

$$C_c = 7.80x = 7.80(277) = 2161$$
$$C_s = 562$$
$$T = 3(500)(0.00021)(200) = -63$$
$$P_n = 2660 \text{ kN}$$
$$\phi P_n = 0.70(2660) = 1862 \text{ kN} \approx P_u = 1861 \text{ kN required} \qquad \text{OK}$$

Check by taking moments about plastic centroid:

$$2660(77.4) = 2161[180 - 0.5(235)] + 562(116) + 63(116)$$
$$205,900 \approx 207,500 \qquad \text{OK}$$

Use 360 mm square column with 6-#25M bars, as shown in Fig. 13.16.5.

13.17 DESIGN FOR STRENGTH—REGION III, TENSION CONTROLS ($e > e_b$)

The "tension controls" region contains the transition in strength from the condition of crushing failure of concrete as a column to that of ductile flexural failure as a beam. It follows that the safety provisions should logically provide for a gradual increase in the strength reduction factor ϕ from that of columns to that of beams.

The ACI Code requires the use of ϕ factors of 0.70 and 0.75 for tied and spirally reinforced columns, respectively, so long as the eccentric compressive load design strength ϕP_n is $0.10 f'_c A_g$ or larger. Then as ϕP_n decreases from $0.10 f'_c A_g$ to zero, the ϕ factor may be linearly increased to 0.90.

In order to be safe, however, it is necessary that the value $\phi P_n = 0.10 f'_c A_g$ is within Region III; in other words, one must be certain that ϕP_n is less than ϕP_b (the value at the balanced strain condition). This will usually be the case when f_y does not exceed 60,000 psi, there is symmetrical reinforcement, and the distance $(h - d' - d_s)$ (see Fig. 13.12.1) between tension and compression reinforcement is not less than $0.70h$. Under these conditions the strength reduction factors ϕ are (ACI-9.3.2) the following:

1. For *tied* compression members:

$$\phi = 0.90 - \frac{2.0\phi P_n}{f'_c A_g} \geq 0.70 \qquad \textbf{(13.17.1)}$$

which gives

$$\phi = \left[\frac{0.90}{1 + \dfrac{2.0 P_n}{f'_c A_g}} \right] \geq 0.70 \qquad \textbf{(13.17.2)}$$

2. For *spirally reinforced* compression members:

$$\phi = 0.90 - \frac{1.5\phi P_n}{f'_c A_g} \geq 0.75 \qquad \textbf{(13.17.3)}$$

which gives

$$\phi = \left[\frac{0.90}{1 + \dfrac{1.5 P_n}{f'_c A_g}} \right] \geq 0.75 \qquad \textbf{(13.17.4)}$$

The relationships expressed by Eqs. (13.17.1) and (13.17.3) are shown in Fig. 13.17.1(a); and those of Eqs. (13.17.2) and (13.17.4), in Fig. 13.17.1(b). The ϕ vs. P_u (or ϕP_n) curves in Fig. 13.17.1(a) are linear, but the ϕ vs. P_n curves in Fig. 13.17.1(b) are slightly nonlinear. The ACI Code rule was made with the intent to provide the *designer* a *practical* transition from the ϕ for a compression member to the ϕ for a beam; there is no theoretical or experimental reason for making ϕ vary linearly with ϕP_n but not with P_n. Equations (13.17.1) and (13.17.3) are useful in preliminary design, because with a trial value of A_g, a trial value of ϕ may be obtained by using the factored force P_u for ϕP_n. Equations (13.17.2) and (13.17.4) are useful for investigating a compression member under bending already designed, because P_n is then available from which ϕ may be computed.

In the relatively few cases where ϕP_b is smaller than $0.10 f'_c A_g$, the increase in the ϕ factor should start from ϕP_b. Using ϕP_b for $0.10 f'_c A_g$ in Eqs. (13.17.1) and (13.17.4) gives

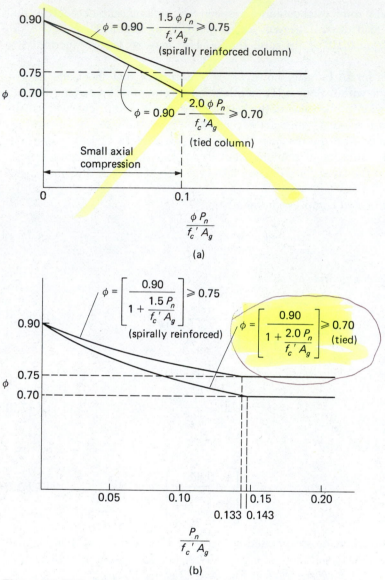

Figure 13.17.1   Variation in ϕ for beam-columns having symmetrical reinforcement, $f_y \leq 60,000$ psi, and $(d - d')/h \geq 0.70$.

1. For *tied* compression members:

$$\phi = 0.90 - \frac{0.20 P_n}{P_b} \geq 0.70 \qquad (13.17.5)$$

$$\phi = \left[\frac{0.90}{1 + \dfrac{0.20 \phi P_n}{P_b}} \right] \geq 0.70 \qquad (13.17.6)$$

2. For *spirally reinforced* members:

$$\phi = 0.90 - \frac{0.15P_n}{P_b} \geq 0.75 \qquad (13.17.7)$$

$$\phi = \left[\frac{0.90}{1 + \dfrac{0.15\phi P_n}{P_b}}\right] \geq 0.75 \qquad (13.17.8)$$

For this situation, Eqs. (13.17.5) and (13.17.7) are useful for investigation of a compression member under bending already designed; and Eqs. (13.17.6) and (13.17.8) are useful in preliminary design because with a trial value of P_b, a trial value of ϕ may be obtained by using the factored force P_u for ϕP_n.

EXAMPLE 13.17.1 Design a rectangular tied column, not over 14 in. wide with about 3% reinforcement, to carry a service dead load of $P = 43$ kips and $M = 96$ ft-kips and a service live load of $P = 32$ kips and $M = 85$ ft-kips. Use symmetrical reinforcement with $f'_c = 4500$ psi, $f_y = 50,000$ psi, and the ACI Code.

Solution: (a) Required nominal strengths and eccentricity. Assuming that $P_u \geq 0.10f'_c A_g$, the value of ϕ is 0.70.

$$\text{required } P_n = \frac{P_u}{\phi} = \frac{1.4(43) + 1.7(32)}{0.70} = \frac{114.6}{0.70} = 164 \text{ kips}$$

$$\text{required } M_n = \frac{M_u}{\phi} = \frac{1.4(96) + 1.7(85)}{0.70} = \frac{278.9}{0.70} = 398 \text{ ft-kips}$$

$$e = \frac{398(12)}{164} = 29.1 \text{ in.}$$

(b) Find the approximate size such that $P_n = P_b$ (see Fig. 13.16.1 and discussion relating to Example 13.16.1).

$$x_b = \frac{0.003d}{0.003 + 0.00172} = 0.635d$$

$$a = \beta_1 x_b = 0.825(0.635)d = 0.524d$$

$$P_b = 0.85(4.5)(0.524)bd + A'_s(f_y - 0.85f'_c) - A_s f_y$$

For $A'_s = A_s$, it is reasonable to assume that the two terms involving steel approximately cancel each other. Therefore

$$P_b \approx 2.00bd$$

$$\text{balanced } bd \approx \frac{164}{2.00} = 82.0 \text{ sq in.} \left(A_g \approx \frac{82.0}{0.8} = 103 \text{ sq in.}\right)$$

It is reasonably certain that an area larger than this must be used; therefore tension would control.

(c) Determine size required for reinforcement ratio ρ_g about 0.03. Using

Eq. (13.14.7) to obtain preliminary size requirement,

$$P_n = 0.85f'_c bd$$

$$\times \left\{ -\rho + 1 - \frac{e'}{d} + \sqrt{\left(1 - \frac{e'}{d}\right)^2 + 2\rho \left[(m-1)\left(1 - \frac{d'}{d}\right) + \frac{e'}{d}\right]} \right\}$$

[13.14.7]

Estimate

$$\rho = 0.015 \text{ (one-half total percentage)}$$

$$\frac{e'}{d} = \frac{d - h/2 + e}{d} \approx 2.0$$

$$\frac{d'}{d} \approx 0.1$$

Also

$$m = \frac{f_y}{0.85f'_c} = \frac{50{,}000}{0.85(4500)} = 13.08$$

Thus

$$P_n = 3.82bd\{-0.015 + 1 - 2 + \sqrt{(-1.0)^2 + 0.03[12.08(0.9) + 2.00]}\}$$

$$= 3.82bd(-1.015 + \sqrt{1.00 + 0.386})$$

$$= 3.82(0.163)bd = 0.623bd$$

$$\text{required } bd = \frac{164}{0.623} = 263 \text{ sq in.}$$

Try 14 × 18 column, $A_g = 252$ sq in. Rechecking several variables,

$$\frac{e'}{d} = \frac{d - h/2 + e}{d} = \frac{15.5 - 9 + 29.1}{15.5} = 2.30$$

$$\frac{d'}{d} = \frac{2.5}{15.5} = 0.16$$

$$P_n = 3.82bd(-1.315 + 1.445)$$

$$\text{required } bd = \frac{164}{0.497} = 330 \text{ sq in.}$$

Try 14 × 20 column, $A_g = 280$ sq in.

$$\frac{e'}{d} = \frac{17.5 - 10.0 + 29.1}{17.5} = 2.09$$

$$\frac{d'}{d} = \frac{2.5}{17.5} = 0.14$$

$$P_n = 3.82bd(-1.105 + 1.252)$$

$$\text{required } bd = \frac{164}{0.562} = 292 \text{ sq in.} \approx 280 \text{ sq in.}$$

The 14 × 20 section will be further investigated, and reinforcement will be determined by statics.

(d) Use approximate statics to determine reinforcement. Assuming that C_s and T are equal in Fig. 13.17.2,

$$P_n = C_c = 0.85 f'_c ab$$

$$164 = 0.85(4.5)a(14)$$

$$a = 3.06 \text{ in.}$$

Equating the moments of the two couples,

$$P_n\left(19.1 + \frac{3.06}{2}\right) = A_s f_y(d - d')$$

$$A_s = \frac{164(20.63)}{50(15)} = 4.51 \text{ sq in.}$$

$$A_{st} = 9.02 \text{ sq in.}$$

$$\rho = \frac{9.02}{14(20)} = 0.0322$$

Try 6-#11 bars, $A_{st} = 9.36$ sq in.

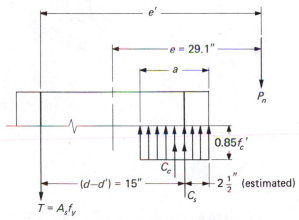

e'

$e = 29.1''$

a

P_n

$0.85 f'_c$

C_c

$(d-d') = 15''$

$2\frac{1}{2}''$ (estimated)

C_s

$T = A_s f_y$

Figure 13.17.2 Statics for Example 13.17.1.

(e) Check by statics (see Fig. 13.17.3). A statics check may be made (1) by determining the neutral axis location to satisfy $\Sigma F_y = 0$ for $P_u = \phi P_n = 114.6$ kips and then comparing the allowable eccentricity of this load with the required eccentricity, or (2) by determining the neutral axis location to satisfy the condition $e = 29.1$ in. and then comparing the design strength ϕP_n at this eccentricity with the factored eccentric load P_u. The first approach is taken here to illustrate an alternative procedure not previously used in this chapter.

For this section, using Eq. (13.17.1) assuming $\phi P_n = P_u$,

$$\phi = 0.90 - \frac{2.0(114.6)}{4.5(280)} = 0.90 - 0.182 = 0.718 > 0.70$$

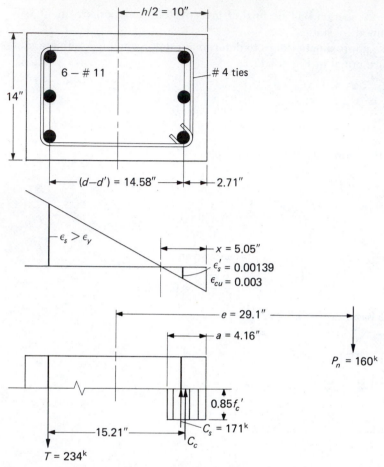

Figure 13.17.3 Section for Example 13.17.1.

Use $\phi = 0.718$.

$$\text{required } P_n = \frac{114.6}{0.718} = 160 \text{ kips}$$

Assuming the neutral axis to be at $x = 4.50$ in. (somewhat larger than $x = a/\beta_1 = 3.06/0.825 = 3.70$ in.) from the extreme compression face,

$$\epsilon_s' \approx 0.003\left(\frac{4.50 - 2.5}{4.50}\right) \approx 0.0012 \qquad \text{say } f_s' = 35 \text{ ksi}$$

$$P_n = 160 \text{ kips}$$

$$T = 4.68(50) = 234 \text{ kips}$$

$$C_s = 4.68[35 - 0.85(4.5)] = 146 \text{ kips}$$

$$C_c = 160 + 234 - 146 = 248 \text{ kips}$$

$$C_c = 0.85f_c'ab$$

$$a = \frac{248}{0.85(4.5)14} = 4.63 \text{ in.}$$

$$\text{revised } x = \frac{a}{0.825} = 5.61 \text{ in.}$$

$$\epsilon'_s = 0.003\left(\frac{5.61 - 2.71}{5.61}\right) = 0.00155, \qquad f'_s = 45 \text{ ksi}$$

At this point it is known that the correct f'_s is between 35 and 45 ksi. The trial and error procedure is to compare a trial f'_s value with the corrected f'_s as follows:

Revise $f'_s = 40$ ksi, $C_s = 169.0$ kips, $C_c = 225$ kips

$a = 4.20$ in., $x = 5.09$ in., $\epsilon'_s = 0.00140$

$f'_s = 40.7$ ksi

Revise $f'_s = 40.4$ ksi, $C_s = 171$ kips, $C_c = 223$ kips

$a = 4.16$ in., $x = 5.05$ in., $\epsilon'_s = 0.00139$

$f'_s = 40.3$ ksi   say OK

Taking summation of moments about the center of the column,

$$160e = 223[10 - 0.5(4.16)] + (234 + 171)(7.29)$$

from which

$$e = 29.5 \text{ in.} > 29.1 \text{ in.} \qquad\qquad \text{OK}$$

Use 14 × 20 column, with 6-#11 bars, 3 bars in each face.

EXAMPLE 13.17.2 Design a rectangular tied column, not over 14 in. wide, to carry a service dead load of $P = 12$ kips and $M = 80$ ft-kips and a service live load of $P = 10$ kips and $M = 85$ ft-kips. Use $f'_c = 4500$ psi, $f_y = 50,000$ psi, and the ACI Code.

Solution: (a) Obtain eccentricity. Because the axial compression is of small magnitude relative to the bending moment, the correct ϕ factor quite likely will be between 0.70 and 0.90. The eccentricity may be obtained from the factored forces P_u and M_u; thus

$$P_u = 1.4(12) + 1.7(10) = 33.8 \text{ kips}$$

$$M_u = 1.4(80) + 1.7(85) = 256 \text{ ft-kips}$$

$$e = \frac{256(12)}{33.8} = 91.0 \text{ in.}$$

With this large eccentricity, the section is clearly near the bottom of the "tension controls" region on the strength interaction diagram.

(b) Approach the design as a compression member. Assuming $\phi \approx 0.80$,

$$\text{required } P_n = \frac{P_u}{\phi} = \frac{33.8}{0.8} = 42.2 \text{ kips}$$

Estimate size required using Eq. (13.14.7) for Region III (tension controls).

$$P_n = 0.85 f'_c bd$$

$$\times \left\{ -\rho + 1 - \frac{e'}{d} + \sqrt{\left(1 - \frac{e'}{d}\right)^2 + 2\rho\left[(m-1)\left(1 - \frac{d'}{d}\right) + \frac{e'}{d}\right]} \right\}$$

Estimate a depth of say 20 in. and $\rho \approx 0.015$ ($\rho_g \approx 0.03$),

$$\frac{e'}{d} = \frac{d - h/2 + e}{d} \approx \frac{17.5 - 10 + 91}{17.5} = 5.6$$

$$\frac{d'}{d} \approx \frac{2.5}{17.5} = 0.143$$

$$m = 13.08 \qquad \text{(see Example 13.17.1)}$$

$$P_n = 0.85(4.5)bd\{-0.015 + 1 - 5.6$$

$$+ \sqrt{(1 - 5.6)^2 + 0.03[(13.08 - 1)(1 - 0.143) + 5.6]}\}$$

$$= 0.134bd$$

$$\text{required } bd = \frac{P_n}{0.134} = \frac{42.2}{0.134} = 314 \text{ sq in.}$$

which would indicate a section about 14×25.

(c) Alternative approach considering the member as a beam. Try about one-half the maximum percentage of reinforcement allowed by ACI Code. From Table 3.6.1, $0.75\rho_b$ is about 0.03. Using Eq. (3.8.4), or Fig. 3.8.1, find

$$R_n \approx 650 \text{ psi}$$

Estimate ϕ again at 0.80.

$$\text{required } M_n = \frac{M_u}{\phi} = \frac{256}{0.80} = 320 \text{ ft-kips}$$

$$\text{required } bd^2 = \frac{\text{required } M_n}{R_n} = \frac{320(12,000)}{650} = 5900 \text{ in.}^3$$

If $b = 14$ in., required $d = 21$ in. and $h = d + 2.5 = 21 + 2.5 = 23.5$, say 24 in. It appears from (b) and (c) that a section about 14×24 will work.

(d) Compute the correct value of ϕ to be used assuming that the 14×24 section is satisfactory.

$$0.10 f'_c A_g = 0.10(4.5)(14)(24) = 151 \text{ kips} > P_u$$

This exceeds ϕP_n, which here is assumed to equal P_u. Assuming one layer of steel in each face,

$$\gamma = \frac{d - d'}{h} \approx \frac{21.5 - 2.5}{24} = 0.79 > 0.70 \qquad\qquad \text{OK}$$

and

$$f_y < 60,000 \text{ psi}$$

Using Eq. (13.17.1) with $\phi P_n = P_u$,

$$\phi = 0.90 - 2.0\left(\frac{P_u}{f_c'A_g}\right)$$

$$= 0.90 - 2.0\left(\frac{33.8}{4.5(14)(24)}\right) = 0.90 - 0.04 = 0.86$$

A statics check may now be made to be certain the nominal strength of the section is at least

$$\text{required } P_n = \frac{P_u}{\phi} = \frac{33.8}{0.86} = 39.3 \text{ kips}$$

$$\text{required } M_n = \frac{M_u}{\phi} = \frac{256}{0.86} = 298 \text{ ft-kips}$$

Selection of bars and the check procedure offer no feature that has not already been treated elsewhere.

13.18 CIRCULAR SECTIONS AS COMPRESSION MEMBERS WITH BENDING

The concepts presented for the calculation of the strength P_n (acting at an eccentricity e from the plastic centroid) for a rectangular section are equally applicable to a circular section. The rectangular stress distribution may be applied to the concrete area under compression according to ACI-10.2.6 and 10.2.7, though the use of statics requires knowing the area and centroid of area for circular segments. This information may be computed from formulas or by the use of coefficients for circular sections, as in Fig. 13.18.1. For a value on the strength interaction diagram in the tension controls zone where the portion of the circular section within the rectangular stress distribution may be small, the relationship between the centroid of a circular segment and its area given by Fig. 13.18.2 may be useful.

When the steel reinforcement is in a circular arrangement, each bar may be located and treated in the same manner as a layer of steel in a rectangular section. The analysis would require first an assumption regarding whether or not the bar (or bars) will yield when $\epsilon_{cu} = 0.003$, as for Example 13.13.2. When about eight or more bars are used the analysis may treat the bars as a steel tube [26, p. 4–8].

Design aids for design of circular columns, both tied and spirally reinforced, are available from the Portland Cement Association [30], American Concrete Institute [26,31], and the Concrete Reinforcing Steel Institute [27]. The strength interaction diagrams of ACI-SP17A (1981) [26] differ from those of ACI-SP17A (1970) [31] in that the more recent edition contains the variable ϕ factor built into the curves whereas the constant ϕ factor was used in the earlier edition. If the actual $P_n - M_n$ interaction is desired in the tension controls region, it is readily obtained from Ref. 31.

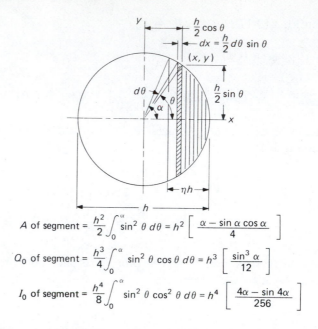

$$A \text{ of segment} = \frac{h^2}{2} \int_0^\alpha \sin^2 \theta \, d\theta = h^2 \left[\frac{\alpha - \sin \alpha \cos \alpha}{4} \right]$$

$$Q_0 \text{ of segment} = \frac{h^3}{4} \int_0^\alpha \sin^2 \theta \cos \theta \, d\theta = h^3 \left[\frac{\sin^3 \alpha}{12} \right]$$

$$I_0 \text{ of segment} = \frac{h^4}{8} \int_0^\alpha \sin^2 \theta \cos^2 \theta \, d\theta = h^4 \left[\frac{4\alpha - \sin 4\alpha}{256} \right]$$

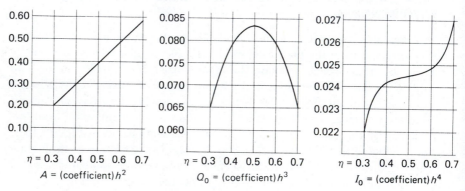

Figure 13.18.1 Properties of circular segments.

13.19 AXIAL TENSION AND BENDING MOMENT

In the relatively uncommon situation of axial tension in combination with bending moment, the strength interaction diagram can be considered to extend on the negative P_n side of the axis, as shown in Fig. 13.19.1. The strength analysis for tension values of P_n is similar to that used for compressive load in the "tension controls" region. When the eccentric load is tensile, the neutral axis distance x will be smaller than it is under pure bending (M_0).

When axial tension and bending moment exist, the usual tendency will be to proportion the section in a manner similar to the design as a beam. The subject of axial tension combined with bending moment has been discussed by Harris [32], Moreadith [33], and Villalta, Carreira, and Erler [34].

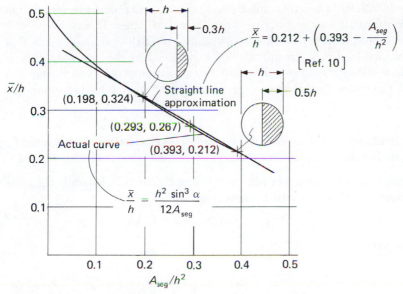

Figure 13.18.2 Centroid of circular segment versus its area.

Since the usual situation will be a section having unsymmetrical reinforcement, the reader is alerted to the fact that the moment used on the interaction diagram must be clearly defined as to whether it is the moment of the tensile force about the centroid of the concrete cross section (this is probably the best choice), or about the "plastic centroid" in compression, or about the "plastic centroid" in tension. Furthermore, the direction of the moment in relation to the unsymmetrical reinforcement must be clearly

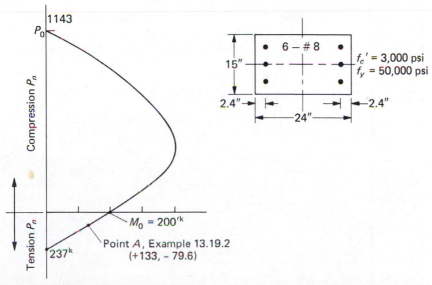

Figure 13.19.1 Strength interaction diagram showing both axial compression and axial tension.

defined; Villalta, Carreira, and Erler [34] have expanded the interaction diagram into all four quadrants of the P_n and M_n axes. In any event, the reader should review a section by using the strain diagram and verify that the external load eccentrically applied is balanced by the internal forces.

The following examples illustrate the calculation of points on the tension side of the interaction diagram (Fig. 13.19.1) for a section having symmetrical reinforcement.

EXAMPLE 13.19.1 Determine the maximum value for the tensile force P_n when no bending moment is acting on the section shown in Fig. 13.19.1.

Solution: Since the concrete will crack before the steel yields, only the steel participates in carrying axial tension. Thus

$$P_n = A_{st}f_y = 6(0.79)50 = 237 \text{ kips}$$

This is plotted on Fig. 13.19.1.

EXAMPLE 13.19.2 Determine the axial tensile strength P_n on the section of Fig. 13.19.1 when the eccentricity is 20 in.

Solution: Referring to Fig. 13.19.2, the eccentric tensile force P_n must be acting on the tension side of the plastic centroid. In this case the neutral axis distance x is smaller than its value for pure bending. For bending alone,

$$C_c = 0.85f_c'ba = 0.85(3)(15)a = 38.3a$$

$$T = A_sf_y = 3(0.79)50 = 118.5 \text{ kips}$$

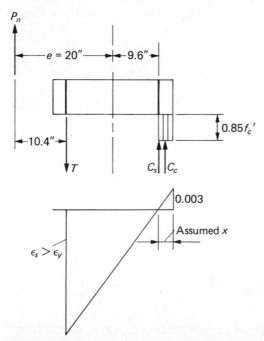

Figure 13.19.2 Free-body diagram and assumed strain diagram for Example 13.19.2.

Assuming the compression steel yields,

$$C_s = 3(0.79)(50 - 2.55) = 112.5 \text{ kips}$$

$$C = T$$

$$a = \frac{118.5 - 112.5}{38.3} = 0.16; \qquad x = \frac{0.16}{0.85} < 2.4 \text{ in.}$$

The above computation shows that compression steel does not yield under bending alone; thus it likely will not yield when P_n is tension at $e = 20$ in.

Thus for axial tension estimate C_s to be in compression but at less than yield stress. Neglecting any displaced concrete effect,

$$C_s = A_s' \epsilon_s' E_s = 3(0.79) \left(\frac{x - 2.4}{x} \right) (0.003)(29{,}000) = \frac{206x - 495}{x}$$

$$C_c = 0.85 f_c' b(0.85x) = 0.85(3)(15)(0.85x) = 32.5x$$

$$T = A_s f_y = 3(0.79)50 = 118.5 \text{ kips}$$

Taking moments about P_n (refer to Fig. 13.19.2) gives

$$T(20 - 12 + 2.4) - C_c(20 + 12 - 0.425x) - C_s(20 + 9.6) = 0$$

$$x^3 - 75.3x^2 - 352x + 1060 = 0$$

$$x = 2.10 \text{ in.}$$

Check assumptions,

$$\epsilon_s' = \left(\frac{2.10 - 2.40}{2.10} \right) 0.003 = -0.000429 < \epsilon_y$$

The steel represented by C_s actually is in tension and does not yield; there is no displaced concrete effect.

$$C_s = 3(0.79)(-0.000429)29{,}000 = -\ 29.4 \text{ kips}$$

$$C_c = 32.5 (2.10) \qquad\qquad = +\ 68.3 \text{ kips}$$

$$T = \qquad\qquad\qquad\quad = \underline{-118.5 \text{ kips}}$$

$$P_n = -\ 79.6 \text{ kips (tension)}$$

$$M_n = 79.6(20)\tfrac{1}{12} = 133 \text{ ft-kips}$$

This value is plotted as point A on Fig. 13.19.1. Check by taking moments about the plastic centroid:

$$133 = [(118.5 - 29.4)(9.6) + 68.3(12 - 0.89)] \tfrac{1}{12}$$

$$133 \approx 134 \qquad\qquad\qquad\qquad\qquad \text{OK}$$

The reader may note that when the axial tension is sufficiently high such that the neutral axis falls beyond the edge of the section, there is no longer a compression face of the member. However, the method illustrated in this example seems sufficient to show how points on the tension–bending moment interaction diagram may be obtained.

Note also that a ϕ factor of 0.90 is to be used for this entire region of axial tension.

13.20 WORKING STRESS METHOD

In the working stress method, or "alternate method" of the ACI Code, the service load moment and axial force are used without applying the overload factors U or the strength reduction factors ϕ.

The allowable service load capacity (ACI-Appendix B.6.1) is taken as 40% of the nominal strength—that is, $0.40M_n$ and $0.40P_n$, subject to possible further reduction due to the effects of slenderness.

Note that limiting the service load capacity to 40% of the nominal strength is equivalent to using a U/ϕ value of 2.5. Even at the higher value of $U = 1.70$ for live load only, combined with the lower value of $\phi = 0.70$ for tied columns, the resulting U/ϕ value is $1.70/0.70 = 2.43$. This shows that design of columns by the ACI "alternate method" will always be more conservative than by the strength method, thus offering no advantage.

13.21 BIAXIAL BENDING AND COMPRESSION

The investigation or design of a square or rectangular section subjected to an axial compression in combination with bending moments about both the x and y axes has received considerable attention [35–53].

One method of analysis is to use the basic principles of equilibrium with the same strength assumptions as were used earlier in this chapter for the case of axial compression and bending about only one axis. This method essentially involves a trial and error process for obtaining the position of an inclined neutral axis; hence any such method is sufficiently complex that no formula may be developed for practical use.

Failure Surfaces. The concept of using failure surfaces has been presented by Bresler [36] and Pannell [38]. The nominal ultimate strength of a section under biaxial bending and compression is a function of three variables, P_n, M_{nx}, and M_{ny}, which may also be expressed in terms of the axial force P_n acting at eccentricities $e_y = M_{nx}/P_n$ and $e_x = M_{ny}/P_n$ with respect to the x and y axes, respectively, as shown in Fig. 13.21.1.

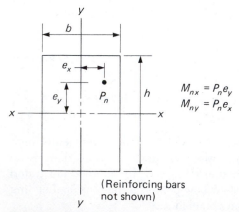

$$M_{nx} = P_n e_y$$
$$M_{ny} = P_n e_x$$

(Reinforcing bars not shown)

Figure 13.21.1 Notation.

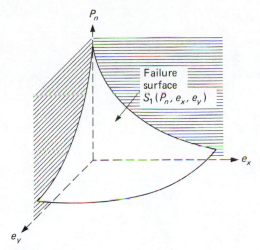

Figure 13.21.2   Failure surfaces S_1 (P_n, e_x, e_y) (from Ref. 36).

Three types of failure surfaces may be defined. In the first type S_1, the variables used along the three orthogonal axes are P_n, e_x, and e_y, as shown in Fig. 13.21.2; in the second type S_2, the variables are $1/P_n$, e_x, and e_y, as shown in Fig. 13.21.3; and in the third type S_3, the variables are P_n, M_{nx}, and M_{ny}, as shown in Fig. 13.21.4. Bresler has developed a very useful analysis procedure [36] using the reciprocal surface S_2. The third type of failure surface S_3 is a three-dimensional extension of the interaction diagram for uniaxial bending and compression as has been used in the earlier part of this chapter. A number of investigators have made approximations to S_3 for use in design and analysis [36,37,39,42,43]. Bresler [36] and Parme, Nieves, and Gouwens [43] have suggested practical approaches to the use of the surface S_3. In the presentation that follows, two analysis methods are presented; the first using the reciprocal $1/P_n - e_x - e_y$ surface S_2 that gives

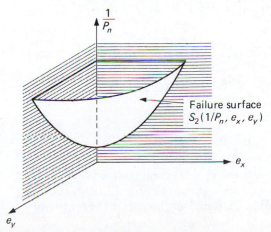

Figure 13.21.3   Reciprocal failure surface S_2 ($1/P_n$, e_x, e_y) (from Ref. 36).

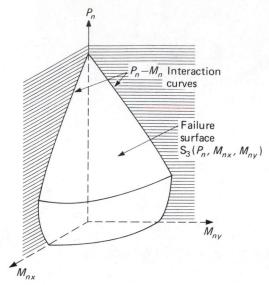

Figure 13.21.4 Failure surface S_3 (P_n, M_{ny}, M_{nx}) (from Ref. 36).

a simple tool for analysis, and the second using the $P_n - M_{ny} - M_{nx}$ surface S_3 which is helpful in design.

Bresler Reciprocal Load Method. Bresler, in an attempt to develop a realistic procedure for analysis, suggested [36] approximating a point $(1/P_{n1}, e_{xA}, e_{yB})$ on the reciprocal failure surface S_2 by a point $(1/P_i, e_{xA}, e_{yB})$ on a *plane S'_2* passing through points A, B, and C (Fig. 13.21.5). Each point on the true surface is approximated by a different plane; that is, the entire failure surface is defined by an infinite number of planes.

The problem then is to determine the strength P_{n1} which exists with biaxial eccentricities e_{xA} and e_{yB} by assuming that P_{n1} equals the value P_i lying on the plane S'_2 specifically established for it. The specific plane is defined by passing it through three points A, B, and C known to lie on the true failure surface S_2,

$$A\left(e_{xA}, \quad 0, \quad \frac{1}{P_y}\right)$$

$$B\left(0, \quad e_{yB}, \quad \frac{1}{P_x}\right)$$

$$C\left(0, \quad 0, \quad \frac{1}{P_0}\right)$$

where P_0 is the nominal strength under axial compression alone without any eccentricity; P_x is the nominal strength at the uniaxial eccentricity e_{yB} $(M_{nx} = P_x e_{yB})$; and P_y is the nominal strength at the uniaxial eccentricity e_{xA} $(M_{ny} = P_y e_{xA})$.

In other words, point A represents a point (P_y, M_{ny}) on the uniaxial P_n–M_n interaction diagram, such as Fig. 13.6.2, for bending about the y

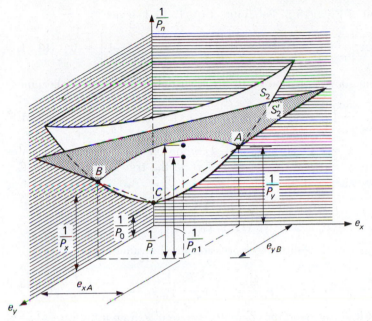

Figure 13.21.5   Graphical representation of the reciprocal load method (from Ref. 36).

axis; point B represents a point (P_x, M_{nx}) on the uniaxial P_n–M_n interaction diagram for bending about the x axis; and point C is a point that is common to both of the uniaxial P_n–M_n interaction diagrams.

The equation of the plane S_2' may be defined in terms of the three points A, B, and C. By letting $x = e_x$, $y = e_y$, and $z = 1/P_n$, the general equation of a plane is

$$A_1 x + A_2 y + A_3 z + A_4 = 0 \tag{13.21.1}$$

Substitution of the coordinates of points A, B, and C (see Fig. 13.21.5) into Eq. (13.21.1) gives

$$A_1 e_{xA} + \quad 0 \quad + A_3 \frac{1}{P_y} + A_4 = 0 \tag{13.21.2a}$$

$$0 \quad + A_2 e_{yB} + A_3 \frac{1}{P_x} + A_4 = 0 \tag{13.21.2b}$$

$$0 \quad + \quad 0 \quad + A_3 \frac{1}{P_0} + A_4 = 0 \tag{13.21.2c}$$

Solving Eqs. (13.21.2abc) for A_1, A_2, and A_3 in terms of A_4,

$$A_1 = \frac{1}{e_{xA}} \left(\frac{P_0}{P_y} - 1 \right) A_4 \tag{13.21.3a}$$

$$A_2 = \frac{1}{e_{yB}} \left(\frac{P_0}{P_x} - 1 \right) A_4 \tag{13.21.3b}$$

$$A_3 = -P_0 A_4 \tag{13.21.3c}$$

Substitution of Eqs. (13.21.3abc) into Eq. (13.21.1) gives

$$A_4\left[\frac{x}{e_{xA}}\left(\frac{P_0}{P_y} - 1\right) + \frac{y}{e_{yB}}\left(\frac{P_0}{P_x} - 1\right) - P_0z + 1\right] = 0 \quad (13.21.4)$$

Dividing the above equation by P_0, the equation of the plane S_2' becomes

$$\frac{x}{e_{xA}}\left(\frac{1}{P_y} - \frac{1}{P_0}\right) + \frac{y}{e_{yB}}\left(\frac{1}{P_x} - \frac{1}{P_0}\right) - z + \frac{1}{P_0} = 0 \quad (13.21.5)$$

At the point $(x = e_{xA}, y = e_{yB}, z = 1/P_i)$ on the plane that approximates the point $(x = e_{xA}, y = e_{yB}, z = 1/P_{n1})$ on the true failure surface, Eq. (13.21.5) becomes

$$\left(\frac{1}{P_y} - \frac{1}{P_0}\right) + \left(\frac{1}{P_x} - \frac{1}{P_0}\right) - \frac{1}{P_i} + \frac{1}{P_0} = 0$$

which reduces to the following expression for the value of P_i:

$$\frac{1}{P_i} = \frac{1}{P_x} + \frac{1}{P_y} - \frac{1}{P_0} \quad (13.21.6)$$

Bresler [36] has found the computed values of P_i from using Eq. (13.21.6) to be "in excellent agreement with test results, the maximum deviation being 9.4%, and the average deviation being 3.3%." Ramamurthy [42] also reported test results and concluded that Eq. (13.21.6) "can be used to predict ultimate loads with reasonable accuracy." Pannell [38] has presented additional test results in his discussion of Ref. 36 and indicated that Eq. (13.21.6) may be inappropriate when small values of axial load are involved, such as when P_n/P_0 is in the range of 0.06 or less. For such cases the member should be designed for flexure only.

EXAMPLE 13.21.1 Determine the adequacy of a 16-in. square tied column section containing 8-#10 bars as shown in Fig. 13.21.6(a). The section is to carry factored loads, $P_u = 144$ kips, $M_{ux} = 120$ ft-kips, and $M_{uy} = 54$ ft-kips. The tied column has $f_c' = 3000$ psi and $f_y = 40,000$ psi.

Solution: The eccentricities are

$$e_y = \frac{M_{ux}}{P_u} = \frac{120(12)}{144} = 10.0 \text{ in.}$$

$$e_x = \frac{M_{uy}}{P_u} = \frac{54(12)}{144} = 4.5 \text{ in.}$$

In order to determine the values of P_x and P_y, the P_n–M_{nx} and P_n–M_{ny} interaction diagrams in uniaxial bending are needed for bending about the x and y axes, respectively. Because of the symmetry in this case, a single P_n–M_{nx} or M_{ny} diagram valid for both bending axes is all that is required, as shown in Fig. 13.21.6(b). The interaction information may be obtained by using equilibrium, by the use of approximate formulas such as developed in Secs. 13.13 and 13.14, or by means of nondimensionalized P_n–M_n diagrams such as provided in *Strength Design Handbook*, Vol. 2, *Columns* [26].

From the $P_n - e_x$ or e_y interaction diagram, Fig. 13.21.6(c), determine P_x and P_y for uniaxial bending,

$$e_x = 4.5 \text{ in.}, \quad \text{find } P_y = 480 \text{ kips}$$
$$e_y = 10.0 \text{ in.}, \quad \text{find } P_x = 260 \text{ kips}$$

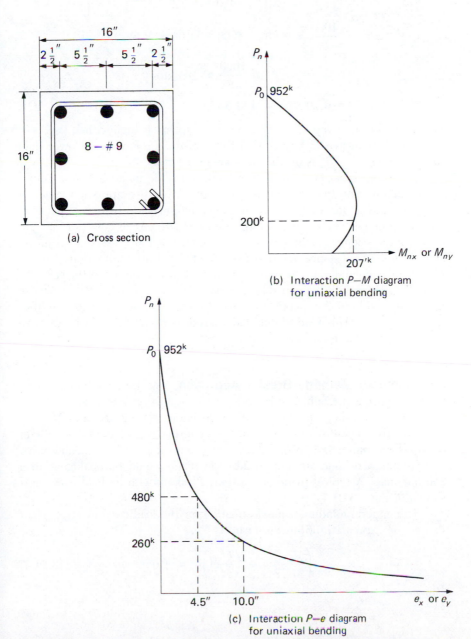

(a) Cross section

(b) Interaction $P-M$ diagram for uniaxial bending

(c) Interaction $P-e$ diagram for uniaxial bending

Figure 13.21.6   Section and interaction diagrams for Example 13.21.1.

Then, using Eq. (13.21.6),

$$\frac{1}{P_n} \approx \frac{1}{P_i} = \frac{1}{P_x} + \frac{1}{P_y} - \frac{1}{P_0}$$

Multiplying by 1000 for convenience,

$$\frac{1000}{P_i} = \frac{1000}{260} + \frac{1000}{480} - \frac{1000}{952}$$

$$\frac{1000}{P_i} = 3.85 + 2.08 - 1.05 = 4.88$$

$$P_n \approx P_i = \frac{1000}{4.88} = 205 \text{ kips}$$

$$[\phi P_n = 0.70(205) = 143.5 \text{ kips}] \approx [P_u = 140 \text{ kips}] \qquad \text{OK}$$

One may note that the $1/P_i$ value on the S_2' plane is higher than the $1/P_n$ value on the concave surface so that the approximate P_i value is lower than the P_n value; hence P_i is on the safe (low) side.

The more exact analysis of this section may be made by using statics on the assumption that there is a uniformly stressed compression zone, in the same way as for uniaxial bending and compression. In biaxial bending and compression, the uniform compressive stress $0.85 f_c'$ is considered to act on a compression zone bounded by the edges of the cross section and a straight line at a distance $a = \beta_1 x$ from the fiber of maximum strain (corner of section) and parallel to the neutral axis. Such an analysis gives the strength $P_n = 216.7$ kips with the neutral axis inclined at an angle of $29.23°$ clockwise with the x axis and the distance a equal to 9.65 in. from the extreme fiber in compression. This kind of detailed analysis is usually practical only with the aid of a computer.

Load Contour Method—Bresler Approach. The load contour method involves cutting the failure surface S_3 (Fig. 13.21.4) at a constant value of P_n to give a so-called "load contour" interaction relating M_{nx} and M_{ny}. In other words, the entire surface S_3 may be considered to include a family of curves (load contours) corresponding to constant values of P_n, which if drawn superimposed on one another in a single plane would be analogous to a contour map. A typical plane at constant P_n along with its load contour is shown in Fig. 13.21.7.

The general nondimensional equation for the load contour at constant P_n may be expressed [36] in the form

$$\left(\frac{M_{nx}}{M_{0x}}\right)^{\alpha_1} + \left(\frac{M_{ny}}{M_{0y}}\right)^{\alpha_2} = 1.0 \qquad (13.21.7)$$

where

$$M_{nx} = P_n e_y; \; M_{ny} = P_n e_x$$

$M_{0x} = M_{nx}$ capacity at axial load P_n when M_{ny} (or e_x) is zero

$M_{0y} = M_{ny}$ capacity at axial load P_n when M_{nx} (or e_y) is zero

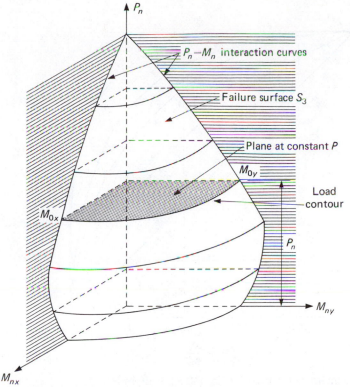

Figure 13.21.7 Load contours for constant P_n on failure surface S_3 (from Ref. 36).

and α_1 and α_2 are exponents that depend on the dimensions of the cross section, the reinforcement amount and location, concrete strength, steel yield stress, and amount of concrete cover.

Bresler [36] suggests that it is acceptable to take $\alpha_1 = \alpha_2 = \alpha$; then

$$\left(\frac{M_{nx}}{M_{0x}}\right)^\alpha + \left(\frac{M_{ny}}{M_{0y}}\right)^\alpha = 1 \qquad (13.21.8)$$

which is shown graphically in Fig. 13.21.8.

In using Eq. (13.21.8) or Fig. 13.21.8, however, it is still necessary to have a value of α that is applicable to the particular column under investigation. Bresler [36] reported the calculated values of α to vary from 1.15 to 1.55.

For practical purposes, it seems satisfactory to take α as 1.5 for rectangular sections and between 1.5 and 2.0 for square sections.

Load Contour Method—Parme Approach.* The approach described herein has been developed by Parme et al. [43], as an extension of the Bresler load contour method. The Bresler interaction equation (13.21.8) is assumed to be the basic strength criterion to define the typical load contour represent-

*This approach may also be referred to as the Portland Cement Association (PCA) method since Ref. 43 is also available as PCA Advanced Engineering Bulletin No. 18.

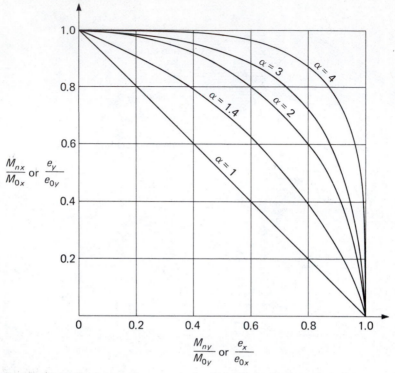

Figure 13.21.8 Interaction curves for Eq. (13.21.8) (from Ref. 36).

ing the intersection of the failure surface S_3 (Fig. 13.21.7) with a horizontal plane at a height P_n. Such a typical load contour is shown in Fig. 13.21.9. A change in the orientation of the M_{nx} and M_{ny} axes has been made in Fig. 13.21.9 to suit the two-dimensional representation.

In the Parme approach, a point B on the load contour is defined such that the biaxial moment strengths M_{nx} and M_{ny} at this point are in the same ratio as the uniaxial moment strengths M_{0x} and M_{0y}; thus at point B

$$\frac{M_{ny}}{M_{nx}} = \frac{M_{0y}}{M_{0x}} \qquad (13.21.9)$$

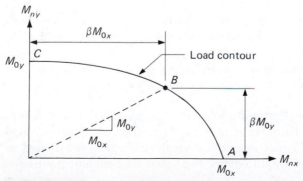

Figure 13.21.9 Load contour at plane of constant P_n cut through failure surface S_3 (Fig. 13.21.7).

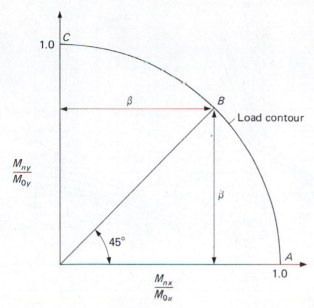

Figure 13.21.10 Nondimensional load contour at constant P_n.

or

$$M_{nx} = \beta M_{0x}; \qquad M_{ny} = \beta M_{0y} \qquad (13.21.10)$$

When the load contour of Fig. 13.21.9 is adjusted to take the nondimensional form of Fig. 13.21.10, the point B will have the ratio β defined by Eq. (13.21.10) as its x and y coordinates. In the physical sense, the ratio β is that constant portion of the uniaxial moment strengths which may be permitted to act simultaneously on the column section. The actual value of β depends on the ratio of P_n to P_0, as well as the material and cross-sectional properties; however, the usual range is between 0.55 and 0.70 [43]. An average value of $\beta = 0.65$ is suggested for design. More accurate values of β have been computed using basic principles of equilibrium and charts for β values have been presented in Ref. 43. These β-value charts appear as Fig. 13.21.11.

Once an empirical value of β has been ascertained for a given cross section and loading, the complete nondimensional load contour is defined if Eq. (13.21.8) is accepted as the correct relationship. The relationship between the α of Eq. (13.21.8) and β is obtained by using the coordinates of point B, which is known to lie on the contour. Thus substituting the coordinates of B into Eq. (13.21.8) gives

$$\left(\frac{\beta M_{0x}}{M_{0x}}\right)^{\alpha} + \left(\frac{\beta M_{0y}}{M_{0y}}\right)^{\alpha} = 1$$

$$\beta^{\alpha} = \tfrac{1}{2}$$

$$\alpha \log \beta = \log 0.5$$

$$\alpha = \frac{\log 0.5}{\log \beta} \qquad (13.21.11)$$

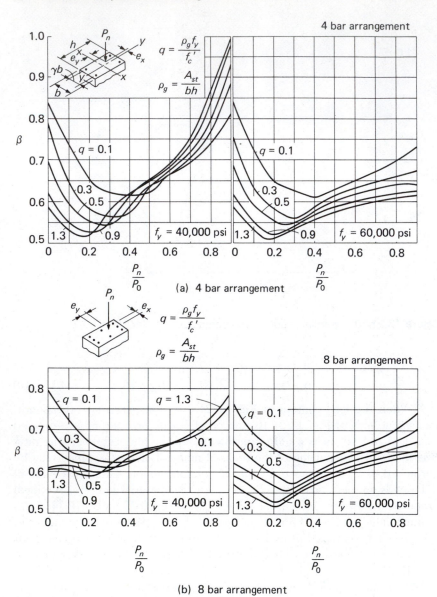

(a) 4 bar arrangement

(b) 8 bar arrangement

Figure 13.21.11 Biaxial bending design constants β (from Ref. 43). For $0.6 \leq \gamma \leq 1.0$; $3000 \leq f'_c \leq 6000$; and $10 \leq h/b \leq 4.0$. (Figure is continued on right page.)

Thus Eq. (13.21.8) may be written

$$\left(\frac{M_{nx}}{M_{0x}}\right)^{\log 0.5/\log \beta} + \left(\frac{M_{ny}}{M_{0y}}\right)^{\log 0.5/\log \beta} = 1 \qquad (13.21.12)$$

Plots of Eq. (13.21.12) for different values of β are shown in Fig. 13.21.12.

Gouwens [50] has presented equations that may be used in place of the curves of Fig. 13.21.12.

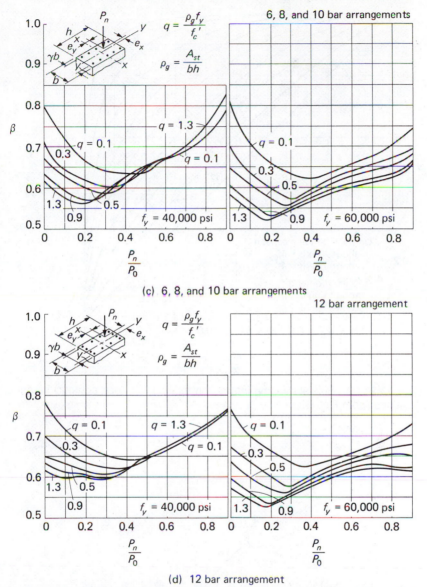

(c) 6, 8, and 10 bar arrangements

(d) 12 bar arrangement

Figure 13.21.11 (continued) Biaxial bending design constants β (from Ref. 43). For $0.6 \leq \gamma \leq 1.0$; $3000 \leq f'_c \leq 6000$; and $10 \leq h/b \leq 4.0$.

For design purposes, the nondimensionalized load contour of Fig. 13.21.10 may be approximated by two straight lines AB and BC as shown in Fig. 13.21.13. When M_{ny}/M_{0y} exceeds M_{nx}/M_{0x}, the straightline approximation equation for BC is

$$\frac{M_{ny}}{M_{0y}} + \frac{M_{nx}}{M_{0x}}\left(\frac{1-\beta}{\beta}\right) = 1 \qquad (13.21.13)$$

When M_{ny}/M_{0y} is less than M_{nx}/M_{0x}, the straightline approximation equa-

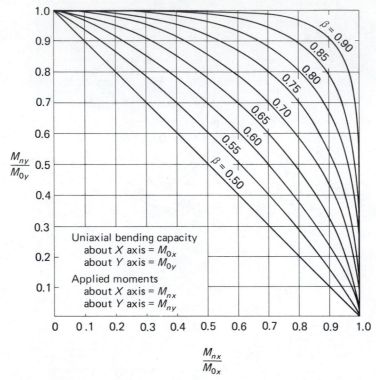

Figure 13.21.12 Biaxial bending interaction relationship (load contour) in terms of β values (from Ref. 43).

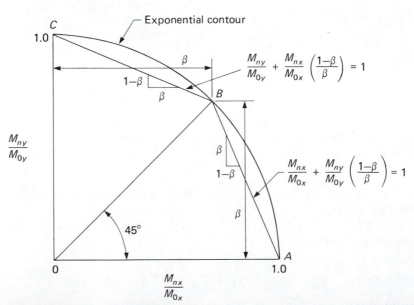

Figure 13.21.13 Straightline approximation of load contour for design (from Ref. 43).

tion for AB is

$$\frac{M_{nx}}{M_{0x}} + \frac{M_{ny}}{M_{0y}}\left(\frac{1-\beta}{\beta}\right) = 1 \qquad (13.21.14)$$

For design purposes, Eqs. (13.21.13) and (13.21.14) may be written

$$M_{ny} + M_{nx}\left(\frac{M_{0y}}{M_{0x}}\right)\left(\frac{1-\beta}{\beta}\right) = M_{0y}; \qquad \left[\text{for } \frac{M_{ny}}{M_{nx}} \geq \frac{M_{0y}}{M_{0x}}\right] \quad (13.21.15)$$

$$M_{nx} + M_{ny}\left(\frac{M_{0x}}{M_{0y}}\right)\left(\frac{1-\beta}{\beta}\right) = M_{0x}; \qquad \left[\text{for } \frac{M_{ny}}{M_{nx}} \leq \frac{M_{0y}}{M_{0x}}\right] \quad (13.21.16)$$

Thus Eqs. (13.21.15) and (13.21.16) represent the alternate algebraic expressions to the exponential equation of Eq. (13.21.12) or Fig. 13.21.12.

When rectangular sections are used with reinforcement distributed uniformly along all faces, the ratio of M_{0y} to M_{0x} (i.e., M_{ny}/M_{nx} of Fig. 13.21.1) will be approximately equal to that of b to h; thus

$$\frac{M_{0y}}{M_{0x}} \approx \frac{b}{h}$$

which gives for Eqs. (13.21.15) and (13.21.16), respectively,

$$M_{ny} + M_{nx}\left(\frac{b}{h}\right)\left(\frac{1-\beta}{\beta}\right) \approx M_{0y} \qquad \left[\text{for } \frac{M_{ny}}{M_{nx}} \geq \frac{b}{h}\right] \quad (13.21.17)$$

$$M_{nx} + M_{ny}\left(\frac{h}{b}\right)\left(\frac{1-\beta}{\beta}\right) \approx M_{0x} \qquad \left[\text{for } \frac{M_{ny}}{M_{nx}} \leq \frac{b}{h}\right] \quad (13.21.18)$$

EXAMPLE 13.21.2 Investigate the section of Example 13.21.1 (Fig. 13.21.6) using the Parme load contour method.

Solution: (a) Use Eq. (13.21.12) or Fig. 13.21.12. In order to make a check by this method, the correct β value must be determined.

$$\frac{\text{required } P_n}{P_0} = \frac{144/0.70}{952} = 0.216 \qquad [P_0 \text{ from Fig. 13.21.6(b)}]$$

$$q = \rho_g \frac{f_y}{f_c'} = \frac{8(1.0)}{16(16)}\left(\frac{40}{3}\right) = 0.417$$

Find $\beta = 0.61$ from Fig. 13.21.11.

$$\frac{\text{required } M_{nx}}{M_{0x}} = \frac{120/0.70}{207} = 0.828$$

where M_{0x} is from Fig. 13.21.6(b).

Using $\beta = 0.61$ and $M_{nx}/M_{0x} = 0.828$, find from Fig. 13.21.12,

$$\frac{M_{ny}}{M_{0y}} = 0.37$$

$$M_{ny} = 0.37(207) = 77 \text{ ft-kips}$$

$$[\phi M_{ny} = 0.70(77) = 54 \text{ ft-kips}] \approx [M_{uy} = 54 \text{ ft-kips}] \qquad \text{OK}$$

(b) Use Eq. (13.21.14). The required biaxial moment strengths M_{nx} and M_{ny} give

$$\frac{\text{required } M_{nx}}{M_{0x}} = \frac{120/0.70}{207} = 0.828;$$

$$\frac{\text{required } M_{ny}}{M_{0y}} = \frac{54/0.70}{207} = 0.373$$

Substituting $\beta = 0.61$ from part (a) in Eq. (13.21.14),

$$\frac{M_{nx}}{M_{0x}} + \frac{M_{ny}}{M_{0y}}\left(\frac{1-\beta}{\beta}\right) \le 1$$

$$0.828 + 0.337 \left(\frac{1 - 0.61}{0.61}\right) = 0.828 + 0.238 = 1.066$$

which would indicate an overstress. As is expected, the straightline relationship is conservative.

Design Procedure. The design procedure may be summarized as follows:

1. Estimate the value of β at 0.65, or use Fig. 13.21.11 to make an estimate.
2. If M_{ny} is larger than M_{nx}, compute the approximate equivalent uniaxial bending moment M_{0y} using Eq. (13.21.17); if M_{nx} is larger than M_{ny}, compute the approximate equivalent uniaxial bending moment M_{0x} using Eq. (13.21.18).
3. Design the section using the methods treated in Secs. 13.15 through 13.17.
4. Check by using any one of the three approaches of this section: (a) the Bresler reciprocal load method, Eq. (13.21.6); (b) the Bresler load contour method, Eq. (13.21.8); (c) the Parme load contour method, Eq. (13.21.12), which is the same as Fig. 13.21.12; or Eqs. (13.21.13) and (13.21.14) which are straightline approximations to the load contour.

Several additional comments in regard to the design of compression members in biaxial bending may be made as follows:

1. Whenever possible, columns subjected to biaxial bending should be circular in cross section.
2. If rectangular or square columns are necessary, the reinforcement should be uniformly spread around the perimeter.
3. Circular or square columns can reasonably be designed to satisfy a given P_u by using the equation [36],

$$\left(\frac{M_{nx}}{M_{0x}}\right)^{1.75} + \left(\frac{M_{ny}}{M_{0y}}\right)^{1.75} = 1.0 \tag{13.21.19}$$

4. Rectangular columns can be approximately designed to satisfy for a given P_u by using the equation [36],

$$\left(\frac{M_{nx}}{M_{0x}}\right)^{1.5} + \left(\frac{M_{ny}}{M_{0y}}\right)^{1.5} = 1.0 \tag{13.21.20}$$

EXAMPLE 13.21.3 Select a rectangular cross section for a compression member subjected to biaxial bending, to take the following service loads: dead load, $P = 80$ kips, $M_x = 40$ ft-kips, and $M_y = 20$ ft-kips; live load, $P = 32$ kips, $M_x = 48$ ft-kips, and $M_y = 16$ ft-kips. Use $f_c' = 4000$ psi and $f_y = 60,000$ psi.

Solution: (a) Determine factored axial force and moments.

$$P_u = 1.4(80) + 1.7(32) = 166 \text{ kips}$$

$$M_{ux} = 1.4(40) + 1.7(48) = 138 \text{ ft-kips}$$

for bending about x axis; and

$$M_{uy} = 1.4(20) + 1.7(16) = 55 \text{ ft-kips}$$

for bending about y axis.

(b) Obtain an approximate equivalent uniaxial factored moment. Estimating β at 0.65 to begin the design, and using the approximate strength relationship, Eq. (13.21.18),

$$\text{equivalent } M_{0x} \approx M_{nx} + M_{ny}\left(\frac{h}{b}\right)\left(\frac{1-\beta}{\beta}\right) \qquad [\textbf{13.21.18}]$$

Using factored moments,

$$\text{equivalent } \phi M_{ox} = M_{ux} + M_{uy}\left(\frac{h}{b}\right)\left(\frac{1-\beta}{\beta}\right)$$

Assume h/b proportional to $M_{ux}/M_{uy} = 138/55 \approx 2.5$; the requirement is

$$\text{equivalent } \phi M_{ox} = 138 + 55(2.5)\left(\frac{1 - 0.65}{0.65}\right) = 212 \text{ ft-kips}$$

(c) Determine equivalent eccentricity for uniaxial bending and compression.

$$P_u = 166 \text{ kips}$$

$$M_u = \text{required equivalent } \phi M_{ox} = 212 \text{ ft-kips}$$

$$\text{equivalent } e_y = \frac{212(12)}{166} = 15.3 \text{ in.}$$

(d) Determine approximate requirements for a section that would have P_u/ϕ equal to the strength P_b at the balanced strain condition (see Sec. 13.16, Example 13.16.1),

$$x_b = \left(\frac{\epsilon_{cu}}{\epsilon_{cu} + \epsilon_y}\right)d = \left[\frac{0.003}{0.003 + (60/29,000)}\right]d = 0.592d$$

If symmetrical reinforcement is used, $\rho = \rho'$,

$$P_b \approx C_c = 0.85 f_c' b \beta_1 x_b$$

$$= 0.85(4)b(0.85)(0.592d) = 1.71bd$$

For $P_b = P_u/\phi = 166/0.70 = 237$ kips,

$$\text{balanced } bd \approx \frac{237}{1.71} = 139 \text{ sq in.}$$

and assuming $d \approx 0.9h$,

$$\text{balanced } A_g \approx \frac{139}{0.9} = 154 \text{ sq in.}$$

If $h/b = 2.5$,

$$\text{balanced } b = \sqrt{\frac{154}{2.5}} = 7.8$$

which gives a section about 8×20. Since an 8-in. section is narrower than generally desired, the section to be chosen will probably be larger than the balanced cross section. Thus, according to Fig. 13.16.1, the section will probably be in Region III where tension controls.

(e) Determine required size assuming tension controls. Use Eq. (13.14.7), and assume $\rho_g \approx 0.04$,

$$P_n = 0.85f_c'bd\left\{- \rho + 1 - \frac{e'}{d}\right.$$
$$\left. + \sqrt{\left(1 - \frac{e'}{d}\right)^2 + 2\rho\left[(m - 1)\left(1 - \frac{d'}{d}\right) + \frac{e'}{d}\right]}\right\}$$

Estimate

$$\rho = 0.02 \text{ (one-half total in each face)}$$

$$\frac{e'}{d} = \frac{d - h/2 + e}{d} \approx 1.3 \quad \text{(assuming } h \text{ as 18 to 22 in.)}$$

$$\frac{d'}{d} \approx 0.14$$

$$m = \frac{f_y}{0.85f_c'} = \frac{60}{0.85(4)} = 17.6$$

Thus

$$P_n = 3.4bd\left\{-0.32 + \sqrt{(0.3)^2 + 0.04[16.6(0.86) + 1.3]}\right\}$$
$$= 1.78bd$$

If $b = 12$ in. and $P_u/\phi = 237$ kips,

$$\text{required } d = \frac{237}{1.78(12)} = 11.1; \quad h \approx 15 \text{ in.}$$

Revise equivalent M_{0x} to assume $h/b \approx 1.33$ (say 12×16),

$$\text{equivalent } \phi M_{0x} = 138 + 55(1.33)\left(\frac{1 - 0.65}{0.65}\right) = 177 \text{ ft-kips}$$

Using the required uniaxial strengths $M_{0x} = 177/0.70 = 253$ ft-kips and $P_n = 237$ kips,

$$e_y = \frac{253(12)}{237} = 12.8 \text{ in.}$$

Revise e'/d to 1.35 (assume $h \approx 16$ in.), and d'/d to 0.185,

$$P_n = 3.4bd\{-0.37 + \sqrt{(-0.35)^2 + 0.04[16.6(0.815) + 1.35]}\}$$

$$= 3.4bd(-0.37 + 0.85) = 1.63bd$$

If $b = 12$ in.,

$$\text{required } d = \frac{237}{1.63(12)} = 12.1 \text{ in.}; \qquad h = 16 \text{ in.}$$

A 12×16 section might work; but since the earlier preliminary indication was for a deeper section, try the deeper one.
Try a section 12×18.

$$A_s \approx 0.04(12)(15.2) = 7.3 \text{ sq in.}$$

(f) Check the section of Fig. 13.21.14(a) using the Bresler reciprocal load method. The eccentricities are

$$e_y = \frac{M_{ux}}{P_u} = \frac{138(12)}{166} = 9.98 \text{ in.}$$

$$e_x = \frac{M_{uy}}{P_u} = \frac{55(12)}{166} = 3.98 \text{ in.}$$

Establish the strength P_n for the required values of e_y and e_x from uniaxial $P_n - M_n$ interaction diagrams as in Fig. 13.21.14. The values of P_x and P_y for $e/h = 0.55$ and $e/b = 0.33$ may also be obtained directly using the *ACI Strength Handbook*, Vol. 2 [26]. Thus

$$\text{for } e_y = 9.98 \text{ in.}, \qquad \text{find } P_x = 400 \text{ kips}$$
$$\text{for } e_x = 3.98 \text{ in.}, \qquad \text{find } P_y = 525 \text{ kips}$$

Using Eq. (13.21.6), with $P_0 = 1271$ kips (Fig. 13.21.14),

$$\frac{1}{P_i} = \frac{1}{P_x} + \frac{1}{P_y} - \frac{1}{P_0}$$

$$\frac{1000}{P_i} = \frac{1000}{400} + \frac{1000}{525} - \frac{1000}{1271}$$

$$= 2.50 + 1.90 - 0.79 = 3.61$$

$$\text{provided } P_n \approx P_i = \frac{1000}{3.61} = 277 \text{ kips}$$

$$[\phi P_n = 0.70(277) = 194 \text{ kips}] > [P_u = 166 \text{ kips}] \qquad \text{OK}$$

Use 12×18 section with 12-#8 bars, as in Fig. 13.21.14.

(g) Check using the Parme load contour method. Using Fig. 13.21.11(d),

$$q = \rho_g \frac{f_y}{f_c'} = \frac{9.48}{12(18)} \left(\frac{60}{4}\right) = 0.66$$

$$\frac{\text{required } P_n}{P_0} = \frac{166/0.70}{1271} = \frac{237}{1271} = 0.19$$

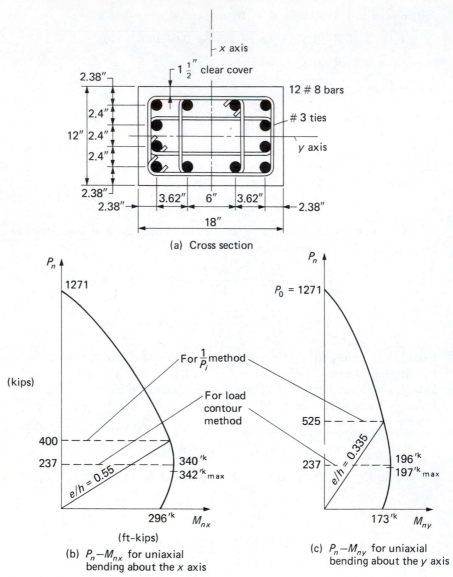

(a) Cross section

(b) $P_n - M_{nx}$ for uniaxial bending about the x axis

(c) $P_n - M_{ny}$ for uniaxial bending about the y axis

Figure 13.21.14 Data for Example 13.21.3.

Find $\beta = 0.56$ from Fig. 13.21.11(d).

Next, enter Fig. 13.21.12 with $M_{nx}/M_{0x} = (138/0.70)/340 = 0.58$ and $\beta = 0.56$. Find

$$\frac{\text{strength } M_{ny}}{M_{0y}} = 0.56$$

$$\text{strength } M_{ny} = 0.56(196) = 110 \text{ ft-kips}$$

$$[\phi M_{ny} = 0.70(110) = 77 \text{ ft-kips}] > [M_{uy} = 55 \text{ ft-kips}] \qquad \text{OK}$$

The basic statics analysis* for this section gives a strength P_n for $e_y = 9.98$ in. and $e_x = 3.98$ in. of 291 kips. From the published computer program of J. C. Smith [54] the statics analysis gives $P_n = 284$ kips. The required strength P_n is 237 kips.

SELECTED REFERENCES

1. A. Considère. "Compressive Resistance of Concrete Steel and Hooped Concrete, Part I," *Engineering Record*, December 20, 1902, 581–583; Part II, December 27, 1902, 605–606.

2. A. Considère. "Concrete-Steel and Hooped Concrete," *Reinforced Concrete*, 1903, p. 119.

3. A. N. Talbot. "Tests of Concrete and Reinforced Concrete Columns." *Bulletin*, No. 10, 1906, and No. 20, 1907, University of Illinois, Urbana.

4. M. O. Withey. "Tests of Plain and Reinforced Concrete Columns," *Engineering Record*, July 1909, 41.

5. M. O. Withey. "Tests on Reinforced Concrete Columns," *Bulletin*, No. 300, 1910, and No. 466, 1911, University of Wisconsin, Madison.

6. Committee 105. "Reinforced Concrete Column Investigation," *ACI Journal, Proceedings*, **26**, April 1930, 601–612; **27**, February 1931, 675–676; **28**, November 1931, 157–158; **29**, September 1932, 53–56; February 1933, 275–284; **30**, September–October 1933, 78–90; November–December 1933, 153–156.

7. Joint ACI-ASCE Committee 441. *Reinforced Concrete Columns* (annotated bibliography), *ACI Bibliography* No. 5. Detroit: American Concrete Institute, 1965.

8. M. J. N. Priestley, R. Park, and R. T. Potangaroa. "Ductility of Spirally-Confined Concrete Columns," *Journal of the Structural Division*, ASCE, **107**, January 1981 (ST1), 181–202.

9. Shamin A. Sheikh and S. M. Uzumeri. "Strength and Ductility of Tied Concrete Columns," *Journal of the Structural Division*, ASCE, **106**, May 1980 (ST5), 1079–1102.

10. E. Hognestad. "A Study of Combined Bending and Axial Load in Reinforced Concrete Members," *Bulletin* No. 399, November 1951, Engineering Experiment Station, University of Illinois, Urbana (117 references).

11. ACI-ASCE Committee 327. "Report on Ultimate Strength Design," *Proceedings ASCE*, **81**, October 1955, Paper No. 809. See also *ACI Journal, Proceedings*, **52**, January 1956, 505–524.

12. A. H. Mattock, L. B. Kriz, and Eivind Hognestad, "Rectangular Concrete Stress Distribution in Ultimate Strength Design," *ACI Journal, Proceedings*, **57**, February 1961, 875–928.

13. E. O. Pfrang, C. P. Siess, and M. A. Sozen. "Load-Moment-Curvature Characteristics of Reinforced Concrete Cross Sections," *ACI Journal, Proceedings*, **61**, July 1964, 763–778. Disc., 1673–1683.

14. German Gurfinkel and Arthur Robinson. "Determination of Strain Distribution and Curvature in a Reinforced Concrete Section Subjected to Bending Moment and Longitudinal Load," *ACI Journal, Proceedings*, **64**, July 1967, 398–403.

15. James G. MacGregor, John E. Breen, and Edward O. Pfrang. "Design of Slender Concrete Columns," *ACI Journal, Proceedings*, **67**, January 1970, 6–28.

16. F. E. Richart, J. O. Draffin, T. A. Olson, and R. H. Heitman. "The Effect of Eccentric Loading, Protective Shells, Slenderness Ratios, and Other Variables

*Unpublished program used by the author in the classroom; originally written by Jagdish Syal.

in Reinforced Concrete Columns," *Bulletin* No. 368, 1947 (130 pp.), Engineering Experiment Station, University of Illinois, Urbana.

17. B. Bresler and P. H. Gilbert. "Tie Requirements for Reinforced Concrete Columns," *ACI Journal, Proceedings*, **58,** November 1961, 555–570; Disc., **58,** 897–907.

18. James F. Pfister. "Influence of Ties on the Behavior of Reinforced Concrete Columns," *ACI Journal, Proceedings*, **61,** May 1964, 521–537.

19. Fred M. Hudson. "Reinforced Concrete Columns: Effects of Lateral Tie Spacing on Ultimate Strength," *Symposium on Reinforced Concrete Columns*, Publication SP-13. Detroit, Michigan: American Concrete Institute, 1966 (pp. 235–244).

20. Edwin G. Burdette and Hubert K. Hilsdorf. "Behavior of Laterally Reinforced Concrete Columns," *Journal of the Structural Division*, ASCE, **97,** February 1971 (ST2), 587–602.

21. N. G. Bunni. "Rectangular Ties in Reinforced Concrete Columns," *Reinforced Concrete Columns* (SP-50). Detroit: American Concrete Institute, 1975 (pp. 193–210).

22. ACI Committee 315. *ACI Detailing Manual—1980* (SP-66). Detroit: American Concrete Institute, 1980.

23. ACI Committee 318. *Commentary on the Building Code Requirements for Reinforced Concrete (ACI 318–83)*, Detroit: American Concrete Institute, 1983.

24. Ti Huang. "On the Formula for Spiral Reinforcement," *ACI Journal, Proceedings*, **61,** March 1964, 351–353. Disc., 1241–1248.

25. Charles S. Whitney. "Plastic Theory of Reinforced Concrete Design," *Transactions ASCE*, **107,** 1942, 251–326.

26. ACI Committee 340 (C. G. Salmon, Chairman). *Design Handbook in Accordance with Strength Design Method of ACI 318-77*, Vol. 2, *Columns* (SP–17A). Detroit: American Concrete Institute, 1978.

27. *CRSI Handbook* (4th ed.). Chicago: Concrete Reinforcing Steel Institute, 1982.

28. *Metric Design Handbook for Reinforced Concrete Elements in Accordance with the Strength Design Methods of CSA Standard CAN3-A23.3-M77*. (Edited by Murat Saatcioglu). Ottawa, Ontario, Canada: Canadian Portland Cement Association, 1980.

29. Joaquin Marin. "Design Aids for L-Shaped Reinforced Concrete Columns," *ACI Journal, Proceedings*, **76,** November 1979, 1197–1216.

30. *Ultimate Load Tables for Circular Columns*. Chicago: Portland Cement Association, 1960.

31. ACI Committee 340. *Ultimate Strength Design Handbook*, Vol. 2, *Columns* (SP-17A). Detroit: American Concrete Institute, 1970.

32. E. C. Harris. "Design of Members Subject to Combined Bending and Tension," *ACI Journal, Proceedings*, **72,** September 1975, 491–495.

33. F. L. Moreadith. "Design of Reinforced Concrete for Combined Bending and Tension," *ACI Journal, Proceedings*, **75,** June 1978, 251–255. Disc., **75,** December 1978, 721–723.

34. Fernando Villalta, Domingo J. Carreira, and Bryan A. Erler. Discussion of "Design of Reinforced Concrete for Combined Bending and Tension," (*ACI Journal, Proceedings*, **75,** June 1978, 251–255), *ACI Journal, Proceedings*, **75,** December 1978, 721–723.

35. K. H. Chu and A. Pabarcius. "Biaxially Loaded Reinforced Concrete Columns," *Proceedings ASCE*, **84,** ST8, December 1958, 1–27.

36. Boris Bresler. "Design Criteria for Reinforced Columns under Axial Load and Biaxial Bending," *ACI Journal, Proceedings*, **57,** November 1960, 481–490. Disc., 1621–1638.

37. Richard W. Furlong. "Ultimate Strength of Square Columns under Biaxially Eccentric Loads," *ACI Journal, Proceedings*, **57,** March 1961, 1129–1140.

38. F. N. Pannell. "Failure Surfaces for Members in Compression and Biaxial Bending," *ACI Journal, Proceedings*, **60,** January 1963, 129–140.

39. J. L. Meek. "Ultimate Strength of Columns With Biaxially Eccentric Loads," *ACI Journal, Proceedings*, **60,** August 1963, 1053–1064.

40. A. Aas-Jakobsen. "Biaxial Eccentricities in Ultimate Load Design," *ACI Journal, Proceedings*, **61,** March 1964, 293–315.

41. John F. Fleming and Stuart D. Werner. "Design of Columns Subjected to Biaxial Bending," *ACI Journal, Proceedings*, **62,** March 1965, 327–342. Disc., 1217–1224.

42. L. N. Ramamurthy. "Investigation of the Ultimate Strength of Square and Rectangular Columns under Biaxially Eccentric Loads," *Symposium on Reinforced Concrete Columns* (SP-13). Detroit: American Concrete Institute, 1966 (pp. 263–298).

43. Alfred L. Parme, Jose M. Nieves, and Albert Gouwens. "Capacity of Reinforced Rectangular Columns Subject to Biaxial Bending," *ACI Journal, Proceedings*, **63,** September 1966, 911–923.

44. Donald C. Weber. "Ultimate Strength Design Charts for Columns with Biaxial Bending." *ACI Journal, Proceedings*, **63,** November 1966, 1205–1320. Disc., 1583–1586.

45. Anis Farah and M. W. Huggins. "Analysis of Reinforced Concrete Columns Subjected to Longitudinal Load and Biaxial Bending," *ACI Journal, Proceedings*, **66,** July 1969, 569–575.

46. K. N. Smith and W. H. Nelles. "Columns Subjected to Biaxial Bending— Preliminary Selection of Reinforcing," *ACI Journal, Proceedings*, **71,** August 1974, 411–413.

47. Peter D. Heimdahl and Albert C. Bianchini. "Ultimate Strength of Biaxially Eccentrically Loaded Concrete Columns Reinforced With High Strength Steel," *Reinforced Concrete Columns* (SP-50). Detroit: American Concrete Institute, 1975 (pp. 93–117).

48. S. I. Abdel-Sayed and N. J. Gardner. "Design of Symmetric Square Slender Reinforced Concrete Columns Under Biaxially Eccentric Loads," *Reinforced Concrete Columns* (SP-50). Detroit: American Concrete Institute, 1975 (pp. 149–164).

49. A. K. Basu and P. Suryanarayana. "Analysis of Restrained Reinforced Concrete Columns Under Biaxial Bending," *Reinforced Concrete Columns* (SP-50). Detroit: American Concrete Institute, 1975 (pp. 211–232).

50. Albert J. Gouwens. "Biaxial Bending Simplified," *Reinforced Concrete Columns* (SP-50). Detroit: American Concrete Institute, 1975 (pp. 233–261).

51. W. F. Chen and M. T. Shoraka. "Tangent Stiffness Method for Biaxial Bending of Reinforced Concrete Columns," *Publications*, International Association for Bridge and Structural Engineering, 35–1, 1975, 23–44.

52. Richard W. Furlong. "Concrete Columns Under Biaxially Eccentric Thrust," *ACI Journal, Proceedings*, **76,** October 1979, 1093–1118.

53. M. Pinto de Magalhaes. "Biaxially Loaded Concrete Sections," *Journal of the Structural Division*, ASCE, **105,** December 1979 (ST12), 2639–2656.

54. J. C. Smith. "Biaxially Loaded Concrete Interaction Curve," *Computers and Structures*, 3(1973), 1461–1464.

55. B. D. Scott, R. Park, and M. J. N. Priestley. "Stress-Strain Behavior of Concrete Confined by Overlapping Hoops at Low and High Strain Rates," *ACI Journal, Proceedings*, **79,** January–February 1982, 13–27.

56. J. S. Ford, D. C. Chang, and J. E. Breen. "Behavior of Concrete Columns Under Controlled Lateral Deformation," *ACI Journal, Proceedings*, **78**, January–February 1981, 3–20.

57. Paul W. Reed. "Simplified Analysis for Thrust and Moment of Concrete Sections," *ACI Journal, Proceedings*, **77**, May–June 1980, 195–200.

58. Curtis J. Young. "Direct Selection of Concrete Dimensions in Columns," *Concrete International*, **3**, October 1981, 27–31.

PROBLEMS

All problems* are to be worked in accordance with the strength method of the ACI Code unless otherwise indicated, and all stated loads are service loads. Note that eccentricities e_x and e_y are measured along the x- and y-axes, respectively (see Fig. 13.21.1). For all design problems a design sketch (to scale) of the cross section is required, showing section dimensions, bars, location of bars, and tie or spiral size and spacing.

Problems on Rectangular Sections Subject to Uniaxial Bending

13.1 Calculate the nominal strength P_n for the section of the accompanying figure for an eccentricity e_y (measured along the y axis) = $0.1h$ = 1.8 in. Use f'_c = 3000 psi and f_y = 40,000 psi. Use basic principles of statics to obtain solution, considering the effect of compression concrete displaced by steel; compare the result (P_n) with that obtained by the Whitney formula (Eq. 13.13.8) and compare with the maximum P_n given by ACI-10.3.5. (e = 46 mm; f'_c = 20 MPa; f_y = 300 MPa.)

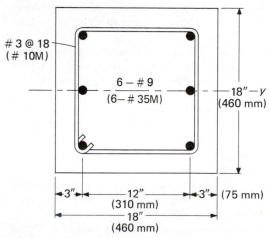

Problems 13.1 and 13.2

---

*Many problems may be solved either as problems stated in Inch-Pound units, or as problems in SI units using quantities in parenthesis at the end of the statement. The SI conversions are approximate to avoid implying higher precision for the given information in metric units than that for the Inch-Pound units.

13.2 Same as Prob. 13.1 except f'_c = 5000 psi and f_y = 60,000 psi. (f'_c = 35 MPa; f_y = 400 MPa.)

13.3 Compute the nominal strength $P_n = P_b$ in the balanced strain condition for bending about the strong axis (measured as e_y) for the column of the accompanying figure. Compute also the eccentricity e_b. Use the basic statics method, including the effect of the compression concrete displaced by steel, f'_c = 3000 psi and f_y = 40,000 psi. (f'_c = 20 MPa; f_y = 300 MPa.)

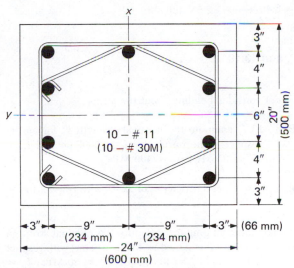

Problems 13.3, 13.4, 13.5, and 13.6

13.4 Using the basic statics method and taking account of the compression concrete displaced by steel, calculate and plot the P_n versus M_n strength interaction diagram for the section of Prob. 13.3. Take bending with respect to the strong axis (h = 24 in.). To obtain points for the diagram, compute P_n for the following cases in addition to e_b (see Prob. 13.3):
(a) $e = 0(P_0)$ (d) $e = 0.7h$
(b) $e = 0.1h$ (e) $e = h$
(c) $e = 0.3h$ (f) $e = \infty(M_0)$

13.5 Same as Prob. 13.3 except use f'_c = 4000 psi and f_y = 60,000 psi. (f'_c = 30 MPa; f_y = 400 MPa.)

13.6 Same as Prob. 13.4 except use f'_c = 4000 psi and f_y = 60,000 psi. (For e_b, see Prob. 13.5).

13.7 For the section of the accompanying figure, using basic statics compute and plot the strength interaction diagram of $P_n - M_n$ for bending about the x axis. Compute the balanced condition in addition to those points indicated for Prob. 13.4. Use f'_c = 3000 psi and f_y = 60,000 psi. (f'_c = 20 MPa; f_y = 400 MPa.)

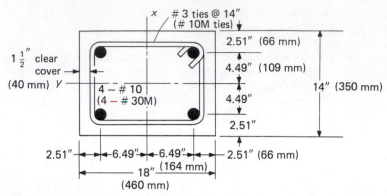

Problems 13.7, 13.8, 13.9, and 13.10

13.8 Repeat Prob. 13.7 except consider bending about the y axis.

13.9 For the section of Prob. 13.7, compute the nominal strength P_n for an eccentricity e_y of 5 in. with respect to the x axis. Use $f'_c = 5000$ psi and $f_y = 60{,}000$ psi. ($e = 125$ mm; $f'_c = 35$ MPa; $f_y = 400$ MPa.)

13.10 Repeat Prob. 13.9, except use an eccentricity e_y of 22 in. with respect to the x axis. ($e = 560$ mm.)

13.11 For the section in the accompanying figure, compute the nominal strength P_n for the eccentricity given below (as assigned by the instructor) with respect to the x axis. Use $f'_c = 3000$ psi and $f_y = 40{,}000$ psi. ($f'_c = 20$ MPa; $f_y = 300$ MPa.)
  (a) $e = 6$ in. (150 mm)   (e) $e = 30$ in. (750 mm)
  (b) $e = 8$ in. (200 mm)   (f) $e = 40$ in. (1000 mm)
  (c) $e = 12$ in. (300 mm)   (g) $e = 200$ in. (5 m)
  (d) $e = 20$ in. (500 mm)   (h) $e = 2000$ in. (50 m)

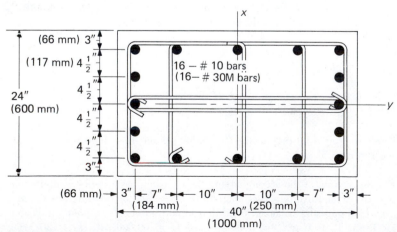

Problem 13.11

13.12 For the section of the accompanying figure, compute the nominal strength P_n for the eccentricity given below (as assigned by the instructor) with respect to the x axis. Use $f'_c = 3000$ psi and $f_y = 40,000$ psi.

(a) $e = 2$ in.   (e) $e = 40$ in.
(b) $e = 3.6$ in.   (f) $e = 70$ in.
(c) $e = 10$ in.   (g) $e = 200$ in.
(d) $e = 15$ in.   (h) $e = 1000$ in.

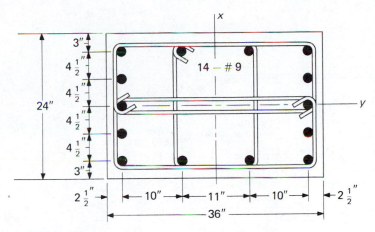

Problem 13.12

13.13 For the section of the accompanying figure, using basic statics compute and plot the strength interaction diagram of $P_n - M_n$ for bending about the x axis. Compute the balanced condition in addition to enough other points to plot the curve. (Hint: By successively setting the neutral axis distance x at specific values the points can be computed without solving a cubic equation.) Use $f'_c = 4000$ psi and $f_y = 60,000$ psi. ($f'_c = 30$ MPa; $f_y = 400$ MPa.)

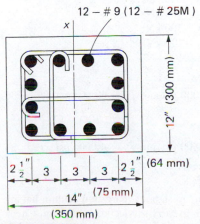

Problem 13.13

13.14 Using basic principles, compute the nominal strength P_n for an eccentricity e_y (as assigned by the instructor) on the column of the accompanying figure. Use $f'_c = 3500$ psi and $f_y = 40,000$ psi. ($f'_c = 25$ MPa; $f_y = 300$ MPa.)

(a) $e = 3$ in. (75 mm)   (f) $e = 30$ in. (750 mm)

(b) $e = 6$ in. (150 mm)   (g) $e = 60$ in. (1.5 m)

(c) $e = 9$ in. (220 mm)   (h) $e = 150$ in. (3.8 m)

(d) $e = 15$ in. (350 mm)   (i) $e = 300$ in. (7.5 m)

(e) $e = 20$ in. (500 mm)   (j) $e = 2000$ in. (50 m)

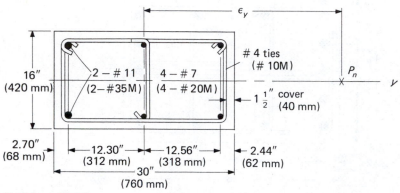

Problem 13.14

13.15 For the case assigned by the instructor, design the smallest (in whole inches) square tied column having not more than 4% reinforcement to carry an axial compression (i.e., $M = 0$ or neglected). Compute the strength P_n when $e = 0.10h$. Comment! Use $f'_c = 3500$ psi and $f_y = 60,000$ psi. ($f'_c = 25$ MPa; $f_y = 400$ MPa.)

	P_D (dead load)	P_L (live load)
(a)	100 kips (450 kN)	80 kips (350 kN)
(b)	200 kips (900 kN)	160 kips (700 kN)
(c)	350 kips (1500 kN)	150 kips (700 kN)
(d)	400 kips (1800 kN)	320 kips (1400 kN)
(e)	600 kips (2700 kN)	480 kips (2100 kN)

13.16 Repeat Prob. 13.15 except use $f'_c = 4000$ psi and $f_y = 60,000$ psi. ($f'_c = 30$ MPa and $f_y = 400$ MPa.)

13.17 Repeat Prob. 13.15 except use $f'_c = 3500$ psi and $f_y = 40,000$ psi. ($f'_c = 25$ MPa and $f_y = 300$ MPa.)

13.18 For the case assigned by the instructor, design a square tied column having symmetrical reinforcement of about $3\frac{1}{2}\%$ to carry service loads as given. Check final answer using basic statics. Use whole inches with $f'_c = 4000$ psi and $f_y = 50,000$ psi.

	P_D (dead load)	P_L (live load)	M_D (dead load)	M_L (live load)
(a)	100 kips	90 kips	40 ft-kips	30 ft-kips
(b)	200 kips	175 kips	75 ft-kips	65 ft-kips
(c)	300 kips	270 kips	120 ft-kips	100 ft-kips
(d)	300 kips	270 kips	250 ft-kips	200 ft-kips
(e)	300 kips	270 kips	500 ft-kips	400 ft-kips
(f)	600 kips	550 kips	125 ft-kips	100 ft-kips
(g)	600 kips	550 kips	250 ft-kips	200 ft-kips
(h)	600 kips	550 kips	500 ft-kips	400 ft-kips

13.19 Redesign the column of Prob. 13.18 using a rectangular section not over 15 in. wide instead of a square section. If the depth h would be greater than 30 in., use $h/b = 2.0$ and exceed the 15-in. limit.

13.20 Design a square tied column with symmetrical reinforcement of about 4% to carry a dead load of $P = 225$ kips and $M = 220$ ft-kips and a live load of $P = 200$ kips and $M = 190$ ft-kips. Check final answer using basic statics. Use whole inches (20 mm for SI) for size with $f'_c = 4000$ psi and $f_y = 40,000$ psi. (DL: $P = 1000$ kN, $M = 300$ kN·m; LL: $P = 900$ kN, $M = 260$ kN·m; $f'_c = 30$ MPa; $f_y = 300$ MPa.)

13.21 Redesign the column of Prob. 13.20 using a rectangular section not over 16 in. (400 mm) wide instead of a square section.

13.22 Design a square tied column with about 4% reinforcement to carry a dead load of $P = 250$ kips and $M = 90$ ft-kips, and a live load of $P = 185$ kips and $M = 70$ ft-kips. Use whole inches (20 mm for SI) for size with $f'_c = 4000$ psi and $f_y = 60,000$ psi. (DL: $P = 1100$ kN, $M = 120$ kN·m; LL: $P = 820$ kN, $M = 95$ kN·m; $f'_c = 30$ MPa; $f_y = 400$ MPa.)

13.23 Design a square tied column with symmetrical reinforcement of about 2% to carry a dead load of $P = 25$ kips and $M = 125$ ft-kips and a live load of $P = 10$ kips and $M = 50$ ft-kips. Use whole inches (20 mm for SI) for size with $f'_c = 4000$ psi and $f_y = 60,000$ psi. (DL: $P = 110$ kN, $M = 170$ kN·m; LL: $P = 45$ kN, $M = 68$ kN·m; $f'_c = 30$ MPa; $f_y = 400$ MPa.)

13.24 Redesign the column of Prob. 13.23 as a rectangular column having a depth-to-width ratio of between 1.5 and 2.0 along with unsymmetrical reinforcement with respect to the bending axis.

13.25 Design a square tied column with symmetrical reinforcement of about $2\frac{1}{2}\%$ to carry a dead load of $P = 75$ kips and $M = 125$ ft-kips and a live load of $P = 30$ kips and $M = 50$ ft-kips. Use whole inches (20 mm for SI) for size with $f'_c = 4000$ psi and $f_y = 60,000$ psi. (DL: $P = 340$ kN, $M = 170$ kN·m; LL: $P = 140$ kN, $M = 70$ kN·m; $f'_c = 30$ MPa; $f_y = 400$ MPa.)

13.26 Redesign the column of Prob. 13.25 using a rectangular member but still using symmetrical reinforcement.

Problems on Circular Sections

13.27 Determine the strength $P_n = P_b$, and the eccentricity e_b, for a balanced strain condition on the section of the accompanying figure by using the method of statics and the rectangular stress distribution as for beams. Use $f'_c = 4000$ psi, $f_y = 50,000$ psi.

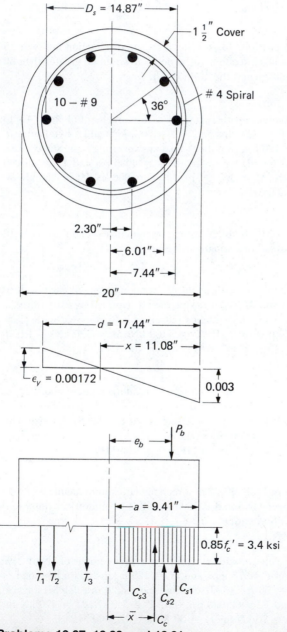

Problems 13.27, 13.28, and 13.31

13.28 Determine the nominal compressive strength P_n for the section of Prob. 13.27 for an eccentricity $e = 5$ in., by the direct application of statics. Use $f'_c = 4000$ psi, $f_y = 50,000$ psi.

13.29 Compute the nominal strength P_n of the spirally reinforced column of the accompanying figure when the loading has an eccentricity $e = 0.05h = 0.9$ in. Use $f'_c = 3000$ psi and $f_y = 40,000$ psi. Use basic statics with the circular section including the effect of compression concrete displaced by steel. Compare with maximum P_n obtained from ACI-10.3.5. ($e = 23$ mm; $f'_c = 20$ MPa; $f_y = 300$ MPa.)

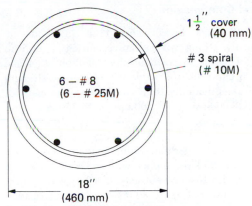

Problems 13.29 and 13.30

13.30 Same as Prob. 13.29 except use $f'_c = 5000$ psi and $f_y = 60,000$ psi. ($f'_c = 35$ MPa; $f_y = 400$ MPa.)

13.31 Determine the nominal compressive strength P_n for the section of Prob. 13.27 for an eccentricity $e = 20$ in., by a direct application of statics. Use $f'_c = 4000$ psi, $f_y = 50,000$ psi, and the ACI Code.

13.32 For the column of the accompanying figure, determine the nominal strength P_n at an eccentricity of 21 in. Use $f'_c = 4000$ psi and $f_y = 50,000$ psi. Apply the basic principles of statics. ($e = 530$ mm; $f'_c = 30$ MPa; $f_y = 350$ MPa.)

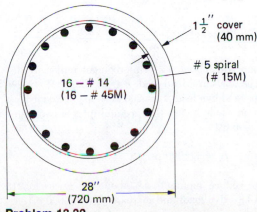

Problem 13.32

13.33 Same as Prob. 13.15, except use a spirally reinforced circular column.

13.34 Redesign the column of Prob. 13.18 as a spirally reinforced circular column.

13.35 Redesign the column of Prob. 13.20 as a spirally reinforced circular column.

13.36 Design a circular spirally reinforced column for the conditions given in Prob. 13.22. Make statics check using the designed circular section.

13.37 Redesign the column of Prob. 13.25 as a spirally reinforced circular column.

Problems on Biaxial Bending and Compression

13.38 Compute the nominal strength P_n for the column of Prob. 13.3 under an eccentricity e_y of 8 in. about the strong axis and an eccentricity e_x of 5 in. about the weak axis. (Hint: Use the interaction curve of Prob. 13.4 and calculate a similar interaction curve for the weak axis.) Compute strength using (a) Bresler reciprocal load method, and (b) Parme load contour method. Use $f'_c = 3000$ psi and $f_y = 40,000$ psi. ($e_y = 200$ mm; $e_x = 130$ mm; $f'_c = 20$ MPa; $f_y = 300$ MPa.)

13.39 Determine the adequacy of the rectangular tied column section of Prob. 13.7 when subjected to an axial load of 90 kips, applied with an eccentricity e_y of 5 in. with respect to the x axis and with an eccentricity e_x of 3 in. with respect to the y axis. Use $f'_c = 3000$ psi and $f_y = 40,000$ psi (use information developed in Probs. 13.7 and 13.8). Compare results using (a) Bresler reciprocal load method and (b) Parme load contour method.

13.40 Repeat Prob. 13.39, except the axial load is 45 kips applied with eccentricities of 10 in. with respect to the x axis and 6 in. with respect to the y axis.

13.41 For the case assigned by the instructor, compute the nominal strength P_n for biaxial bending and compression for the section of Prob. 13.7. Use both the Bresler reciprocal load and the Parme load contour methods. (If available, the strength interaction diagrams from Probs. 13.7 and 13.8 may be used.) Use $f'_c = 3000$ psi and $f_y = 60,000$ psi.
(a) $e_y = 1.8$ in. $e_x = 1.4$ in.
(b) $e_y = 5.4$ in. $e_x = 1.4$ in.
(c) $e_y = 15$ in. $e_x = 2$ in.
(d) $e_y = 30$ in. $e_x = 3$ in.
(e). $e_y = 40$ in. $e_x = 6$ in.
(f) $e_y = 40$ in. $e_x = 20$ in.

13.42 Design a square tied column with about 4% reinforcement uniformly distributed around its sides. The loads are dead load, $P = 225$ kips, $M_x = 236$ ft-kips, and $M_y = 108$ ft-kips, and live load, $P = 200$ kips, $M_x = 204$ ft-kips, $M_y = 64$ ft-kips. Select size in whole inches that are multiples of two, using $f'_c = 4000$ psi and $f_y = 40,000$ psi.

13.43 Repeat Prob. 13.42, except use $f_y = 60,000$ psi.

13.44 Repeat Prob. 13.42, except reduce moments M_x about the x axis to 215 ft-kips dead load and 162 ft-kips live load, and increase f_y to 60,000 psi.

14

Deflections

14.1 DEFLECTIONS—GENERAL

Throughout the period from 1910 to 1956, while the working stress method was used nearly exclusively, concrete with a compressive strength f'_c from 1500 to 3000 psi (approximately 10.5 to 21 MPa) and reinforcement with a yield stress from 33 to 40 ksi (approximately 230 to 280 MPa) were predominant. The use of these materials with conservative allowable stresses, along with the straightline working stress method, resulted in large stiff sections having small deflections. Ordinary reinforced concrete design involved little concern for deflections.

With the widespread use of the strength method and the realistic acceptance of the additional strength of concrete in compression according to the nonlinear relationship between stress and strain, sections could be made smaller. Such smaller sections deflect a greater amount than those designed under the working stress method.

The common use of 60,000 psi (420 MPa) yield strength steel and of concrete having strengths f'_c ranging from the ordinary value of 4000 psi (28 MPa) up to 9000 psi (63 MPa) permits smaller sections than those resulting from the use of lower strength materials.

The permissible deflection is governed by the serviceability requirements for the structure, such as the amount of deformation that can be tolerated by the interacting components of the structure. Excessive deflection of the member may not in itself be detrimental, but the effect on structural components that are supported by the deflecting member frequently determines the acceptable amount. Both the short-time (instantaneous or immediate) and the long-time effects must be considered. The acceptable deflection depends on many factors, among which are the type of building (warehouse, school, factory, residence, etc.), the presence of plastered ceilings, the type and arrangement of partitions, the sensitivity of equipment to deflection, and the magnitude and duration of live load. Vibration and noise transmission are also serviceability concerns that depend on the stiffness of flexural members just as does deflection.

The maximum acceptable immediate deflections due to live load, for flat roofs or floors that do not support and are not attached to nonstructural elements such as plastered ceilings or frangible partitions likely to be damaged by large deflections, are prescribed by ACI-Table 9.5(b) to be (a) for flat roofs, $L/180$ and (b) for floors, $L/360$. Further, in recognition of the increase of deflection with time, ACI-Table 9.5(b) also prescribes limits for the sum of the *creep and shrinkage deflection due to all sustained loads plus any additional live load deflection*. The two stated limits for this deflection combination are (a) when nonstructural elements are likely to be damaged, $L/480$, and (b) when no damage to nonstructural elements is likely, $L/240$.

Limitations on deflection are somewhat arbitrary; historically, $L/360$ has been the accepted limit to prevent the cracking of plastered ceilings. Other limits should be considered as guidelines, with the designer having the responsibility for evaluating the possible adverse effects of excessive deflection in any given situation.

ACI Committee 435 has reported recommendations for allowable deflections for a great variety of situations [1].

The general concepts dealt with in this chapter are applicable to both one-way (beams and slabs) and two-way systems. Specific examples are applied only to one-way systems. The reader is referred to Sec. 16.20 for a brief discussion of two-way system deflections, along with specific references.

14.2 DEFLECTIONS FOR ELASTIC SECTIONS

Various methods are available in structural analysis for computing deflections on uniform and variable moment of inertia sections in statically determinate and indeterminate structures. In general, using any of the several methods, the maximum deflection in an elastic member may be expressed as

$$\Delta_{\max} = \beta_a \frac{ML^2}{EI_c} \tag{14.2.1}$$

where

M = a reference value of bending moment such as the maximum positive value

L = span length

E = modulus of elasticity

I_c = moment of inertia of a standard section in the member

β_a = a coefficient that depends on the degree of fixity at supports, the variation in moment of inertia along the span, and the distribution of loading; such as $\frac{5}{48}$ for simply supported uniformly loaded beam; $\frac{1}{4}$ for uniformly loaded cantilever.

The deflection coefficient β_a for simple cases may be found in handbooks, such as *ACI Design Handbook*, Vol. 1 [2], or *Handbook of Concrete Engineering* [3].

The following example will demonstrate one elastic method, the conjugate beam method, while also deriving one of the most useful deflection expressions.

EXAMPLE 14.2.1 Using the conjugate beam method, derive the general expression for the elastic midspan deflection for a uniformly loaded span with unequal end moments, as shown in Fig. 14.2.1.

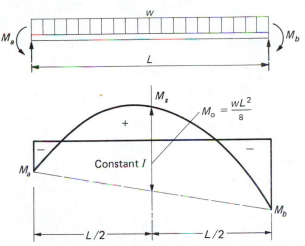

Figure 14.2.1 Typical bending moment diagram for a uniformly loaded span of a continuous beam.

Solution: In the conjugate beam method, the deflection at a given point equals the bending moment at that point for a beam loaded with the M/EI diagram. Thus the system of Fig. 14.2.1 may be regarded as being composed of three separate conjugate beams as shown in Fig. 14.2.2.

The midspan deflections, or the bending moments on the three conjugate beams, are

$$\text{uniform load, } \Delta_s = \frac{2}{3}\left(\frac{L}{2}\right)\left(\frac{M_0}{EI}\right)\left[\frac{L}{2} - \frac{3}{8}\left(\frac{L}{2}\right)\right] = \frac{5M_0 L^2}{48EI}$$

$$\text{left end moment, } \Delta_a = \frac{1}{3}\left(\frac{-M_a}{EI}\right)\left(\frac{L}{2}\right)\left(\frac{L}{2}\right) - \frac{1}{2}\left(\frac{-M_a}{2EI}\right)\left(\frac{L}{2}\right)\left(\frac{1}{3}\right)\left(\frac{L}{2}\right)$$

$$= \frac{-M_a L^2}{16EI}$$

$$\text{right end moment, } \Delta_b = \frac{-M_b L^2}{16EI}$$

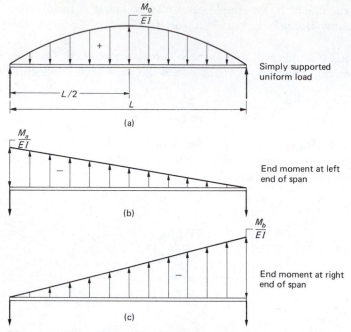

Figure 14.2.2 Component conjugate beams.

The total midspan deflection Δ_m is

$$\Delta_m = \Delta_s + \Delta_a + \Delta_b$$

$$= \frac{5M_0L^2}{48EI} - \frac{M_aL^2}{16EI} - \frac{M_bL^2}{16EI}$$

$$= \frac{L^2}{48EI} [5M_0 - 3(M_a + M_b)] \qquad \textbf{(14.2.2)}$$

The net positive midspan moment M_s is

$$M_s = M_0 - \tfrac{1}{2}(M_a + M_b) \qquad \textbf{(14.2.3)}$$

Then, upon substituting Eq. (14.2.3) in Eq. (14.2.2), one obtains

$$\Delta_m = \frac{L^2}{48EI}\left[5M_s + \frac{5}{2}(M_a + M_b) - 3(M_a + M_b) \right]$$

$$= \frac{5L^2}{48EI}\left[M_s - \frac{1}{10}(M_a + M_b) \right] \qquad \textbf{(14.2.4)}$$

Equation (14.2.4) may be used with satisfactory results for practically all prismatic (i.e., constant EI_g) beams, even though the absolute maximum deflection will be obtained only when the loading is uniform and the end moments are equal.

Continuous Beams Having Variable Flexural Rigidity.

When significant changes of gross cross section along the span length are involved, such changes should be included when performing the statically indeterminate analysis for end moments. When deflections are desired for such cases, the best approach is to account for the variable flexural rigidity EI exactly, either mathematically in the component conjugate beams (see Fig. 14.2.2), or by the use of numerical integration. Section 14.4 gives further treatment of this subject when discussion is made on what should be the value of the moment of inertia to be used for reinforced concrete sections.

14.3 MODULUS OF ELASTICITY

As discussed in Sec. 1.9, variance in interpretation is encountered when referring to the modulus of elasticity of concrete. Ordinary beam theory presumes the modulus in tension to be the same as in compression for a homogeneous material. In a reinforced concrete section, creep affects the apparent modulus primarily on the compression side. Even the short-time loading modulus, measured as the secant modulus, is considerably more variant than the compressive strength f'_c. On the tension side there are cracks in the concrete in regions of high bending moment, while at low moment sections the concrete may not crack. The secant modulus in tension is essentially the same as in compression when the stress magnitude is low, but the modulus reduces markedly as the stress nears the cracking level. In other words, in both tension and compression the actual modulus of elasticity varies not only with the magnitude of stress from top to bottom at a section, but also along the span.

Further, creep and shrinkage over a period of time effectively reduce the modulus in compression and will generally magnify deflections by a factor of two or three. Referring to Fig. 1.10.1 one may note that the true elastic strain decreases with time, indicating that the modulus of elasticity increases. The modulus increases because it is dependent on f'_c which increases with age. In design practice, usually the apparent elastic strain is used as computed with a code-specified E_c, presumably corresponding to the 28-day age at loading. Beyond 28 days, however, the increase in modulus of elasticity is relatively slight.

14.4 MOMENT OF INERTIA

In addition to the modulus of elasticity, the flexural rigidity includes the cross-sectional property, moment of inertia. The moment of inertia, even for sections whose external dimensions are constant, varies considerably along a span. Consider a span with a T-shaped section in a continuous beam or frame, as shown in Fig. 14.4.1. In such a continuous structure, at or near the support the concrete slab is cracked, so that the effective section is section A–A; while in the positive moment zone the stem is cracked, so that the effective section is section B–B. Furthermore, near the points of inflection the entire section may be uncracked and fully effective, as shown in section C–C.

In an elastic continuity analysis as discussed in Chaps. 7 and 10, only relative stiffness values are required; but in deflection computations, the absolute magnitudes for E_c and I must first be determined or assumed. Moreover, the amount of deflection under each increment of load is not

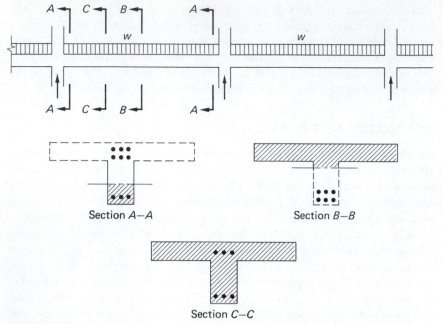

Figure 14.4.1 Effective moment of inertia for continuous T-shaped sections.

constant. The flexural rigidity E_cI is greater at a low load level, since the fully uncracked section provides the greatest effective moment of inertia.

The subject of effective moment of inertia for computing deflections due to short-term loading has received significant study by Yu and Winter [4], and by Branson [5,6]. It is known that the flexural rigidity E_cI varies with the magnitude of bending moment in the general manner shown in Fig. 14.4.2. Of course, the moment of inertia I_{cr} of the transformed cracked section* increases roughly in proportion with an increase in the percentage

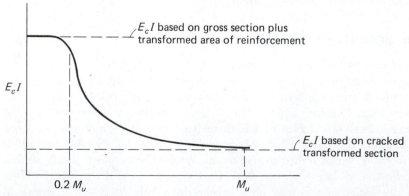

Figure 14.4.2 Typical variation of flexural rigidity with applied bending moment.

*See Secs. 4.5, 4.6, and 4.11.

of reinforcement. Sections with higher percentages of reinforcement exhibit less change in rigidity under increasing load than those with low percentages of reinforcement.

For loads below the cracking load (see Fig. 14.4.3), deflections may be based on the gross concrete section, with generally a small difference arising from whether or not the transformed area of reinforcement is also included. However, as the load increases above the cracking load, the moment of inertia approaches that of the cracked transformed section, although it may be greater between cracks.

Generally, as shown by the typical load deflection curve in Fig. 14.4.3, the use of gross section underestimates the deflection, and the use of transformed cracked section overestimates the deflection. However, the degree of accuracy is affected by the magnitude of the service load compared to the load to cause cracking.

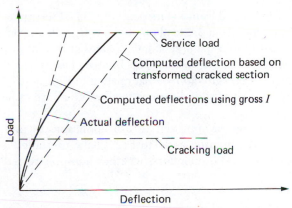

Figure 14.4.3 Typical load-deflection curve for reinforced concrete beams.

ACI Effective Moment of Inertia I_e. In order to provide a smooth transition between the moment of inertia I_{cr} of the transformed cracked section and the moment of inertia I_g of the gross uncracked concrete section, since 1971 the ACI Code used the expression developed by Branson [5],

$$I_e = \left(\frac{M_{cr}}{M_{max}}\right)^3 I_g + \left[1 - \left(\frac{M_{cr}}{M_{max}}\right)^3\right] I_{cr} \leq I_g \qquad (14.4.1)$$

where

$M_{cr} = f_r I_g / y_t$ = cracking moment

M_{max} = maximum service load moment acting at the condition under which deflection is computed

I_g = moment of inertia of gross uncracked concrete section about centroidal axis, neglecting reinforcement

I_{cr} = moment of inertia of transformed cracked section (see Secs. 4.5 and 4.11)

f_r = modulus of rupture of concrete (see Sec. 1.8), taken by ACI Code as $7.5 \sqrt{f'_c}$ for normal-weight concrete; in general may be taken [7,8,9] as $0.65 \sqrt{w_c f'_c}$, where w_c is the unit weight of concrete

y_t = distance from neutral axis to extreme fiber of concrete in tension

Equation (14.4.1) is intended to be calculated at the location of maximum moment as a single value for the entire span in the case of simply supported beams, or as a single value between points of inflection in continuous beams. If one wishes to recognize the continuous variation of the moment of inertia along the span, Branson proposed using the fourth power instead of the third power in Eq. (14.4.1). In this case M_{cr} and M_{max} are the cracking moment and the applied moment, respectively, at each section along the span. The span may be broken into segments with each segment having a different moment of inertia and numerical integration used to compute the deflection.

Equation (14.4.1) was developed from a statistical study of 54 test specimens which had M_{max}/M_{cr} values from 2.2 to about 4 and I_g/I_{cr} values from 1.3 to 3.5. The study included simple-span rectangular beams [10], T-beams [4], and two-span continuous rectangular beams [11]. Branson has provided an excellent summary [6,8,12] of background material relating to Eq. (14.4.1). Lutz [34] has provided an ingenious set of diagrams to allow graphical evaluation of Eq. (14.4.1).

The effective moment of inertia approach is also applicable to prestressed concrete (see Chap. 21); Branson and Trost [29] have presented a unified procedure for partially cracked members, whether non-prestressed or prestressed.

Single Value of Effective Moment of Inertia for Practical Use. As an *approximation*, a single value of effective moment of inertia is suggested for practical use when the variable I results from the variation in the extent of tension concrete cracking. Three methods have been suggested [13].

1. *Midspan value alone.* Recognized by the 1983 ACI Code (ACI-9.5.2.4), this assumption is

$$I_e = I_m \tag{14.4.2}$$

where I_m is the effective moment of inertia at midspan for simply supported and continuous spans, and at the support section for cantilevers. This is the simplest method and when the inflection point is between about $0.2L$ and the support, the results are within $\pm 20\%$ of those obtained using variable I, as long as $0.33 \leq \alpha \leq 1.0$ [where $\alpha = (I_m$ at midspan$)/(I_e$ at end$)$]. When α lies between 0.50 and 1.0, the results are within $\pm 5\%$ of those obtained using variable I. Zuraski, Salmon, and Shaikh [14] have suggested that this method is satisfactory for ordinary design situations, and it is endorsed by ACI Committee 435 [9].

2. *Weighted average.* In this method the adjusted I is obtained by weighting the moments of inertia in accordance with the magnitudes of the

end moments [13]. The following weighted average expression has been recommended by ACI Committee 435 [9] as giving a somewhat improved result over the use of the midspan value alone. For spans with *both* ends continuous,

$$\text{average } I_e = 0.70I_m + 0.15\,[I_{e1} + I_{e2}] \tag{14.4.3}$$

For spans with *one* end continuous,

$$\text{average } I_e = 0.85I_m + 0.15I_{e1} \tag{14.4.4}$$

3. *Simple average*. With this assumption the I to be used is the average I

$$\text{average } I_e = \frac{\frac{1}{2}(I_{e1} + I_{e2}) + I_m}{2} \tag{14.4.5}$$

where I_{e1} and I_{e2} are the effective moments of inertia at the two ends of the span. The use of both I_{e1} and I_{e2} is appropriate only when there are end moments at both ends.

For uniform loading on continuous spans Eq. (14.4.3) is slightly more accurate (say on the order of 5%) than using only the midspan value, but for concentrated loads it is less accurate [14]. When an average value is used as *permitted* by ACI-9.5.2.4 it should be done in accordance with Eq. (14.4.5), rather than taking the sum of I_m, I_{e1}, and I_{e2} and dividing by three [13].

For a single heavy concentrated load, averaging *reduces* accuracy [14]; Eq. (14.4.2) should be used in such cases (ACI Commentary-9.5.2.4).

14.5 INSTANTANEOUS DEFLECTIONS IN DESIGN

Throughout the history of reinforced concrete construction, computation of instantaneous (short-time) deflection has usually involved using either transformed cracked section or gross uncracked section. In either case, Eq. (14.2.1) is suitable after it is slightly rewritten using E_c and I_e

$$\Delta = \beta_a \left(\frac{ML^2}{E_c I_e}\right) \tag{14.5.1}$$

where

β_a = coefficient based on load and support conditions
I_e = effective moment of inertia
E_c = modulus of elasticity of concrete

For instantaneous (elastic) deflection and also generally for long-time deflection under sustained loads, the basic value of modulus of elasticity to be used for concrete is (ACI-8.5.1)

$$E_c = 33w_c^{1.5}\sqrt{f_c'}$$

for concrete having a unit weight between 90 and 155 pcf. For normal-weight concrete,

$$E_c = 57{,}000\sqrt{f_c'}$$

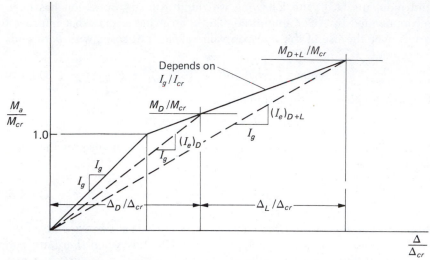

Figure 14.5.1 Typical idealized moment-deflection diagram for short-time loading (from Refs. 8, 9, and 15).

The generally accepted effective moment of inertia for Eq. (14.5.1) is Eq. (14.4.1), using a single value in accordance with Eqs. (14.4.2) through (14.4.5).

In order to compute deflection at different load levels, such as dead load or dead load plus live load, the effective moment of inertia I_e should be computed using Eq. (14.4.1) for that total load level in each case. The incremental deflection, such as for live load only, is then computed as the difference between the deflections due to dead plus live load and dead load only. It should be assumed that the live load cannot act in the absence of dead load.

Computation of the live load deflection as $\Delta_{D+L} - \Delta_D$ gives the live load deflection occurring during the *first* application of live load. Figure 14.5.1 shows the typical idealized load or moment versus deflection relationship [15]. For repeated loadings, the upper envelope of load-deflection curves is nearly the same as the single-loading curve for both reinforced and prestressed concrete members, even though increasing residual deflections occur due to creep and cracking effects [9,28]. Thus it seems reasonable to compute short-time deflections using I_e as described above and the residual deflection separately as discussed in the sections on creep and shrinkage.

A comparison of the measured short-time deflections with the deflections computed by Eq. (14.4.1) [16] is shown in Fig. 14.5.2. A study by Committee 435 [33] indicates that by using the present ACI Code criteria for deflection for simply supported beams under *controlled laboratory conditions* "there is approximately a 90% chance that the deflections of a particular beam will be within the range of 20% less than to 30% more than the calculated value."

The following two examples demonstrate the computation of instantaneous deflection.

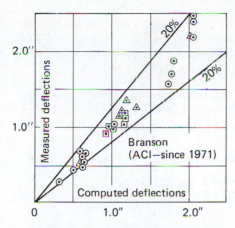

- ⊙ Simply supported rectangular beams
- △ Simply supported T beams
- ⊡ 2 span continuous rectangular beams

Figure 14.5.2 Comparison of computed and measured short-time deflections (from Ref. 16).

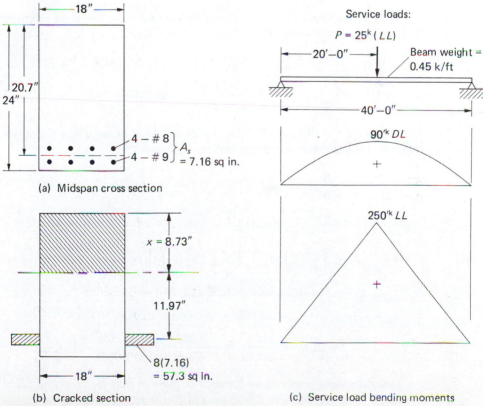

(a) Midspan cross section

(b) Cracked section

(c) Service load bending moments

Figure 14.5.3 Beam for Example 14.5.1.

EXAMPLE 14.5.1 Investigate the instantaneous (short-time loading) deflection for the simply supported beam of Fig. 14.5.3 over a span of 40 ft. Assume that the member has been designed by the strength method using $f'_c = 4000$ psi, $f_y = 60,000$ psi, and the ACI Code.

Solution: According to ACI-Table 9.5(a) the minimum depth unless deflection is computed is $L/16 = 40(12)/16 = 30$ in. Thus a deflection computation is required regardless of whether or not excessive deflection is of concern.

(a) Dead load short-time deflection. The gross moment of inertia is

$$I_g = \tfrac{1}{12}(18)(24)^3 = 20,700 \text{ in.}^4$$

$$M_{max} = \frac{0.45(40)^2}{8} = 90 \text{ ft-kips}$$

For the transformed cracked section, as shown in Fig. 14.5.3(b), the neutral axis position is

$$\frac{18x^2}{2} = 57.3 \,(20.7 - x)$$

$$x^2 + 6.37x = 131.8$$

$$x = 8.73$$

$$I_{cr} = \tfrac{1}{3} (18)(8.73)^3 + 57.3(20.7 - 8.73)^2 = 12,200 \text{ in.}^4$$

The effective moment of inertia I_e is dependent on the bending moment M_{cr} that causes cracking at the extreme tension fiber.

$$f_r = 7.5 \sqrt{f'_c} = 7.5 \sqrt{4000} = 474 \text{ psi} \qquad \text{(ACI-9.5.2.3)}$$

$$M_{cr} = \frac{f_r I_g}{y_t} = \frac{0.474(20,700)}{12}\left(\frac{1}{12}\right) = 68 \text{ ft-kips}$$

Note that y_t is the distance $h/2$ for the 24-in. deep beam when the gross section is used and the reinforcement is neglected.

$$\frac{M_{cr}}{M_{max}} = \frac{68}{90} \text{ (dead load only)} = 0.756; \qquad \left(\frac{M_{cr}}{M_{max}}\right)^3 = 0.431$$

From ACI Formula (9-7), Eq. (14.4.1), the effective moment of inertia is

$$I_e = \left(\frac{M_{cr}}{M_{max}}\right)^3 I_g + \left[1 - \left(\frac{M_{cr}}{M_{max}}\right)^3\right] I_{cr}$$

$$= 0.431(20,700) + 0.569(12,200) = 15,860 \text{ in.}^4$$

$$E_c = 33w_c^{1.5} \sqrt{f'_c} = 33(145)^{1.5} \sqrt{4000} = 3.64 \times 10^6 \text{ psi}$$

$$(\Delta_i)_D = \frac{5WL^3}{384EI} = \frac{5(0.45)(40)(40)^3(1728)}{384(3.64)(10^3)(15,860)} = 0.45 \text{ in.}$$

This deflection may not be harmful because it may be accommodated by using a negative deflection (camber) in construction. Even if camber is not used, such dead load instantaneous deflection will not affect plastered ceilings or other items that are put into place after the immediate dead load deflection has taken place. Concern is primarily with the instantaneous deflection from live load and the long-term creep and shrinkage deflection from sustained loads.

(b) Dead load plus live load short-time deflection. The maximum service load moment at this load level is

$$M_{max} = \frac{25(40)}{4} + 90 = 340 \text{ ft-kips (for dead load only)}$$

$$\frac{M_{cr}}{M_{max}} = \frac{68}{340}\left(\frac{\text{dead load}}{+ \text{ live load}}\right) = 0.200; \qquad \left(\frac{M_{cr}}{M_{max}}\right)^3 = 0.008$$

$$I_e = (0.008)(20,700) + (0.992)(12,200) = 12,270 \text{ in.}^4$$

At this higher load level, I_e is only slightly larger than I_{cr}. Using $I_e = I_{cr} = 12,200$ in.^{4},

$$(\Delta_i)_{\substack{\text{beam} \\ \text{weight}}} = \frac{5(0.45)(40)(40)^3 1728}{384(3640)(12,200)} = 0.58 \text{ in.}$$

$$(\Delta_i)_{\substack{\text{conc.} \\ \text{load}}} = \frac{25(40)^3 1728}{48(3640)(12,200)} = 1.30 \text{ in.}$$

$$(\Delta_i)_{D+L} = 0.58 + 1.30 = 1.88 \text{ in.}$$

(c) Live load short-time deflection. Consistent logic dictates that live load deflection must be obtained indirectly as

$$(\Delta_i)_L = (\Delta_i)_{D+L} - (\Delta_i)_D$$
$$= 1.88 - 0.45 = 1.43 \text{ in.}$$

It is assumed that live load cannot act in the absence of dead load. Thus if the effective moment of inertia when dead load alone is acting is considerably different from that when dead load plus live load is acting, the live load deflection is properly obtained only by subtracting Δ_D from Δ_{D+L}. Even if the service live load has been preceded by a construction load of equal magnitude, Refs. 8, 9, and 15 as shown in Fig. 14.5.1 indicate that the procedure is proper for repeated loadings.

From a practical viewpoint, $I_e = I_{cr}$ may be used whenever $(M_{cr}/M_{max})^3$ is less than about 0.1. Assuming this to be a floor beam not supporting partitions, the acceptable immediate live load deflection from ACI-Table 9.5(b) is

$$\text{allowable } (\Delta_i)_L = \frac{L}{360} = \frac{40(12)}{360} = 1.33 \text{ in.} < 1.43 \text{ in.} \qquad \text{NG}$$

If plastered ceilings or frangible partitions are to be supported, the deflection due to long-time creep and shrinkage must be added to that due to live loads; the acceptable limit for such deflection is $L/480$ (ACI-Table 9.5b). It might have been expected that the deflection of this beam would be excessive since the reinforcement ratio ρ is 0.0192, which is $0.67\rho_b$. This exceeds the guideline value of about $0.5\rho_{max}$, or $0.375\rho_b$, suggested in Chap. 3 as the maximum reinforcement ratio for deflection control.

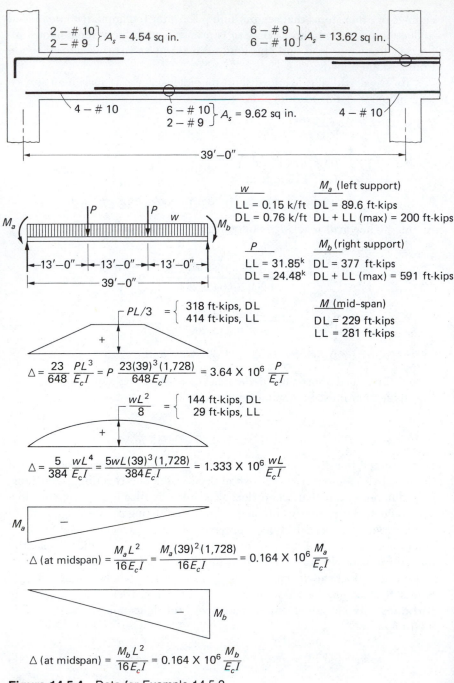

$$\Delta = \frac{23}{648} \frac{PL^3}{E_c I} = P \frac{23(39)^3(1,728)}{648 E_c I} = 3.64 \times 10^6 \frac{P}{E_c I}$$

$$\Delta = \frac{5}{384} \frac{wL^4}{E_c I} = \frac{5wL(39)^3(1,728)}{384 E_c I} = 1.333 \times 10^6 \frac{wL}{E_c I}$$

$$\Delta \text{ (at midspan)} = \frac{M_a L^2}{16 E_c I} = \frac{M_a(39)^2(1,728)}{16 E_c I} = 0.164 \times 10^6 \frac{M_a}{E_c I}$$

$$\Delta \text{ (at midspan)} = \frac{M_b L^2}{16 E_c I} = 0.164 \times 10^6 \frac{M_b}{E_c I}$$

Figure 14.5.4 Data for Example 14.5.2.

EXAMPLE 14.5.2 Investigate the instantaneous deflection on the continuous beam of Fig. 14.5.4. Use $f'_c = 3000$ psi, $f_y = 40,000$ psi, and the ACI Code.

Solution: This continuous girder supports two smaller beams that frame to it and some uniform loading that comes directly to it. ACI-9.5.2.4 indicates that I_e may be computed as an average value for the positive and negative moment regions. For prismatic members, the effective moment of inertia I_e at the positive moment region may be used instead of the average value. In fact, for loading that is largely concentrated load, as in the situation for this example, ACI Committee 435 recommends [9] *against* using an average value. The results will be compared using the various assumptions for the single value of I_e, as discussed in Sec. 14.4.

(a) Section at the left support. For the gross section neglecting reinforcement (see Fig. 14.5.5),

$$x = \frac{36(18)18 + 90(4.5)38.25}{36(18) + 90(4.5)} = \frac{27,160}{1053} = 25.79 \text{ in.}$$

$$I_g = \tfrac{1}{3}(18)[(25.79)^3 + (14.71)^3] + \tfrac{1}{12}(72)(4.5)^3 + 72(4.5)(12.46)^2$$

$$= 173,000 \text{ in.}^4$$

Had reinforcement been included, $x = 25.29$ in. and $I_g = 200,000$ in.[4] It is generally accepted that I_g for use in ACI Formula (9-7) should not include steel reinforcement.

For the transformed cracked section [see Fig. 14.5.6(a)],

$$18x \left(\frac{x}{2}\right) + 40.6(x - 2.6) = 40.9(36.7 - x)$$

$$x^2 + 9.05x = 178.2$$

$$x = 9.56 \text{ in.}$$

$$I_{cr} = \tfrac{1}{3}(18)(9.56)^3 + 40.6(9.56 - 2.6)^2 + 40.9(36.7 - 9.56)^2$$

$$= 37,200 \text{ in.}^4$$

It is noted that the modulus of elasticity ratio n is used to transform the steel in the compression zone, which agrees with the concept of computing an elastic (short time) deflection.

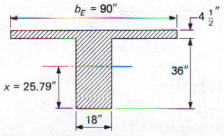

Figure 14.5.5 Gross section of beam for Example 14.5.2.

Compute the cracking moment for the beam with tension on the flange.

$$f_r = 7.5\sqrt{f_c'} = 7.5\sqrt{3000} = 411 \text{ psi}$$

$$M_{cr} = \frac{f_r I_g}{y_t} = \frac{0.411(173,000)}{14.71}\left(\frac{1}{12}\right) = 402 \text{ ft-kips}$$

$$\frac{M_{cr}}{M_{max}} = \frac{402}{89.6}\left(\begin{array}{c}\text{dead}\\\text{load}\end{array}\right) > 1; \qquad \text{use } I_e = I_g = 173,000 \text{ in.}^4$$

$$\frac{M_{cr}}{M_{max}} = \frac{402}{200}\left(\begin{array}{c}\text{dead load}\\+ \text{ live load}\end{array}\right) > 1; \qquad \text{use } I_e = I_g = 173,000 \text{ in.}^4$$

(b) Midspan section [Fig. 14.5.6(b)]. Determine whether or not the neutral axis occurs in the flange by taking moments about the bottom of the flange,

$$90(4.5)(2.25) < 86.6(32.3)$$

Locate the neutral axis, including the effect of compression in the stem.

$$90(4.5)(x - 2.25) + 18(x - 4.5)^2(\tfrac{1}{2}) = 86.6(36.8 - x)$$

$$x^2 + 45.6x = 435$$

$$x = 8.09 \text{ in.}$$

$$I_{cr} = \tfrac{1}{12}(72)(4.5)^3 + 72(4.5)(5.84)^2 + \tfrac{1}{3}(18)(8.09)^3 + 86.6(36.8 - 8.09)^2$$
$$= 86,200 \text{ in.}^4$$

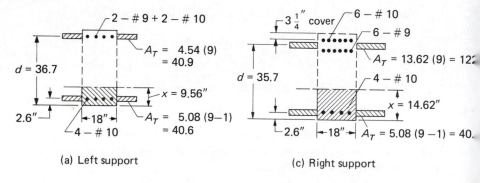

(a) Left support

2 – # 9 + 2 – # 10

$A_T = 4.54 (9) = 40.9$

$d = 36.7$

$x = 9.56''$

$A_T = 5.08 (9–1) = 40.6$

2.6''   ├─18''─┤   4 – # 10

(c) Right support

$3\frac{1}{4}''$ cover   6 – # 10

6 – # 9

$A_T = 13.62 (9) = 12?$

$d = 35.7$

4 – # 10

$x = 14.62''$

2.6''   ├─18''─┤ $A_T = 5.08 (9 –1) = 40.$

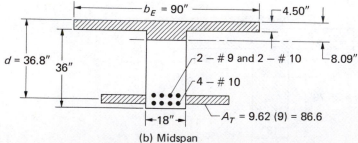

$b_E = 90''$ 4.50''

$d = 36.8''$ 36''

2 – # 9 and 2 – # 10   8.09''

4 – # 10

├─18''─┤ $A_T = 9.62 (9) = 86.6$

(b) Midspan

Figure 14.5.6 Transformed cracked sections for Example 14.5.2.

The cracking moment for the beam with tension in the stem is

$$M_{cr} = \frac{f_r I_g}{y_t} = \frac{0.411(173,000)}{25.79}\left(\frac{1}{12}\right) = 229.5 \text{ ft-kips}$$

$$\frac{M_{cr}}{M_{max}} = \frac{229.5}{229}\left(\frac{\text{dead}}{\text{load}}\right) > 1; \qquad \text{use } I_e = I_g = 173,000 \text{ in.}^4$$

$$\frac{M_{cr}}{M_{max}} = \frac{229.5}{510}\left(\frac{\text{dead load}}{+ \text{ live load}}\right) = 0.45; \qquad \left(\frac{M_{cr}}{M_{max}}\right)^3 = 0.091$$

$$I_e = \left(\frac{M_{cr}}{M_{max}}\right)^3 I_g + \left[1 - \left(\frac{M_{cr}}{M_{max}}\right)^3\right] I_{cr}$$

$$= 0.091(173,000) + 0.909(86,200) = 94,000 \text{ in.}^4$$

(c) Section at the right support [Fig. 14.5.6(c)]. Use transformed cracked section and locate the neutral axis,

$$18x\left(\frac{x}{2}\right) + 40.6(x - 2.6) = 122.6(35.7 - x)$$

$$x^2 + 18.13x = 498$$

$$x = 15.02 \text{ in.}$$

$$I_{cr} = \tfrac{1}{3}(18)(15.02)^3 + 122.6(35.7 - 15.02)^2 + 40.6(15.02 - 2.60)^2$$

$$= 79,000 \text{ in.}^4$$

$M_{cr} = 402$ ft-kips (same as left support)

$$\frac{M_{cr}}{M_{max}} = \frac{402}{377}\left(\frac{\text{dead}}{\text{load}}\right) > 1; \qquad \text{use } I_e = I_g = 173,000 \text{ in.}^4$$

$$\frac{M_{cr}}{M_{max}} = \frac{402}{591}\left(\frac{\text{dead load}}{+ \text{ live load}}\right) = 0.68; \qquad \left(\frac{M_{cr}}{M_{max}}\right)^3 = 0.314$$

$$I_e = 0.314(173,000) + 0.686(79,000) = 109,000 \text{ in.}^4$$

(d) Summary of values for I_e. The values of effective moment of inertia are

	For DL	For DL + LL
Left end	$I_e = 173,000$ in.^{4}	$I_e = 173,000$ in.^{4}
Midspan	$I_e = 173,000$ in.^{4}	$I_e = 94,000$ in.^{4}
Right end	$I_e = 173,000$ in.^{4}	$I_e = 109,000$ in.^{4}

Having obtained the above values, usual practice is to use a single adjusted value of I_e as discussed in Sec. 14.4.

(e) Compute a single adjusted value for I_e using the various procedures of Sec. 14.4.

1. Midspan value:

$$I_e = 173,000 \text{ in.}^4 \text{ (for DL only)}$$

$$I_e = 94,000 \text{ in.}^4 \text{ (for DL + LL)}$$

2. Weighted average:

$$I_e = 0.70I_m + 0.15(I_{e1} + I_{e2})$$

$$I_e = 173,000 \text{ in.}^4 \text{ (for DL only)}$$

$$I_e = 0.70(94,000) + 0.15(173,000 + 109,000)$$

$$= 108,100 \text{ in.}^4 \text{ (for DL + LL)}$$

3. Simple average:

$$I_e = 173,000 \text{ in.}^4 \text{ (for DL only)}$$

$$I_e = \frac{1}{2}\left(\frac{173,000 + 109,000}{2}\right) + 94,000$$

$$= 118,000 \text{ in.}^4 \text{ (for DL + LL)}$$

The task of determining the immediate deflection is actually one of analyzing a continuous beam with variable moment of inertia. Because the most "exact" computations at best give deflections within probably $\pm 20\%$, procedures more complex than those illustrated here are not justified.

(f) Immediate dead load deflection. I_e for dead load only is 173,000 in.^{4} regardless of whether midspan, weighted average, or simple average is used.

$$M_a = 89.6 \text{ ft-kips} \qquad M_b = 377 \text{ ft-kips}$$
$$w = 0.76 \text{ kip/ft} \qquad P = 24.48 \text{ kips}$$

It is to be noted that the span is taken as that measured between the centerlines of supports, and the end moments are those computed for the same locations. Equally acceptable results are obtained by using the clear span and the face-of-support moments. Referring to Fig. 14.5.4 the total midspan deflection is

$$\Delta = \frac{10^6}{E_c I_e} (3.64P + 1.333wL - 0.164M_a - 0.164M_b)$$

$$E_c = 33w_c^{1.5} \sqrt{f_c'} = 57,000 \sqrt{3000} = 3.15 \times 10^6 \text{ psi (ACI-8.5.1)}$$

$$(\Delta_i)_D = \frac{1}{3.15(173)} [3.64(24.48) + 1.333(0.76)(39) - 0.164(89.6 + 377.0)]$$

$$= \frac{89.2 + 39.5 - 76.4}{545} = \frac{52.3}{545} = 0.10 \text{ in.}$$

As previously mentioned, dead load deflection will usually cause no difficulty, as most of it may be compensated for by the construction. However, it is used as a basis for determining the long-time creep and shrinkage deflection which is discussed in the next section.

(g) Immediate live load deflection using the midspan value of I_e. The midspan value is 94,000 in.^{4} as computed in part (e).

$$(\Delta_i)_D = 0.10 \text{ in.} \qquad [\text{see part (f)}]$$

$$M_a = 200 \text{ ft-kips} \qquad M_b = 591 \text{ ft-kips}$$
$$w = 0.91 \text{ kip/ft} \qquad P = 56.33 \text{ kips}$$

$$(\Delta_i)_{D+L} = \frac{1}{3.15(94)} [3.64(56.33) + 1.333(0.91)(39) - 0.164(200 + 591)]$$

$$= \frac{205.0 + 47.0 - 129.7}{296} = \frac{122.3}{296} = 0.41 \text{ in.}$$

$$(\Delta_i)_L = 0.41 - 0.10 = 0.31 \text{ in.}$$

(h) Immediate live load deflection using the weighted average value of I_e.

$$(\Delta_i)_D = 0.10 \text{ in.} \qquad [\text{see part (f)}]$$

weighted average value of $I_e = 108,100 \text{ in.}^4$

$$(\Delta_i)_{D+L} = 0.41\left(\frac{94,000}{108,100}\right) = 0.36 \text{ in.}$$

$$(\Delta_i)_L = 0.36 - 0.10 = 0.26 \text{ in.}$$

(i) Immediate live load deflection using the simple average value of I_e.

$$(\Delta_i)_D = 0.10 \text{ in.} \qquad [\text{see part (f)}]$$

simple average value of $I_e = 118,000 \text{ in.}^4$

$$(\Delta_i)_{D+L} = 0.41\left(\frac{94,000}{118,000}\right) = 0.33 \text{ in.}$$

$$(\Delta_i)_L = (\Delta_i)_{D+L} - (\Delta_i)_D = 0.33 - 0.10 = 0.23 \text{ in.}$$

(j) Conclusion. The immediate live load deflection is computed to be 0.31 in., 0.26 in., and 0.23 in., respectively, depending on whether midspan, weighted average, or simple average value of I_e is used for dead and live load deflection. The use of midspan I_e seems appropriate; for this example one can conclude that the dead load deflection is about 0.1 in. and the live load deflection is about 0.3 in., using *one* significant figure.

If this is a floor that does not support frangible partitions, the limiting permissible deflection is $L/360$, or

$$\frac{L}{360} = \frac{39(12)}{360} = 1.3 \text{ in.} > 0.3 \text{ in. (computed)} \qquad \text{OK}$$

The reinforcement ratios used ($0.14\rho_b$ for the positive moment region and $0.37\rho_b$ for the negative moment regions) were below the guideline value suggested in Chap. 3 for deflection control; thus excessive deflection was not expected.

Recommended Values for Maximum Reinforcement Ratio ρ for Deflection Control.

ACI Committee 435 [9] has recommended the following values of the maximum reinforcement ratio ρ to be used in the positive moment zone for deflection control:

1. For members of normal weight concrete *not supporting* or *not attached* to nonstructural elements likely to be damaged by large deflections,

Rectangular beams $\qquad$ $0.35\rho_b$

T-beams or box beams $\qquad$ $0.40\rho_b$

2. For members of normal weight concrete *supporting* or *attached* to nonstructural elements likely to be damaged by large deflections,

Rectangular beams $\qquad$ $0.25\rho_b$

T-beams or box beams $\qquad$ $0.30\rho_b$

3. For members of lightweight concrete, use $0.05\rho_b$ *less* than that indicated in 1 and 2.

14.6 CREEP EFFECT ON DEFLECTIONS UNDER SUSTAINED LOAD

The total long-term deflection includes the instantaneous elastic deflection plus the contributions from creep and shrinkage. Creep is inelastic deformation with time under sustained loads at unit stresses within the accepted elastic range (say, below $0.5f_c'$), as shown in Fig. 14.6.1. This inelastic deformation increases at a decreasing rate during the time of loading.

Factors that affect the magnitude of creep deformation [8,17,18] are (1) the constituents—such as the composition and fineness of the cement, the admixtures, and the size, grading, and mineral content of the aggregates; (2) proportions, such as water content and water–cement ratio; (3) curing temperature and humidity; (4) relative humidity during storage; (5) size of the concrete member, particularly the thickness and the volume to surface ratio; (6) age at loading; (7) duration of loading; and (8) magnitude of stress.

Since, as seen from Fig. 14.6.1, the result of creep is an increase in strain with constant stress, one of the ways of accounting for it is by the use

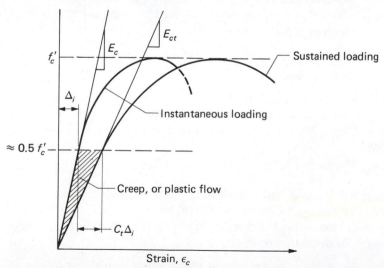

Figure 14.6.1 Typical stress-strain curves for instantaneous and long-time loading.

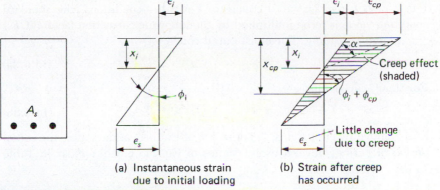

Figure 14.6.2 Creep effect on beam curvature.

of a modified modulus of elasticity E_{ct}. An alternative and more recently preferred procedure is to apply a multiplier C_t to the elastic deflection Δ_i.

In order to help understand the qualitative effect of creep on beam deformation, consider the singly reinforced beam of Fig. 14.6.2. It is noted that the strain at the tension steel is essentially unchanged because the concrete contributes little in taking tension and ordinary deformed steel reinforcement exhibits little creep. Since the neutral axis moves down, two observations may be made: (1) The concrete stress reduces at the compression face (i.e., same compressive force acting and x_{cp} exceeds x_i); and (2) the increase in compressive strain is much greater than the increase in curvature ϕ.

For deflection purposes it is frequently desirable to use a creep coefficient C_t defined as the ratio of creep strain to elastic strain,

$$C_t = \frac{\epsilon_{cp}}{\epsilon_i} \qquad (14.6.1)$$

ACI Committee 209 has recommended [7] the hyperbolic-type equation of Branson et al. [7,8,19] for the creep coefficient, as follows:

$$C_t = \left(\frac{t^{0.60}}{10 + t^{0.60}}\right)C_u \qquad (14.6.2)$$

where

C_t = ratio of creep strain to elastic strain at any time t after a basic curing period
t = time in days after loading
C_u = ultimate creep coefficient: recommended *average* value is 2.35 for 40% humidity.

The general relationship of C_t/C_u appears in Fig. 1.10.2.

Equation (14.6.2) applies to the standard condition of 40% ambient relative humidity, 4 in. (100 mm) or less slump, average thickness of member 6 in. (150 mm), and loading age of 7 days for moist-cured concrete or

1 to 3 days for steam-cured concrete. For other conditions, the standard condition value is to be multiplied by the following correction factors (CF):

1. Age at loading. For moist-cured concrete,

$$(CF)_a = 1.25t_a^{-0.118} \tag{14.6.3a}$$

For steam-cured concrete,

$$(CF)_a = 1.13t_a^{-0.095} \tag{14.6.3b}$$

In the above two equations, t_a is the age at loading, in days after the initial period of curing. Several useful values of Eq. (14.6.3ab) appear in Table 14.6.1.

Table 14.6.1 CREEP CORRECTION FACTOR, $(CF)_a$ FOR AGE AT LOADING, EQ. (14.6.3ab)

	CORRECTION FACTOR, $(CF)_a$	
t_a, AGE IN DAYS AFTER INITIAL CURING PERIOD	MOIST CURED FOR 7 DAYS INITIAL CURING PERIOD	STEAM CURED FOR 1–3 DAYS INITIAL CURING PERIOD
10	0.95	0.90
20	0.87	0.85
30	0.83	0.82
60	0.77	0.76
90	0.74	0.74

2. Humidity. For $H \geq 40\%$,

$$(CF)_h = 1.27 - 0.0067H \tag{14.6.4}$$

where H is the ambient relative humidity in percent. Values for this correction factor appear in Table 14.6.2.

Table 14.6.2 CREEP CORRECTION FACTOR, $(CF)_h$ FOR HUMIDITY, EQ. (14.6.4)

AMBIENT RELATIVE HUMIDITY, H, PERCENT	CORRECTION FACTOR $(CF)_h$
40 or less	1.00
50	0.94
60	0.87
70	0.80
80	0.73
90	0.67
100	0.60

3. Average thickness of member. Where the average thickness of the member in inches exceeds 6 in. (150 mm), a correction factor (reduction factor) may be applied. However, for most design purposes such a correction may be neglected. For members whose average thickness greatly exceeds 12 in. (300 mm), Ref. 20 provides a chart that may be used to correct for the effect of average thickness.

4. Other correction factors. Additional correction factors are available to account for variations in thickness (pp. 22–23 of Branson [8]), slump greater than 4 in. [8, p. 45], cement content [8, p. 45], percent of fine aggregate [8, p. 45], and air content [8, p. 47]; however, these tend to be either small or offset one another and may generally be neglected. Meyers and Branson [20] provide a simple chart for these additional correction factors if they are desired.

Compression Steel Effect on Creep. The presence of compression steel decreases the deformation due to creep (and shrinkage as discussed in the next section). Evaluation of the effect of compression steel has been reported by Washa and Fluck [10], Yu and Winter [4], and Hollington [21], and a multiplier factor has been given by Branson [22] and recommended by Committee 435 [9], as follows:

$$k_r = \frac{0.85}{1 + 50\rho'} \qquad (14.6.5)$$

where ρ' is A_s'/bd, the compression steel reinforcement ratio.

Thus from Fig. 14.6.1, $C_t \Delta_i$ would become $k_r C_t \Delta_i$ when the compression steel effect is included.

14.7 SHRINKAGE EFFECT ON DEFLECTIONS UNDER SUSTAINED LOAD

Shrinkage of concrete in beams may have a similar effect on the deflection as creep. Shrinkage of an isolated plain concrete member would merely shorten it without causing curvature. When steel reinforcement is added, however, bond between concrete and steel restrains the shrinkage. Thus, a singly reinforced beam, having its shrinkage restrained at the reinforced face and unrestrained at the unreinforced face, will have considerable curvature. Generally it is difficult to separate the effects of creep and shrinkage. Shrinkage occurs more pronounced during the first few months than does creep. Typically, 90% of the shrinkage will have occurred at the end of 1 year, whereas not until the end of 5 years will 90% of the creep have occurred. A number of investigators have studied shrinkage effects separately from those of creep [7,8,18–20,23–25].

If the free shrinkage strain is known, shrinkage curvature ϕ_{sh} must be determined as a function of shrinkage strain. Such curvature will be dependent on the relative amounts of compression and tension steel just as creep is so affected. Finally, the shrinkage deflection will involve the geometry of the support system. Shrinkage deflection Δ_{sh} may be expressed [7,16] as

$$\Delta_{sh} = \alpha_1 \phi_{sh} L^2 \qquad (14.7.1)$$

where α_1 is a factor relating to the geometry of the support system and may be taken as the following:

$$\alpha_1 = 0.50 \text{ cantilever beams}$$

$$= 0.125 \text{ simply supported beams}$$

$$= 0.086 \text{ beams continuous at one end only}$$

$$= 0.063 \text{ beams continuous at both ends}$$

and L is the span length of the beam.

Shrinkage Strain, ϵ_{sh}. ACI Committee 209 has recommended [7] that the following expressions by Branson et al. [8, 19] may be used for shrinkage strain ϵ_{sh}:

For any time t after age 7 days for moist-cured concrete,

$$\epsilon_{sh} = \frac{t}{35 + t} (\epsilon_{sh})_u \tag{14.7.2a}$$

For any time t after age 1 to 3 days for steam-cured concrete,

$$\epsilon_{sh} = \frac{t}{55 + t} (\epsilon_{sh})_u \tag{14.7.2b}$$

where

ϵ_{sh} = shrinkage strain at any time t after initial curing
t = time in days after initial curing
$(\epsilon_{sh})_u$ = ultimate shrinkage strain: average value suggested is 800×10^{-6} in./in. for 40% humidity.

Equation (14.7.2a) is shown graphically in Fig. 1.10.4. For conditions other than 40% ambient relative humidity, the standard condition value of Eqs. (14.7.2ab) is to be multiplied by the following correction factor (CF):

$$(CF)_h = 1.40 - 0.010H, \qquad 40 \le H \le 80\% \tag{14.7.3}$$

$$(CF)_h = 3.00 - 0.030H, \qquad H \ge 80\% \tag{14.7.4}$$

where H is the relative humidity in percent. Values for $(CF)_h$ appear in Table 14.7.1.

Table 14.7.1 SHRINKAGE CORRECTION FACTOR $(CF)_h$ FOR HUMIDITY, EQS. (14.7.3) AND (14.7.4)

AMBIENT RELATIVE HUMIDITY, H, PERCENT	CORRECTION FACTOR $(CF)_h$
40 or less	1.00
50	0.90
60	0.80
70	0.70
80	0.60
90	0.30
100	0

Other correction factors may normally be neglected. Should such factors be desired, corrections for average thickness other than 6 in. [8, pp. 22–23], slump greater than 4 in. [8, p. 45], cement content [8, p. 45], percentage of fines [8, p. 45], and air content [8, p. 47] are available.

Shrinkage Curvature, ϕ_{sh}. Several investigators [5,7,16,25] have developed expressions for curvature due to warping that arises from nonuniform shrinkage. Reinforcement of different amounts in the two faces of a beam is the principal cause of shrinkage warping.

Miller [25] established the following relationship for the singly reinforced beam. Referring to Fig. 14.7.1, by straightline proportion,

$$\phi_{sh} = \frac{\epsilon_{sh} - \epsilon_s}{d} = \frac{\epsilon_{sh}}{d}\left(1 - \frac{\epsilon_s}{\epsilon_{sh}}\right) \qquad (14.7.5)$$

where ϵ_s is the compressive strain induced in the steel from shrinkage; ϵ_{sh} is the free shrinkage strain at the unreinforced face. Miller established empirically values for ϵ_s/ϵ_{sh} as a function of the percentage of reinforcement ρ.

Branson [5] modified Miller's equation and empirically extended the results to give equations including the effects of compression steel.

$$\phi_{sh} = 0.7 \frac{\epsilon_{sh}}{h} (\rho - \rho')^{1/3}\left(\frac{\rho - \rho'}{\rho}\right)^{1/2} \qquad \text{for } (\rho - \rho') \le 3\% \quad (14.7.6)$$

$$\phi_{sh} = \frac{\epsilon_{sh}}{h} \qquad \text{for } (\rho - \rho') > 3\% \qquad (14.7.7)$$

Note that ρ and ρ' are in percent, $100\,(A_s$ or $A_s')/(bd)$. Equations (14.7.6) and (14.7.7) are recommended [7,8,16] as the most accurate relationships available.

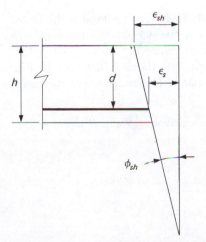

Figure 14.7.1   Shrinkage strain related to beam curvature for a singly reinforced beam (after Miller, Ref. 25).

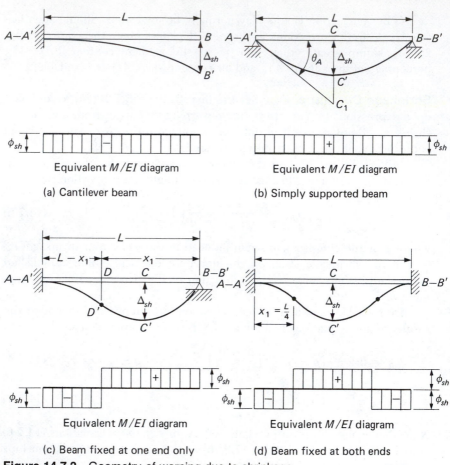

Figure 14.7.2 Geometry of warping due to shrinkage.

Geometry of Warping from Shrinkage. In order to establish the factor α_1 in Eq. (14.7.1) for the four typical cases, the well-known moment area theorems for beam deflections may be used. Since the quantity $M/(EI)$ is in fact the curvature due to bending moment, the ϕ_{sh} diagrams in Fig. 14.7.2 may be regarded as the equivalent $M/(EI)$ diagrams. For the cantilever beam [Fig. 14.7.2(a)],

$$\Delta_{sh} = BB' = \text{moment of } \phi_{sh} \text{ diagram between } A \text{ and } B \text{ about } B$$

$$= (\phi_{sh}L)\left(\frac{L}{2}\right) = 0.50\phi_{sh}L^2$$

For the simply supported beam [Fig. 14.7.2(b)],

$$\theta_A = \text{area of } \phi_{sh} \text{ diagram between } A \text{ and } C$$

$$= \phi_{sh}\left(\frac{L}{2}\right)$$

$$\Delta_{sh} = CC' = CC_1 - C_1C'$$

$$= \theta_A\left(\frac{L}{2}\right) - (\text{moment of } \phi_{sh} \text{ diagram between } A \text{ and } C \text{ about } C)$$

$$= \phi_{sh}\left(\frac{L}{2}\right)\left(\frac{L}{2}\right) - \phi_{sh}\left(\frac{L}{2}\right)\left(\frac{L}{4}\right) = 0.125\phi_{sh}L^2$$

For the beam fixed at one end only [Fig. 14.7.2(c)],

deflection of B from tangent at A

= moment of ϕ_{sh} diagram between A and B about B

$$= -\phi_{sh}(L - x_1)\left(x_1 + \frac{L - x_1}{2}\right) + \phi_{sh}\left(\frac{x_1^2}{2}\right) = 0$$

from which

$$x_1 = \frac{\sqrt{2}L}{2}$$

In order that the tangent at C' be horizontal, the distances AD and DC must be equal; thus

$$AD = DC = L - x_1 = \left(1 - \frac{\sqrt{2}}{2}\right)L$$

$$\Delta_{sh} = CC' = \text{moment of } \phi_{sh} \text{ diagram between } A \text{ and } C \text{ about } C$$

$$= \text{moment of a couple} = \phi_{sh}(L - x_1)^2$$

$$= \phi_{sh}L^2(1 - \tfrac{1}{2}\sqrt{2})^2 = \phi_{sh}L^2(1 - \sqrt{2} + \tfrac{1}{2})$$

$$= 0.086\phi_{sh}L^2$$

For the beam fixed at both ends [Fig. 14.7.2(d)], in order that the slope be horizontal at midspan for symmetry,

$$x_1 = \frac{L}{4}$$

Then

$$\Delta_{sh} = CC' = \text{moment of } \phi_{sh} \text{ diagram between } A \text{ and } C \text{ about } C$$

$$= \text{moment of a couple} = \phi_{sh}\left(\frac{L}{4}\right)^2$$

$$= 0.063\phi_{sh}L^2$$

Compression Steel Effect on Combined Shrinkage and Creep. Whenever shrinkage is to be considered *in combination with creep* in the computation for the deflection due to sustained load, a multiplier similar to that given by Eq. (14.6.5) may be applied [7] to the short-time deflection. The general relationship $k_r T \Delta_i$ to replace $C_t \Delta_i$ in Fig. 14.6.1 when compression steel is present has been discussed by Branson [22] and Shaikh [26], and the following has been recommended by ACI Committee 435 [9] as most appropriate,

$$k_r = \frac{1}{1 + 50\rho'} \qquad (14.7.8)$$

where $\rho' = A_s'/bd$, and k_r includes the effects of both creep and shrinkage. Equation (14.7.8) has been adopted by the 1983 ACI Code.

14.8 CREEP AND SHRINKAGE DEFLECTION—ACI CODE METHOD

In the ACI Code method, the creep and shrinkage deflection due to sustained load is obtained by multiplying the short-time (instantaneous) deflection by a factor λ (ACI-9.5.2.5). Thus

$$\Delta_{cp+sh} = k_r T(\Delta_i)_D = \lambda(\Delta_i)_D \qquad (14.8.1)$$

where

$$\lambda = k_r T = \frac{T}{1 + 50\rho'} \qquad (14.8.2)$$

and $(\Delta_i)_D$ is the instantaneous deflection due to all sustained loads (usually dead load).

The value of T (ACI uses symbol ξ) may be taken as the following in accordance with duration of sustained load:

5 years or more	2.0
1 year	1.4
6 months	1.2
3 months	1.0

EXAMPLE 14.8.1 For the beam of Example 14.5.1 determine the creep and shrinkage deflection according to the ACI Code. Assume only the dead load is sustained.

Solution: First it is necessary to compute the short-time (instantaneous) deflection due to all sustained loads, in this case the dead load. From Example 14.5.1, part (b) of the solution,

$$(\Delta_i)_D = 0.45 \text{ in.}$$

Since no compression steel is used, Eq. (14.8.2) gives for 5 years or more load duration,

$$\lambda = k_r T = 2.0$$

Then from Eq. (14.8.1)

$$\Delta_{cp+sh} = \lambda(\Delta_i)_D = 2.0(0.45) = 0.90 \text{ in.}$$

If part of the live load were considered as sustained, such as certain types of equipment whose placement or installation is not expected to change for a period of 5 years or more, then it would be necessary to compute Δ_i for the dead load plus the sustained live load. Under the ACI Code, an additional effective moment of inertia I_e would be computed using M_{cr}/M_{max} where M_{max} is due to dead load plus sustained live load.

14.9 CREEP AND SHRINKAGE DEFLECTION— ALTERNATE PROCEDURES

Separate Creep and Shrinkage Multiplier Procedure. A procedure for computing the deflections due to creep and shrinkage separately was recommended by ACI Committee 435 [16,27], based on the work of Branson [5], as modfied by recent improvement in the prediction of creep and shrinkage [7]. Thus

$$\Delta_{cp+sh} = \Delta_{cp} + \Delta_{sh} \tag{14.9.1}$$

where, using Eq. (14.7.1),

$$\Delta_{sh} = \alpha_1 \phi_{sh} L^2 \tag{14.9.2}$$

and

$$\Delta_{cp} = k_r C_t (\Delta_i)_D \tag{14.9.3}$$

For evaluation of Eqs. (14.9.2) and (14.9.3),

α_1 = constant (see Eq. 14.7.1)
ϕ_{sh} = shrinkage curvature [Eqs. (14.7.6) and (14.7.7)]
L = span length
k_r = compression steel factor [Eq. (14.6.5)]
C_t = creep coefficient, originally given in tabular form [16], but more recently [7] and more accurately as Eq. (14.6.2) with correction factors of Eqs. (14.6.3) and (14.6.4)
$(\Delta_i)_D$ = instantaneous deflection due to all sustained loads

Combined Creep and Shrinkage Multiplier Procedure. This method is similar to the ACI Code method, except the time-dependent factor T can be more accurately evaluated [7,8,22,27]. It may be stated as

$$\Delta_{cp+sh} = k_r T (\Delta_i)_D \tag{14.9.4}$$

where

$k_r = 1/(1 + 50\rho')$ (same as ACI) $\tag{14.9.5}$
T = time-dependent coefficient (creep plus shrinkage), which may be taken from Table 14.9.1

EXAMPLE 14.9.1 For the beam of Example 14.5.1 determine the ultimate (i.e., 5 years duration of load) creep and shrinkage deflection using methods more accurate than the basic ACI method. Assume only the dead load is sustained, the ambient relative humidity is 70%, the age at loading is 20 days after the initial moist-curing period.

Solution: It is noted that ACI-9.5.2.5 permits computation of long-time deflection by a "more comprehensive analysis," which could include either of these alternate methods.

(a) Separate creep and shrinkage multiplier procedure.

$$\Delta_{cp+sh} = \Delta_{cp} + \Delta_{sh}$$

Table 14.9.1 TIME-DEPENDENT COEFFICIENT T INCLUDING BOTH CREEP AND SHRINKAGE EFFECTS, FOR BOTH NORMAL WEIGHT AND LIGHTWEIGHT CONCRETE MEMBERS OF COMMON TYPES, SIZES, AND COMPOSITION [FROM BRANSON (27)][a]

CONCRETE STRENGTH f'_c AT 28 DAYS	AVERAGE RELATIVE HUMIDITY, AGE WHEN LOADED								
	100%			70%			50%		
	$\leq 7d$	$14d$	$\geq 28d$	$\leq 7d$	$14d$	$\geq 28d$	$\leq 7d$	$14d$	$\geq 28d$
2500 to 4000 psi (17 to 28 MPa)	2.0	1.5	1.0	3.0	2.0	1.5	4.0	3.0	2.0
>4,000 psi (28 MPa)	1.5	1.0	0.7	2.5	1.8	1.2	3.5	2.5	1.5

[a]It is suggested that the following percentages of the values in the table be used for sustained loads that are maintained for the periods indicated:

> 25% for 1 month or less
> 50% for 3 months
> 75% for 1 year
> 100% for 5 years or more

The 50% values may normally be used for average relative humidities lower than 50%, which might be the case in heated buildings, for example.

Using Eq. (14.9.3),

$$\Delta_{cp} = k_r C_t (\Delta_i)_D$$

$$k_r = \frac{0.85}{1 + 50\rho'} = 0.85 \qquad \text{for } \rho' = 0$$

and from Eq. (14.6.2), for $t = 5(365)$ days,

$$C_t = \left(\frac{t^{0.60}}{10 + t^{0.60}} \right) C_u = 0.90 C_u$$

which for $C_u = 2.35$ as recommended for average conditions gives the basic value of C_t as

$$C_t = 2.12$$

Adjusting for 70% humidity, $(CF)_h = 0.80$ from Eq. (14.6.4) or Table 14.6.2, and for 20-day age of loading after initial moist-curing period, $(CF)_a = 0.87$. Thus the adjusted C_t is

$$C_t = 2.12(0.80)(0.87) = 1.47$$

From Example 14.5.1 using the ACI Code method,

$$(\Delta_i)_D = 0.45 \text{ in.}$$

Then

$$\Delta_{cp} = k_r C_t (\Delta_i)_D = 0.85(1.47)0.45 = 0.56 \text{ in.}$$

For shrinkage, from Eq. (14.7.1),

$$\Delta_{sh} = \alpha_1 \phi_{sh} L^2$$

where

$$\alpha_1 = 0.125 \quad \text{for simply supported beams}$$

For this beam, since $\rho = 1.92\%$ and $\rho' = 0$, Eq. (14.7.6) gives

$$\phi_{sh} = 0.7\left(\frac{\epsilon_{sh}}{h}\right)\sqrt[3]{\rho}$$

Using ϵ_{sh} from Eq. (14.7.2a),

$$\epsilon_{sh} = \frac{t}{35 + t}(\epsilon_{sh})_u$$

which for $t = 5(365)$ days is, for average conditions,

$$\epsilon_{sh} \approx (\epsilon_{sh})_u = 800 \times 10^{-6} \text{ in./in.}$$

Adjusting for 70% humidity, $(CF)_h = 0.70$, from Eq. (14.7.3) or Table 14.7.1, the adjusted ϵ_{sh} is

$$\epsilon_{sh} = (800 \times 10^{-6})0.70 = 560 \times 10^{-6} \text{ in./in.}$$

Then,

$$\phi_{sh} = 0.7\left(\frac{560 \times 10^{-6}}{24}\right)\sqrt[3]{1.92} = 20.3 \times 10^{-6} \text{ in.}$$

$$\Delta_{sh} = \alpha_1\phi_{sh}L^2 = 0.125(20.3 \times 10^{-6})(480)^2 = 0.58 \text{ in.}$$

$$\Delta_{cp+sh} = 0.56 + 0.58 = 1.14 \text{ in.}$$

(b) Combined creep and shrinkage multiplier procedure. Using Eq. (14.9.4),

$$\Delta_{cp+sh} = k_r T(\Delta_i)_D$$

$$k_r = \frac{1}{1 + 50\rho'} = 1.0$$

$$T = \text{value from Table 14.9.1} \approx 1.8$$

Note that age at loading is *after* initial curing period.

$$(\Delta_i)_D = 0.45 \text{ in.} \quad \text{(from Example 14.5.1)}$$

$$\Delta_{cp+sh} = 1.0(1.8)0.45 = 0.81 \text{ in.}$$

A comparison of computation methods for creep and shrinkage deflection may be obtained from the following summary.

METHOD	Δ_{sh+cp}
1. ACI, using $\lambda = k_r T = 2.0$	0.90 in. (Example 14.8.1)
2. Separate creep and shrinkage as per Eqs. (14.9.1) to (14.9.3), etc.	1.14 in. (Example 14.9.1a)
3. Combined creep and shrinkage using $k_r T = 1.8$	0.81 in. (Example 14.9.1b)

14.10 ACI MINIMUM DEPTH OF FLEXURAL MEMBERS

According to the ACI Code the minimum depths specified in ACI-9.5.2.1 shall apply to all cases of "one-way construction . . . unless computation of deflection indicates a lesser thickness may be used without adverse effects." The minimum depth (thickness) values apply to members where large deflection is *not* likely to damage partitions, ceilings, or other frangible attachments. When large deflection may cause such damage, deflections must be computed whether or not the minimum thickness requirement is satisfied.

The minimum depths prescribed by any such table are arbitrary and not necessarily conservative. The following logic may be used to explain the limitation on span–depth ratio as an attempt to control deflection.

The deflection at midspan of a simply supported beam is

$$\Delta = \frac{5wL^4}{384EI} \tag{14.10.1}$$

The maximum bending moment is

$$M = \frac{wL^2}{8} = \frac{fI}{c} \tag{14.10.2}$$

Substituting Eq. (14.10.2) in Eq. (14.10.1) gives

$$\Delta = \frac{5L^2 f}{48Ec} \tag{14.10.3}$$

Assuming that cracked section is effective at service load conditions,

$$\frac{f}{c} = \frac{f_s}{n(d - x)} = \frac{f_c}{x} \tag{14.10.4}$$

where x is the distance between the neutral axis and the extreme compression fiber. Assume the beam is fully stressed at $f_s = 24{,}000$ psi for Grade 60 steel, and that $x \approx 0.4h$.

$$\frac{f}{c} \approx \frac{24{,}000}{(E_s/E_c)(0.6h)} = \frac{40{,}000 E_c}{h E_s} = \frac{E_c}{725h} \tag{14.10.5}$$

for $E_s = 29 \times 10^6$ psi. Substituting Eq. (14.10.5) in Eq. (14.10.3) and calling $E = E_c$.

$$\frac{\Delta}{L} = \frac{5}{48}\left(\frac{1}{725}\right)\frac{L}{h}$$

$$\min h = \frac{1}{6950}\left(\frac{L}{\Delta}\right)L \tag{14.10.6}$$

Equation (14.10.6) represents an approximate relationship between depth, span, and span-to-deflection ratio for a fully stressed section under *short-time loading*. If the member is under a reduced stress, the depth may be decreased proportionally to give the same short-time deflection. To account

Table 14.10.1 MINIMUM DEPTHS FOR VARIOUS EQUIVALENT IMMEDIATE DEFLECTIONS AND PERCENTAGE STRESSED[a]

PERCENT STRESSED	$\Delta = L/300$	$\Delta = L/360$	$\Delta = L/480$	$2\Delta = L/300$	$2\Delta = L/360$
100	$L/23.2$	$L/19.3$	$L/14.5$	$L/11.5$	$L/9.7$
67	$L/35$	$L/29$	$L/21.5$	$L/17.5$	$L/14.5$
60	$L/39$	$L/32$	$L/24$	$L/19.5$	$L/16$
50	$L/46.5$	$L/38.5$	$L/29$	$L/23$	$L/19.5$

[a]Assumes $f_s = 24,000$ psi at 100% stressed.

for the sustained load creep and shrinkage deflection, the depth must be increased. Table 14.10.1 shows the evaluation of Eq. (14.10.6) to give the minimum depth required for various deflection limitations under fully and partially stressed conditions. The last two columns in Table 14.10.1 assume that total deflection including creep and shrinkage effects is twice the immediate deflection.

Any selection of limiting values for minimum depth from Table 14.10.1 can only be a crude attempt to control deflection. Table 14.10.2 (ACI-Table 9.5a) is the result of a compromise between the relative conservative recommendations of ACI Committee 435 and the values practicing engineers believe suitable on the basis of experience.

When large deflections may cause cracking of partitions and other frangible attachments, the total deflection (ACI-9.5.2.6) that occurs after installation of such elements is limited to $L/480$. This shows that the minimum depths of ACI-Table 9.5a are likely to be too low; hence that table *does not apply for such cases*, and deflection computations must be made.

In general, minimum depth as a proportion of span is an inadequate criterion for controlling deflection; computation of deflection should be made whenever deflection is of concern.

Table 14.10.2 MINIMUM DEPTH h FOR BEAMS AND ONE-WAY SLABS, FOR MEMBERS *NOT* SUPPORTING OR ATTACHED TO PARTITIONS OR OTHER CONSTRUCTION LIKELY TO BE DAMAGED BY LARGE DEFLECTIONS (FROM ACI-TABLE 9.5a)

TYPE OF MEMBERS		SIMPLE SUPPORT	ONE END CONTINUOUS	BOTH ENDS CONTINUOUS	CANTILEVER
Beams	$f_y = 60$ ksi	$L/16$	$L/18.5$	$L/21$	$L/8$
	$f_y = 40$ ksi	$L/20$	$L/23$	$L/26$	$L/10$
One-way slabs (solid)	$f_y = 60$ ksi	$L/20$	$L/24$	$L/28$	$L/10$
	$f_y = 40$ ksi	$L/25$	$L/30$	$L/35$	$L/12.5$

NOTE: For structural lightweight concrete having unit weights w_c from $90 - 120$ pcf, multiply table values by $1.65 - 0.005\ w_c$ but not less than 1.09. (60 ksi = 420 MPa, 40 ksi = 280 MPa, approximately.)

14.11 SPAN-TO-DEPTH RATIO TO ACCOUNT FOR CRACKING AND SUSTAINED LOAD EFFECTS

In order to illustrate the effects of the many variables on the span-to-depth ratio, the following general development, similar to that of Branson [27], is presented.

Grossman [30] has also provided a procedure for determining the minimum thickness that would approximately satisfy any deflection limitation given by ACI-Table 9.5b. The Grossman procedure has approximated the effective moment of inertia, thus eliminating the need to compute the moment of inertia I_{cr} of the cracked section. ACI Committee 318 seriously considered adopting the Grossman approach for the 1983 ACI Code; the approach is still being considered for a future code revision. The development in this section may be less practical as a means to obtain an answer than the Grossman method, but it is intended to illustrate how the variables interrelate, and to show why it is impossible to have a simple table of minimum thicknesses. The Grossman work [30] along with Branson's discussion [31] and Grossman's closure [30] provides the reader an interesting and useful discourse on the subject of deflection control. Rangan and Donaghy [32] and Rangan [35] have also presented a minimum thickness approach similar to that of Branson [27] and to the following development.

The short-time deflection Δ_i may be expressed, according to Eq. (14.5.1), as

$$\Delta_i = \beta_a \left(\frac{M_{\max} L^2}{E_c I_e} \right) \tag{14.11.1}$$

where

$M_{\max}$ = maximum moment at the stage for which deflection is desired

I_e = Eq. (14.4.1), [which is ACI Code Formula (9-7)]
= $(M_{cr}/M_{\max})^3 I_g + [1 - (M_{cr}/M_{\max})^3] I_{cr} \le I_g$

M_{cr} = maximum moment to cause a beam to crack at the extreme fiber in tension = $f_r I_g / y_t$

f_r = modulus of rupture = $0.65 \sqrt{w_c f_c'}$ (ACI uses $7.5 \sqrt{f_c'}$ for normal weight concrete)

y_t = distance from neutral axis to extreme fiber in tension

Multiplying Eq. (14.11.1) by $f_r I_g / (y_t M_{cr})$, which is equal to unity, gives

$$\Delta_i = \beta_a \left(\frac{M_{\max} L^2}{E_c I_e} \right) \frac{f_r I_g}{y_t M_{cr}} \tag{14.11.2}$$

Solving Eq. (14.11.2) for L/y_t gives

$$\frac{L}{y_t} = \frac{\Delta_i}{L} \left(\frac{E_c}{f_r \beta_a} \right) \left(\frac{M_{cr}}{M_{\max}} \right) \frac{I_e}{I_g} \tag{14.11.3}$$

Since both E_c and f_r are proportional to $\sqrt{f_c'}$, let

$$\beta_w = \frac{E_c}{f_r} = \frac{33 w_c^{1.5} \sqrt{f_c'}}{0.65 w_c^{0.5} \sqrt{f_c'}} = 50.8 w_c \tag{14.11.4}$$

For normal-weight concrete

$$\beta_w = 50.8 w_c = 50.8(145) = 7370$$

Conversion from normal-weight concrete to lightweight concrete may be made by multiplying L/y_t by the ratio of the unit weight of lightweight concrete to 145 pcf (2330 kg/m³).

Letting $\gamma = I_e/I_g$ and $\beta_w = E_c/f_r$ in Eq. (14.11.3),

$$\frac{L}{y_t} = \frac{\Delta_i}{L}\left(\frac{\beta_w}{\beta_a}\right)\left(\frac{M_{cr}}{M_{max}}\right)\gamma \tag{14.11.5}$$

where

$$\gamma = \frac{I_e}{I_g} = \left(\frac{M_{cr}}{M_{max}}\right)^3 + \left[1 - \left(\frac{M_{cr}}{M_{max}}\right)^3\right]\frac{I_{cr}}{I_g} \tag{14.11.6}$$

Using the methods of Chap. 4, the ratio of the moment of inertia of the cracked section to that of the gross section may be computed for various shapes of beams. As an example, for a singly reinforced rectangular beam assuming $d = 0.9h$,

$$\frac{I_{cr}}{I_g} = \frac{bx^3/3 + nA_s(d-x)^2}{bh^3/12}$$

$$= 8.75\left[\frac{(x/d)^3}{3} + n\rho\left(1 - \frac{x}{d}\right)^2\right] \tag{14.11.7}$$

where

$$x/d = \sqrt{(\rho n)^2 + 2\rho n} - \rho n \tag{14.11.8}$$

Approximate expressions of $\gamma = I_e/I_g$ for the positive and negative moment regions of a T-shaped beam are given by Branson [27].

For dead load deflection, $M_{max} = M_D$, which makes $\gamma = \gamma_D$; Eq. (14.11.5) then becomes

$$\frac{(\Delta_i)_D}{L} = \frac{\beta_a}{\beta_w}\left(\frac{M_D}{M_{cr}}\right)\left(\frac{L}{y_t}\right)\left(\frac{1}{\gamma_D}\right) \tag{14.11.9}$$

For dead load plus live load,

$$\frac{(\Delta_i)_{D+L}}{L} = \frac{\beta_a}{\beta_w}\left(\frac{M_{D+L}}{M_{cr}}\right)\left(\frac{L}{y_t}\right)\left(\frac{1}{\gamma_{D+L}}\right) \tag{14.11.10}$$

As discussed in Sec. 14.5, since live load cannot act in the absence of dead load, the live load deflection must be obtained indirectly,

$$\frac{(\Delta_i)_L}{L} = \frac{(\Delta_i)_{D+L}}{L} - \frac{(\Delta_i)_D}{L} \tag{14.11.11}$$

which, using Eqs. (14.11.9) and (14.11.10), and letting $C_L = M_L/M_D$, gives

$$\frac{(\Delta_i)_L}{L} = \frac{\beta_a}{\beta_w}\left(\frac{L}{y_t}\right)\left(\frac{M_{D+L}}{M_{cr}}\right)\left[\frac{1}{\gamma_{D+L}} - \frac{1}{\gamma_D(1 + C_L)}\right] \tag{14.11.12}$$

When excessive deflection may cause damage to partitions and other nonstructural construction, it is the sum of deflections due to live load plus creep and shrinkage that is of concern. The instantaneous (short-time) dead load deflection will have occurred when forms are removed and before any breakable attachments are put in place. Thus using Eq. (14.8.1),

$$\Delta_{cp+sh} = k_r T(\Delta_i)_D$$

Finally, the deflection to be controlled to minimize possible damage is

$$
\frac{(\Delta_i)_L}{L} + \frac{\Delta_{cp+sh}}{L} = \frac{\beta_a}{\beta_w}\left(\frac{L}{y_t}\right)\left(\frac{M_{D+L}}{M_{cr}}\right)\left[\frac{1}{\gamma_{D+L}} + \frac{(k_rT-1)}{\gamma_D(1+C_L)}\right]
$$

$$
= \frac{\beta_a}{\beta_w}\left(\frac{L}{y_t}\right)\left(\frac{M_{D+L}}{M_{cr}}\right)\left[\frac{1+C_L + (k_rT-1)(\gamma_{D+L}/\gamma_D)}{1+C_L}\right]\frac{1}{\gamma_{D+L}}
$$

$$(14.11.13)$$

Solving for L/y_t gives

$$
\left(\frac{L}{y_t}\right)_{\substack{\text{limit for} \\ L+cp+sh}} = \left(\frac{\Delta}{L}\right)\left(\frac{\beta_w}{\beta_a}\right)\left(\frac{M_{cr}}{M_{D+L}}\right)\left[\frac{C_L+1}{C_L+1+(k_rT-1)(\gamma_{D+L}/\gamma_D)}\right]\gamma_{D+L}
$$

$$(14.11.14)$$

For instantaneous dead load plus live load, Eq. (14.11.10) gives

$$
\left(\frac{L}{y_t}\right)_{\substack{\text{limit for} \\ D+L}} = \left(\frac{\Delta}{L}\right)\left(\frac{\beta_w}{\beta_a}\right)\left(\frac{M_{cr}}{M_{D+L}}\right)\gamma_{D+L} \qquad (14.11.15)
$$

As shown below if the span-to-depth ratio limit is available for short-time deflection under dead load plus live load, such as from Table 14.10.1, the effect of creep and shrinkage can be obtained by the use of a multiplier. Comparing Eqs. (14.11.14) and (14.11.15), in which Δ/L may be taken as a stated limit,

$$
\left(\frac{L}{y_t}\right)_{L+cp+sh} = \left(\frac{L}{y_t}\right)_{D+L}\left[\frac{C_L + 1}{C_L + 1 + (k_rT - 1)(\gamma_{D+L}/\gamma_D)}\right] \qquad (14.11.16)
$$

For the situation in which the I_e under dead load only is approximately the same as I_e under dead plus live load, $\gamma_{D+L} \approx \gamma_D$, in which case Eq. (14.11.16) becomes

$$
\left(\frac{L}{y_t}\right)_{L+cp+sh} = \left(\frac{L}{y_t}\right)_{D+L}\left(\frac{C_L + 1}{C_L + k_rT}\right) \qquad (14.11.17)
$$

Charts are available [27] for the L/h ratio for short-time effects of dead load plus live load, Eq. (14.11.10). The chart for singly reinforced rectangular beams ($y_t = 0.5h$) is given in Fig. 14.11.1.

EXAMPLE 14.11.1 Determine the depth of beam required for the loading conditions of Example 14.5.1 (Fig. 14.5.3) if the sum of the immediate live load plus creep and shrinkage deflection must not exceed $L/480$. Assume only the dead load is sustained and the sustained load factor $k_rT = 2$ as given by ACI-9.5.2.5.

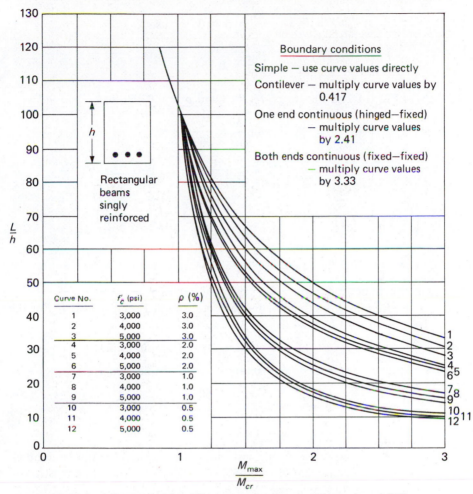

Figure 14.11.1 L/h versus M_{max}/M_{cr} curves for the conditions: $\Delta = L/360$, normal-weight concrete, and uniformly distributed short-time loading, and for different boundary conditions, steel percentages, and concrete strengths (from Ref. 27).

Solution: (a) Use ACI-Table 9.5a.

$$\frac{L}{h} = 16$$

This considers average conditions and includes some effect of sustained load deflection.

$$\min h = \frac{480}{16} = 30 \text{ in.}$$

(b) Use more accurate procedure with Fig. 14.11.1.

$$\rho = 0.0192$$

$$M_{cr} = \frac{f_r I_g}{y_t} = \frac{7.5 \sqrt{4000}(20,700)}{12(12)} = 68 \text{ ft-kips}$$

$$\frac{M_{max}}{M_{cr}} = \frac{250 + 90}{68} = 5 > 3; \qquad \text{use } 3$$

$$C_L = \frac{M_L}{M_D} = \frac{250}{90} = 2.78$$

From Fig. 14.11.1 for $M_{max}/M_{cr} = 3$ and $\rho = 0.0192$, find

$$\left(\frac{L}{h}\right)_{D+L} = 23, \qquad \text{for } \frac{\Delta}{L} = \frac{1}{360}$$

$$\left(\frac{L}{h}\right)_{D+L} = 23\left(\frac{360}{480}\right) = 17.3, \qquad \text{for } \frac{\Delta}{L} = \frac{1}{480}$$

$$\left(\frac{L}{h}\right)_{L+cp+sh} = 17.3\left(\frac{C_L + 1}{C_L + k_rT}\right) = 17.3\left(\frac{2.78 + 1}{2.78 + 2}\right) = 13.7$$

$$\min h = \frac{480}{13.7} = 35 \text{ in.}$$

which is somewhat more severe than ACI-Table 9.5a.

(c) Required depth and adequacy of beam in Example 14.5.1. For the given beam with $h = 24$ in.,

$$(\Delta_i)_L = 1.43 \text{ in.} \qquad \text{(Example 14.5.1)}$$

$$\Delta_{cp+sh} = 0.90 \text{ in.} \qquad \text{(Example 14.8.1)}$$

$$\text{allowable } \Delta = \frac{L}{480} = \frac{480}{480} = 1 \text{ in.}$$

$$\text{actual } \Delta = 1.43 + 0.90 = 2.23 \text{ in.}$$

If the beam were 28 in. deep, with the same loading it would be approximately carrying only $(24/28 = 0.855)$ 85.5% of its capacity, that is, stressed to only 85.5%. In which case,

$$\min h = 35(0.855) = 29.9 \text{ in.}$$

It would appear that 29 or 30 in. is required to satisfy deflection limit. Unless the width is reduced, which seems undesirable here, increasing the depth will increase dead load moment, so the 30-in. minimum given by ACI-Table 9.5a seems in this case to be about right.

14.12 ACI CODE DEFLECTION PROVISIONS—BEAM EXAMPLES

The ACI Code provisions (ACI-9.5) regarding deflection computations may be summarized as follows. In both the strength method and the working stress method (ACI "alternate" method), deflections *under service load conditions* must be computed

1. Whenever excessive deflection may adversely affect the strength or serviceability of the structure at service loads (ACI-9.5.1).
2. When minimum thickness used is *less* than that given by ACI-Table 9.5a for beams and one-way slabs (see also Table 14.10.2 in this chapter).

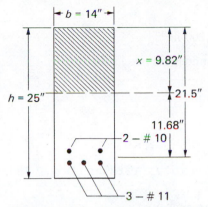

Figure 14.12.1   Section for Example 14.12.1.

The use of minimum thicknesses from ACI-Table 9.5a eliminates the need for computation of deflection *only for those cases where partitions, ceilings, and other nonstructural elements are not being supported.* Beams or one-way slabs supporting such frangible elements are in category (1) above, and deflections must be computed whether or not the minimum thickness limits of Table 9.5a have been met.

EXAMPLE 14.12.1 Investigate the deflection for the beam of Fig. 14.12.1 (same as obtained in Example 3.8.1) used on a simple span of 25 ft. The maximum bending moments under service load are 158 ft-kips dead load and 105 ft-kips live load, which correspond to M_u = 400 ft-kips as used for design by the strength method. Assume that all loading is uniformly distributed and that none of the live load is sustained. The beam supports partitions and other construction likely to be damaged by large deflections. Use f'_c = 4000 psi, f_y = 40,000 psi, and the ACI Code.

Solution: (a) Check minimum thickness (ACI-Table 9.5a), or from Table 14.10.2 for f_y = 40 ksi,

$$\text{min } h = \frac{L}{20} = \frac{25(12)}{20} = 15 \text{ in.} < 25 \text{ in. used}$$

This would make it seem that deflection is not likely to be excessive; however, deflection computations must be made because of the frangible items the beam supports.

(b) Examine reinforcement ratio ρ

$$\rho = \frac{A_s}{bd} = \frac{7.22}{14(21.5)} = 0.024$$

This exceeds $0.5\rho_{\text{max}}$ = $0.375\rho_b$ = 0.0185 suggested in Chap. 3 as a guideline for deflection control; thus from this check one may predict that deflection may be a problem.

(c) Determine the moment of inertia I_g for gross uncracked section without steel and I_{cr} for the cracked transformed section.

For gross uncracked section,

$$I_g = \tfrac{1}{12}(14)(25)^3 = 18,200 \text{ in.}^4$$

For cracked section, locate neutral axis under service loads,

$$\frac{14x^2}{2} = 7.22(8)(21.5 - x)$$

$$x = 9.82 \text{ in.}$$

$$I_{cr} = \tfrac{1}{3}(14)(9.82)^3 + 7.22(8)(11.68)^2 = 12,300 \text{ in.}^4$$

(d) Determine effective moment of inertia I_e (ACI Formula 9-7). The cracking moment is

$$M_{cr} = \frac{f_r I_g}{y_t} = \frac{7.5 \sqrt{4000}(18,200)}{12.5(12)} = 57.6 \text{ ft-kips}$$

For dead load deflection,

$$\frac{M_{cr}}{M_{max}} = \frac{57.6}{158} = 0.365; \qquad \left(\frac{M_{cr}}{M_{max}}\right)^3 = 0.05$$

$$I_e = \left(\frac{M_{cr}}{M_{max}}\right)^3 I_g + \left[1 - \left(\frac{M_{cr}}{M_{max}}\right)^3\right] I_{cr}$$

$$= 0.05(18,200) + 0.95(12,300) = 12,600 \text{ in.}^4$$

For dead load plus live load deflection,

$$\frac{M_{cr}}{M_{max}} = \frac{57.6}{263} = 0.22; \qquad \left(\frac{M_{cr}}{M_{max}}\right)^3 = 0.01$$

$$I_e \approx I_{cr} = 12,300 \text{ in.}^4$$

(e) Compute immediate deflections.

$$E_c = 57,000 \sqrt{f_c'} = 57,000 \sqrt{4000} = 3.6 \times 10^6 \text{ psi}$$

For dead load,

$$(\Delta_i)_D = \frac{5wL^4}{384EI} = \frac{5ML^2}{48EI} = \frac{5(158)(12)(300)^2}{48(3.6)(10^3)(12,600)} = 0.39 \text{ in.}$$

For dead load plus live load,

$$(\Delta_i)_{D+L} = \frac{5(263)(12)(300)^2}{48(3.6)(10^3)(12,300)} = 0.67 \text{ in.}$$

Then immediate live load deflection is

$$(\Delta_i)_L = (\Delta_i)_{D+L} - (\Delta_i)_D = 0.67 - 0.39 = 0.28 \text{ in.}$$

(f) Compute creep and shrinkage deflection. From ACI-9.5.2.5 the multiplier is

$$\lambda = k_r T = \frac{2.0}{1 + 50\rho'} = 2.0$$

for sustained load at 5 years or more.

$$\Delta_{cp+sh} = \lambda(\Delta_i)_D = 2.0(0.39) = 0.78 \text{ in.}$$

(g) Check limitation of ACI-Table 9.5b. For roof or floor construction supporting or attached to nonstructural elements likely to be damaged by large deflection,

$$(\Delta_i)_L + \Delta_{cp+sh} \leq \frac{L}{480}$$

This limit is for the deflection that is estimated to occur *after* the nonstructural elements are put in place. Whatever portion, if any, of the live load or creep and shrinkage deflection has occurred prior to the placement of nonstructural elements may be excluded from the $L/480$ limitation. Further, if adequate measures are taken to prevent damage to supported or attached elements, the $L/480$ limit may be exceeded (footnote, ACI-Table 9.5b).

For this example,

$$\left[0.28 + 0.78 = 1.06 \text{ in.}\right] > \left[\frac{L}{480} = \frac{300}{480} = 0.63 \text{ in.}\right]$$

which is not acceptable. This shows that the minimum thickness indicated by ACI-Table 9.5a was not even close.

EXAMPLE 14.12.2 Investigate the deflection for the one-way continuous slab shown in Fig. 14.12.2. The slab is $4\frac{1}{2}$ in. thick. Assume that 60% of the 100 psf live load is sustained. Use $f'_c = 3000$ psi and $f_y = 40,000$ psi.

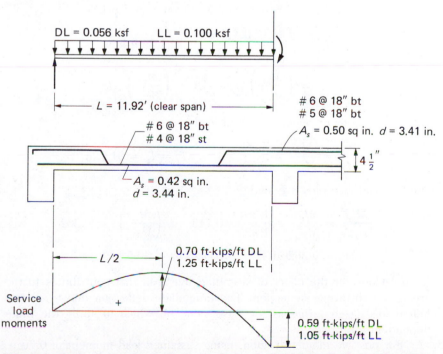

Figure 14.12.2 End-span details for continuous slab of Example 14.12.2. [See also Fig. 7.5.1(a).]

Solution: Use may be made of Eq. (14.2.4) to compute midspan deflection:

$$\Delta_m = \frac{5L^2}{48EI}\left[M_s - \frac{1}{10}(M_a + M_b)\right] \qquad [14.2.4]$$

Note that clear span is used with face-of-support moments as recommended by ACI Committee 435 [9]; further, a conservative assumption of zero moment at the exterior support is used. Thus

M_s = net midspan moment

$$= \tfrac{1}{8}(0.156)(11.92)^2 - \tfrac{1}{2}(1.64) = 2.77 - 0.82 = 1.95 \text{ ft-kips/ft}$$
$$E_c = 57{,}000 \sqrt{f_c'} = 57{,}000 \sqrt{3000} = 3.12 \times 10^6 \text{ psi}$$

(a) Determine I_g and I_{cr}, using a 1-ft width of section.

$$I_g = \tfrac{1}{12}(12)(4.5)^3 = 91.1 \text{ in.}^4$$

For positive moment region, $d = 3.44$ in.,

$$\frac{12x^2}{2} = 9(0.42)(3.44 - x)$$

$$x = 1.19 \text{ in.}$$
$$I_{cr} = \tfrac{1}{3}(12)(1.19)^3 + 9(0.42)(2.44 - 1.19)^2 = 25.9 \text{ in.}^4$$

Note that only the positive moment region properties need to be used according to ACI-9.5.2.4.

(b) Determine effective moment of inertia I_e.

$$f_r = 7.5 \sqrt{f_c'} = 7.5 \sqrt{3000} = 411 \text{ psi}$$

$$M_{cr} = \frac{f_r I_g}{y_t} = \frac{0.411(91.1)}{2.25(12)} = 1.39 \text{ ft-kips}$$

$$I_e = \left(\frac{M_{cr}}{M_{max}}\right)^3 I_g + \left[1 - \left(\frac{M_{cr}}{M_{max}}\right)^3\right]I_{cr}$$

In positive moment region, for dead load

$$\frac{M_{cr}}{M_D} = \frac{M_{cr}}{M_{max}} = \frac{1.39}{0.70} > 1; \qquad I_e = I_g = 91.1 \text{ in.}^4$$

and for dead load plus live load,

$$\frac{M_{cr}}{M_{D+L}} = \frac{M_{cr}}{M_{max}} = \frac{1.39}{1.95} = 0.713; \qquad \left(\frac{M_{cr}}{M_{max}}\right)^3 = 0.362$$

$$I_e = 0.362(91.1) + 0.638(25.9) = 49.5 \text{ in.}^4$$

(c) Consider the effect of sustained live load that contributes to the creep and shrinkage deflection. The immediate deflection due to all sustained loads is required·as the base value on which to apply the time-dependent multiplier.

For positive moment region, using sustained load moment of $0.70 + 0.6(1.25) = 1.46$ ft-kips/ft,

$$\frac{M_{cr}}{M_{max}} = \frac{1.39}{1.46} = 0.952; \qquad \left(\frac{M_{cr}}{M_{max}}\right)^3 = 0.863$$

$$I_e = 0.863(91.1) + 0.137(25.9) = 82.2 \text{ in.}^4$$

(d) Immediate live load deflection.

$$(\Delta_i)_L = (\Delta_i)_{D+L} - (\Delta_i)_D$$

$$(\Delta_i)_{D+L} = \frac{5(11.92)^2 144}{48(3.12)(10^3)49.5}\left(1.95 - \frac{1.64}{10}\right)(12) = 0.30 \text{ in.}$$

$$(\Delta_i)_D = \frac{5(11.92)^2 144}{48(3.12)(10^3)91.1}\left(0.70 - \frac{0.59}{10}\right)(12) = 0.06 \text{ in.}$$

$$(\Delta_i)_L = 0.30 - 0.06 = 0.24 \text{ in.}$$

(e) Creep and shrinkage deflection. The immediate deflection due to sustained loads,

$$(\Delta_i)_{D+\text{sust } L} = \frac{5(11.92)^2 144}{48(3.12)(10^3)82.2}\left(1.45 - \frac{1.22}{10}\right)(12) = 0.13 \text{ in.}$$

Considering 5 years or more load duration, and with no compression steel, $\lambda = 2.0$

$$(\Delta_i)_{cp+sh} = \lambda(\Delta_i)_{D+\text{sust } L} = 2.0(0.13) = 0.26 \text{ in.}$$

(f) Check deflection if the limit of $L/480$ in ACI-Table 9.5b applies,

$$(\Delta_i)_L + \Delta_{cp+sh} = 0.24 + 0.26 = 0.50 \text{ in.}$$

$$\text{allowable } \Delta = \frac{L}{480} = \frac{11.92(12)}{480} = 0.30 \text{ in.}$$

which appears unacceptable. If none of the live load was considered sustained,

$$(\Delta_i)_L + \Delta_{cp+sh} = 0.24 + 2(0.06) = 0.36 \text{ in.} > 0.30 \text{ in.}$$

The result is closer but still may not be considered acceptable. Note that if I_e is taken as the average for the midspan and support regions (as shown in Example 14.5.2), as is also permitted by ACI-9.5.2.4, the live load plus creep and shrinkage deflection would have been computed to be 0.31 in., very close to the 0.30-in. limit.

The decision regarding whether or not part of the live load is sustained should be made by considering the actual loading and its duration. For instance, a floor system supporting library stacks as the live load might well be considered to have a significant part of the live load treated as being sustained. In most situations it is acceptable to consider that only the dead load is sustained and thereby affects the magnitude of creep and shrinkage deflection.

In spite of the fact the deflection calculations have been illustrated throughout this chaper without shortcutting any steps in the formal procedure, deflection computations should be made keeping practicality in mind.

The effective moment of inertia I_e theoretically should be computed at each total load level for which deflection is of concern: usually dead load, dead load plus live load, and dead load plus sustained live load. I_e should be computed at each end and at the midspan region in order to obtain an average as is encouraged by the ACI Code.

However, when the true accuracy obtainable from a deflection computation is recognized, the designer should use the I_e equation only when the result will be significantly different from using either I_{cr} or I_g. Further, for most situations only the midspan I_e need be computed; an average does not measurably improve accuracy. The authors believe these practical suggestions satisfy the spirit of the ACI Code.

For additional practical deflection examples based on the ACI Code and ACI Committee 435 recommendations, including composite beams and two-way systems, see chapters by Branson in *Handbook of Concrete Engineering* [36] and the *PCA Notes* [37].

SELECTED REFERENCES

1. ACI Committee 435, Subcommittee 1. "Allowable Deflections," *ACI Journal, Proceedings,* **65,** June 1968, 433–444. Disc., 1037–1038.
2. ACI Committee 340 (C. G. Salmon, Chairman). *Design Handbook In Accordance With the Strength Design Method of ACI 318-77, Vol. 1—Beams, Slabs, Brackets, Footings, and Pile Caps* [SP-17(81)]. Detroit: American Concrete Institute, 1981 (p. 306).
3. Mark Fintel (Ed.). *Handbook of Concrete Engineering.* New York: Van Nostrand, 1974 (801 pp.) (Chap. 2, "Deflections," p. 49).
4. Wei-Wen Yu and George Winter. "Instantaneous and Long-Time Deflections of Reinforced Concrete Beams Under Working Loads," *ACI Journal, Proceedings,* **57,** July 1960, 29–50. Disc., 1165–1171.
5. Dan E. Branson. "Instantaneous and Time-Dependent Deflections of Simple and Continuous Reinforced Concrete Beams," Part 1, Report No. 7. Alabama Highway Research Report, Bureau of Public Roads, August 1963 (1965) (pp. 1–78).
6. Dan E. Branson. Discussion of "Variability of Deflections of Simply Supported Reinforced Concrete Beams," by ACI Committee 435, *ACI Journal, Proceedings,* **69,** July 1972, 449–451.
7. ACI Committee 209. "Prediction of Creep, Shrinkage, and Temperature Effects in Concrete Structures," *Designing for Effects of Creep, Shrinkage, and Temperature in Concrete Structures* (SP-27). Detroit: American Concrete Institute, 1971 (pp. 51–93).
8. Dan E. Branson. *Deformation of Concrete Structures.* New York: McGraw-Hill, 1977.
9. ACI Committee 435. "Proposed Revisions By Committee 435 to ACI Building Code and Commentary Provisions on Deflections," *ACI Journal, Proceedings,* **75,** June 1978, 229–238.
10. G. W. Washa and P. G. Fluck. "Effect of Compressive Reinforcement on the Plastic Flow of Reinforced Concrete Beams," *ACI Journal, Proceedings,* **49,** October 1952, 89–108.
11. G. W. Washa and P. G. Fluck. "Plastic Flow (Creep) of Reinforced Concrete Continuous Beams," *ACI Journal, Proceedings,* **52,** January 1956, 549–561.

12. Dan E. Branson. Discussion of "Proposed Revision of ACI 318-63 Building Code Requirements for Reinforced Concrete," *ACI Journal, Proceedings*, **67**, September 1970, 692–693.

13. ACI Committee 435, Subcommittee 7. "Deflections of Continuous Beams," *ACI Journal, Proceedings*, **70**, December 1973, 781–787.

14. P. D. Zuraski, C. G. Salmon, and A. Fattah Shaikh. "Calculation of Instantaneous Deflections for Continuous Reinforced Concrete Beams," *Deflections of Concrete Structures* (SP-43). Detroit: American Concrete Institute, 1974 (pp. 315–331).

15. N. H. Burns and C. P. Siess. "Repeated and Reverse Loading in Reinforced Concrete," *Journal of Structural Division*, ASCE, **92**, October 1966 (ST5), 65–78.

16. ACI Committee 435. "Deflections of Reinforced Concrete Flexural Members," *ACI Journal, Proceedings*, **63**, June 1966, 637–674.

17. Adam Neville and Bernard Meyers. "Creep of Concrete: Influencing Factors and Prediction," *Symposium on Creep of Concrete* (SP-9). Detroit: American Concrete Institute, 1964 (pp. 1–33).

18. ACI Committee 209. "Effects of Concrete Constituents, Environment, and Stress on the Creep and Shrinkage of Concrete," *Designing for Effects of Creep, Shrinkage, and Temperature in Concrete Structures* (SP-27). Detroit: American Concrete Institute, 1971 (pp. 1–42).

19. D. E. Branson and M. L. Christiason. "Time Dependent Concrete Properties Related to Design—Strength and Elastic Properties, Creep, and Shrinkage," *Designing for Effects of Creep, Shrinkage, and Temperature in Concrete Structures* (SP-27). Detroit: American Concrete Institute, 1971 (pp. 257–277).

20. B. L. Meyers and D. E. Branson. "Design Aid for Predicting Creep and Shrinkage Properties of Concrete," *ACI Journal, Proceedings*, **69**, September 1972, 551–555.

21. M. R. Hollington. *A Series of Long-Term Tests to Investigate the Deflection of a Representative Precast Concrete Floor Component*, Technical Report TRA 442. London: Cement and Concrete Association, April 1970 (43 pp.).

22. Dan E. Branson. "Compression Steel Effect on Long-Time Deflection," *ACI Journal, Proceedings*, **68**, August 1971, 555–559.

23. T. C. Hansen and A. H. Mattock. "Influence of Size and Shape of Member on Shrinkage and Creep of Concrete," *ACI Journal, Proceedings*, **63**, February 1966, 267–289.

24. Hans Gesund. "Shrinkage and Creep Influence on Deflections and Moments of Reinforced Concrete Beams," *ACI Journal, Proceedings*, **59**, May 1962, 689–704.

25. Alfred L. Miller. "Warping of Reinforced Concrete Due to Shrinkage," *ACI Journal, Proceedings*, **54**, May 1958, 939–950.

26. A. F. Shaikh. Discussion of "Proposed Revision of ACI 318–63: Building Code Requirements for Reinforced Concrete," *ACI Journal, Proceedings*, **67**, September 1970, 722–723.

27. Dan E. Branson. "Design Procedures for Computing Deflections," *ACI Journal, Proceedings*, **65**, September 1968, 730–742.

28. Koladi M. Kripanarayanan and Dan E. Branson. "Short-Time Deflections of Beams Under Single and Repeated Load Cycles," *ACI Journal, Proceedings*, **69**, February 1972, 110–117.

29. Dan E. Branson and Heinrich Trost. "Unified Procedures for Predicting the Deflection and Centroidal Axis Location of Partially Cracked Nonprestressed and Prestressed Concrete Members," *ACI Journal, Proceedings*, **79**, March–April 1982, 119–130.

30. Jacob S. Grossman. "Simplified Computations for Effective Moment of Inertia I_e and Minimum Thickness to Avoid Deflection Computations," *ACI Journal, Proceedings*, **78**, November–December 1981, 423–439. Disc., **79**, September–October 1982, 413–419.

31. Dan E. Branson. Discussion of "Simplified Computations for Effective Moment of Inertia I_e and Minimum Thickness to Avoid Deflection Computations," (*ACI Journal, Proceedings*, **78**, November–December 1981, 423–439), *ACI Journal, Proceedings*, **79**, September–October 1982, 413–414.

32. B. V. Rangan and T. Donaghy. *Deflection Design of Flexural Concrete Members by Allowable Span/Depth Ratios* (UNICIV Report No. R-166). Kensington, N. S. W., Australia: University of New South Wales, December 1976.

33. ACI Committee 435. "Variability of Deflections of Simply Supported Reinforced Concrete Beams," *ACI Journal, Proceedings*, **69**, January 1972, 29–35.

34. LeRoy A. Lutz. "Graphical Evaluation of the Effective Moment of Inertia for Deflection," *ACI Journal, Proceedings*, **70**, March 1973, 207–213. Disc., **70**, September 1973, 662–663.

35. B. Vijaya Rangan. "Control of Beam Deflections by Allowable Span-Depth Ratios," *ACI Journal, Proceedings*, **79**, September–October 1982, 372–377.

36. Dan E. Branson. "Deflections," *Handbook of Concrete Engineering*, 2nd ed., ed. by Mark Fintel. New York: Van Nostrand Reinhold, 1984, Chapter 2.

37. Dan E. Branson. "Deflections," *Notes on ACI 318-83 Building Code Requirements for Reinforced Concrete*, 4th ed., ed. by G. B. Neville. Skokie, Illinois: Portland Cement Association, 1984, Chapter 7.

PROBLEMS

All problems* are to be done in accordance with the ACI Code unless otherwise indicated.

14.1 Compute the immediate deflections due to dead load and live load on the beam of the accompanying figure. The span of the uniformly loaded, simply supported beam is 30 ft, and the maximum service load moments are 80 ft-kips for dead load and 95 ft-kips for live load. Use $f'_c = 3500$ psi ($n = 8.5$) and $f_y = 60,000$ psi. (Span = 9.2 m; $M_D = 110$ kN·m; $M_L = 130$ kN·m; $f'_c = 25$ MPa; $f_y = 400$ MPa.)

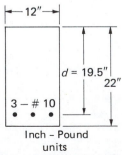

Inch – Pound
units

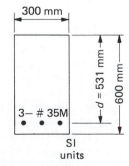

SI
units

Problems 14.1, 14.7, and 14.14

---

*Most problems may be solved as problems stated in Inch-Pound units, or as problems in SI units using quantities in parenthesis at the end of the statement. The SI conversions are approximate to avoid implying higher precision for given information in metric units than that for the Inch-Pound units.

14.2   Compute the immediate deflections due to dead load and live load on the beam of the accompanying figure. The uniformly loaded, simply supported beam on a span of 29 ft must resist maximum service load moments of 300 ft-kips dead load and 500 ft-kips live load. Use $f_c' = 4000$ psi ($n = 8$) and $f_y = 40,000$ psi. (Span $= 8.9$ m; $M_D = 407$ kN·m; $M_L = 680$ kN·m; $f_c' = 30$ MPa; $f_y = 300$ MPa.)

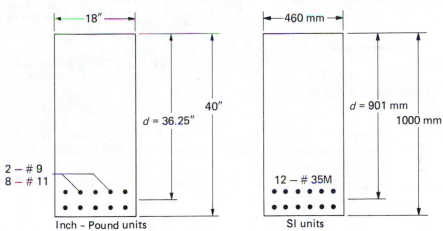

Inch – Pound units   SI units

Problems 14.2, 14.15, and 14.16

14.3   Compute the immediate deflections due to dead load and live load on the beam of the accompanying figure. The uniformly loaded, simply supported beam on a span of 30 ft must resist maximum service load moments of 20 ft-kips dead load and 14 ft-kips live load. Use $f_c' = 3000$ psi ($n = 9$) and $f_y = 40,000$ psi. (Span $= 9.1$ m; $M_D = 27$ kN·m; $M_L = 19$ kN·m; $f_c' = 20$ MPa; $f_y = 300$ MPa.)

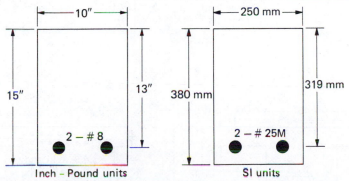

Inch – Pound units   SI units

Problems 14.3, 14.4, 14.17, and 14.18

14.4   Repeat Prob. 14.3 except the service load moments are 29 ft-kips dead load and 19 ft-kips live load, and $f_y = 60,000$ psi. ($M_D = 40$ kN·m; $M_L = 26$ kN·m; $f_y = 400$ MPa.)

14.5   Investigate the acceptability of the beam of the accompanying figure for immediate live load deflection if the limitation is $L/360$. Use $f_c' = 3500$ psi ($n = 8.5$) and $f_y = 40,000$ psi.

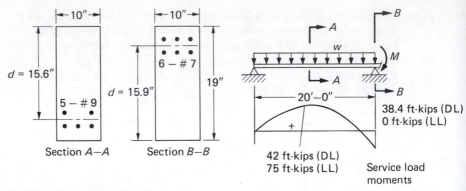

Problems 14.5 and 14.19

14.6 Investigate the acceptability of the beam of the accompanying figure for immediate live load deflection if the limitation is $L/360$. Use $f'_c = 3000$ psi $(n = 9)$ and $f_y = 60,000$ psi.

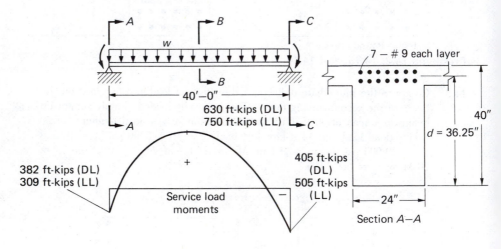

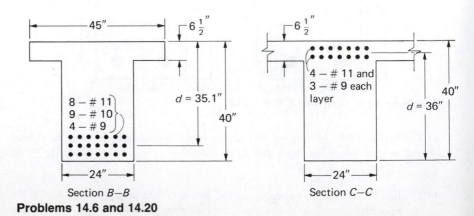

Problems 14.6 and 14.20

14.7 Investigate the acceptability of the beam of Prob. 14.1, if the limit for live load plus creep and shrinkage deflection is $L/360$. Consider that none of the live load is sustained. Use data computed in Prob. 14.1 if that problem was previously assigned.

14.8 Investigate the acceptability of the deflection due to live load plus creep and shrinkage on the 24-ft span for the beam described in Prob. 3.13 if the cross section is 12×24 with 6-# 10 (3 in each of two layers) in the positive moment zone and 3-# 9 (one layer) in the negative moment zone. Assume # 3 stirrups are used. Only the dead load is sustained. Use $f'_c = 3500$ psi and $f_y = 40,000$ psi.

14.9 Investigate the acceptability of the deflection for the cantilever portion of the beam of Prob. 14.8.

14.10 Investigate the acceptability of the deflection due to live load plus creep and shrinkage on the 30-ft span of the beam described in Prob. 3.14 if the cross section is 14×24 with 4-# 10 (one layer) as positive moment reinforcement and 6-# 6 (one layer) as negative moment reinforcement. Assume # 3 stirrups are used. Assume also only the dead load is sustained. Use $f'_c = 3500$ psi and $f_y = 60,000$ psi.

14.11 Investigate the acceptability of the deflection for the cantilever portions of the beam of Prob. 14.10.

14.12 Repeat Prob. 14.10 if the beam cross section is 16×28 having 4-#9 for positive moment reinforcement and 5-#6 for negative moment reinforcement.

14.13 Investigate the acceptability of the deflection for the cantilever portions of the beam of Prob. 14.12.

14.14 Repeat Prob. 14.7 except use the separate creep and shrinkage multiplier procedure instead of ACI method. Assume humidity is 80%, age at loading is 28 days after initial curing period, and duration of sustained load is 5 years.

14.15 Investigate the acceptability of the beam of Prob. 14.2 if the beam supports partitions and other nonstructural construction likely to be damaged by large deflection. Consider that 20% of the live load is sustained.

14.16 Repeat Prob. 14.15 except use the separate creep and shrinkage multiplier procedure instead of the ACI method. The deflection limit is the same as for ACI Code, however. Assume humidity is 90%, age at loading is 60 days after initial curing period, and duration of sustained load is 5 years.

14.17 Investigate the acceptability of the beam of Prob. 14.3 if the beam supports partitions, etc., which limit the maximum deflection due to live load plus creep and shrinkage to $L/360$. Assume none of the live load is sustained, the relative humidity is 50%, age at loading is 20 days after initial curing, and sustained load will be in place for 1 year.
(a) Use ACI method.
(b) Use separate creep and shrinkage multiplier procedure.
(c) Use combined creep and shrinkage multiplier procedure.

14.18 Investigate the acceptability of the beam of Prob. 14.4 if the beam supports partitions, etc., which limit the maximum deflection due to live load plus creep and shrinkage to $L/480$. Assume none of the live load is sustained, the relative humidity is 90%, age at loading is 30 days after initial curing period, and duration of sustained load is 5 years or more.
(a) Use ACI method.
(b) Use separate creep and shrinkage multiplier procedure.

14.19 Investigate the acceptability of the beam of Prob. 14.5 if the live load plus creep and shrinkage deflection is limited to $L/250$ and 10% of the live load is considered sustained. Assume relative humidity is 50%, age at loading is 10 days after initial curing period, and duration of sustained loading is 1 year.
(a) Use ACI Code method.
(b) Use separate creep and shrinkage multiplier procedure.

14.20 Investigate the acceptability of the beam of Prob. 14.6 if the beam supports partitions and other nonstructural construction likely to be damaged by large deflections. Assume that 10% of the live load is sustained.

14.21 Investigate the acceptability of an interior 45-ft span (see the accompanying figure) of a continuous T-section with regard to deflection. The service load moments at midspan are 40 ft-kips dead load and 100 ft-kips live load; and at both supports are 50 ft-kips dead load and 114 ft-kips live load. Assume that the $L/480$ limit of ACI-Table 9.5b applies and that none of the live load should be considered as sustained. Use $f'_c = 3500$ psi ($n = 8.5$) and $f_y = 60,000$ psi.

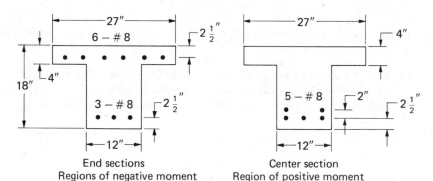

End sections
Regions of negative moment

Center section
Region of positive moment

Problem 14.21

14.22 Use the span-to-depth ratio approach described in Sec. 14.11 to check the deflection for Prob. 14.8.

14.23 Use the span-to-depth ratio approach described in Sec. 14.11 to check the deflection for Prob. 14.10.

15

Length Effects on Columns

15.1 GENERAL

In the basic treatment of compression members in Chap. 13, the assumption was made that the effects of buckling and lateral deflection on strength were small enough to be neglected. Short compression members—that is, those having a low slenderness ratio L/r (L = column height and r = radius of gyration = $\sqrt{I/A}$)—when overloaded experience a material failure (crushing of concrete) prior to reaching a buckling mode of failure. Further, the lateral deflections of short compression members subjected to bending moments are small, thus contributing little secondary bending moment $P\Delta$ as shown in Fig. 15.1.1. It is these buckling and deflection effects that reduce the strength of a compression member below the value computed according to the principles of Chap. 13.

The adoption of the strength method for design, along with the use of higher strength steel and concrete, has led to the increased use of slender members. A stocky member having an L/r less than 20 will achieve essentially the strength discussed in Chap. 13, whereas a member having L/r greater than about 70 will have a considerable reduction in strength, both due to the likelihood of buckling and because of secondary bending moment. To permit the greatest flexibility in structural design, specifications must provide for adequate determination of strength with any slenderness ratio. Thus the provisions of ACI-10.10 and 10.11 take into account the length effects on long compression members.

In computing r, ACI-10.11.3 endorses a simple value for the

Hotel Americana, Houston, Texas. (Courtesy of Portland Cement Association.)

rectangular column having b as the width and h as the depth

$$r = \sqrt{\frac{I}{A}} = \sqrt{\frac{\frac{1}{12}bh^3}{bh}} = 0.288h \approx 0.30h$$

and for the circular column having h as the diameter

$$r = \sqrt{\frac{I}{A}} = \sqrt{\frac{\pi h^4(4)}{64\pi h^2}} = 0.25h$$

These values actually should be slightly larger due to the effect of the reinforcement.

In the ACI Code, the evaluation of the effect of slenderness may be approximated by using the moment magnifier approach, whereby the sum of the primary and secondary moments (Fig. 15.1.1) is treated as being equal to the product of the primary moment and a magnification factor δ. The general idea relating to this approach is derivable from the differential equation of the beam-column.

In the next several sections, the general concepts relating to the effect of slenderness on the strength of compression members are presented.

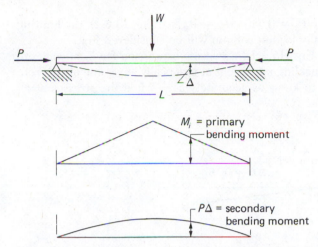

Figure 15.1.1 Primary and secondary moment for beam-columns.

15.2 BUCKLING OF CONCENTRICALLY LOADED COLUMNS

Over 200 years ago Leonhard Euler derived the well-known Euler formula [1] for concentrically loaded columns stressed below the proportional limit. Engesser [2] in 1889 proposed the tangent modulus modification of the Euler formula. In 1910 von Kármán [3] performed a series of careful tests verifying Engesser's assumptions. The tangent modulus formula has now been accepted as representing the lower bound for buckling strength of concentrically loaded columns,

$$P_c = \frac{\pi^2 E_t I}{(kL_u)^2} \tag{15.2.1}$$

where

P_c = buckling load

E_t = tangent modulus of elasticity of concrete at the buckling load

I = moment of inertia of the effective section

kL_u = equivalent pin-end length (L_u = actual unbraced length)

Although von Kármán was the first to use a rational analytical method for the inelastic buckling of long slender columns, his work did not consider reinforced concrete.

The fact that concentrically loaded columns rarely, if ever, exist in reinforced concrete structures led investigators [4–25] during the past 20 years to focus attention on the interaction of long columns with beams in frame structures, resulting in the more rational provisions for length effects on compression members beginning with the 1971 ACI Code.

The basic design limitation of a maximum axial strength equal to 80 or 85% of the concentric capacity P_0 (see Fig. 13.11.2) means, of course, that from a practical viewpoint the concentrically loaded column is considered not to exist. However, to help the reader understand the effect of the slenderness ratio on the behavior of beam-columns over the entire range from

$P_n = P_0$ with $M_n = 0$ to $P_n = 0$ with $M_n = M_0$ (see Fig. 13.6.2), the limiting case of the concentrically loaded column will be considered first.

In order to apply Eq. (15.2.1), a realistic expression for E_t of concrete must be used. Since buckling may occur at practically any value of concrete strain, it is necessary to know as accurately as possible the stresses at all strain levels.

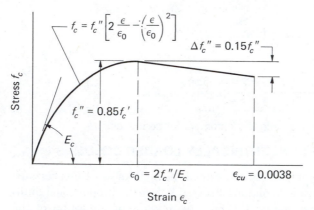

Figure 15.2.1 Hognestad's stress-strain diagram for flexure (Ref. 26).

The idealized stress-strain diagram for steel was shown in Chap. 3 (Fig. 3.2.2), where the modulus of elasticity is taken at 29,000,000 psi. One of the more realistic stress-strain diagrams for concrete in compression is that of Hognestad [26] shown in Fig. 15.2.1 in which the initial modulus of elasticity for concrete is taken as

$$E_c = 1,800,000 + 500f_c'' \qquad \text{psi} \qquad (15.2.2)$$

in which $f_c'' = 0.85f_c'$. Such a stress-strain relationship may be used along with the assumptions that (1) concrete resists no tensile stress, (2) linear strain exists across the section, and (3) the deflected shape is part of a sine wave. On the basis of these assumptions, evaluation of Eq. (15.2.1) gives the typical column strength curves for concentric loading, such as those of Fig. 15.2.2.

A study of Fig. 15.2.2 shows that in the curves for $f_y = 40,000$ psi there occurs a flat leveling off (such as portion BC of Fig. 15.2.2) of the curve, indicating yielding of the steel with a sudden drop in E_s from 29×10^6 psi to zero. In such cases where $\epsilon_y < \epsilon_0$, the concrete may still increase the capacity (such as portion AB in Fig. 15.2.2) with an increased strain up to ϵ_0. For $f_y = 50,000$ psi and $f_c' = 4000$ psi, the strains $\epsilon_y = 0.00173$ and $\epsilon_0 = 0.00194$ are nearly equal so that little increase above the plateau of yielding in the steel can occur. The effect of creep on long-time loading may be noted by the crosshatched region.

EXAMPLE 15.2.1 Calculate the ordinate and abscissa of points A, B, and C of Fig. 15.2.2. Use $f_c' = 4000$ psi, $f_y = 40,000$ psi, and the steel area of

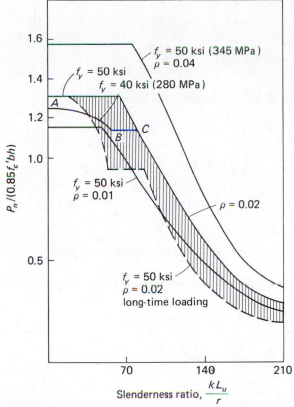

Figure 15.2.2   Strength curves for reinforced concrete (f'_c = 4000 psi) (28 MPa) concentrically loaded pin-end columns (adapted from Ref. 27).

$A_s = 0.02bh$ ($0.01bh$ in each face located at $0.45h$ from the center). Use the stress-strain diagrams of Figs. 15.2.1 and 3.2.2.

Solution: The basic quantities of the concrete stress-strain diagram are computed first:

$$f''_c = 0.85f'_c = 0.85(4) = 3.4 \text{ ksi}$$

$$E_c = 1800 + 500f''_c = 1800 + 500(3.4) = 3500 \text{ ksi}$$

$$\epsilon_0 = \frac{2f''_c}{E_c} = 2\left(\frac{3.4}{3500}\right) = 0.00194$$

$$\epsilon_y = \frac{f_y}{E_s} = \frac{40}{29,000} = 0.00138$$

(a) Point A, maximum strength of section according to the principles of Chap. 13 (upper limit, $\epsilon_c = \epsilon_0 > \epsilon_y$),

$$P_n = 0.85f'_c bh + A_s f_y$$

without correcting for the displaced concrete.

$$P_n = 0.85f'_c bh + 0.02bh(40)$$

$$= 0.85f'_c bh \left[1 + \frac{0.02(40)}{0.85(4)} \right]$$

$$\frac{P_n}{0.85f'_c bh} = 1.235 \qquad \text{(ordinate of point A)}$$

(b) Point B, $\epsilon_c = \epsilon_y = 0.00138 < \epsilon_0$, $E_s = 0$.

$$f_c = f''_c \left[2\left(\frac{\epsilon}{\epsilon_0}\right) - \left(\frac{\epsilon}{\epsilon_0}\right)^2 \right]$$

$$= 3.4 \left[2\left(\frac{1.38}{1.94}\right) - \left(\frac{1.38}{1.94}\right)^2 \right]$$

$$= 3.4 \,[2(0.712) - 0.508] = 3.12 \text{ ksi}$$

$$P_n = f_c bh + A_s f_y = 3.12bh + 0.02bh(40) = 3.92bh$$

$$\frac{P_n}{0.85f'_c bh} = \frac{3.92}{3.4} = 1.153 \qquad \text{(ordinate of point B)}$$

Applying Eq. (15.2.1) using for I the uncracked transformed section,

$$E_t = \frac{df_c}{d\epsilon} = f''_c \left(\frac{2}{\epsilon_0} - \frac{2\epsilon}{\epsilon_0^2} \right) = E_c - \frac{E_c \epsilon}{\epsilon_0}$$

$$= 3500 \left(1 - \frac{1.38}{1.94} \right) = 1008 \text{ ksi}$$

$$I = \frac{bh^3}{12} + 0.02bh\left(\frac{E_s}{E_t}\right)(0.45h)^2$$

$$= bh^3 \left[\frac{1}{12} + 0.02 \left(\frac{0}{1008}\right)(0.45)^2 \right] = 0.0833bh^3$$

$$(kL_u)^2 = \frac{\pi^2 E_t I}{P_c} \qquad\qquad\qquad\qquad [15.2.1]$$

$$(kL_u)^2 = \frac{\pi^2 (1008)(0.0833bh^3)}{3.92bh} = 212h^2$$

$$\frac{kL_u}{h} = 14.55$$

For a rectangular section, $r = h/\sqrt{12}$,

$$\frac{kL_u}{r} = 14.55\sqrt{12} = 50.3 \qquad \text{(abscissa of point B)}$$

(c) Point C, ϵ_c at an infinitesimal amount less than ϵ_y; $E = 29,000$ ksi.

$$I = bh^3 \left[\frac{1}{12} + 0.02 \left(\frac{29,000}{1008}\right)(0.45)^2 \right]$$

$$= bh^3(0.0833 + 0.1167) = 0.2000bh^3$$

$$(kL_u)^2 = \frac{\pi^2(1008)(0.200bh^3)}{3.92bh} = 507h^2$$

$$\frac{kL_u}{h} = 22.5$$

$$\frac{kL_u}{r} = 22.5\sqrt{12} = 78 \qquad \text{(abscissa of point } C)$$

15.3 EQUIVALENT PIN-END LENGTHS

For conditions other than pin ends where the factor k in Eq. (15.2.1) is 1.0, the equivalent pin-end length (also called *effective* length) factor k must be determined for various rotational and translational end restraint conditions. Where translation at both ends is adequately prevented, the distance between points of inflection is shown in Fig. 15.3.1. For all such cases the equivalent pin-end length is less than the actual unbraced length (i.e., k is less than one).

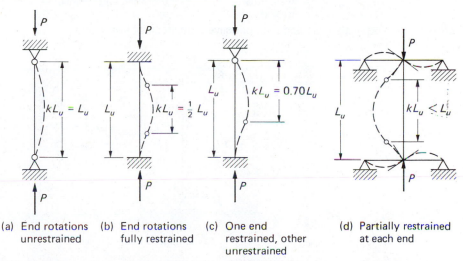

(a) End rotations unrestrained   (b) End rotations fully restrained   (c) One end restrained, other unrestrained   (d) Partially restrained at each end

Figure 15.3.1   Equivalent pin-end (i.e., effective) lengths; no joint translation.

If sidesway or joint translation is possible, as in the case of the unbraced frame, the equivalent pin-end length exceeds the actual unbraced length (i.e., k is greater than one), as shown in Fig. 15.3.2.

As reinforced concrete columns are in general part of a larger frame, it is necessary to understand the concepts of a *braced frame* (where joint translation is prevented by rigid bracing, shear walls, or attachment to an adjoining structure) and the *unbraced frame* (where buckling stability is dependent on the stiffness of the beams and columns that constitute the frame). As shown in Figs. 15.3.3(a) and (c), the effective length kL_u for cases where joint translation is prevented may never exceed the actual length L_u.

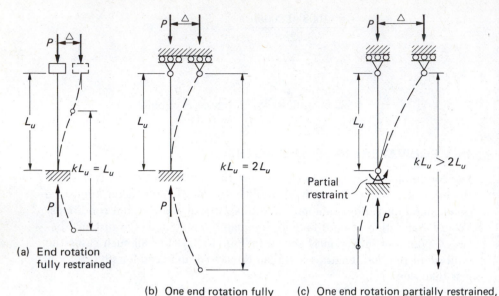

(a) End rotation
fully restrained

(b) One end rotation fully
restrained, other
unrestrained

(c) One end rotation partially restrained,
other end unrestrained

Figure 15.3.2 Equivalent pin-end (i.e., effective) lengths; joint translation
possible.

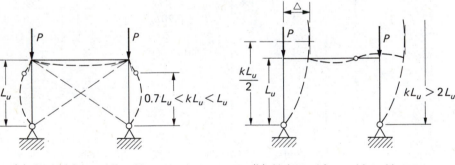

(a) Braced frame, hinged base

(b) Unbraced frame, hinged base

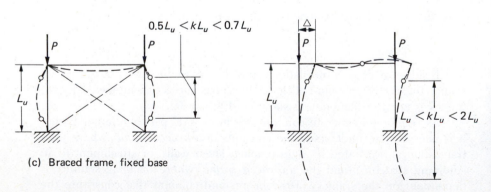

(c) Braced frame, fixed base

(d) Unbraced frame, fixed base

Figure 15.3.3 Equivalent pin-end (i.e., effective) lengths for frames.

In an unbraced frame [Fig. 15.3.3(b) and (d)] instability results in a sidesway type of buckling with the effective length kL_u always exceeding the actual length L_u. Frame stability is discussed further in Secs. 15.7 and 15.8.

15.4 MOMENT MAGNIFICATION—SIMPLIFIED TREATMENT FOR MEMBERS IN SINGLE CURVATURE WITHOUT END TRANSLATION (i.e., NO SIDESWAY)

As stated previously, nearly all compression members are simultaneously subjected to some bending moment that causes lateral deflections. Any deflected compression member is further subjected to a secondary bending moment $P\Delta$, as shown in Fig. 15.1.1. One may consider this as a magnification of the applied bending moment. An approximate determination of the magnifying effect may be made by considering the member as finally achieving a deflection Δ_{max}, which is composed of the deflection Δ_0 due to the primary applied bending moment and the additional deflection Δ_1 due to the secondary moment from axial compression (see Fig. 15.4.1). It may be assumed that the secondary bending moment takes the shape of a sine curve (very nearly exact for members with no end restraint and whose primary bending moment and deflection are both maximum at midspan). The midspan deflection Δ_1 equals the moment of the $M/(EI)$ diagram (for secondary bending moment) between the support and midspan taken about the support, according to the moment area principle. Thus

$$\Delta_1 = \frac{P}{EI}(\Delta_0 + \Delta_1)\left(\frac{L}{2}\right)\frac{2}{\pi}\left(\frac{L}{\pi}\right) = (\Delta_0 + \Delta_1)\frac{PL^2}{\pi^2 EI} \qquad (15.4.1)$$

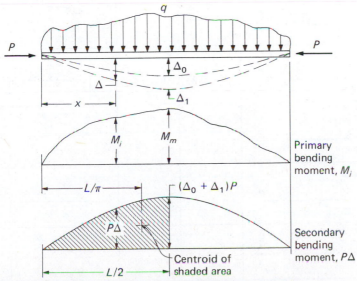

Figure 15.4.1 Primary and secondary bending moment.

from which

$$\Delta_1 = \Delta_0 \left[\frac{PL^2/(\pi^2 EI)}{1 - PL^2/(\pi^2 EI)} \right] = \Delta_0 \left(\frac{\alpha}{1 - \alpha} \right) \qquad (15.4.2)$$

where $\alpha = PL^2/(\pi^2 EI)$. Since $\Delta_{\max}$ is the sum of Δ_0 and Δ_1,

$$\Delta_{\max} = \Delta_0 + \Delta_1 = \Delta_0 + \Delta_0 \left(\frac{\alpha}{1 - \alpha} \right) = \frac{\Delta_0}{1 - \alpha} \qquad (15.4.3)$$

The maximum bending moment, including the effect of axial load, becomes

$$M_{\max} = M_m + P\Delta_{\max} \qquad (15.4.4)$$

Table 15.4.1 SUGGESTED VALUES OF C_m FOR COMMON SITUATIONS WITH NO JOINT TRANSLATION[a]

	Case	C_m (positive moment)	C_m (negative moment)	Primary Bending Moment
1		$1.0 + 0.2\alpha$	—	
2		1.0	—	
3		$1.0 - 0.2\alpha$	—	
4		$1.0 - 0.3\alpha$	$1 - 0.4\alpha$	
5		$1.0 - 0.4\alpha$	$1 - 0.4\alpha$	
6		$1.0 - 0.4\alpha$	$1 - 0.3\alpha$	
7		$1.0 - 0.6\alpha$	$1 - 0.2\alpha$	
8		Eq.(15.4.7)	not available	

[a]Adapted from Ref. 28.

Substituting the expression for Δ_{max} of Eq. (15.4.3) and making $P = \alpha\pi^2 EI/L^2$, Eq. (15.4.4) becomes

$$M_{max} = M_m\left(\frac{C_m}{1 - \alpha}\right) = M_m\delta \tag{15.4.5}$$

where

$$\delta = \frac{C_m}{1 - \alpha} = \text{magnification factor} \tag{15.4.6}$$

and

$$C_m = 1 + \left(\frac{\pi^2 EI\Delta_0}{M_m L^2} - 1\right)\alpha \tag{15.4.7}$$

Thus, for common cases of single curvature deflection, the magnification factor to be applied to the primary bending moment is equal to $C_m/(1 - \alpha)$. For Cases 1 to 7 shown in Table 15.4.1, rigorous solution of the differential equation may be made including the term Py in the expression for $EI (d^2y/dz^2)$. The approximate values of C_m for positive moment shown in this table are computed using Eq. (15.4.7) and they are in general agreement with theoretical results even though Eq. (15.4.7) is derived using a sine curve deflection. Approximate values of C_m for negative moment are given in the AISC Commentary [28]. One may note that for all cases this C_m value will be close to 1.0, because in actual concrete structures α rarely exceeds about 0.3. The approximate treatment of slenderness in ACI-10.11.5 conservatively requires that C_m shall be taken as 1.0 for all cases with transverse loading between supports.

15.5 MOMENT MAGNIFICATION—MEMBERS SUBJECT TO END MOMENTS ONLY; NO JOINT TRANSLATION

Consider the general case shown in Fig. 15.5.1 wherein the end moments M_1 and M_2 constitute the primary bending moment M_i which is a function of z. The sum of primary and secondary moments causes the member to have a deflection y which gives rise to the secondary moment Py. Stating the total moment M_z at the section z of Fig. 15.5.1 gives

$$M_z = M_i + Py = -EI\frac{d^2y}{dz^2} \tag{15.5.1}$$

for sections with constant EI; and dividing by EI gives

$$\frac{d^2y}{dz^2} + \frac{P}{EI}y = -\frac{M_i}{EI} \tag{15.5.2}$$

For design purposes, the general expression for moment M_z is of greater importance than the deflection y. Differentiating Eq. (15.5.2) twice gives

$$\frac{d^4y}{dz^4} + \frac{P}{EI}\frac{d^2y}{dz^2} = -\frac{1}{EI}\frac{d^2M_i}{dz^2} \tag{15.5.3}$$

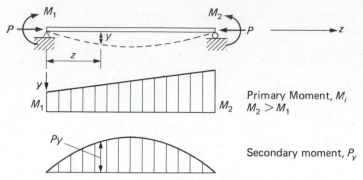

Figure 15.5.1 Beam-column having end moments without transverse loading.

From Eq. (15.5.1),

$$\frac{d^2y}{dz^2} = -\frac{M_z}{EI} \quad \text{and} \quad \frac{d^4y}{dz^4} = -\frac{1}{EI}\frac{d^2M_z}{dz^2}$$

Substitution into Eq. (15.5.3) gives

$$-\frac{1}{EI}\frac{d^2M_z}{dz^2} + \frac{P}{EI}\left(-\frac{M_z}{EI}\right) = -\frac{1}{EI}\frac{d^2M_i}{dz^2}$$

Simplifying and letting $\lambda^2 = P/EI$,

$$\frac{d^2M_z}{dz^2} + \lambda^2 M_z = \frac{d^2M_i}{dz^2} \tag{15.5.4}$$

which is the same form as the deflection differential equation, Eq. (15.5.2).
The homogeneous solution for Eq. (15.5.4) is

$$M_z = A \sin \lambda z + B \cos \lambda z \tag{15.5.5}$$

To this must be added the particular solution that will satisfy the right-hand side of the differential equation. In the special case of unequal end moments acting without transverse loading,

$$M_i = M_1 + \frac{M_2 - M_1}{L} z \tag{15.5.6}$$

Since

$$\frac{d^2M_i}{dz^2} = 0$$

Eq. (15.5.4) becomes a homogeneous equation, in which case Eq. (15.5.5) represents the entire solution.
In order to determine the maximum moment,

$$\frac{dM_z}{dz} = 0 = A\lambda \cos \lambda z - B\lambda \sin \lambda z \tag{15.5.7}$$

or

$$\tan \lambda z = \frac{A}{B} \tag{15.5.8}$$

At maximum M_z,

$$\sin \lambda z = \frac{A}{\sqrt{A^2 + B^2}} \qquad \cos \lambda z = \frac{B}{\sqrt{A^2 + B^2}} \tag{15.5.9}$$

Substitution of Eq. (15.5.9) in Eq. (15.5.5) gives

$$M_{\max} = \frac{A^2}{\sqrt{A^2 + B^2}} + \frac{B^2}{\sqrt{A^2 + B^2}}$$

$$= \sqrt{A^2 + B^2} \tag{15.5.10}$$

Now the constants A and B are evaluated by applying the boundary conditions to Eq. (15.5.5). The conditions are

(1) at $z = 0$, $\quad M_z = M_1$

$$\therefore B = M_1$$

(2) at $z = L$, $\quad M_z = M_2$

$$M_z = A \sin \lambda L + M_1 \cos \lambda L$$

$$\therefore A = \frac{M_2 - M_1 \cos \lambda L}{\sin \lambda L}$$

so that

$$M_z = \left(\frac{M_2 - M_1 \cos \lambda L}{\sin \lambda L} \right) \sin \lambda z + M_1 \cos \lambda z \tag{15.5.11}$$

and

$$M_{\max} = \sqrt{\left(\frac{M_2 - M_1 \cos \lambda L}{\sin \lambda L} \right)^2 + M_1^2}$$

$$= M_2 \sqrt{\frac{1 - 2(M_1/M_2) \cos \lambda L + (M_1/M_2)^2}{\sin^2 \lambda L}} \tag{15.5.12}$$

For the general case of a beam-column subject to end moments, the maximum moment may be either (1) the larger end moment M_2 at the braced (supported) location, [Fig. 15.5.2(a)] or (2) the magnified moment given by Eq. (15.5.12) that occurs at a variable location out along the span [Fig. 15.5.2(b)], depending on the ratio M_1/M_2 and the value of α, since $\lambda L = \pi \sqrt{\alpha}$. In order to investigate the strength of a beam-column, one needs to know whether the maximum moment occurs at a location between the supports, and if so, the correct *distance*. To eliminate the need for such information, the concept of equivalent uniform moment [Fig. 15.5.2(c)] is used. Thus, for the case with unequal end moments, use of the equivalent moment assumes $M_{\max}$ to be at midspan.

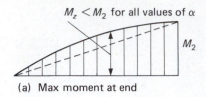

(a) Max moment at end

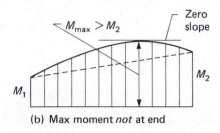

(b) Max moment *not* at end

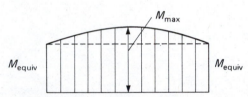

(c) Equivalent uniform moment with max magnified moment at midspan

Figure 15.5.2 Combined primary and secondary bending moment diagrams for beam-columns having end moments without transverse loading.

To establish the equivalent moment, let $M_1 = M_2 = M_{equiv}$ in Eq. (15.5.12),

$$M_{max} = M_{equiv} \sqrt{\frac{2(1 - \cos \lambda L)}{\sin^2 \lambda L}} \qquad (15.5.13)$$

Equating Eq. (15.5.12) and Eq. (15.5.13) gives

$$M_{equiv} = M_2 \sqrt{\frac{(M_1/M_2)^2 - 2(M_1/M_2) \cos \lambda L + 1}{2(1 - \cos \lambda L)}} \qquad (15.5.14)$$

According to the procedure of Sec. 15.4, the approximate expression for maximum moment was shown to be

$$M_{max} = M_m \delta = M_m \left(\frac{C_m}{1 - \alpha} \right) \qquad (15.5.15)$$

For the case of uniform moment ($M_1 = M_2 = M_{equiv}$),

$$\Delta_0 = \frac{M_{equiv}L^2}{8EI}$$

$$M_m = M_{equiv}$$

$$C_m = 1 + \left[\left(\frac{\pi^2 EI}{L^2} \right) \frac{M_{equiv}L^2}{8EIM_{equiv}} - 1 \right] \alpha \approx 1$$

Thus

$$M_{max} = M_{equiv} \left(\frac{1}{1 - \alpha} \right) \tag{15.5.16}$$

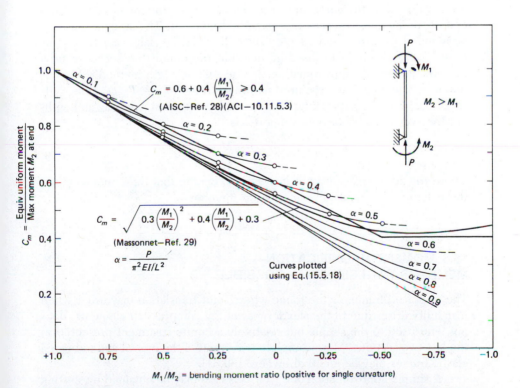

Figure 15.5.3 Comparison of theoretical C_m with design recommendations for members subject to end moments only, without joint translation (from Salmon and Johnson [30]).

Substitution of Eq. (15.5.14) into Eq. (15.5.16) gives

$$M_{\max} = M_2\left(\frac{C_m}{1 - \alpha}\right) \tag{15.5.17}$$

in which

$$C_m = \sqrt{\frac{(M_1/M_2)^2 - 2(M_1/M_2)\cos \lambda L + 1}{2(1 - \cos \lambda L)}} \tag{15.5.18}$$

Comparing Eq. (15.5.17) with Eq. (15.5.16), $C_m M_2$ may be considered to be the equivalent uniform moment along the span.

Equation (15.5.18) assumes that the strength of the beam-column is limited by excessive deflection *in the plane of bending*. Also, it does not fully cover the double-curvature cases where M_1/M_2 lies between -0.5 and -1.0. The actual failure mode of members bent in double curvature with such bending moment ratios is generally one of "unwinding" from double to single curvature in a sudden type of buckling.

Massonnet [29] and the AISC steel design specification [28] have suggested expressions for C_m to be used for design in place of Eq. (15.5.18). The comparison of Eq. (15.5.18) with the recommendations of Massonnet and the AISC is shown in Fig. 15.5.3. The reader should note that for a given value of α, the curve shown terminates when the moment M_2 at the end of the member exceeds the magnified moment. The straight line recommended by AISC, and adopted by ACI-10.11.5.3, falls near the upper limit for C_m at any given bending moment ratio, and thus seems to be a realistic and simple approximation. In the ACI strength method α is computed by using for P the required nominal strength $P_n = P_u/\phi$. Thus ACI-10.11.5.3, Formula (10-11), for members braced against sidesway and without transverse loads between supports, is

$$C_m = 0.6 + 0.4\frac{M_{1b}}{M_{2b}} \geq 0.4 \tag{15.5.19}$$

where the additional subscript b is used to denote that these moments are those acting on a "braced" compression member.

15.6 MOMENT MAGNIFICATION— MEMBERS WITH SIDESWAY POSSIBLE

The unbraced frame, or the frame where joint translation may occur when instability arises due to the slenderness of the compression elements, does not lend itself to the simple but relatively accurate treatment presented in the last two sections. More complete treatment of the braced and unbraced elastic frames may be found elsewhere [30,31].

A simple approximation of C_m for this case may be obtained by starting with Eq. (15.4.5) which applies for the single-curvature case,

$$M_{\max} = M_m\left(\frac{C_m}{1 - \alpha}\right) \tag{15.6.1}$$

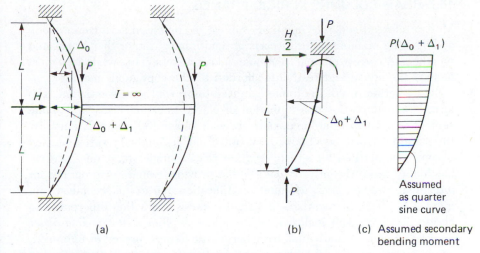

(a)   (b)   (c) Assumed secondary
                                                          bending moment

Figure 15.6.1   Beam-column with sidesway instability.

Next consider the situation of Fig. 15.6.1. Whatever the degree of restraint at the top and bottom of the two-story member, the deflection curve, and therefore the secondary bending moment (P times deflection), may be reasonably assumed to be a sine curve, in which case the development used when no sidesway occurs (Fig. 15.4.1) is also valid here. Since $2L$ from Fig. 15.6.1 equals L for Fig. 15.4.1, Eq. (15.4.7) for C_m becomes

$$C_m = 1 + \left(\frac{\pi^2 E I \Delta_0}{4L^2 M_m} - 1 \right) \alpha \qquad (15.6.2)$$

The larger effective length ($2L$ instead of L) is also used in the computation of α. Next, referring to Fig. 15.6.1,

$$\Delta_0 = \frac{(H/2)L^3}{3EI} \qquad (15.6.3)$$

$$M_m = \frac{HL}{2} \qquad (15.6.4)$$

Substitution of Eqs. (15.6.3) and (15.6.4) into Eq. (15.6.2) gives

$$C_m = 1 + \left[\frac{\pi^2 EI}{4L^2} \left(\frac{HL^3}{6EI} \right) \left(\frac{2}{HL} \right) - 1 \right] \alpha$$

$$C_m = 1 + \left(\frac{\pi^2}{12} - 1 \right) \alpha = 1 - 0.18\alpha \qquad (15.6.5)$$

which is suggested for the unbraced frame by the AISC Commentary [28]. Again, it will be conservative to take $C_m = 1$ for unbraced frames, as required in the approximate evaluation of slenderness effects in ACI-10.11.5.3.

15.7 BEAM-COLUMNS IN RIGID FRAMES

Most reinforced concrete members subject to combined axial compression and bending moment occur as parts of rigid frames, rather than as isolated members. A correct rational treatment of long members in such frames must include the actual end restraints afforded by the contiguous members, as well as whether or not the frame is braced to prevent joint translation. The formal procedures for determination of elastic buckling loads in a frame subject to primary axial forces only are well known [30,32–34]. Generally, the solutions must be obtained by trial from an implicit expression and become complicated for situations other than a single story, one- or two-bay frame. More recently, Wang [35] has provided computer programs for individual case solutions. Moment magnification due to the secondary moment ($P\Delta$) effect for members of a rigid frame is a situation different from elastic buckling. The inclusion of $P\Delta$ effect is often called *second-order analysis*. Again, for individual irregular elastic frames, braced or unbraced, computer programs exist to give maximum bending moment everywhere in the structure [35]. Practical design procedures for frames are discussed in more detail in Secs. 15.10 through 15.15.

Design of the compression members in rigid frames has traditionally been done by estimating the relative sizes of members and performing a nominal elastic analysis to determine moments, shears, and axial loads. The nominal analysis is one using any elastic method wherein stiffnesses are determined from gross uncracked sections and the secondary effects of deflection and reduction in stiffness due to axial load are neglected. Each member of the frame is then designed or investigated individually using the loading from the nominal analysis.

In the absence of a convenient second-order analysis for obtaining the total magnified moments, a compression member restrained by adjoining members at its ends may be considered removed from the frame and replaced by an equivalent pin-end column (see Fig. 15.3.3) whose length is equal to the effective length kL_u for axial compression on the real column. The equivalent column is then analyzed for compression plus the end moments carried by the member. The effective length kL_u would be used both for the axial effect as in Eq. (15.2.1) and for determining any magnified bending moment by Eq. (15.4.5).

One of the difficulties that arises in using the nominal analysis for the reinforced concrete frame is that when the gross moment of inertia is used for the beam stiffness, the restraining effect of the beams on the columns is overestimated [18,19]. If a beam is cracked under service load but assumed uncracked, the amount of moment transmitted to the column will be underestimated. This is the common situation.

15.8 ALIGNMENT CHARTS FOR EFFECTIVE LENGTH FACTOR *k*

The most commonly used procedure for obtaining the equivalent pin-end (effective) length is to use the alignment charts from the Structural Stability Research Council Guide [31], originally developed by O. J. Julian and L. S. Lawrence, and presented in detail by T. C. Kavanagh [36]. The charts are shown in Fig. 15.8.1, the complete derivations of which appear in Chap. 14 of Salmon and Johnson [30].

The effective length factor k is a function of the end restraint factors ψ_A and ψ_B, for the top and bottom joints at the ends of the member, respectively, defined as

$$\psi = \frac{\Sigma EI/L \text{ for column members in the plane of bending}}{\Sigma EI/L \text{ for beam members in the plane of bending}} \qquad (15.8.1)$$

which for a hinged end gives $\psi = \infty$ and for a fixed end, $\psi = 0$. Since a frictionless hinge cannot exist in practical construction, ψ is to be taken equal to 10 for an end assumed as hinged in the analysis.

One nomogram (or alignment chart), Fig. 15.8.1(a), is for braced frames where sidesway (joint translation) is prevented, and the other, Fig. 15.8.1(b), is for the unbraced frame where sidesway is possible, being restrained only by the stiffness of interacting beams and columns.

An effective length procedure has been adopted by ACI-10.11 in the approximate evaluation of slenderness effects. The alignment charts are implicitly endorsed for determining the k factor by their inclusion in the ACI Commentary [37].

The assumptions inherent in the development of the alignment chart for the braced frame [Fig. 15.8.1(a)] are as follows [30]:

1. All columns reach their respective critical loads simultaneously.
2. The structure is assumed to consist of symmetrical rectangular frames.
3. At any joint, the restraining moment provided by the beams is distributed among the columns in proportion to their stiffnesses.
4. The beams are elastically restrained at their ends by the columns, and at the onset of buckling the rotations of the beam at its ends are equal and opposite (i.e., the beams are deflected in single curvature).
5. The beams carry no axial loads.

For the unbraced frame alignment chart [Fig. 15.8.1(b)] the assumptions (1) through (3) and (5) are unchanged; however, the beams are assumed to be deflected in double curvature, where the rotations of the ends are equal in magnitude and direction.

By means of the alignment charts one may determine the k factor for a column of constant cross section in a multistory, multibay frame. With steel frames where the material is homogeneous and isotropic, the modulus of elasticity E is constant for all members, and the moment of inertia I is computed for the gross cross section. In reinforced concrete, E varies with concrete strength and magnitude of loading, while I also varies depending on the degree of cracking and the reinforcement percentage. ACI-10.11.2.2 requires that the effective length factor k for the *unbraced* frame "shall be determined with due consideration of effects of cracking and reinforcement on relative stiffness." For the *braced* frame, ACI-10.11.2.1 merely says the effective length factor "shall be taken as 1.0, unless analysis shows that a lower value may be used." It is believed that an appropriate use of the alignment chart would constitute an "analysis" as required by the ACI Code.

Thus, for the purpose of evaluating the end restraint factor ψ, cracked section moment of inertia should be used for the beams whereas gross moment of inertia is probably satisfactory for the columns [38] (or ACI Formula 10-9 with $\beta_d = 0$, as discussed in Sec. 15.10). Particularly recognition of different behavior in beams and columns is necessary when the reinforcement ratio ρ is significantly different, such as with $\rho_g = 0.08$ in the columns and $\rho = 0.005$ in the beams. The ACI Commentary-10.11.2 indicates that when kL_u/r is less than 60, the use of $0.5I_g$ instead of the cracked transformed section moment of inertia for the beam ". . . will usually result in reasonable member sizes for columns . . ."

As an alternative to actually using the nomograms of Fig. 15.8.1 some approximate formulas for the effective length factor k have been proposed and are endorsed by the ACI Commentary.

For members in *braced* frames, the 1972 British Code of Standard Practice [39] suggests that an upper bound for k is obtained by using the *smaller* of the following two equations:

$$k = 0.7 + 0.05(\psi_A + \psi_B) \le 1.0 \tag{15.8.2}$$

$$k = 0.85 + 0.05\psi_{min} \le 1.0 \tag{15.8.3}$$

where ψ_A and ψ_B are the ψ values at the two ends of the member, and ψ_{min} is the smaller of the two values.

For members in *unbraced* frames, Furlong [40] proposed for members restrained at both ends,

when $\psi_{avg} < 2$ (i.e., high end restraint),

$$k = \frac{20 - \psi_{avg}}{20} \sqrt{1 + \psi_{avg}} \tag{15.8.4}$$

when $\psi_{avg} > 2$ (i.e., moderate to low end restraint),

$$k = 0.9\sqrt{1 + \psi_{avg}} \tag{15.8.5}$$

where ψ_{avg} is the average of the ψ values at the two ends of the member. Equations (15.8.4) and (15.8.5) give k values that are within 2% of those obtained by the nomograms.

For members in *unbraced* frames, when hinged at one end, the British Code of Standard Practice [39] proposes

$$k = 2.0 + 0.3\psi \tag{15.8.6}$$

where ψ is the value at the restrained end.

15.9 INTERACTION DIAGRAMS—EFFECT OF SLENDERNESS

In order to understand the ACI Code procedure and its approximations as discussed in Secs. 15.10 through 15.15, the general approach is described for determining a point on the P_n-M_n interaction diagram for kL/r not equal to zero. In Chap. 13 the basic strength of a section with zero kL/r was treated, giving an interaction diagram such as Fig. 13.6.2, and designated by $kL/r = 0$ in Fig. 15.9.1.

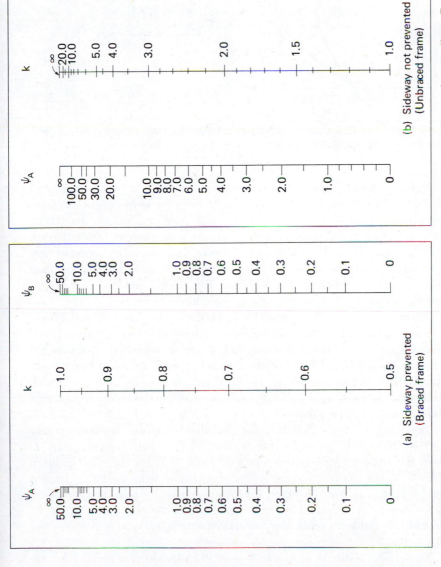

Figure 15.8.1 Alignment charts for effective length factor for columns in continuous frames (from Ref. 36), where

$$\psi = \frac{\Sigma EI/L, \text{ columns}}{\Sigma EI/L, \text{ beams}}$$

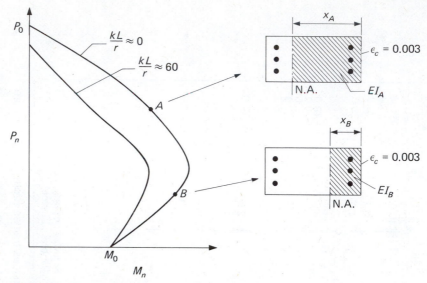

Figure 15.9.1 Beam-column strength interaction diagram, including effective cross section for moment of inertia.

Points A and B represent combinations of P_n and M_n with the neutral axes (NA) located at x_A and x_B, respectively, from the extreme compression fiber whose crushing strain is taken as the ACI Code prescribed value 0.003. In the following development it will be shown that when kL/r is not zero, the total nominal strength M_n to resist primary and secondary moment may be achieved when the strain at the extreme compression fiber is less than 0.003. The curve labeled $kL/r \approx 60$ in Fig. 15.9.1 represents a typical strength interaction curve that includes slenderness effects.

Pfrang and Siess [8], Pfrang [15], and MacGregor, Breen, and Pfrang [41] have provided excellent discussions of the slenderness effects on interaction diagrams. A long column may fail in one of two ways: (1) it may fail by reaching a combined P_n–M_n that exceeds the cross-section strength computed by the methods of Chap. 13; (2) it may fail by instability when an infinitesimal increase in axial load results in additional deflection such that equilibrium cannot be achieved.

Referring to Fig. 15.9.2(a), the member of large slenderness, say $kL/r = 100$, will generally follow the loading path up to point D where the material strength is reached. Point D is on the short column $(kL/r = 0)$ interaction diagram but is at a smaller axial load P_{long} than it would be [P_{short} in Fig. 15.9.2(a)] if kL/r were actually zero. If the column fails by instability, it would follow the path (dashed) up to point E; that is, it would be unable to reach the material strength interaction diagram (for $kL/r = 0$).

Generally, columns in braced frames are capable of achieving a "material failure," while the "instability failure" is not common but may occur in unbraced frames.

In the construction of interaction diagrams, as shown in Fig. 15.9.2b, a material failure occurring at D on the $kL/r = 0$ curve due to an axial load

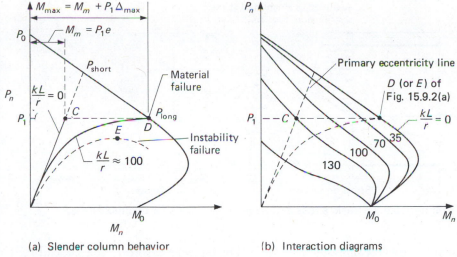

(a) Slender column behavior   (b) Interaction diagrams

Figure 15.9.2   Slender column interaction diagrams (adapted from Ref. 41).

P_1 plus a magnified moment $M_m\delta$ (equal to $M_m + P_1\Delta_{max}$) may be plotted *for the particular loading arrangement* at point C on the primary moment radial line. Whatever the primary moment loading arrangement—such as equal end eccentricities of axial load, unequal end eccentricities of axial load, or lateral transverse loading—the deflection of the member will differ so that the secondary bending moment $P\Delta$ will differ. This means that different types of primary moments will cause a member of $kL/r = 100$ to follow a curved path to intersect the $kL/r = 0$ material strength interaction diagram at different locations such as point D. The radial line through point C is a function of only the primary moment, which is the same for all slenderness ratios.

The development of a correct interaction diagram for members of large slenderness, such as Fig. 15.9.2(b), would require an elaborate analysis for each structure taking into account such factors as the following: (1) a realistic moment-curvature relationship; (2) the time-dependent and cracking effects on deflections; and (3) the influence of axial load on the flexural stiffness of members. Mockry and Darwin [52,53] have made such an analysis and prepared a practical set of design charts. Parme [14] has treated the subject in a similar manner and the Portland Cement Association has published design aids [55].

EXAMPLE 15.9.1 Illustrate qualitatively the determination of the long column strength, indicating the process of computing point D (material failure) or point E (instability failure) of Fig. 15.9.2(a). Assume that the loading condition is a simply supported member with axial load P_1 and subjected to a primary bending moment M_i which arises from uniform lateral loading. The Hognestad stress-strain curve for concrete (Fig. 15.2.1) is to be used (see also Fig. 15.9.3).

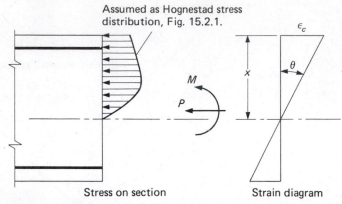

Figure 15.9.3 Basic stress and strain relationships for beam-columns.

Solution: (a) Compute *M*–θ curve. For the *constant* value of axial load P_1, a series of moment capacities *M* may be determined for various values of the extreme fiber strain ϵ_c. The principles of Chap. 13 are used except *M* is computed for other values of ϵ_c in addition to 0.003, which is the ACI prescribed maximum. Points *A* and *B* of Fig. 15.9.1 correspond to using 0.003. When ϵ_c is less than 0.003, the neutral axis distances *x* will also change. The determination of *M* involves solving also for *x*. Then, using the strain geometry relationship,

$$\theta = \frac{\epsilon_c}{x} = \frac{M}{EI} \tag{15.9.1}$$

the *M*–θ curve may be plotted as in Fig. 15.9.4. For determining values of *M* and *x* for a given *P*, the Hognestad compressive stress distribution should be used for good results, and this will be more complicated than using the Whitney rectangular stress distribution though the principles are the same.

(b) Compute deflection under primary bending moment. For the constant P_1 select a value of maximum primary moment M_1 that is not expected to cause failure. For the uniform lateral loading condition (parabolic variation) the values of θ for values of *M* (such as θ_1 for M_1 in Fig. 15.9.5) are

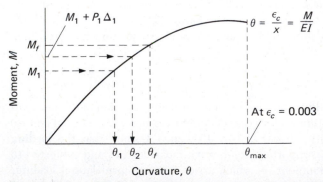

Figure 15.9.4 Moment-rotation characteristics for a section.

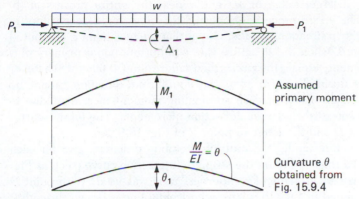

Figure 15.9.5   Loading and first approximation for Δ.

obtained from the relationship of Fig. 15.9.4. Using the conjugate beam with the θ loading, compute the deflections (including Δ_1 at midspan).

(c) Compute deflection under primary plus secondary bending moment. The secondary bending moment $P\Delta$ is computed and added to the primary moment, as in Fig. 15.9.6. For this first approximation of total moment, the θ_2 variation is determined from Fig. 15.9.4. The conjugate beam with the θ_2 loading is used to compute the second approximation of deflection.

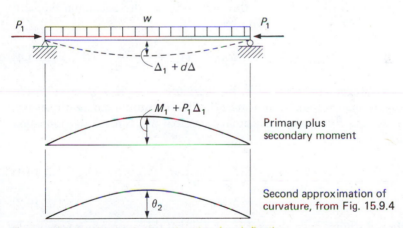

Figure 15.9.6   Second approximation for deflection.

(d) Iteration. The newly computed deflection means increased secondary moment which will result in increased deflection. The process in (c) is repeated until the newly computed deflection agrees with that used for computing the secondary bending moment. These iterations then yield one combination of P_1 and M for which the beam-column is stable. If the maximum value of ϵ_c computed from Eq. (15.9.1) exceeds 0.003, the applied moment is too large, resulting in a material failure. Generally the initial value of M applied will result in maximum ϵ_c well below 0.003.

(e) Apply a larger value of M_1 and repeat the entire process of (b) through (d). The entire process is repeated for larger and larger values of M_1 until one of two situations results. One is that the maximum ϵ_c computed from maximum θ, when the structure is in equilibrium under primary plus secondary moment, equals the prescribed failure limit 0.003 (or whatever other value is decided on); or two, the equilibrium situation cannot be achieved due to the fact that each succeeding moment increment due to $P_1\Delta$ results in successively larger deflection increments. The former corresponds to point D and the latter to point E of Fig. 15.9.2(a).

In either failure mode, the multiple iteration processes give for each axial load level a *single* point on the strength interaction curve [such as Fig. 15.9.2(b)] for whatever value of slenderness ratio was used. With point D (or E) thus obtained in Fig. 15.9.2(b), a horizontal line is drawn through it to intersect the primary eccentricity line at point C. Point C is one point on the interaction diagram containing slenderness effect. In a design situation, the maximum primary moment (point C) is known. The magnification factor δ is the ratio of the moment at point D to the moment at point C. If this concept is to be used in practical design, simpler approaches, such as presented in Secs. 15.10 through 15.15, are necessary.

15.10 ACI CODE—MOMENT MAGNIFIER METHOD FOR BRACED FRAMES

It has been shown in Secs. 15.4 through 15.6 that the maximum moment in an elastic beam-column is given by Eq. (15.4.4)

$$M_{max} = M_m + P\Delta_{max} = M_m + \frac{P\Delta_0}{1 - \alpha} \qquad [15.4.4]$$

where $\alpha = P/P_c$ and $P_c = \pi^2 EI/L^2$. Furthermore, Eqs. (15.4.5) and (15.5.17), applicable to braced frames, indicated that the maximum moment may also be expressed as the maximum primary moment M_m times a magnification factor δ_b

$$M_{max} = \left(\frac{C_m}{1 - \alpha}\right) M_m \qquad (15.10.1)$$

$$= \delta_b M_m \qquad (15.10.2)$$

A number of studies have shown [21,41,42] that this approach is acceptable for reinforced concrete compression members using the strength design method. Design of the members is based on the factored axial load P_u combined with the corresponding factored moment M_m from an elastic frame analysis magnified by the factor δ_b. M_m is the maximum moment acting on the member and it may occur at either end, or if there is transverse loading, in the midspan region.

The required nominal strength of the designed section must be

$$P_n = \frac{P_u}{\phi} \quad \text{and} \quad M_n = \frac{M_m}{\phi}\left(\frac{C_m}{1 - P_n/P_c}\right) \qquad (15.10.3)$$

For practical purposes the strength reduction factor ϕ in the denominator may be treated as shown, or as a multiplier on the other side of the equation. However, since P_n appears in the magnifier, the expression for δ_b contains ϕ in any case,

$$\delta_b = \frac{C_m}{1 - \dfrac{P_u}{\phi P_c}} \tag{15.10.4}$$

which is ACI Formula (10-7).

Factor C_m. The quantity C_m has two basic meanings: (1) for braced frames with transverse loading and single-curvature deflection, it is truely a part of the moment magnifier; (2) for braced frames with end moments alone acting, the factor C_m is really not part of the magnifier, rather $C_m M_m$ gives an equivalent uniform moment which is then magnified by multiplying by $1/[1 - P_u/(\phi P_c)]$. For the first meaning of C_m, given by Eq. (15.4.6), ACI-10.11.5.3 states "C_m shall be taken as 1.0." This is a conservative approach since the correct C_m will usually be between 0.9 and 1.0. Thus the use of C_m under ACI-10.11.5 may be summarized as follows for *braced* frames:

1. Transverse loading,

$$C_m = 1.0 \tag{15.10.5}$$

2. End moments only, use Eq. (15.5.19)

$$C_m = 0.6 + 0.4\frac{M_{1b}}{M_{2b}} \geq 0.4 \tag{15.10.6}$$

which is ACI Formula (10-11). Note that M_{2b} is larger than M_{1b} and the ratio M_{1b}/M_{2b} is positive when the member is bent in single curvature.

Stiffness Parameter EI. The other quantity required for evaluating the moment magnifier δ_b is

$$P_c = \frac{\pi^2 EI}{(kL_u)^2} \tag{15.10.7}$$

which is slightly modified from the form used in Secs. 15.4 through 15.6 by the term kL_u, which is the unsupported length L_u of the member in a reinforced concrete frame modified for the pin-end condition by the effective length factor k.

The principal difficulty with the magnifier method is that it requires a value for EI, as illustrated by its general development in Sec. 15.9. EI correctly varies due to cracking, time-dependent effects, and nonlinearity of the concrete stress-strain curve. MacGregor, Breen, and Pfrang [41] proposed to use the larger of two simple expressions *when more precise values are not available*. These appear in ACI-10.11.5.2 as Formulas (10-9) and (10-10), respectively,

$$EI = \frac{0.2E_c I_g + E_s I_s}{1 + \beta_d} \tag{15.10.8}$$

or

$$EI = \frac{0.4E_cI_g}{1 + \beta_d} \tag{15.10.9}$$

where

E_c = concrete modulus of elasticity = 57,000 $\sqrt{f_c'}$ for normal-weight concrete (ACI-8.5)

I_g = gross moment of inertia of concrete section, neglecting reinforcement

I_s = moment of inertia of reinforcement

β_d = proportion of the factored load *moment* that is considered sustained so as to contribute to time-dependent deformations (usually the factored dead load moment to total factored load moment)

The larger of Eqs. (15.10.8) and (15.10.9) is appropriate for use and is still an underestimate of the correct EI.

The relative accuracy of these EI expressions is shown in Fig. 15.10.1 from Ref. 41. The theoretical values were for the case of no sustained load ($\beta_d = 0$) [41]. In the study of Eqs. (15.10.8) and (15.10.9), Design Subcommittee of ASCE-ACI Committee 441 [41] estimated EI values for about 100 cases using theoretical load–moment–curvature diagrams computed in a manner similar to that discussed in Sec. 15.9, considering columns of various dimensions, strengths, and steel percentages. Effective EI values were also computed for the University of Texas frame tests [6,10,11], and for a series of frames simulated by the computer.

Alternate expressions for EI have been proposed by MacGregor, Oelhafen, and Hage [44] and Medland and Taylor [45]. As a result of these studies it seemed appropriate to divide only the concrete term in Eq. (15.10.8) by the factor $(1 + \beta_d)$. However, when this is done, the number of cases where the ratio of theoretical EI to formula computed EI is less than 1.0 becomes three times more than otherwise, particularly when the reinforce-

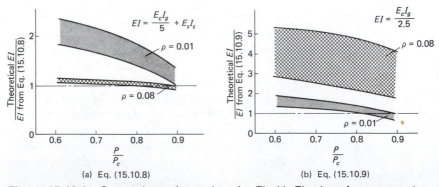

(a) Eq. (15.10.8) (b) Eq. (15.10.9)

Figure 15.10.1 Comparison of equations for EI with EI values from moment-curvature diagrams, for short-duration loading ($\beta_d = 0$) (adapted from Ref. 41).

ment ratios are low. Thus, ACI Committee 318 concluded that the $(1 + \beta_d)$ factor should be retained as the divisor for both the steel and concrete terms.

The sustained-load effect on slender columns has also been studied by Goyal and Jackson [46] and Drysdale and Huggins [47].

15.11 GENERAL CONCEPT FOR DESIGN OF UNBRACED FRAMES

As shown in Figs. 15.3.2 and 15.3.3(b) and (d), an unbraced frame must rely on the interaction of its beams and columns to limit the horizontal displacement when subject to lateral loads, such as wind or earthquake. Under lateral loads, a "braced" frame will resist the lateral force by such components as diagonal bracing or shear walls so that any lateral distortion will be of small magnitude. Thus secondary bending moments $P\Delta$ from sidesway (the P–Δ effect) may ordinarily be neglected. However, for "unbraced" frames, the relatively larger sidesway deflection Δ due to lateral load will give rise to secondary moments $P\Delta$ (P is the gravity load) that must be provided for in design. Thus, an unbraced concrete frame requires an analysis to accomplish the following tasks:

1. Provide strength under factored loads to resist gravity load, neglecting any sidesway effect except in rare cases of unbalanced loading or unsymmetrical structural configuration as to cause "appreciable" sidesway. ACI Commentary-10.11.5.1 indicates that "appreciable" would be where the sway Δ divided by the clear height L_u exceeds 1/1500. Randomly occurring construction inaccuracies cause columns to be "out-of-plumb"; that is, the centroid of the top of the column is not directly over the centroid of the bottom of the column. MacGregor [56] indicates that Δ/h of 1/450 may be expected due to out-of-plumbness, and further that out-of-plumbness typically increases first-order moments by about $6\frac{1}{2}\%$; whereas a second-order analysis performed on a structure where out-of-plumbness was neglected increases first-order moments by only about $4\frac{1}{2}\%$. In other words, out-of-plumbness typically has more effect than the P–Δ effect from a second-order analysis.
2. Provide strength under factored loads to resist lateral load (i.e., factored wind or earthquake load). The moments under lateral load include the primary moments from first-order analysis plus secondary moments due to the P–Δ effect.
3. Provide stiffness such that the relative horizontal deflection between adjacent floors, and for the entire frame, is within specified limits (usually, say, equal to the clear height L_u divided by 400 or 500).

MacGregor and Hage [43] have provided an excellent summary of the various methods and have suggested practical approaches; much of what follows is based on their work. Referring to Fig. 15.11.1, the basic first-

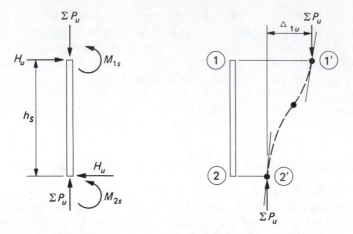

(a) First-order analysis: For equilibrium, $M_{1s} + M_{2s} = H_u h_s$

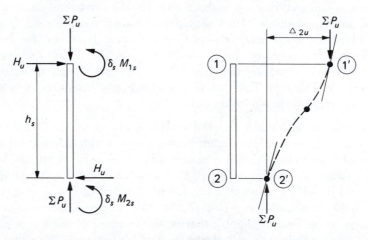

(b) Second-order analysis: For equilibrium, $\delta_s (M_{1s} + M_{2s}) = H_u h_s + \Sigma P_u \Delta_{2u}$

Figure 15.11.1 Summation of forces acting on all columns in one story of a multistory building frame.

order value for the sum of moments at the column ends is

$$M_{1s} + M_{2s} = H_u h_s \tag{15.11.1}$$

in which the ACI Code has taken M_{2s} to be the larger end moment. However, the model of Fig. 15.11.1(a) acted upon by H_u deflects relatively by the amount Δ_{1u}, which means that the gravity load ΣP_u is acting with an eccentricity Δ_{1u}, thus increasing the lateral load moment $H_u h_s$ by the amount $\Sigma P_u \Delta_{1u}$. Since the total moment now acting is $H_u h_s + \Sigma P_u \Delta_{1u}$ the relative lateral deflection will finally be Δ_{2u} (larger than Δ_{1u}) when the structure reaches equilibrium in the final displaced position, as shown in Fig. 15.11.1(b). The total sum of moments at the column ends (primary plus secondary) then may be expressed

$$\delta_s(M_{1s} + M_{2s}) = H_u h_s + \Sigma P_u \Delta_{2u} \tag{15.11.2}$$

in which the ACI Code has defined δ_s to be the magnification factor.

In general, the primary moments M_{1s} and M_{2s} are obtained from a first-order elastic analysis of the unbraced frame. ACI-10.11.5.1 uses M_{2s} to be the "larger factored end moment on a compression member due to loads which result in appreciable sidesway . . ."

There are two ways by which magnified column end moments $\delta_s M_{1s}$ and $\delta_s M_{2s}$ may be obtained:

1. Moment magnifier method. As previously described for braced frames, the magnified moment $\delta_s M_{2s}$ may be treated as the primary moment M_{2s} multiplied by a magnification factor δ_s in accordance with ACI-10.11.5.1. This method is described in Secs. 15.12 and 15.13.

2. Second-order analysis. Alternatively, the magnified moments $\delta_s M_{1s}$ and $\delta_s M_{2s}$ may be computed directly from a second-order analysis of the structure in accordance with ACI-10.10.1. This procedure involves utilization of the final displaced position [the Δ_{2u} position of Fig. 15.11.1(b)] of the structure in the frame analysis, and for a rigorous solution requires an iterative procedure.

No matter which way the "P–Δ effect" is accounted for, a lateral load analysis must be performed to obtain at least the primary moments.

15.12 ACI CODE—MOMENT MAGNIFIER METHOD FOR UNBRACED FRAMES

The ACI Code gives the highest priority to a general frame analysis taking into account "influence of axial loads and variable moment of inertia on member stiffness and fixed-end moments, effect of deflections on moments and forces, and the effects of duration of loads" (ACI-10.10.1). In recognition of the fact that this general analysis is practical only with aid of a sophisticated computer program, the ACI Code prescribes an approximate method. ACI Committee 441, Columns, does recommend [37], however, that improved structural analysis should be used to overcome weaknesses in the conventional methods of nominal analysis. Many analysis shortcomings affect not only long columns but in fact affect short columns as much or more [37].

The following requirements are suggested by the ACI Commentary-10.10.1 [37] as minimum for an adequate rational analysis.

1. Realistic moment–curvature or moment–end rotation relationships should be used to provide accurate values of deflections and secondary moments (P–Δ moments). Since column design is based on strength and stability under factored loads, the stiffnesses used, even though in an elastic analysis, should be consistent with the strength limit state. ACI Commentary accepts as satisfactory the following:

$$\text{column } EI = E_c I_g (0.2 + 1.2 \rho_g E_s / E_c) \qquad \textbf{(15.12.1)}$$

$$\text{beam } EI = 0.5 E_c I_g \qquad \textbf{(15.12.2)}$$

These are as suggested by MacGregor, Oelhafen, and Hage [44].

2. The effect of foundation rotations on the lateral deformations should be considered.

3. The influence of axial load on the flexural stiffness must be considered for very slender columns when L_u/r exceeds 45. For instance, the flexural stiffness of a prismatic member without axial load is $4EI/L$. When axial compression is present, the 4 becomes reduced. The general idea relating to the effect of axial compression on flexural stiffness of elastic members can be found in Salmon and Johnson [30].

4. The effects of creep should be considered, particularly in frames subject to sustained lateral loads, such as a building resisting a horizontal reaction from an arch or an unbalanced horizontal earth force. Creep must also be considered in frames where unbalanced dead load gives rise to differential shortening of the two sides of a building resulting in lateral deflection.

5. The maximum moment in the compression member should be determined considering effects of the lateral deflections of the frame and the deflection of the compression member itself. This means that a second-order analysis must be made to include the P–Δ effects. The methods of MacGregor and Hage [43] and Wood, Beaulieu, and Adams [48,49] all are directed at making the second-order analysis, using various degrees of precision.

ACI Code Moment Magnifier Method. ACI-10.11.5.1 prescribes the use of the factored axial load P_u in combination with a magnified factored moment M_c defined by ACI Formula (10-6),

$$M_c = \delta_b M_{2b} + \delta_s M_{2s} \qquad (15.12.3)$$

where the subscript 2 refers to the larger of the two end moments on the compression member, the subscript b refers to "braced" or moments resulting from "no appreciable" sidesway, and the subscript s refers to "sway" or "sidesway" related moments. This separate magnifier approach for the braced and unbraced portions of the frame action is based on the recommendation of Ford, Chang, and Breen [54]. As defined in ACI-10.0,

δ_b = moment magnification factor for frames braced against sidesway to reflect effects of member curvature between ends of compression members

δ_s = moment magnification factor for frames not braced against sidesway to reflect lateral drift resulting from lateral and gravity loads

M_{2b} = value of larger factored end moment on compression members due to loads which result in no appreciable sidesway, calculated by conventional elastic frame analysis

M_{2s} = value of larger factored end moment on compression member due to loads which result in appreciable sidesway, calculated by conventional elastic frame analysis

For a typical relatively symmetrical frame subject only to gravity loading, the sidesway effect may be considered negligible. According to ACI Commentary-10.11.5.1, "All supportive evidence in the frame tests and accompanying theoretical analyses indicated that sway magnification of the gravity moment by the sway magnifier is unwarranted."

The reader is reminded that in the *braced* system, the magnified moment *may* occur at the end of the member or at an unknown location away from the ends. The gravity moment M_g must therefore be magnified by the braced frame magnifier δ_b. To conservatively allow for the separate magnification effects in the case of combined gravity and lateral loads, ACI Formula (10-6), Eq. (15.12.3), may for practical purposes be written as

$$M_c = \delta_b M_g + \delta_s M_{2s} \qquad (15.12.4)$$

according to ACI Commentary-10.11.5.1, Eq. (5).

The magnifier δ_b for the braced frame has been described as Eq. (15.10.4); the magnifier δ_s for the unbraced frame is

$$\delta_s = \frac{1}{1 - \dfrac{\Sigma P_u}{\phi \Sigma P_c}} \geq 1.0 \qquad (15.12.5)$$

in which

$$P_c = \frac{\pi^2 EI}{(kL_u)^2} \qquad (15.12.6)$$

and ΣP_u and ΣP_c are the summations for all columns in a story.

An approximate procedure in which a quantity Q_u is used in place of $(\Sigma P_u/\phi)/\Sigma P_c$ is suggested by MacGregor and Hage [43] and described in Sec. 15.13.

15.13 MacGREGOR–HAGE MOMENT MAGNIFIER METHOD

MacGregor and Hage [43] have shown that the expression $(\Sigma P_u/\phi)/\Sigma P_c$ in Eq. (15.12.5) may be substituted by the quantity Q_u, where

$$Q_u = \frac{\Sigma P_u \, \Delta_{1u}}{H_u h_s} \qquad (15.13.1)$$

and therefore

$$\delta_s = \frac{1}{1 - Q_u} \qquad (15.13.2)$$

The derivation of Eqs. (15.13.1) and (15.13.2) may be made as follows. Referring to Figs. 15.11.1(a) and (b) and Eqs. (15.11.1) and (15.11.2), and using a proportionality factor η, let

$$\Delta_{1u} = \eta H_u \qquad (15.13.3)$$

which is valid for linear first-order analysis. The equivalent magnified lateral load in Fig. 15.11.1(b) may be taken as

$$\text{equivalent lateral load} = H_u + \frac{\Sigma P_u \, \Delta_{2u}}{h_s} \qquad (15.13.4)$$

Thus,

$$\Delta_{2u} = \eta \text{ (equivalent lateral load)}$$

$$= \eta \left(H_u + \frac{\Sigma P_u \, \Delta_{2u}}{h_s} \right)$$

$$= \Delta_{1u} + \frac{\Delta_{1u} \Sigma P_u \, \Delta_{2u}}{H_u h_s}$$

$$= \Delta_{1u} + Q_u \, \Delta_{2u}$$

from which

$$\Delta_{2u} = \frac{\Delta_{1u}}{1 - Q_u} \qquad (15.13.5)$$

Comparing Eq. (15.11.2) with Eq. (15.11.1),

$$\delta_s = \frac{H_u h_s + \Sigma P_u \, \Delta_{2u}}{H_u h_s} \qquad (15.13.6)$$

Substituting Eq. (15.13.5) into Eq. (15.13.6) gives

$$\delta_s = 1 + \frac{\Sigma P_u}{H_u h_s} \frac{\Delta_{1u}}{1 - Q_u} = 1 + \frac{Q_u}{1 - Q_u} = \frac{1}{1 - Q_u} \qquad (15.13.7)$$

Using Eqs. (15.13.1) and (15.13.2) as the basic formulas, MacGregor and Hage have suggested the following steps in the application of the ACI moment magnifier method:

1. Perform a first-order analysis for service lateral loads to obtain un-factored sway moments. Apply factors from ACI-9.2 appropriate for lateral load to obtain factored lateral load moments M_s. In this step no gravity loads are acting. Compute Δ_{1s}/h_s due to service load for each story and for the overall frame, and compare with the desired limit if any is desired.
2. Perform a factored gravity load analysis to obtain moments M_g.
3. Compute Q_u for each story.

$$Q_u = \frac{\Sigma P_u \, \Delta_{1u}}{H_u h_s} \qquad [15.13.1]$$

in which ΣP_u is the sum of the factored axial loads for all columns in the story, H_u is the factored lateral force at the story, and Δ_{1u} is the relative factored load deflection in the story. Note that Δ_{1u} and H_u are *both* for the same load level, either service load or factored load. For Q_u exceeding 0.04 but less than 0.2,

$$\delta_s = \frac{1}{1 - Q_u} \qquad [15.13.2]$$

4. Combine the effects of gravity and lateral loads for the design of the member.

P_u = factored gravity load on member

$M_c = \delta_b M_g + \delta_s M_{2s}$

where

δ_b = Eq. (15.10.4)

M_g = moments from step 2

δ_s = Eq. (15.13.2)

M_{2s} = larger of the end moments on the member, computed in step 1 and increased by the appropriate factors from ACI-9.2

15.14 MacGREGOR–HAGE AMPLIFIED LATERAL LOAD METHOD

In this method, the lateral loads are amplified (magnified) before the first-order structural analysis is performed, and the resulting moments are used directly in the design. The following steps may be followed:

1. Establish whether the frame is to be treated as *braced* or *unbraced*. According to ACI Commentary-10.11.2, par. 4, a compression member in a story may be considered braced when the following is satisfied within the story:

$$\frac{\Sigma K \text{ of walls or lateral bracing}}{\Sigma K \text{ of columns in parallel frames}} \geq 6 \qquad \textbf{(15.14.1)}$$

where K = stiffness of element.

2. Establish the maximum Δ_{2s}/h_s (say 1/400 or 1/500) that can be tolerated in any story under service load. The deflection Δ_{2s} is the total including second-order deflection. [MacGregor and Hage [43] used Δ_s/h. It was not clear whether Δ_s meant Δ_{1s} or Δ_{2s}; however, any limitation Δ_s/h established by the designer would likely be one related to the *total* (including second-order) deflection.] There are no prescribed ACI Code limits on Δ_s/h_s.

3. Estimate the drift index Δ_{2u}/h_s at factored loads. Δ_{2u} would be the total primary plus secondary deflection (Δ_{2u} of Fig. 15.11.1). MacGregor and Hage [43] have indicated that if $Q_u \leq 0.2$ the service load deflections (Δ_{1s} or Δ_{2s}) will be about 40% of the factored load deflections (Δ_{1u} or Δ_{2u}). Further, Δ/h_s (either Δ_{1s} or Δ_{2s}) for an entire multistory frame will be 85% or less of the maximum sway (first- or second-order, respectively) in one story. Thus, the Δ_{2u}/h_s at factored load may be estimated for the overall frame to be

$$\text{estimated } \frac{\Delta_{2u}}{h_s} \approx \left[\left(\frac{\Delta_{2s}}{h_s}\right)^{\text{service}}_{\substack{\text{load} \\ \text{limit}}}\right]\left(\frac{0.85}{0.40}\right) = 2.125\left[\left(\frac{\Delta_{2s}}{h_s}\right)^{\text{service}}_{\substack{\text{load} \\ \text{limit}}}\right]$$

$$\textbf{(15.14.2)}$$

4. Compute the estimated Q_u for the overall frame, and for each story,

$$Q_u = \frac{\Sigma P_u}{H_u}\left(\frac{\Delta_{2u}}{h_s}\right) \qquad \textbf{(15.14.3)}$$

where

Δ_{2u}/h_s = the value from Eq. (15.14.2) for the overall frame and 1/0.85 of that value for each story
H_u = factored lateral force within the story
h_s = story height
ΣP_u = sum of the factored gravity loads acting on the columns of the story

The value of Q_u must not exceed 0.2.

5. Compute the approximate magnified lateral loads. The magnified lateral forces H_{2u} to use in the lateral load analysis are, following the concept of Eq. (15.13.7),

$$H_{2u} = H_u \left(\frac{1}{1 - Q_u} \right) \qquad (15.14.4)$$

6. Apply the magnified factored lateral forces H_{2u} (omitting all gravity load) and perform a first-order analysis. For moment of inertia I, the use of $0.4I_g$ for beams and $0.8I_g$ for columns is suggested [43]. Obtain the moments $\delta_s M_s$ directly from the analysis, as well as the total deflection Δ_{2u} at each story level.

7. Check that Δ_{2u}/h_s computed for the entire frame does not exceed the limit given by Eq. (15.14.2), and that Δ_{2u}/h_s for any story does not exceed 1/0.85 times the value given by Eq. (15.14.2). If any of the limits is exceeded the structure must be stiffened.

8. Perform an elastic analysis under factored gravity load. Obtain the moments M_g.

9. Obtain the factored moments for the design of the member.

$$M_c = \delta_b M_g + \delta_s M_{2s} \qquad [15.12.4]$$

where

$$\delta_b = \frac{C_m}{1 - \dfrac{P_u}{\phi P_c}} \geq 1.0$$

$\delta_s M_{2s}$ = values from analysis in step 6

M_g = factored gravity load moments from step 8

P_u = factored gravity axial compression in member

15.15 FURLONG'S RATIONAL METHOD FOR UNBRACED FRAMES

Furlong [50] has proposed guidelines for a rational analysis that is intended to satisfy the intent of ACI-10.10.1 within the complexity implied by the five requirements of the ACI Commentary referred to previously in Sec. 15.12 for the general analysis. The recommended procedure applies to *unbraced* frames. For such frames, the slenderness effects are the products of the column load P_u times the sway deflection Δ.

Furlong indicates that lateral displacements may be obtained from either a second-order analysis or an iterative first-order analysis that uses a "dummy horizontal load" to give the second-order result, provided the appropriate "model" for the structure is used. The effects of the actual nonlinear moment–curvature behavior may be accounted for by the use in the model of prismatic member EI values that give moment–rotation characteristics *no stiffer* than real members after concrete cracks and steel yields prior to nominal strength being reached. This will be accomplished by following ACI Commentary recommendations for EI values, as given by Eqs. (15.12.1) and (15.12.2).

The Furlong recommended [50] procedure for *unbraced* frames is as follows:

1. Estimate member sizes and compute EI values for each component of the frame, using $0.5E_cI_g$ for beams and $0.3E_cI_g$ for columns [Eq. (15.12.1) gives about $0.3E_cI_g$ for $\rho_g = 0.01$].
2. Perform factored gravity load analyses for the following cases:
   (a) $1.4D + 1.7L$ in every span of every floor
   (b) $1.4D + 1.7L$ in alternate spans on every floor, beginning with the end span
   (c) $1.4D + 1.7L$ in alternate spans on every floor, beginning with the second span from end
   When live load reductions are appropriate for multistory frames or for large floor areas (see for example, ANSI [10, Chap. 1], Sec. 4.7), the reduced live load is to be used in loading cases (a), (b), and (c).
3. Compute factored axial loads acting on all columns at each floor level using $1.4D + 1.7L$, where L is the *total* live load tributary to each column.
4. Perform lateral load analysis, including second-order analysis, using factored wind ($1.7W$) for horizontal forces and nodal point factored gravity loads obtained in Step 3. There are iterative methods (such as that of Wood, Beaulieu, and Adams [48]) for determining the cumulative amount of additional moments so that a first-order analysis procedure could be used.
5. Select beam reinforcement for the more severe of the following:
   (a) Factored gravity load moments from Step 2
   (b) 0.75 (factored gravity load moments from Step 2 + factored lateral load moments from Step 4)
   This latter combination corresponds to ACI Formula (9-2).
6. Select column reinforcement for the more severe of the following:
   (a) Factored axial forces from Step 3 + factored gravity load moments from the most severe condition of Step 2
   (b) 0.75 (Step 3 factored axial forces + Step 4 lateral load factored moments + Step 2 gravity load factored moments)
7. Examine individual columns for possible slenderness effect strength reduction as a *braced* member. For every interior column having a ratio L_u/h greater than 8, and every exterior column having a ratio L_u/h greater than 12, compute and use moment magnification factor δ_b for the factored gravity load moments of Step 2.

8. The overall stiffness of the unbraced frame may be evaluated by *estimating* that 25% of the factored load horizontal deflection (lateral drift) computed in Step 4 will be the traditional service load drift computed using gross properties. If 25% of the factored load drift exceeds the desired limit of 1/400 to 1/500, the members should probably be stiffened. There are no ACI Code serviceability limits on drift.

15.16 MINIMUM ECCENTRICITY IN DESIGN

Braced Frames. When computations indicate only a small moment to be acting on a member such that the eccentricity M_u/P_u is less than $(0.6 + 0.03h)$ in., the primary moment M_m in Eq. (15.10.2) is to be computed as $P_u(0.6 + 0.03h)^*$ (ACI-10.11.5.4). Although ACI-10.11.5.4 refers only to *end* eccentricities, the authors believe the intent is to use this code-stated minimum primary moment as the basis for applying the magnifier whether the small primary moment actually acts at the end of a member or at some point between the ends.

Further, when computed *end* moments are small (or zero) for a *braced* frame member having only end moments acting, ACI-10.11.5.4 specifies determining the ratio M_{1b}/M_{2b} for that C_m equation by either of the following:

1. When actual computed end moments give eccentricities less than a minimum of $(0.6 + 0.03h)$ in., such actual end moments may be used for M_{1b}/M_{2b} in Eq. (15.10.6).
2. If computations indicate essentially no moment at both ends—that is, the member is considered to be axially loaded—single curvature is to be assumed with $C_m = 1$.

Unbraced Frames. When computations indicate that both end moments M_{1s} and M_{2s} are less than those based on an end eccentricity e_{min} of $(0.6 + 0.03h)^*$ in., M_{2s} in ACI Formula (10-6), Eqs. (15.12.3) or (15.12.4), must be based on $e_{min} = (0.6 + 0.03h)^*$ in.

When the minimum eccentricity requirements control, the provisions are intended to be applied to bending about only one axis at a time, *not* as a case of biaxial bending.

15.17 BIAXIAL BENDING AND AXIAL COMPRESSION

For compression members subject to bending about both principal axes, the moment about *each* axis is to be magnified by the factor δ which is computed from the restraint conditions for that axis (ACI-10.11.7). In gen-

---

*For SI, ACI 318M-83 gives
$$e_{min} = (15 + 0.03h) \text{ mm}$$

eral, the effective length factor k, the stiffness factor EI, and C_m may differ for each bending axis. The member may also be considered as part of a braced system in one direction and unbraced in the other. Additional data on biaxial bending of slender members may be found in Refs. 47 and 51. See also Sec. 13.21 and Refs. 35 to 50 of Chap. 13.

15.18 ACI CODE—SLENDERNESS RATIO LIMITATIONS

Although the slenderness ratio is never zero in actual structures, there are certain limits for kL_u/r below which the reduction in strength may reasonably be neglected. The ACI Code provisions are based on the assumption that a strength loss of up to 5% can be tolerated without the designer having to consider the slenderness effect; thus a significant number of ordinary columns can be designed considering only the provisions of Chap 13.

ACI-ASCE Committee 441 surveyed typical reinforced concrete buildings to determine the normal range of variables found in columns of such buildings [41]. A great variety of buildings were studied, including towers (braced frames) as high as 33 stories and an unbraced frame 20 stories high. The total number of columns exceeded 20,000. The following results were reported [41]. For braced frames, 98% of the columns had L/h less than 12.5 ($L/r \approx 42$) and e/h less than 0.64. For unbraced frames, 98% of the columns had L/h less than 18 ($L/r \approx 60$) and e/h less than 0.84. Further, it was found that the practical upper limit on the slenderness ratio kL/r is about 70 in building columns. In general, these limits for the variables provide a guide to the range of variables that are used in any approximate method.

Taking the idea that attainment of at least 95% of the material strength of a short column is acceptable, the effects of slenderness may be neglected when:

For *braced* frame members (ACI-10.11.4.1),

$$\frac{kL_u}{r} < 34 - 12\,\frac{M_{1b}}{M_{2b}} \tag{15.18.1}$$

where M_{1b} is the smaller and M_{2b} the larger of end moments on the member.

For single curvature cases the ratio M_{1b}/M_{2b} is positive, while for double curvature the ratio is negative. When the member is subject to large transverse loading (other than end moments alone), the ratio M_{1b}/M_{2b} should probably be taken as $+1$.

For *unbraced* frame members (ACI-10.11.4.2),

$$\frac{kL_u}{r} < 22 \tag{15.18.2}$$

For all compression members with kL_u/r exceeding 100, a more elaborate analysis as discussed in Secs. 15.11 to 15.15 is required (ACI-10.11.4.3).

Comparison of the slenderness limits of Eqs. (15.18.1) and (15.18.2) with actual columns in existing buildings indicates [41] that over 90% of the columns in braced frames and over 40% of the columns in unbraced frames will fall within the limits of those equations and allow neglect of the slenderness effect.

15.19 RESTRAINING EFFECT OF BEAMS

The restraining effect of beams has a major effect on column behavior. The problem is discussed in detail by Pagay, Ferguson, and Breen [18]; Okamura, Pagay, Breen, and Ferguson [19]; and Breen, MacGregor, and Pfrang [41].

Braced Frames. Consider the portion of a braced frame shown in Fig. 15.19.1. The primary moment M_c on the column, computed from a nominal first-order analysis using gross moment of inertia I_g, depends on the relative stiffnesses of beam and column. The moment M_{ext} applied to the cantilever at A is resisted by the beam and the column; thus

$$M_{\text{ext}} = M_b + M_c \qquad (15.19.1)$$

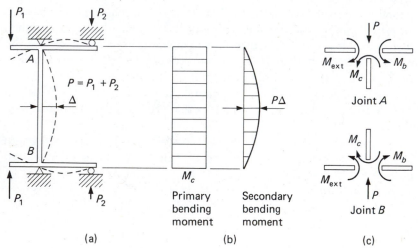

Primary bending moment
Secondary bending moment

Figure 15.19.1 Braced frame—restraining effect of beam.

Since the column deflects, there will be an additional moment $P\Delta$ on it, such that

$$M_{\text{max}} = M_c + P\Delta \qquad (15.19.2)$$

Solving for M_c in Eq. (15.19.1) and substituting into Eq. (15.19.2) gives

$$M_{\text{max}} = M_{\text{ext}} - M_b + P\Delta \qquad (15.19.3)$$

As the column deflects laterally due to $M_c + P\Delta$, joint A rotates forcing more and more of the applied moment to be resisted by the beam. This effect is further increased by the reduction in column stiffness due to axial load. However, beam deflection due to creep and shrinkage gives the reverse effect by putting moment back on the column.

When the beam is relatively stiff (high reinforcement ratio ρ) compared to the column (i.e., slender column), the primary moment on the column is overestimated by the nominal analysis and the moment on the beam is underestimated. This is not a problem because the slender column will be conservatively designed, and underestimating the beam end moment will increase the positive moment and also the chance that beam deflection will control.

When the beam is relatively flexible (low reinforcement ratio ρ) compared to the column (i.e., short stiff column), the primary moment on the column is underestimated by the nominal analysis and the beam end moment overestimated. In this case the column could be significantly underdesigned. For the beam design, it makes little difference because the beam having low ρ is ductile and redistribution of moments (see Sec. 10.12) can occur. In other words, on the ductile beam it is not too essential whether the moment strength is somewhat larger at the support or at midspan as long as the total load is capable of being carried.

A multiplier to increase the design moment on short columns has been suggested by Okamura, Pagay, Breen, and Ferguson [19]. For the *single-curvature* case in *braced* frames, the nominal e/h should be multiplied by

$$\text{multiplier} = 1.38 + 5.5(\rho_g - 5.5\rho) \qquad (15.19.4)$$

where

ρ_g = ratio of column steel to gross area bh

ρ = ratio of beam tension steel to effective area bd

The relationship of Eq. (15.19.4) was developed for an average end restraint factor ψ of Eq. (15.8.1) for the column equal to 1.0, which corresponds to an effective length factor k about 0.8. It would seem Eq. (15.19.4) could be used for any short column in a braced frame.

Unbraced Frames. In unbraced frames, the beam may be inadequately designed for moment at its junction with the column, when a lateral load such as wind is applied. The beam moment must be equal to the magnified moment on the column. When the moment magnifier method or the general analysis is used, no problem arises as long as one recognizes that the moment, $M_m + P_u\Delta$, or $\delta_b M_{2b} + \delta_s M_{2s}$, acting at the end of the column must be carried by the beams in a manner such that equilibrium of the joint is maintained.

Moment of Inertia for Restraining Beam. Even though gross section is used for a nominal elastic frame analysis, the adjustment to the nominal e/h may be eliminated if cracked transformed section is used for the moment of inertia in computing the end restraint factor ψ of Eq. (15.8.1) [18]. In general, more accurate results are obtained by using Eq. (15.10.8) or (15.10.9) with $\beta_d = 0$ for the moment of inertia of the column members and transformed cracked section for the moment of inertia of the beam members in determining the effective length factor k.

15.20 EXAMPLES

In the preceding sections of this chapter, basic concepts underlying the ACI Code provisions relating to length effects on columns have been discussed. Details are given for the moment magnifier method as well as the suggestions for a second-order analysis of the unbraced frame. In order to facilitate easy reference to the actual quantitative procedures or formulas, most of which appear in the ACI Code or Commentary, Table 15.20.1 is presented summarizing the information needed for solving practical problems. For illustration, the following six examples are presented.

EXAMPLE 15.20.1 Determine the adequacy of the interior top floor column (column A) of the *braced* frame of Fig. 15.20.1. The column is 10 × 10 with 4-#8 and 4-#9 bars (f_y = 50 ksi and f_c' = 3 ksi) and is to carry a service axial compression of 108 kips live load and 36 kips dead load. The bending moments that may act in combination with the axial load have been computed and found to be negligible. If the member is not adequate, revise the design so that it satisfies the moment magnifier method of the ACI Code.

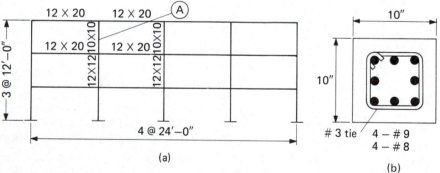

Figure 15.20.1 Rigid frame for Examples 15.20.1 and 15.20.2.

Solution: (a) Determine slenderness ratio. Unless a rational evaluation of end restraint is made, ACI-10.11.2.1 requires taking the effective length factor k for a braced frame equal to 1.0. The radius of gyration may be taken as $0.3h$ according to ACI-10.11.3. The clear height L_u is

$$L_u = 12 - \frac{20}{12} = 10.33 \text{ ft}$$

Then

$$\frac{kL_u}{r} = \frac{1.0(10.33)(12)}{0.3(10)} = 41.3$$

(b) Slenderness ratio limits. Since the end moments are negligible, the minimum eccentricity provisions of ACI-10.11.5.4 govern the design. Accordingly, the deformation should be considered as single curvature with M_{1b}/M_{2b} = 1.0. The slenderness limit is

$$\left(\frac{kL_u}{r}\right)_{\text{limit}} = 34 - 12\frac{M_{1b}}{M_{2b}} = 22 < 41.3$$

Thus slenderness effects must be considered.

(c) Moment magnifier δ_b.

$$\delta_b = \frac{C_m}{1 - \frac{P_u}{\phi P_c}}$$

where

$C_m = 1.0$ for single-curvature member in braced frame

$P_u = 1.4(36) + 1.7(108) = 234$ kips

$\phi = 0.7$ for this tied column

$$P_c = \frac{\pi^2 EI}{(kL_u)^2}$$

For the stiffness parameter EI using Eqs. (15.10.8) or (15.10.9),

$$E_c = 57,000 \sqrt{f_c'} = 3120 \text{ ksi}$$

$$I_g = \frac{10(10)^3}{12} = 833 \text{ in.}^4$$

$$E_s = 29,000 \text{ ksi}$$

$$I_s = 2(2.79)(2.59)^2 = 37.4 \text{ in.}^4$$

$$0.2E_c I_g + E_s I_s = 0.2(3120)(833) + 29,000(37.4)$$

$$= 520,000 + 1,090,000 = 1,610,000 \text{ kip in.}^2$$

$$0.4E_c I_g = 1,040,000 \text{ kip in.}^2$$

The EI values are to be divided by $(1 + \beta_d)$ to account for time-dependent deflection due to creep and shrinkage. Use the larger of the two values of EI, and divide by $(1 + \beta_d)$, where β_d is the proportion of factored load moment that is sustained. In this case, since moments are assumed proportional to axial load, β_d is the sustained proportion of factored axial load,

$$\beta_d = \frac{36(1.4)}{108(1.7) + 36(1.4)} = 0.215$$

$$EI = \frac{1,610,000}{1 + 0.215} = 1,330,000 \text{ kips in.}^2$$

$$P_c = \frac{\pi^2(1,330,000)}{[1.0(10.33)(12)]^2} = 854 \text{ kips}$$

$$\frac{P_u}{\phi P_c} = \frac{234}{0.7(854)} = 0.391$$

$$\delta_b = \frac{1.0}{1 - 0.391} = 1.64$$

In this case, the minimum eccentricity $(0.6 + 0.03h)$ should be magnified so that

required $e = 1.64(e_{\text{min}}) = 1.69(0.6 + 0.03h) = 0.98 + 0.049h$ in.

Table 15.20.1 LENGTH EFFECTS ON COLUMNS—SUMMARY OF USEFUL FORMULAS

1. Definitions

$$\psi = \frac{\Sigma EI/L \text{ of columns}}{\Sigma EI/L \text{ of beams}}; \qquad \begin{aligned} \psi &= 0 \text{ (fixed end)} \\ \psi &= 10 \text{ (column end supported on footing)} \\ \psi &= \infty \text{ (theoretical hinged end)} \end{aligned} \qquad \text{[15.8.1]}$$

$$P_c = \frac{\pi^2 EI}{(kL_u)^2} \qquad \text{[15.10.7]}$$

2. Slenderness Ratio Limitation (No Length Effect)

Braced frames:
$$\frac{kL_u}{r} < 34 - 12\frac{M_{1b}}{M_{2b}} \qquad \text{[15.18.1]}$$

$$|M_{1b}| < |M_{2b}|; \qquad M_{1b}/M_{2b} \text{ positive for single curvature}$$

Unbraced frames:
$$\frac{kL_u}{r} < 22 \qquad \text{[15.18.2]}$$

Approximate values of r: $0.30h$ for rectangular columns [Sec. 15.1]
$0.25h$ for circular columns [Sec. 15.1]

3. Stiffness Factor EI in Moment Magnifier Method

Larger of
$$\begin{cases} EI = \dfrac{0.2E_c I_g + E_s I_s}{1 + \beta_d} & \text{ACI Formula (10-9)} \qquad \text{[15.10.8]} \\[2mm] EI = \dfrac{0.4E_c I_g}{1 + \beta_d} & \text{ACI Formula (10-10)} \qquad \text{[15.10.9]} \end{cases}$$

$$E_c = 57{,}000\sqrt{f'_c} \qquad \beta_d = \frac{\text{factored sustained load moment}}{\text{factored total load moment}}$$

$$EI_{cr} \text{ of beam} \approx \frac{EI_g}{2}\left(\text{for }\frac{kL_u}{r} < 60; \text{ ACI Commentary-10.11.2}\right) \qquad \text{[Sec. 15.8]}$$

4. Effective Length Factor k in Moment Magnifier Method
Braced frames:

(a) Nomogram (alignment chart) [Fig. 15.8.1(a)]
(b) 1972 British Code $\left.\begin{aligned} k &= 0.7 + 0.05(\psi_A + \psi_B) \le 1.0 \\ k &= 0.85 + 0.05\psi_{\min} \le 1.0 \end{aligned}\right\}$ (use smaller) [15.8.2] [15.8.3]

(c) ACI Code $k = 1.0$ (on safe side) [Sec. 15.8]

Unbraced frames:

(a) Nomogram (alignment chart) [Fig. 15.8.1(b)]

(b) Furlong
$$k = \frac{20 - \psi_{avg}}{20}\sqrt{1 + \psi_{avg}}; \qquad \psi_{avg} < 2 \qquad \text{[15.8.4]}$$
$$k = 0.9\sqrt{1 + \psi_{avg}}; \qquad \psi_{avg} > 2 \qquad \text{[15.8.5]}$$

(c) 1972 British Code $k = 2.0 + 0.3\psi;$ one hinged end [15.8.6]

5. C_m Factor in Moment Magnifier Method

Braced frames

Transverse loading, $C_m = 1.0$ [15.11.5]

End moments only, $C_m = 0.6 + 0.4\dfrac{M_{1b}}{M_{2b}} \ge 0.4$ [15.11.6]

$$|M_{1b}| < |M_{2b}|; \qquad M_{1b}/M_{2b} \text{ positive for single curvature}$$

Unbraced frames $C_m = 1.0$ [15.11.7]

Table 15.20.1 (*CONT.*)

6. Moment Magnifier Method

Braced frames: $M_c = \delta_b M_{2b}$ (M_{2b} = larger end moment)   **[15.10.6]**

$M_c = \delta_b M_m$ (M_m = primary bending moment   **[15.10.2]**
due to transverse loading)

$$\delta_b = \frac{C_m}{1 - \dfrac{P_u}{\phi P_c}} \geq 1.0 \qquad \textbf{[15.10.4]}$$

Unbraced frames: $M_c = \delta_b M_{2b} + \delta_s M_{2s}$ **[15.12.3]**

(M_{2b} = larger end moment in braced case)
(M_{2s} = larger end moment in sway case)

$M_c = \delta_b M_g + \delta_s M_{2s}$ **[15.12.4]**

(M_g = gravity load end moment)
(M_{2s} = larger end moment in sway case)

$$\delta_s = \frac{1}{1 - \dfrac{\Sigma P_u}{\phi \Sigma P_c}} \geq 1.0 \qquad \textbf{[15.12.5]}$$

---

7. Second-Order Analysis Method

Unbraced frames:

$$Q_u = \text{stability index} \approx \frac{\Sigma P_u}{\phi \Sigma P_c}$$

(a) $Q_u = \dfrac{\Sigma P_u \, \Delta_1}{H_1 h_s} \leq 0.04;$ $\left(\dfrac{\Delta_1}{h_s} = \dfrac{1}{400} \text{ or } \dfrac{1}{500} \right)$ **[15.13.2]**

Second-order analysis may be neglected.

(b) $0.04 < Q_u \leq 0.2$

Second-order analysis must be performed.

$$\delta_s = \frac{1}{1 - Q_u} \qquad \textbf{[15.14.2]}$$

(c)   Stiffness factor EI

From ACI Commentary:

    column $EI = E_c I_g (0.2 + 1.2 \rho_g E_s / E_c)$ **[15.12.1]**
    beam $EI = 0.5 E_c I_g$ **[15.12.2]**

From MacGregor and Hage:

    column $EI = 0.8 E_c I_g$ **[Sec. 15.13]**
    beam $EI = 0.4 E_c I_g$ **[Sec. 15.13]**

---

8. Restraining Effect of Beams

Braced frames: multiplier for $\dfrac{e}{h} = 1.38 + 5.5(\rho_g - 5.5\rho)$ **[15.19.4]**

$\left(\begin{array}{c} \text{single curvature} \\ \text{short columns} \end{array} \right)$

    $\rho_g(\text{column}) = A_{st}/(bh);$ ρ (beam) = $A_s/(bd)$

Unbraced frames: magnified moment at beam end
    = magnified moment at column end

Note that even the magnified eccentricity might not exceed the eccentricity corresponding to the maximum axial compressive strength of $0.80P_0$ ($P_{n(\text{max})}$ of Fig. 13.11.2). If this happens as in Fig. 13.11.2(b), there is still no reduction in strength due to the slenderness effect.

(d) Rational analysis for effective length factor k. For the beam the cracked section moment of inertia is recommended. An approximation is $I_{cr} = I_g/2 = 4000$ in.[4]

$$(EI)_{\text{bm}} = E_c I_{cr} = 3120(4000) = 12,500,000 \text{ kip in.}^2$$
$$(EI)_{\text{col}} = 1,610,000 \text{ kip in.}^2$$

End restraint factors,

$$\psi_A(\text{top}) = \frac{\Sigma EI/L \text{ for cols}}{\Sigma EI/L \text{ for beams}} = \frac{1610/12}{2(12,500)/24} = 0.13$$

$$\psi_B(\text{bottom}) = \frac{(1610 + 3230)/12}{2(12,500)/24} = 0.39$$

Since the 12×12 column below has not been designed, its EI value is taken as $0.6E_c I_g$, which is approximately the general expression obtained for the 10×10 column by the ACI formula. From Fig. 15.8.1(a), $k = 0.62$. The more correct effective slenderness ratio is

$$\frac{kL_u}{r} = \frac{0.62(10.33)12}{0.3(10)} = 25.6$$

The magnification factor is also affected,

$$P_c = \frac{\pi^2(1,330,000)}{[0.62(10.33)(12)]^2} = 2220 \text{ kips}$$

$$\frac{P_u}{\phi P_c} = \frac{234}{0.7(2220)} = 0.151$$

$$\delta_b = \frac{1.0}{1 - 0.151} = 1.18$$

$$\text{required } e = 1.18(e_{\text{min}}) = 0.71 + 0.035h \text{ in.}$$

In this case the beams are very stiff compared to the columns. Using cracked section for the beams and the ACI EI formula for the columns gave little different result than would be obtained by using gross section.

(e) Check capacity. The strength of the section may be checked by the methods of Chap. 13.

$$\text{required } P_n = \frac{P_u}{\phi} = \frac{234}{0.7} = 334 \text{ kips}$$

$$\text{required } e = 0.71 + 0.035(10) = 1.06 \text{ in. } (0.106h)$$
$$[\text{according to (d) above}]$$

The actual nominal strength P_n at $e = 1.06$ in. is 431 kips (a convenient source is Ref. 26 of Chap. 13). Even when $e = 0.98 + 0.049h = 1.47$ in. as obtained in (c), the nominal strength P_n is 382 kips. So this section is adequate as a braced frame column.

Note that the strength P_n may not be taken in design greater than $0.80P_0$ according to ACI-10.3.5,

$$P_{n(max)} = 0.80[0.85f'_c(A_g - A_{st}) + f_y A_{st}]$$
$$= 0.80[0.85(3)(100 - 7.16) + 50(7.16)] = 476 \text{ kips} > 431 \text{ kips OK}$$

Thus, $P_n = 431$ kips at $e = 1.06$ in. is the correct strength, a situation corresponding to point A of Fig. 13.11.2(b).

EXAMPLE 15.20.2 Repeat Example 15.20.1 except consider the frame as unbraced.

Solution: In general, members in unbraced frames will have end moments on the members. One might assume that this case was the result of gravity load analysis that happened to give negligible column moment. In some cases where beams are unusually stiff, the behavior of the unbraced frame is little different from that of the braced frame.

(a) Effective pin-end length. From part (d) of Example 15.20.1 the end restraint factors are

$$\psi_A(\text{top}) = 0.13 \qquad \psi_B(\text{bottom}) = 0.39$$

From Fig. 15.8.1(b), $k = 1.07$.

(b) Compute magnification factor δ_s for sidesway. From part (c) of Example 15.20.1,

$$EI = 1,330,000 \text{ kip in.}^2 \qquad \text{(for column)}$$

which includes the effect of 21.5% sustained factored load.

$$P_c = \frac{\pi^2 EI}{(kL_u)^2} = \frac{\pi^2(1,330,000)}{[1.07(10.33)(12)]^2} = 746 \text{ kips}$$

Assuming P_u/P_c is the same for all columns in the story, $\Sigma P_u/\Sigma P_c = P_u/P_c$,

$$\frac{P_u}{\phi P_c} = \frac{234}{0.7(746)} = 0.448$$

For C_m, ACI-10.11.5.3 says to use 1.0.

$$\delta_s = \frac{C_m}{1 - \Sigma P_u/(\phi \Sigma P_c)} = \frac{1.0}{1 - 0.448} = 1.81$$

(c) Compute magnified factored moment M_c. ACI-10.11.5.1 prescribes

$$M_c = \delta_b M_{2b} + \delta_s M_{2s}$$

Assuming this is the gravity dead and live load case where the moment $M_{2b} = M_g$ and is negligible; M_{2s} may also be considered negligible, and design is nominally based on minimum eccentricity e_{min}. The magnified factored moment M_c is

$$M_c = \delta_s M_{2s} = \delta_s P_u e_{min}$$
$$= 1.81 P_u (0.6 + 0.03h)$$

which for $h = 10$ in. gives

$$M_c = 1.63 P_u$$

$$\text{required } e \text{ (for short column)} = \frac{M_c}{P_u} = 1.63 \text{ in.}$$

The strength ϕP_n for the member with $e = 1.63$ in. is 255 kips, which exceeds the requirement of $P_u = 234$ kips, and is acceptable.

EXAMPLE 15.20.3 Determine the adequacy of the square tied column (17 in. square, with 10-#9 bars, $f_c' = 3000$ psi, $f_y = 40,000$ psi) which is an exterior first-floor column in the frame of Fig. 15.20.2. Assume that this frame is braced sufficiently to prevent relative translation of its joints. Also assume 40% of the factored load moment is sustained.

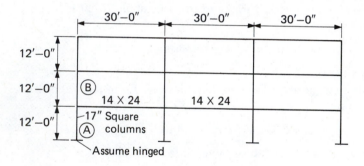

(a) Frame dimensions

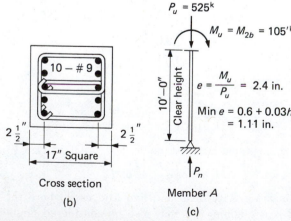

Cross section

(b)

Member A

(c)

Figure 15.20.2 Rigid frame for Example 15.20.3.

Solution: (a) Effective length. In accordance with ACI-10.11.2.1, neglecting an analysis to determine a k value less than 1.0 for a braced frame, use

$$kL_u = L_u = 10 \text{ ft}$$

(b) Slenderness ratio limits. The actual column slenderness ratio is

$$\frac{kL_u}{r} = \frac{120}{0.3(17)} = 23.6$$

Slenderness effects may be neglected when

$$\frac{kL_u}{r} < 34 - 12 \frac{M_{1b}}{M_{2b}}$$

In this case, $M_{1b}/M_{2b} = 0$,

$$\left(\frac{kL_u}{r}\right)_{\text{limit}} = 34 > 23.6$$

Slenderness effects may be neglected. In the following sections, the method will be illustrated even though it would not be required by the ACI Code.

(c) Moment magnifier δ_b.

$$E_c = 57,000 \sqrt{f'_c} = 3120 \text{ ksi}$$

$$I_g = \tfrac{1}{12}(17)(17)^3 = 6950 \text{ in.}^4$$

$$I_s = 2(5)(6)^2 = 360 \text{ in.}^4$$

Using ACI Formulas (10-10) and (10-11), Eqs. (15.10.8) and (15.10.9),

$$EI = 0.2E_cI_g + E_sI_s$$

$$= 0.2(3120)(6950) + 29,000(360)$$

$$= 4,340,000 + 10,440,000 = 14,800,000 \text{ kip in.}^2$$

or

$$EI = 0.4E_cI_g = 8,680,000 \text{ kip in.}^2$$

Using the larger value of EI and applying the factor $(1 + \beta_d)$ to account for sustained load,

$$\frac{EI}{1 + \beta_d} = \frac{14,800,000}{1.40} = 10,500,000 \text{ kip in.}^2$$

$$P_c = \frac{\pi^2 EI}{(kL_u)^2} = \frac{\pi^2(10,500,000)}{[1.0(10.0)(12)]^2} = 7200 \text{ kips}$$

$$P_u = 525 \text{ kips} \qquad [\text{from Fig. 15.20.2(c)}]$$

$$\frac{P_u}{\phi P_c} = \frac{525}{0.70(7200)} = 0.104$$

$$C_m = 0.6 + 0.4 \frac{M_{1b}}{M_{2b}} = 0.6$$

$$\delta_b = \frac{C_m}{1 - P_u/\phi P_c} = \frac{0.6}{1 - 0.104} = 0.67 < 1.0$$

In this case the magnified moment out in the span is less than that at the braced point. The factored loads to be carried are $P_u = 525$ kips and $M_u = 105$ ft-kips at the top of the column. The strength ϕP_n is found to be 566 kips at $e = 2.4$ in.

EXAMPLE 15.20.4 Determine the adequacy of the square tied column (20 in. square, with 12-#10 bars, $f'_c = 4000$ psi, $f_y = 60,000$ psi) which is an exterior first-floor column in the unbraced frame of Fig. 15.20.3. Assume 40% of the factored load moment is sustained. The ΣP_u on the four columns in the lowest story is 3150 kips.

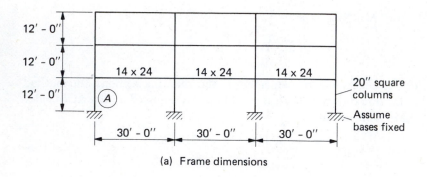

(a) Frame dimensions

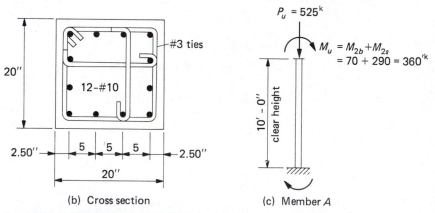

(b) Cross section   (c) Member A

Figure 15.20.3 Rigid frame for Example 15.20.4.

Solution: (a) Effective length and slenderness ratio. The end restraint factors ψ must be determined. Assuming the cracked section moment of inertia for the beam to be half of the gross moment of inertia.

$$I_{cr} \approx \frac{I_g}{2} = \frac{14(24)^3/12}{2} = 8070 \text{ in.}^4$$

When the beam reinforcement is known, the actual I_{cr} for the beam should be used here. For the column, either I_g or the larger EI from ACI Formulas (10-10) or (10-11) should be used.

For the 20-in. square column,

$$I_g = \tfrac{1}{12}(20)(20)^3 = 13,333 \text{ in.}^4$$

$$I_s = 1.27(2)[4(7.5)^2 + 2(2.5)^2] = 603 \text{ in.}^4$$

$$E_c = 57,000 \sqrt{f'_c} = 57,000 \sqrt{4000} = 3605 \text{ ksi}$$

$$EI = 0.2E_cI_g + E_sI_s$$

$$= 0.2(3605)(13,333) + 29,000(603)$$

$$= 9,610,000 + 17,500,000$$

$$= 27,110,000 \text{ kip in.}^2 \text{ (larger than } 0.4E_cI_g)$$

The EI for obtaining k is taken *without* the sustained load effect.

For the unbraced frame, the magnifier δ_s involves the *sum* of the Euler loads P_c for all columns participating in the sidesway resistance at the story level. Thus, the effective length factor k is needed for each of these columns. In this example,

$$\psi_A\text{(top of exterior column)} = \frac{\Sigma EI/L \text{ for columns}}{\Sigma EI/L \text{ for beams}}$$

$$= \frac{2(27,110,000)/12}{3605(8070)/30} = 4.66$$

$$\psi_A\text{(top of interior column)} = 2.33$$

For this calculation use of center-to-center span distances is recommended as being consistent with the nominal frame analysis using those distances. At the fixed end theoretically ψ equals zero; however, the Structural Stability Research Council [31] recommends that for practical purposes ψ should not be taken smaller than 1.0. Thus,

$$\psi_B\text{(bottom)} = 1.0$$

Using Fig. 15.8.1(b), find

$$k \text{ of exterior column} = 1.68$$

$$k \text{ of interior column} = 1.50$$

The effective slenderness ratio for the exterior column being investigated is

$$\frac{kL_u}{r} = \frac{1.68(10.0)(12)}{0.3(20)} = 33.6$$

which exceeds the limit of 22 given by ACI-10.11.4.2 for unbraced frames. Slenderness effects must be considered.

(b) Magnification factors. For the unbraced frame, the ACI Code conservatively requires adding the magnified moments *that would exist if the frame were braced* to the frame sidesway moments magnified by the P–Δ effect. Thus, ACI Formula (10-6) states

$$M_c = \delta_b M_{2b} + \delta_s M_{2s} \qquad\qquad [15.12.3]$$

For this example the end moment M_{2b} under factored gravity load is given as 70 ft-kips and the end moment M_{2s} caused by factored lateral load is given as 290 ft-kips.

1. For the *braced frame magnifier* δ_b:

$$\delta_b = \frac{C_m}{1 - \dfrac{P_u}{\phi P_c}}$$

$$C_m = 0.6 + 0.4 \frac{M_{1b}}{M_{2b}} = 0.6 - 0.4\left(\frac{35}{70}\right) = 0.4$$

Note that 0.4 is the minimum permissible value for C_m.

Next, the Euler load P_c must be computed.

$$P_c = \frac{\pi^2 EI}{(kL_u/r)^2}$$

and for the 20-in. square column,

$$EI = \frac{27,110,000}{1 + \beta_d}$$

Since β_d is the proportion of factored load *moment* that is sustained, β_d will be the sustained portion (say the dead load portion of M_{2b}) of the total factored moment M_u (that is, $M_{2b} + M_{2s}$) that acts on the frame. For this example, assuming M_{2b} is the factored dead load (say, $0.9D$ in accordance with ACI-9.2) and M_{2s} is the factored wind load (say, $1.3W$ in accordance with ACI-9.2.2), then

$$\beta_d = \frac{70}{360} = 0.194, \qquad \text{say } 0.2$$

Assume for this example that β_d is the same for the interior columns as for the exterior columns. Thus,

$$EI = \frac{27,110,000}{1 + 0.2} = 22,600,000 \text{ kip in.}^2$$

$$P_c = \frac{\pi^2(22,600,000)}{[0.85(10)(12)]^2} = 21,400 \text{ kips}$$

Note that for the braced frame multiplier, the k is for a braced frame; for $\psi_A = 4.66$ and $\psi_B = 1.0$ Fig. 15.8.1(a) gives $k = 0.85$. Then,

$$\delta_b = \frac{C_m}{1 - \dfrac{P_u}{\phi P_c}} = \frac{0.4}{1 - \dfrac{525}{0.70(21,400)}} = 0.41 < 1.0$$
$$(\text{Use } \delta_b = 1.0)$$

2. For the *unbraced frame magnifier* δ_s:

$$\delta_s = \frac{1}{1 - \dfrac{\Sigma P_u}{\phi \Sigma P_c}}$$

For the 20-in. square exterior column,

$$P_c = \frac{\pi^2(22,600,000)}{[(1.68)(10)(12)]^2} = 5490 \text{ kips}$$

For the 20-in. square interior column,

$$P_c = \frac{\pi^2(22,600,000)}{[(1.50)(10)(12)]^2} = 6880 \text{ kips}$$

The P_c is needed for each of the first-story columns; in this case

$$\Sigma P_c = 2(5490 + 6880) = 24,740 \text{ kips}$$

The total factored load ΣP_u coming to the four columns is given as 3150 kips. Thus,

$$\frac{\Sigma P_u}{\Sigma P_c} = \frac{3150}{24,740} = 0.127$$

$$\frac{\Sigma P_u}{\phi \Sigma P_c} = \frac{0.127}{0.70} = 0.181$$

This value is the value of Q_u (see Sec. 15.14) and was recommended by MacGregor and Hage to not be permitted to exceed 0.2, ensuring overall frame stability. The magnifier δ_s for the P–Δ effect is

$$\delta_s = \frac{1}{1 - \dfrac{\Sigma P_u}{\phi \Sigma P_c}} = \frac{1}{1 - 0.181} = 1.22$$

(c) Compute the required short column strength P_n and M_n for the member. Using ACI Formula (10-6), Eq. (15.12.3),

$$M_c = \delta_b M_{2b} + \delta_s M_{2s}$$
$$= 1.0(70) + 1.22(290) = 423.8 \text{ ft-kips}$$

The strength required from the 20-in. exterior square column having 12-#10 is

$$\text{required } P_n = \frac{P_u}{\phi} = \frac{525}{0.70} = 750 \text{ kips}$$

$$\text{required } M_n = \frac{M_c}{\phi} = \frac{423.8}{0.70} = 605 \text{ ft-kips}$$

$$\text{required } e = \frac{M_n}{P_n} = \frac{605(12)}{750} = 9.68 \text{ in.}$$

Using the principles of Chap. 13, the nominal strength $P_n = 750$ kips at $e = 9.68$ in.

EXAMPLE 15.20.5 Determine the adequacy of the 14 × 20 in. compression member designed without regard to length effects in Sec. 13.17, Example 13.17.1 (6-#11 bars, $f_c' = 4500$ psi, and $f_y = 50,000$ psi). The member serves as an exterior column in a braced frame, with loading as shown in Fig. 15.20.4, and having a clear height of 22 ft 6 in.

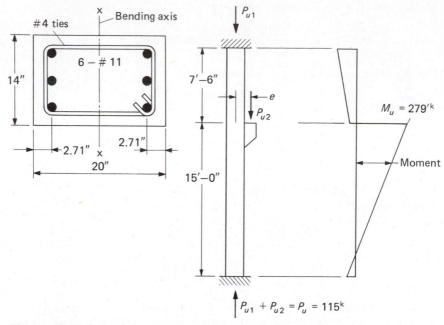

Figure 15.20.4   Member and loading for Example 15.20.5.

Solution: It is logical and proper to determine primary bending moment by any elastic method such as moment distribution. Assume that the maximum factored moment M_u at an intermediate point is 279 ft-kips. Design as a short column neglecting slenderness, as demonstrated in Example 13.17.1, has resulted in a member that is in the tension controls zone and has ϕ slightly greater than 0.70. For this braced frame member, $kL_u = L_u = 22.5$ ft.

(a) For possible instability in the plane of the frame,

$$\frac{kL_u}{r_x} = \frac{1.0(22.5)12}{0.3(20)} = 45$$

which exceeds the maximum value of 22 (used conservatively for braced frame having this irregular moment diagram) for which slenderness effects may be neglected according to ACI-10.11.4. When the bending moment diagram has the largest moment at a location other than at an end, M_{1b}/M_{2b} should be taken conservatively as 1.0. Note that the moment diagram is similar to the case of transverse loading and should be similarly treated.

$$I_g = \tfrac{1}{12}(14)(20)^3 = 9330 \text{ in.}^4$$

$$I_s = 2(3)(1.56)(7.29)^2 = 497 \text{ in.}^4$$

$$E_c = 57,000 \sqrt{4500} = 3820 \text{ ksi}$$

Assuming no sustained load, $\beta_d = 0$,

$$EI = 0.2E_cI_g + E_sI_s$$

$$= 0.2(3820)(9330) + 29,000(497) = 21,500,000 \text{ kip in.}^2$$

or

$$EI = 0.4E_cI_g = 0.4(3820)(9330) = 14,300,000 \text{ kip in.}^2$$

$$C_m = 1.0$$

$$P_c = \frac{\pi^2 EI}{(kL_u)^2} = \frac{\pi^2(21,500,000)}{[1.0(22.5)(12)]^2} = 2910 \text{ kips}$$

$$\delta_b = \frac{C_m}{1 - (P_u/\phi)/P_c} = \frac{1.0}{1 - (115/0.70)/2910} = 1.06$$

The required capacity of the member *in the plane of the frame* (i.e., the strong orientation of the member), using the $\phi = 0.718$ computed in Example 13.17.1, is

$$\text{required } P_n = \frac{P_u}{\phi} = \frac{115}{0.718} = 160 \text{ kips}$$

$$\text{required } M_n = \frac{\delta_b M_u}{\phi} = \frac{1.06(279)}{0.718} = 412 \text{ ft-kips}$$

A statics analysis of this section using an eccentricity of

$$e = \frac{412(12)}{160} = 30.9 \text{ in.}$$

gives $P_n = 151$ kips, which is probably not close enough to 160 kips to be acceptable.

(b) Buckling transverse to the plane of the frame. The slenderness ratio is

$$\frac{kL_u}{r_y} = \frac{22.5(12)}{0.3(14)} = 64.4$$

which exceeds the limiting value of 22 for which the effect of slenderness may be neglected. Again, as in part (a), M_{1b}/M_{2b} should be conservatively taken as 1.0. Since slenderness effects must be considered, the minimum $e = 0.6 + 0.03h$ must be magnified by the factor δ_b.

$$I_g = \tfrac{1}{12}(20)(14)^3 = 4570 \text{ in.}^4$$

$$I_s = 2(2)(1.56)(4.29)^2 = 115 \text{ in.}^4$$

$$EI = 0.2E_cI_g + E_sI_s$$

$$= 0.2(3820)(4570) + 29,000(115) = 6,830,000 \text{ kip in.}^2$$

or

$$EI = 0.4E_cI_g = 0.4(3820)(4570) = 6,980,000 \text{ kip in.}^2$$

$$C_m = 1.0$$

$$P_c = \frac{\pi^2 EI}{(kL_u)^2} = \frac{\pi^2(6,980,000)}{[1.0(22.5)(12)]^2} = 945 \text{ kips}$$

For this axis, small eccentricity makes $\phi = 0.70$.

$$\delta_b = \frac{C_m}{1 - (P_u/\phi)/P_c} = \frac{1.0}{1 - (115/0.70)/945} = 1.21$$

Thus, in the weaker direction the member must have the strength $P_n = P_u/0.70 = 164$ kips at an eccentricity

$$e = 1.21(0.6 + 0.03h) = 0.72 + 0.036h$$
$$= 0.72 + 0.036(14) = 1.22 \text{ in.}$$

A statics analysis indicates that the nominal strength P_n is 1205 kips at this eccentricity with respect to the weaker axis.

The strength P_n may not be taken greater than $0.80P_0$,

$$P_{n(\max)} = 0.80[0.85(4.5)(280 - 9.36) + 50(9.36)] = 1200 \text{ kips}$$

which is approximately the same as the capacity at the minimum eccentricity of 1.22 in.

EXAMPLE 15.20.6 Design column A for the unbraced frame of Fig. 15.20.5 for the dead load plus live load plus wind load case of ACI-9.2.2. The service axial compression is 106 kips dead load and 106 kips live load, and service bending moment is 32 ft-kips dead load, 32 ft-kips live load, and 150 ft-kips wind load. Consider only the dead load to be sustained. The factored axial load P_u on the interior columns is 1.8 times that on the exterior columns. Select a square member to contain approximately $2\frac{1}{2}\%$ reinforcement. Use $f_c' = 5000$ psi, $f_y = 60,000$ psi, and the strength method of the ACI Code.

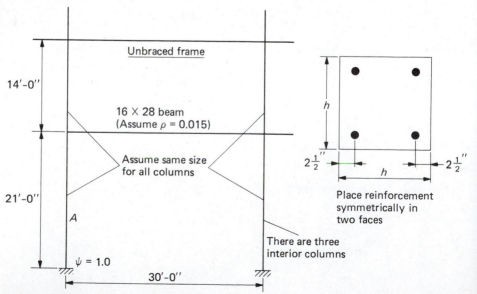

Figure 15.20.5   Data for Example 15.20.6.

Solution: (a) Factored loads. According to ACI-9.2.2,

$$P_u = 0.75[1.4(106) + 1.7(106)] = 246 \text{ kips}$$

$$M_u = 0.75[1.4(32) + 1.7(32) + 1.7(150)] = 266 \text{ ft-kips}$$

(b) Estimate size at balanced condition. Using the procedure described in Secs. 13.16 and 13.17, assume that the axial force contributions of the steel in the two faces are approximately equal, $T \approx C_s$; then

$$C_c \approx P_b = \text{required } P_n$$

$$0.85 f'_c \beta_1 x_b b = \frac{246}{\phi}$$

$$x_b = \left(\frac{0.003}{0.003 + 60/29{,}000} \right) d = 0.592d$$

$$0.85(5)(0.8)(0.592d)b = \frac{246}{0.7} = 351$$

$$\text{balanced } bd = \frac{351}{2.01} = 175 \text{ sq in.}$$

Assume $d \approx 0.85h$

$$\text{balanced } bh = \frac{175}{0.85} = 206 \text{ sq in.}, \qquad \text{say 15 in. square}$$

The moment magnifier requirement is

$$M_c = \delta_b M_{2b} + \delta_s M_{2s}$$

The magnifier δ_b is assumed to be 1.0; and since the column is relatively long a magnifier δ_s of 1.2 to 1.5 may be expected. Thus, if the factored moments are separated into the moment M_{2b} causing no sidesway and the moment M_{2s} causing sidesway,

$$M_{2b} = 0.75[1.4(32) + 1.7(32)] = 75 \text{ ft-kips}$$

$$M_{2s} = 0.75(1.7)(150) = 191 \text{ ft-kips}$$

$$\text{estimated } M_c = 1.0(75) + 1.3(191) = 323 \text{ ft-kips}$$

$$\text{required } e \text{ (estimated)} = \frac{323(12)}{246} = 15.8 \text{ in.}$$

For this large an eccentricity, it is likely the design will be in the tension controls region and the section will be larger than the 15-in. section for the balanced condition. Try an 18-in. section to obtain a first approximation for the magnifier δ_s. The magnifier δ_b is likely to be 1.0.

(c) Estimate effective length factor k for the unbraced frame to be used in computing the sidesway magnifier δ_s.

$$I_g \text{ for column} = \tfrac{1}{12}(18)(18)^3 = 8750 \text{ in.}^4$$

$$I_g \text{ for beam} = \tfrac{1}{12}(16)(28)^3 = 29{,}200 \text{ in.}^4$$

Since cracked section should be used for the beam moment of inertia, assume I_{cr} for the beam to be about 15,000 in.[4]

$$\psi \text{ (top of exterior column)} = \frac{\Sigma EI/L, \text{ columns}}{\Sigma EI/L, \text{ beams}}$$

$$= \frac{8750/14 + 8750/21}{15,000/30} = \frac{1042}{500} \approx 2.0$$

$$\psi \text{ (top of interior column)} = \frac{1042}{1000} \approx 1.0$$

Using the alignment chart of Fig. 15.8.1(b), with ψ (bottom) = 1.0 (given), find k of exterior column = 1.45 and k of interior column = 1.30.

(d) Estimate magnification factor δ_s. Since reinforcement is not yet selected, the following approximation is used,

$$EI \approx \frac{0.5E_c I_g}{1 + \beta_d}$$

$$= \frac{0.5(57 \sqrt{5000})8750}{1.13} = 15,605,000 \text{ kip in.}^2$$

where

$$\beta_d = \frac{0.75(1.4)(32)}{266} = 0.13$$

assuming only dead load moment is sustained. $\beta_d = 0$ might be used since the sway is caused by nonsustained load (wind).

$$P_c = \frac{\pi^2 EI}{(kL_u)^2} = \frac{\pi^2(15,605,000)}{[1.45(18.67)(12)]^2} = 1460 \text{ kips} \quad \text{(exterior column)}$$

$$P_c = \frac{\pi^2(15,605,000)}{[1.30(18.67)(12)]^2} = 1820 \text{ kips} \quad \text{(interior column)}$$

where $L_u = 21.0 - 2.33 = 18.67$ ft.

$$\delta_s = \frac{1.0}{1 - \dfrac{\Sigma P_u}{\phi \Sigma P_c}}$$

$$\Sigma P_c = 2(1460) + 3(1820) = 8380 \text{ kips}$$

$$\Sigma P_u = 246[2 + 3(1.8)] = 1820 \text{ kips}$$

$$\delta_s = \frac{1}{1 - \dfrac{1820}{0.70(8380)}} = 1.45$$

$$\text{estimated } M_c = 1.0(75) + 1.45(191) = 352 \text{ ft-kips}$$

$$\text{required } e = \frac{352(12)}{246} = 17.2 \text{ in.}$$

(e) Determine column size considering $e = 17.2$ in. and $\rho_g = 0.025$, and assuming tension controls. Use Eq. (13.14.7),

$$P_n = 0.85f'_cbd\left\{-\rho + 1 - \frac{e'}{d}\right.$$

$$\left. + \sqrt{\left(1 - \frac{e'}{d}\right)^2 + 2\rho\left[(m - 1)\left(1 - \frac{d'}{d}\right) + \frac{e'}{d}\right]}\right\}$$

Estimate (see Fig. 13.14.1 for e') using $h = 18$ in.,

$$\frac{e'}{d} = \frac{17.2 + 6.5}{15.5} = 1.53$$

$$\frac{d'}{d} = \frac{2.5}{15.5} = 0.16$$

$$m = \frac{f_y}{0.85f'_c} = \frac{60}{0.85(5)} = 14.1$$

$$\rho = 0.5\rho_g = 0.0125$$

$$P_n = 0.85(5)bd\{-0.0125 + 1 - 1.53$$

$$+ \sqrt{(1 - 1.53)^2 + 2(0.0125)[13.1(1 - 0.16) + 1.53]}\}$$

$$= 4.25(0.228)bd = 0.970bd$$

Assuming $d \approx 0.85h$

$$\text{required } bh = \frac{246/0.70}{0.970(0.85)} = 426 \text{ sq in.}$$

An 18-in. square section is not acceptable. A second iteration of the formula with $h = 20$ in. gives required $bh = 345$ sq in. Try 20 in. square with 10-#9 bars ($\rho_g = 0.025$).

(f) Recheck effective length factor k. Use ACI Formula (10-10) without sustained load factor β_d for EI.

EI for column $= 0.2E_cI_g + E_sI_s$

$$I_g = \tfrac{1}{12}(20)(20)^3 = 13,300 \text{ in.}^4$$

$$E_c = 57\sqrt{5000} = 4030 \text{ ksi}$$

$$I_s = 2(4)(7.5)^2 = 450 \text{ in.}^4$$

$$EI = 0.2(4030)(13,300) + 29,000(450) = 23,800,000 \text{ kips in.}^2$$

For the beam, cracked transformed section should be used. In lieu of using the general method of Chap. 4, Eqs. (14.11.7) and (14.11.8) may be used with $n = 7$ for $f'_c = 5000$ psi.

$$\frac{x}{d} = \sqrt{(\rho n)^2 + 2\rho n} - \rho n$$

$$\rho n = 0.015(7) = 0.105$$

$$\frac{x}{d} = \sqrt{(0.105)^2 + 0.21} - 0.105 = 0.365$$

$$\frac{I_{cr}}{I_g} = 8.75\left[\frac{(x/d)^3}{3} + pn\left(1 - \frac{x}{d}\right)^2\right]$$

$$= 8.75\left[\frac{(0.365)^3}{3} + 0.105(0.635)^2\right] = 0.512$$

$$EI \text{ for beam } = 0.512E_cI_g$$

$$= 0.512(4030)(29,200) = 60,300,000 \text{ kip in.}^2$$

$$\psi \text{ (top of exterior column)} = \frac{23.8/14 + 23.8/21}{60.3/30} = \frac{2.833}{2.01} = 1.4$$

$$\psi \text{ (top of interior column)} = \frac{2.833}{4.02} = 0.7$$

Using Fig. 15.8.1(b), find $k = 1.37$ for exterior column and $k = 1.27$ for interior column.

(g) Recompute magnification factor δ_s.

$$\frac{EI}{1 + \beta_d} = \frac{23,800,000}{1.13} = 21,060,000 \text{ kip in.}^2$$

$$P_c = \frac{\pi^2 EI}{(kL_u)^2} = \frac{\pi^2(21,060,000)}{[1.37(18.67)(12)]^2} = 2210 \text{ kips} \quad \text{(exterior column)}$$

$$P_c = \frac{\pi^2(21,060,000)}{[1.27(18.67)(12)]^2} = 2570 \text{ kips} \quad \text{(interior column)}$$

$$\Sigma P_c = 2(2210) + 3(2570) = 12,130 \text{ kips}$$

$$\Sigma P_u = 1820 \text{ kips} \quad \text{[from (d) above]}$$

$$\frac{\Sigma P_u}{\phi \Sigma P_c} = \frac{1820}{0.70(12,130)} = 0.214$$

$$\delta_s = \frac{1.0}{1 - \dfrac{\Sigma P_u}{\phi \Sigma P_c}} = \frac{1.0}{1 - 0.214} = 1.27$$

(h) Recheck capacity by approximate formula, Eq. (13.14.7). Using the 20-in. square section with 10-#9 (4 in each face), the more accurate distance from face of concrete to center of bars is

$$d' = d_s = 1.5 \text{ (cover)} + 0.375(\#3 \text{ tie}) + 0.564 \text{ (bar radius)} = 2.44 \text{ in.}$$

instead of 2.5 in. used in the preliminary computations.

$$\text{required } e = \frac{\delta_b M_{2b} + \delta_s M_{2s}}{P_u}$$

$$= \frac{[1.0(75) + 1.27(191)]12}{246} = 15.5 \text{ in.}$$

$$\frac{e'}{d} = \frac{15.5 + 7.56}{17.56} = 1.31$$

$$\frac{d'}{d} = \frac{2.44}{17.56} = 0.139$$

$$\rho = 0.01 \qquad \text{(using only the bars in one face)}$$

$$m = 14.1$$

$$m' = m - 1 = 13.1$$

$$P_n = 4.25bd\{-0.01 + 1 - 1.31$$
$$+ \sqrt{(1 - 1.31)^2 + 2(0.01)[13.1(1 - 0.139) + 1.31]}\}$$
$$= 4.25(0.270)bd = 1.148bd$$

$$P_n = 1.148(20)(17.56) = 403 \text{ kips}$$

$$\phi P_n = 0.70(403) = 282 \text{ kips} > [P_u = 246 \text{ kips}] \qquad \text{OK}$$

A statics analysis of the selected section indicates a strength $P_n = 479$ kips at $e = 15.5$ in.

(i) Check magnification factor δ_b. For a fixed base member having no transverse load,

$$C_m = 0.6 + 0.4\left(\frac{M_{1b}}{M_{2b}}\right)$$

$$= 0.6 + 0.4\left(\frac{-37.5}{75}\right) = 0.4 \qquad \text{(the minimum value)}$$

For the braced frame $k = 1.0$ (conservatively); thus

$$P_c = \frac{\pi^2 EI}{(kL_u)^2} = \frac{\pi^2(20,900,000)}{[1.0(18.67)(12)]^2} = 4110 \text{ kips}$$

$$\delta_b = \frac{C_m}{1 - \dfrac{P_u}{\phi P_c}} = \frac{0.4}{1 - \dfrac{246}{0.70(4110)}} = 0.44 < 1.0 \qquad \text{(Use } \delta_b = 1.0)$$

This confirms the assumption of $\delta_b = 1.0$ made earlier.

Use 20-in. square section with 10-#9 bars as shown in Fig. 15.20.6.

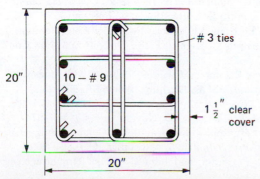

Figure 15.20.6 Section selected for Example 15.20.6.

SELECTED REFERENCES

1. L. Euler, *DeCurvis Elasticis, Additamentum I, Methodus Inveniendi Lineas Curvas Maximi Minimive Proprietate Gaudentes.* Lausanne and Geneva: 1744, (pp. 267–268); and "Sur la Force des Colonnes," *Mémoires de l'Académie de Berlin*, Vol. 13. Berlin: 1759 (pp. 252–282).

2. F. Engesser. "Ueber die Knickfestigkeit gerader Stäbe," *Zeitschrift des Architekten- und Ingenieur-Vereins zu Hannover*, 35(1889), 455–462. Also "Die Knickfestigkeit gerader Stäbe," *Zentralblatt der Bauverwaltung*, Berlin, Dec. 5, 1891, p. 483.

3. T. von Kármán. "Die Knickfestigkeit gerader Stäbe," *Physikalische Zeitschrift*, Vol. 9, 1908 (p. 136). Also, "Untersuchungen über Knickfestigkeit," *Mitteilungen über Forschungsarbeiten auf dem Gebiete des Ingenieurwesens*, No. 81. Berlin: 1910.

4. Wen F. Chang and Phil M. Ferguson. "Long Hinged Reinforced Concrete Columns," *ACI Journal, Proceedings*, **60**, January 1963, 1–25.

5. Luis P. Sáenz and Ignacio Martín. "Test of Reinforced Concrete Columns with High Slenderness Ratios," *ACI Journal, Proceedings*, **60,** May 1963, 589–616.

6. John E. Breen and Phil M. Ferguson. "The Restrained Long Concrete Column As a Part of A Rectangular Frame," *ACI Journal, Proceedings*, **61,** May 1964, 563–587.

7. John E. Breen. "Computer Use In Studies of Frames with Long Columns," *International Symposium on Flexural Mechanics of Reinforced Concrete* (SP-12). Detroit: American Concrete Institute, 1965 (pp. 535–556).

8. Edward O. Pfrang and Chester P. Siess. "Behavior of Restrained Reinforced Concrete Columns," *Journal of the Structural Division*, ASCE, **90,** (ST5) October 1964, 113–135; Disc., **91,** June 1965 (ST3), 280–287.

9. J. G. MacGregor. Discussion of "Behavior of Restrained Reinforced Concrete Columns," by E. O. Pfrang and C. P. Siess, *Journal of the Structural Division*, ASCE, **91,** June 1965 (ST3), 280–287.

10. Richard W. Furlong and Phil M. Ferguson. "Tests of Frames with Columns in Single Curvature," *Symposium on Reinforced Concrete Columns* (SP-13). Detroit: American Concrete Institute, 1966 (pp. 55–73).

11. Phil M. Ferguson and John E. Breen. "Investigation of the Long Column in a Frame Subject to Lateral Loads," *Symposium on Reinforced Concrete Columns* (SP-13). Detroit: American Concrete Institute, 1966 (pp. 75–119).

12. Ignacio Martín and Elmer Olivieri. "Test of Slender Reinforced Concrete Columns Bent in Double Curvature," *Symposium on Reinforced Concrete Columns* (SP-13). Detroit: American Concrete Institute, 1966 (pp. 121–138).

13. J. G. MacGregor and S. L. Barter. "Long Eccentrically Loaded Concrete Columns Bent in Double Curvature," *Symposium on Reinforced Concrete Columns* (SP-13). Detroit: American Concrete Institute, 1966 (pp. 139–156).

14. Alfred L. Parme. "Capacity of Restrained Eccentrically Loaded Long Columns," *Symposium on Reinforced Concrete Columns* (SP-13). Detroit: American Concrete Institute, 1966 (pp. 325–367).

15. Edward O. Pfrang. "Behavior of Reinforced Concrete Columns with Sidesway," *Journal of the Structural Division*, ASCE, **92,** June 1966 (ST3), 225–252.

16. R. Green and John E. Breen. "Eccentrically Loaded Concrete Columns Under Sustained Load," *ACI Journal, Proceedings*, **66,** November 1969, 866–874.

17. John E. Breen and Phil M. Ferguson. "Long Cantilever Columns Subject to Lateral Forces," *ACI Journal, Proceedings*, **66,** November 1969, 884–893.

18. Shriniwas N. Pagay, Phil M. Ferguson, and John E. Breen. "Importance of

Beam Properties on Concrete Column Behavior," *ACI Journal, Proceedings,* **67,** October 1970, 808–815.

19. Hajime Okamura, Shriniwas N. Pagay, John E. Breen, and Phil M. Ferguson. "Elastic Frame Analysis—Corrections Necessary for Design of Short Concrete Columns in Braced Frames," *ACI Journal, Proceedings,* **67,** November 1970, 894–897.

20. Phil M. Ferguson, Hajime Okamura, and Shriniwas N. Pagay. "Computer Study of Long Columns in Frames," *ACI Journal, Proceedings,* **67,** December 1970, 955–958.

21. G. A. Blomier and J. E. Breen. "Effect of Yielding of Restraints on Slender Concrete Columns With Sidesway Prevented," *Reinforced Concrete Columns* (SP-50). Detroit: American Concrete Institute, 1975 (pp. 41–65).

22. F. N. Rad and R. W. Furlong. "Behavior of Unbraced Reinforced Concrete Frames," *ACI Journal, Proceedings,* **77,** July–August 1980, 269–278.

23. J. S. Ford, D. C. Chang, and J. E. Breen. "Behavior of Concrete Columns Under Controlled Lateral Deformation," *ACI Journal, Proceedings,* **78,** January–February 1981, 3–20.

24. J. S. Ford, D. C. Chang, and J. E. Breen. "Experimental and Analytical Modeling of Unbraced Multipanel Concrete Frames," *ACI Journal, Proceedings,* **78,** January–February 1981, 21–35.

25. J. S. Ford, D. C. Chang, and J. E. Breen. "Behavior of Unbraced Multipanel Concrete Frames," *ACI Journal, Proceedings,* **78,** March–April 1981, 99–115.

26. E. Hognestad. *A Study of Combined Bending and Axial Load in Reinforced Concrete Members* (Bulletin No. 399). Urbana: University of Illinois Engineering Experiment Station, November 1951, (128 pp.).

27. Bengt Broms and I. M. Viest. "Ultimate Strength Analysis of Long Hinged Reinforced Concrete Columns," *Journal of Structural Division,* ASCE, **84,** January 1958 (ST1) (Paper No. 1510).

28. *Specification for the Design, Fabrication and Erection of Structural Steel for Buildings* (effective November 1, 1978). Also, *Commentary* on 1978 Specification. Chicago: American Institute of Steel Construction, 1978.

29. Charles Massonnet. "Stability Considerations in the Design of Steel Columns," *Journal of Structural Division,* ASCE, **85,** September 1959 (ST7), 75–111.

30. Charles G. Salmon and John E. Johnson. *Steel Structures: Design and Behavior* (2nd ed.). New York: Harper & Row, 1980 (Chap. 14).

31. Bruce G. Johnston (Ed.). *Guide to Stability Design Criteria for Metal Structures* (3rd ed.) (Structural Stability Research Council). New York: Wiley, 1976 (Chap. 15).

32. S. P. Timoshenko and J. M. Gere. *Theory of Elastic Stability* (2nd ed.). New York: McGraw-Hill, 1961 (pp. 17–19, 59–70).

33. Friedrich Bleich. *Buckling Strength of Metal Structures.* New York: McGraw-Hill, 1952 (Chaps. 6 and 7).

34. William McGuire. *Steel Structures.* Englewood Cliffs, N.J.: Prentice-Hall, 1968 (pp. 424–520).

35. C. K. Wang. *Computer Methods in Advanced Structural Analysis.* New York: Intext Educational Publishers (Harper & Row), 1973.

36. Thomas C. Kavanagh. "Effective Length of Framed Columns," *Transactions ASCE,* **127,** Part II, 1962, 81–101.

37. *Commentary on Building Code Requirements for Reinforced Concrete* (ACI 318-83). Detroit: American Concrete Institute, 1983 (Sections 10.8–10.11).

38. John E. Breen, James G. MacGregor, and Edward O. Pfrang. "Determination

of Effective Length Factors for Slender Concrete Columns," *ACI Journal, Proceedings,* **69,** November 1972, 669–672.

39. *Code of Practice for the Structural Use of Concrete,* Part I. *Design, Materials and Workmanship.* London: British Standards Institute, 1972.

40. Richard W. Furlong. "Column Slenderness and Charts for Design," *ACI Journal, Proceedings,* **68,** January 1971, 9–18.

41. James G. MacGregor, John E. Breen, and Edward O. Pfrang. "Design of Slender Concrete Columns," *ACI Journal, Proceedings,* **67,** January 1970, 6–28.

42. James Colville. "Slenderness Effects in Reinforced Concrete Square Columns," *Reinforced Concrete Columns* (SP-50). Detroit: American Concrete Institute, 1975 (pp. 165–191).

43. James G. MacGregor and Sven E. Hage. "Stability Analysis and Design of Concrete Frames," *Journal of the Structural Division,* ASCE, **103,** October 1977 (ST10), 1953–1970.

44. J. G. MacGregor, U. H. Oelhafen, and S. E. Hage. "A Reexamination of the *EI* Value for Slender Columns," *Reinforced Concrete Columns* (SP-50). Detroit: American Concrete Institute, 1975 (pp. 1–40).

45. Ian C. Medland and Donald A. Taylor. "Flexural Rigidity of Concrete Column Sections," *Journal of the Structural Division,* ASCE, **97,** February 1971 (ST2), 573–586.

46. Brij B. Goyal and Neil Jackson. "Slender Concrete Columns under Sustained Load," *Journal of the Structural Division,* ASCE, **97,** November 1971 (ST11), 2729–2750.

47. Robert G. Drysdale and Mark W. Huggins. "Sustained Biaxial Load on Slender Concrete Columns," *Journal of the Structural Division,* ASCE, **97,** May 1971 (ST5), 1423–1443.

48. Brian R. Wood, Denis Beaulieu, and Peter F. Adams. "Column Design by P Delta Method," *Journal of the Structural Division,* ASCE, **102,** February 1976 (ST2), 411–427.

49. Brian R. Wood, Denis Beaulieu, and Peter F. Adams. "Further Aspects of Design by P-Delta Method," *Journal of the Structural Division,* ASCE, **102,** March 1976 (ST3), 487–500.

50. Richard W. Furlong. "Rational Analysis of Multistory Concrete Structures," *Concrete International,* **3,** June 1981, 29–35.

51. S. I. Abdel-Sayed and N. J. Gardner. "Design of Symmetric Square Slender Reinforced Concrete Columns under Biaxially Eccentric Loads," *Reinforced Concrete Columns* (SP-50). Detroit: American Concrete Institute, 1975 (pp. 149–164).

52. Eldon F. Mockry and David Darwin. "Slender Column Interaction Diagrams," *Concrete International,* **4,** June 1982, 44–50.

53. Eldon F. Mockry and David Darwin. "Simplified Design of Slender Reinforced Concrete Columns," SM Report No. 2, Structural Engineering and Engineering Materials, University of Kansas Center for Research, Lawrence, Kansas, July 1980 (312 pp).

54. J. S. Ford, D. C. Chang, and J. E. Breen. "Design Indications from Tests of Unbraced Multipanel Concrete Frames," *Concrete International,* **3,** March 1981, 37–47.

55. *Design Constants for Rectangular Long Columns.* Advanced Engineering Bulletin No. 12, Portland Cement Association, Skokie, Illinois, 1964 (36 pp).

56. J. G. MacGregor. "Out-of-Plumb Columns in Concrete Structures," *Concrete International,* **1,** June 1979, 26–31.

PROBLEMS

All problems* are to be done in accordance with the strength method of the ACI
Code unless otherwise specified. All loads given are service loads unless otherwise
indicated.

15.1 Determine the adequacy of the section including length effect, for a 16-in.
square tied column that has a clear height of 18 ft and serves as an interior
member of a braced frame. The member is designed as axially loaded with
the following service loads: live load, 130 kips; dead load, 200 kips. Use
$f'_c = 4000$ psi and $f_y = 60,000$ psi. Without actually selecting bars, assume
about $2\frac{1}{2}\%$ total reinforcement equally divided in the opposite faces of the
member. Use ACI moment magnifier method. (400 mm square section; 5.5
m clear height; LL $= 580$ kN; DL $= 890$ kN; $f'_c = 30$ MPa; $f_y = 400$ MPa.)

15.2 Determine the adequacy, including length effects, for a 14-in. diameter spirally reinforced column (assume about $2\frac{1}{2}\%$ reinforcement) which is an interior second-floor column (column A) in the braced frame of the accompanying
figure. The member is to carry an axial service load of 87 kips dead load and

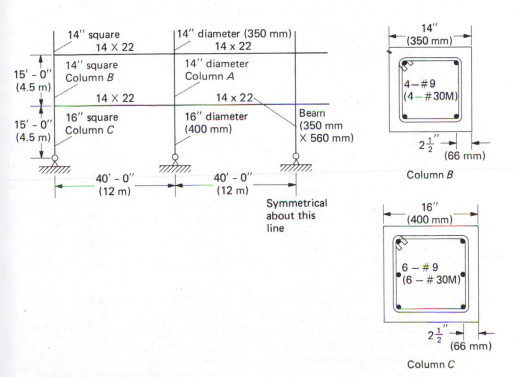

Problems 15.2 to 15.8

*Most problems may be solved as problems stated in Inch-Pound units, or as problems in SI
units using quantities in parenthesis at the end of the statement. The SI conversions are
approximate to avoid implying higher precision for the given information in metric units than
that given for the Inch-Pound units.

133 kips live load, with only the dead load considered sustained. The 14 × 22 beams contain $\rho = 0.015$ for negative moment tension steel. Use $f'_c = 5000$ psi and $f_y = 60,000$ psi. Use the ACI moment magnifier method. (350 mm diameter; $P_n = 2200$ kN; 350 mm × 560 mm beams; $f'_c = 35$ MPa; $f_y = 300$ MPa.)

15.3 Assume the frame of Prob. 15.2 is *unbraced* and consists of 4 bays symmetrical about the middle. The gravity axial loads are 44 kips dead load and 66 kips live load. The largest moments in column A are 12 ft-kips dead load, 10 ft-kips live load, and 58 ft-kips wind load. An exterior column carries one-half the factored axial load of an interior column. Investigate the adequacy of the interior column A. If not adequate, redesign it to make it satisfactory according to the moment magnifier method of the ACI Code.

15.4 Determine the adequacy of the exterior square column (column B) of the figure for Prob. 15.2, if the member is carrying a gravity axial compression of 217 kips dead load and 145 kips live load and primary bending moments of 27 ft-kips dead load and 18 ft-kips live load. Assume that the primary bending moment varies from $+M$ at the top of the member linearly to $-M/2$ at the bottom, with joint translation adequately prevented (i.e., a braced frame). Use $f'_c = 5000$ psi and $f_y = 60,000$ psi. The 14 × 22 beams contain $\rho = 0.015$ for negative moment tension steel. Use ACI moment magnifier method. ($P_D = 965$ kN; $P_L = 645$ kN; $M_D = 37$ kN·m; $M_L = 24$ kN·m; $f'_c = 35$ MPa; $f_y = 400$ MPa, 350 mm × 560 mm beams.)

15.5 Repeat Prob. 15.4, except consider the primary bending moment constant over the height of the column.

15.6 Assume the member (column B) in Prob. 15.4 is part of an *unbraced* frame, and that maximum moments M_2 of 28 ft-kips dead load, 20 ft-kips live load, and 85 ft-kips wind load must be carried. The axial compression loads are 30 kips dead load and 40 kips live load. The total factored load ΣP_u to all five columns in the story is 590 kips for ACI Formula (9-2). Investigate the adequacy of the 14-in. square exterior column (column B). If found to be inadequate, redesign the member assuming the loadings are not affected by a change in member stiffness.

15.7 Determine the adequacy of the 16-in. square first-floor column (column C) of the figure for Prob. 15.2, which is carrying service axial load of 370 kips and a primary bending moment of 121 ft-kips (assume loads 70% dead load). The primary bending moment varies from a maximum at the top to zero at the bottom and joint translation is adequately prevented. Use $f'_c = 5000$ psi and $f_y = 60,000$ psi. Use ACI moment magnifier method. ($P = 1645$ kN; $M = 164$ kN·m; $f'_c = 35$ MPa; $f_y = 400$ MPa.)

15.8 Assume the member (column C) of Prob. 15.7 is part of an *unbraced* frame. The axial loads are 45 kips dead load and 60 kips live load. Assume the 14 × 22 beams contain negative moment reinforcement $\rho = 0.015$. The bending moments are 30 ft-kips dead load, 22 ft-kips live load, and 110 ft-kips wind load. Investigate the adequacy according to the moment magnifier method of the ACI Code. If not adequate, redesign assuming the loadings are not affected by changes in member stiffnesses. The total factored load ΣP_u to all columns in the lower story is 885 kips for ACI Formula (9-2).

15.9 Determine the adequacy of an 18-in. square tied member in a *braced* frame where the member has end flexibilities $\psi = 2.0$ at its top and 10.0 at its bottom and has a clear height of 16 ft. The member is required to carry a factored axial load $P_u = 70$ kips, and the maximum factored primary bending moment $M_u = 168$ ft-kips. Assume primary bending moment is constant over the height of the member, and that 70% of the loads are dead load, and remainder live load. Use $f'_c = 3000$ psi and $f_y = 40,000$ psi. The member reinforcement consists of 8-#9 bars, four in each face centered $2\frac{1}{2}$ in. (or 64 mm) from edge of member. Use ACI moment magnifier method. (460 mm square member; clear height = 5 m; $P_u = 312$ kN; $M_u = 228$ kN·m; $f'_c = 20$ MPa; $f_y = 300$ MPa; reinforcement, 4-#25M and 4-#35M bars.)

15.10 Repeat Prob. 15.9 if bending moment arises from uniform lateral load, and apply rational analysis to determine k (less than 1.0) and for determination of C_m (assume the member fixed at one end and simply supported at the other). Use moment magnifier method.

15.11 (a) Redesign the column used in Example 15.20.6 except for service loads use an axial compression of 132 kips dead load and 132 kips live load, and bending moments M_2 of 26 ft-kips dead load, 26 ft-kips live load, and 123 ft-kips wind load. The length for column A is 20 ft instead of 21 ft. ($P_D = 580$ kN; $P_L = 580$ kN; $M_D = 35$ kN·m; $M_L = 35$ kN·m; $M_W = 167$ kN·m; length of column above = 4.3 m; length of column A = 6.1 m; beam = 400 mm × 700 mm on 9-m span; $f'_c = 35$ MPa; $f_y = 400$ MPa.)
(b) How much smaller could the member have been if the frame were adequately braced to prevent joint translation?

15.12 (a) Redesign the columns used in Example 15.20.6 except use 18 ft (5.5 m) instead of 21 ft for the length of column A.
(b) How much smaller could the member have been if the frame were adequately braced to prevent joint translation?

15.13 through 15.19. An unbraced multistory frame has been analyzed using a conventional first-order elastic analysis. The service load results of a gravity load analysis to get dead load and live load values, and the separately computed lateral load analysis to obtain the wind load moments (M_W) are given in the accompanying table. The rotational moments given are in the same rotational direction at each end of the member. For each design assume the column being designed is one of five identical ones resisting sidesway in that story. For the beams, assume the ratio of moment of inertia I to span L is 30 in.[3] for each beam.

For the loading condition given in the accompanying table, design a square tied column having symmetrical reinforcement with an approximate gross reinforcement ratio ρ_g of 0.010. Design formulas from Chap. 13 may be used to size the section but the final check of strength is to be done using statics with the rectangular stress distribution. Use $f'_c = 4000$ psi and $f_y = 60,000$ psi.

TABLE FOR PROBS. 15.13 THROUGH 15.19

PROB. NO.	L_u (ft)	P_D (kips)	P_L (kips)	M_D (ft-kips)	M_L (ft-kips)	M_W (ft-kips)	LOCATION
15.13	14	48	13	Negl.	32	7.5	Top of column
				Negl.	46	2.0	Bottom of column
$\Sigma P_D = 250$; $\Sigma P_L = 75$ kips							
15.14	14	423	234	Negl.	15	99	Top of column
				Negl.	30	80	Bottom of column
$\Sigma P_D = 2120$ kips; $\Sigma P_L = 1100$ kips							
15.15	14	352	202	39	26	62	Top of column
				39	26	32	Bottom of column
$\Sigma P_D = 1050$ kips; $\Sigma P_L = 600$ kips							
15.16	14	352	202	Negl.	13	83	Top of column
				Negl.	13	62	Bottom of column
$\Sigma P_D = 1050$ kips; $\Sigma P_L = 600$ kips							
15.17	14	176	133	18	13	59	Top of column
				18	13	37	Bottom of column
$\Sigma P_D = 880$ kips; $\Sigma P_L = 650$ kips							
15.18	14	176	154	20	13	26	Top of column
				20	13	10	Bottom of column
$\Sigma P_D = 880$ kips; $\Sigma P_L = 650$ kips							
15.19	22	500	320	15	13	41	Top of column
				30	26	316	Bottom of column
$\Sigma P_D = 2800$ kips; $\Sigma P_L = 1760$ kips							

16

Design of Two-Way Floor Systems

16.1 GENERAL DESCRIPTION

In reinforced concrete buildings, a basic and common type of floor is the slab–beam–girder construction, which has been treated in Chaps. 8, 9, and 10. As shown in Fig. 16.1.1(a) the shaded slab area is bounded by the two adjacent beams on the sides and portions of the two girders at the ends. When the length of this area is two or more times its width, almost all of the floor load goes to the beams, and very little, except some near the edge of the girders, goes directly to the girders. Thus the slab may be designed as a one-way slab, with the main reinforcement parallel to the girder and

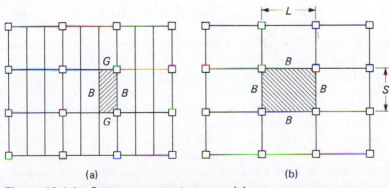

Figure 16.1.1 One-way versus two-way slabs.

Flat slab (waffle slab) with capitals in the Fisher Cleveland Plant. (Courtesy of Portland Cement Association.)

the shrinkage and temperature reinforcement parallel to the beams. The deflected surface of a one-way slab is primarily one of single curvature.

When the ratio of the long span L to the short span S as shown in Fig. 16.1.1(b) is less than about 2, the deflected surface of the shaded area becomes one of double curvature. The floor load is carried in both directions to the four supporting beams around the panel, hence the panel is a *two-way slab*. Obviously, when S is equal to L, the four beams around a typical interior panel should be identical; for other cases the long beams take more load than the short beams.

Both the *flat-slab* and *flat-plate* floors shown in Figs. 16.1.2(a) and (b) are characterized by the absence of beams along the interior column lines, but edge beams may or may not be used at the exterior edges of the floor. Flat-slab floors differ from flat-plate floors in that flat-slab floors provide adequate shear strength by having either or both of the following: (a) drop panels (i.e., increased thickness of slab) in the region of the columns; or (b) column capitals (i.e., tapered enlargement of the upper ends of columns). In flat-plate floors a uniform slab thickness is used and the shear strength is obtained by the embedment of multiple-U stirrups or structural steel devices known as *shearhead reinforcement* [22] within the slab of uniform thickness. Relatively speaking, flat slabs are more suitable for larger panel size or heavier loading than flat plates.

Historically, flat slabs predate both two-way slabs on beams and flat plates. Flat-slab floors were originally patented by O. W. Norcross [1] in the United States on April 29, 1902. Several systems of placing reinforcement have been developed and patented since then—the four-way system,

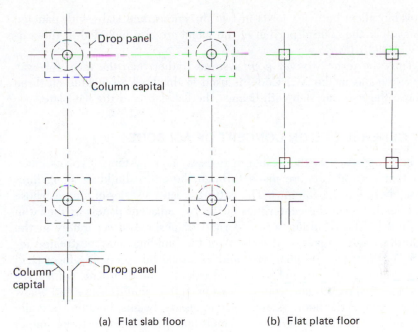

(a) Flat slab floor   (b) Flat plate floor

Figure 16.1.2   Flat slab and flat plate floor construction.

the two-way system, the three-way system, and the circumferential system. C. A. P. Turner [1] was one of the early advocates of a flat-slab system known as the "mushroom" system. About 1908 the flat slab began being recognized as an acceptable floor system, but for many years designers were confronted with difficulties of patient infringements.

Actually the terms *two-way slab* [Fig. 16.1.1(b)], *flat slab* [Fig. 16.1.2(a)], and *flat plate* [Fig. 16.1.2(b)] are arbitrary, because there is in fact two-way action in all three types and a flat (usually nearly square) ceiling area usually exists within the panel in all three types. Following tradition, the implication is that there are beams between columns in two-way slabs; but no such beams, except perhaps edge beams along the exterior sides of the entire floor area, are used in flat slabs or flat plates. From the viewpoint of structural analysis, however, the distinction as to whether or not there are beams between columns is not pertinent, because if beams of any relative size could be designed to interact with the slab, use of beams of zero size would be only the limit condition.

If methods of structural analysis and design are developed for two-way slabs with beams, many of these general provisions should apply equally well to flat slabs or flat plates. Until 1971 the design of two-way slabs supported on beams has, historically, been treated separately from the flat slabs or flat plates without beams. Various empirical procedures have been proposed and used [2–4]. The present ACI Code takes an integrated view and Chap. 13 of the Code refers to two-way slab *systems* with or without beams. In addition to solid slabs, hollow slabs with interior voids to reduce dead weight, slabs (such as waffle slabs) with recesses made by permanent or removable fillers between joists in two directions, and slabs with paneled ceilings near the central portion of the panel are also included in this category (ACI-13.1.3).

Thus the term two-way *floor* systems (rather than the term two-way *slab* systems as in the ACI Code) is used in this book to include all three systems: the two-way slab with beams, the flat slab, and the flat plate.

16.2 GENERAL DESIGN CONCEPT OF ACI CODE

The basic approach to the design of two-way floor systems involves imagining that vertical cuts are made through the entire building along lines midway between the columns. The cutting creates a series of frames whose width lies between the centerlines of the two adjacent panels as shown in Fig. 16.2.1. The resulting series of rigid frames, taken separately in the longitudinal and transverse directions of the building, may be treated for gravity loading floor by floor as would generally be acceptable for a rigid frame structure consisting of beams and columns, in accordance with ACI-8.9.1. A typical rigid frame would consist of (1) the columns above and below the floor, and (2) the floor system, with or without beams, bounded laterally between the centerlines of the two panels (one panel for an exterior line of columns) adjacent to the line of columns.

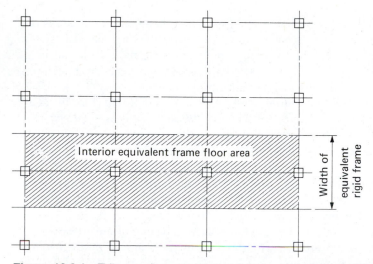

Figure 16.2.1 Tributary floor area for an interior equivalent rigid frame of a two-way floor system.

Thus the design of a two-way floor system (including two-way slab, flat slab, and flat plate) is reduced to that of a rigid frame; hence the name "equivalent frame method."

As in the case of design of actual rigid frames consisting of beams and columns, approximate methods of analysis may be suitable for many usual floor systems, spans, and story heights. As treated in Chap. 7 the analysis for actual frames could be (a) approximate using the moment and shear coefficients of ACI-8.3, or (b) more accurate using structural analysis after assuming the relative stiffnesses of the members. For gravity load only and for floor systems within the specified limitations, the moments and shears on these equivalent frames may be determined (a) approximately using moment and shear coefficients prescribed by the "direct design method" of ACI-Chapter 13, or (b) by structural analysis in a manner similar to that for actual frames using the special provisions of the "equivalent frame method" of ACI-Chapter 13. An elastic analysis (such as by the equivalent frame method) must be used for lateral load even if the floor system meets the limitations of the direct design method for gravity load.

The equivalent rigid frame is the structure being dealt with whether the moments are determined by the "direct design method (DDM)" or by the "equivalent frame method (EFM)." These two ACI Code terms describe two ways of obtaining the longitudinal variation of bending moments and shears.

When the "equivalent frame method" is used for obtaining the longitudinal variation of moments and shears, the relative stiffness of the columns, as well as that of the floor system, can be assumed in the preliminary analysis and then reviewed, as is the case for design of any statically indeterminate structure. Design moment envelopes may be obtained for dead load in combination with various patterns of live load, as described in Chap. 7 (Sec. 7.2). In lateral load analysis, moment magnification in columns due to sidesway of vertical loads must be taken into account as prescribed in ACI-10.11.5 and 10.12.

Once the longitudinal variation in factored moments and shears has been obtained, whether by ACI "DDM" or "EFM," the moment across the entire width of the floor system being considered is distributed laterally to the beam, if used, and to the slab. The lateral distribution procedure and the remainder of the design is essentially the same whether "DDM" or "EFM" has been used.

The accuracy of analysis methods utilizing the concept of dividing the structure into equivalent frames has been verified for *gravity load* analysis by tests [5–16] and analytical studies [17,18,53,58,64–67]. For *lateral load* analysis, where there is less agreement on procedure, proposals have been made by Pecknold [68], Allen and Darvall [69], Vanderbilt [58,70], Elias [71–73], and Vanderbilt and Corley [74].

16.3 TOTAL FACTORED STATIC MOMENT

Consider two typical interior panels $ABCD$ and $CDEF$ in a two-way floor system, as shown in Fig. 16.3.1(a). Let L_1 and L_2 be the panel size in the longitudinal and transverse directions, respectively. Let lines 1–2 and 3–4 be centerlines of panels $ABCD$ and $CDEF$, both parallel to the longitudinal direction. Isolate as a free body [see Fig. 16.3.1(b)] the floor slab and the included beam bounded by the lines 1–2 and 3–4 in the longitudinal direc-

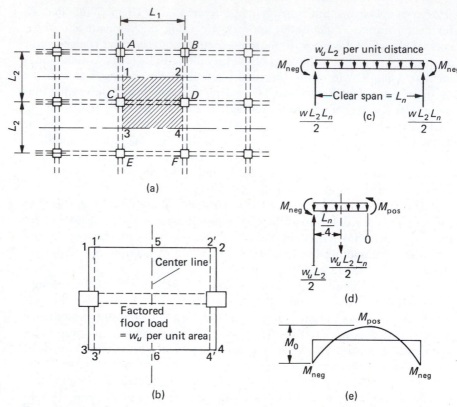

Figure 16.3.1 Statics of a typical interior panel in a two-way floor system.

tion and the transverse lines $1'$–$3'$ and $2'$–$4'$ at the faces of the beams in the transverse direction. The load acting on this free body [see Fig. 16.3.1(c)] is $w_u L_2$ per unit distance in the longitudinal direction. The total upward force acting on lines $1'$–$3'$ or $2'$–$4'$ is $w_u L_2 L_n/2$, where w_u is the factored load per unit area and L_n is the clear span in the longitudinal direction between faces of supports (as defined by ACI-13.6.2.5).

If M_{neg} and M_{pos} are the numerical values of the total negative and positive bending moments along lines $1'$–$3'$ and 5–6, then moment equilibrium of the free body of Fig. 16.3.1(d) requires

$$M_{\text{neg}} + M_{\text{pos}} = \frac{w_u L_2 L_n^2}{8} \qquad (16.3.1)$$

For a typical exterior panel, the negative moment at the interior support would be larger than that at the exterior support, as has been shown in Sec. 7.5. The maximum positive moment would occur at a section to the left of the midspan, as shown in Fig. 16.3.2(c). In practical design, it is customary to use M_{pos} at midspan for determining the required positive moment reinforcement. For this case,

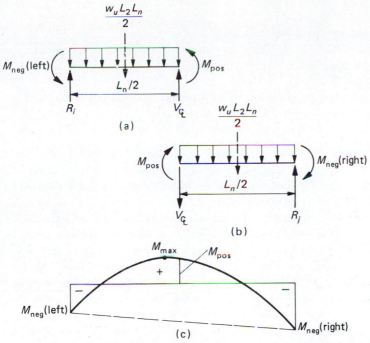

Figure 16.3.2   Statics of a typical exterior panel in a two-way floor system.

$$\frac{M_{\text{neg}}(\text{left}) + M_{\text{neg}}(\text{right})}{2} + M_{\text{pos}} = \frac{w_u L_2 L_n^2}{8} \qquad \textbf{(16.3.2)}$$

A proof for Eq. (16.3.2) can be obtained by writing the moment equilibrium equation about the left end of the free body shown in Fig. 16.3.2(a),

$$M_{\text{neg}}(\text{left}) + M_{\text{pos}} = \frac{w_u L_2 L_n}{2}\left(\frac{L_n}{4}\right) - V_{\mathbb{C}}\left(\frac{L_n}{2}\right)$$

and, by writing the moment equilibrium equation about the right end of the free body shown in Fig. 16.3.2(b),

$$M_{\text{neg}}(\text{right}) + M_{\text{pos}} = \frac{w_u L_2 L_n}{2}\left(\frac{L_n}{4}\right) + V_{\mathbb{C}}\left(\frac{L_n}{2}\right)$$

Equation (16.3.2) is arrived at by adding the two preceding equations and dividing by 2 on each side. Note that Eq. (16.3.2) may also be obtained, as shown in Fig. 16.3.2(c), by the superposition of the simple span uniform loading parabolic positive moment diagram over the trapezoidal negative moment diagram due to end moments.

ACI-13.6.2 uses the symbol M_0 to mean $w_u L_2 L_n^2/8$ and calls M_0 the *total factored static moment*. It states, "Absolute sum of positive and average negative factored moments in each direction shall not be less than" M_0; or

$$\frac{M_{\text{neg}}(\text{left}) + M_{\text{neg}}(\text{right})}{2} + M_{\text{pos}} \geq M_0 = \frac{w_u L_2 L_n^2}{8} \qquad \textbf{(16.3.3)}$$

in which

w_u = factored load per unit area

L_n = clear span in the direction moments are being determined, measured face to face of supports (ACI-13.6.2.5), but not less than $0.65L_1$

L_1 = span length in the direction moments are being determined, measured center to center of supports

L_2 = transverse span length, measured center to center of supports

Equations (16.3.1) and (16.3.2) are theoretically derived on the basis that M_{neg}(left), M_{pos}, and M_{neg}(right) occur simultaneously for the same live load pattern on the adjacent panels of the equivalent rigid frame defined in Fig. 16.2.1. If the live load is relatively heavy compared with dead load, then different live load patterns should be used to obtain the critical positive moment at midspan and the critical negative moments at the supports. In such a case, the "equal" sign in Eqs. (16.3.1) and (16.3.2) becomes the "greater" sign. This is the reason why ACI-13.6.2 states "absolute sum . . . shall not be *less* than" M_0 as the design requirement. The designer should keep this in mind when steel reinforcement is selected for positive and negative bending moment in two-way floors when the direct design method is used for gravity load. To avoid the use of excessively small values of M_0 in the case of short spans and large columns or column capitals, the clear span L_n to be used in Eq. (16.3.3) is not to be less than $0.65L_1$ (ACI-13.6.2.5).

When the limitations for using the direct design method are met, it is customary to divide the value of M_0 into M_{neg} and M_{pos}, if the restraints at each end of the span are identical; or into $[M_{neg}$ (left) $+ M_{neg}$ (right)$]/2$ and M_{pos} if the span end restraints are different. Then the moments M_{neg} (left), M_{neg} (right), and M_{pos} must be distributed transversely along the lines $1'–3'$, $2'–4'$, and $5–6$, respectively. This last distribution is a function of the relative flexural stiffness between the slab and the included beam.

Total Factored Static Moment in Flat Slabs. The ability of flat-slab floor systems to carry load has been substantiated by numerous tests of actual structures [1]. However, the amount of reinforcement used, say, in a typical interior panel, was less than what it should be to satisfy an analysis by statics, as is demonstrated in this section. This led to some controversy [19], but after studies by Westergaard and Slater [20], a provision was adopted (about 1921) into the code that a reduction of moment coefficient from the statically required value of 0.125 to 0.09 may be made. This reduction was not regarded as a violation of statics but it was used as a way of permitting an increase in the allowable unit stresses. The reduction, moreover, was applicable only to flat slabs that satisfied the limitations then specified in the code. Over the years these limitations had been liberalized, but at the same time the moment coefficient was raised to values closer to 0.125. The present ACI Code logically stipulates the use of the full statically required coefficient of 0.125.

The statical analysis of a typical interior panel was first made in 1914 by Nichols [19] and further developed later by Westergaard and others [10,20,21].

Consider the typical interior panel of a flat-slab floor subjected to a factored load of w_u per unit area, as shown in Fig. 16.3.3(a). The total load on the panel area (rectangle minus four quadrantal areas) is supported by the vertical shears at the four quandrantal arcs. Let M_{neg} and M_{pos} be the total negative and positive moments about a horizontal axis in the L_2 direction along the edges of $ABCD$ and EF, respectively. Then

$$\text{load on area } ABCDEF = \text{sum of reactions at arcs } AB \text{ and } CD$$
$$= w_u\left(\frac{L_1 L_2}{2} - \frac{\pi c^2}{8}\right)$$

Considering the half-panel $ABCDEF$ as a free body, recognizing that there is no shear at the edges BC, DE, EF, and FA, and taking moments about axis 1-1,

$$M_{neg} + M_{pos} + w_u\left(\frac{L_1 L_2}{2} - \frac{\pi c^2}{8}\right)\left(\frac{c}{\pi}\right) - \frac{w_u L_1 L_2}{2}\left(\frac{L_1}{4}\right) + \frac{w_u \pi c^2}{8}\left(\frac{2c}{3\pi}\right) = 0$$

Letting $M_0 = M_{neg} + M_{pos}$,

$$M_0 = \frac{1}{8}w_u L_2 L_1^2\left(1 - \frac{4c}{\pi L} + \frac{c^3}{3 L_2 L_1^2}\right) \approx \frac{1}{8}w_u L_2 L_1^2\left(1 - \frac{2c}{3 L_1}\right)^2 \quad (16.3.4)$$

Actually Eq. (16.3.4) may be more easily visualized by inspecting the equivalent interior span as shown in Fig. 16.3.3(b).

ACI-13.6.2.5 states that circular or regular polygon shaped supports shall be treated as square supports having the same area. For flat slabs,

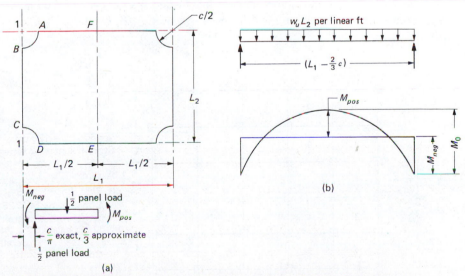

Figure 16.3.3 Statics of a typical interior panel in a flat slab floor system.

particularly with column capitals, the clear span L_n computed from using equivalent square supports should be compared with that indicated by Eq. (16.3.4), which is L_1 minus $2c/3$. In some cases the latter value is larger and should be used, consistent with the fact that ACI-13.6.2.2 does express its intent in an inequality.

Design Examples. In an effort to present, explain, and illustrate the design procedure for the three types of two-way floor systems, identified in this chapter as two-way slabs (with beams), flat slabs, and flat plates, it will be necessary to assume that preliminary dimensions and sizes of the slab (and drop, if any), beams, and columns (and column capitals, if any) are available. In the usual design processes, not only the preliminary sizes may need to be revised as they are found unsuitable, but also designs based on two or three different relative beam sizes (when used) to slab thickness should be made and compared. Preliminary data for the three types of two-way floor systems to be illustrated are as follows:

Data for Two-Way Slab (with Beams) Design Example. Figure 16.3.4 shows a two-way slab floor with a total area of 12,500 sq ft. It is divided into 25 panels with a panel size of 25 ft × 20 ft. Concrete strength is $f'_c = 3000$ psi and steel yield strength is $f_y = 40,000$ psi. Service live load is to be taken as 120 psf. Story height is 12 ft. The preliminary sizes are as follows: slab thickness is $6\frac{1}{2}$ in.; long beams are 14 × 28 in. overall; short beams are 12 × 24 in. overall; upper and lower columns are 15 × 15 in. The four kinds of panels (corner, long-sided edge, short-sided edge, and interior) are numbered 1, 2, 3, and 4 in Fig. 16.3.4.

EXAMPLE 16.3.1 For the two-way slab (with beams) design example, determine the total factored static moment in a loaded span in each of the four equivalent rigid frames whose widths are designated A, B, C, and D in Fig. 16.3.5.

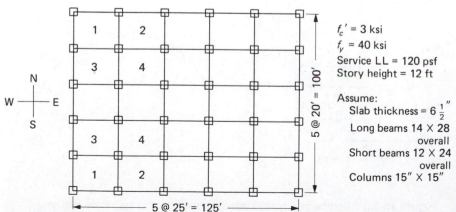

Figure 16.3.4   Floor plan in design example for two-way slab with beams.

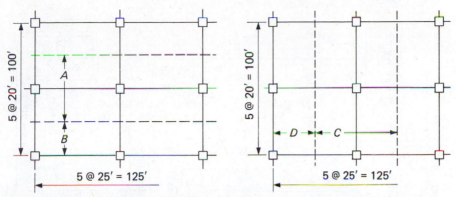

Figure 16.3.5 Equivalent rigid frame notations in the two-way slab (with beams) design example.

Solution: The factored load w_u per unit floor area is

$$w_u = w_D + w_L = 1.4(6.5)\left(\frac{150}{12}\right) + 1.7(120)$$

$$= 114 + 204 = 318 \text{ psf}$$

Note that for two-way slabs supported on beams the definition of clear span L_n is ambiguous. In ACI-13.6.2.5 it says ". . . shall extend from face to face of columns, capitals, brackets, or walls." but in ACI-9.0 L_n is defined as measured face-to-face of *beams* for two-way slabs supported on beams. The authors will conservatively use L_n measured to the face of beams.

for frame *A*, $M_0 = \frac{1}{8}w_u L_2 L_n^2 = \frac{1}{8}(0.318)(20)(24)^2 = 458$ ft-kips

for frame *B*, $M_0 = 229$ ft-kips

for frame *C*, $M_0 = \frac{1}{8}w_u L_2 L_n^2 = \frac{1}{8}(0.318)(25)(18.83)^2 = 352$ ft-kips

for frame *D*, $M_0 = 176$ ft-kips

Data for Flat Slab Design Example. Figure 16.3.6 shows a flat slab floor with a total area of 12,500 sq ft. It is divided into 25 panels with a panel size of 25 × 20 ft. Concrete strength is $f'_c = 3000$ psi and steel yield strength is $f_y = 40,000$ psi. Service live load is 120 psf. Story height is 10 ft. Exterior columns are 16 in. square and interior columns are 18 in. round. Edge beams are 14 × 24 in. overall. Thickness of slab is $7\frac{1}{2}$ in. outside of drop panel and $10\frac{1}{2}$ in. through the drop panel. Sizes of column capitals and drop panels are as shown in Fig. 16.3.6.

EXAMPLE 16.3.2 Compute the total factored static moment in the long and short directions of an interior panel in the flat-slab design example as shown in Fig. 16.3.6. Compare the results obtained by using Eqs. (16.3.3) and (16.3.4).

Solution: Neglecting the weight of the drop panel, the service dead load is $(150/12)(7.5) = 94$ psf; thus

$$w_u = 1.4w_D + 1.7w_L = 1.4(94) + 1.7(120) = 132 + 204 = 336 \text{ psf}$$

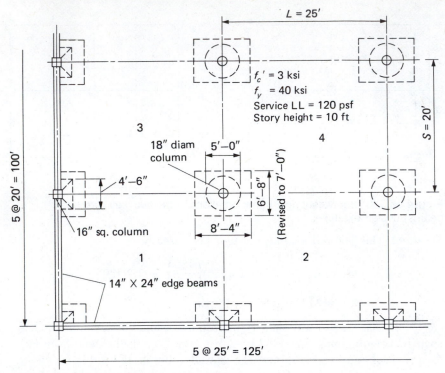

Figure 16.3.6 Flat slab design example.

Using Eq. (16.3.4),

$$M_0 = \frac{1}{8} w_u L_2 L_1^2 \left(1 - \frac{2c}{3L_1} \right)^2 = \frac{1}{8}(0.336)(20)(25)^2 \left[1 - \frac{2(5)}{3(25)} \right]^2 = 395 \text{ ft-kips}$$

(in long direction)

$$M_0 = \frac{1}{8} w_u L_2 L_1^2 \left(1 - \frac{2c}{3L_1} \right)^2 = \frac{1}{8}(0.336)(25)(20)^2 \left[1 - \frac{2(5)}{3(20)} \right]^2 = 292 \text{ ft-kips}$$

(in short direction)

The equivalent square area for the column capital (ACI-13.6.2.5) has its side equal to 4.43 ft; then, using Eq. (16.3.3), with L_n measured to the face of capital (i.e., equivalent square),

$$M_0 = \tfrac{1}{8} w_u L_2 L_n^2 = \tfrac{1}{8}(0.336)(20)(25 - 4.43)^2 = 356 \text{ ft-kips}$$

(in long direction)

$$M_0 = \tfrac{1}{8} w_u L_2 L_n^2 = \tfrac{1}{8}(0.336)(25)(20 - 4.43)^2 = 255 \text{ ft-kips}$$

(in short direction)

Insofar as flat slabs with column capitals are concerned, it appears that the larger values of 395 ft-kips and 292 ft-kips should be used because Eq. (16.3.4) is specially suitable; in particular, ACI-13.6.2.2 states that the total factored static moment shall *not be less than* that given by Eq. (16.3.3).

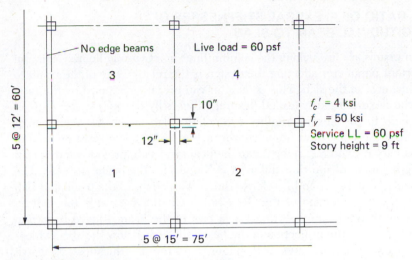

Figure 16.3.7 Flat plate design example.

Data for Flat Plate Design Example. Figure 16.3.7 shows a flat plate floor with a total area of 4500 sq ft. It is divided into 25 panels with a panel size of 15 × 12 ft. Concrete strength is $f'_c = 4000$ psi and steel yield strength is $f_y = 50,000$ psi. Service live load is 60 psf. Story height is 9 ft. All columns are rectangular, 12 in. in the long direction and 10 in. in the short direction. Preliminary slab thickness is set at $5\frac{1}{2}$ in. No edge beams are used along the exterior edges of the floor.

EXAMPLE 16.3.3 Compute the total factored static moment in the long and short directions of a typical panel in the flat plate design example as shown in Fig. 16.3.7.

Solution: The dead load for a $5\frac{1}{2}$ in. slab is

$$w_D = (5.5/12)(150) = 69 \text{ psf}$$

The factored load per unit area is

$$w_u = 1.4w_D + 1.7w_L = 1.4(69) + 1.7(60) = 96 + 102 = 198 \text{ psf}$$

Using Eq. (16.3.3),

$$M_0 = \tfrac{1}{8}(0.198)(12)(15 - 1)^2 = 58.2 \text{ ft-kips}$$

(in long direction)

$$M_0 = \tfrac{1}{8}(0.198)(15)(12 - 0.83)^2 = 46.3 \text{ ft-kips}$$

(in short direction)

These moment values obtained on the basis of the clear spans of 14 ft and 11 ft-2 in. are appropriate for flat plate floors.

16.4 RATIO OF FLEXURAL STIFFNESSES OF LONGITUDINAL BEAM TO SLAB

When beams are used along the column lines in a two-way floor system, an important parameter affecting the design is the relative size of the beam to the thickness of the slab. This parameter can best be measured by the ratio α of the flexural rigidity (called flexural stiffness by the ACI Code) $E_{cb}I_b$ of the beam to the flexural rigidity $E_{cs}I_s$ of the slab in the transverse cross section of the equivalent frame shown in Fig. 16.4.1. The separate moduli of elasticity E_{cb} and E_{cs}, referring to the beam and slab, provide for different strength concrete (and thus different E_c values) for the beam and slab. The moments of inertia I_b and I_s refer to the gross sections of the beam and slab within the cross section of Fig. 16.4.1(c). ACI-13.2.4 permits the slab on each side of the beam web to act as a part of the beam, this slab portion being limited to the projection of the beam above or below the slab, whichever is greater, but not greater than four times the slab thickness, as shown in Fig. 16.4.2. More accurately, the small portion of the slab already counted in the beam should not be used in I_s, but ACI permits the use of the total width of the equivalent frame in computing I_s. Thus,

$$\alpha = \frac{E_{cb}I_b}{E_{cs}I_s} \tag{16.4.1}$$

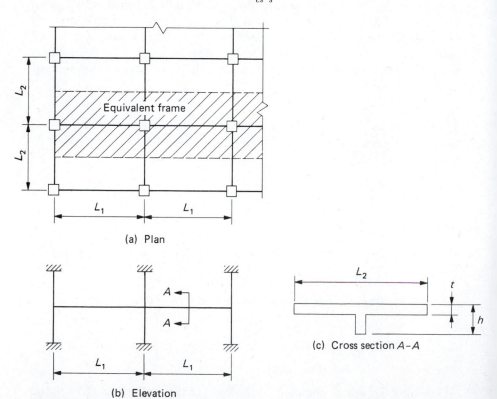

(a) Plan

(b) Elevation

(c) Cross section A–A

Figure 16.4.1 Plan, elevation, and cross section of equivalent frame in a two-way floor system.

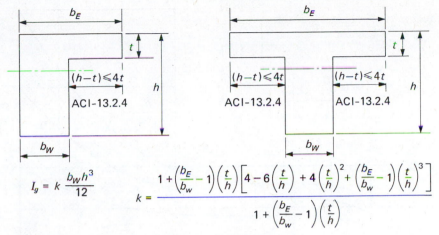

Figure 16.4.2 Moment of inertia of a flanged section.

The moment of inertia of a flanged beam section about its own centroidal axis (Fig. 16.4.2) may be shown to be

$$I_b = k\,\frac{b_w h^3}{12} \qquad (16.4.2a)$$

in which

$$k = \frac{1 + \left(\dfrac{b_E}{b_w} - 1\right)\left(\dfrac{t}{h}\right)\left[4 - 6\left(\dfrac{t}{h}\right) + 4\left(\dfrac{t}{h}\right)^2 + \left(\dfrac{b_E}{b_w} - 1\right)\left(\dfrac{t}{h}\right)^3\right]}{1 + \left(\dfrac{b_E}{b_w} - 1\right)\left(\dfrac{t}{h}\right)}$$

$$(16.4.2b)$$

where

h = overall beam depth
t = overall slab thickness
b_E = effective width of flange
b_w = width of web

Equation (16.4.2b) expresses the nondimensional constant k in terms of (b_E/b_w) and (t/h). Typical values of k are tabulated in Table 16.4.1 and three curves are plotted in Fig. 16.4.3. The values of k are about 1.4, 1.6,

Table 16.4.1 VALUES OF k IN TERMS OF (b_E/b_w) AND (t/h)
IN EQ. (16.4.2b)

b_E/b_w					t/h					
	0.1	0.2	0.3	0.4	0.5	0.6	0.7	0.8	0.9	1.0
2	1.222	1.328	1.366	1.372	1.375	1.396	1.454	1.565	1.743	2.000
3	1.407	1.564	1.605	1.608	1.625	1.694	1.844	2.098	2.477	3.000
4	1.564	1.744	1.777	1.781	1.825	1.956	2.212	2.621	3.209	4.000

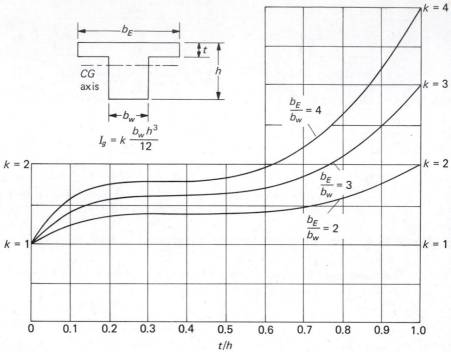

Figure 16.4.3 Values of k in terms of b_E/b_w and t/h.

and 1.8, respectively, for b_E/b_w values of 2, 3, and 4, when t/h values are between 0.2 and 0.5. Thus

$$k \approx 1.0 + 0.2\left(\frac{b_E}{b_w}\right) \quad \text{for} \quad 2 < \frac{b_E}{b_w} < 4 \quad \text{and} \quad 0.2 < \frac{t}{h} < 0.5 \quad \textbf{(16.4.2c)}$$

EXAMPLE 16.4.1 For the two-way slab (with beams) design example described in Sec. 16.3, compute the ratio α of the flexural stiffness of the longitudinal beam to that of the slab in the equivalent rigid frame, for all the beams around panels 1, 2, 3, and 4 in Fig. 16.4.4.

Solution: (a) *B1-B2.* Referring to Fig. 16.4.4, the effective width b_E for B1-B2 is the smaller of $14 + 2(21.5) = 57$ in. or $14 + 8(6.5) = 66$ in.; thus $b_E = 57$ in. Using Eq. (16.4.2b),

$$\frac{b_E}{b_w} = \frac{57}{14} = 4.07, \qquad \frac{t}{h} = \frac{6.5}{28} = 0.232$$

$$k = 1.774, \qquad I_b = 1.774\,\frac{14(28)^3}{12} = 45,400 \text{ in.}^4$$

A slightly higher value of k would have been obtained using Eq. (16.4.2c). Using Eq. (16.4.1), where $E_{cb} = E_{cs}$,

$$I_s = \tfrac{1}{12}(240)(6.5)^3 = 5490 \text{ in.}^4, \qquad \alpha = \frac{E_{cb}I_b}{E_{cs}I_s} = \frac{45,400}{5490} = 8.27$$

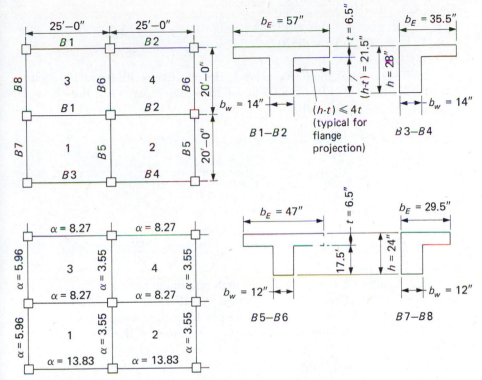

Figure 16.4.4   Computation of α values in Example 16.4.1.

(b) *B3-B4*. Referring to Fig. 16.4.4, the effective width b_E for *B3-B4* is the smaller of $14 + 21.5 = 35.5$ in. or $14 + 4(6.5) = 40$ in.; thus $b_E = 35.5$ in. Using Eq. (16.4.2b),

$$\frac{b_E}{b_w} = \frac{35.5}{14} = 2.54, \qquad \frac{t}{h} = \frac{6.5}{28} = 0.232$$

$$k = 1.484, \qquad I_b = 1.484 \frac{14(28)^3}{12} = 38,000 \text{ in.}^4$$

Using Eq. (16.4.1),

$$I_s = \tfrac{1}{12}(120)(6.5)^3 = 2745 \text{ in.}^4, \qquad \alpha = \frac{E_{cb}I_b}{E_{cs}I_s} = \frac{38,000}{2745} = 13.83$$

(c) *B5-B6*. Referring to Fig. 16.4.4, the effective width b_E for *B5-B6* is the smaller of $12 + 2(17.5) = 47$ in. or $12 + 8(6.5) = 64$ in.; thus $b_E = 47$ in. Using Eq. (16.4.2b),

$$\frac{b_E}{b_w} = \frac{47}{12} = 3.92, \qquad \frac{t}{h} = \frac{6.5}{24} = 0.271$$

$$k = 1.762, \qquad I_b = 1.762 \frac{12(24)^3}{12} = 24,400 \text{ in.}^4$$

Using Eq. (16.4.1),

$$I_s = \tfrac{1}{12}(300)(6.5)^3 = 6870 \text{ in.}^4, \qquad \alpha = \frac{E_{cb}I_b}{E_{cs}I_s} = \frac{24,400}{6870} = 3.55$$

(d) *B7-B8*. Referring to Fig. 16.4.4, the effective width b_E for *B7-B8* is the smaller of $12 + 17.5 = 29.5$ in. or $12 + 4(6.5) = 38$ in.; thus $b_E = 29.5$ in. Using Eq. (16.4.2b),

$$\frac{b_E}{b_w} = \frac{29.5}{12} = 2.46, \qquad \frac{t}{h} = 0.271$$

$$k = 1.480, \qquad I_b = 1.480 \frac{12(24)^3}{12} = 20,500 \text{ in.}^4$$

Using Eq. (16.4.1),

$$I_s = \tfrac{1}{12}(150)(6.5)^3 = 3435 \text{ in.}^4, \qquad \alpha = \frac{E_{cb}I_b}{E_{cs}I_s} = \frac{20,500}{3435} = 5.96$$

The resulting α values for *B1* through *B8* around panels 1, 2, 3, and 4 are shown in Fig. 16.4.4. For this design, the α values vary between 3.55 and 13.83; thus the equivalent rigid frames have their substantial portion along or close to the column lines, even though their widths vary from 10 to 25 ft.

16.5 MINIMUM SLAB THICKNESS FOR DEFLECTION CONTROL

Control of deflection in two-way roof and floor systems is dealt with in ACI-9.5.3. If deflections are computed according to ACI-9.5.3.4, then ACI-Table 9.5(b) gives the maximum permissible computed deflection. To compute the deflections, the designer may apply one of the various methods proposed [59–63,75,76]; however, no specific method is endorsed by the ACI Code or Commentary. Computation of deflections in two-way floor systems is outside the scope of this book.

To aid the designer, semiempirical equations are prescribed by the ACI Code (ACI-9.5.3.1) to give a minimum thickness for use in the design of two-way floor systems in order to control deflections. These equations, applicable to situations wherein the ratio of long to short span does not exceed 2, give a minimum thickness that from experience has been found to be satisfactory. Five parameters enter into these equations; they are (1) L_n, the longer clear span of a panel, (2) L_n/S_n (ACI calls it β), the ratio of longer clear span to shorter clear span of a panel, (3) f_y, the yield strength of the reinforcement, (4) α_m, average of the four α values for the beams around the panel, as described in Sec. 16.4, and (5) β_s, the ratio of length of continuous edges to total perimeter of the panel. In terms of these parameters, ACI-9.5.3.1 requires for two-way floor systems the following:

$$\min t = \frac{L_n(0.8 + 0.2f_y/40,000)}{36 - 2.5(1 - \beta_s)(1 + L_n/S_n) + 5\alpha_m L_n/S_n}, \qquad \text{ACI Formula (9-11)}$$

$$(16.5.1)$$

or

$$\min t = \frac{L_n(0.8 + 0.2f_y/40{,}000)}{36 + 5(1 + \beta_s)L_n/S_n}, \quad \text{ACI Formula (9-12)} \quad \textbf{(16.5.2)}$$

but it need not be more than

$$t = \frac{L_n(0.8 + 0.2f_y/40{,}000)}{36}, \quad \text{ACI Formula (9-13)} \quad \textbf{(16.5.3)}$$

The rationale for the ACI formulas may be examined by observing the plot of the formulas in Fig. 16.5.1. The quantity obtained is the long direction clear span to thickness ratio L_n/t, similar to the span to depth ratio limitations that are used for beams (see Secs. 14.10 and 14.11).

Figure 16.5.1 includes the full feasible range of parameters: (1) the panel proportions L_n/S_n ranging from square to two-to-one rectangular, (2) the β_s from a maximum of 1.0 for an interior panel continuous over all four edges to a minimum of 0.5 for a corner panel continuous over only two sides, and (3) the α_m ranging from zero with no edge beams to 2.5 or so for very stiff edge beams. The parameter f_y is included in the footnote that the allowable L_n/t must be divided by 1.1 for $f_y = 60{,}000$ psi, or by 1.05 for

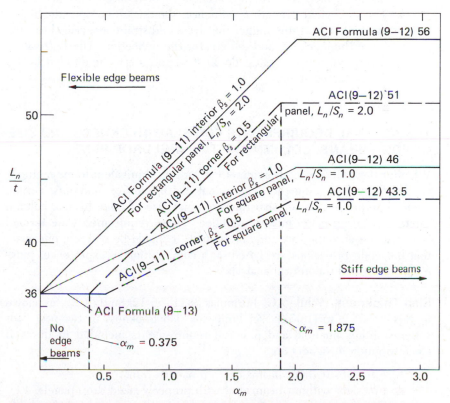

Figure 16.5.1 ACI minimum slab thickness formulas (for Grade 40 steel). For Grade 60, divide L_n/t by 1.1; for Grade 50, divide L_n/t by 1.05 (adapted from Ref. 23).

$f_y = 50,000$ psi, which means that the required minimum slab thickness is larger for a given L_n when f_y exceeds 40,000 psi. The parameter L_n is included in the diagram of Fig. 16.5.1 for maximum L_n/t that may be allowed.

The following logical results are indicated from a study of Fig. 16.5.1:

1. The stiffness of edge beams is the predominant factor; the no-edge-beam condition requires the thickest slab ($L_n/t = 36$ max), from which the thickness required decreases until the edge beams provide an average stiffness corresponding to $\alpha_m = 1.50 + 0.50 S_n/L_n + 0.50\beta_s(1 - S_n/L_n)$, which has a maximum value of 2.0 for combinations of variables within the applicable range.

2. The proportion β_s of the perimeter of the panel that is continuous affects the required thickness; an interior panel with β_s at maximum of 1.0 allows a higher L_n/t value, 5% higher for a square panel and 10% higher for a two-to-one rectangular panel.

3. The ratio L_n/S_n of the plan dimensions of the panel affects the thickness by requiring the thickest slab (lowest L_n/t) for the square panel. As the panel becomes more rectangular, more load is transferred to the supports in the short direction; thus with the shorter span becoming more effective the thickness can be reduced (L_n/t increased).

4. The steel yield strength f_y influences the thickness, since the higher the yield stress, the higher the stress and strain at service load, the more the cracking, and the greater the deflection. The L_n/t values of Fig. 16.5.1 are for Grade 40 steel and are to be divided by 1.1 for Grade 60 steel.

16.6 NOMINAL REQUIREMENTS FOR SLAB THICKNESS AND SIZE OF EDGE BEAMS, COLUMN CAPITAL, AND DROP PANEL

Whether the ACI "direct design method" or the "equivalent frame method" is used for determining the longitudinal distribution of moments, certain nominal requirements for slab thickness and size of edge beams, column capital, and drop panel must be fulfilled. These requirements are termed "nominal" because they are code prescribed. It should be realized, of course, that the code provisions are based on a combination of experience, judgment, tests, and theoretical analysis.

Slab Thickness. While ACI Formulas (9-11), (9-12), and (9-13), as shown in Eqs. (16.5.1) to (16.5.3), set limits for slab thickness in two-way floor systems, other nominal and practical requirements included in the ACI Code to guide designers are:

1. For slabs without beams or drop panels, 5 in.
2. For slabs without beams but with properly sized drop panels, 4 in.
3. For slabs with beams on all four sides with α_m equal to at least 2.0, $3\frac{1}{2}$ in.

Edge Beams. When edge beams are not used, or if the edge beam is so small that α is less than 0.80, ACI-9.5.3.3 states that the requirement for slab thickness expressed by Eqs. (16.5.1) to (16.5.3) must be increased by 10% in the panel having the discontinuous edge.

Column Capital. Used in flat slab construction, the column capital (Fig. 16.6.1) is an enlargement of the top of the column as it meets the floor slab or drop panel. Since no beams are used, the purpose of the capital is to gain increased perimeter around the column to transmit shear from the floor loading and to provide increasing thickness as the perimeter decreases near the column. Assuming a maximum 45° line for distribution of the shear into the column, ACI-13.1.2 requires that the effective column capital for strength considerations be within the largest circular cone, right pyramid, or tapered wedge with a 90° vertex that can be included within the outlines of the actual supporting element (see Fig. 16.6.1). The diameter of the column capital is usually about 20 to 25% of the average span length between columns.

Drop Panel. Used in flat slab and flat plate construction, the drop panel (Fig. 16.1.2) is an area of increased slab thickness surrounding the column. When drop panels extend from the centerline of supports a minimum distance of one-sixth of the span length measured from center to center in each direction, and when the projection below the slab is at least one-fourth of the slab thickness outside of the drop, ACI-9.5.3.2 permits the minimum slab thickness required by Eqs. (16.5.1) to (16.5.3) to be reduced by 10%. For determining reinforcement, ACI-13.4.7.3 requires that the thickness of the drop below the slab be assumed at a value no larger than one fourth of the distance between the edge of the drop panel and the edge of the column capital. Because of this requirement, there is little reason to use a drop panel of larger thickness.

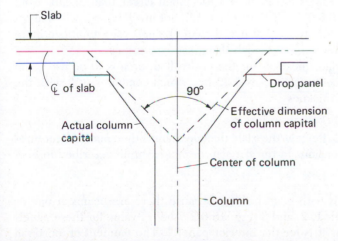

Figure 16.6.1   Effective dimension of column capital.

EXAMPLE 16.6.1 For the two-way slab (with beams) design example described in Sec. 16.3, determine the minimum thickness requirement for deflection control; and compare it with the preliminary thickness of $6\frac{1}{2}$ in.

Solution: The average ratios α_m for panels 1, 2, 3, and 4 may be computed from the α values shown in Fig. 16.4.4; thus

$$\alpha_m \text{ for panel } 1 = \tfrac{1}{4}(5.96 + 8.27 + 3.55 + 13.83) = 7.90$$

$$\alpha_m \text{ for panel } 2 = \tfrac{1}{4}(3.55 + 8.27 + 3.55 + 13.83) = 7.30$$

$$\alpha_m \text{ for panel } 3 = \tfrac{1}{4}(5.96 + 8.27 + 3.55 + 8.27) = 6.51$$

$$\alpha_m \text{ for panel } 4 = \tfrac{1}{4}(3.55 + 8.27 + 3.55 + 8.27) = 5.91$$

The continuity fractions β_s of the panel perimeters are

$$\beta_s \text{ for panel } 1 = 0.50$$

$$\beta_s \text{ for panel } 2 = \frac{65}{90}$$

$$\beta_s \text{ for panel } 3 = \frac{70}{90}$$

$$\beta_s \text{ for panel } 4 = 1.0$$

Using $L_n = 24$ ft, $S_n = 18.83$ ft, and $f_y = 40,000$ psi, the minimum slab thickness requirements are as follows:

PANEL	1	2	3	4
Eq. (16.5.1), minimum thickness	3.45	3.56	3.78	3.91
Eq. (16.5.2), minimum thickness	6.32	6.13	6.09	5.91
Eq. (16.5.3), need not be more than	8	8	8	8

Note that the α_m values for all panels are much larger than 2; thus from Fig. 16.5.1, Eq. (16.5.2) [ACI Formula (9-12)] not involving α_m will be most critical. Figure 16.5.1 shows that it will rarely be necessary to utilize more than *one* of the three minimum slab thickness expressions.

If a uniform slab thickness for the entire floor area is to be used, the minimum for deflection control is 6.32 in., which compares well with the $6\frac{1}{2}$-in. preliminary thickness.

EXAMPLE 16.6.2 Review the slab thickness and other nominal requirements for the dimensions in the flat slab design example described in Sec. 16.3.

Solution: (a) Panels with edge beams. Because there are beams at one or more sides of panels 1, 2, and 3 (Fig. 16.6.2), the α_m value for these panels will not be zero, as it is for the interior panel 4. The moment of inertia of the edge beam section shown in Fig. 16.6.2(b) is 22,900 in.[4] Thus the α value for the long edge beam is

$$\alpha = \frac{I_b}{I_s} = \frac{22{,}900}{120(7.5)^3/12} = \frac{22{,}900}{4220} = 5.42$$

and for the short edge beam, it is

$$\alpha = \frac{I_b}{I_s} = \frac{22{,}900}{150(7.5)^3/12} = \frac{22{,}900}{5270} = 4.34$$

These α values are entered on Fig. 16.6.2(a). The average α (i.e., α_m) for panels 1, 2, and 3 are, respectively, 2.44, 1.36, and 1.08.

An examination of Fig. 16.5.1 showing the ACI minimum thickness

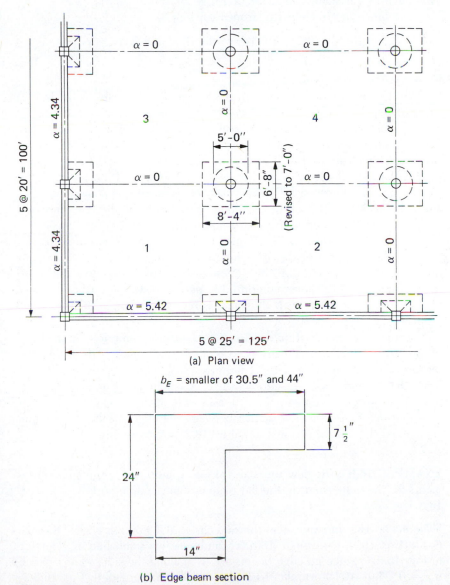

(a) Plan view

(b) Edge beam section

Figure 16.6.2   Computation of α values in Example 16.6.2.

equations reveals that ACI Formula (9-12) [Eq. (16.5.2)] governs for panel 1, whereas Formula (9-11) [Eq. (16.5.1)] governs for panels 2 and 3.

(b) Interior panel. Panel 4 having no beams along its perimeter has $\alpha_m = 0$. Figure 16.5.1 shows for $\alpha_m = 0$ ACI Formula (9-13) [Eq. (16.5.3)] controls.

(c) Minimum slab thickness. Assuming the entire slab is to have a uniform thickness, ACI Formula (9-13) applied to the interior panel gives $L_n/t = 36$, a lower value than obtained for panels 1, 2, or 3. This, of course, is the primary reason for using edge beams—that is, to stiffen the edges of exterior panels and thus make the interior panel govern the slab thickness. The longest clear span L_n is $(25 - 4.43) = 20.57$ ft.

The minimum thickness is, from Eq. (16.5.3),

$$\min t = \frac{L_n(0.8 + 0.2f_y/40,000)}{36} = \frac{20.57(12)(0.8 + 0.2)}{36} = 6.86 \text{ in.}$$

Thus the $7\frac{1}{2}$-in. slab as used is ample.

(d) The minimum slab thickness of 4 in. for flat slabs with drop is satisfied by the $7\frac{1}{2}$-in. slab used here.

(e) The ratio α of edge-beam stiffness to slab stiffness is 4.34 in the short direction and 5.42 in the long direction, as shown in Fig. 16.6.2. These α values are well above 0.8. Thus the minimum slab thickness requirements of ACI Formula (9-12) for panel 1 and ACI Formula (9-11) for panels 2 and 3 need not be increased by 10%. When α_m is small enough so that ACI Formula (9-13) would govern, then that value would be the one increased by 10%, as is illustrated in the next example, Example 16.6.3.

(f) Since the size of the drop panel (see Fig. 16.6.2) is equal in each direction to one-third of the span length, and the 3-in. projection of the drop below the slab is more than one-fourth of the $7\frac{1}{2}$-in. slab thickness, the minimum slab thickness computed in part (a) could be further reduced by 10%. Thus the $7\frac{1}{2}$-in. slab thickness is more than ample; $6\frac{1}{2}$-in. should probably have been used.

(g) Reinforcement within the drop panel must be computed on the basis of the $10\frac{1}{2}$-in. thickness actually used or $7\frac{1}{2}$-in. plus one-fourth of the projection of the drop beyond the column capital, whichever is smaller. In order that the full 3-in. projection of the drop below the $7\frac{1}{2}$ in. slab is usable in computing reinforcement, the 6 ft–8 in. side of the drop is revised to 7 ft so that one-fourth of the distance between the edges of the 5-ft column capital and the 7-ft drop is just equal to $(10.5 - 7.5) = 3$ in.

EXAMPLE 16.6.3 Review the slab thickness and other nominal requirements for the dimensions in the flat plate design example described in Sec. 16.3.

Solution: (a) The minimum slab thickness from ACI Formulas (9-11) through (9-13) [Eqs. (16.5.1) through (16.5.3)] can be easily established by referring to Fig. 16.5.1. Since no edge beams are used along the sides of any panel, $\alpha_m = 0$. Thus ACI Formula (9-13) governs. From the data in Fig. 16.3.7, the longest clear span L_n is $(15 - 1) = 14$ ft.

$$\min t = \frac{L_n(0.8 + 0.2f_y/40{,}000)}{36}$$

$$= \frac{14(1.05)(12)}{36} = 4.90 \text{ in.} < 5\tfrac{1}{2} \text{ in. used} \qquad\qquad \text{OK}$$

(b) The minimum slab thickness of 5 in. for slabs without beams or drop panels is satisfied by the $5\tfrac{1}{2}$-in. slab used here.

(c) Since no edge beams are used, $\alpha = E_{cb}I_b/(E_{cs}I_s) = 0$, which is less than 0.8. Thus the minimum slab thickness becomes 1.1 times the controlling 4.90 in. computed from ACI Formula (9-13) [Eq. (16.5.3)] in part (a); or

$$\min t = 1.10(4.90) = 5.39 \text{ in.} < 5\tfrac{1}{2} \text{ in. used}$$

Thus the $5\tfrac{1}{2}$-in. slab is appropriate.

16.7 LIMITATIONS OF DIRECT DESIGN METHOD

Over the years the use of two-way floor systems has been extended from one-story or low-rise to medium or high-rise buildings. For the common cases of one-story or low-rise buildings, lateral load (wind or earthquake) is of lesser concern; thus most of the ACI Code refers only to gravity load (dead and live uniform load). In particular when the dimensions of the floor system are quite regular and when the live load is not excessively large compared to the dead load, the use of a set of prescribed coefficients to distribute longitudinally the total factored static moment M_0 seems reasonable. As shown in Figs. 16.3.1 and 16.3.2, for each clear span in the equivalent rigid frame, the equation

$$\frac{M_{\text{neg}} \text{ (left)} + M_{\text{neg}} \text{ (right)}}{2} + M_{\text{pos}} \geq \left[M_0 = \frac{w_u L_2 L_n^2}{8} \right] \qquad [16.3.3]$$

is to be satisfied.

In order that the designer may use the direct design method, in which a set of prescribed coefficients give the negative end moments and the positive moment within the span of the equivalent rigid frame, ACI-13.6.1 imposes the following limitations:

1. There is a minimum of three continuous spans in each direction.
2. Panels must be rectangular with the ratio of longer to shorter span within a panel not greater than 2.0.
3. The successive span lengths (center-to-center of supports) in each direction do not differ by more than one-third of the longer span.
4. Columns are not offset more than 10% of the span in the direction of the offset.
5. The load is due to gravity only and is uniformly distributed over an entire panel, and the service live load does not exceed 3 times the service dead load.
6. The relative stiffness ratio of L_1^2/α_1 to L_2^2/α_2 must lie between 0.2 and 5.0, where α is the ratio of the flexural stiffness of the included beam to that of the slab.

Though the design of two-way floor systems is to a large extent empirical, the ACI limitations conform to the experimental results that are available [11–16] and to many years of experience with slabs in actual structures. The "direct design method" can also be used when it can be demonstrated that variations from any of the six limitations will still produce a slab system that satisfies the conditions of equilibrium and geometric compatibility and provides adequate strength and serviceability. Van Buren [52] has provided such an analysis for staggered columns in flat plates.

EXAMPLE 16.7.1 Show that for the two-way slab (with beams) design example described in Sec. 16.3 the six limitations of the direct design method are satisfied.

Solution: The first four limitations are satisfied by inspection. For the fifth limitation,

$$\text{service dead load } w_D = 6.5\left(\frac{150}{12}\right) = 81 \text{ psf}$$

$$\text{service live load } w_L = 120 \text{ psf}$$

$$\frac{w_L}{w_D} = \frac{120}{81} < 3 \qquad\qquad \text{OK}$$

For the sixth limitation, referring to Fig. 16.4.4 and taking L_1 and L_2 in the long and short directions, respectively,

Panel 1,
$$\frac{L_1^2}{\alpha_1} = \frac{625}{0.5(13.83 + 8.27)} = 56.6$$

$$\frac{L_2^2}{\alpha_2} = \frac{400}{0.5(5.96 + 3.55)} = 84.0$$

Panel 2,
$$\frac{L_1^2}{\alpha_1} = \frac{625}{0.5(13.83 + 8.27)} = 56.6$$

$$\frac{L_2^2}{\alpha_2} = \frac{400}{3.55} = 112.7$$

Panel 3,
$$\frac{L_1^2}{\alpha_1} = \frac{625}{8.27} = 75.6$$

$$\frac{L_2^2}{\alpha_2} = \frac{400}{0.5(5.96 + 3.55)} = 84.0$$

Panel 4,
$$\frac{L_1^2}{\alpha_1} = \frac{625}{8.27} = 75.6$$

$$\frac{L_2^2}{\alpha_2} = \frac{400}{3.55} = 112.7$$

All ratios of L_1^2/α_1 to L_2^2/α_2 lie between 0.2 and 5.

16.8 DIRECT DESIGN METHOD—LONGITUDINAL DISTRIBUTION OF MOMENTS

In the "direct design method," moment curves in the direction of span length need not be computed by an elastic analysis (such as the moment distribution method) of the equivalent rigid frame subjected to various pattern loadings; instead they are nominally defined for regular situations, with additional prescribed adjustments for pattern loading effects.

Figure 16.8.1 shows the longitudinal moment diagram for the typical interior span of the equivalent rigid frame in a two-way floor system, as prescribed by ACI-13.6.3.2. Later in Sec. 16.12, the positive moment $0.35M_0$ or the negative moment $0.65M_0$ is to be distributed transversely to the slab having width L_2 and to the included beam (if any) having clear span L_n. Note that

$$M_0 = \tfrac{1}{8}w_u L_2 L_n^2 \qquad\qquad [16.3.3]$$

For a span that is completely fixed at both ends, the negative moment at the fixed end is twice as large as the positive moment at midspan. For a typical interior span satisfying the limitations for the direct design method, the specified negative moment of $0.65M_0$ is a little less than twice the specified positive moment of $0.35M_0$, which is fairly reasonable because the restraining effect of the columns and adjacent panels is definitely less than that of a completely fixed-ended beam.

For the exterior span, ACI-13.6.3.3 provides the longitudinal moment diagram for each of the five categories as described in Fig. 16.8.2. Upon examination of these diagrams, one sees that the negative moment at the exterior support increases from 0 to $0.65M_0$, the positive moment within the span decreases from $0.63M_0$ to $0.35M_0$, and the negative moment at the interior support decreases from $0.75M_0$ to $0.65M_0$, all gradually as the restraint at the exterior support increases from the case of a masonry wall support to that of a reinforced concrete wall built monolithically with the slab. ACI Commentary-13.6.3.3 states that high positive moments are purposely assigned into the span since design for exterior negative moment will be governed by minimum reinforcement to control cracking.

Regarding the ACI Code suggested moment diagrams of Figs. 16.8.1 and 16.8.2, ACI-13.6.7 permits these moments to be modified by 10% pro-

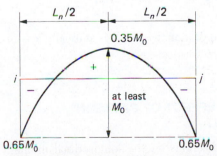

Figure 16.8.1 Longitudinal moment diagram for interior span.

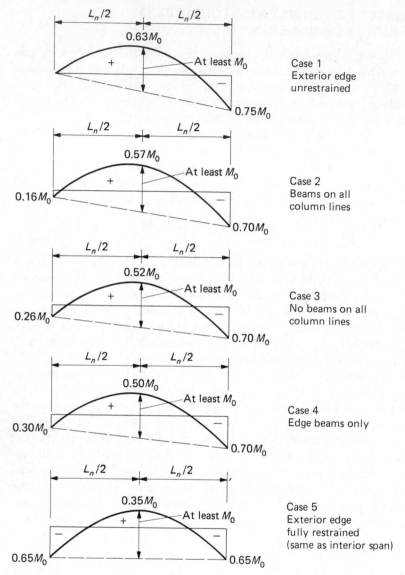

Figure 16.8.2 Longitudinal moment diagram for exterior span.

vided the total factored static moment M_0 for the panel is statically accommodated.

16.9 DIRECT DESIGN METHOD—EFFECT OF PATTERN LOADINGS ON POSITIVE MOMENT

To understand the effect of pattern loadings on the longitudinal moment values in multiple panel two-way floor systems, it is convenient to review some aspects of the continuity analysis of the usual column-beam type of rigid frames discussed earlier in Chap. 7. Some of the findings, which might be visualized by those knowledgeable in influence lines and maximum moment envelopes due to dead and live load combinations, are as follows: (1) the higher the ratio of column stiffness to beam stiffness, the smaller the effect of pattern loadings, because the ends of the span are closer to the fixed condition and less effect is exerted on the span by loading patterns on adjacent spans; (2) the lower the ratio of dead load to live load, the larger the effect of pattern loadings, because dead load exists constantly on all spans and the pattern is related to live load only; and (3) maximum negative moments at supports are less affected by pattern loadings than maximum positive moments within the span.

Based on studies by Jirsa, Sozen, and Siess [24], it is recommended that in order to limit the increase in positive moment within the two-way floor system caused by pattern loadings to such a tolerance (33% according to ACI Commentary-13.6.10) that the moment values in Fig. 16.8.2 (ACI-13.6.3) may be used, the ratio of column stiffness to slab stiffness must have at least the values shown in Table 16.9.1 (from ACI-Table 13.6.10). If the ratio α_c of column stiffness to slab and beam stiffness is less than the minimum value α_{min} shown in Table 16.9.1, then the positive moments in Figs. 16.8.1 and 16.8.2 must be multiplied by the coefficient δ_s given by the following formula (ACI Formula 13-5),

Table 16.9.1 MINIMUM RATIO α_{min} OF COLUMN STIFFNESS TO COMBINED BEAM AND SLAB STIFFNESS TO QUALIFY FOR POSITIVE MOMENT MULTIPLIER $\delta_s = 1$ (α = RATIO OF BEAM TO SLAB STIFFNESS)[a]

$\beta_a = \dfrac{DL}{LL}$	ASPECT RATIO L_2/L_1	RATIO OF BEAM TO SLAB STIFFNESS, α				
		0	0.5	1.0	2.0	4.0
2.00	0.50–2.00	0	0	0	0	0
1.00	0.50	0.6	0	0	0	0
	0.80	0.7	0	0	0	0
	1.00	0.7	0.1	0	0	0
	1.25	0.8	0.4	0	0	0
	2.00	1.2	0.5	0.2	0	0
0.50	0.50	1.3	0.3	0	0	0
	0.80	1.5	0.5	0.2	0	0
	1.00	1.6	0.6	0.2	0	0
	1.25	1.9	1.0	0.5	0	0
	2.00	4.9	1.6	0.8	0.3	0
0.33	0.50	1.8	0.5	0.1	0	0
	0.80	2.0	0.9	0.3	0	0
	1.00	2.3	0.9	0.4	0	0
	1.25	2.8	1.5	0.8	0.2	0
	2.00	13.0	2.6	1.2	0.5	0.3

[a]From ACI-Table 13.6.10.

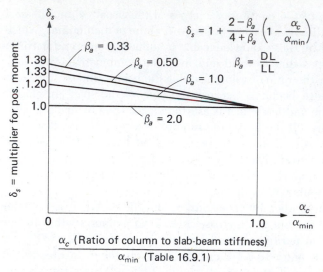

Figure 16.9.1   Direct design method—effect of pattern loadings.

$$\delta_s = 1 + \frac{2 - \beta_a}{4 + \beta_a}\left(1 - \frac{\alpha_c}{\alpha_{\min}}\right) \qquad (16.9.1)$$

in which

> β_a = ratio of service dead load to service live load
> α_c = ratio of column stiffness to slab and beam stiffness
> $\quad = \Sigma K_c/(\Sigma K_s + \Sigma K_b)$
> $\alpha_{\min}$ = value shown in Table 16.9.1

For the half column strip parallel to an exterior edge, it is conservative, and therefore permissible to use the value of α_c computed for the adjacent interior column if it is equal in size to the exterior column. The graph shown in Fig. 16.9.1 is not so much to save the arithmetic involved in Eq. (16.9.1) as to show the variation of δ_s with respect to $\alpha_c/\alpha_{\min}$ and β_a.

16.10 DIRECT DESIGN METHOD—PROCEDURE FOR COMPUTATION OF LONGITUDINAL MOMENTS

The background explanation for the distribution of the total static moment M_0 in the longitudinal direction and for the modification of positive moment due to the effect of pattern loadings has been presented in the two preceding sections. Utilizing this information the procedure for the computation of longitudinal moments by the "direct design method" may be summarized as follows:

1. Check if the six limitations for the "direct design method" listed in Sec. 16.7 are satisfied.

2. Compute the total static moment $M_0 = w_u L_2 L_n^2/8$ as stated by Eq. (16.3.2). Note that L_n is not to be taken less than $0.65L_1$.

3. Compute the slab stiffness

$$K_s = \frac{4EI_s}{L_1}, \qquad I_s = \sum L_2 \left(\frac{t^3}{12} \right)$$

4. Compute the moment of inertia I_b of the longitudinal beam (if any) (ACI-13.2.4) and its flexural stiffness by

$$K_b = \frac{4E_{cb}I_b}{L_1}$$

5. Compute the column stiffness

$$\sum K_c = K_{c1} + K_{c2} = \frac{4EI_{c1}}{L_{c1}} + \frac{4EI_{c2}}{L_{c2}}$$

6. Compute the ratio α of the longitudinal beam (ACI-13.2.4) stiffness to that of the slab by $\alpha = (E_{cb}I_b)/(E_{cs}I_s)$ and with the value of $\beta_a = $ (service DL/service LL), obtain the value of α_{min} from Table 16.9.1 or ACI-Table 13.6.10.

7. Compute for the exterior and interior spans the ratio α_c of the flexural stiffnesses of the upper and lower columns to the combined flexural stiffnesses of the slab and beam (if any) on both sides of the columns by the following:

$$\alpha_c = \frac{K_{c1} + K_{c2}}{\sum K_s + \sum K_b}$$

8. If the value of α_c computed in step 7 does not reach the α_{min} shown in Table 16.9.1, the positive moments M_{pos} must be increased for pattern loading effects by the multiplier δ_s, Eq. (16.9.1). If α_c/α_{min} exceeds one, $\delta_s = 1$ (i.e., no pattern loading effect).

9. Obtain the three critical ordinates on the longitudinal moment diagrams for the exterior and interior spans using Figs. 16.8.1 and 16.8.2, but multiplying the positive moments by the δ_s for that span if δ_s is larger than 1.

EXAMPLE 16.10.1 For the two-way slab (with beams) design example described in Sec. 16.3, determine the longitudinal moments in frames A, B, C, and D, as shown in Figs. 16.3.4 and 16.10.1.

Solution: (a) Check the six limitations for the direct design method. These limitations have been checked in Example 16.7.1.

(b) Total factored static moment M_0. The total factored static moments M_0 for the equivalent rigid frames A, B, C, and D have been computed in Example 16.3.1; they are

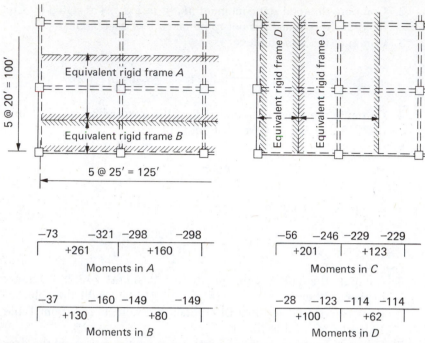

Figure 16.10.1 Longitudinal moments for two-way slab (with beams) design example.

$$M_0 \text{ (frame } A) = 458 \text{ ft-kips}$$
$$M_0 \text{ (frame } B) = 229 \text{ ft-kips}$$
$$M_0 \text{ (frame } C) = 352 \text{ ft-kips}$$
$$M_0 \text{ (frame } D) = 176 \text{ ft-kips}$$

(c) Compute the slab stiffness, beam stiffness, and column stiffness. Since all columns are of the same size, 15 by 15 in. and 12 ft long,

$$K_{c1} = K_{c2} = \frac{4E(15)(15)^3/12}{144} = 117E$$

For frame A,

$$K_s = \frac{4E(240)(6.5)^3/12}{300} = 73.2E$$

$$K_b = \frac{4EI_b \text{ of B1-B2(Example 16.4.1)}}{300} = \frac{4E(45,400)}{300} = 605E$$

$$\alpha = \frac{K_b}{K_s} = \frac{605E}{73.2E} = 8.27 \quad \text{(same as in Fig. 16.4.4)}$$

$$\alpha_c = \frac{234E}{605E + 73.2E} = 0.344 \quad \text{(at exterior column)}$$

$$\alpha_c = 0.177 \quad \text{(at interior column)}$$

For frame B,

$$K_s = 36.6E$$

$$K_b = \frac{4EI_b \text{ of } B3\text{-}B4 \text{ (Example 16.4.1)}}{300} = \frac{4E(38,000)}{300} = 507E$$

$$\alpha = \frac{507E}{36.6E} \approx 13.83 \qquad \text{(same as in Fig. 16.4.4)}$$

$$\alpha_c = \frac{234E}{507E + 36.6E} = 0.430 \qquad \text{(at exterior column)}$$

$$\alpha_c = 0.215 \qquad \text{(at interior column)}$$

For frame C,

$$K_s = \frac{4E(300)(6.5)^3/12}{240} = 114E$$

$$K_b = \frac{4EI_b \text{ of } B5\text{-}B6 \text{ (Example 16.4.1)}}{240} = \frac{4E(24,400)}{240} = 406E$$

$$\alpha = \frac{406E}{114E} \approx 3.55 \qquad \text{(same as in Fig. 16.4.4)}$$

$$\alpha_c = \frac{234E}{406E + 114E} = 0.450 \qquad \text{(at exterior column)}$$

$$\alpha_c = 0.225 \qquad \text{(at interior column)}$$

For frame D,

$$K_s = 57E$$

$$K_b = \frac{4EI_b \text{ of } B7\text{-}B8 \text{ (Example 16.4.1)}}{240} = \frac{4E(20,500)}{240} = 341E$$

$$\alpha = \frac{341E}{57E} \approx 5.96 \qquad \text{(same as in Fig. 16.4.4)}$$

$$\alpha_c = \frac{234E}{341E + 57E} = 0.587 \qquad \text{(at exterior column)}$$

$$\alpha_c = 0.294 \qquad \text{(at interior column)}$$

(d) Minimum ratio α_{min}. The ratio β_a of service dead to live load is

$$\beta_a = \frac{6.5(150/12)}{120} = \frac{81}{120} = 0.67$$

The minimum ratios α_{min} required for making positive moment multiplier $\delta_s = 1.0$ are tabulated below.

FRAME	A	B	C	D
$\alpha = \dfrac{E_{cb}I_b}{E_{cs}I_s}$ from part (c) above or Fig. 16.4.4	8.27	13.83	3.55	5.96
L_2/L_1	0.80	0.80	1.25	1.25
α_{min} from Table 16.9.1	0	0	0	0

Note that when α is larger than 4, α_m is zero (thus the positive moment multiplier δ_s is 1.0) except when $L_2/L_1 = 2.00$ and $\beta_a = 0.33$.

(e) Positive moment multiplier δ_s. Since there is sufficient stiffness in the columns to minimize the pattern loading effects for all frames, $\delta_s = 1.0$ in all cases.

(f) Longitudinal moments in the frames. The longitudinal moments in frames A, B, C, and D are computed using Case 2 of Fig. 16.8.2 for the exterior span and Fig. 16.8.1 for the interior span. The computations are as shown below, and the results are summarized in Fig. 16.10.1.

For frame A, $\qquad\qquad\qquad M_0 = 458$ ft-kips

M_{neg} at exterior support	$= 0.16(458) = 73$ ft-kips
M_{pos} in exterior span	$= 1.0(0.57)(458) = 261$ ft-kips
M_{neg} at first interior support	$= 0.70(458) = 321$ ft-kips
M_{neg} at typical interior support	$= 0.65(458) = 298$ ft-kips
M_{pos} in typical interior span	$= 1.0(0.35)(458) = 160$ ft-kips

For frame B, $\qquad\qquad\qquad M_0 = 229$ ft-kips

M_{neg} at exterior support	$= 0.16(229) = 37$ ft-kips
M_{pos} in exterior span	$= 1.0(0.57)(229) = 130$ ft-kips
M_{neg} at first interior support	$= 0.70(229) = 160$ ft-kips
M_{neg} at typical interior support	$= 0.65(229) = 149$ ft-kips
M_{pos} in typical interior span	$= 1.0(0.35)(229) = 80$ ft-kips

For frame C, $\qquad\qquad\qquad M_0 = 352$ ft-kips

M_{neg} at exterior support	$= 0.16(352) = 56$ ft-kips
M_{pos} in exterior span	$= 1.0(0.57)(352) = 201$ ft-kips
M_{neg} at first interior support	$= 0.70(352) = 246$ ft-kips
M_{neg} at typical interior support	$= 0.65(352) = 229$ ft-kips
M_{pos} in typical interior span	$= 1.0(0.35)(352) = 123$ ft-kips

For frame D, $\qquad\qquad\qquad M_0 = 176$ ft-kips

M_{neg} at exterior support	$= 0.16(176) = 28$ ft-kips
M_{pos} in exterior span	$= 1.0(0.57)(176) = 100$ ft-kips
M_{neg} at first interior support	$= 0.70(176) = 123$ ft-kips
M_{neg} at typical interior support	$= 0.65(176) = 114$ ft-kips
M_{pos} in typical interior span	$= 1.0(0.35)(176) = 62$ ft-kips

EXAMPLE 16.10.2 For the flat slab design example described in Sec. 16.3, compute the longitudinal moments in frames A, B, C, and D as shown in Figs. 16.3.6 and 16.10.2.

Solution: (a) Check the five limitations (the sixth limitation does not apply here) for the direct design method. These five limitations are all satisfied.

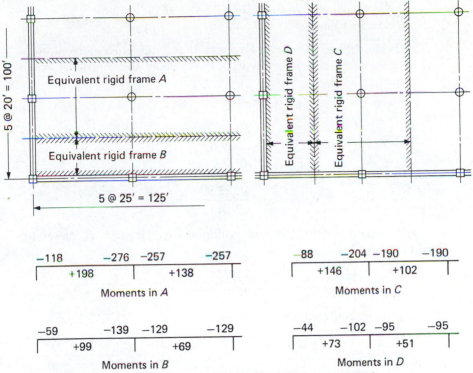

Figure 16.10.2 Longitudinal moments for flat slab design example.

(b) Total factored static moment M_0. Referring to the equivalent rigid frames A, B, C, and D in Fig. 16.10.2, the total factored static moment may be taken from the results of Example 16.3.2; thus

$$M_0 \text{ for } A = 395 \text{ ft-kips}$$

$$M_0 \text{ for } B = \tfrac{1}{2}(395) = 198 \text{ ft-kips}$$

$$M_0 \text{ for } C = 292 \text{ ft-kips}$$

$$M_0 \text{ for } D = \tfrac{1}{2}(292) = 146 \text{ ft-kips}$$

(c) Slab stiffness K_s.

$$K_s \text{ for } A = \frac{4EI_s}{L_1} = \frac{4E(240)(7.5)^3/12}{25(12)} = 112.5E$$

$$K_s \text{ for } B = 56.25E$$

$$K_s \text{ for } C = \frac{4EI_s}{L_1} = \frac{4E(300)(7.5)^3/12}{20(12)} = 176E$$

$$K_s \text{ for } D = 88E$$

(d) Column stiffness K_c.

$$\Sigma K_c \text{ (upper and lower} \atop \text{interior columns)} = \frac{2(4EI_c)}{L} = \frac{2(4E)\pi(18)^4/64}{10(12)} = 343E$$

$$\Sigma K_c \text{ (upper and lower}\atop \text{exterior columns)} = \frac{2(4EI_c)}{L} = \frac{2(4E)(16)^4/12}{10(12)} = 364E$$

(e) Beam stiffness K_b. From Example 16.6.2, the gross moment of inertia of the edge beam shown in Fig. 16.6.2 about the centroidal axis is 22,900 in.[4]

$$K_b \text{ for } A = 0$$

$$K_b \text{ for } B = \frac{4E(22,900)}{300} = 305E$$

$$K_b \text{ for } C = 0$$

$$K_b \text{ for } D = \frac{4E(22,900)}{240} = 381E$$

(f) Compute positive moment multipliers δ_s. The ratio β_a of service dead to live load is

$$\beta_a = \frac{7.5(150/12)}{120} = 0.78$$

The α_{min} values required for $\delta_s = 1.0$ are

FRAME	A	B	C	D
α (from Fig. 16.6.2)	0	5.42	0	4.34
L_2/L_1	0.80	0.80	1.25	1.25
α_{min} from Table 16.9.1	1.05	0	1.28	0

For frame A,

$$\alpha_c \text{ (exterior)} = \frac{364E}{0 + 112.5E} = 3.24 > \alpha_{min} = 1.05; \qquad \delta_s = 1.0$$

$$\alpha_c \text{ (interior)} = \frac{343E}{0 + 225E} = 1.52 > \alpha_{min} = 1.05; \qquad \delta_s = 1.0$$

For frame B,

$$\alpha_c \text{ (exterior)} = \frac{364E}{305E + 56.25E} = 1.01 > \alpha_{min} = 0; \qquad \delta_s = 1.0$$

$$\alpha_c \text{ (interior)} = \frac{364E}{610E + 112.5E} = 0.50 > \alpha_{min} = 0; \qquad \delta_s = 1.0$$

For frame C,

$$\alpha_c \text{ (exterior)} = \frac{364E}{0 + 176E} = 2.07 > \alpha_{min} = 1.28; \qquad \delta_s = 1.0$$

$$\alpha_c \text{ (interior)} = \frac{343E}{0 + 352E} = 0.97 < \alpha_{min} = 1.28; \qquad \delta_s > 1.0$$

$$\delta_s \text{ (interior)} = \frac{2 - 0.78}{4 + 0.78}\left(1 - \frac{0.97}{1.28}\right) = 1.06$$

Table 16.10.1 LONGITUDINAL MOMENTS (FT-KIPS) FOR THE
FLAT SLAB DESIGN EXAMPLE

FRAME	A	B	C	D
M_0	395	198	292	146
M_{neg} at exterior support, $0.30M_0$	118	59	88	44
M_{pos} in exterior span, $\delta_s\,(0.50M_0)$	198 ($\delta_s = 1.0$)	99 ($\delta_s = 1.0$)	146 ($\delta_s = 1.0$)	73 ($\delta_s = 1.0$)
M_{neg} at first interior support, $0.70M_0$	276	139	204	102
M_{neg} at typical interior support, $0.65M_0$	257	129	190	95
M_{pos} in typical interior span, $\delta_s\,(0.35M_0)$	138 ($\delta_s = 1.0$)	69 ($\delta_s = 1.0$)	102 ($\delta_s = 1.06$)	51 ($\delta_s = 1.0$)

For frame D,

$$\alpha_c \text{ (exterior)} = \frac{364E}{381E + 88E} = 0.78 > \alpha_{min} = 0; \qquad \delta_s = 1.0$$

$$\alpha_c \text{ (interior)} = \frac{364E}{762E + 176E} = 0.39 > \alpha_{min} = 0; \qquad \delta_s = 1.0$$

Only the positive moment in the interior span· of frame C needs to be
multiplied by 1.06.

(g) Longitudinal moments in the frames. The longitudinal moments in
frames A, B, C, and D are computed using Case 4 of Fig. 16.8.2 for the
exterior span and Fig. 16.8.1 for the interior span. The computations are
shown in Table 16.10.1 and the results are summarized in Fig. 16.10.2.

EXAMPLE 16.10.3 For the flat plate design example described in Sec.
16.3, compute the longitudinal moments in frames A, B, C, and D as shown
in Figs. 16.3.7 and 16.10.3.

Solution: (a) Check the five limitations (the sixth limitation does not apply
here) for the direct design method. These limitations are all satisfied.

(b) Total factored static moment M_0 from the results of Example 16.3.3.

$$M_0 \text{ for } A = 58.2 \text{ ft-kips}$$

$$M_0 \text{ for } B = \tfrac{1}{2}(58.2) = 29.1 \text{ ft-kips}$$

$$M_0 \text{ for } C = 46.3 \text{ ft-kips}$$

$$M_0 \text{ for } D = 23.1 \text{ ft-kips}$$

(c) Slab stiffness K_s.

$$K_s \text{ for } A = \frac{4EI_s}{L_1} = \frac{4E(144)(5.5)^3/12}{15(12)} = 44.4E$$

$$K_s \text{ for } B = 22.2E$$

$$K_s \text{ for } C = \frac{4EI_s}{L_1} = \frac{4E(180)(5.5)^3/12}{12(12)} = 69.4E$$

$$K_s \text{ for } D = 34.7E$$

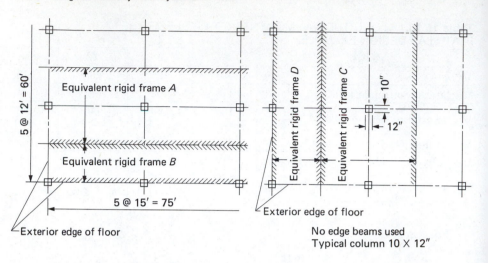

-15.1   -40.7 -37.8   -37.8   -12.0   -32.4 -30.1   -30.1
      +30.3   +20.4   +24.1   +16.9

        Moments in A   Moments in C

-7.6   -20.4 -18.9   -18.9   -6.0   -16.2 -15.0   -15.0
      +15.1   +10.2   +12.0   +8.1

        Moments in B   Moments in D

Figure 16.10.3 Longitudinal moments for flat plate design example.

(d) Column stiffness K_c.

$$\Sigma K_c \text{ for } A \text{ and } B = \frac{2(4EI_c)}{L} = \frac{2(4E)(10)(12)^3/12}{9(12)} = 106.7E$$

(upper and lower columns)

$$\Sigma K_c \text{ for } C \text{ and } D = \frac{2(4EI_c)}{L} = \frac{2(4E)(12)(10)^3/12}{9(12)} = 74.1E$$

(upper and lower columns)

(e) Compute positive moment multipliers δ_s. The ACI Code will be followed to obtain δ_s. Alternate equations for the multiplier δ_s have been proposed by Jofriet and McNeice [25]. The ratio β_a of service dead to live load is

$$\beta_a = \frac{5.5(150/12)}{60} = 1.15$$

The $\alpha_{\min}$ values required for $\delta_s = 1.0$ are, at $\alpha = E_{cb}I_b/E_{cs}I_s = 0$,

FRAME	A	B	C	D
L_2/L_1	0.80	0.80	1.25	1.25
$\alpha_{\min}$ from Table 16.9.1	0.60	0.60	0.68	0.68

For frame A,

$$\alpha_c \text{ (exterior)} = \frac{106.7E}{0 + 44.4E} = 2.40 > 0.60; \qquad \delta_s = 1.0$$

$$\alpha_c \text{ (interior)} = \frac{106.7E}{0 + 88.8E} = 1.20 > 0.60; \qquad \delta_s = 1.0$$

For frame B,

$$\alpha_c \text{ (exterior)} = \frac{106.7E}{0 + 22.2E} = 4.81 > 0.60; \qquad \delta_s = 1.0$$

$$\alpha_c \text{ (interior)} = \frac{106.7E}{0 + 44.4E} = 2.40 > 0.60; \qquad \delta_s = 1.0$$

For frame C,

$$\alpha_c \text{ (exterior)} = \frac{74.1E}{0 + 69.4E} = 1.07 > 0.68; \qquad \delta_s = 1.0$$

$$\alpha_c \text{ (interior)} = \frac{74.1E}{0 + 138.8E} = 0.53 < 0.68; \qquad \delta_s > 1.0$$

$$\delta_s \text{ (interior)} = 1 + \frac{2 - 1.15}{4 + 1.15}\left(1 - \frac{0.53}{0.68}\right) = 1.04$$

For frame D,

$$\alpha_c \text{ (exterior)} = \frac{74.1E}{0 + 34.7E} = 2.14 > 0.68; \qquad \delta_s = 1.0$$

$$\alpha_c \text{ (interior)} = \frac{74.1E}{0 + 69.4E} = 1.07 > 0.68; \qquad \delta_s = 1.0$$

Only the positive moment in the interior spans of frame C needs to be multiplied by 1.04.

(f) Longitudinal moments in the frames. The longitudinal moments in frames A, B, C, and D are computed using Case 3 of Fig. 16.8.2 for the exterior span and Fig. 16.8.1 for the interior span. The computations are as shown in Table 16.10.2 and the results are summarized in Fig. 16.10.3.

Table 16.10.2 LONGITUDINAL MOMENTS (FT-KIPS) FOR THE FLAT PLATE DESIGN EXAMPLE

FRAME	A	B	C	D
M_0	58.2	29.1	46.3	23.1
M_{neg} at exterior support, $0.26M_0$	15.1	7.6	12.0	6.0
M_{pos} in exterior span, $\delta_s (0.52M_0)$	30.3 ($\delta_s = 1.0$)	15.1 ($\delta_s = 1.0$)	24.1 ($\delta_s = 1.0$)	12.0 ($\delta_s = 1.0$)
M_{neg} at first interior support, $0.70M_0$	40.7	20.4	32.4	16.2
M_{neg} at typical interior support, $0.65M_0$	37.8	18.9	30.1	15.0
M_{pos} in typical interior span, $\delta_s (0.35M_0)$	20.4 ($\delta_s = 1.0$)	10.2 ($\delta_s = 1.0$)	16.9 ($\delta_s = 1.04$)	8.1 ($\delta_s = 1.0$)

16.11 TORSION CONSTANT C OF THE TRANSVERSE BEAM

One important parameter useful in the analysis (that is, the application of the equivalent frame method described in Chap. 17) and design (that is, for the transverse distribution of the longitudinal moment) is the torsional constant C of the transverse beam spanning from column to column. Even if there is no such beam (as defined by projection above or below the slab) actually visible, for the present use one still should imagine that there is a beam made of a portion of the slab having a width equal to that of the column, bracket, or capital in the longitudinal direction (ACI-13.7.5.1a). When there is actually a transverse beam web above or below the slab, the cross section of the transverse beam should include the portion of slab within the width of column, bracket, or capital described above plus the projection of beam web above or below the slab (ACI-13.7.5.1b). As a third possibility the transverse beam may include that portion of slab on each side of the beam web equal to its projection above or below the slab, whichever is greater, but not greater than four times the slab thickness (ACI-13.7.5.1c). The largest of the three definitions as shown in Fig. 16.11.1 may be used.

The torsional constant C of the transverse beam equals (ACI Formula 13-7),

$$C = \sum\left(1 - 0.63\frac{x}{y}\right)\left(\frac{x^3 y}{3}\right)$$

(16.11.1)

where

x = shorter dimension of a component rectangle

y = longer dimension of a component rectangle

and the component rectangles should be taken in such a way that the largest value of C is obtained (ACI Commentary-13.7.5). Equation (16.11.1) is identical to Eq. (19.3.5), for which there is additional discussion in Chap. 19.

EXAMPLE 16.11.1 For the two-way slab (with beams) design example, compute the torsional constant C for the edge and interior beams in the short and long directions.

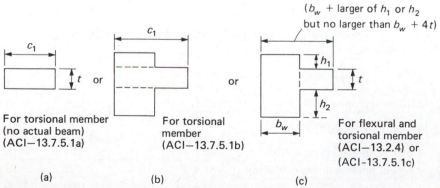

Figure 16.11.1 Definition of cross sections for transverse beams in torsion. [Projection of slab beyond beam in Case (c) is allowed on each side for interior beam.]

Solution: Each cross section shown in Fig. 16.11.2 may be divided into component rectangles in two different ways and the larger value of C is to be used. For the short direction,

$$C\left(\begin{array}{c}\text{edge}\\\text{beam}\end{array}\right) = \left[1 - \frac{0.63(6.5)}{29.5}\right]\frac{29.5(6.5)^3}{3} + \left[1 - \frac{0.63(12)}{17.5}\right]\frac{17.5(12)^3}{3}$$

$$= 2325 + 5725 = 8050 \text{ in.}^4$$

or

$$C\left(\begin{array}{c}\text{edge}\\\text{beam}\end{array}\right) = \left[1 - \frac{0.63(6.5)}{17.5}\right]\frac{17.5(6.5)^3}{3} + \left[1 - \frac{0.63(12)}{24}\right]\frac{24(12)^3}{3}$$

$$= 1230 + 9470 = 10,700 \text{ in.}^4 \qquad\qquad Use$$

$$C\left(\begin{array}{c}\text{interior}\\\text{beam}\end{array}\right) = \left[1 - \frac{0.63(6.5)}{47}\right]\frac{47(6.5)^3}{3} + \left[1 - \frac{0.63(12)}{17.5}\right]\frac{17.5(12)^3}{3}$$

$$= 3925 + 5725 = 9650 \text{ in.}^4$$

or

$$C\left(\begin{array}{c}\text{interior}\\\text{beam}\end{array}\right) = 2(1230) + 9470 = 11,930 \text{ in.}^4 \qquad\qquad Use$$

For the long direction,

$$C\left(\begin{array}{c}\text{edge}\\\text{beam}\end{array}\right) = \left[1 - \frac{0.63(6.5)}{35.5}\right]\frac{35.5(6.5)^3}{3} + \left[1 - \frac{0.63(14)}{21.5}\right]\frac{21.5(14)^3}{3}$$

$$= 2900 + 11,600 = 14,500 \text{ in.}^4$$

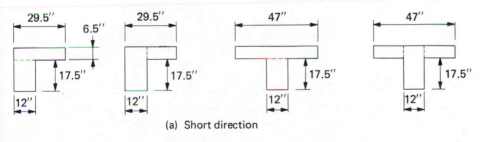

(a) Short direction

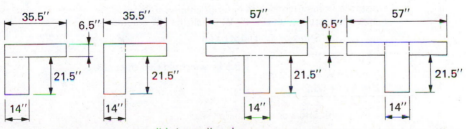

(b) Long direction

Figure 16.11.2 Effective cross sections of transverse beams resisting torsion, in two-way slab (with beams) design example.

or

$$C\left(\frac{\text{edge}}{\text{beam}}\right) = \left[1 - \frac{0.63(6.5)}{21.5}\right]\frac{21.5(6.5)^3}{3} + \left[1 - \frac{0.63(14)}{28}\right]\frac{28(14)^3}{3}$$

$$= 1600 + 17,500 = 19,100 \text{ in.}^4 \qquad \textit{Use}$$

$$C\left(\frac{\text{interior}}{\text{beam}}\right) = \left[1 - \frac{0.63(6.5)}{57}\right]\frac{57(6.5)^3}{3} + \left[1 - \frac{0.63(14)}{28}\right]\frac{28(14)^3}{3}$$

$$= 4800 + 11,600 = 16,400 \text{ in.}^4$$

or

$$C\left(\frac{\text{interior}}{\text{beam}}\right) = 2(1600) + 17,500 = 20,700 \text{ in.}^4 \qquad \textit{Use}$$

EXAMPLE 16.11.2 For the flat slab design example, compute the torsional constant C for the edge beam and the interior beam in the short and long directions.

Solution: For the short or long edge beam [Fig. 16.11.3(a)], the torsional constant C is computed on the basis of the cross section shown in Fig. 16.11.3(a).

$$C = \left[1 - \frac{0.63(7.5)}{51.5}\right]\frac{(7.5)^3(51.5)}{3} + \left[1 - \frac{0.63(14)}{16.5}\right]\frac{(14)^3(16.5)}{3}$$

$$= 6575 + 7025 = 13,600 \text{ in.}^4$$

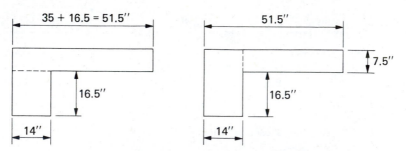

35" = distance from outer edge of exterior column to inner edge of square capital (i.e. 2'-3" + half the 16" column)

(a) Short or long edge beam

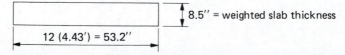

(b) Short or long interior beam

Figure 16.11.3 Cross sections of torsional transverse beams in flat slab design example.

or

$$C = \left[1 - \frac{0.63(14)}{24}\right]\frac{(14)^3(24)}{3} + \left[1 - \frac{0.63(7.5)}{37.5}\right]\frac{(7.5)^3(37.5)}{3}$$

$$= 13{,}890 + 4610 = 18{,}500 \text{ in.}^4 \qquad\qquad Use$$

For the short or long interior beam [Fig. 16.11.3(b)], a weighted slab thickness of 8.5 in. is used, on the assumption that one-third of the span has a $10\frac{1}{2}$-in. thickness and the remainder has a $7\frac{1}{2}$-in. thickness. (Actually the ratio is not exactly so because the drop width has been revised from 6 ft–8 in. to 7 ft.)

$$C = \left(1 - 0.63\frac{x}{y}\right)\frac{x^3 y}{3} = \left[1 - \frac{0.63(8.5)}{12(4.43)}\right]\left[\frac{(8.5)^3(12)(4.43)}{3}\right] = 9800 \text{ in.}^4$$

EXAMPLE 16.11.3 For the flat plate design example, compute the torsional constant C for the short and long beams.

Solution: Since no actual edge beams are used, the torsional member is, according to Fig. 16.11.4, equal to the slab thickness t by the column width c_1

$$C \text{ for short beams} = \left[1 - \frac{0.63(5.5)}{12}\right]\frac{(5.5)^3(12)}{3} = 474 \text{ in.}^4$$

$$C \text{ for long beams} = \left[1 - \frac{0.63(5.5)}{10}\right]\frac{(5.5)^3(10)}{3} = 363 \text{ in.}^4$$

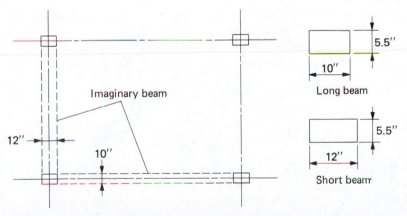

Figure 16.11.4 Cross sections of torsional transverse beams in flat plate design example.

16.12 TRANSVERSE DISTRIBUTION OF LONGITUDINAL MOMENT

The longitudinal moment values, whether those of the "direct design method" shown in Figs. 16.8.1 and 16.8.2 or those obtained by structural analysis using the "equivalent frame method" (Chap. 17), are for the entire width (sum of the two half panel widths in the transverse direction, for an interior column line) of the equivalent rigid frame. Each of these moments is to be divided, on the basis of studies by Gamble, Sozen, and Siess [5], between the column strip and the two half middle strips as defined in Fig. 16.12.1. If the two adjacent transverse spans are each equal to L_2, the width of the column strip is then equal to one-half of L_2, or one-half of the longitudinal span L_1, whichever is smaller (ACI-13.2.1). This seems reasonable, since when the longitudinal span is shorter than the transverse span, a larger portion of the moment across the width of the equivalent frame might be expected to concentrate near the column centerline.

The transverse distribution of the longitudinal moment to column and middle strips is a function of three parameters, using L_1 and L_2 for the longitudinal and transverse spans, respectively: (1) the aspect ratio L_2/L_1; (2) the ratio $\alpha_1 = E_{cb}I_b/(E_{cs}I_s)$ of the longitudinal beam stiffness to slab stiffness; and (3) the ratio $\beta_t = E_{cb}C/(2E_{cs}I_s)$ of the torsional rigidity of edge beam section to the flexural rigidity of a width of slab equal to the span length of the edge beam. According to ACI-13.6.4, the column strip is to take the percentage of the longitudinal moment as shown in Table 16.12.1. As may be seen from Table 16.12.1, only the first two parameters affect the transverse distribution of the negative moments at the first and typical interior supports as well as the positive moments in exterior and interior spans; but all three parameters are involved in the transverse distribution of the negative moment at the exterior support.

Regarding the distributing percentages shown in Table 16.12.1, the following observations may be made:

1. In general, the column strip takes more than 50% of the longitudinal moment.
2. The column strip takes a larger share of the negative longitudinal moment than the positive longitudinal moment.
3. When no longitudinal beams are present, the column strip takes the same share of the longitudinal moment, irrespective of the as-

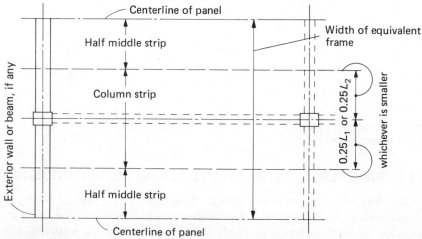

Figure 16.12.1   Definition of column and middle strips.

Table 16.12.1 PERCENTAGE OF LONGITUDINAL MOMENT IN COLUMN STRIP (ACI-13.6.4)

ASPECT RATIO L_2/L_1			0.5	1.0	2.0
Negative moment at exterior support	$\alpha_1 L_2/L_1 = 0$	$\beta_t = 0$	100	100	100
		$\beta_t \geq 2.5$	75	75	75
	$\alpha_1 L_2/L_1 \geq 1.0$	$\beta_t = 0$	100	100	100
		$\beta_t \geq 2.5$	90	75	45
Positive moment	$\alpha_1 L_2/L_1 = 0$		60	60	60
	$\alpha_1 L_2/L_1 \geq 1.0$		90	75	45
Negative moment at interior support	$\alpha_1 L_2/L_1 = 0$		75	75	75
	$\alpha_1 L_2/L_1 \geq 1.0$		90	75	45

pect ratio. The reader may note, however, that the column strip width is a fraction of L_1 or L_2 ($0.25L_1$ or $0.25L_2$ on each side of column line), whichever is smaller.

4. In the presence of longitudinal beams, the larger the aspect ratio, the smaller the distribution to the column strip. This seems consistent because the same reduction in the portion of moment going into the slab is achieved by restricting the column strip width to a fraction of L_1 when L_2/L_1 is greater than one.

5. The column strip takes a smaller share of the exterior moment as the torsional rigidity of the edge beam section increases.

When the exterior support consists of a column or wall extending for a distance equal to or greater than three-fourths of the transverse width, the exterior negative moment is to be uniformly distributed over the transverse width (ACI-13.6.4.3).

The procedure for distributing the longitudinal moment across a transverse width to the column and middle strips may be summarized as follows:

1. Divide the total transverse width applicable to the longitudinal moment into a column strip width and two half middle strip widths, one adjacent to each side of the column strip. For an exterior column line, the column strip width is $\frac{1}{4}L_1$ or $\frac{1}{4}L_2$, whichever is smaller; for an interior column line, the column strip width is $\Sigma(\frac{1}{4}L_1$ or $\frac{1}{4}L_2$, whichever is smaller, of the panels on both sides).

2. Determine the ratio $\beta_t = E_{cb}C/(2E_{cs}I_s)$ of edge beam torsional rigidity to slab flexural rigidity. (Note: The 2 arises from approximating the shear modulus of elasticity in the numerator as $E_{cb}/2$.)

3. Determine the ratio $\alpha_1 = E_{cb}I_b/(E_{cs}I_s)$ of longitudinal beam flexural stiffness to slab flexural stiffness.

4. Divide the longitudinal moment at each critical section into two parts according to the percentage shown in Table 16.12.1: one part to the column strip width; and the remainder to the half middle strip for an exterior column line, or to the half middle strips on each side of an interior column line.

5. If there is an exterior wall instead of an exterior column line, the strip ordinarily called the exterior column strip will not deflect and therefore no moments act. In this case there can be no longitudinal

distribution of moments; thus there is no computed moment to distribute laterally to the half middle strip adjacent to the wall. This half middle strip should be combined with the next adjacent half middle strip, which itself receives a lateral distribution in the frame of the first interior column line. The total middle strip in this situation is designed for twice the moment in the half middle strip from the first interior column line (ACI-13.6.6.3).

Distribution of Moment in Column Strip to Beam and Slab. When a longitudinal beam exists in the column strip along the column centerline, the column strip moment as determined by the percentages in Table 16.12.1 (ACI-13.6.4) should be divided to the beam and the slab. ACI-13.6.5 states that 85% of this moment be taken by the beam if $\alpha L_2/L_1$ is equal to or greater than 1.0, and for values of $\alpha L_2/L_1$ between 1.0 and 0, the proportion of moment to be resisted by the beam is to be obtained by linear interpolation between 85 and 0%. In addition, any beam must be designed to carry its own weight (projection above and below the slab), and any concentrated or linear loads applied directly on it (ACI-13.6.5.3).

EXAMPLE 16.12.1 For the two-way slab (with beams) design example described in Sec. 16.3, distribute the longitudinal moments computed for frames A, B, C, and D (see Fig. 16.10.1) into three parts—namely, for the longitudinal beam, for the column strip slab, and for the middle strip slab.

Solution: The values for the total longitudinal moments in frames A, B, C, and D at the five critical sections are taken from Example 16.10.1 and shown again in Table 16.12.2. The results of transverse distribution of these moments are also shown in this table.

(a) Negative moment at face of exterior support. For frame A, L_2/L_1 = 0.80; α_1 = 8.27 (Fig. 16.4.4); $\alpha_1 L_2/L_1$ = 6.61; C = 10,700 in.^{4} (Example 16.11.1); I_s = 240(6.5)^3/12 = 5490 in.^{4}; and β_t = $C/(2I_s)$ = 10,700/[2(5490)] = 0.98. Table 16.12.3 shows the linear interpolation for obtaining the column strip percentage from the prescribed limits of Table 16.12.1. The total moment of 73 ft-kips is divided into three parts, 92.6% to column strip (of which 85% goes to the beam and 15% to the slab since $\alpha_1 L_2/L_1$ = 6.61 ≥ 1.0) and 7.4% to the middle strip slab. The results are shown in Table 16.12.2.

For frame B, L_2/L_1 = 0.80; α_1 = 13.83 (Fig. 16.4.4); $\alpha_1 L_2/L_1$ = 11.1; β_t = 0.98 same as for frame A; and column strip moment percentage = 92.6%, the same as for frame A.

For frame C, L_2/L_1 = 1.25; α_1 = 3.55 (Fig. 16.4.4); $\alpha_1 L_2/L_1$ = 4.44; C = 19,100 in.^{4} (Example 16.11.1); I_s = 300(6.5)^3/12 = 6870 in.^{4}; and β_t = $C/(2I_s)$ = 19,100/[2(6870)] = 1.39. Table 16.12.4 shows the linear interpolation for obtaining the column strip percentage from the prescribed limits of Table 16.12.1. The total moment of 56 ft-kips is divided into three parts, 81.9% to column strip (of which 85% goes to the beam and 15% to the slab since $\alpha_1 L_2/L_1$ = 4.44 ≥ 1.0) and 18.1% to the middle strip slab.

For frame D, L_2/L_1 = 1.25; α_1 = 5.96 (Fig. 16.4.4), $\alpha_1 L_2/L_1$ = 7.45;

Table 16.12.2 (also see Fig. 16.10.1)   TRANSVERSE DISTRIBUTION OF LONGITUDINAL MOMENTS IN TWO-WAY SLAB (WITH BEAMS) DESIGN EXAMPLE

FRAME A

TOTAL WIDTH = 20 ft, COLUMN STRIP WIDTH = 10 ft, MIDDLE STRIP WIDTH = 10 ft

	EXTERIOR SPAN			INTERIOR SPAN	
	EXTERIOR NEGATIVE	POSITIVE	INTERIOR NEGATIVE	NEGATIVE	POSITIVE
Total moment	−73	+261	−321	−298	+160
Moment in beam	−57	+179	−221	−205	+110
Moment in column strip slab	−10	+32	−39	−36	+20
Moment in middle strip slab	−6	+50	−61	−57	+30

FRAME B

TOTAL WIDTH = 10 ft, COLUMN STRIP WIDTH = 5 ft, HALF MIDDLE STRIP WIDTH = 5 ft

	EXTERIOR SPAN			INTERIOR SPAN	
	EXTERIOR NEGATIVE	POSITIVE	INTERIOR NEGATIVE	NEGATIVE	POSITIVE
Total moment	−37	+130	−160	−149	+80
Moment in beam	−29	+89	−110	−103	+55
Moment in column strip slab	−5	+16	−20	−18	+10
Moment in middle strip slab	−3	+25	−30	−28	+15

FRAME C

TOTAL WIDTH = 25 ft, COLUMN STRIP WIDTH = 10 ft, MIDDLE STRIP WIDTH = 15 ft

	EXTERIOR SPAN			INTERIOR SPAN	
	EXTERIOR NEGATIVE	POSITIVE	INTERIOR NEGATIVE	NEGATIVE	POSITIVE
Total moment	−56	+201	−246	−229	+123
Moment in beam	−39	+112	−141	−132	+70
Moment in column strip slab	−7	+20	−25	−23	+13
Moment in middle strip slab	−10	+63	−80	−74	+40

FRAME D

TOTAL WIDTH = 12.5 ft, COLUMN STRIP WIDTH = 5 ft, HALF MIDDLE STRIP WIDTH = 7.5 ft

	EXTERIOR SPAN			INTERIOR SPAN	
	EXTERIOR NEGATIVE	POSITIVE	INTERIOR NEGATIVE	NEGATIVE	POSITIVE
Total moment	−28	+100	−123	−114	+62
Moment in beam	−20	+55	−71	−65	+36
Moment in column strip slab	−3	+10	−12	−12	+6
Moment in middle strip slab	−5	+31	−40	−37	+20

Table 16.12.3   LINEAR INTERPOLATION FOR COLUMN STRIP PERCENTAGE OF EXTERIOR NEGATIVE MOMENT—FRAME A

	L_2/L_1	0.5	0.8	1.0
$\alpha_1 L_2/L_1 = 6.61$	$\beta_t = 0$	100%	100%	100%
	$\beta_t = 0.98$	96.1%	92.6%	90.2%
	$\beta_t \geq 2.50$	90%	81%	75%

Table 16.12.4   LINEAR INTERPOLATION FOR COLUMN STRIP
PERCENTAGE OF EXTERIOR NEGATIVE MOMENT—FRAME C

L_2/L_1		1.0	1.25	2.0
	$\beta_t = 0$	100%	100%	100%
$\alpha_1 L_2/L_1 = 4.44$	$\beta_t = 1.39$	86.1%	81.9%	69.4%
	$\beta_t \geq 2.50$	75%	67.5%	45%

$\beta_t = 1.39$ same as for frame C; and column strip moment percentage $=$ 81.9%, the same as for frame C.

(b) Negative moments at exterior face of first interior support and at face of typical interior support. For frame A, $L_2/L_1 = 0.80$ and $\alpha_1 L_2/L_1 = 6.61 > 1.0$. Using the prescribed values in Table 16.12.1, the proportion of moment going to the column strip is determined to be 81% by linear interpolation:

L_2/L_1	0.5	0.8	1.0
$\alpha_1 L_2/L_1 = 6.61$	90%	81%	75%

For frame B, $L_2/L_1 = 0.80$ and $\alpha_1 L_2/L_1 = 11.1$. The proportion of moment is again 81% for the column strip, the same as for frame A.

For frame C, $L_2/L_1 = 1.25$ and $\alpha_1 L_2/L_1 = 4.44$. Using the prescribed values in Table 16.12.1, the proportion of moment going to the column strip is determined to be 67.5% by linear interpolation:

L_2/L_1	1.0	1.25	2.0
$\alpha_1 L_2/L_1 = 4.44$	75%	67.5%	45%

For frame D, $L_2/L_1 = 1.25$ and $\alpha_1 L_2/L_1 = 7.47$. The proportion of moment is again 67.5% for the column strip, the same as for frame C.

(c) Positive moments in exterior and interior spans. Since the prescribed limits for $\alpha_1 L_2/L_1 \geq 1.0$ are the same for positive moment and for negative moment at interior support, the percentages of column strip moment for positive moments in exterior and interior spans are identical to those for negative moments as determined in part (b) of this example.

EXAMPLE 16.12.2   Divide the five critical moments in each of the equivalent rigid frames A, B, C, and D in the flat slab design example, as shown in Fig. 16.10.2, into two parts: one for the half column strip (for frames B and D) or the full column strip (for frames A and C), and the other for the half middle strip (for frames B and D) or the two half middle strips on both sides of the column line (for frames A and C).

Solution: The percentages of the longitudinal moments going into the column strip width are shown in lines 10 to 12 of Table 16.12.5. Note that the column strip width shown in line 2 is one-half of the shorter panel dimension for both frames A and C, and one-fourth of this value for frames B and D. Note also the sum of the values on lines 2 and 3 should be equal to that on line 1, for each respective frame.

Table 16.12.5 TRANSVERSE DISTRIBUTION OF LONGITUDINAL MOMENT FOR FLAT SLAB DESIGN EXAMPLE

LINE NUMBER	EQUIVALENT RIGID FRAME	A	B	C	D
1	Total transverse width (in.)	240	120	300	150
2	Column strip width (in.)	120	60	120	60
3	Half middle strip width (in.)	2@60	60	2@90	90
4	C (in.^{4}) from Example 16.11.2	18,500	18,500	18,500	18,500
5	I_s (in.^{4}) in β_t	8,440	8,440	10,600	10,600
6	$\beta_t = E_{cb}C/(2E_{cs}I_s)$	1.10	1.10	0.87	0.87
7	α_1 from Example 16.10.2(f)	0	5.42	0	4.34
8	L_2/L_1	0.80	0.80	1.25	1.25
9	$\alpha_1 L_2/L_1$	0	4.33	0	5.43
10	Exterior negative moment, percent to column strip	89.0%	91.6%	91.3%	88.7%
11	Positive moment, percent to column strip	60.0%	81.0%	60.0%	67.5%
12	Interior negative moment, percent to column strip	75.0%	81.0%	75.0%	67.5%

The moment of inertia of the slab equal in width to the transverse span of the edge beam is

$$I_s \text{ in } \beta_t \text{ for } A \text{ and } B = \frac{240(7.5)^3}{12} = 8440 \text{ in.}^4$$

and

$$I_s \text{ in } \beta_t \text{ for } C \text{ and } D = \frac{300(7.5)^3}{12} = 10,600 \text{ in.}^4$$

These values are shown in line 5 of Table 16.12.5.

The percentages shown in lines 10 to 12 are obtained from Table 16.12.1, by interpolation (as illustrated in Tables 16.12.3 and 16.12.4) if necessary.

Table 16.12.6 TRANSVERSE DISTRIBUTION OF LONGITUDINAL MOMENT FOR FLAT PLATE DESIGN EXAMPLE

LINE NUMBER	EQUIVALENT RIGID FRAME	A	B	C	D
1	Total transverse width (in.)	144	72	180	90
2	Column strip width (in.)	72	36	72	36
3	Half middle strip width (in.)	2@36	36	2@54	54
4	C(in.^{4}) from Example 16.11.3	474	474	363	363
5	I_s(in.^{4}) in β_t	2,000	2,000	2,500	2,500
6	$\beta_t = E_{cb}C/(2E_{cs}I_s)$	0.118	0.118	0.073	0.073
7	α_1	0	0	0	0
8	L_2/L_1	0.80	0.80	1.25	1.25
9	$\alpha_1 L_2/L_1$	0	0	0	0
10	Exterior negative moment, percent to column strip	98.8%	98.8%	99.3%	99.3%
11	Positive moment, percent to column strip	60%	60%	60%	60%
12	Interior negative moment, percent to column strip	75%	75%	75%	75%

Having had these percentages, the separation of each of the longitudinal moment values shown in Fig. 16.10.2 into two parts is a simple matter and thus is not shown further.

EXAMPLE 16.12.3 Divide the five critical moments in each of the equivalent rigid frames A, B, C, and D in the flat plate design example, as shown in Fig. 16.10.3, into two parts: one for the half column strip (for frames B and D) or the full column strip (for frames A and C), and the other for the half middle strip (for frames B and D) or the two half middle strips on both sides of the column line (for frames A and C).

Solution: The percentages of the longitudinal moments going into the column strip width are shown in lines 10 to 12 of Table 16.12.6. Explanations are identical to those for the preceding example.

16.13 DESIGN OF SLAB THICKNESS AND REINFORCEMENT

Slab Thickness. The slab thickness must be sufficient for resisting the bending moment and shear at the critical sections. First it is necessary to search for the location where the intensity of bending moment is the largest per unit width of slab. With the preliminary thickness already assumed, the percentage of tension reinforcement required may be compared with $0.75\rho_b$. Deflection control is usually satisfactory if the percentage of reinforcement is less than one-half of $0.75\rho_b$. For flat slabs, the slab thickness outside of the drop panel, if any, and the total thickness through the drop must both be investigated so that under the most severe bending moment the maximum ratio of tension reinforcement satisfies ACI-10.3.3 (i.e., less than $0.75\rho_b$ for flexural members with tension steel only). With the minimum thickness requirements of ACI-9.5.3 already satisfied, however, the maximum reinforcement ratio is more likely to be well below $0.375\rho_b$. In evaluating the drop-panel thickness for flexure, the drop width should be used as the transverse width of the compression area, since the drop is usually narrower than the width of the column strip. Also, the effective depth to be used for computing the reinforcement should not be more than what would be furnished by a drop thickness below the slab equal to one-fourth the distance from the edge of drop to the edge of column capital.

The shear requirement for two-way slabs (with beams) may be investigated by observing strips 1-1 and 2-2 in Fig. 16.13.1. Beams with $\alpha_1 L_2/L_1$ values larger than 1.0 are assumed to carry the loads acting on the tributary floor areas bounded by 45° lines drawn from the corners of the panel and the centerline of the panel parallel to the long side (ACI-13.6.8.1). If this is the case, the loads on the trapezoidal areas E and F of Fig. 16.13.1 go to the long beams; and those on the triangular areas G and H go to the short beams. The shear per unit width of slab along the beam is highest at ends of slab strips 1-1 and 2-2, which, considering the increased shear at the exterior face of the first interior support, is approximately equal to

$$V_u = 1.15\left(\frac{w_u S}{2}\right) \tag{16.13.1}$$

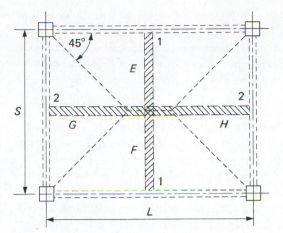

Figure 16.13.1 Load transfer from floor area to beams.

If $\alpha_1 L_2/L_1$ is equal to zero, there is, of course, no load on the beams (because there are no beams). When the value of $\alpha_1 L_2/L_1$ is between 0 and 1.0, the percentage of the floor load going to the beams should be obtained by linear interpolation. In such a case, the shear expressed by Eq. (16.13.1) would be reduced, but the shear around the column due to the portion of the floor load going directly to the columns by two-way action should be investigated as for flat plate floors.

The shear strength requirement for flat slab and flat plate systems (without beams) is treated separately in Secs. 16.15, 16.16, and 16.18.

Reinforcement. When the nominal requirements for slab thickness as discussed in Sec. 16.6 are satisfied, no compression reinforcement will likely be required. The tension steel area required within the strip being considered can then be obtained by the following steps:

1. required $M_n = \dfrac{\text{factored moment } M_u \text{ in the strip}}{(\phi = 0.90)}$

2. $m = \dfrac{f_y}{0.85 f'_c}$, $R_n = \dfrac{M_n}{bd^2}$,

$$\rho = \frac{1}{m}\left(1 - \sqrt{1 - \frac{2mR_n}{f_y}}\right), \qquad A_s = \rho bd$$

Instead of using the equation for ρ in step 2, the curves in Fig. 3.8.1 may be used. Note also that the values of b and d to be used in step 2 for negative moment in a column strip with drop are the drop width for b, and for d the smaller of the actual effective depth through the drop and that provided by a drop thickness below the slab at no more than one-fourth the distance between the edges of the column capital and the drop. For positive moment computation the full column strip width should be used for b, and the effective depth in the slab for d. After obtaining the steel area A_s required within the strip, a number of bars may be chosen so that they provide either the area required for strength or the area required for shrinkage and temperature reinforcement, which is $0.002bt$ for Grades 40 and 50 steel, but somewhat less for higher grades (see ACI-7.12). The spacing of reinforcing bars must not exceed 2 times the slab thickness (ACI-13.4.2), except in slabs of cellular or ribbed construction where the requirement for shrinkage and temperature reinforcement governs (i.e., 5 times the slab thickness or 18 in.).

Corner Reinforcement for Two-Way Slab (With Beams). It is well known from plate bending theory that a transversely loaded slab simply supported along four edges will tend to develop corner reactions as shown in Fig. 16.13.2, for which reinforcement must be provided. Thus in slabs supported on beams having a value of α greater than 1.0, special reinforcement (Fig. 16.13.3) shall be provided at exterior corners in both the bottom and top of the slab. This reinforcement (ACI-13.4.6) is to be provided for a distance in each direction from the corner equal to one-fifth the longer span. The reinforcement in both the top and bottom of the slab must be sufficient to resist a moment equal to the maximum positive moment per foot of width in the slab, and it may be placed in a single band parallel to the diagonal in the top of the slab and perpendicular to the diagonal in the bottom of the slab, or in two bands parallel to the sides of the slab.

Crack Control. In addition to deflection control, crack control is the other major serviceability requirement usually considered in the design of flexural members. ACI-10.6 gives criteria for beams and one-way slabs to ensure distribution of flexural reinforcement to minimize crack width under service loads. No ACI Code provisions are given for two-way floor (or roof) systems; however, ACI Committee 224, Cracking, has suggested a formula to predict the possible crack width in two-way acting slabs, flat slabs, and flat plates. The recommendations are based on the work of Nawy et al. [26–29]. When the predicted crack width is considered excessive (there are no ACI Code limits for slabs), the distribution (size and spacing) of flexural reinforcement may be adjusted [29] to decrease predicted crack width. Ordinarily crack width is not a problem on two-way acting slabs, but when steel with f_y equal to 60,000 psi or higher is used, crack control should be considered.

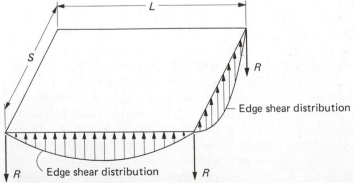

Figure 16.13.2 Edge reactions for simply supported two-way slab.

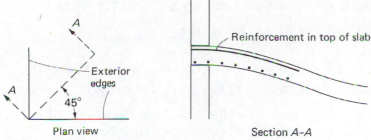

Figure 16.13.3   Corner reinforcement in two-way slab.

Development Lengths and Bar Cutoffs and Bends. General guidance for laying out the reinforcement within the slab is provided by ACI-Fig. 13.4.8, which shows in detail the minimum lengths of straight and bent bars in column as well as middle strips. The determination of minimum lengths is similar to that for bars in one-way slabs, as discussed in Chap. 8. Specifically all positive moment reinforcement perpendicular to a discontinuous edge should extend at least 6 in. into the edge beam, wall, or column; and all negative moment reinforcement perpendicular to such an edge should be bent, hooked, or in some way anchored in the edge beam, wall, or column. Development lengths required for all negative moment reinforcement should be provided on either side of the face of support, in the manner treated in Chap. 6.

EXAMPLE 16.13.1 Investigate if the preliminary slab thickness of $6\frac{1}{2}$ in. in the two-way slab (with beams) design example described in Sec. 16.3 is sufficient for resisting flexure and shear.

Solution: For each of the equivalent frames A, B, C, and D, the largest bending moment in the slab occurs at the exterior face of the first interior support in the middle strip slab. From Table 16.12.1, this moment is observed to be 61/10, 30/5, 80/15, or 40/7.5 ft-kips per ft of width in frames A, B, C, and D, respectively. Taking the effective depth to the contact level between the reinforcing bars in the two directions, and assuming #5 bars,

$$\text{average } d = 6.50 - 0.75 - 0.63 = 5.12 \text{ in.}$$

The largest R_n required is

$$R_n = \frac{M_u}{\phi bd^2} = \frac{6100(12)}{0.90(12)(5.12)^2} = 260 \text{ psi}$$

From Fig. 3.8.1, the reinforcement ratio ρ for this value of R_n is 0.007, which is well below $0.375\rho_b = 0.0139$. Hence excessive deflection should not be expected; this is further verification of the minimum thickness formulas given in ACI-9.5.3.

The factored floor load w_u is

$$w_u = 1.4w_D + 1.7w_L = 318 \text{ psf}$$

Since all $\alpha L_2/L_1$ values are well over 1.0, take V from Eq. (16.13.1) as

$$V_u = \frac{1.15 w_u S}{2} = \frac{1.15(0.318)(20)}{2} = 3.66 \text{ kips}$$

$$V_c = 2\sqrt{f'_c} b_w d = 2\sqrt{3000}(12)(5.12)\tfrac{1}{1000} = 6.73 \text{ kips}$$

$$\phi V_c = 0.85(6.73) = 5.72 \text{ kips} > [V_u = 3.66 \text{ kips}] \qquad\qquad \text{OK}$$

Note that the factored shear 3.66 kips is the maximum at strip 1-1 of Fig. 16.13.1; actually the average for all such strips will be lower.

EXAMPLE 16.13.2 Design the reinforcement in the exterior and interior spans of a typical column strip and a typical middle strip in the short direction of the flat slab design example. As described earlier in Sec. 16.3, $f'_c = 3000$ psi and $f_y = 40,000$ psi.

Solution: (a) Moments in column and middle strips. The typical column strip is the column strip of equivalent rigid frame C of Fig. 16.10.2; but the typical middle strip is the sum of two half middle strips, taken from each of the two adjacent equivalent rigid frames C. The factored moments in the typical column and middle strips are shown in Table 16.13.1.

(b) Slab thickness for flexure. For $f'_c = 3000$ psi and $f_y = 40,000$ psi, the maximum percentage for tension reinforcement only is $0.75\rho_b = 0.0278$ (Table 3.6.1). The actual percentages used (line 6 of Tables 16.13.2 and 16.13.3) are nowhere near this maximum. Thus there is ample compressive strength in the slab. This phenomenon is usual because of the deflection control exerted by the minimum slab thickness requirements.

(c) Design of reinforcement. The design of reinforcement for the typical column strip is shown in Table 16.13.2; for the typical middle strip, it is shown in Table 16.13.3. Because the moments in the long direction are larger than those in the short direction, the larger effective depth is assigned to the long direction wherever the two layers of steel are in contact. This contact at crossing occurs in the top steel at the intersection of column strips

Table 16.13.1 FACTORED MOMENTS IN A TYPICAL COLUMN STRIP AND MIDDLE STRIP, EXAMPLE 16.13.2 (FLAT SLAB)

LINE NUMBER	MOMENTS AT CRITICAL SECTION (*ft-kips*)	EXTERIOR SPAN			INTERIOR SPAN		
		NEGATIVE MOMENT	POSITIVE MOMENT	NEGATIVE MOMENT	NEGATIVE MOMENT	POSITIVE MOMENT	NEGATIVE MOMENT
1	Total M in column and middle strips (Fig. 16.10.2) (rigid frame C)	−88	+146	−204	−190	+102	−190
2	Percentage to column strip (Table 16.12.5)	91.3%	60%	75%	75%	60%	75%
3	Moment in column strip	−80	+88	−153	−142	+61	−142
4	Moment in middle strip	−8	+58	−51	−48	+41	−48

Table 16.13.2 DESIGN OF REINFORCEMENT IN COLUMN STRIP, EXAMPLE 16.13.2 (FLAT SLAB) (f_y = 40,000 psi, f_c' = 3000 psi)

LINE NUMBER	CRITICAL SECTION	EXTERIOR SPAN			INTERIOR SPAN		
		NEGATIVE MOMENT	POSITIVE MOMENT	NEGATIVE MOMENT	NEGATIVE MOMENT	POSITIVE MOMENT	NEGATIVE MOMENT
1	Moment, Table 16.13.1, line 3 (ft-kips)	-80	$+88$	-153	-142	$+61$	-142
2	Width b of drop or strip (in.)	100	120	100	100	120	100
3	Effective depth d (in.)	8.81	6.44	8.81	8.81	6.44	8.81
4	M_u/ϕ (ft-kips)	-89	$+98$	-170	-158	$+68$	-158
5	R_n(psi) = $M_u/(\phi bd^2)$	138	236	263	244	164	244
6	ρ, Eq. (3.8.5) or Fig. 3.8.1	0.35%	0.62%	0.70%	0.64%	0.42%	0.64%
7	$A_s = \rho bd$	3.08	4.79	6.17	5.63	3.24	5.63
8	$A_s = 0.002bt^*$	2.40	1.80	2.40	2.40	1.80	2.40
9	N = larger of (7) or (8)/0.31	9.9	15.5	19.9	18.2	10.5	18.2
10	N = width of strip/(2t)	5	8	5	5	8	5
11	N required, larger of (9) or (10)	10	16	20	19	11	19
12	Use #5 bars	10 $st^\dagger$	6 st 10 bt	3 st 17 bt	3 st 17 bt	5 st 7 bt	5 st 14 bt

*$bt = 100(10.5) + 20(7.5) = 1200$ in.² for negative moment region.
†Bent bars at exterior supports may be used if a general analysis is made (ACI-Fig. 13.4.8).

Table 16.13.3 DESIGN OF REINFORCEMENT IN MIDDLE STRIP, EXAMPLE 16.13.2 (FLAT SLAB) (f_y = 40,000 psi, f_c' = 3000 psi)

LINE NUMBER	CRITICAL SECTION	EXTERIOR SPAN			INTERIOR SPAN		
		NEGATIVE MOMENT	POSITIVE MOMENT	NEGATIVE MOMENT	NEGATIVE MOMENT	POSITIVE MOMENT	NEGATIVE MOMENT
1	Moment, Table 16.13.1, line 4 (ft-kips)	-8	$+58$	-51	-48	$+41$	-48
2	Width of strip, b (in.)	180	180	180	180	180	180
3	Effective depth d (in.)	6.44	5.81	6.44	6.44	5.81	6.44
4	M_u/ϕ (ft-kips)	-9	$+64$	-57	-53	$+46$	-53
5	R_n (psi) = $M_u/(\phi bd^2)$	14	126	92	85	91	85
6	ρ, Eq. (3.8.5) or Fig. 3.8.1	0.04%	0.32%	0.23%	0.22%	0.23%	0.22%
7	$A_s = \rho bd$	0.46	3.30	2.67	2.53	2.41	2.53
8	$A_s = 0.002bt$	2.70	2.70	2.70	2.70	2.70	2.70
9	N = larger of (7) or (8)/0.31*	8.7	10.6	8.7	8.7	8.7	8.7
10	N = width of strip/(2t)	12	12	12	12	12	12
11	N required, larger of (9) or (10)	12	12	12	12	12	12
12	Use of #5 bars	12 $st^\dagger$	6 st 6 bt	12 bt	12 bt	6 st 6 bt	12 bt

*A mixture of #5 and #4 bars could have been selected.
†Bent bars at exterior supports may be used if a general analysis is made (ACI-Fig. 13.4.8).

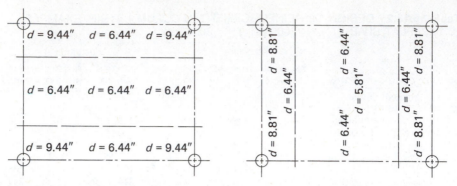

Figure 16.13.4 Effective depths provided at critical sections in flat slab design example.

and in the bottom steel at the intersection of middle strips. Assuming #5 bars and $\frac{3}{4}$ in. clear cover, the effective depths provided at various critical sections of the long and short directions are shown in Fig. 16.13.4.

16.14 BEAM (IF USED) SIZE REQUIREMENT IN FLEXURE AND SHEAR

The size of the beams along the column centerlines in a two-way slab (with beams) should be sufficient to take the bending moments and shears at the critical sections.

For approximately equal spans, the largest bending moment should occur at the exterior face of the first interior column where the available section for strength computation is rectangular in nature because the effective slab projection is on the tension side. Then with the preliminary beam size the required reinforcement ratio ρ may be determined and compared with $0.75\rho_b$, the maximum value permitted. Deflection is unlikely to be a problem with T-sections, but must be investigated if excessive deflection may cause difficulty.

The maximum shear in the beam should also occur at the exterior face of the first interior column. The shear diagram for the exterior span may be obtained by placing the negative moments already computed for the beam at the face of the column at each end and loading the span with the percentage of floor load interpolated between $\alpha_1 L_2/L_1 = 0$ and $\alpha_1 L_2/L_1 \geq 1.0$. The maximum nominal shear stress v_n should stay below, say, $6\sqrt{f_c'}$, as discussed in Sec. 10.2.

EXAMPLE 16.14.1 Investigate if the preliminary overall sizes of 14×28 in. for the long beam and 12×24 in. for the short beam are suitable for the two-way slab (with beams) design example.

Solution: Since the values of α, or of $\alpha_1 L_2/L_1$, are considerably larger than 1.0 for all beam spans, there is to be no reduction of the floor load going into the beams from the tributary areas (ACI-13.6.8). As shown in Fig. 16.14.1, the most critical span is $B1$ for the long direction and $B5$ for the

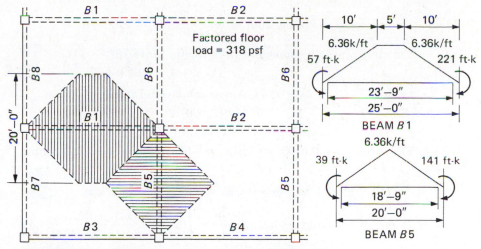

Figure 16.14.1 Beams around the two-way slab panel.

short direction. Actually the load acting on the clear span of the beam should include the floor load (including the weight of the beam stem itself or any other load) directly over the beam stem width plus the floor load on the tributary areas bounded by the 45° lines from the corner of the panel. Also for practical purposes it is acceptable to consider the shear due to floor load at the face of column equal to one half of the floor load on the tributary areas between column centerlines, as shown in Fig. 16.14.1.

(a) Size of long beam $B1$. The negative moments at the face of supports, 57 and 221 ft-kips, are taken from Table 16.12.2, frame A.

$$\text{weight of beam stem} = \frac{14(21.5)}{144}(150) = 314 \text{ lb/ft}$$

$$\text{maximum negative moment} = \tfrac{1}{10}(1.4)(0.314)(23.75)^2 + 221$$
$$= 25 + 221 = 246 \text{ ft-kips}$$

$b = 14$ in. $d = 28 - 2.5$ (assume one layer of steel) $= 25.5$ in.

$$R_n = \frac{M_u}{\phi b_w d^2} = \frac{246(12{,}000)}{0.90(14)(25.5)^2} = 360 \text{ psi}$$

From Fig. 3.8.1, $\rho = 0.010$, which is well below $0.75\rho_b = 0.027$. Perhaps the beam size should be reduced. From Fig. 16.14.1,

$$\text{total factored floor load on } B1 = 6.36(15) = 95.4 \text{ kips}$$

$$\text{max } V_u = 1.15(1.4)(0.314)\frac{23.75}{2} + \frac{1}{2}(95.4) + \frac{221 - 57}{23.75}$$

$$= 6.0 + 47.7 + 6.9 = 60.6 \text{ kips}$$

$$v_n = \frac{V_u}{\phi b_w d} = \frac{60{,}600}{0.85(14)(25.5)} = 200 \text{ psi} = 3.7\sqrt{f_c'} \qquad \text{OK}$$

(b) Size of short beam $B5$. The negative moments at the face of supports, 39 and 141 ft-kips, are taken from Table 16.12.2, frame C.

$$\text{weight of beam stem} = \frac{12(17.5)}{144}(150) = 219 \text{ lb/ft}$$

$$\text{maximum negative moment} = \tfrac{1}{10}(1.4)(0.219)(18.75)^2 + 141$$
$$= 11 + 141 = 152 \text{ ft-kips}$$

$$b = 12 \text{ in.} \qquad d = 24 - 2.5 \text{ (assume one layer of steel)} = 21.5 \text{ in.}$$

$$R_n = \frac{M_u}{\phi b_w d^2} = \frac{152(12,000)}{0.90(12)(21.5)^2} = 365 \text{ psi}$$

From Fig. 3.8.1, $\rho = 0.0105$, which is well below $0.75\rho_b = 0.027$. From Fig. 16.14.1,

$$\text{total factored floor load on } B5 = 6.36(10) = 63.6 \text{ kips}$$

$$\text{max } V_u = 1.15(1.4)(0.219)\frac{18.75}{2} + \frac{1}{2}(63.6) + \frac{141 - 39}{18.75}$$

$$= 3.3 + 31.8 + 5.4 = 40.5 \text{ kips}$$

$$v_n = \frac{V_u}{\phi b_w d} = \frac{40,500}{0.85(12)(21.5)} = 185 \text{ psi} = 3.4\sqrt{f_c'} \qquad \text{OK}$$

Both beams probably should have been made smaller.

16.15 SHEAR STRENGTH IN TWO-WAY FLOOR SYSTEMS

The shear strength of a flat slab or flat plate floor around a typical interior column under dead and full live loads is analogous to that of a square or rectangular spread footing subjected to a concentrated column load, except each is an inverted situation of the other. The area enclosed between the parallel pairs of centerlines of the adjacent panels of the floor is like the area of the footing, because there is no shear force along the panel centerline of a typical interior panel in a floor system. Consequently the discussion here is essentially identical to what is included in Chap. 20 on footings.

The shear strength of two-way slab systems without shear reinforcement has been studied by many investigators [30–36]. An excellent summary is provided by ASCE-ACI Task Committee 426 under Chairman N. M. Hawkins [38].

The shear strength of the flat slab or flat plate should be first investigated for wide-beam action and then for two-way action (ACI-11.11). In the wide beam action, the critical section is parallel to the panel centerline in the transverse direction and extends across the full distance between two adjacent longitudinal panel centerlines. As in one-way beams, this critical section of width b_w times the effective depth d is located at a distance d from the face of the equivalent square column capital or from the face of the drop panel, if any. The nominal strength in usual cases where no shear reinforcement is used is

$$V_n = V_c = 2\sqrt{f_c'}b_w d \qquad (16.15.1)$$

according to the simplified method of ACI-11.3.1.1. Alternatively, V_c may be determined using the more detailed expression (ACI-11.3.2.1) involving $\rho V_u d/M_u$, Eq. (5.10.5).

In the two-way action, potential diagonal cracking may occur along a truncated cone or pyramid around the column. Thus the critical section is located so that its periphery b_0 is at a distance equal to one-half of the effective depth through the drop from the periphery of the column capital, and also at a distance equal to one-half of the effective depth outside of the drop from the periphery of the drop. Where no drop is used, of course, there would be only one critical section for two-way action. If shear reinforcement is not used, the nominal shear strength is (ACI-11.11.2)

$$V_n = V_c = \left(2 + \frac{4}{\beta_c}\right)\sqrt{f'_c}\,b_0 d \leq 4\sqrt{f'_c}\,b_0 d \qquad \textbf{(16.15.2)}$$

where for a rectangular column capital or drop, β_c is the ratio of the long side to the short side of the rectangle. Unless β_c is larger than 2.0, the expression involving β_c does not control and V_c is limited to $4\sqrt{f'_c}\,b_0 d$.

Even when shear reinforcement is used (ACI-11.11.3.2), the nominal strength is limited to a maximum of

$$V_n = V_c + V_s \leq 6\sqrt{f'_c}\,b_0 d \qquad \textbf{(16.15.3)}$$

Further, in the design of any shear reinforcement, the portion of the strength V_c may not exceed $2\sqrt{f'_c}\,b_0 d$ (ACI-11.11.3.4). If shearhead reinforcement such as described in Sec. 16.16 is used (ACI-11.11.4), the maximum V_n in Eq. (16.15.3) is $7\sqrt{f'_c}\,b_0 d$.

The investigation for concentric shear (without moment) transfer from slab to column is shown in the following two examples, for the flat slab and flat plate design examples, respectively. When there must be transfer of both shear and moment from the slab to the column, ACI-11.12.2 applies, as will be discussed in Sec. 16.18.

EXAMPLE 16.15.1 Investigate the shear strength in wide-beam and two-way actions in the flat slab design example. Note that $f'_c = 3000$ psi.

Solution: (a) Wide-beam action. Investigation for the wide-beam action is made for sections 1-1 and 2-2 in the long direction, as shown in Fig. 16.15.1(a). The short direction has a wider critical section and shorter span; thus it does not control. For section 1-1, if the entire width of 20 ft is conservatively assumed to have an effective depth of 6.12 in.,

$$V_u = 0.336(20)(9.52) = 64 \text{ kips} \qquad \text{(section 1-1)}$$
$$V_n = V_c = 2\sqrt{f'_c}(240)(6.12)\tfrac{1}{1000} = 161 \text{ kips}$$
$$\phi V_n = 0.85(161) = 137 \text{ kips} > V_u \qquad\qquad\qquad \text{OK}$$

If, however, b_w is taken as 84 in. and d as 9.12 in. on the contention that the increased depth d is only over a width of 84 in.,

$$V_n = V_c = 2\sqrt{f'_c}(84)(9.12)\tfrac{1}{1000} = 84 \text{ kips}$$

This latter value is probably unrealistically low. For section 2-2, the shear resisting section has a constant d of 6.12 in.; thus

$$V_u = 0.336(20)(7.82) = 53 \text{ kips} \qquad \text{(section 2-2)}$$
$$\phi V_n = 137 \text{ kips} > V_u \qquad\qquad\qquad\qquad\qquad\qquad \text{OK}$$

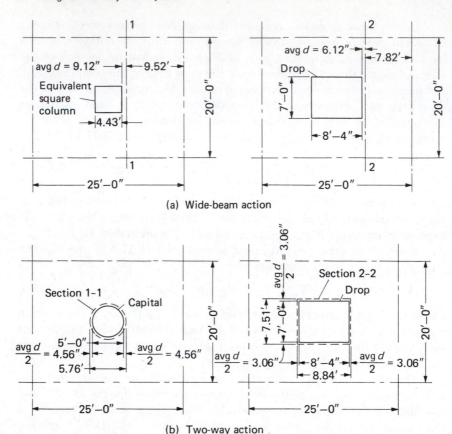

(a) Wide-beam action

(b) Two-way action

Figure 16.15.1 Critical sections for shear in flat slab design example.

It will be rare that wide-beam (one-way) action will govern.

(b) Two-way action. The critical sections for two-way action are the circular section 1-1 at $d/2 = 4.56$ in. from the edge of the column capital and the rectangular section 2-2 at $d/2 = 3.06$ in. from the edge of the drop, as shown in Fig. 16.15.1(b). Since there are no shearing forces at the centerlines of the four adjacent panels, the shear forces around the critical sections 1-1 and 2-2 in Fig. 16.15.1(b) are

$$V_u = 0.336 \left[500 - \frac{\pi(5.76)^2}{4} \right] + 1.4(0.038) \left[7(8.33) - \frac{\pi(5.76)^2}{4} \right]$$

$$= 159.2 + 1.7 = 161 \text{ kips} \quad \text{(section 1-1)}$$

In the second term, the 0.038 is the weight of the 3-in. drop in ksf.

$$V_u = 0.336 \left[500 - 8.84(7.51) \right] = 146 \text{ kips} \quad \text{(section 2-2)}$$

The corresponding shear strengths are

$$\phi V_n = \phi V_c = \phi(4\sqrt{f_c'})b_0 d$$

$$= 0.85(4\sqrt{f_c'})(69.12)(9.12)\tfrac{1}{1000} = 369 \text{ kips} \quad \text{(section 1-1)}$$

$$= 0.85(4\sqrt{f_c'})(2)(106.12 + 90.12)(6.12)\tfrac{1}{1000} = 447 \text{ kips (section 2-2)}$$

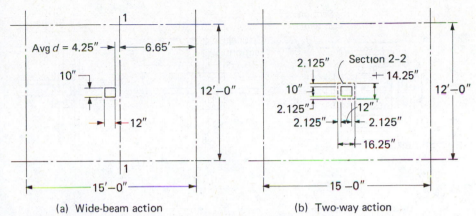

(a) Wide-beam action   (b) Two-way action

Figure 16.15.2 Critical sections for shear in flat plate design example.

The value of V_c is $4\sqrt{f'_c}\, b_0 d$ since β_c is less than 2.0. For both sections, ϕV_n is well above V_u; thus shear reinforcement is not required.

EXAMPLE 16.15.2 Investigate nominal shear stresses in wide-beam and two-way actions in the flat plate design example. Note that $f'_c = 4000$ psi.

Solution: (a) Wide-beam action. Assuming $\frac{3}{4}$ in. clear cover and #4 bars, the average effective depth when bars in two directions are in contact is

$$\text{avg } d = 5.50 - 0.75 - 0.50 = 4.25 \text{ in.}$$

Referring to Fig. 16.15.2(a),

$$v_n = \frac{V_u}{\phi b_w d} = \frac{198(12)(6.65)}{0.85(144)(4.25)} = 30.4 \text{ psi} < (2\sqrt{f'_c} = 126 \text{ psi}) \quad \text{(section 1-1)}$$

(b) Two-way action. Referring to Fig. 16.15.2(b),

$$v_n = \frac{V_u}{\phi b_0 d} = \frac{198[180 - 1.35(1.19)]}{0.85(61)(4.25)} = 160 \text{ psi} < (4\sqrt{f'_c} = 253 \text{ psi}) \quad \text{(section 2-2)}$$

Thus shear reinforcement is not required for this flat plate floor.

16.16 SHEAR REINFORCEMENT IN FLAT PLATE FLOORS

In flat plate floors where neither column capitals nor drop panels are used, shear reinforcement is frequently necessary. In such cases two-way action usually controls. The shear reinforcement may take the form of properly anchored bars or wires placed in vertical sections around the column [Fig. 16.16.1(a)], or consist of shearheads, which are steel I- or channel-shaped sections fabricated by welding into four (or three for an exterior column) identical arms at right angles and uninterrupted within the column section [Fig. 16.16.1(b)]. A summary of shear reinforcement for flat plates has been provided by Dilger and Ghali [77]. The strength of two-way slab systems with shear reinforcement has been summarized by Hawkins [39]. Corley and Hawkins [40,46] have studied shearhead reinforcement. Other studies

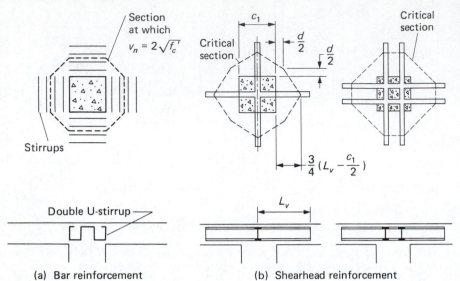

Figure 16.16.1 Bar and shearhead reinforcement in flat plate floors.

of shear reinforcement in flat plates have been made by Ghali, Dilger et al. [37,77–80], and Pillai, Kirk, and Scavuzzo [81].

When *bar or wire shear reinforcement* is used, the nominal strength is

$$V_n = V_c + V_s = 2\sqrt{f'_c}b_0d + \frac{A_v f_y d}{s} \tag{16.16.1}$$

where b_0 is the periphery around the critical section for two-way shear action and A_v is the total stirrup bar area around b_0. Such bar or wire reinforcement is required wherever V_u exceeds ϕV_c based on V_c of $(2 + 4/\beta_c)\sqrt{f'_c}b_0d$ (or $4\sqrt{f'_c}b_0d$ maximum); however, in the design of shear reinforcement V_c for Eq. (16.16.1) equals $2\sqrt{f'_c}b_0d$, and the maximum strength V_s from the shear reinforcement may not exceed $4\sqrt{f'_c}b_0d$ according to ACI-11.11.3.2.

Shear strength may be provided by *shearheads* under ACI-11.11.4 whenever V_u/ϕ at the critical section is between $(2 + 4/\beta_c)\sqrt{f'_c}b_0d$ and $7\sqrt{f'_c}b_0d$. These provisions, based on the tests of Corley and Hawkins [40], apply only where shear alone (i.e., no bending moment) is transferred at an interior column. When there is moment transfer to columns, ACI-11.12.2.5 applies, as is discussed in Sec. 16.18.

With regard to the size of the shearhead, it must furnish a ratio α_v of 0.15 or larger (ACI-11.11.4.5) between the stiffness for each shearhead arm $(E_s I_x)$ and that for the surrounding composite cracked slab section of width $(c_2 + d)$; or

$$\min \alpha_v = \frac{E_s I_x}{E_c (\text{composite } I_s)} = 0.15 \tag{16.16.2}$$

The steel shape used must not be deeper than 70 times its web thickness; and the compression flange must be located within $0.3d$ of the compression surface of the slab (ACI-11.11.4.2 and 11.11.4.4). In addition, the plastic moment capacity M_p of the shearhead arm must be at least (ACI-11.11.4.6).

$$\min M_p = \frac{V_u}{2\eta\phi}\left[h_v + \alpha_v\left(L_v - \frac{c_1}{2}\right)\right]$$ (16.16.3)

where

> η = number (usually 4) of identical shearhead arms
> V_u = factored shear around the periphery of column face
> h_v = depth of shearhead
> L_v = length of shearhead measured from column centerline
> ϕ = 0.90, the strength reduction factor for flexure

Equation (16.16.3) is to assure that the required shear strength of the slab is reached before the flexural strength of the shearhead is exceeded.

The length of the shearhead should be such that the nominal shear strength V_n will not exceed $4\sqrt{f_c'}b_0d$ computed at a peripheral section located at $\frac{3}{4}(L_v - c_1/2)$ along the shearhead but no closer elsewhere than $d/2$ from the column face (ACI-11.11.4.7 and 11.11.4.8). This length requirement is shown in Fig. 16.16.1(b).

When a shearhead is used, it may be considered to contribute a resisting moment

$$M_v = \frac{\phi\alpha_v V_u}{2\eta}\left(L_v - \frac{c_1}{2}\right)$$ (16.16.4)

to each column strip, but not more than 30% of the total moment resistance required in the column strip, nor the change in column strip moment over the length L_v, nor the required M_p given by Eq. (16.16.3).

EXAMPLE 16.16.1 Using the dimensions of the flat plate design example but changing the live load to 200 psf, investigate the shear strength for wide-beam and two-way actions around an interior column. If the required nominal shear strength V_n for two-way action is between $4\sqrt{f_c'}b_0d$ (which controls since β_c is less than 2.0) and $6\sqrt{f_c'}b_0d$, determine the A_v/s requirement for shear reinforcement at the peripheral critical section and show the nominal shear stress (which is factored shear V_u divided by $\phi b_0 d$) variation from the critical section to the panel centerline. Use $f_c' = 4000$ psi and $f_y = 50,000$ psi; assume # 5 slab reinforcement.

Solution: (a) Wide-beam action.

$$w_u = 1.4w_D + 1.7w_L = 1.4(150)(5.5/12) + 1.7(200)$$

$$= 96 + 340 = 436 \text{ psf}$$

$$\text{avg } d \text{ in column strip} = 5.50 - 0.75 - 0.63 = 4.12 \text{ in.}$$

For a 12-in. wide strip along section 1-1 of Fig. 16.16.2,

$$v_n = \frac{V_u}{\phi b_w d} = \frac{436(6.66)}{0.85(12)(4.12)} = 69 \text{ psi} < (2\sqrt{f_c'} = 126 \text{ psi}) \quad \text{OK}$$

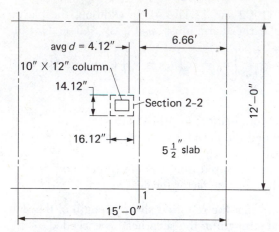

Figure 16.16.2 Critical sections for shear, Example 16.16.1.

(b) Two-way action. Referring to section 2-2 of Fig. 16.16.2,

$$v_n = \frac{V_u}{\phi b_o d} = \frac{436[180 - 1.34(1.18)]}{0.85(60.48)(4.12)} = 368 \text{ psi}$$

Since the maximum nominal shear stress of 368 psi is between $4\sqrt{f'_c} = 253$ psi and $6\sqrt{f'_c} = 380$ psi, shear reinforcement is required to take the excess stress v_n which exceeds $2\sqrt{f'_c} = 126$ psi. The shear reinforcement in this case may consist of properly anchored bars or wires and need not be a shearhead. The variation of the nominal shear stress from the maximum value of 368 psi to zero at the panel centerline over the equally spaced points 1 to 5 is shown in Fig. 16.16.3. The A_v/s requirement around the critical section of 60.48 in. periphery is, from applying Eq. (16.16.1) with required $V_n = V_u/\phi$,

$$\frac{A_v}{s} = \frac{V_u/\phi - (2\sqrt{f'_c})b_o d}{f_y d} = \frac{(v_n - 2\sqrt{f'_c})b_o}{f_y}$$

$$= \frac{(368 - 126)(60.48)}{50,000} = 0.292 \text{ in.}$$

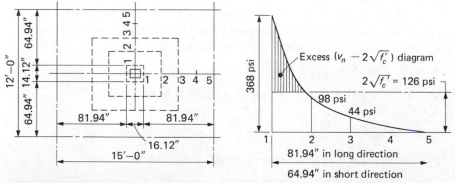

Figure 16.16.3 Variation of two-way nominal shear stress ($V_u/\phi b_o d$), Example 16.16.1.

Assuming $s = d/2 \approx 2$ in. spacing,

$$A_v = 0.58 \text{ sq in.}$$

If two double #3 U stirrups are used at each of the four sides,

$$\text{provided } A_v = 4(4)(0.11) = 1.76 \text{ sq in.}$$

As shown in Fig. 16.16.3, the nominal shear stress $(V_u/\phi b_0 d)$ drops to 126 psi in a rather steep manner so that the number and spacing of these U stirrips can be laid out by the aid of the excess $(v_n - 2\sqrt{f_c'})$ diagram.

EXAMPLE 16.16.2 Design the shearhead reinforcement for the two-way shear action of Example 16.16.1.

Solution: (a) Two-way action nominal shear stress. Since the maximum nominal shear stress $v_n = V_u/\phi b_0 d$ of 368 psi is between $4\sqrt{f_c'} = 253$ psi and $7\sqrt{f_c'} = 442$ psi, shearhead reinforcement for the interior column (having no moment transfer to column) may be designed according to ACI-11.11.4.

(b) Length of shearhead. The length of shearhead should be such that the nominal shear stress be less than $4\sqrt{f_c'}$, computed around a periphery passing through points at $\frac{3}{4}(L_v - c_1/2)$ from but no closer than $d/2$ to the column faces. Assuming a square as the critical periphery since the shearhead is to have four identical arms (ACI-11.11.4.1), the required b_0 (ft) may be computed from Fig. 16.16.4; thus,

$$4\sqrt{4000} = \frac{436\,[180 - (b_0/4)^2]}{0.85(12b_0)(4.12)}$$

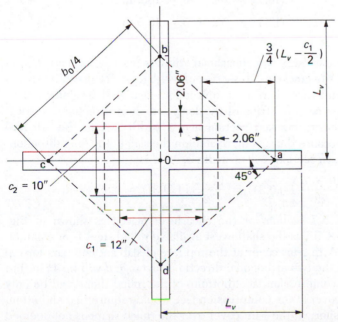

Figure 16.16.4 Required length of shearhead, Example 16.16.2.

Neglecting the $(b_0/4)^2$ in the numerator,

$$b_0 = \frac{436(180)}{253(0.85)(4.12)(12)} = 7.4 \text{ ft (88.5 in.)}$$

The required distance L_v may be computed from the following geometric considerations. From right triangle oab,

$$\left[\frac{3}{4}\left(L_v - \frac{c}{2}\right) + \frac{c}{2}\right]\sqrt{2} = \frac{b_0}{4}$$

which gives, based on leg ob,

$$L_v = \left(\frac{88.5}{4\sqrt{2}} - 5\right)\frac{4}{3} + 5 = 19.2 \text{ in.}$$

and, based on leg oa,

$$L_v = \left(\frac{88.5}{4\sqrt{2}} - 6\right)\frac{4}{3} + 6 = 18.8 \text{ in.}$$

For the periphery $abcd$ not to approach closer than $d/2$ to the periphery of the column section,

$$oa = ob = 15.12 \text{ in.}$$

But,

$$oa = 6 + \tfrac{3}{4}(L_v - 6) \quad \text{and} \quad ob = 5 + \tfrac{3}{4}(L_v - 5)$$

which gives

$$L_v = (15.12 - 6)\tfrac{4}{3} + 6 = 18.2 \text{ in.}$$
$$L_v = (15.12 - 5)\tfrac{4}{3} + 5 = 18.5 \text{ in.}$$

Use $L_v = 20$ in.

(c) Size of shearhead. The shearhead stiffness must be at least 0.15 of that of the composite cracked slab section of width $(c_2 + d)$. It can be shown that 14-#5 bars and 10-#5 bars are required for negative slab reinforcement in the 72-in. wide column strips of the long and short directions, respectively. The composite cracked section across width A-A in Fig. 16.16.5 should be used because there is more steel in the slab in the long direction. The steel area A_s in section A-A of width $c_2 + d$ is

$$A_s = \frac{10 + d}{72}(14)(0.31) = \frac{14.12}{72}(14)(0.31) = 0.85 \text{ sq in.}$$

Assume a S3 × 5.7 section for the shearhead placed as shown in Fig. 16.16.5. The S3 × 5.7 is the shallowest available rolled steel I- or channel-shaped section. With $\tfrac{3}{4}$-in. cover at the top face of slab and #5 bars for top reinforcement in the two orthogonal directions, average d will be $4\tfrac{1}{8}$ in. but the cover to the compression face (bottom) of the rolled shape will be only $\tfrac{1}{2}$ in. Even $\tfrac{3}{4}$-in. cover at the compression face would require that all bottom slab steel be cut short. If the $\tfrac{1}{2}$-in. cover over the rolled shape is not deemed adequate either a thicker slab must be used or a shallower shearhead fabricated (welded) from three plates would have to be used.

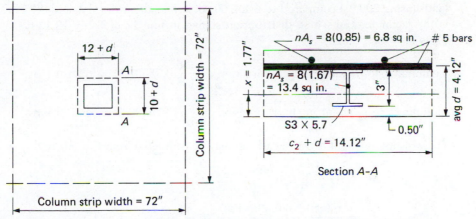

Figure 16.16.5 Cracked slab section of width ($c_2 + d$), Example 16.16.2.

The centroidal axis of the composite cracked section may be obtained by equating the static moments of the compression and tension transformed areas,

$$\frac{14.12x^2}{2} = 13.4(2.0 - x) + 6.8(4.12 - x)$$

$$x = 1.70 \text{ in.}$$

$$\text{composite } I_s = \frac{14.12(1.70)^3}{3} + n(I_x \text{ of steel section})$$

$$+ 13.4(0.30)^2 + 6.8(2.42)^2$$

$$= 23.1 + 8(2.52) + 1.2 + 39.8 = 87.3 \text{ in.}^3$$

$$\text{provided } \alpha_v = \frac{E_s(2.52)}{E_c \text{ (composite } I_s)} = \frac{8(2.52)}{87.3} = 0.23 > 0.15 \quad \text{OK}$$

The plastic section modulus of the S3 × 5.7 is given by the *AISC Manual** as 1.95 in.[3]. Using A36 steel, the provided M_p is

$$\text{provided } M_p = 36(1.95) = 70.2 \text{ in.-kips}$$

The required M_p is computed from using Eq. (16.16.4) as

$$\text{required } M_p = \frac{V_u}{8\phi}\left[h_v + \alpha_v\left(\text{required } L_v - \frac{c_1}{2}\right)\right]$$

$$= \frac{0.436(180)}{8(0.90)}[3 + 0.23(19.2 - 5)]$$

$$= 68.3 \text{ in.-kips} < 70.2 \text{ in-kips} \quad \text{OK}$$

(d) Shearhead contribution to resist negative moment in slab. The negative moments at the face of column in the 72-in. column strip width in the long and short directions are $(436/198)(0.75)$ times those for equivalent rigid frames A and C in Fig. 16.10.3, wherein $(436/198)$ is the ratio of factored

---

*See *Manual of Steel Construction* (8th ed.). Chicago: American Institute of Steel Construction, 1980.

loads (using 200 psf compared to using 60 psf live load) on the slab and 0.75 is the factor for transverse distribution shown in line 12 of Table 16.12.6. Thus

$$\text{column strip moment in long direction} = \frac{436}{198}(0.75)(37.8) = 62.4 \text{ ft-kips}$$

$$\text{column strip moment in short direction} = \frac{436}{198}(0.75)(30.1) = 49.7 \text{ ft-kips}$$

The resisting moment of the shearhead may be computed from Eq. (16.16.4),

$$M_v = \frac{\phi \alpha_v V_u}{8}\left(L_v - \frac{c_1}{2}\right)$$

$$= \frac{0.90(0.23)(0.436)(180)}{8(12)}[(20 - 6) \text{ or } (20 - 5)]$$

$$= 2.37 \text{ or } 2.54 \text{ ft-kips}$$

Thus the contribution is rather small and the revision of slab reinforcement is unnecessary.

16.17 DIRECT DESIGN METHOD—MOMENTS IN COLUMNS

The moments in columns due to unbalanced loads on adjacent panels are readily available when an elastic analysis is performed on the equivalent rigid frame for the various pattern loadings. In the "direct design method," wherein the six limitations listed in Sec. 16.7 are satisfied, the longitudinal moments in the slab are prescribed by the provisions of the Code (ACI-13.6.3). In a similar manner, the Code prescribes the unbalanced moment at an interior column as follows [ACI Formula (13-4)]:

$$M = 0.07\left[\left(w_D + \frac{1}{2}w_L\right)L_2 L_n^2 - w_D' L_2'(L_n')^2\right] \qquad \textbf{(16.17.1)}$$

where

$$w_D = \text{factored dead load per unit area}$$
$$w_L = \text{factored live load per unit area}$$
$$w_D', L_2', L_n' = \text{quantities referring to shorter span}$$

The moment is yet to be distributed between the two ends of the upper and lower columns meeting at the joint.

The rationale for Eq. (16.17.1) may be observed from the stiffness ratios at a typical interior joint shown in Fig. 16.17.1(a), wherein the distribution factor for the sum of the column end moments is taken as $\frac{7}{8}$ and the unbalanced moment in the column strip is taken to be 0.080/0.125 times the difference in the total static moments due to dead plus half live load on the longer span and dead load only on the shorter span.

For the edge column, ACI-13.6.3.6 requires using the column strip nominal moment strength *provided* as the moment to be transferred. Alternatively, one might apply Eq. (16.17.1) taking the shorter span as zero, as shown in Fig. 16.17.1(b).

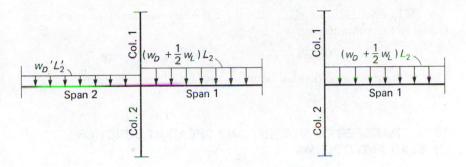

(a) Interior joint   (b) Exterior joint

Figure 16.17.1   Direct design method—moments in columns.

EXAMPLE 16.17.1 Obtain the moments in the interior and exterior columns in each direction for the flat plate design example.

Solution: (a) Exterior column, long direction.

$$w_D = 1.4(150)(5.5/12) = 1.4(69) = 96 \text{ psf}$$

$$w_L = 1.7(60) = 102 \text{ psf}$$

The factored moment M_u to be transferred to the columns is, according to Eq. (16.17.1),

$$M_u = 0.07(0.096 + 0.051)(12)(15 - 1)^2 = 24.2 \text{ ft-kips}$$

This moment is to be divided between upper and lower columns in proportion to their stiffnesses (in this case, equally).

The column strip nominal moment strength *provided* (ACI-13.6.3.6) is not available (i.e., reinforcement not selected) in this example. However, the required amount would be based on M_u equal to 98.8% (Table 16.12.6 for frame A) of the 15.1 ft-kips from Table 16.10.2 for frame A. This moment is much lower than the 24.2 ft-kips from ACI Formula (13-4). Conservatively, the larger of the slab requirement or ACI Formula (13-4) should be used.

(b) Interior column, long direction. The factored moment to be transferred to the columns is

$$M_u = 0.07[(0.096 + 0.051)(12)(15 - 1)^2 - 0.096(12)(15 - 1)^2]$$

$$= 0.07(0.051)(12)(14)^2 = 8.4 \text{ ft-kips}$$

This moment is to be divided between upper and lower columns.

(c) Exterior column, short direction. The factored moment to be transferred to the columns is, according to Eq. (16.17.1),

$$M_u = 0.07(0.096 + 0.051)(15)(12 - 0.83)^2 = 19.3 \text{ ft-kips}$$

This moment is to be divided between upper and lower columns. The 19.3 ft-kips computed above is more conservative than the column strip moment M_u (99.3% of 12.0 ft-kips) indicated for the exterior support for frame C (Tables 16.10.2 and 16.12.6).

(d) Interior column, short direction. The factored moment to be transferred to the columns is

$$M_u = 0.07(0.051)(15)(12 - 0.83)^2 = 6.7 \text{ ft-kips}$$

This moment is to be divided between upper and lower columns.

16.18 TRANSFER OF MOMENT AND SHEAR AT JUNCTION OF SLAB AND COLUMN

Inasmuch as the columns meet the slab at monolithic joints, there must be moment as well as shear transfer between the slab and the column ends. The moments may arise out of lateral loads due to wind or earthquake effects acting on the multistory frame, or they may be due to unbalanced gravity loads as considered in Sec. 16.17. In addition, the shear forces at the column ends and throughout the columns must be considered in the design of lateral reinforcement (ties or spiral) in the columns (ACI-11.12.1.1). The transfer of moment and shear at the slab-column interface is extremely important in the design of flat plates and has been the subject of numerous research studies [41–48,82–86].

Let M_u be the total factored moment that is to be transferred to both ends of the columns meeting at an exterior or an interior joint. Test results by Hanson and Hanson [42] have shown that about 60% of the moment is transferred by flexure and the remainder by unbalanced shear stresses around the critical periphery located at $d/2$ from the column faces. The ACI Code requires the division of the total factored moment M_u into M_b transferred by flexure (ACI-13.3.3) and M_v transferred by shear (ACI-11.12.2.3) such that

$$M_b = \frac{M_u}{1 + \dfrac{2}{3}\sqrt{\dfrac{c_1 + d}{c_2 + d}}} \tag{16.18.1}$$

and

$$M_v = M_u - M_b \tag{16.18.2}$$

The moment M_b is considered to be transferred through a slab width equal to $(c_2 + 3t)$ at the column (ACI-13.3.3), where t is the slab or drop panel thickness. Concentration of reinforcement in this width by closer spacing or additional reinforcement may be used to resist this moment.

Note that the quantities $(c_1 + d)$ and $(c_2 + d)$, appearing as numerator and denominator, of Eq. (16.18.1) really are the dimensions of the critical sections (Fig. 16.18.1) in the longitudinal and transverse directions, respectively. Thus, Eq. (16.18.1) is applicable as stated only for an interior column. For an exterior column the numerator $(c_1 + d)$ should be taken as $(c_1 + d/2)$.

If $c_1 = c_2$, Eq. (16.18.1) becomes

$$M_b = 0.60M_u$$

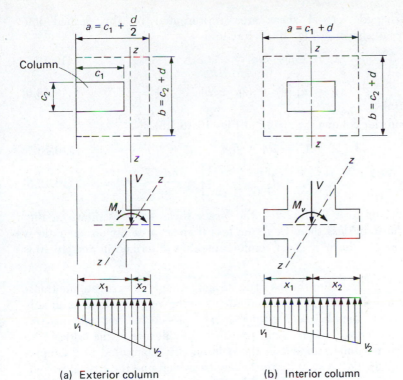

(a) Exterior column   (b) Interior column

Figure 16.18.1 Shear transfer of moment to columns.

If $c_2 = 2c_1$ and $c_1 = d$, Eq. (16.18.1) becomes

$$M_b = 0.648 M_u$$

It appears reasonable that when c_2 is larger than c_1, the moment transferred by flexure is greater because the effective slab width $(c_2 + 3t)$ resisting the moment is relatively larger.

The moment M_v transferred by shear acts with the associated shear force V_u at the centroid of the shear area around the critical periphery located at $d/2$ from the column faces, as shown in Fig. 16.18.1. Referring to Fig. 16.18.1,

$$v_1 = \frac{V_u}{\phi A_c} - \frac{M_v x_1}{\phi J_c} \qquad (16.18.3)$$

$$v_2 = \frac{V_u}{\phi A_c} + \frac{M_v x_2}{\phi J_c} \qquad (16.18.4)$$

It is noted that in order to be consistent with the strength design method, the strength reduction factor ϕ is included in the denominators of Eqs. (16.18.3) and (16.18.4). By using a section property J_c analogous to the polar moment of inertia about the z-z axis (perpendicular to the column and located at the centroid of the shear area) of the shear areas around the critical periphery, it is assumed that there are both horizontal and vertical shear stresses on the shear areas with dimensions a by d in Fig. 16.18.1. For an exterior column, x_1 and x_2 are obtained by locating the centroid of the channel-shaped vertical shear area represented by the dashed line $(a + b + a)$ shown in Fig. 16.18.1(a), and

$$A_c = (2a + b)d \qquad (16.18.5)$$

$$J_c = d\left[\frac{2a^3}{3} - (2a + b)(x_2)^2\right] + \frac{ad^3}{6} \qquad (16.18.6)$$

For an interior column, referring to Fig. 16.18.1(b),

$$A_c = 2(a + b)d \qquad (16.18.7)$$

$$J_c = d\left[\frac{a^3}{6} + \frac{ba^2}{2}\right] + \frac{ad^3}{6} \qquad (16.18.8)$$

According to ACI-11.12.2.4, the larger shear stress v_2 shown in Fig. 16.18.1 must be less than $4\sqrt{f_c'}$ (or less if shear area dimension ratio is larger than 2), otherwise shear reinforcement as described in Sec. 16.16 is required.

EXAMPLE 16.18.1 For the flat plate design example, investigate the transfer of the unbalanced gravity load moments in the long direction, as already computed in Example 16.17.1, to the exterior and interior columns, respectively. Compare the moment transferred by flexure in the critical slab width with the total moment in the column strip computed in Examples 16.10.3 and 16.12.3. Compute the shear stresses around the critical periphery due to the moment transferred by shear.

Solution: (a) Exterior column (long direction), transfer by flexure. From Example 16.17.1, the moment to be transferred is

$$M_u = 24.2 \text{ ft-kips}$$

From Eq. (16.18.1), using the average effective depth for #4 slab reinforcement of 4.25 in.,

$$M_b = \frac{M_u}{1 + \dfrac{2}{3}\sqrt{\dfrac{c_1 + d/2}{c_2 + d}}} = \frac{24.2}{1 + \dfrac{2}{3}\sqrt{\dfrac{12 + 2.125}{10 + 4.25}}} = 0.60(24.2)$$

$$= 14.5 \text{ ft-kips}$$

As shown by Fig. 16.18.2, this moment is in a critical slab width of 26.5 in. From Table 16.12.6 and Fig. 16.10.3, the total amount in the 72-in. wide column strip is

$$M \text{ in column strip} = 0.988(15.1) = 15 \text{ ft-kips}$$

If the slab reinforcement is placed at equal spacing in the column strip, additional reinforcement is needed in the 26.5 in. width for a moment of

$$M_b - 15\left(\frac{26.5}{72}\right) = 14.5 - 5.5 = 9 \text{ ft-kips}$$

(b) Exterior column (long direction), transfer by shear. The shear force V is taken as $(w_D + w_L)$ times the floor area of 12×7.5 ft around the exterior column.

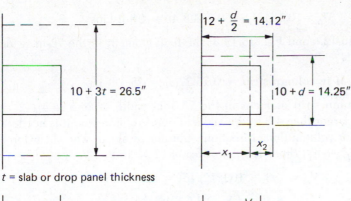

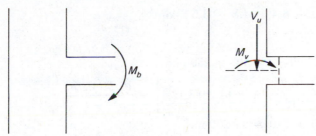

t = slab or drop panel thickness

Figure 16.18.2   Transfer of moments at exterior column, Example 16.18.1.

$$V_u = (96 + 102)(12)(7.5) = 17{,}800 \text{ lb}$$

$$M_v = M_u - M_b = 24.2 - 14.5 = 9.7 \text{ ft-kips}$$

From Fig. 16.18.2,

$$x_2 = \frac{2(14.12)(7.06)}{28.24 + 14.25} = 4.70 \text{ in.}$$

$$A_c = 4.25(28.24 + 14.25) = 180 \text{ in.}^2$$

$$J_c = 4.25\left[\frac{2(14.12)^3}{3} - 42.49(4.70)^2\right] + \frac{14.12(4.25)^3}{6}$$

$$= 3990 + 180 = 4170 \text{ in.}^4$$

$$v_1 = \frac{17{,}800}{0.85(180)} - \frac{9700(12)(9.42)}{0.85(4170)} = 116 - 309 = -193 \text{ psi}$$

$$v_2 = \frac{17{,}800}{0.85(180)} + \frac{9700(12)(4.70)}{0.85(4170)} = 116 + 154 = 270 \text{ psi}$$

The capacity v_c is $4\sqrt{f_c'} = 253$ psi when no shear reinforcement is provided. Additional shear reinforcement must be provided as done in Example 16.16.1.

(c) Interior column (long direction), transfer by flexure. From Example 16.17.1, the moment to be transferred is

$$M_u = 8.4 \text{ ft-kips}$$

$$M_b = \frac{M_u}{1 + \dfrac{2}{3}\sqrt{\dfrac{12 + 4.25}{10 + 4.25}}} = 0.584 M_u$$

$$M_b = 0.584M_u = 0.584(8.4) = 4.8 \text{ ft-kips}$$

From Table 16.12.6 and Fig. 16.10.3, the total moment in the 72-in. wide column strip is

$$M \text{ in column strip} = 0.75(37.8) = 28.4 \text{ ft-kips}$$

Since the column strip moment in the 26.5-in. width of $26.5(28.4)/72 = 10.5$ ft-kips is larger than 4.8 ft-kips, no additional reinforcement is needed.

(d) Interior column (long direction), transfer by shear. The shear force is taken as $(w_D + w_L)$ times the floor area of 12×15 ft.

$$V_u = (96 + 102)(12)(15) = 35,600 \text{ lb}$$

$$M_v = 8.4 - 4.8 = 3.6 \text{ ft-kips}$$

From Fig. 16.18.3,

$$A_c = 4.25(32.50 + 28.50) = 259 \text{ in.}^2$$

$$J_c = 4.25\left[\frac{2(16.25)^3}{12} + 2(14.25)(8.12)^2\right] + \frac{16.25(4.25)^3}{6}$$

$$= 11,040 + 210 = 11,250 \text{ in.}^4$$

$$v_1 = \frac{35,600}{0.85(259)} - \frac{3600(12)(8.12)}{0.85(11,250)} = 162 - 37 = 125 \text{ psi}$$

$$v_2 = \frac{35,600}{0.85(259)} + \frac{3600(12)(8.12)}{0.85(11,250)} = 162 + 37 = 199 \text{ psi}$$

The capacity v_c is $4\sqrt{f_c'} = 253$ psi when no shear reinforcement is provided.

Moment Transfer from Flat Plate to Column When Shearheads Are Used.

Tests [40,46] have indicated that shear stresses computed for factored loads at the critical section distance $d/2$ from the column face are appropriate for transfer of $M_v = M_u - M_b$ as described above, even when shearheads are used. However, the critical section for V_u is at a periphery passing through points at $\frac{3}{4}(L_v - c_1/2)$ from but no closer than $d/2$ to the column faces. When there are both V_u and M_u to be transferred, ACI-11.12.2.5 requires that the sum of the shear stresses computed for M_v and V_u at their respective locations not exceed $\phi(4\sqrt{f_c'})$. The reason for this apparent inconsistency (ACI Commentary-11.12.2.5) is that these two critical sections are in close proximity at the column corners where the failures initiate (under factored loads, of course).

EXAMPLE 16.18.2 Recompute the periphery length b_0 required, located at $\frac{3}{4}(L_v - c/2)$ but no closer than $d/2$ from the column face for the shearhead in Example 16.16.2 when there is unbalanced moment at the interior column equal to that of Eq. (16.17.1), ACI Formula (13-4).

Solution: (a) Determine whether or not additional reinforcement is necessary for moment transfer. Referring to Example 16.16.1, part (a) of solution,

$$w_D = 1.4(5.5)(150/12) = 96 \text{ psf}$$

$$w_L = 1.7(200) = 340 \text{ psf}$$

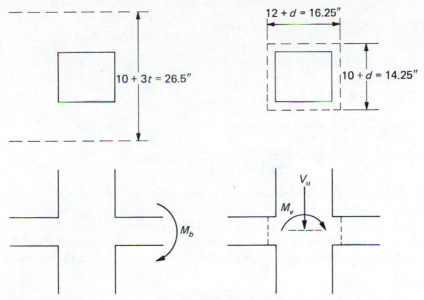

Figure 16.18.3 Transfer of moments in the long direction at interior column, Example 16.18.1.

Referring to Example 16.16.2, part (d) of solution,

$$\text{column strip moment in long direction} = 62.4 \text{ ft-kips}$$

The moment to be transferred to the column is, using Eq. (16.17.1),

$$M_u = 0.07[(0.096 + 0.170)(12)(15 - 1)^2 - 0.096(12)(15 - 1)^2]$$

$$= 0.07(0.170)(12)(14)^2 = 28.0 \text{ ft-kips}$$

Referring to Example 16.18.1, part (c) of solution,

$$M_b = 0.584M_u = 0.584(28.0) = 16.4 \text{ ft-kips}$$

Column strip moment in 26.5-in. width of Fig. 16.18.3 is

$$62.4\left(\frac{26.5}{72}\right) = 23.0 \text{ ft-kips}$$

Additional reinforcement is needed to take $(28.0 - 23.0) = 5.0$ ft-kips within the 26.5-in. width unless $28.0/62.4 = 44.9\%$ of the total column strip reinforcement is concentrated in the 26.5-in. width.

(b) Compute nominal shear stress at critical section of Fig. 16.18.3 due to M_v only.

$$M_v = 28.0 - 16.4 = 11.6 \text{ ft-kips}$$

Using critical section properties in Example 16.18.1, part (d) of solution,

$$v_n = \frac{11,600(12)(8.12)}{0.85(11,250)} = 118 \text{ psi}$$

(c) Compute the required periphery b_0. Referring to Example 16.16.2, part (b) of solution,

$$4\sqrt{4000} - [118 \text{ from part (b)}] = \frac{436[180 - (b_0/4)^2]}{0.85(12b_0)(4.12)}$$

Neglecting the $(b_0/4)^2$ in the numerator,

$$b_0 = \frac{436(180)}{135(0.85)(12)(4.12)} = 13.83 \text{ ft}$$

Placing $b_0 = 13.83$ ft in the numerator and solving for b_0 again

$$b_0 = \frac{436[180 - (3.46)^2]}{135(0.85)(12)(4.12)} = 12.91 \text{ ft}$$

(d) Discussion. This example is for illustration of the procedure only. Actually when the service live load is increased from 60 psf in the original flat plate design example to 200 psf, the slab thickness of $5\frac{1}{2}$ in. would have to be increased if the spans are not reduced. The requirement of ACI-11.12.2.5 is expected to be more controlling for the shearhead in the exterior column.

16.19 OPENINGS AND CORNER CONNECTIONS IN FLAT SLABS

When openings and corner connections are present in flat slab floors, designers must make sure that adequate provisions for them are made. The ASCE-ACI Joint Task Committee [38] has summarized available information. Recent tests by Roll et al. [49] have provided additional data for treating openings, while Zaghlool et al. [50,51] have provided data for corner connections.

ACI-13.5.1 first prescribes in general that openings of any size may be provided if it can be shown by analysis that all strength and serviceability conditions including the limits on the deflections are satisfied. However, in common situations (ACI-13.5.2) a special analysis need not be made for slab systems not having beams when (1) openings are within the middle half of the span in each direction, provided the total amount of reinforcement required for the panel without the opening is maintained; (2) openings in the area common to two column strips do not interrupt more than one-eighth of the column strip width in either span, and the equivalent of reinforcement interrupted is added on all sides of the openings; (3) openings in the area common to one column strip and one middle strip do not interrupt more than one-fourth of the reinforcement in either strip, and the equivalent of reinforcement interrupted be added on all sides of the openings.

In regard to two-way action nominal shear stress, the critical section for slabs without shearhead is not to include that part of the periphery which is enclosed by radial projections of the openings to the center of the column (ACI-11.11.5). For slabs with shearhead, the critical periphery is to be reduced only by one-half of what is cut away by the radial lines from the center of the column to the edges of the opening. Some critical sections with cutaways by openings are shown in Fig. 16.19.1.

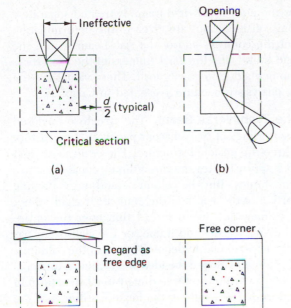

Figure 16.19.1 Effect of openings and free edges on critical periphery of two-way shear action (from ACI Commentary-11.11.5).

16.20 EQUIVALENT FRAME METHOD FOR GRAVITY AND LATERAL LOAD ANALYSIS

For gravity load analysis the "equivalent frame method" prescribed by the ACI Code differs from the "direct design method" *only* in the way by which the longitudinal moments along the spans of the equivalent rigid frame (as defined in Sec. 16.2) are obtained. When lateral (wind) load needs to be considered, an elastic analysis must be made of the structure under lateral load and the results combined with those due to gravity load. Consistent with the tradition under ACI Code of using an equivalent frame for gravity load analysis, a logical extension is to use an equivalent frame approach to lateral load analysis. ACI-13.3.1.2 does not prescribe an equivalent frame method for lateral load analysis, but rather only refers to "unbraced frames" and requires taking into account "effects of cracking and reinforcement on stiffness of frame members."

The maximum positive moments (and reversals) within the span and negative moments at the supports should be obtained for various combinations of gravity load patterns with lateral load as indicated by ACI-9.2. When the equivalent frame method *is* used for gravity load analysis of two-way floor systems meeting the limitations of the direct design method, the resulting factored moments may be reduced in a proportion such that the absolute sum of positive and average negative moments used in design need not exceed $w_u L_2 L_n^2 / 8$ (ACI-13.7.7.4).

The elastic analysis for the equivalent rigid frame is treated separately in Chap. 17. In this section, considerations are given to the determination of the flexural stiffnesses of the columns and of the slab-beam within the width of the equivalent rigid frame, the torsional stiffness of the transverse beam, and the fixed-end moments due to gravity load. These values are the required input data for the analysis procedure presented in Chap. 17.

Torsional Stiffness of the Transverse Beam. The structure enclosed between the two parallel centerlines of two adjacent panels in a multistory two-way floor system is a three-dimensional structure. The equivalent rigid frame described in Sec. 16.2 approximates the three-dimensional structure by a series of two-dimensional ones. But the columns stand on or provide support for only a small portion of the width of the equivalent rigid frame. Hence either the column stiffness has to be spread thin over the entire width (denoted by L_2) of the equivalent rigid frame, or the slab-beam has to be shrunk to the narrow transverse width (denoted by c_2) of the columns. Corley, Sozen, and Siess [53] first developed the idea of attaching a torsional member to the column (a cutaway from the three-dimensional structure) in the transverse direction and in essence shifting the flexural stiffness of the slab-beam to the end of the torsional member away from the columns (see Fig. 17.3.1). Thus the effectiveness of the column to restrain the ends of the slab-beam is reduced; hence the name of a less effective "equivalent column." As shown by Fig. 16.20.1, under gravity loading, the restraint at the column is more like a fixed end to the slab-beam but the restraint away from the column tends to approach that of a simple support.

Corley and Jirsa [17] developed a formula for the torsional stiffness K_t of the attached torsional member so that results of the equivalent frame analysis are close to those of tests, as follows [ACI Formula (13-6)]:

$$K_t = \sum \frac{9E_{cs}C}{L_2\left(1 - \dfrac{c_2}{L_2}\right)^3}\left(\frac{I_{sb}}{I_s}\right) \qquad (16.20.1)$$

in which

$\quad\quad C$ = torsional constant of the transverse beam (see Sec. 16.11)
$\quad E_{cs}$ = modulus of elasticity of slab concrete
$\quad\quad I_s$ = moment of inertia of slab over width of equivalent frame
$\quad I_{sb}$ = moment of inertia of entire T-section (if so) within the width of the equivalent rigid frame
$\quad\quad L_2$ = span of member subject to torsion

The summation sign is for the transverse spans (denoted by L_2) on each side of the column.

This parameter K_t is the most influential parameter to relate results of the theoretical elastic analysis to those of tests for gravity (or lateral) load. As test results become more available, especially for lateral load, the coefficient "9" in Eq. (16.20.1) might be adjusted in the future.

Treatment of Flexural Element Having Variable Moment of Inertia. The flexural stiffness of a flexural element ij having variable moment of

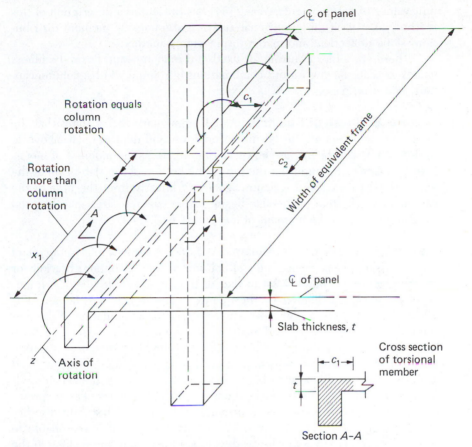

Figure 16.20.1 Attached torsional member for the columns.

inertia can be expressed by two near-end stiffnesses S_{ii} and S_{jj} and a cross stiffness S_{ij}. In applying the moment distribution method, the carry-over factor from i to j is S_{ji}/S_{ii} and the carry-over factor from j to i is S_{ij}/S_{jj}. In applying the matrix displacement method the element stiffness matrix $[S]$ is

$$[S] = \begin{bmatrix} S_{ii} & S_{ij} \\ S_{ji} & S_{jj} \end{bmatrix} \quad \text{(in which } S_{ij} = S_{ji}) \quad (16.20.2)$$

In applying the slope deflection method [55],

$$\left. \begin{array}{l} M_i = M_{0i} + S_{ii}\theta_i + S_{ij}\theta_j - (S_{ii} + S_{ij})R \\ M_j = M_{0j} + S_{ji}\theta_i + S_{jj}\theta_j - (S_{ii} + S_{ij})R \end{array} \right\} \quad (16.20.3)$$

in which M_{0i} and M_{0j} are the fixed-end moments, θ_i and θ_j are the slopes at ends i and j of the flexural element, and R is the clockwise rotation of the element axis.

The column analogy method may be the best way to compute the fixed-end moments, the near-end stiffnesses S_{ii} and S_{jj}, and the cross stiffness S_{ij} or S_{ji} of a flexural element having variable moment of inertia, thus avoiding using the flexibility analysis for pin-end rotations due to applied loads or unit values of end moments. Wang [54] has presented a description of this method, as well as the elastic analysis of rigid frames by moment distribution, slope deflection, and matrix displacement methods.

There are tables available for fixed-end moments and flexural stiffness values for slabs having various combinations of column widths, column capitals, and drop panels [55–57].

Flexural Stiffness of Columns. ACI Commentary-13.7.4 states that the height of the column is to be measured from mid-depth of slab above to mid-depth of slab below, as shown in Fig. 16.20.2. The moment of inertia is to be taken as infinite from the top to the bottom of the slab-beam at the joint (ACI-13.7.4.3), and is assumed to vary linearly from the base of the capital to the bottom of the slab-beam. Gross concrete area may be used for computation of the moment of inertia (ACI-13.7.4.1).

Flexural Stiffness of Slab-Beam. ACI-13.7.3.3 states that moment of inertia of slab-beams from center of column to face of column, bracket, or capital shall be assumed equal to that of the slab-beam at face of column, bracket, or capital divided by the quantity $(1 - c_2/L_2)^2$. Although ACI-13.7.3.1 permits the use of gross concrete area for computation of the moment of inertia in gravity load analysis, at the same time ACI-13.3.1.2 requires taking into account the effects of cracking and reinforcement in lateral load analysis. When computers are used, loadings for different gravity load patterns to be combined with lateral load can be listed together in an input load matrix; thus, it would be convenient to use only one set of assumptions for flexural stiffness properties, especially in the final analysis after the finished design. ACI Commentary-13.3.1.2 suggests that it is reasonable to use ACI Formula (9-7), which was developed to account for cracking in the moment of inertia used for deflection computations.

There is not definitive agreement on what constitutes the appropriate assumptions for stiffness, either for gravity load analysis or lateral load analysis. For gravity load analysis the use of gross section is reasonable because it is the simplest assumption and the results are acceptable. For lateral load analysis, particularly for the unbraced frame where the entire lateral re-

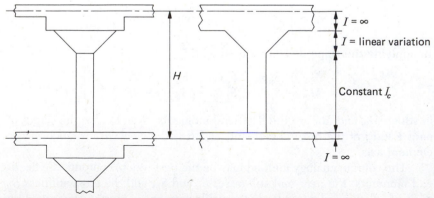

Figure 16.20.2   Basis of calculation of column stiffness.

sistance is provided by the flexural stiffnesses of the slab-beams and columns, the use of gross section for stiffness overemphasizes the resistance to lateral loads. In Chap. 17, devoted to the analysis of two-way floor systems, several analytical models are presented for possible use in lateral load analysis, or in combined gravity and lateral load analysis. For an extended discussion of lateral load analysis, the reader should consult the study by Vanderbilt [58]. Perhaps the acceptance of an equivalent beam or reduced beam method of analysis would make possible the use of gross section in lateral load analysis.

Arrangement of Live Load. When there is a definitely known load pattern, of course, analysis should be made for it (ACI-13.7.6.1). When service live load does not exceed $\frac{3}{4}$ of the service dead load, analysis needs to be made only for full factored dead and live load on all spans (ACI-13.7.6.2). When load patterns in accordance with influence line concepts are used, only $\frac{3}{4}$ of the full factored live load needs to be used (ACI-13.7.6.3); however, factored moments used in design should not be less than those due to full factored dead and live loads on all panels (ACI-13.7.6.4).

Reduction of Negative Moments Obtained at Column Centerlines from Structural Analysis. Negative moments obtained at interior column centerlines may be reduced to the face of rectilinear or equivalent square (for circular or polygon-shaped supports) supports, but not greater than $0.175L_1$ from the column centerline (ACI-13.7.7). For exterior columns, having capitals or brackets, reduction of negative moments can be made only to a section no greater than halfway between the face of column and edge of the capital or bracket (ACI-13.7.7.2).

Deflections. When the deflection must be calculated for a two-way slab system, the ACI Code (ACI-9.5.3.4) provides little guidance other than that one should take into account "size and shape of panel, conditions of support, and nature of restraints at panel edges." The effective moment of inertia I_e [Eq. (14.4.1)] is required to be used in such calculations. Although a number of techniques have been proposed [59–62], adaption of the equivalent frame concept seems to have the most promise of being relatively simple to apply and giving reasonable results. This equivalent frame application has been developed by Nilson and Walters [61] for essentially uncracked systems and extended by Kripanarayanan and Branson [63] for partially cracked load ranges.

EXAMPLE 16.20.1 Assuming the equivalent frame method is to be applied to the two-way slab (with beams) design example described in Sec. 16.3, obtain the stiffnesses and carry-over factors necessary for the analysis of the equivalent rigid frames A, B, C, and D as shown by the notations in Fig. 16.3.5. Also obtain the fixed-end moments for gravity load. Note that with the information obtained here, the matrix displacement method as described in Chap. 17 can be applied as an alternative (perhaps in preference to) the method of moment distribution.

Solution: (a) Compute flexure properties of slab-beam. The variations in the moment of inertia of the slab-beam in the long and short directions are shown in Fig. 16.20.3. For the long slab-beam, the ratio of moment of inertia between the center and the face of the column to the moments of inertia of the rest of the span is $1.0/(1 - 15/240)^2 = 1.138$; and it is $1.0/(1 - 15/300)^2 = 1.108$ for the short slab-beam (ACI-13.7.3.3). The stiffness K, carry-over factor COF, and fixed-end moment FEM coefficients may be computed by the column analogy method [55].

For the long direction, referring to Fig. 16.20.3(a),

$$A = 23.75 + 2(0.879)(0.625) = 23.75 + 1.10 = 24.85$$
$$I = \tfrac{1}{12}(23.75)^3 + 1.10(12.1875)^2 = 1117 + 164 = 1281$$

stiffness factor $s_{ii} = L\left(\dfrac{1}{A} + \dfrac{Mc}{I}\right)$

$$s_{ii} = \dfrac{25}{24.85} + \dfrac{25(12.5)^2}{1281} = 1.005 + 3.045 = 4.05$$

stiffness factor $s_{ij} = L\left(\dfrac{Mc}{I} - \dfrac{1}{A}\right)$

$$s_{ij} = -(1.005 - 3.045) = 2.04$$

$$\text{COF} = \dfrac{2.04}{4.05} = 0.503$$

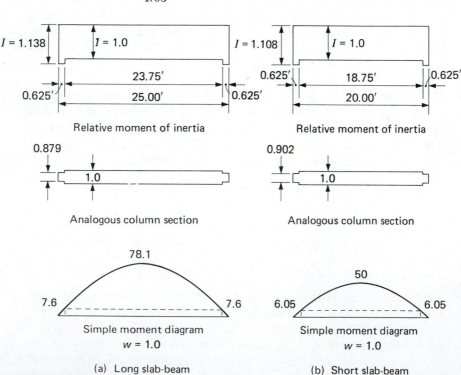

(a) Long slab-beam   (b) Short slab-beam

Figure 16.20.3 Flexure properties of slab-beam in two-way slab (with beams) design example.

load on analogous column for uniform load ($w = 1.0$)

$$= \tfrac{2}{3}(78.1 - 7.6)(23.75) + 7.6(23.75) + 0.879(7.6)(0.625)$$
$$= 1117 + 180 + 4 = 1301$$

$$\text{FEM coefficient} = \frac{1301}{24.85L_1^2} = \frac{1301}{24.85(625)} = 0.084$$

For the short direction, referring to Fig. 16.20.3(b),

$$A = 18.75 + 2(0.902)(0.625) = 18.75 + 1.13 = 19.88$$
$$I = \tfrac{1}{12}(18.75)^3 + 1.13(9.6875)^2 = 550 + 106 = 656$$
$$s_{ii} = \frac{20}{19.88} + \frac{20(10)^2}{656} = 1.01 + 3.05 = 4.06$$
$$s_{ij} = -(1.01 - 3.05) = 2.04$$

$$\text{COF} = \frac{2.04}{4.06} = 0.502$$

load on analogous column for uniform load ($w = 1.0$)

$$= \tfrac{2}{3}(50 - 6.05)(18.75) + 6.05(18.75) + 0.902(6.05)(0.625)$$
$$= 550 + 113.5 + 3.4 = 666.9$$

$$\text{FEM coefficient} = \frac{666.9}{19.88L_1^2} = \frac{666.9}{19.88(400)} = 0.084$$

The flexural stiffnesses of the slab-beams in frames A, B, C, and D are, using the I_{sb} values shown in Fig. 16.20.4,

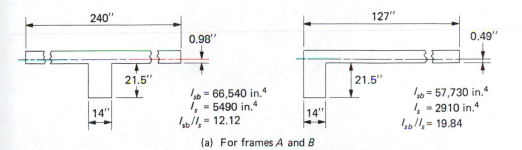

(a) For frames A and B

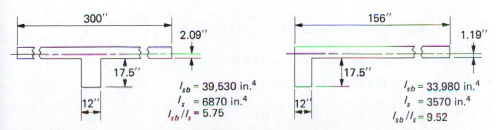

(b) For frames C and D

Figure 16.20.4 Slab-beam cross sections in two-way slab (with beams) design example.

Frame A, $$K_{sb} = \frac{4.05E(66,540)}{300} = 898E$$

Frame B, $$K_{sb} = \frac{4.05E(57,730)}{300} = 779E$$

Frame C, $$K_{sb} = \frac{4.06E(39,530)}{240} = 669E$$

Frame D, $$K_{sb} = \frac{4.06E(33,980)}{240} = 575E$$

Note that the moment of inertia values used above are based on the gross cross sections as shown in Fig. 16.20.4, an acceptable procedure for gravity load analysis. These stiffness values are likely to be too high for lateral load analysis, since cracking reduces flexural stiffness. ACI-13.3.1.2 states that for lateral load analysis, effects of cracking and reinforcement must be taken into account.

(b) Compute flexure properties of columns. The variations in the moment of inertia of the column section in the long and short directions are shown in Fig. 16.20.5. The stiffness coefficients and carry-over factors may be computed by the column analogy method.

For the short direction, referring to Fig. 16.20.5(a),

$$A = 9.67, \qquad I = \tfrac{1}{12}(9.67)^3 = 75.3$$

$$s_{TT} = \frac{12}{9.67} + \frac{12(6.90)^2}{75.3} = 1.24 + 7.60 = 8.84$$

$$s_{BB} = \frac{12}{9.67} + \frac{12(5.10)^2}{75.3} = 1.24 + 4.15 = 5.39$$

$$s_{TB} = s_{BT} = -\left[\frac{12}{9.67} - \frac{12(6.90)(5.10)}{75.3} \right] = -1.24 + 5.60 = 4.36$$

$$(\text{COF})_{TB} = \frac{4.36}{8.84} = 0.493$$

$$(\text{COF})_{BT} = \frac{4.36}{5.39} = 0.809$$

Stiffness at top, $$K_{cT} = s_{TT}\frac{EI}{L} = \frac{8.84E(15)^4/12}{144} = 259E$$

Stiffness at bottom, $$K_{cB} = s_{BB}\frac{EI}{L} = \frac{5.39E(15)^4/12}{144} = 158E$$

For the long direction, referring to Fig. 16.20.5(b),

$$A = 10.00, \qquad I = \tfrac{1}{12}(10)^3 = 83.3$$

$$s_{TT} = \frac{12}{10} + \frac{12(6.73)^2}{83.3} = 1.20 + 6.53 = 7.73$$

$$s_{BB} = \frac{12}{10} + \frac{12(5.27)^2}{83.3} = 1.20 + 4.01 = 5.21$$

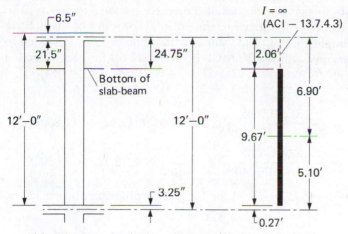

(a) Column section in short direction (for frames C and D)

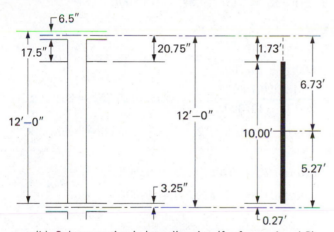

(b) Column section in long direction (for frames A and B)

Figure 16.20.5 Flexure properties of columns in two-way slab (with beams) design example.

$$s_{TB} = s_{BT} = -\left[\frac{12}{10} - \frac{12(6.73)(5.27)}{83.3}\right] = -1.20 + 5.12 = 3.92$$

$$(\text{COF})_{TB} = \frac{3.92}{7.73} = 0.507$$

$$(\text{COF})_{BT} = \frac{3.92}{5.21} = 0.752$$

Stiffness at top, $\quad K_{cT} = \dfrac{7.73E(15)^4/12}{144} = 226E$

Stiffness at bottom, $\quad K_{cB} = \dfrac{5.21(15)^4/12}{144} = 153E$

(c) Compute torsional stiffnesses of transverse torsional members. The torsional constants C for the transverse members shown in Fig. 16.20.6 are taken from Example 16.11.1. The values for the ratio of I_{sb} to I_s needed to increase the torsional stiffness K_t [ACI Formula (13-6)] for each direction are shown in Fig. 16.20.6.

For frame A,

$$\text{exterior } K_t = \frac{18E(10,700)}{240(1 - 15/240)^3}(12.12) = 974E(12.12) = 11,800E$$

$$\text{interior } K_t = \frac{18E(11,930)}{240(1 - 15/240)^3}(12.12) = 1086E(12.12) = 13,200E$$

For frame B, using $I_{sb}/I_s = 19.84$ for 14×21.5 projection below 127×6.5 slab,

$$\text{exterior } K_t = 487E(19.84) = 9660E$$

$$\text{interior } K_t = 543E(19.84) = 10,800E$$

For frame C,

$$\text{exterior } K_t = \frac{18E(19,100)}{300(1 - 15/300)^3}(5.75) = 1340E(5.75) = 7700E$$

$$\text{interior } K_t = \frac{18E(20,700)}{300(1 - 15/300)^3}(5.75) = 1450E(5.75) = 8340E$$

For frame D, using $I_{sb}/I_s = 9.52$ for 12×17.5 projection below 156×6.5 slab,

$$\text{exterior } K_t = 670E(9.52) = 6380E$$

$$\text{interior } K_t = 725E(9.52) = 6900E$$

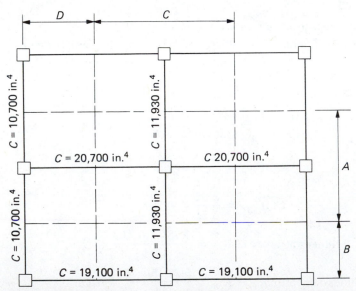

Figure 16.20.6 Torsional constants in two-way slab (with beams) design example (from Example 16.11.1).

EXAMPLE 16.20.2 Assuming the equivalent frame method is to be applied to the flat slab design example described in Sec. 16.3, obtain the stiffnesses and carry-over factors necessary for the analysis of equivalent rigid frame A in the long direction. Also obtain the fixed-end moments for gravity load.

Solution: (a) Compute flexure properties of slab strip. The stiffnesses, carry-over factors, and fixed-end moments may be determined by various analysis methods. The column analogy method [54] is used in this example. Simmonds and Misic [56] have provided design aids to meet the ACI Code assumptions of the "equivalent frame method."

The variation in the moment of inertia along an exterior span of the slab strip is shown in Fig. 16.20.7(a). Taking the moment of inertia through the $7\frac{1}{2}$-in. slab as the reference value of 1, the moment of inertia through the drop is $(10.5/7.5)^3 = 2.742$, and the moment of inertia between the column centerline and the face of the equivalent square column capital is $2.742/(1 - 4.43/20)^2 = 2.742/0.607 = 4.52$.

The variation in the width of the analogous column section is $1/I$, which is shown in Fig. 16.20.7(b). The area of the analogous column section is

$$A = 16.66 + 2(0.364)(1.96) + 2(0.221)(2.21)$$
$$= 16.66 + 1.43 + 0.98 = 19.07$$

The moment of inertia about the midspan, neglecting the moments of inertia of the short segments about their own centroidal axes, is

$$I = \tfrac{1}{12}(16.66)^3 + 1.43(9.31)^2 + 0.98(11.40)^2 = 385 + 124 + 128 = 637$$

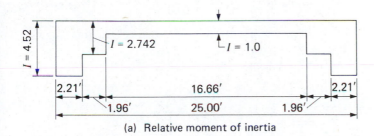

(a) Relative moment of inertia

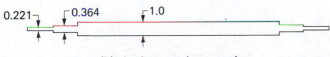

(b) Analogous column section

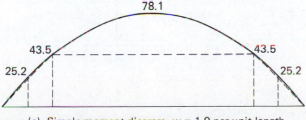

(c) Simple moment diagram, w = 1.0 per unit length

Figure 16.20.7 Flexure properties of slab strip in flat slab design example.

$$\text{stiffness factor } s_{ii} = L\left(\frac{1}{A} + \frac{Mc}{I}\right) = 25\left[\frac{1}{19.07} + \frac{12.5(12.5)}{637}\right]$$

$$s_{ii} = 1.31 + 6.14 = 7.45$$

$$s_{ij} = -1.31 + 6.14 = 4.83$$

$$\text{COF} = \frac{4.83}{7.45} = 0.649$$

$$\text{stiffness } K \text{ at end of 20-ft wide slab strip} = \frac{7.45E(7.5)^3/12}{300}(240) = 210E$$

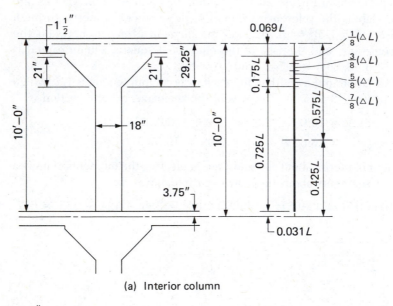

(a) Interior column

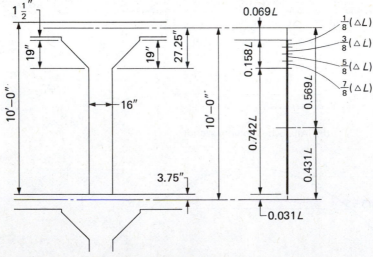

(b) Exterior column

Figure 16.20.8 Flexure properties of columns in flat slab design example.

The load on the analogous column is equal to the summation of the product of the width of the analogous column section and the area of the simple beam moment diagram of Fig. 16.20.7(c). Considering the moment areas over the short segments as being trapezoidal, the load on the analogous column is

$$P = \tfrac{2}{3}(78.1 - 43.5)(16.66) + 43.5(16.66)$$
$$+ 2(0.364)(\tfrac{1}{2})(43.5 + 25.2)(1.96)$$
$$+ 2(0.221)(\tfrac{1}{2})(25.2)(2.21)$$
$$= 385 + 725 + 49 + 12 = 1171$$

$$\text{FEM coefficient} = \frac{P}{AL^2} = \frac{1171}{19.07(25)^2} = 0.0985$$

Since the edge column capital is almost equal in size to the equivalent square of the interior column capital, the FEM coefficient, stiffness, and carry-over factor obtained above for the interior span will also be used for the exterior span.

(b) Compute flexure properties of columns. The length of the column is measured between the centerlines of slab thickness, as shown in Fig. 16.20.8. The moment of inertia is assumed to be infinite from the top of the slab to the bottom of the drop panel and then linearly to the base of the column capital.

The $0.175L$ of the analogous column representing the column capital is divided into four parts, $\Delta L = 0.175L/4$, with $1/I$ of $1/8$, $3/8$, $5/8$, and $7/8$.

$$A = 0.725L + \left(\frac{1}{8} + \frac{3}{8} + \frac{5}{8} + \frac{7}{8}\right)\left(\frac{0.175L}{4}\right) = 0.8125L$$

$$\Sigma Ay \text{ from top} = \frac{0.175L}{4}(0.091L + 0.135L + 0.178L + 0.222L)$$
$$+ 0.725L(0.6065L) = 0.4671L^2$$

$$\bar{y} \text{ from top} = \frac{0.4671L^2}{0.8125L} = 0.575L$$

$$I = \frac{1}{12}(0.725L)^3 + 0.725L(0.0315L)^2$$

$$+ \frac{0.175L}{4}[(0.484L)^2 + (0.440L)^2 + (0.397L)^2 + (0.353L)^2]$$

$$= 0.06354L^3$$

The stiffness factors at the top and bottom are

$$s_{TT} = \frac{1}{0.8125} + \frac{(0.575)^2}{0.06354} = 1.23 + 5.20 = 6.43$$

$$s_{BB} = \frac{1}{0.8125} + \frac{(0.425)^2}{0.06354} = 1.23 + 2.84 = 4.07$$

for which the stiffnesses are

$$K_{cT} = \frac{s_{TT}EI}{L} = \frac{6.43E\pi(9)^4/4}{120} = 276E$$

$$K_{cB} = \frac{s_{BB}EI}{L} = \frac{4.07E\pi(9)^4/4}{120} = 175E$$

The carry-over factors are

$$(COF)_{TB} = \frac{0.575(0.425)/(0.06354) - 1.23}{6.43} = \frac{2.62}{6.43} = 0.407$$

$$(COF)_{BT} = \frac{2.62}{4.07} = 0.644$$

For the exterior column [Fig. 16.20.8(b)],

$$A = 0.742L + \left(\frac{1}{8} + \frac{3}{8} + \frac{5}{8} + \frac{7}{8}\right)\left(\frac{0.158L}{4}\right) = 0.821L$$

$$\Sigma A_y \text{ from top} = \frac{0.158L}{4}(0.089L + 0.128L + 0.168L + 0.207L)$$

$$+ 0.742L(0.598L) = 0.4671L^2$$

$$\bar{y} \text{ from top} = \frac{0.4671L^2}{0.821L} = 0.569L$$

$$I = \frac{1}{12}(0.742L)^3 + 0.742L(0.0296L)^2$$

$$+ \frac{0.158L}{4}[(0.480L)^2 + (0.441L)^2 + (0.401L)^2 + (0.362L)^2]$$

$$= 0.06298L^3$$

$$s_{TT} = \frac{1}{0.821} + \frac{(0.569)^2}{0.06298} = 1.22 + 5.14 = 6.36$$

$$s_{BB} = \frac{1}{0.821} + \frac{(0.431)^2}{0.06298} = 1.22 + 2.95 = 4.17$$

$$K_{cT} = \frac{s_{TT}EI}{L} = \frac{6.36E(16)^4/12}{120} = 289E$$

$$K_{cB} = \frac{s_{BB}EI}{L} = \frac{4.17E(16)^4/12}{120} = 190E$$

(c) Compute torsional stiffness of transverse torsional members.

$$C = 9800 \text{ in.}^4 \qquad \text{(from Example 16.11.2)}$$

For the two members, one framing in from each side,

$$K_t = \frac{2(9E)C}{L_2\left(1 - \dfrac{c_2}{L_2}\right)^3} = \frac{2(9E)(9800)}{240\left(1 - \dfrac{4.43}{20}\right)^3} = 1560E$$

SELECTED REFERENCES

1. George A. Hool and Nathan C. Johnson. *Concrete Engineers Handbook*. New York: McGraw-Hill, 1918 (pp. 457–486).
2. Joseph DiStasio and M. P. Van Buren. "Slabs Supported on Four Sides," *ACI Journal, Proceedings*, **32**, January–February 1936, 350–364.
3. R. L. Bertin, Joseph DiStasio, and M. P. Van Buren. "Slabs Supported on Four Sides," *ACI Journal, Proceedings*, **41**, June 1945, 537–556.
4. C. P. Siess and N. M. Newmark. "Rational Analysis and Design of Two-Way Concrete Slabs," *ACI Journal, Proceedings*, **45**, December 1948, 273–316.
5. W. L. Gamble, M. A. Sozen, and C. P. Siess. "Measured and Theoretical Bending Moments in Reinforced Concrete Floor Slabs," Civil Engineering Structural Research Series No. 246. Urbana: University of Illinois, June 1962.
6. M. A. Sozen and C. P. Siess. "Investigation of Multi-Panel Reinforced Concrete Floor Slabs," *ACI Journal, Proceedings*, **60**, August 1963, 999–1028.
7. S. A. Guralnick and R. W. LaFraugh. "Laboratory Study of a 45-Foot Square Flat Plate Structure," *ACI Journal, Proceedings*, **60**, September 1963, 1107–1185.
8. W. L. Gamble, M. A. Sozen, and C. P. Siess. "Tests of a Two-Way Reinforced Concrete Floor Slab," *Journal of the Structural Division*, ASCE, **95**, June 1969 (ST6), 1073–1096.
9. M. Daniel Vanderbilt, Mete A. Sozen, and Chester P. Siess. "Tests of a Modified Reinforced Concrete Two-Way Slab," *Journal of the Structural Division*, ASCE, **95**, June 1969 (ST6), 1097–1116.
10. H. M. Westergaard. "Formulas for the Design of Rectangular Floor Slabs and Supporting Girders," *ACI Proceedings*, **22**, 1926, 26.
11. David S. Hatcher, Mete A. Sozen, and Chester P. Siess. "Test of a Reinforced Concrete Flat Plate," *Journal of the Structural Division*, ASCE, **91**, October 1965 (ST5), 205–231.
12. James O. Jirsa, Mete A. Sozen, and Chester P. Siess. "Test of a Flat Slab Reinforced with Welded Wire Fabric," *Journal of the Structural Division*, ASCE, **92**, June 1966 (ST3), 199–224.
13. D. S. Hatcher, Mete A. Sozen, and Chester P. Siess. "Test of a Reinforced Concrete Flat Slab," *Journal of the Structural Division*, ASCE, **95**, June 1969 (ST6), 1051–1072.
14. E. Ramzy F. Zaghlool, H. A. Rawdon de Paiva, and Peter G. Glockner. "Tests of Reinforced Concrete Flat Plate Floors," *Journal of the Structural Division*, ASCE, **96**, March 1970 (ST3), 487–507.
15. Alex E. Cardenas and Paul H. Kaar. "Field Test of a Flat Plate Structure," *ACI Journal, Proceedings*, **68**, January 1971, 50–58.
16. Donald D. Magura and W. Gene Corley. "Tests to Destruction of a Multipanel Waffle Slab Structure—1964–1965 New York World's Fair." *ACI Journal, Proceedings*, **68**, September 1971, 699–703.
17. W. G. Corley and J. O. Jirsa. "Equivalent Frame Analysis for Slab Design," *ACI Journal, Proceedings*, **67**, November 1970, 875–884.
18. William L. Gamble. "Moments in Beam Supported Slabs," *ACI Journal, Proceedings*, **69**, March 1972, 149–157.
19. J. R. Nichols. "Statical Limitations Upon the Steel Requirement in Reinforced Concrete Flat Slab Floors," *Transactions ASCE*, **77**, 1914, 1670–1736.
20. H. M. Westergaard and W. A. Slater. "Moments and Stresses in Slabs," *ACI Proceedings*, **17**, 1921, 415.
21. Joseph A. Wise. "Design of Reinforced Concrete Slabs," *ACI Proceedings*, **25**, 1929, 712.

22. "Shearhead Reinforcement for Flat-Plate Floors," *Modern Developments in Reinforced Concrete* (No. 22). Chicago: Portland Cement Association, 1948.

23. *Notes on ACI 318-83 Building Code Requirements with Design Applications.* Skokie, Illinois: Portland Cement Association, 1984.

24. J. O. Jirsa, M. A. Sozen, and C. P. Siess. "Pattern Loadings on Reinforced Concrete Floor Slabs," *Journal of the Structural Division*, ASCE, **95,** June 1969 (ST6), 1117–1137.

25. Jan C. Jofriet and Gregory M. McNeice. "Pattern Loading on Reinforced Concrete Flat Plates," *ACI Journal, Proceedings*, **68,** December 1971, 968–972.

26. Edward G. Nawy. "Crack Width Control in Welded Fabric Reinforced Centrally Loaded Two-Way Concrete Slabs," *Causes, Mechanism, and Control of Cracking in Concrete* (SP-20). Detroit: American Concrete Institute, 1968 (pp. 211–235).

27. Edward G. Nawy and G. S. Orenstein. "Crack Width Control in Reinforced Concrete Two-Way Slabs," *Journal of the Structural Division*, ASCE, **96,** March 1970 (ST3), 701–721.

28. Edward G. Nawy and Kenneth W. Blair. "Further Studies on Flexural Crack Control in Structural Slab Systems," *Cracking, Deflection, and Ultimate Load of Concrete Slab Systems* (SP-30). Detroit: American Concrete Institute, 1971, (pp. 1–41).

29. Edward G. Nawy. "Crack Control Through Reinforcement Distribution in Two-Way Acting Slabs and Plates," *ACI Journal, Proceedings*, **69,** April 1972, 217–219.

30. Johannes Moe. *Shearing Strength of Reinforced Concrete Slabs and Footings Under Concentrated Loads*, Development Department Bulletin D47. Chicago: Portland Cement Association, April 1961 (130 pp.).

31. ACI-ASCE Committee 326. "Report on Shear and Diagonal Tension," *ACI Journal, Proceedings*, **59,** January, February, and March, 1962, 1–30, 277–344, and 352–396.

32. Neil M. Hawkins, H. B. Fallsen, and R. C. Hinojosa. "Influence of Column Rectangularity on the Behavior of Flat Plate Structures," *Cracking, Deflection, and Ultimate Load of Concrete Slab Systems* (SP-30). Detroit: American Concrete Institute, 1971 (pp. 127–146).

33. M. Daniel Vanderbilt. "Shear Strength of Continuous Plates," *Journal of the Structural Division*, ASCE, **98,** May 1972 (ST5), 961–973.

34. M. E. Criswell and N. M. Hawkins. "Shear Strength of Slabs: Basic Principles and Their Relation to Current Methods of Analysis," *Shear in Reinforced Concrete*, Vol. 2 (SP-42). Detroit: American Concrete Institute, 1974 (pp. 641–676).

35. N. M. Hawkins, M. E. Criswell, and F. Roll. "Shear Strength of Slabs Without Shear Reinforcement," *Shear in Reinforced Concrete*, Vol. 2 (SP-42). Detroit: American Concrete Institute, 1974 (pp. 677–720).

36. Brian E. Hewitt and Bärrington de V. Batchelor. "Punching Shear Strength of Restrained Slabs," *Journal of the Structural Division*, ASCE, **101,** September 1975 (ST9), 1837–1853.

37. Paul H. Langohr, Amin Ghali, and Walter H. Dilger. "Special Shear Reinforcement for Concrete Flat Plates," *ACI Journal, Proceedings*, **73,** March 1976, 141–146.

38. Neil M. Hawkins, (Chmn.). "The Shear Strength of Reinforced Concrete Members—Slabs, by the Joint ASCE-ACI Task Committee 426 on Shear and Diagonal Tension of the Committee on Masonry and Reinforced Concrete of the Structural Division," *Journal of the Structural Division*, ASCE, **100,** August 1974 (ST8), 1543–1591.

39. N. M. Hawkins. "Shear Strength of Slabs With Shear Reinforcement," *Shear in Reinforced Concrete*, Vol. 2 (SP-42). Detroit: American Concrete Institute, 1974 (pp. 785–816).

40. W. G. Corley and N. M. Hawkins. "Shearhead Reinforcement for Slabs," *ACI Journal, Proceedings,* **65,** October 1968, 811–824.

41. Joseph DiStasio and M. P. Van Buren. "Transfer of Bending Moment between Flat Plate Floor and Column," *ACI Journal, Proceedings,* **57,** September 1960, 299–314.

42. N. W. Hanson and J. M. Hanson. "Shear and Moment Transfer Between Concrete Slabs and Columns." *Journal PCA Research and Development Laboratories,* **10,** (1) January 1968, 2–16.

43. Neil M. Hawkins and W. Gene Corley. "Transfer of Unbalanced Moment and Shear from Flat Plates to Columns," *Cracking, Deflection, and Ultimate Load of Concrete Slab Systems* (SP-30). Detroit: American Concrete Institute, 1971 (pp. 147–176).

44. Dieter D. Pfaffinger. "Column-Plate Interaction in Flat Slab Structures," *Journal of the Structural Division,* ASCE, **98,** January 1972 (ST1), 307–326.

45. N. M. Hawkins. "Shear Strength of Slabs With Moments Transferred to Columns," *Shear in Reinforced Concrete,* Vol. 2 (SP-42). Detroit: American Concrete Institute, 1974 (pp. 817–846).

46. N. M. Hawkins and W. G. Corley. "Moment Transfer to Columns in Slabs With Shearhead Reinforcement," *Shear in Reinforced Concrete,* Vol. 2 (SP-42). Detroit: American Concrete Institute, 1974 (pp. 847–880).

47. Shafiqul Islam and Robert Park. "Tests on Slab-Column Connections with Shear and Unbalanced Flexure," *Journal of the Structural Division,* ASCE, **102,** March 1976 (ST3), 549–568.

48. Robert Park and Shafiqul Islam. "Strength of Slab-Column Connections with Shear and Unbalanced Flexure," *Journal of the Structural Division,* ASCE, **102,** September 1976 (ST9), 1879–1901.

49. Frederic Roll, S. T. H. Zaidi, Gajanan Sabnis, and Kuang Chuang. "Shear Resistance of Perforated Reinforced Concrete Slabs," *Cracking, Deflection, and Ultimate Load of Concrete Slab Systems* (SP-30). Detroit: American Concrete Institute, 1971 (pp. 77–100).

50. E. Ramzy F. Zaghlool and H. A. Rawdon de Paiva. "Strength Analysis of Corner Column-Slab Connections," *Journal of the Structural Division,* ASCE, **99,** January 1973 (ST1), 53–70.

51. E. Ramzy F. Zaghlool and H. A. Rawdon de Paiva. "Tests of Flat-Plate Corner Column-Slab Connections," *Journal of the Structural Division,* ASCE, **99,** March 1973 (ST3), 551–572.

52. Maurice P. Van Buren. "Staggered Columns in Flat Plates," *Journal of the Structural Division,* ASCE, **97,** June 1971 (ST6), 1791–1797.

53. W. G. Corley, M. A. Sozen, and C. P. Siess. "The Equivalent Frame Analysis for Reinforced Concrete Slabs," Structural Research Series No. 218. Urbana: University of Illinois, Civil Engineering Department, June 1961.

54. C. K. Wang. *Intermediate Structural Analysis.* New York: McGraw-Hill, 1983.

55. Alex E. Cardenas, Rolf J. Lenschow, and Mete A. Sozen. "Stiffness of Reinforced Concrete Plates," *Journal of the Structural Division,* ASCE, **98,** November 1972 (ST11), 2587–2603.

56. Sidney H. Simmonds and Janko Misic. "Design Factors for the Equivalent Frame Method." *ACI Journal, Proceedings,* **68,** November 1971, 825–831.

57. ACI Committee 340 (C. G. Salmon, Chairman). *Design Handbook In Accordance with the Strength Design Method of ACI 318-77. Vol. 1—Beams, Slabs, Brackets, Footings, and Pile Caps* [SP-17(81)] (3rd ed.). Detroit, Michigan: American Concrete Institute, 1981.

58. M. D. Vanderbilt. *Equivalent Frame Analysis of Unbraced Reinforced Concrete*

Buildings for Static Lateral Loads. Structural Research Report No. 36, Civil Engineering Department, Colorado State University, Fort Collins, June 1981.

59. Mortimer D. Vanderbilt, Mete A. Sozen, and Chester P. Siess. "Deflections of Multiple-Panel Reinforced Concrete Floor Slabs," *Journal of the Structural Division*, ASCE, **91**, August 1965 (ST4), 77–101.

60. ACI Committee 435. "State-of-the-Art Report, Deflection of Two-Way Reinforced Concrete Floor Systems," *Deflections of Concrete Structures* (SP-43). Detroit: American Concrete Institute, 1974 (pp. 55–81).

61. Arthur H. Nilson and Donald B. Walters, Jr. "Deflection of Two-Way Floor Systems by the Equivalent Frame Method," *ACI Journal, Proceedings*, **72**, May 1975, 210–218.

62. B. Vijaya Rangan. "Prediction of Long-Term Deflections of Flat Plates and Slabs," *ACI Journal, Proceedings*, **73**, April 1976, 223–226.

63. K. M. Kripanarayanan and D. E. Branson. "Short Time Deflections of Flat Plates, Flat Slabs, and Two-Way Slabs," *ACI Journal, Proceedings*, **73**, December 1976, 686–690.

64. Donald J. Fraser. "Equivalent Frame Method for Beam-Slab Structures," *ACI Journal, Proceedings*, **74**, May 1977, 223–228.

65. D. J. Fraser. "The Equivalent Frame Method Simplified for Beam and Slab Construction," *Concrete International*, **4**, April 1982, 66–73.

66. S. K. Sharan, D. Clyde, and D. Turcke. "Equivalent Frame Analysis Improvements for Slab Design," *ACI Journal, Proceedings*, **75**, February 1978, 55–59.

67. J. F. Mulcahy and J. M. Rotter. "Moment Rotation Characteristics of Flat Plate and Column Systems," *ACI Journal, Proceedings*, **80**, March–April 1983, 85–92.

68. David A. Pecknold. "Slab Effective Width for Equivalent Frame Analysis," *ACI Journal, Proceedings*, **72**, April 1975, 135–137.

69. Fred Allen and Peter Darvall. "Lateral Load Equivalent Frame," *ACI Journal, Proceedings*, **74**, July 1977, 294–299.

70. M. Daniel Vanderbilt. "Equivalent Frame Analysis for Lateral Loads," *Journal of the Structural Division*, ASCE, **105**, October 1979 (ST10), 1981–1998.

71. Ziad M. Elias. "Sidesway Analysis of Flat Plate Structures," *ACI Journal, Proceedings*, **76**, March 1979, 421–442.

72. Ziad M. Elias and Constantinos Georgiadis. "Flat Slab Analysis Using Equivalent Beams," *ACI Journal, Proceedings*, **76**, October 1979, 1063–1078.

73. Ziad M. Elias. "Lateral Stiffness of Flat Plate Structures," *ACI Journal, Proceedings*, **80**, January–February 1983, 50–54.

74. M. Daniel Vanderbilt and W. Gene Corley. "Frame Analysis of Concrete Buildings," *Concrete International: Design and Construction*, **5**, December 1983, 33–43.

75. P. J. Taylor and J. L. Heiman. "Long-Term Deflection of Reinforced Concrete Flat Slabs and Plates," *ACI Journal, Proceedings*, **74**, November 1977, 556–561.

76. B. Vijaya Rangan and Arthur E. McMullen. "A Rational Approach to Control of Slab Deflections," *ACI Journal, Proceedings*, **75**, June 1978, 256–262.

77. Walter H. Dilger and Amin Ghali. "Shear Reinforcement for Concrete Slabs," *Journal of the Structural Division*, ASCE, **107**, December 1981 (ST12), 2403–2420.

78. Amin Ghali, Mahmoud Z. Elmasri, and Walter Dilger. "Punching of Flat Plates Under Static and Dynamic Horizontal Forces," *ACI Journal, Proceedings*, **73**, October 1976, 566–576.

79. Walter Dilger, Mahmoud Z. Elmasri, and Amin Ghali. "Flat Plates with Special Shear Reinforcement Subjected to Static Dynamic Moment Transfer," *ACI Journal, Proceedings*, **75**, October 1978, 543–549.

80. Frieder Seible, Amin Ghali, and Walter H. Dilger. "Preassembled Shear Rein-

forcing Units for Flat Plates," *ACI Journal, Proceedings,* **77,** January–February 1980, 28–35.

81. S. Unnikrishna Pillai, Wayne Kirk, and Leonard Scavuzzo. "Shear Reinforcement at Slab-Column Connections in a Reinforced Concrete Flat Plate Structure," *ACI Journal, Proceedings,* **79,** January–February 1982, 36–42.

82. Adrian E. Long. "Punching Failure of Slabs—Transfer of Moment and Shear," *Journal of the Structural Division,* ASCE, **99,** April 1973 (ST4), 665–685.

83. V. W. Neth, H. A. R. de Paiva, and A. E. Long. "Behavior of Models of a Reinforced Concrete Flat Plate Edge-Column Connection," *ACI Journal, Proceedings,* **78,** July–August 1981, 269–275.

84. Hans Gesund and Harinatha B. Goli. "Local Flexural Strength of Slabs at Interior Columns," *Journal of the Structural Division,* ASCE, **106,** May 1980 (ST5), 1063–1078.

85. Harinatha B. Goli and Hans Gesund. "Flexural Strength of Flat Slabs at Exterior Columns," *Journal of the Structural Division,* ASCE, **108,** November 1982 (ST11), 2479–2495.

86. B. Vijaya Rangan and A. S. Hall. "Moment and Shear Transfer Between Slab and Edge Column," *ACI Journal, Proceedings,* **80,** May–June 1983, 183–191.

PROBLEMS

Two-Way Slabs (with Beams)

16.1 Design the typical interior frame along columns 2-5-7 for the two-way slab system shown. The 13-ft long columns are connected by beams, and no column capitals or drop panels are used. As an initial trial, assume all beams (interior) are 12 × 24 in. overall. Revise beam size as necessary during the

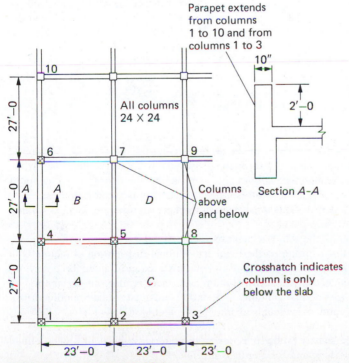

Problems 16.1 through 16.4

design. Determine slab thickness based on ACI-9.5.3, then use the "direct design method" for longitudinal distribution of moments. Show design sketch giving all your decisions, including dimensions, bar sizes, bar lengths, and stirrups for the two spans from column 2 to column 7. The live load is 150 psf, $f'_c = 4000$ psi, and $f_y = 60,000$ psi.

16.2   Design the interior frame of Prob. 16.1, except that a 12-in. wall exists at the lower-story level and contains the 24-in. square columns at locations 1, 2, 3, 4, 6, and 10.

16.3   Design the typical interior frame along column lines 4-5-8 for the two-way slab system of Prob. 16.1.

16.4   Design the exterior half-frame along column lines 1-2-3 for the two-way slab system of Prob. 16.1.

16.5   Design an interior frame in the long direction for a floor system of slabs supported on beams which has 5 panels at 21 ft in one direction and 5 panels at 27 ft in the other direction. The live load is 175 psf and the dead load is 40 psf in addition to the slab weight. Assume that all panels are bounded by beams that are 14 in. wide. Columns 15 in. square and 13 ft long are located at the corners of all panels. Use $f'_c = 4000$ psi, $f_y = 50,000$ psi, and the "direct design method" of the ACI Code.

16.6   Design a floor system of slabs supported on beams which has two panels at 16 ft in one direction and two panels at 21 ft in the other direction. Assume that all panels are bounded by beams 12 in. wide and the columns are 14 in. square and 11 ft long. The live load is 200 psf, and the dead load is 50 psf in addition to the slab weight. Use $f'_c = 3000$ psi, $f_y = 60,000$ psi, and the ACI Code.

16.7   Design a simply supported sidewalk slab for an 18-ft square panel to carry a live load of 250 psf. The panel is supported by beams 12 in. wide on all four sides. There are no walls or columns above the slab. Use $f'_c = 4000$ psi, $f_y = 60,000$ psi, and the ACI Code.

Flat Slabs

16.8   Given the flat slab shown in the accompanying figure. The columns are 24 in. square with columns 1 through 6 existing only below the floor slab, while columns 7 through 9 exist both above and below the floor slab. All columns are 13 ft long center to center of floor slabs. The live load is 150 psf, $f'_c = 4000$ psi, and $f_y = 60,000$ psi. Design the flat slab using rectangular column capitals and drop panels.
(a) Determine the slab thickness based on ACI-9.5.3.
(b) Use "direct design method" for longitudinal distribution of moments in interior equivalent frame defined by columns 2, 5, and 7 along its centerline.
(c) Determine transverse distribution and select reinforcement for the column strip (defined by columns 2, 5, and 7) and adjacent half middle strips.
(d) Specify and show details giving lengths and locations of bars.

16.9   For the flat slab of Prob. 16.8, design the interior equivalent frame defined by columns 4, 5, and 8 along its centerline.

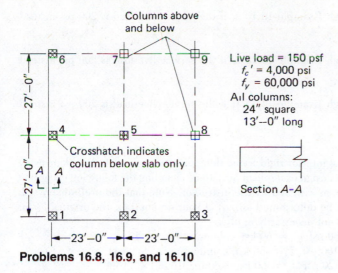

Columns above
and below

Live load = 150 psf
f_c' = 4,000 psi
f_y = 60,000 psi
All columns:
24" square
13'–0" long

Section A-A

Crosshatch indicates
column below slab only

27'–0"
27'–0"

23'–0" — 23'–0"

Problems 16.8, 16.9, and 16.10

16.10 For the flat slab of Prob. 16.8, redesign the interior equivalent frame defined by column line 2-5-7, but consider that a 12-in. wall 13 ft high exists along column line 1-2-3 at the story below the slab to be designed.

16.11 Investigate the moment and shear transfer at the exterior support of a flat-slab structure as detailed in the accompanying figure. The exterior support has a 5-ft flat-sided column capital on a 24-in. square column, along with a 7 ft-8 in. width of drop panel that is 11 in. thick. The slab is $8\frac{1}{2}$ in. thick. Assume there is no edge beam or wall at the exterior support location. The factored moment M_u to be transferred is 340 ft-kips and the factored shear V_u is 115 kips. The negative moment reinforcement provided in the column strip is #5 at 10 in. spacing. Use f_c' = 4000 psi and f_y = 60,000 psi.

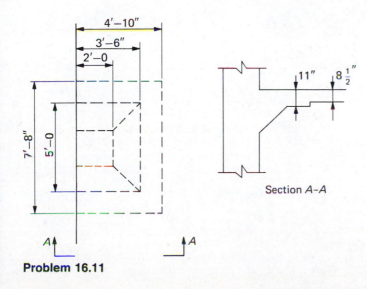

4'–10"
3'–6"
2'–0
7'–8"
5'–0

11" $8\frac{1}{2}''$

Section A-A

Problem 16.11

16.12 Rework through Example 16.10.2, if the service live load is 200 psf instead of 120 psf.

16.13 Rework through Example 16.13.2, if the service live load is 200 psf instead of 120 psf.

16.14 Rework through Example 16.15.1, if the service live load is 200 psf instead of 120 psf.

Flat Plates

16.15 Design a typical interior rigid frame (long direction) of a flat plate floor using the data for the design example of Sec. 16.3, including the following exception to the example, as assigned by the instructor. Note that the preliminary slab thickness must be determined as part of your design (i.e., the original $5\frac{1}{2}$-in. thickness does not necessarily apply).
  (a) Live load 80 psf; $f_y = 60$ ksi; columns 14 in. $\times$ 12 in.
  (b) Live load 100 psf; $f_y = 60$ ksi; columns 16 in. $\times$ 14 in.
  (c) Live load 120 psf; $f_y = 60$ ksi; columns 16 in. $\times$ 14 in.
  (d) Live load 120 psf; $f_y = 60$ ksi; columns 18 in. $\times$ 18 in.; panel size 18 ft $\times$ 15 ft

16.16 Assuming the slab thickness is $6\frac{1}{2}$ in., investigate the moment and shear transfer in the flat plate design example (see Examples 16.17.1 and 16.18.1) for a service live load of 150 psf instead of 60 psf.

16.17 Rework through Example 16.10.3, if the service live load is 150 psf instead of 60 psf.

16.18 Rework through Example 16.15.2, if the service live load is 150 psf instead of 60 psf.

17

Equivalent Frame Analysis of Two-Way Floor Systems in Unbraced Frames

17.1 GENERAL INTRODUCTION

The design of two-way floor systems has been presented in Chap. 16, wherein the longitudinal distribution of moments in the equivalent rigid frame* follows the coefficients of the ACI Code, within the limitations set forth in the direct design method. Beyond these limitations, an elastic analysis of the equivalent rigid frame must be made for various gravity load patterns to obtain the longitudinal moment and shear envelopes. For lateral load, however, elastic analysis of the equivalent rigid frame has to be made, whether or not the system is within the limitations of the direct design method for gravity load. Note that ACI-13.3.1 states "A slab system may be designed by *any procedure* satisfying conditions of equilibrium and geometric compatibility if shown that the design strength at every section is at least equal to the required strength . . . , and that all serviceability conditions . . . are met."

For gravity load analysis of a typical multibay, multistory frame as shown in Fig. 17.1.1, ACI-13.7.2.5 permits the separate analysis of each floor, with far ends of columns considered fixed in rotation and in sidesway. For lateral load analysis, the implication is that the entire frame should be analyzed as an unbraced frame. The authors belive that the preliminary design, say, of the third floor in Fig. 17.1.1 may be based on results of analysis of a subassembly as shown in Fig. 17.1.2 for both gravity load and lateral load. In this subassembly, the joint rotations at D, E, F, G, H, and K are zero, but

*Note that the concept of "equivalent rigid frame" is applicable in both the direct design method and the equivalent frame method.

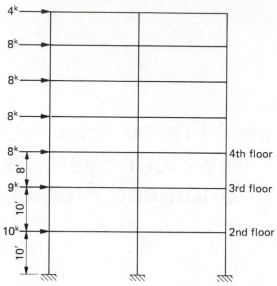

Figure 17.1.1 A typical 2-bay, 7-story equivalent rigid frame.

sideways at GHK and ABC are permitted. Thus the degree of freedom, in this particular case, is 5 (two sideways plus rotations at A, B, and C); and the analysis is manageable on a microcomputer.

At this point of presentation, complicating factors such as axial deformation of the columns, reduction of column stiffness due to heavy gravity load, and increase of column moments and shears due to "P-delta"* effects are neglected in order that attention may be focused on characteristics of the equivalent rigid frame in a two-way floor system.

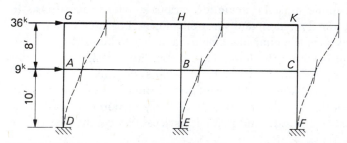

Figure 17.1.2 A subassembly in the 2-bay, 7-story equivalent rigid frame of Fig. 17.1.1.

17.2 ANALYTICAL MODELS FOR ELASTIC FRAME ANALYSIS

The elevation and plan of the subassembly containing a floor with the attached upper and lower columns in a two-way floor system are shown in Figs. 17.2.1(a) and (b). Had the elastic section properties been constant

---

*In the horizontally deflected position, the gravity load (P) times the horizontal deflection (delta) gives so-called secondary bending moments; hence the term "P-delta."

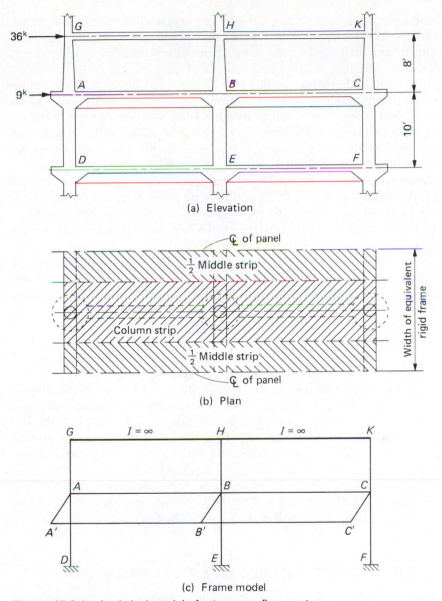

Figure 17.2.1   Analytical model of a two-way floor system.

across the width of the equivalent rigid frame, the structure could be modeled simply as a plane frame with two beam elements and six column elements, at a degree of freedom of 5. But, since the columns are concentrated along the column line, their restraining effect on the horizontal elements has to be spread in some way across the entire width of the equivalent rigid frame. Furthermore, in the general case, there may be transverse and longitudinal beams through the columns, in addition to the monolithic slab. The most sophisticated model would be to divide the slab into a mesh of finite elements in bending and install "nodes" all over the floor area, with each node having two rotational degrees of freedom and one vertical degree of freedom. (The sidesway degrees of freedom will still need to be included.) This technique has been used in some recent investigations; a summary of these investigations is given by Vanderbilt [1].

To model the structure of Figs. 17.2.1(a) and (b) as an elastic frame with straight elements only, it is necessary to treat the slab area as an equivalent beam. One way is to imagine a beam having moment of inertia equal to that of the entire cross section (ACI-Chap. 13, with modifications for cracking and reinforcement) across the full width of the equivalent rigid frame but located at $A'B'C'$ shown in Fig. 17.2.1(c) (i.e., offset from the plane defined by the columns). In such a case, the beam ABC located on the column line in Fig. 17.2.1(c) will have no stiffness; but the beam $A'B'C'$ will be connected to the columns by torsion elements AA', BB', and CC'. This is the model implied by the ACI Code and Commentary, and for purpose of identification will be called Model 1.

Investigations by the authors using computer programs written for microcomputers seem to indicate a more rational model, in which the moment of inertia of the cross section in the column strip is placed in beam ABC and that in the middle strip, in beam $A'B'C'$. Again, for purpose of identification, this model will be called Model 2.

It may be noted that the behavior (from tests) of the actual structure and the solution from the analytical model (Models 1 or 2) can be adjusted to correspond through the quantification and classification of the following parameters: (1) moment of inertia (with or without the effect of cracking and reinforcement) and its variation along the length of the beam ABC and $A'B'C'$; (2) moment of inertia and its variation along the length of the columns; and (3) torsional stiffness K_t of the transverse beams as now defined by the ACI Code. As understanding and use of personal computers in structural analysis become more popular and as more correlations are obtained between analytical solutions and test data, there is little doubt that an appropriate method of gravity and lateral load analysis of unbraced building frames containing two-way floor systems will be achieved.

17.3 TREATMENT OF MODEL 1 AS A PLANE FRAME

Although the limited investigation by the authors indicates that Model 2 (placing column strip stiffness in ABC and middle strip stiffness in $A'B'C'$ of Fig. 17.2.1) gives more rational results than the ACI implied Model 1 (placing all stiffness within width of equivalent rigid frame in $A'B'C'$, leaving no stiffness in ABC of Fig. 17.2.1), there is an advantage in using Model 1 over Model 2. Model 1 may be "approximately" analyzed as a plane frame having a much smaller degree of freedom than if it is analyzed as a space frame,[*] as will be shown later. Perhaps it is for this reason that Corley et al. [2,3] developed the equivalent column concept, since at that time the

---

[*]The term "space frame" is used for simple identification only. In fact the torsion elements projecting out of the plane of the columns are rigid except in torsional rotation.

moment distribution method was the usual method for structural analysis of plane frames.

As shown in Fig. 17.3.1, when the three torsion elements are treated as such, the degree of freedom of the subassembly (defined as Model 1) is 8.

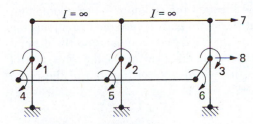

Degree of freedom = 8
Number of beam elements = 2
Number of column elements = 6
Number of torsion elements = 3

Figure 17.3.1   Model 1 as a space frame. (The term *space frame* is used for simple identification only. In fact the torsion elements projecting out of the plane of the columns are rigid except in torsional rotations.)

In Fig. 17.3.2(a), the torsion element at each column is arbitrarily divided into two such elements in parallel, one joined to the lower end of the upper column and the other joined to the upper end of the lower column, with the stiffness of the original element divided proportionally according to the column stiffness at that end. If the resulting frame, called Model 1A, is still treated as a space frame, its degree of freedom is 11. However, if the six torsion elements and the two beam elements are compressed into the plane of the columns, the torsion elements become six rotational springs set internally at the column ends *before* the column joins the "node," as shown in Fig. 17.3.2(b). The flexural properties of the columns can then be modified owing to the presence of rotational springs at the ends so that the degree of freedom of the plane frame in Fig. 17.3.2(b) is 5.

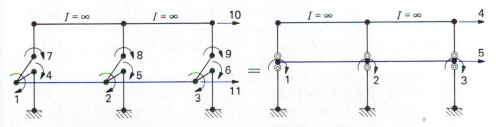

Degree of freedom = 11
Number of beam elements = 2
Number of column elements = 6
Number of torsion elements = 6

(a) As a space frame

Degree of freedom = 5
Number of beam elements = 2
Number of equivalent column elements = 6
(No torsion element)

(b) As a plane frame

Figure 17.3.2   Model 1A: Replacing torsion elements by rotational springs at column ends, distributed to upper and lower columns in proportion to the column flexural stiffnesses (same as K_{ec} model of Vanderbilt [1]).

As an alternative, the torsion element at each interior column may be divided into two such elements in parallel [as shown in Fig. 17.3.3(a)], one joined to the right end of the left beam and the other to the left end of the right beam, with the total torsional stiffness proportioned according to the beam stiffness at that end. The degree of freedom of the resulting frame, called Model 1B, is 9. Again, if the four torsion elements and the two beam elements are compressed into the plane of the columns, they become four rotational springs set internally at the beam ends *before* the beam joins the "node," as shown in Fig. 17.3.3(b). The fixed-end moments and the flexural properties of the beams can then be modifed owing to the presence of rotational springs at the ends so that the degree of freedom of the plane frame in Fig. 17.3.3(b) is still 5.

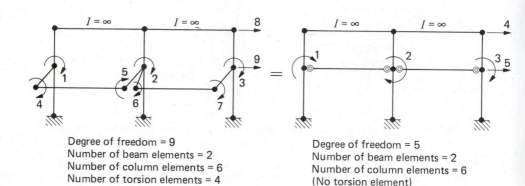

Degree of freedom = 9
Number of beam elements = 2
Number of column elements = 6
Number of torsion elements = 4

(a) As a space frame

Degree of freedom = 5
Number of beam elements = 2
Number of column elements = 6
(No torsion element)

(b) As a plane frame

Figure 17.3.3 Model 1B: Replacing torsion elements by rotational springs at slab-beam ends (same as K_{eb} model of Vanderbilt [1]).

The use of Model 1A by modifying the flexural properties of the columns is the "equivalent column method" of Corley [3]; and the use of Model 1B by modifying the fixed-end moments and the flexural properties of the beams is the "equivalent beam method" of Vanderbilt [1]. Other models specifically for flat plate structures using equivalent beams without torsional springs have been used by Pecknold [68, Chap. 16], Allen and Darvall [69, Chap. 16], and Elias [71–73, Chap. 16].

Whether the results of the equivalent column method [Model 1A of Fig. 17.3.2(b)] are good approximations to those of actually analyzing Model 1 of Fig. 17.3.1 as a space frame can be examined by comparing the slopes $\theta_{AU}-\theta_{BU}-\theta_{CU}$ (subscript U means upper) with the slopes $\theta_{AL}-\theta_{BL}-\theta_{CL}$ (subscript L means lower) of Fig. 17.3.4(a). Likewise, the results of the equivalent beam method [Model 1B of Fig. 17.3.3(b)] may be examined by comparing the slope θ_{BL} (subscript L here means left) with slope θ_{BR} (subscript R means right) of Fig. 17.3.4(b).

In the remainder of this chapter, procedures of analysis for the four models (Model 1, Model 1A, Model 1B, and Model 2) will be described and numerical examples shown.

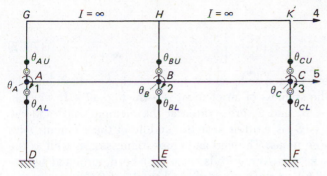

(a) Model 1A: Equivalent column method

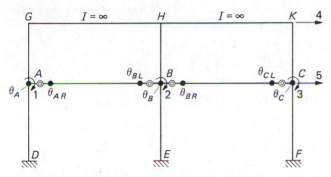

(b) Model 1B: Equivalent beam method

Figure 17.3.4 Plane frame analysis of equivalent rigid frame in a two-way floor system.

17.4 REDUCED STIFFNESS MATRIX AND REDUCED FIXED-END MOMENTS FOR A FLEXURAL ELEMENT WITH ROTATIONAL SPRINGS AT ENDS

The presence of a rotational spring at the end of a flexural element is identical to the situation in which the member is attached to a rigid gusset plate at the joint by a semirigid connection [4]. Methods of obtaining the reduced fixed-end moments and the reduced stiffness matrix for a member with constant moment of inertia have been treated by Wang [4].

For reinforced concrete buildings with two-way floor systems, the beams (whether for the combined column and middle strip as for Model 1, or for the column strip and middle strip separately as for Model 2) and the columns all have variations in moment of inertia along their lengths. Even for such an element with rigid end connection, the element stiffness matrix is no longer

$$[S] = \begin{bmatrix} \dfrac{4EI}{L} & \dfrac{2EI}{L} \\[2mm] \dfrac{2EI}{L} & \dfrac{4EI}{L} \end{bmatrix} \tag{17.4.1}$$

wherein EI is the constant moment of inertia for the entire length L, but it becomes

$$[S] = \begin{bmatrix} S_{ii} & S_{ij} \\ S_{ji} & S_{jj} \end{bmatrix} \tag{17.4.2}$$

in which $S_{ij} = S_{ji}$ but written as S_{ij} and S_{ji} for theoretical clarification. The best way to show the stiffness of an element is as in Fig. 17.4.1, wherein the "end stiffnesses" S_{ii} and S_{jj} are written at the element ends and the "cross stiffness" $S_{ij} = S_{ji}$ is written near the middle of the element, with lines linking the three items. The end and cross stiffnesses, as well as the fixed-end moments for transverse loads, may best be determined by the column-analogy method, as demonstrated in Chap. 16 of Ref. 4.

The presence of rotational springs at the member ends, with stiffnesses K_{ti} and K_{tj}, will cause slip angles ψ_i and ψ_j, both in the *counterclockwise* direction as shown in Fig. 17.4.1(b). By the definition of K_{ti} and K_{tj},

$$\begin{Bmatrix} \psi_i \\ \psi_j \end{Bmatrix} = \begin{bmatrix} \dfrac{1}{K_{ti}} & 0 \\ 0 & \dfrac{1}{K_{tj}} \end{bmatrix} \begin{Bmatrix} M_i \\ M_j \end{Bmatrix} = [B] \begin{Bmatrix} M_i \\ M_j \end{Bmatrix} \tag{17.4.3}$$

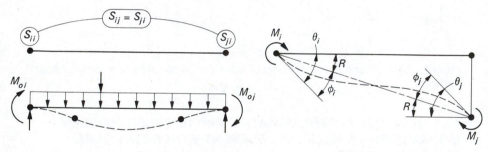

(a) Flexural element with rigid end connection

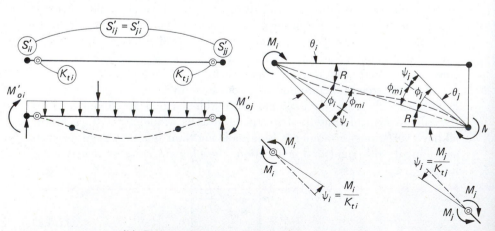

(b) Flexural element with semirigid end connection
(same as with rotational springs at ends)

Figure 17.4.1 Flexural element with rigid and semirigid end connections.

Applying the definition of the regular stiffness matrix $[S]$ to the member itself inside the rotational springs in Fig. 17.4.1(b),

$$\begin{Bmatrix} M_i \\ M_j \end{Bmatrix} = \begin{bmatrix} S_{ii} & S_{ij} \\ S_{ji} & S_{jj} \end{bmatrix} \begin{Bmatrix} \phi_{mi} \\ \phi_{mj} \end{Bmatrix} = [S]\{\phi_m\} \qquad (17.4.4)$$

But, from the geometry shown in Fig. 17.4.1(b),

$$\{\phi_m\} = \{\phi\} - \{\psi\} \qquad (17.4.5)$$

Substituting Eq. (17.4.5) in Eq. (17.4.4),

$$\{M\} = [S]\{\phi\} - [S]\{\psi\} \qquad (17.4.6)$$

Substituting Eq. (17.4.3) in Eq. (17.4.6),

$$\{M\} = [S]\{\phi\} - [S][B]\{M\}$$

from which

$$[I + SB]\{M\} = [S]\{\phi\}$$

and

$$\{M\} = [I + SB]^{-1}[S]\{\phi\} \qquad (17.4.7)$$

By definition in Fig. 17.4.1(b),

$$\{M\} = [S']\{\phi\} \text{ for semirigid end connection} \qquad (17.4.8)$$

Comparing Eq. (17.4.8) with Eq. (17.4.7),

$$[S'] = [I + SB]^{-1}[S] \qquad (17.4.9)$$

Equation (17.4.9) gives the reduced stiffness matrix of a flexural element with rotational springs at its ends. That the $[S']$ matrix is still symmetric can be proved as follows:

To Prove: $[S'] = [I + SB]^{-1}[S]$ is symmetric.

Proof:
1. Taking the inverse of each side of the above equation, $[S']^{-1} = [S^{-1}][I + SB]$, because in general $[CD]^{-1} = [D^{-1}][C^{-1}]$.
2. Expanding the equation in step 1, $[S']^{-1} = [S^{-1}I] + [S^{-1}SB] = [S^{-1}] + [B]$.
3. Observing the equation in step 2, $[S']^{-1}$ is symmetric because $[S^{-1}]$ and $[B]$ are both symmetric.
4. From step 3, $[S']$ is symmetric, because $[S']^{-1}$ is symmetric.

Next, to obtain formulas for the reduced fixed-end moments under transverse loads, observe first that the rotation R of the member axis is zero in the fixed condition and thus θ's and ϕ's are also zero. There are two ways of reasoning by which the formulas for M'_{0i} and M'_{0j} in Fig. 17.4.1(b) may be obtained. In the first way, the beginning premise is that M'_{0i} and M'_{0j} in Fig. 17.4.1(b) are there to keep $\theta_i = \theta_j = 0$. If the unreduced fixed end moments M_{0i} and M_{0j} were there instead, then $\theta_{mi} = \theta_{mj} = 0$ (note that when R is zero, θ_{mi} and θ_{mj} are the same as ϕ_{mi} and ϕ_{mj}) but θ_i and θ_j would be, using Eq. (17.4.5),

$$\begin{Bmatrix} \text{unwanted } \theta_i \\[2mm] \text{unwanted } \theta_j \end{Bmatrix} = \begin{Bmatrix} \theta_{mi} = 0 \\[2mm] \theta_{mj} = 0 \end{Bmatrix} + \begin{Bmatrix} \dfrac{M_{0i}}{K_{ti}} \\[4mm] \dfrac{M_{0j}}{K_{tj}} \end{Bmatrix} = [B] \begin{Bmatrix} M_{0i} \\[2mm] M_{0j} \end{Bmatrix} \quad \textbf{(17.4.10)}$$

The M'_{0i} and M'_{0j} must then be the sum of the regular fixed-end moments and those to cancel out [thus for the negative sign in Eq. (17.4.11)] the unwanted slopes in Eq. (17.4.10); or

$$\begin{aligned} \{M'_0\} &= \{M_0\} - [S'][B]\{M_0\} \\ &= [I - S'B]\{M_0\} \end{aligned} \quad \textbf{(17.4.11)}$$

Substituting Eq. (17.4.9) for $[S']$ in Eq. (17.4.11),

$$\{M'_0\} = [I - [I + SB]^{-1}[S][B]]\{M_0\}$$

Premultiplying each side of the above equation by $[I + SB]$,

$$[I + SB]\{M'_0\} = [[I + SB] - [SB]]\{M_0\} = \{M_0\}$$

from which

$$\{M'_0\} = [I + SB]^{-1}\{M_0\} \quad \textbf{(17.4.12)}$$

The second way of reasoning is more straightforward because the formula for $[S']$ is not required in the process. The beginning premise is the same, in that M'_{0i} and M'_{0j} are there to keep $\theta_i = \theta_j = 0$, which means that θ_{mi} and θ_{mj} (or ϕ_{mi} and ϕ_{mj}) will not be zero but equal to, using Eq. (17.4.5),

$$\{\theta_m\} = \{\theta = 0\} - [B]\{M'_0\} \quad \textbf{(17.4.13)}$$

The excess of $\{M'_0\}$ over $\{M_0\}$ acting on the member ends between the rotational springs should be consistent with $\{\theta_m\}$ in Eq. (17.4.13) through the regular stiffness matrix $[S]$; or

$$\{M'_0 - M_0\} = -[S][B]\{M'_0\}$$
$$\{M'_0\} + [SB]\{M'_0\} = \{M_0\}$$

from which

$$\{M'_0\} = [I + SB]^{-1}\{M_0\}$$

same as Eq. (17.4.12).

In a computer program, the formulas for $\{M'_0\}$ and $[S']$, as expressed by Eqs. (17.4.12) and (17.4.9), may be written as

$$
\left.
\begin{aligned}
&1.\ F_{ti} = \frac{1}{K_{ti}}; \qquad F_{tj} = \frac{1}{K_{tj}} \\
&2.\ T_1 = (S_{ii}S_{jj}) - (S_{ij})^2 \\
&3.\ T_2 = 1 + F_{ti}S_{ii} + F_{tj}S_{jj} + F_{ti}F_{tj}T_1 \\
&4.\ M'_{0i} = [(1 + F_{tj}S_{jj})M_{0i} - (F_{tj}S_{ij})M_{0j}]/T_2 \\
&\quad M'_{0j} = [(1 + F_{ti}S_{ii})M_{0j} - (F_{ti}S_{ij})M_{0i}]/T_2 \\
&5.\ S'_{ii} = (S_{ii} + F_{tj}T_1)/T_2 \\
&6.\ S'_{ij} = S_{ij}/T_2 \\
&7.\ S'_{jj} = (S_{jj} + F_{ti}T_1)/T_2
\end{aligned}
\right\} \qquad \textbf{(17.4.14)}
$$

The necessary algebraic manipulation to obtain Eqs. (17.4.14) is not shown here but left as an exercise for the reader.

17.5 THE ACI EQUIVALENT COLUMN

In the 1977 ACI Code, the equivalent column was one having a stiffness obtained by ACI Formula (13-6), defined in words as follows: "Flexibility (inverse of stiffness) of an equivalent column shall be taken as the sum of the flexibilities of the actual columns above and below the slab-beam and the flexibility of the attached torsion member . . ." This formula is removed from the 1983 ACI Code and placed in Commentary-13.7.4. The formula is

$$
\frac{1}{K_{ec}} = \frac{1}{\Sigma K_c} + \frac{1}{K_t} \qquad \textbf{(17.5.1)}
$$

from which

$$
K_{ec} = \frac{\Sigma K_c}{1 + \dfrac{\Sigma K_c}{K_t}} \qquad \textbf{(17.5.2)}
$$

It can be shown that S'_{ii} of Eq. (17.4.14) is identical to K_{ec} of Eq. (17.5.2), for one column with stiffness K_c. From Eq. (17.4.14), if K_{tj} at the far end is infinity, then

$$
T_2 = 1 + F_{ti}S_{ii}
$$

and

$$
S'_{ii} = \frac{S_{ii}}{T_2} = \frac{S_{ii}}{1 + \dfrac{S_{ii}}{K_{ti}}} \qquad \textbf{(17.5.3)}
$$

which is identical with Eq. (17.5.2) for $S_{ii} = \Sigma K_c$. So Eq. (17.5.2) is basically correct for one-story buildings in gravity load analysis, as it will be shown later that the equivalent column model (Model 1A) is good for gravity load analysis, although the equivalent beam model (Model 1B) is somewhat better for lateral load analysis. The original idea of Corley for developing the attached transverse torsion element is noteworthy.

17.6 EQUIVALENT COLUMN METHOD (MODEL 1A) vs EQUIVALENT BEAM METHOD (MODEL 1B)

As explained in Sec. 17.3, the equivalent column method (Model 1A) and the equivalent beam method (Model 1B) are both approximations of the space frame model (Model 1). It turns out that for lateral load analysis both Models 1A and 1B are good approximations with preference for Model 1B, but for gravity load analysis only Model 1A is good. The reason is that for both gravity and lateral loads, the columns are subjected only to end moments causing reverse curvature (that is, they have an inflection point in the midheight region) and the elastic curves in the upper and lower columns are geometrically similar. Hence, the $\theta_{AU}-\theta_{BU}-\theta_{CU}$ are nearly equal to $\theta_{AL}-\theta_{BL}-\theta_{CL}$ in Fig. 17.3.4(a). On the other hand, the moment diagrams for the beams are parabolic for gravity load. Consequently θ_{BL} and θ_{BR} in Fig. 17.3.4(b) are no longer nearly equal. In fact, they are of opposite sign so that there is no continuity of slab-beam at the column line under gravity load.

The authors favor of Model 1 for use on the microcomputer with little additional programming effort; in fact, there is no need of using the reduced stiffness matrix and the reduced fixed-end moments. If this is the choice, the analyst might just as well go further to use Model 2.

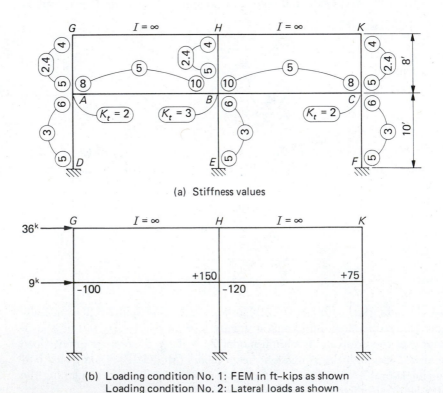

(a) Stiffness values

(b) Loading condition No. 1: FEM in ft-kips as shown
     Loading condition No. 2: Lateral loads as shown

Figure 17.6.1   Details of floor assembly.

EXAMPLE 17.6.1 By the equivalent column method analyze the floor sub-assembly in a building frame with two-way floor systems as detailed in Fig. 17.6.1.

Solution: (a) Choice of method. With the popularly available microcomputer, the matrix displacement method [4,5,6] is the method that should be used, not to mention its expandability to analysis of the whole building frame and to consideration of axial deformation, stability functions, and *P*-delta effects.

(b) Global degree of freedom. The degree of freedom is $NP = 5$, assigned as shown in Fig. 17.3.4(a).

(c) Local stiffness matrix of the beam elements. The formulas shown in Fig. 17.6.2(a) are used to enter the contributions of beams *AB* and *BC* to the global stiffness matrix. Input data include

BEAM	NP1	NP2	S_{ii}	S_{ij}	S_{jj}	M_{oi}	M_{ij}
AB	1	2	8	5	10	-100	$+150$
BC	2	3	10	5	8	-120	$+75$

P $\diagdown$ X	NP1	NP2
NP1	S_{ii}	S_{ij}
NP2	S_{ji}	S_{jj}

(a) Typical beam element

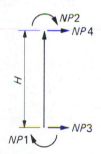

P $\diagdown$ X	NP1	NP2	NP3	NP4
NP1	S_{ii}	S_{ij}	$+T1$	$-T1$
NP2	S_{ji}	S_{jj}	$+T2$	$-T2$
NP3	$+T1$	$+T2$	$+T3$	$-T3$
NP4	$-T1$	$-T2$	$-T3$	$+T3$

$$T1 = \frac{S_{ii} + S_{ij}}{H} \qquad T2 = \frac{S_{jj} + S_{ij}}{H} \qquad T3 = \frac{S_{ii} + S_{jj} + 2S_{ij}}{H^2}$$

(b) Typical column element

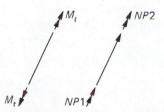

P $\diagdown$ X	NP1	NP2
NP1	$+K_t$	$-K_t$
NP2	$-K_t$	$+K_t$

(c) Typical torsion element

Figure 17.6.2 Local stiffness matrix of typical beam, column, and torsion elements.

(d) Local stiffness matrix of the column elements. Equations (17.4.14) are used first to compute the reduced stiffness matrix of each column and then the formulas in Fig. 17.6.2(b) are used to enter the contributions of the six equivalent columns to the global stiffness matrix. Input data include

COLUMN	NP1	NP2	NP3	NP4	S_{ii}	S_{ij}	S_{jj}	H	K_{ti}	K_{tj}
DA	6	1	6	5	5	3	6	10	∞	1.0909
EB	6	2	6	5	5	3	6	10	∞	1.6364
FC	6	3	6	5	5	3	6	10	∞	1.0909
AG	1	6	5	4	5	2.4	4	8	0.9091	∞
BH	2	6	5	4	5	2.4	4	8	1.3636	∞
CK	3	6	5	4	5	2.4	4	8	0.9091	∞

Note that the local degree of freedom is assigned the value $NP + 1 = 6$ wherever there is restraint so the $(NP + 1)$th row and column are never used. The torsional stiffness at each column intersection is divided into two parts in the ratio of the column stiffnesses; for example, $K_{tj} = [(S_{ii})_{\text{lower}}/(S_{ii})_{\text{upper}}]K_t$. The modified stiffness values are

COLUMN	S'_{ii}	S'_{ij}	S'_{jj}
DA	3.7308	0.4615	0.9231
EB	3.8214	0.6428	1.2857
FC	3.7308	0.4615	0.9231
AG	0.7692	0.3692	3.0252
BH	1.0714	0.5143	3.0948
CK	0.7692	0.3692	3.0252

(e) The load matrix. The loads that are applied directly at the joints are entered first. Those arising from the fixed-end moments are programmed for internal calculation and compilation. The resulting load matrix is

$[P]_{5 \times 2} =$

P \ LC	1	2
1	+100	0
2	−30	0
3	−75	0
4	0	+36
5	0	+9

(f) Displacements and member end moments. The displacements in terms of kip-ft^2/EI and kip-ft^3/EI are

	GRAVITY LOAD	LATERAL LOAD
θ_A	+11.636	+5.644
θ_B	−2.487	+1.283
θ_C	−6.419	+5.644
Δ_{GHK}	+2.504	+436.69
Δ_{ABC}	+1.384	+266.78

The member end moments in ft-kips are

	GRAVITY LOAD		LATERAL LOAD	
	M_i	M_j	M_i	M_j
Beam AB	− 19.34	+ 183.31	+ 51.57	+ 41.05
BC	− 176.96	+ 11.21	+ 41.05	+ 51.57
Column DA	+ 4.79	+ 10.55	− 109.24	− 31.73
AG	+ 8.79	+ 3.82	− 19.84	− 70.01
EB	− 2.22	− 3.46	− 118.27	− 49.80
BH	− 2.89	− 1.78	− 32.30	− 76.00
FC	− 3.54	− 6.12	− 109.24	− 31.73
CK	− 5.10	− 2.84	− 19.84	− 70.01

(g) Column end slopes computed from beam end slopes. The θ's in the output are the slopes of beams at the ends of the torsion element. The slopes at the column ends can be computed from

$$\theta_{mi} = \theta_i - \frac{M_i}{K_{ti}}$$

The closeness or divergence of the two upper and lower column slopes indicates the degree of approximation of Model 1A to Model 1. The results are

	INTERSECTION	A	B	C	
Gravity load	θ (upper)	+ 1.966	− 0.370	− 0.812	(very good)
	θ (lower)	+ 1.966	− 0.370	− 0.812	
Lateral load	θ (upper)	+ 27.466	+ 24.973	+ 27.466	(not too good)
	θ (lower)	+ 34.728	+ 31.716	+ 34.728	

EXAMPLE 17.6.2 Rework Example 17.6.1 except use the equivalent beam method.

Solution: (a) Global degree of freedom. The degree of freedom is $NP = 5$, assigned as shown in Fig. 17.3.4(b).

(b) Local stiffness matrix of the beam elements. Equations (17.4.14) are used first to compute the reduced fixed-end moments and the reduced stiffness matrix of each beam before the formulas in Fig. 17.6.2(a) for the local stiffness matrix are used. Input data include

BEAM	NP1	NP2	S_{ii}	S_{ij}	S_{jj}	K_{ti}	K_{tj}	M_{0i}	M_{0j}
AB	1	2	8	5	10	2	1.5	− 100	+ 150
BC	2	3	10	5	8	1.5	2	− 120	+ 75

The reduced fixed-end moments and the reduced stiffnesses are

BEAM	M'_{0i}	M'_{0j}	S'_{ii}	S'_{ij}	S'_{jj}
AB	− 42.22	+ 33.33	1.4889	0.1667	1.2500
BC	− 26.25	+ 32.50	1.2500	0.1667	1.4889

Note that the carry-over factors are $0.1667/1.4889 = 0.112$ from A to B and $0.1667/1.2500 = 0.133$ from B to A, which means that convergence in moment distribution is extremely fast for gravity load analysis.

(c) Local stiffness matrix of the column elements. The column stiffnesses do not need modification. The input data are the same as in part (d) of the previous example, except there are no rotational springs at the ends.

(d) The load matrix. The load matrix is

$$[P]_{5\times 2} =$$

P \ LC	1	2
1	$+42.22$	0
2	-7.08	0
3	-32.50	0
4	0	$+36$
5	0	$+9$

(e) Displacements and member end moments. The displacements in terms of kip-ft$^2/EI$ and kip-ft$^3/EI$ are

	GRAVITY LOAD	LATERAL LOAD
θ_A	$+3.568$	$+31.425$
θ_B	-0.370	$+28.649$
θ_C	-2.415	$+31.425$
Δ_{GHK}	$+2.504$	$+436.20$
Δ_{ABC}	$+1.383$	$+249.70$

The member end moments in ft-kips are

		GRAVITY LOAD		LATERAL LOAD	
		M_i	M_j	M_i	M_j
Beam	AB	-36.97	$+33.46$	$+51.56$	$+41.05$
	BC	-27.11	$+28.84$	$+41.05$	$+51.56$
Column	DA	$+9.60$	$+20.16$	-105.49	-36.18
	AG	$+16.80$	$+7.67$	-15.38	-73.77
	EB	-2.22	-3.46	-113.82	-52.84
	BH	-2.89	-1.78	-29.26	-80.44
	FC	-8.35	-15.73	-105.49	-36.18
	CK	-13.11	-6.69	-15.38	-73.77

Note that the sideways in this Model 1B are almost identical to those in Model 1A. So are the beam and column moments for lateral load. But the beam moments at the ends for gravity load are clearly too small in this Model 1B and should not be relied upon; furthermore, as will be seen below in part (f), slope continuity is *not* satisfied at the junction of the two elastic curves of the beam.

(f) Beam end slopes computed from column end slopes. The θ's in the output are the slopes of the columns at their intersections with the torsion elements. The slopes at the beam ends of the torsion elements can be computed from

$$\theta_{mi} = \theta_i - \frac{M_i}{K_{ti}}$$

The closeness or divergence of the left and right slopes of the beams at the interior support indicates the degree of approximation of Model 1B to Model 1. The results are

	INTERSECTION	A	B	C
Gravity load	θ (left)	(none)	− 22.680	− 16.836
	θ (right)	+ 22.054	+ 17.707	(none)
			(no good)	
Lateral load	θ (left)	(none)	+ 1.283	+ 5.643
	θ (right)	+ 5.643	+ 1.283	(none)
			(very good)	

The slopes of the slab-beams are $22.680/EI$ counterclockwise on one side and $17.707/EI$ clockwise on the other side of the interior support, indicating that Model 1B should not be used for gravity load. For lateral load analysis, Model 1B gives even better results than Model 1A because there is slope compatibility at the columns as well as at the beam junctions (such as A', B', and C' of Fig. 17.3.1) occurring away from the columns.

17.7 TREATMENT OF MODEL 1 AS A SPACE FRAME

The equivalent column method (Model 1A, or K_{ec} model) and the equivalent beam method (Model 1B, or K_{eb} model) are, after all, approximations to the analysis of Model 1 of Fig. 17.3.1 as a space frame. When sidesway is ignored in gravity load analysis, Model 1A (not Model 1B) has merit because then the usual moment distribution method can be applied to the subassembly. This was exactly the ingenious idea of Corley et al. [2,3] at a time when electronic computers were not conveniently available. With sidesways as unknowns in lateral load analysis, the indirect moment distribution method [4,5] involving the solution of as many simultaneous equations as there are unknown sidesways has to be used. While this approach can itself be programmed on the computer, it is mainly taught in schools for the purpose of making students better understand the matrix displacement method.

With an isolated floor assembly as shown in Fig. 17.3.1, the degree of freedom of Model 1 would be larger than that of Models 1A or 1B, by the number of column lines. This is not a large deterrent, even on the microcomputer. For an entire tall building frame, say, of 15 stories and 4 column lines, the degree of freedom of Model 1 would be 15 times 4, or 60, larger than that of Models 1A and 1B. In this case, the use of Model 1A for gravity load and Model 1B for lateral load might be justifiable. But with a computer programming technique such as the triband elimination method for tall building frame analysis [6, Chap. 19], there is no need of a matrix inversion larger than number of column lines plus one. In time, it is hoped that Model 1, or Model 2, as advocated in Sec. 17.8, will be used.

The local stiffness matrix of a typical torsion element is shown in Fig. 17.6.2(c). Here, M_t is taken positive if its rotational vectors act "tension-like" on the ends of the element. Taking the local degree of freedom numbers NP1 and NP2 at the near and far ends of the element,

$$M_t = K_t(X_2 - X_1) \tag{17.7.1}$$

wherein X_1 and X_2 are the clockwise slopes at the near and far ends of the torsion element, respectively.

EXAMPLE 17.7.1 Rework Examples 17.6.1 and 17.6.2 except use the torsion elements as such in the space frame (Model 1) of Fig. 17.3.1.

Solution: (a) Global degree of freedom. The degree of freedom is $NP = 8$, assigned as shown in Fig. 17.3.1.

(b) Local stiffness matrix of the beam elements. Input data include

BEAM	NP1	NP2	S_{ii}	S_{ij}	S_{jj}	M_{0i}	M_{0j}
AB	4	5	8	5	10	− 100	+ 150
BC	5	6	10	5	8	− 120	+ 75

(c) Local stiffness matrix of the column elements. Input data include

COLUMN	NP1	NP2	NP3	NP4	S_{ii}	S_{ij}	S_{jj}	H
DA	9	1	9	8	5	3	6	10
EB	9	2	9	8	5	3	6	10
FC	9	3	9	8	5	3	6	10
AG	1	9	8	7	5	2.4	4	8
BH	2	9	8	7	5	2.4	4	8
CK	3	9	8	7	5	2.4	4	8

(d) Local stiffness matrix of the torsion elements. Input data include

TORSION ELEMENT	NP1	NP2	K_t
A'A	4	1	2
B'B	5	2	3
C'C	6	3	2

(e) The load matrix. The load matrix is

$$[P]_{8 \times 2} =$$

P \ LC	1	2
1	0	0
2	0	0
3	0	0
4	+100	0
5	−30	0
6	−75	0
7	0	+36
8	0	+9

(f) Displacements and member end moments from the output. The displacements in terms of kip-ft^2/EI and kip-ft^3/EI are

	GRAVITY LOAD	LATERAL LOAD
θ_A	+1.966	+31.425
θ_B	−0.370	+28.649
θ_C	−0.812	+31.425
θ'_A	+11.636	+5.643
θ'_B	−2.487	+1.283
θ'_C	−6.419	+5.643
Δ_{CHK}	+2.504	+436.20
Δ_{ABC}	+1.383	+249.70

The member end moments in ft-kips are

		GRAVITY LOAD		LATERAL LOAD	
		M_i	M_j	M_i	M_j
Beam	A'B'	−19.34	+183.31	+51.56	+41.05
	B'C'	−176.96	+11.21	+41.05	+51.56
Column	DA	+4.79	+10.55	−105.49	−36.18
	AG	+8.79	+3.82	−15.38	−73.77
	EB	−2.22	−3.46	−113.82	−52.84
	BH	−2.89	−1.78	−29.26	−80.44
	FC	−3.54	−6.12	−105.49	−36.18
	CK	−5.10	−2.84	−15.38	−73.77
Torsion	A'A	−19.34	—	+51.56	—
	B'B	+6.35	—	+82.10	—
	C'C	+11.21	—	+51.56	—

The results for the gravity load are exactly the same as those of Model 1A, and the results for the lateral load are exactly the same as those of Model 1B. This exact coincidence is due to the fact that the isolated floor subassembly has all far ends of columns fixed against rotation. If the entire structural frame is analyzed, there will be some rippling effect. However, if the analyst wishes to use an equivalent plane frame, Models 1A and 1B should be used for gravity and lateral load, respectively.

17.8 MODEL 2 WITH TORSION ELEMENT CONNECTING COLUMN STRIP TO MIDDLE STRIP

The basic frame model for the structure enclosed between the center lines of adjacent panels of a building with two-way floor systems has been shown in Fig. 17.2.1(c). This basic model contains beams ABC and $A'B'C'$, as well as torsion elements $A'A$, $B'B$, and $C'C$. Model 1 is derived from this basic model by placing all longitudinal flexural stiffness within the width of the equivalent rigid frame in beam $A'B'C'$, with none in beam ABC, in an attempt to analyze the space frame as a plane frame, by, say, the moment distribution method. Consequently, the moments in beam $A'B'C'$ obtained from the analysis have yet to be distributed transversely into the column and middle strips, for the purpose of allocating the reinforcement. By placing all of the longitudinal stiffness away from the plane of the columns (granting that the parameter K_t may be adjusted for doing so), not much unwanted effect is produced in the design moments for gravity load; but the model would certainly underestimate the effectiveness of the actual structure to resist lateral loads applied directly in the plane of the columns.

Considering the current capability of the computer to solve mathematics, one might just as well use the basic model, placing the longitudinal stiffness of the column strip in beam ABC and that of the middle strip in beam $A'B'C'$ (granting again the torsional stiffness K_t might need adjustment to correlate with test data). This model, called Model 2, is believed to be more rational than Model 1. The following example shows the results of a Model 2 analysis.

EXAMPLE 17.8.1 Rework Examples 17.6.1, 17.6.2, and 17.7.1, except place two-thirds of the total longitudinal stiffness and fixed-end moments in the plane of the columns and the remaining one-third at the far ends of the torsion elements (Model 2 analysis).

Solution: The degree of freedom numbers and input data are mostly identical to those in Example 17.7.1 except there are now four beam elements instead of two. The displacements in terms of kip-ft^2/EI and kip-ft^3/EI in the output are

	GRAVITY LOAD	LATERAL LOAD
θ_A	$+5.090$	$+11.058$
θ_B	-1.113	$+4.964$
θ_C	-3.016	$+11.058$
$\theta_{A'}$	$+10.074$	$+4.777$
$\theta_{B'}$	-2.100	-0.107
$\theta_{C'}$	-5.900	$+4.777$
Δ_{GHK}	$+3.073$	$+230.40$
Δ_{ABC}	$+1.697$	$+136.02$

The member end moments in ft-kips are

		GRAVITY LOAD		LATERAL LOAD	
		M_i	M_j	M_i	M_j
Beam	AB	−43.23	+109.55	+75.52	+69.95
	A'B'	−9.97	+59.79	+12.56	+7.60
	BC	−97.47	+30.21	+69.95	+75.52
	B'C'	−56.83	+5.77	+7.60	+12.56
Column	DA	+13.91	+29.01	−75.64	−56.07
	AG	+24.18	+11.12	−32.01	−48.96
	EB	−4.70	−8.20	−93.93	−92.94
	BH	−6.84	−3.77	−62.48	−63.59
	FC	−10.40	−19.62	−75.64	−56.07
	CK	−16.35	−8.34	−32.01	−48.96
Torsion	A'A	−9.97	—	+12.56	—
	B'B	+2.96	—	+15.21	—
	C'C	+5.77	—	+12.56	—

Comparing the results of the Model 2 with those of Model 1, it is seen that, for gravity load there are higher negative moments at the exterior supports and lower negative moments at the interior support, the column moments are somewhat larger, and the torsion elements transmit less torsional moment. For lateral load, the sidesway is reduced to about 53% of that obtained for Model 1, the column moments will still have to satisfy statics but they are more equalized between the two ends, and the torsional moments are reduced to 20–25% of those from the Model 1 analysis. Although Model 2 is preferred, its use should be deferred, especially for estimates of sidesway, until the torsional stiffness K_t is adjusted by further studies and tests.

17.9 CONCLUDING REMARKS

At the present time (1983), analysis of concrete frames (including those with two-way floor systems) having their lateral resistance provided only by the stiffnesses of interconnected beams (or slab-beams) and columns should be separated into two parts: (1) analysis for gravity loads only, and (2) analysis for lateral loads plus appropriate simultaneous gravity load. Analytical models that do not allow sidesway should be used for gravity load only, although when the structure or the loading is unsymmetrical a model allowing sidesway should be considered. Simplified assemblies for isolated floors with far ends of columns fixed against rotation but permitted to deflect horizontally may be used for lateral load, although final review analysis should be made for the whole frame in action.

Gravity Load Analysis. The plane frame analysis using the equivalent column model (Model 1A, Fig. 17.3.2) with or without the sidesway degree of freedom, seems appropriate as a mathematical model. This model, developed by Corley et al. [2,3], has been the one endorsed by the ACI Code since 1971. Both the equivalent frame method and the direct design method as prescribed by ACI-13.3.1.1 are equivalent column models. Vanderbilt [1] also indicated that this model is appropriate for gravity load analysis.

Lateral Load Analysis. For lateral load, ACI-13.3.1.2 states that ". . . analysis of unbraced frames shall take into account effects of cracking and reinforcement on stiffness of frame members." During deliberations of ACI Committee 318 regarding the 1983 ACI Code, strong consideration was given to endorsing the equivalent frame model (Models 1A and 1B, Figs. 17.3.2 and 17.3.3, necessarily with the sidesway degree of freedom included) for lateral load analysis primarily based on the work of Vanderbilt [1,7]. According to Vanderbilt [1], although there is little difference in the magnitudes of the sidesway obtained by the two equivalent models, the preferred model for lateral load is the equivalent beam model (Model 1B, Fig. 17.3.2), which agrees with the authors' analysis and results presented in this chapter.

Further, the Vanderbilt report [1] indicates that for stiffness computations in lateral load analysis, a beam-slab stiffness equal to $\frac{1}{4}$ to $\frac{1}{3}$ of that obtained for gross uncracked concrete may be used in lieu of a calculation taking cracking of the beam-slab into account. This reduction would achieve the result of increasing the computed sidesway. The Model 2 method of analysis is advocated by the authors, wherein the longitudinal stiffnesses of the column strip and the middle strip are placed in beams ABC and $A'B'C'$, respectively, of Fig. 17.2.1(c). Indeed, a major difficulty of any lateral load analysis arises in the stiffness determination; the flexural stiffnesses when Models 1A (Fig. 17.3.2) and 1B (Fig. 17.3.3) are to be used, and additionally an appropriate torsional stiffness when Models 1 (Fig. 17.3.1) or 2 (Fig. 17.2.1) are to be used. For the space frame Models 1A and 1B, the torsional stiffness has been calibrated for the gravity load analysis and then used in Model 1B for the lateral load analysis. With a space frame model (especially the general model of Fig. 17.2.1), use of the proper torsional stiffness is more important, and recommendations based on calibration with actual structures are not available.

SELECTED REFERENCES

1. M. Daniel Vanderbilt. "Equivalent Frame Analysis of Unbraced Reinforced Concrete Buildings for Static Lateral Loads," *Structural Research Report No. 36*, Civil Engineering Department, Colorado State University, Fort Collins, June 1981.
2. W. G. Corley, M. A. Sozen, and C. P. Siess. "The Equivalent Frame Analysis for Reinforced Concrete Slabs," *Structural Research Series No. 218*, Department of Civil Engineering, University of Illinois, June 1961, 168 pp.
3. W. G. Corley and J. O. Jirsa. "Equivalent Frame Analysis for Slab Design," *ACI Journal, Proceedings*, **67,** November 1970, 875–884.
4. C. K. Wang. *Intermediate Structural Analysis*. New York: McGraw-Hill, 1983.
5. Chu-Kia Wang and Charles G. Salmon. *Introductory Structural Analysis*. Englewood Cliffs, N.J.:Prentice-Hall, 1984.
6. C. K. Wang. *Matrix Methods of Structural Analysis* (2nd ed.). Madison, Wisconsin: American Publishing Company, 1970.
7. M. Daniel Vanderbilt. "Equivalent Frame Analysis for Lateral Loads," *Journal of the Structural Division*, ASCE, **105,** October 1979 (ST10), 1981–1998. Disc., **106,** July 1980 (ST7), 1671–1672; **107,** January 1981 (ST1), 245.
8. M. Daniel Vanderbilt and W. Gene Corley. "Frame Analysis of Concrete Buildings," *Concrete International: Design and Construction*, 5, December 1983, 33–43.

18

Yield Line Theory of Slabs

18.1 INTRODUCTION

Reinforced concrete design methods under the present ACI Code are based on the results of an elastic analysis of the structure as a whole, when subjected to the action of factored loads (ACI-9.2), such as $1.4D + 1.7L$ where D and L refer to service dead and live loads. Actually the behavior of a statically indeterminate structure is such that after the moment strengths at one or more points have been reached, discontinuities develop in the elastic curve at those points, and the results of an elastic analysis are no longer valid. If there is sufficient ductility, redistribution of bending moments will occur until a sufficient number of sections of discontinuity, commonly called "plastic hinges," form to change the structure into a mechanism, at which time the structure collapses or fails. The term "ultimate load analysis," as opposed to "elastic analysis," relates to the use of the bending moment diagram at the verge of collapse as the basis for design. Other than the provisions for redistribution of moments at the supports of continuous flexural members (ACI-8.4), the present ACI Code has made no allowance for ultimate load analysis. The redistribution as described in ACI-8.4 has been presented and illustrated in Sec. 10.12.

The design and analysis of two-way slab systems has been treated in Chaps. 16 and 17. The factored moments are based on the elastic analysis of an equivalent frame, which has been devised as a simple substitute for the elastic analysis of a three-dimensional system.

The chief concern of this chapter is to develop the yield line theory for two-way slabs. Although not included in the ACI Code, slab analysis by yield line theory may be useful in providing the needed information for

Flat slabs; Marina City, Chicago. (Courtesy of Portland Cement Association.)

understanding the behavior of irregular or single-panel slabs with various boundary conditions.

18.2 GENERAL CONCEPT

Although the study of flexural behavior of plates up to the ultimate load may date back to the 1920s [1], the fundamental concept of the yield line theory for the ultimate load design of slabs has been expanded considerably by K. W. Johansen [2,3]. In this theory the strength of a slab is assumed to be governed by flexure alone; other effects such as shear and deflection are to be separately considered. The steel reinforcement is assumed to be fully yielded along the yield lines at collapse and the bending and twisting moments are assumed to be uniformly distributed along the yield lines.

Yield line theory for one-way slabs is not much different from the limit analysis of continuous beams. On a continuous beam the achievement of flexural strength at one location, say in the negative moment region over a support, does not necessarily constitute reaching the ultimate load on the beam. If the section having reached its flexural strength can continue to provide a constant resistance while undergoing further rotation, then the flexural strength may be reached at additional locations. Complete failure theoretically cannot occur until yielding has occurred at several locations (or along several lines in case of one-way slabs) so that a mechanism forms giving a condition of unstable equilibrium.

Consider for example, the one-way slab of finite width shown in Fig. 18.2.1. A uniform loading on the slab will cause uniform maximum negative bending moment along *AB* and *EF* and uniform positive bending moment along *CD*, which is parallel to the supports. When the uniform load is increased until the moments along *AB*, *CD*, and *EF* reach their respective ultimate moment capacities, rotation of the slab segments will occur with the yield lines acting as axes of rotation. Once the ultimate moment capacity is achieved, angle change can occur without additional resisting moment being developed. Thus, under the limiting condition with the slab segments able to rotate with no change in resisting moment, the slab system is geometrically unstable. This condition is known as a "collapse mechanism."

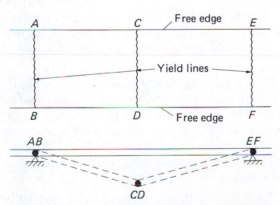

Figure 18.2.1 Collapse mechanism of a one-way slab.

Yield line theory for two-way slabs requires a different treatment from limit analysis of continuous beams, because in this case the yield lines will not in general be parallel to each other but instead form a yield line pattern. The entire slab area will be divided into several segments which can rotate along the yield lines as rigid bodies at the condition of collapse or unstable equilibrium. Some yield line patterns for typical situations are shown in Fig. 18.2.2.

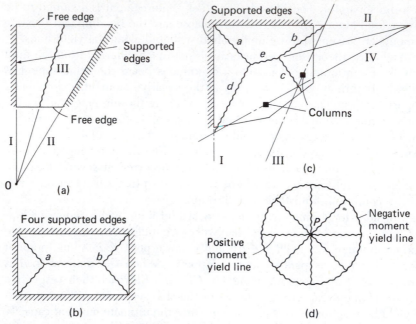

Figure 18.2.2 Typical yield line patterns.

The slab of Fig. 18.2.2(a) has nonparallel supports. At the collapse condition this slab will break into two segments; one segment will have an edge rotating about I and the other will have an edge rotating about II. The positive moment yield line must then intersect lines I and II at their intersection, point 0. The exact position of yield line III will depend on the reinforcement amount and direction, both in the positive and negative moment regions.

For the case of Fig. 18.2.2(b) where a rectangular panel is either simply supported or continuous over four linear supports, the collapse mechanism consists of four slab segments. The exact locations of points a and b will depend on the moment strengths at the supports and the positive moment reinforcement in each direction.

The slab in Fig. 18.2.2(c) is supported along two edges and in addition is supported by two isolated columns. The rotational axes for the slab segments at collapse must occur along the supports (lines I and II), and additional rotational axes must pass through the isolated columns. The critical position of the positive moment yield lines a, b, c, d, and e is a function of the reinforcement amount and direction; in the meantime, compatibility of deflection along the yield lines must be maintained during the rigid body rotations of the slab segments.

For a concentrated load at a significant distance from a supported edge, the yield line pattern will be circular as shown in Fig. 18.2.2(d). The circle pattern will be a yield line of negative bending moment, while the radial yield lines are due to positive bending moment. For concentrated loads near a free edge, a fan or partial circular pattern is typical.

18.3 FUNDAMENTAL ASSUMPTIONS

In applying the yield line theory to the ultimate load analysis of reinforced concrete slabs, the following fundamental assumptions are made:

1. The steel reinforcement is fully yielded along the yield lines at failure. In the usual case, when the slab reinforcement is well below that in the balanced condition, the moment-curvature relationship [4] is as shown in Fig. 18.3.1.

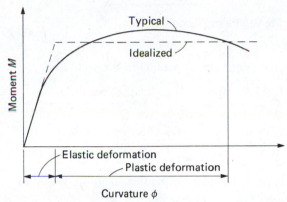

Figure 18.3.1 Typical and idealized M–ϕ relationship for reinforced concrete slab.

2. The slab deforms plastically at failure and is separated into segments by the yield lines.
3. The bending and twisting moments are uniformly distributed along the yield line and they are the maximum values provided by the moment strengths in two orthogonal directions (for two-way slabs).
4. The elastic deformations are negligible compared with the plastic deformations; thus the slab parts rotate as *plane* segments in the collapse condition.

Assumption No. 3 may be considered to be the yield criterion of orthotropic reinforced concrete slabs. It means that along a yield line as shown in Fig. 18.3.2, the bending moment strength M_{nb} and twisting moment strength M_{nt}, each per unit distance along the yield line, are exactly equal to the moment strengths M_{nx} and M_{ny} per unit distance in the y and x directions, respectively. It may be noted that M_{nx} is the strength contributed by the reinforcement in the x direction, and M_{ny} is the strength contributed by the reinforcement in the y direction. Also the sign convention is that the bending moments M_{nx}, M_{ny}, and M_{nb} are positive for tension in the lower portion of the slab and the twisting moment M_{nt} is positive if its vector is directed away from the free body on which it acts.

The bending moment strength M_{nb} and twisting moment strength M_{nt} along the yield line in Fig. 18.3.2 may be expressed in terms of M_{nx} and

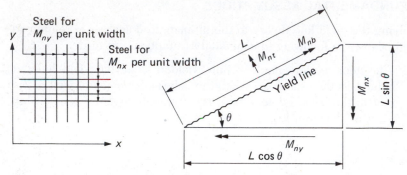

Figure 18.3.2   Bending and twisting moments on yield line.

M_{ny}. Taking equilibrium of moment vectors parallel to the yield line,

$$M_{nb}(L) = M_{nx}(L \sin \theta) \sin \theta + M_{ny} (L \cos \theta) \cos \theta$$

$$M_{nb} = M_{nx} \sin^2 \theta + M_{ny} \cos^2 \theta$$

$$M_{nb} = \frac{M_{nx} + M_{ny}}{2} - \frac{M_{nx} - M_{ny}}{2} \cos 2\theta \qquad (18.3.1)$$

and, taking equilibrium of moment vectors perpendicular to the yield line,

$$M_{nt}(L) = M_{nx} (L \sin \theta) \cos \theta - M_{ny} (L \cos \theta) \sin \theta$$

$$M_{nt} = (M_{nx} - M_{ny}) \sin \theta \cos \theta$$

$$= \frac{(M_{nx} - M_{ny})}{2} \sin 2\theta \qquad (18.3.2)$$

In using Eqs. (18.3.1) and (18.3.2), it is important to note that θ is the counterclockwise angle measured from the positive x axis to the yield line.

18.4 METHODS OF ANALYSIS

There are two methods of yield line analysis of slabs: the virtual work method and the equilibrium method. Based on the same fundamental assumptions, the two methods should give exactly the same results. In either method, a yield line pattern must be first assumed so that a collapse mechanism is produced. For a collapse mechanism, rigid body movements of the slab segments are possible by rotation along the yield lines while maintaining deflection compatibility at the yield lines between slab segments. There may be more than one possible yield line pattern, in which case solutions to all possible yield line patterns must be sought and the one giving the smallest ultimate load would actually happen and thus should be used in design. For instance the failure pattern of the simply supported rectangular slab subjected to uniform load may be that shown either in Fig. 18.4.1(a) or in Fig. 18.4.1(b), depending on the aspect ratio of the rectangular panel and the moment strengths M_{nx} and M_{ny}.

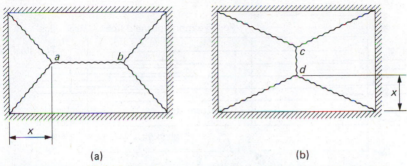

Figure 18.4.1   Yield line patterns of a simply supported rectangular slab.

After the yield line pattern has been assumed, the next step is to determine the position of the yield lines, such as defined by the unknown x in Fig. 18.4.1(a) or (b). It is at this point that one may choose to use the virtual work method or the equilibrium method. In the virtual work method, an equation containing the unknown x is established by equating the total positive work done by the ultimate load during simultaneous rigid body rotations of the slab segments (while maintaining deflection compatibility) to the total negative work done by the bending and twisting moments on all the yield lines. Then that value of x which gives the smallest ultimate load is found by means of differential calculus. In the equilibrium method, the value of x is obtained by applying the usual equations of statical equilibrium to the slab segments, but the optimal position x is defined by the placement of predetermined nodal forces at the intersection of yield lines. Expressions for the nodal forces in typical situations, once derived, can be conveniently used to avoid the necessity of mathematical differentiation as required in the virtual work method.

In the following sections, yield line analysis for one-way slabs is dealt with first in a manner similar to limit analysis of continuous beams. Then both the virtual work method and the equilibrium method are presented and illustrated for two-way slabs.

18.5 YIELD LINE ANALYSIS OF ONE-WAY SLABS

The continuous slab span shown in Fig. 18.5.1(a) has nominal moment strengths of M_{ni} and M_{nj} provided by top reinforcement at the supports and a nominal strength M_{np} provided by bottom reinforcement within the span. The nominal strengths M_{ni}, M_{nj}, and M_{np} are absolute values of the moment strength per unit slab width. For uniform loading the only possible yield pattern consists of three parallel yield lines as shown in Fig. 18.5.1(a), one each along the left and right supports and one at a distance x from the left support. The moment strength per unit slab width when collapse is imminent should be as shown in Fig. 18.5.1(b). The problem is to determine the collapse load w_u/ϕ per unit slab area in terms of M_{ni}, M_{nj}, and M_{np}, and the span length L. It will be shown that the virtual work and equilibrium methods will give exactly the same results.

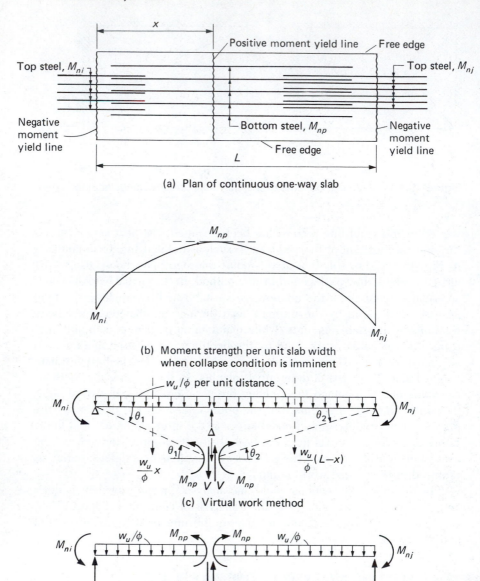

(a) Plan of continuous one-way slab

(b) Moment strength per unit slab width
when collapse condition is imminent

(c) Virtual work method

(d) Equilibrium method

Figure 18.5.1 Yield line analysis of one-way slabs.

Referring to Fig. 18.5.1(c), the rigid body rotations of the slab segments at collapse are measured from the original horizontal positions to those of the dashed lines, where the rotation of the left segment is θ_1 in the clockwise direction and that of the right segment is θ_2 in the counterclockwise direction while maintaining the compatible deflection Δ at the positive yield line. The total positive work done by the uniform load on the left and right segments of unit slab width is

$$\text{total positive work} = (w_u/\phi)x\frac{\Delta}{2} + (w_u/\phi)(L - x)\frac{\Delta}{2}$$

The total negative work done by M_{ni} and M_{np} on the left segment is $(M_{ni} + M_{np})\theta_1$ and that done by M_{nj} and M_{np} on the right segment is $(M_{nj} + M_{np})\theta_2$, both in absolute quantities. Thus

$$\text{total negative work} = (M_{ni} + M_{np})\theta_1 + (M_{nj} + M_{np})\theta_2$$

in which

$$\theta_1 = \frac{\Delta}{x}, \qquad \theta_2 = \frac{\Delta}{L - x}$$

The principle of virtual work states that the total work done by a force system in equilibrium in going through a virtual rigid body displacement is zero. By means of this principle, one can equate the total positive work to the absolute value of the total negative work; or

$$(w_u/\phi)x\frac{\Delta}{2} + (w_u/\phi)(L - x)\frac{\Delta}{2} = (M_{ni} + M_{np})\frac{\Delta}{x} + (M_{nj} + M_{np})\frac{\Delta}{(L - x)}$$

Dividing out the compatible deflection Δ from the above equation and solving for w_u/ϕ,

$$\frac{w_u}{\phi} = \frac{2M_{ni}}{Lx} + \frac{2M_{np}}{x(L - x)} + \frac{2M_{nj}}{L(L - x)} \qquad (18.5.1)$$

Differentiating Eq. (18.5.1) for w_u with respect to x and setting the derivative to zero, one obtains the quadratic equation,

$$(M_{nj} - M_{ni})x^2 + 2(M_{ni} + M_{np})Lx - (M_{ni} + M_{np})L^2 = 0 \quad (18.5.2)$$

In the virtual work method, then, first the value of x is found by solving the quadratic equation (18.5.2) and the collapse load w_u/ϕ is determined from Eq. (18.5.1).

In the equilibrium method the value of x is not to be obtained by differential calculus but defined by the magnitude of the nodal forces [V and V shown in Fig. 18.5.1(d)] acting on the slab segments on either side of the positive yield line. In this particular instance, it is known from the elementary theory of bending of beams that the shear at a section of maximum positive bending moment should be zero. From this point on, the unknown values of R_i, R_j, w_u, and x in Fig. 18.5.1(d) are obtained from the four independent equilibrium equations for the entire slab or either of the two slab segments. The left and right reactions on the free bodies of Fig. 18.5.1(d) are found by applying the equilibrium equations to the entire slab; thus

$$R_i = \left[\frac{1}{2}\left(\frac{w_u}{\phi}\right)L - \frac{M_{nj} - M_{ni}}{L}\right], \qquad R_j = \left[\frac{1}{2}\left(\frac{w_u}{\phi}\right)L + \frac{M_{nj} - M_{ni}}{L}\right]$$

Then, summing the vertical forces on the left slab segment,

$$R_i = \frac{1}{2}\left(\frac{w_u}{\phi}\right)L - \frac{M_{nj} - M_{ni}}{L} = \frac{w_u}{\phi}x$$

from which

$$\frac{w_u}{\phi} = \frac{(M_{nj} - M_{ni})}{L(L/2 - x)} \tag{18.5.3}$$

and, taking moments about the positive moment yield line on the left segment,

$$M_{np} = -M_{ni} + \left[\frac{1}{2}\left(\frac{w_u}{\phi}\right)L - \frac{M_{nj} - M_{ni}}{L}\right]x - \frac{1}{2}\left(\frac{w_u}{\phi}\right)x^2 \tag{18.5.4}$$

Substitution of Eq. (18.5.3) into Eq. (18.5.4) results in the same quadratic equation as Eq. (18.5.2). This shows that the two methods, virtual work and equilibrium, give exactly the same results.

In the solution of a numerical problem the quadratic equation (18.5.2) is first solved for x. Although w_u can then be computed from either Eq. (18.5.1) or Eq. (18.5.3), using Eq. (18.5.1) is far superior to Eq. (18.5.3), because Eq. (18.5.1) contains the sum of three absolute quantities but Eq. (18.5.3) involves the division by a sensitive quantity $(L/2 - x)$.

EXAMPLE 18.5.1 Given the nominal moment strengths $M_{ni} = 14$ ft-kips, $M_{np} = 16$ ft-kips, and $M_{nj} = 22$ ft-kips per ft width of a 20-ft continuous slab span as shown in Fig. 18.5.2, determine the location x of the positive yield line and the collapse load w_u/ϕ in kips per ft per ft width of slab.

Solution: Using the quadratic equation (18.5.2),

$$(M_{nj} - M_{ni})x^2 + 2(M_{ni} + M_{np})Lx - (M_{ni} + M_{np})L^2 = 0$$
$$(22 - 14)x^2 + 2(14 + 16)(20x) - (14 + 16)(20)^2 = 0$$
$$8x^2 + 1200x - 12{,}000 = 0$$
$$x^2 + 150x + 75^2 = 1500 + 5625$$
$$x = \sqrt{7125} - 75 = 84.41 - 75 = 9.41 \text{ ft}$$

Substituting the value of $x = 9.41$ ft in Eq. (18.5.1),

$$\frac{w_u}{\phi} = \frac{2M_{ni}}{Lx} + \frac{2M_{np}}{x(L - x)} + \frac{2M_{nj}}{L(L - x)}$$

$$= \frac{2(14)}{20(9.41)} + \frac{2(16)}{9.41(10.59)} + \frac{2(22)}{20(10.59)}$$

$$= 0.149 + 0.321 + 0.208 = 0.678 \text{ kip/ft}$$

Using $w_u/\phi = 0.678$ kip/ft and the end moments of 14 ft-kips and 22 ft-kips, the end reactions, the shear diagram, and the moment diagram are computed and shown in Fig. 18.5.2.

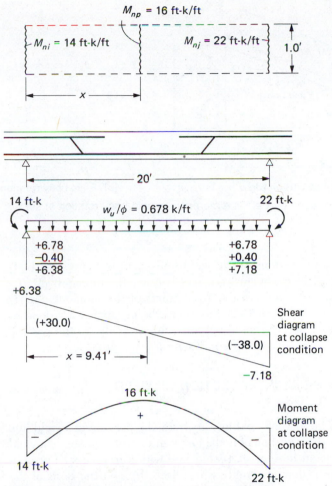

Figure 18.5.2   One-way slab of Example 18.5.1.

18.6 WORK DONE BY YIELD LINE MOMENTS IN RIGID BODY ROTATION OF SLAB SEGMENT

Before taking up the virtual work method of yield line analysis of two-way slabs, it is desirable to derive a general procedure for obtaining the absolute value of the negative work done by the bending and twisting moments acting on the yield line in going through a rigid body rotation of the slab segment. In Fig. 18.6.1(a) are shown the moments acting on the edges of a slab segment having yield line of length L with horizontal and vertical projections of L_x and L_y, respectively. Let this slab undergo a rigid body rotation, whose components are θ_x and θ_y, shown in vector notations in Fig. 18.6.1(b). It can be shown algebraically that the absolute value of the negative work done by the moments $(M_{nb}L)$ and $(M_{nt}L)$ acting on the yield line is equal to the positive work done by the moments $(M_{nx}L_y)$ and $(M_{ny}L_x)$ acting on the hor-

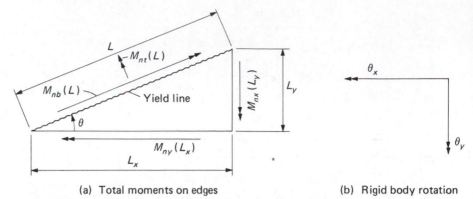

(a) Total moments on edges   (b) Rigid body rotation

Figure 18.6.1 Work done by yield line moments in rigid body rotation of slab segment.

izontal and vertical projections of the yield line—for the same rigid body rotation, of course. This is obviously correct because the moments $(M_{nx}L_y)$ and $(M_{ny}L_x)$ on Fig. 18.6.1(a) are the equilibrants of the moments $(M_{nb}L)$ and $(M_{nt}L)$ on the same figure. Certainly then, the negative work done by one set of generalized forces should be equal, numerically, to the positive work done by the alternative set of equilibrating generalized forces.

18.7 NODAL FORCE AT INTERSECTION OF YIELD LINE WITH FREE EDGE

In the equilibrium method of yield line analysis, the position of a yield line is characterized by the insertion of nodal forces at the intersection of a yield line with another yield line, or of a yield line with a free edge. For one-way slabs it has been demonstrated in Sec. 18.5 that, on the basis of the elementary theory of bending of beams, the nodal force on either side of a positive moment yield line is equal to zero. In this section the expression for the pair of equal and opposite nodal forces V acting on each side of the intersection of a yield line with a free edge in a two-way slab is derived.

Shown in Fig. 18.7.1(a) is a two-way slab with nominal moment strengths of M_{nx} and M_{ny} provided by the bottom reinforcement in the horizontal and vertical directions, respectively. A positive moment yield line is assumed to intersect the free edge at an angle α. The upward nodal force V acts on the left segment and the downward nodal force V acts on the right segment, shown by a dot and a cross for the upward and downward forces in Fig. 18.7.1(a) according to the convention used by Johansen [2,3]. For equilibrium the upward and downward nodal forces must be equal in magnitude.

Consider the equilibrium of an infinitesimal slab element shown in either Fig. 18.7.1(b) or (c). By its own definition a yield line is always at the optimal position. Inasmuch as the edge ac in Fig. 18.7.1(b), or the edge ad in Fig. 18.7.1(c), is at an angle $\Delta\alpha$ from the yield line, the same yield line moments as those on the yield line ab act on edges ac or ad. By applying the principle of equivalent force systems as indicated by Fig. 18.6.1(a), the

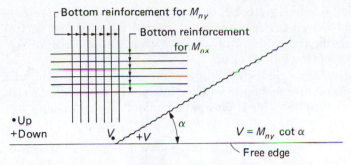

(a) Two-way slab with free edge

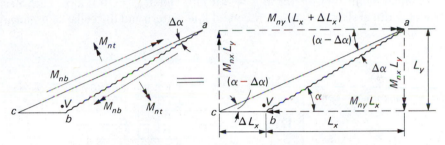

(b) Free-body diagram of infinitesimal piece to left of yield line

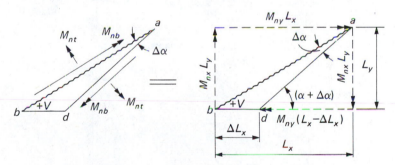

(c) Free-body diagram of infinitesimal piece to right of yield line

Figure 18.7.1   Nodal force at intersection of yield line with free edge.

free body diagram shown on the left side of the equal sign in Fig. 18.7.1(b) or (c) is transformed to the equivalent free body diagram on the right side of the equal sign. Using the equivalent free body and summing the moments about the line ac in Fig. 18.7.1(b),

$$M_{ny}(\Delta L_x) \cos(\alpha - \Delta\alpha) = V(\Delta L_x) \sin(\alpha - \Delta\alpha) \qquad (18.7.1)$$

and summing the moments about the line ad in Fig. 18.7.1(c),

$$M_{ny}(\Delta L_x) \cos(\alpha + \Delta\alpha) = V(\Delta L_x) \sin(\alpha + \Delta\alpha) \qquad (18.7.2)$$

Solving for V by using either Eq. (18.7.1) or Eq. (18.7.2),

$$V = M_{ny} \cot \alpha \qquad (18.7.3)$$

Note that the left side of Eq. (18.7.1) or Eq. (18.7.2), $M_{ny} (\Delta L_x)$, represents the net moment along the edges ca and ab, or the edges ba and ad, respectively; while the right side involves the moment of the nodal force V about an axis coincident with ac or ad.

Equation (18.7.3) is used when applying the equilibrium method to a situation where a positive moment yield line intersects a free edge of a slab at an angle other than 90°. This is illustrated by the following example.

EXAMPLE 18.7.1 The triangular slab ABC shown in Fig. 18.7.2(a) is simply supported along edges AC and BC but has a free edge along AB. The horizontal and vertical reinforcement in the lower face of the slab provides nominal moment strengths $M_{nx} = 8$ ft-kips and $M_{ny} = 10$ ft-kips, each per foot width of slab. Determine the yield line pattern and the collapse uniform load w_u/ϕ.

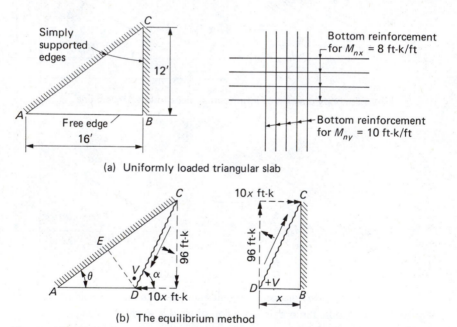

(a) Uniformly loaded triangular slab

(b) The equilibrium method

Figure 18.7.2 Yield line analysis of triangular slab in Example 18.7.1.

Solution: (a) Yield line pattern. It has been demonstrated in Fig. 18.2.2(a) that a yield line should pass through the point of intersection of two non-parallel supported edges. In this example a positive moment yield line CD will break the slab into two segments ACD and BCD, wherein a common compatible deflection Δ of point D can be effected by rigid body rotations of the slab segment ACD about the supported edge AC and of the slab segment BCD about the supported edge BC.

(b) Equilibrium method of finding $BD = x$. The nodal forces V and $-V$ acting on the left and right slab segments of Fig. 18.7.2(b) are computed from Eq. (18.7.3); thus

$$V = M_{ny} \cot \alpha = (+10)\left(\frac{x}{12}\right) = \frac{5}{6} x \text{ kips}$$

Note that the nodal force acts upward in the obtuse angle and it acts downward in the acute angle. The equilibrium of moments about the edge AC of the left segment requires

$$\tfrac{1}{2}(w_u/\phi)(16 - x)(12)\tfrac{1}{3}(DE) = V(DE) + 96 \sin\theta + 10x \cos\theta$$

$$\tfrac{1}{2}(w_u/\phi)(16 - x)(12)(\tfrac{1}{3})(16 - x)(0.6) = \tfrac{5}{6}x(16 - x)(0.6) + 96(0.6) + 10x(0.8)$$

from which

$$\frac{w_u}{\phi} = \frac{576 + 160x - 5x^2}{12(16 - x)^2}$$

Similarly, the equilibrium of moments about the edge BC of the right segment requires

$$\frac{1}{2}(w_u/\phi)(x)(12)\left(\frac{BD}{3}\right) + V(BD) = 96$$

$$\frac{1}{2}\left(\frac{w_u}{\phi}\right)(x)(12)\left(\frac{x}{3}\right) + \frac{5x^2}{6} = 96$$

from which

$$\frac{w_u}{\phi} = \frac{576 - 5x^2}{12x^2}$$

Equating the two expressions for w_u/ϕ,

$$\frac{576 + 160x - 5x^2}{12(16 - x)^2} = \frac{576 - 5x^2}{12x^2}$$

$$x^2 + 14.4x - 115.2 = 0$$

$$x = 5.72 \text{ ft}$$

With the known value of x, the value of w_u/ϕ may be computed,

$$\frac{w_u}{\phi} = 1.048 \text{ kips/sq ft}$$

It can be shown that the same quadratic equation in x may be obtained using the virtual work method by following the procedure as illustrated in Sec. 18.5.

18.8 NODAL FORCES AT INTERSECTION OF THREE YIELD LINES

In Fig. 18.8.1 are shown three possible yield line patterns for an irregular quadrilateral slab with four simply supported edges. The optimum positions of the yield lines in each pattern should be such as to give the lowest collapse load. These positions are defined by the locations of points a and b in Fig. 18.8.1(a), of points c and d in Fig. 18.8.1(b), and of point e in Fig. 18.8.1(c).

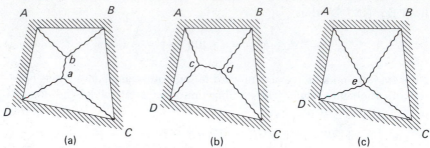

Figure 18.8.1 Yield line patterns of a simply supported quadrilateral slab.

In the equilibrium method the yield lines are characterized by the insertion of predetermined nodal forces. It is the object of this section to derive the expressions for the nodal forces at the intersection of three yield lines, such as at points a, b, c, or d in Fig. 18.8.1. The derivation shown below follows the works of Johansen [2,3], and Jones and Wood [4,5].

Assume that three yield lines 1–2–3 intersecting at a common point 0 are situated at angles ϕ_1–ϕ_2–ϕ_3 measured counterclockwise from the positive x axis, as shown in Fig. 18.8.2(a). The nominal moment strengths under the yield lines 1–2–3 are, as shown in Figs. 18.8.2(b) and (c), $M_{nx1} - M_{nx2} - M_{nx3}$ provided by the reinforcement shown horizontally and $M_{ny1} - M_{ny2} - M_{ny3}$ provided by the reinforcement shown vertically, all reinforcement being near the lower face of the slab. The nodal forces V_{12}, V_{23}, and V_{31} are shown by the dots (which mean upward nodal forces) in Fig. 18.8.2(a). Note that for vertical equilibrium at the point of intersection, the sum of V_{12}, V_{23}, and V_{31} must be zero.

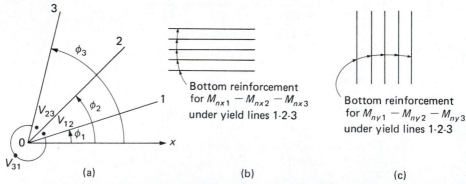

Figure 18.8.2 Nodal forces at intersection of three yield lines.

It is important to emphasize at the outset that by definition the yield lines 1–2–3 are all situated at their respective optimal positions. The bending and twisting moments on any line passing through the point of intersection and deviating by an infinitesimal angle from a particular yield line should be equal to the bending and twisting moments on that line as provided by the orthogonal moment strengths.

Consider the equilibrium of an infinitesimal slab segment $0AB$ [Fig. 18.8.3(a)], bounded by a differential length $0A$ on yield line 3 and an arbitrary length $0B$ on yield line 1. Since BA is at a differential angle $\Delta\alpha$ from $B0$, the bending and twisting moments on both $B0$ and BA are those provided by M_{ny1} and M_{nx1} on their respective horizontal and vertical projections. Likewise the bending and twisting moments on the differential length $0A$ are those provided by M_{ny3} and M_{nx3} on its horizontal and vertical projections. Call the upward nodal force at point 0 inside triangle $B0A$ and

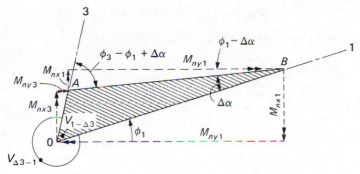

(a) Nodal force $V_{1-\Delta3}$ between yield line 1 and $0A$

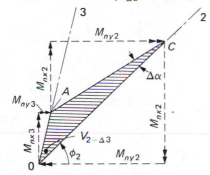

(b) Nodal force $V_{2-\Delta3}$ between yield line 2 and $0A$

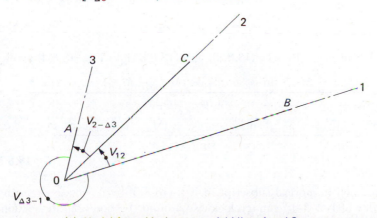

(c) Nodal force V_{12} between yield lines 1 and 2

Figure 18.8.3 Determination of nodal force V_{12} between yield lines 1 and 2.

bounded by yield line 1 and a differential length on yield line 3 by the name $V_{1-\Delta 3}$.

Next write the equation of equilibrium for moments about AB as the axis of rotation for the slab segment $0AB$ in Fig. 18.8.3(a), noting that the moment of the uniform load on $0AB$ about any axis is a differential of the second order and may be neglected.

$$-V_{1-\Delta 3} \, 0A \sin (\phi_3 - \phi_1 + \Delta\alpha)$$
$$+ (M_{ny3} - M_{ny1}) \, 0A \cos \phi_3 \cos(\phi_1 - \Delta\alpha)$$
$$+ (M_{nx3} - M_{nx1}) \, 0A \sin \phi_3 \sin (\phi_1 - \Delta\alpha) = 0$$

Solving the above equation for $V_{1-\Delta 3}$ and letting $\Delta\alpha$ approach zero at the limit,

$$V_{1-\Delta 3} = \frac{(M_{nx3} - M_{nx1}) \sin \phi_3 \sin \phi_1 + (M_{ny3} - M_{ny1}) \cos \phi_3 \cos \phi_1}{\sin (\phi_3 - \phi_1)}$$

$$\text{(18.8.1)}$$

Making a similar analysis for the infinitesimal slab segment $0AC$ [Fig. 18.8.3(b)] bounded by a differential length $0A$ on yield line 3 and an arbitrary length $0C$ on yield line 2 gives the expression for the upward nodal force $V_{2-\Delta 3}$ in Fig. 18.8.3(b) as

$$V_{2-\Delta 3} = \frac{(M_{nx3} - M_{nx2}) \sin \phi_3 \sin \phi_2 + (M_{ny3} - M_{ny2}) \cos \phi_3 \cos \phi_2}{\sin (\phi_3 - \phi_2)}$$

$$\text{(18.8.2)}$$

For vertical equilibrium at the point of intersection, the upward nodal force $V_{\Delta 3-1}$ in Fig. 18.8.3(a) in the zone going counterclockwise from the differential length on yield line 3 to yield line 1 is

$$V_{\Delta 3-1} = -V_{1-\Delta 3} \qquad \text{(18.8.3)}$$

and, for the same reason, the upward nodal force V_{12} in Fig. 18.8.3(c) is

$$V_{12} = -V_{2-\Delta 3} - V_{\Delta 3-1} \qquad \text{(18.8.4)}$$

Substitution of Eqs. (18.8.1), (18.8.2), and (18.8.3) into Eq. (18.8.4) gives

$$V_{12} = \frac{(M_{nx3} - M_{nx1}) \sin \phi_3 \sin \phi_1 + (M_{ny3} - M_{ny1}) \cos \phi_3 \cos \phi_1}{\sin (\phi_3 - \phi_1)}$$
$$- \frac{(M_{nx3} - M_{nx2}) \sin \phi_3 \sin \phi_2 + (M_{ny3} - M_{ny2}) \cos \phi_3 \cos \phi_2}{\sin (\phi_3 - \phi_2)}$$

$$\text{(18.8.5)}$$

Replacing each numerical subscript in Eq. (18.8.5) by its successor in the cyclic order of 1–2–3–1 (counterclockwise around the point of intersection) and then once more in the same manner, the following expressions for the upward nodal forces V_{23} and V_{31} as shown in Fig. 18.8.2(a) are obtained.

$$V_{23} = \frac{(M_{nx1} - M_{nx2}) \sin \phi_1 \sin\phi_2 + (M_{ny1} - M_{ny2}) \cos \phi_1 \cos \phi_2}{\sin (\phi_1 - \phi_2)}$$
$$- \frac{(M_{nx1} - M_{nx3}) \sin \phi_1 \sin \phi_3 + (M_{ny1} - M_{ny3}) \cos \phi_1 \cos \phi_3}{\sin (\phi_1 - \phi_3)}$$

$$(18.8.6)$$

$$V_{31} = \frac{(M_{nx2} - M_{nx3}) \sin \phi_2 \sin \phi_3 + (M_{ny2} - M_{ny3}) \cos \phi_2 \cos \phi_3}{\sin (\phi_2 - \phi_3)}$$
$$- \frac{(M_{nx2} - M_{nx1}) \sin \phi_2 \sin \phi_1 + (M_{ny2} - M_{ny1}) \cos \phi_2 \cos \phi_1}{\sin (\phi_2 - \phi_1)}$$

$$(18.8.7)$$

Equations (18.8.5), (18.8.6), and (18.8.7) are expressions for the upward nodal forces at the intersection of three yield lines.

Nodal Force at Intersection of Yield Line with Free Edge. The nodal forces at the intersection of a yield line with a free edge, as shown by Fig. 18.8.4, may be obtained by substituting $\phi_1 = 0$, $\phi_2 = \alpha$, $\phi_3 = \pi$, $M_{nx1} = M_{nx3} = M_{ny1} = M_{ny3} = 0$, $M_{nx2} = M_{nx}$, and $M_{ny2} = M_{ny}$ in Eqs. (18.8.5) to (18.8.7); thus

$$V_{12} = -\frac{(-M_{ny2})(-1) \cos \alpha}{\sin (\pi - \alpha)} = -M_{ny} \cot \alpha$$

$$V_{23} = \frac{(-M_{ny2})(+1) \cos \alpha}{\sin (0 - \alpha)} = +M_{ny} \cot \alpha$$

$$V_{31} = \frac{(M_{ny2}) \cos \alpha \, (-1)}{\sin (\alpha - \pi)} - \frac{(M_{ny2}) \cos \alpha \, (+1)}{\sin (\alpha - 0)} = 0$$

The above results check with the findings in Sec. 18.7.

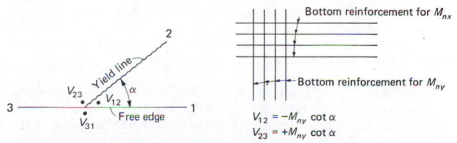

Figure 18.8.4   Nodal forces at intersection of yield line with free edge.

Nodal Forces at Intersection of Three Yield Lines Having Identical M_{nx} and M_{ny} Nominal Moment Strengths. From Eqs. (18.8.5) to (18.8.7) it can be observed that wherever the nominal moment strengths under three intersecting yield lines are identical—that is, $M_{nx1} = M_{nx2} = M_{nx3} = M_{nx}$ and $M_{ny1} = M_{ny2} = M_{ny3} = M_{ny}$—the nodal forces at the intersection are zero. This fact is of great convenience when using the equilibrium method for the yield line analysis of two-way rectangular slabs.

18.9 YIELD LINE ANALYSIS OF RECTANGULAR TWO-WAY SLABS

A typical rectangular two-way slab panel shown in Fig. 18.9.1 has two-way reinforcement within the panel near the bottom face providing positive moment nominal strengths M_{npx} and M_{npy}, and it also has two-way reinforcement along the edges near the top face providing negative moment nominal strengths M_{nnx} and M_{nny}; these strengths are per unit width of slab. The uniform load to give the collapse condition based on the yield line theory may be determined in terms of the sides a and b, and the absolute values of M_{npx}, M_{npy}, M_{nnx}, and M_{nny}.

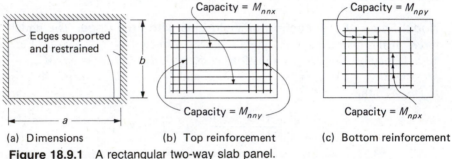

(a) Dimensions   (b) Top reinforcement   (c) Bottom reinforcement

Figure 18.9.1   A rectangular two-way slab panel.

Yield Line Pattern. Three possible yield line patterns are shown in Fig. 18.9.2. There is no unknown position in yield line pattern No. 1 of Fig. 18.9.2(a); consequently the nodal forces V need not be predetermined and their value is dictated by statics alone. The unknowns x and y in yield line patterns Nos. 2 and 3 of Figs. 18.9.2(b) and (c) must be determined by means of differential calculus in the virtual work method; but for the equilibrium method, in this particular case the nodal forces to define the yield lines are all zero because the moment strengths under a set of three intersecting yield lines are identical.

Analysis for Yield Pattern No. 1. Assuming a vertical deflection of Δ at the intersection of the diagonal yield lines in Fig. 18.9.3, the deflection at the centroids of the four triangles A–B–C–D is $\Delta/3$. The work done at the

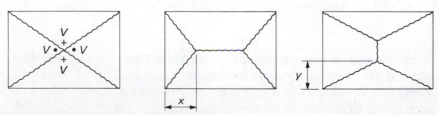

(a) Yield pattern No. 1   (b) Yield pattern No. 2   (c) Yield pattern No. 3

Figure 18.9.2   Yield line patterns for a rectangular two-way slab panel.

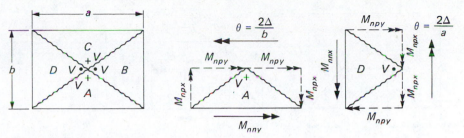

Figure 18.9.3 Analysis for yield pattern No. 1.

collapse condition by the uniform load is the product of the total load on the entire panel and $\Delta/3$; thus

$$W = \frac{w_u}{\phi} ab \left(\frac{\Delta}{3}\right) \tag{18.9.1}$$

The work done by the yield moments on the boundaries of all four slab segments is, referring to Fig. 18.9.3,

$$W = 2(M_{nny} + M_{npy})(a)\left(\frac{2\Delta}{b}\right) + 2(M_{nnx} + M_{npx})(b)\left(\frac{2\Delta}{a}\right) \tag{18.9.2}$$

Equating Eq. (18.9.1) to Eq. (18.9.2) and solving for w_u,

$$\frac{w_u}{\phi} = 12 \left(\frac{M_{nnx} + M_{npx}}{a^2} + \frac{M_{nny} + M_{npy}}{b^2}\right) \tag{18.9.3}$$

Taking moments about the lower edge of slab segment A in Fig. 18.9.3,

$$\frac{1}{2}\left(\frac{w_u}{\phi}\right)a\left(\frac{b}{2}\right)\left(\frac{b}{6}\right) + V\left(\frac{b}{2}\right) = (M_{nny} + M_{npy})(a) \tag{18.9.4}$$

Taking moments about the left edge of slab segment D in Fig. 18.9.3,

$$\frac{1}{2}\left(\frac{w_u}{\phi}\right)b\left(\frac{a}{2}\right)\left(\frac{a}{6}\right) = (M_{nnx} + M_{npx})(b) + V\left(\frac{a}{2}\right) \tag{18.9.5}$$

Eliminating V between Eqs. (18.9.4) and (18.9.5) and solving for w_u/ϕ, the same expression for w_u/ϕ as Eq. (18.9.3) is obtained.

Analysis for Yield Pattern No. 2. Assuming a vertical deflection of Δ at the two points of intersection of the yield lines in Fig. 18.9.4, the work done at the collapse condition by the uniform load on the entire panel is

$$W = 2W_D + 2W_{A1} + 4W_{A2}$$

$$= 2\left[\frac{1}{2}\left(\frac{w_u}{\phi}\right)bx\right]\left(\frac{\Delta}{3}\right) + 2\left(\frac{w_u}{\phi}\right)(a - 2x)\left(\frac{b}{2}\right)\left(\frac{\Delta}{2}\right) + 4\left[\frac{1}{2}\left(\frac{w_u}{\phi}\right)x\frac{b}{2}\right]\left(\frac{\Delta}{3}\right)$$

$$= \frac{w_u}{\phi}\left(\frac{\Delta}{6}\right)(3ab - 2bx) \tag{18.9.6}$$

The work done by the yield moments on the boundaries of all four slab

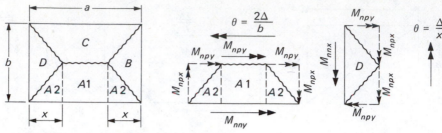

Figure 18.9.4 Analysis for yield pattern No. 2.

segments is, referring to Fig. 18.9.4,

$$W = 2(M_{nny} + M_{npy})(a)\left(\frac{2\Delta}{b}\right) + 2(M_{nnx} + M_{npx})(b)\left(\frac{\Delta}{x}\right) \quad (18.9.7)$$

Equating Eq. (18.9.6) to Eq. (18.9.7) and solving for w_u/ϕ,

$$\frac{w_u}{\phi} = \frac{12[b^2(M_{nnx} + M_{npx}) + 2ax(M_{nny} + M_{npy})]}{b^2(3ax - 2x^2)} \quad (18.9.8)$$

Setting to zero the derivative of Eq. (18.9.8) with respect to x gives the quadratic equation in x,

$$4a(M_{nny} + M_{npy})x^2 + 4b^2(M_{nnx} + M_{npx})x - [3ab^2(M_{nnx} + M_{npx})] = 0 \quad (18.9.9)$$

Taking moments about the lower edge of slab segment A in Fig. 18.9.4,

$$2\left[\frac{1}{2}\left(\frac{w_u}{\phi}\right)x\frac{b}{2}\right]\left(\frac{b}{6}\right) + \frac{w_u}{\phi}(a - 2x)\left(\frac{b}{2}\right)\left(\frac{b}{4}\right) = (M_{nny} + M_{npy})(a)$$

$$\frac{w_u}{\phi} = \frac{24a(M_{nny} + M_{npy})}{2b^2x + 3b^2(a - 2x)} \quad (18.9.10)$$

Taking moments about the left edge of slab segment D in Fig. 18.9.4,

$$\frac{1}{2}\left(\frac{w_u}{\phi}\right)bx\left(\frac{x}{3}\right) = (M_{nnx} + M_{npx})(b)$$

$$\frac{w_u}{\phi} = \frac{6(M_{nnx} + M_{npx})}{x^2} \quad (18.9.11)$$

Equating Eq. (18.9.10) to Eq. (18.9.11) gives the same quadratic equation in x as Eq. (18.9.9).

The conditon for $x = a/2$ in Eq. (18.9.9) can be shown to be

$$\frac{M_{nnx} + M_{npx}}{M_{nny} + M_{npy}} = \frac{a^2}{b^2} \quad \text{for } x = \frac{a}{2} \quad (18.9.12)$$

which means that if the sum of positive and negative moment reinforcement in the a direction, each per unit width of slab, is equal to (a^2/b^2) times the sum of positive and negative moment reinforcement in the b direction, each per unit width of slab, yield pattern No. 1 prevails.

The condition for $x < a/2$ in Eq. (18.9.9) can be shown to be

$$\frac{M_{nnx} + M_{npx}}{M_{nny} + M_{npy}} < \frac{a^2}{b^2} \qquad \text{for } x < \frac{a}{2} \qquad (18.9.13)$$

which means that in order for yield pattern No. 2 to prevail, the reinforcement in the a direction is less than that for yield pattern No. 1 to control.

Analysis for Yield Pattern No. 3. By interchanging the subscripts x and y as well as the quantities a and b in Eqs. (18.9.8), (18.9.9), (18.9.10), and (18.9.11), the following equations applicable to yield line pattern No. 3 are obtained. The quadratic equation in y (Fig. 18.9.2) is

$$4b(M_{nnx} + M_{npx})y^2 + 4a^2(M_{nny} + M_{npy})y - [3ba^2(M_{nny} + M_{npy})] = 0 \qquad (18.9.14)$$

Three expressions for w_u/ϕ in terms of y are

$$\frac{w_u}{\phi} = \frac{12[a^2(M_{nny} + M_{npy}) + 2by(M_{nnx} + M_{npx})]}{a^2(3by - 2y^2)} \qquad (18.9.15)$$

$$\frac{w_u}{\phi} = \frac{24b(M_{nnx} + M_{npx})}{2a^2y + 3a^2(b - 2y)} \qquad (18.9.16)$$

$$\frac{w_u}{\phi} = \frac{6(M_{nny} + M_{npy})}{y^2} \qquad (18.9.17)$$

The condition for $y < b/2$ in Eq. (18.9.14) can be shown to be

$$\frac{M_{nnx} + M_{npx}}{M_{nny} + M_{npy}} > \frac{a^2}{b^2} \qquad \text{for } y < \frac{b}{2} \qquad (18.9.18)$$

which means that in order for yield pattern No. 3 to prevail, the reinforcement in the a direction is more than that for yield pattern No. 1 to control.

EXAMPLE 18.9.1 Determine the controlling yield line pattern and the corresponding collapse condition uniform load for a rectangular two-way slab panel with dimensions as shown in Fig. 18.9.5(a). The slab has reinforcement in the top near the edges and in the bottom within the panel. Obtain solutions for the following three cases:

1. $M_{nnx} + M_{npx} = 6.25$ ft-kips/ft $M_{nny} + M_{npy} = 4$ ft-kips/ft
2. $M_{nnx} + M_{npx} = 2$ ft-kips/ft $M_{nny} + M_{npy} = 4$ ft-kips/ft
3. $M_{nnx} + M_{npx} = 8$ ft-kips/ft $M_{nny} + M_{npy} = 4$ ft-kips/ft

Solution: (a) Case 1. The applicable yield line pattern may be determined by comparing the ratio of $(M_{nnx} + M_{npx})$ to $(M_{nny} + M_{npy})$ with the ratio of a^2 to b^2. In this case

$$\frac{(M_{nnx} + M_{npx})}{(M_{nny} + M_{npy})} = \frac{6.25}{4} = 1.5625, \qquad \frac{a^2}{b^2} = \frac{625}{400} = 1.5625$$

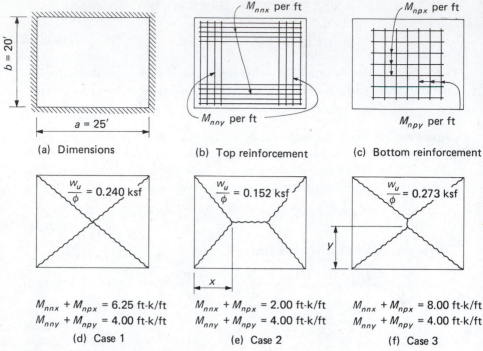

Figure 18.9.5 Rectangular two-way slab of Example 18.9.1.

Since the ratio of reinforcement in direction a to that in direction b per foot slab width is equal to the ratio of a^2 to b^2, the yield pattern is shown in Fig. 18.9.5(d). Then from Eq. (18.9.3),

$$\frac{w_u}{\phi} = 12\left(\frac{M_{nnx} + M_{npx}}{a^2} + \frac{M_{nny} + M_{npy}}{b^2}\right)$$

$$= 12\left(\frac{6.25}{625} + \frac{4}{400}\right) = 0.240 \text{ ksf}$$

(b) Case 2. The ratio of $(M_{nnx} + M_{npx})$ to $(M_{nny} + M_{npy})$ is, in this case,

$$\frac{(M_{nnx} + M_{npx})}{(M_{nny} + M_{npy})} = \frac{2}{4} = 0.5 < \frac{a^2}{b^2} = 1.5625$$

The yield line pattern is as shown in Fig. 18.9.5(e). The quadratic equation (18.9.9) is used to solve for x.

$$4a(M_{nny} + M_{npy})x^2 + 4b^2(M_{nnx} + M_{npx})x - 3ab^2(M_{nnx} + M_{npx}) = 0$$

$$4(25)(4)x^2 + 4(400)(2)x - 3(25)(400)(2) = 0$$

$$x^2 + 8x - 150 = 0$$

$$x = \sqrt{166} - 4 = 8.884 \text{ ft}$$

The same uniform load w_u/ϕ is obtained from Eqs. (18.9.8), (18.9.10), or

(18.9.11); the fact that it is so serves as a check on the numerical computation.

$$\frac{w_u}{\phi} = \frac{12[b^2(M_{nnx} + M_{npx}) + 2ax(M_{nny} + M_{npy})]}{b^2(3ax - 2x^2)}$$

$$= \frac{12[400(2) + 2(25)(8.884)(4)]}{400[3(25)(8.884) - 2(8.884)^2]} = 0.152 \text{ ksf}$$

$$\frac{w_u}{\phi} = \frac{24a(M_{nny} + M_{npy})}{2b^2x + 3b^2(a - 2x)}$$

$$= \frac{24(25)(4)}{2(400)(8.884) + 3(400)[25 - 2(8.884)]} = 0.152 \text{ ksf}$$

$$\frac{w_u}{\phi} = \frac{6(M_{nnx} + M_{npx})}{x^2} = \frac{6(2)}{(8.884)^2} = 0.152 \text{ ksf}$$

(c) Case 3. The ratio of $(M_{nnx} + M_{npx})$ to $(M_{nny} + M_{npy})$ is, in this case,

$$\frac{(M_{nnx} + M_{npx})}{(M_{nny} + M_{npy})} = \frac{8}{4} = 2 > \frac{a^2}{b^2} = 1.5625$$

The yield line pattern is as shown in Fig. 18.9.5(f). The quadratic equation (18.9.14) is used to solve for y.

$$4b(M_{nnx} + M_{npx})y^2 + 4a^2(M_{nny} + M_{npy})y - 3ba^2(M_{nny} + M_{npy}) = 0$$

$$4(20)(8)y^2 + 4(625)(4)y - 3(20)(625)(4) = 0$$

$$8y^2 + 125y - 1875 = 0$$

$$y = 9.375 \text{ ft}$$

The same uniform load w_u/ϕ is obtained from Eqs. (18.9.15), (18.9.16), or (18.9.17); the fact that it is so serves as a check on the numerical computation.

$$\frac{w_u}{\phi} = \frac{12[a^2(M_{nny} + M_{npy}) + 2by(M_{nnx} + M_{npx})]}{a^2(3by - 2y^2)}$$

$$= \frac{12[625(4) + 2(20)(9.375)(8)]}{625[3(20)(9.375) - 2(9.375)^2]} = 0.273 \text{ ksf}$$

$$\frac{w_u}{\phi} = \frac{24b(M_{nnx} + M_{npx})}{2a^2y + 3a^2(b - 2y)}$$

$$= \frac{24(20)(8)}{2(625)(9.375) + 3(625)[20 - 2(9.375)]} = 0.273 \text{ ksf}$$

$$\frac{w_u}{\phi} = \frac{6(M_{nny} + M_{npy})}{y^2} = \frac{6(4)}{(9.375)^2} = 0.273 \text{ ksf}$$

18.10 CORNER EFFECTS IN RECTANGULAR SLABS

In Sec. 18.4 on method of yield line analysis, it has been stated that there may be more than one possible yield line pattern, in which case solutions to all possible yield line patterns must be sought and the one giving the smallest collapse load would actually happen and thus should be used in design. Although the three typical yield line patterns for a rectangular two-way slab panel have been shown in Fig. 18.9.2 and their analysis has been completely treated in Sec. 18.9, it can be demonstrated that the corner yield patterns 4–5–6 shown in Fig. 18.10.1—in one-to-one correspondence to yield patterns 1–2–3 of Fig. 18.9.2—may indeed give a smaller collapse load and therefore control. These corner patterns are complicated to analyze, either by virtual work method or by equilibrium method. For instance, there are three unknowns E–F–G for the yield line positions; then once the expression for w_u/ϕ is obtained from the virtual work equation as a function of three independent variables, the partial derivative of w_u/ϕ with respect to each of the three unknown variables can be equated to zero. In the equilibrium method the same set of equations for the positions of points E–F–G may be obtained by inserting the predetermined zero or nonzero nodal forces and applying the moment equation of equilibrium to each of the slab segments.

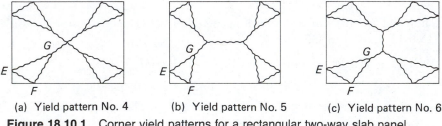

(a) Yield pattern No. 4   (b) Yield pattern No. 5   (c) Yield pattern No. 6

Figure 18.10.1 Corner yield patterns for a rectangular two-way slab panel.

An analysis [6] of a square slab with equal reinforcement in the x and y directions will show that the corner yield pattern No. 4 of Fig. 18.10.2(b) [see also Fig. 18.10.1(a)] results in $w_u/\phi = 22(M_{nn} + M_{np})/a^2$, whereas the regular yield pattern of Fig. 18.10.2(a) indicates $w_u/\phi = 24(M_{nn} + M_{np})/a^2$. M_{nn} and M_{np} are the nominal moment strengths per unit slab width for the negative moment and positive moment regions, respectively, in each direction, and a is the side of the square. Thus the corner pattern is more critical by approximately $(24 - 22)/24 = 8.3\%$. It may be proper then to discount the results of a regular yield pattern analysis as made in Sec. 8.9 for most rectangular slabs by 8 to 10% for reason of corner effects.

It may be pointed out that the yield line EF in Fig. 18.10.2(b) is a negative moment yield line; thus when there is no negative reinforcement, the moment strength along EF is zero. In this case the crack or yield line EF will not form if the corner A is not held down because the corner would simply lift up. ACI-13.4.6 requires the provision of special reinforcement at exterior corners in both top and bottom of the slab, for a distance in each direction from the corner equal to one-fifth the longer span. The use of negative moment reinforcement near the corner tends to move the point G farther away from the corner and thus such reinforcement helps to increase the load that will cause a collapse mechanism.

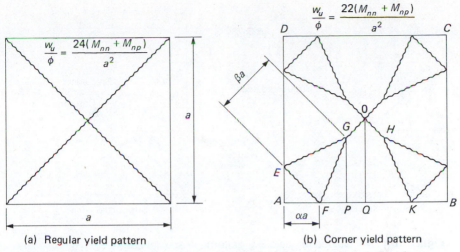

$$\frac{w_u}{\phi} = \frac{24(M_{nn} + M_{np})}{a^2}$$

$$\frac{w_u}{\phi} = \frac{22(M_{nn} + M_{np})}{a^2}$$

(a) Regular yield pattern

(b) Corner yield pattern

Figure 18.10.2   Square slab panel with equal reinforcement in two directions.

18.11 APPLICATION OF YIELD LINE ANALYSIS TO SPECIAL CASES

The yield line theory of slabs, as has been developed and illustrated in the preceding sections, is particularly suitable for special cases involving irregular shapes or irregular boundary conditions. Prerequisite to the analysis of these cases is the picturing of an applicable yield line pattern. The governing concept here is that rigid body plane rotations of slab segments separated at yield lines are possible under compatible deflection conditions. To this end the following guides may be provided:

1. Yield lines end at a slab boundary.
2. A yield line (or its prolongation) between two slab segments passes through the intersection of the axes of rotation of the two adjacent slab segments.
3. The axes of rotation lie along lines of supports or pass over column supports.

In addition to those already described, two other yield line patterns to further illustrate the use of the above guides are shown in Fig. 18.11.1.

Special Case. Shown in Fig. 18.11.2 is a rectangular slab simply supported at three edges and free at the upper edge. The positive moment reinforcement parallel to the a dimension provides a nominal moment strength of M_{nx} per unit of the b distance; and the positive moment reinforcement parallel to the b dimension provides strength M_{ny} per unit of the a distance. Two possible yield patterns are shown in Figs. 18.11.2(c) and (d); the unknown is x in yield pattern No. 1 and it is y in yield pattern No. 2.

For yield pattern No. 1, referring to Fig. 18.11.2(c) and letting Δ be the deflection where the yield line meets the free edge, the virtual work

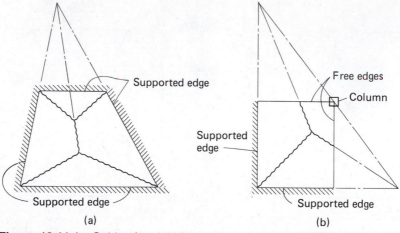

Figure 18.11.1 Guides for yield line patterns in special cases.

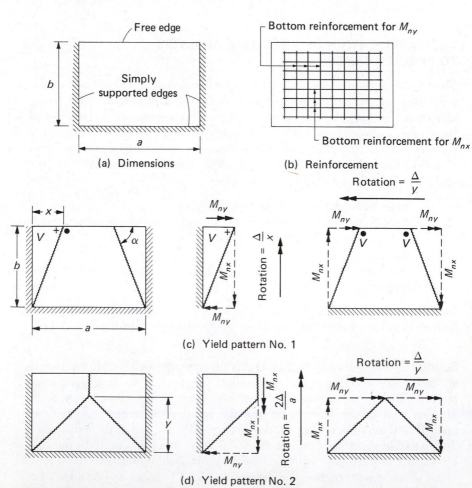

Figure 18.11.2 Rectangular slab simply supported at three edges and free at one edge.

done by the uniform load is

$$W = \frac{1}{2}\left(\frac{w_u}{\phi}\right)bx\left(\frac{\Delta}{3}\right)2 + \frac{1}{2}\left(\frac{w_u}{\phi}\right)xb\left(\frac{\Delta}{3}\right)2 + \frac{w_u}{\phi}(a - 2x)b\frac{\Delta}{2}$$

$$= \frac{w_u}{\phi}(\Delta)\left(\frac{b}{6}\right)(3a - 2x)$$

The virtual work done by the yield line moments is

$$W = 2M_{nx}\, b\left(\frac{\Delta}{x}\right) + 2M_{ny}\, x\left(\frac{\Delta}{b}\right)$$

Equating the two expressions for W and solving for w_u/ϕ,

$$\frac{w_u}{\phi} = \frac{12(b^2 M_{nx} + x^2 M_{ny})}{b^2 x(3a - 2x)} \tag{18.11.1}$$

Setting to zero the derivative of Eq. (18.11.1) with respect to x,

$$3a\left(\frac{M_{ny}}{M_{nx}}\right)x^2 + 4b^2 x - 3ab^2 = 0 \tag{18.11.2a}$$

In order that the root of Eq. (18.11.2a) be less than $a/2$,

$$\left(\frac{M_{ny}}{M_{nx}}\right) > \frac{4b^2}{3a^2} \tag{18.11.2b}$$

The nodal force V in Fig. 18.11.2(c) may be predetermined by use of Eq. (18.7.3); or

$$V = M_{ny} \cot \alpha = M_{ny}\left(\frac{x}{b}\right)$$

For equilibrium of the triangular segment in Fig. 18.11.2(c),

$$\frac{1}{2}\left(\frac{w_u}{\phi}\right)bx\left(\frac{x}{3}\right) + M_{ny}\left(\frac{x}{b}\right)x = M_{nx}b$$

from which

$$\frac{w_u}{\phi} = \frac{6(b^2 M_{nx} - x^2 M_{ny})}{b^2 x^2} \tag{18.11.3}$$

For equilibrium of the trapezoidal segment in Fig. 18.11.2(c),

$$\frac{1}{2}\left(\frac{w_u}{\phi}\right)xb\left(\frac{b}{3}\right)2 + \frac{w_u}{\phi}(a - 2x)b\left(\frac{b}{2}\right) = 2M_{ny}x + 2M_{ny}\left(\frac{x}{b}\right)b$$

from which

$$\frac{w_u}{\phi} = \frac{24x M_{ny}}{b^2(3a - 4x)} \tag{18.11.4}$$

The same quadratic equation in x as Eq. (18.11.2) is obtained from equating Eq. (18.11.3) to Eq. (18.11.4).

For yield pattern No. 2, referring to Fig. 18.11.2(d) and letting Δ be the deflection at the yield line perpendicular to the free edge, the virtual work done by the uniform load is

$$W = \frac{1}{2}\left(\frac{w_u}{\phi}\right)y\left(\frac{a}{2}\right)\left(\frac{\Delta}{3}\right)2 + \frac{w_u}{\phi}(b - y)\left(\frac{a}{2}\right)\left(\frac{\Delta}{2}\right)2 + \frac{1}{2}\left(\frac{w_u}{\phi}\right)ay\left(\frac{\Delta}{3}\right)$$

$$= \frac{w_u}{\phi}(\Delta)\left(\frac{a}{6}\right)(3b - y)$$

The virtual work done by the yield line moments is

$$W = 2M_{nx}b\left(\frac{2\Delta}{a}\right) + M_{ny}a\left(\frac{\Delta}{y}\right)$$

Equating the two expressions for W and solving for w_u/ϕ,

$$\frac{w_u}{\phi} = \frac{6(4byM_{nx} + a^2M_{ny})}{a^2y(3b - y)} \tag{18.11.5}$$

Setting to zero the derivative of Eq. (18.11.5) with respect to y,

$$4by^2 + 2a^2\left(\frac{M_{ny}}{M_{nx}}\right)y - 3a^2b\left(\frac{M_{ny}}{M_{nx}}\right) = 0 \tag{18.11.6a}$$

In order that the root of Eq. (18.11.6a) be less than b,

$$\left(\frac{M_{ny}}{M_{nx}}\right) < \frac{4b^2}{a^2} \tag{18.11.6b}$$

Since the moment strengths at the three intersecting yield lines in Fig. 18.11.2(d) are identical, the nodal forces are all zero, based on the treatment presented in Sec. 18.8. For equilibrium of the trapezoidal segment in Fig. 18.11.2(d),

$$\frac{1}{2}\left(\frac{w_u}{\phi}\right)y\left(\frac{a}{2}\right)\frac{a}{6} + \frac{w_u}{\phi}(b - y)\left(\frac{a}{2}\right)\frac{a}{4} = M_{nx}b$$

from which

$$\frac{w_u}{\phi} = \frac{24bM_{nx}}{a^2(3b - 2y)} \tag{18.11.7}$$

For equilibrium of the triangular segment in Fig. 18.11.2(d),

$$\frac{1}{2}\left(\frac{w_u}{\phi}\right)ay\frac{y}{3} = M_{ny}a$$

from which

$$\frac{w_u}{\phi} = \frac{6M_{ny}}{y^2} \tag{18.11.8}$$

The same quadratic equation in y as Eq. (18.11.6) is obtained from equating Eq. (18.11.7) to Eq. (18.11.8).

Thus Eqs. (18.11.1) to (18.11.4) apply to yield pattern No. 1 which cannot happen if (M_{ny}/M_{nx}) is smaller than $4b^2/(3a^2)$ and Eqs. (18.11.5) to (18.11.8) apply to yield pattern No. 2 which cannot happen if (M_{ny}/M_{nx}) is larger than $4b^2/a^2$. When (M_{ny}/M_{nx}) is between $4b^2/(3a^2)$ and $4b^2/a^2$, analysis for both yield patterns should be made and the one giving the smaller collapse load controls. Although an exact value of (M_{ny}/M_{nx}) in terms of b^2/a^2 (between 1.33 and 4.00) at the transition point where yield pattern No. 1 begins to control over yield pattern No. 2 may be determined, the complication in the algebra does not seem to warrant the effort. Table 18.11.1 shows the results of analysis for six different cases in which the values of (M_{ny}/M_{nx}) become progressively larger while keeping M_{ny} constant.

Table 18.11.1 YIELD LINE ANALYSIS OF A RECTANGULAR SLAB WITH THREE SUPPORTED EDGES AND ONE FREE EDGE (FIG. 18.11.2); a = 25 FT, b = 20 FT

CASE	1	2	3	4	5	6
M_{ny} (ft-kips/ft)	16	16	16	16	16	16
M_{nx} (ft-kips/ft)	24	18.75	16	8	6.25	4
Yield pattern No. 1						
x (ft), Eq. (18.11.2)	—	12.5	12	9.78	9.01	7.68
w (ksf), Eqs. (18.11.1), (18.11.3), or (18.11.4)	—	0.480	0.427	0.262	0.222	0.167
Yield pattern No. 2						
y (ft), Eq. (18.11.6)	13.22	14.41	15.20	18.75	20	—
w (ksf), Eqs. (18.11.5), (18.11.7), or (18.11.8)	0.549	0.462	0.415	0.273	0.240	—
Yield pattern controlling	No. 2	No. 2	No. 2	No. 1	No. 1	No. 1
Collapse load w_u/ϕ (ksf)	0.549	0.462	0.415	0.262	0.222	0.167

SELECTED REFERENCES

1. A. Ingerslev. "The Strength of Rectangular Slabs," *Journal of Institute of Structural Engineers*, London, **1**, (1) January 1923, 3–14; Disc., 14–19.
2. K. W. Johansen. "The Ultimate Strength of Reinforced Concrete Slabs," *Final Report*. Third Congress, International Association for Bridge and Structural Engineering, Liege. September 1948 (pp. 565–570).
3. K. W. Johansen. *Yield Line Theory*. London: Cement and Concrete Association, 1962.
4. L. L. Jones and R. H. Wood. *Yield Line Analysis of Slabs*. New York: Elsevier, 1967.
5. R. H. Wood. "Plastic Design of Slabs using Equilibrium Methods," *Flexural Mechanics of Reinforced Concrete*. Proceedings of the International Symposium, Miami, 1964 (pp. 319–336).
6. E. Hognestad. "Yield Line Theory for the Ultimate Flexural Strength of Reinforced Concrete Slabs," *ACI Journal, Proceedings*, **49**, March 1953, 637–656.

PROBLEMS

18.1 Assuming a 6-in. slab thickness for the continuous slab span described in Example 18.5.1, the uniform load w_u due to the weight of slab itself may be considered to be $1.4w_D = 1.4(75) = 105$ psf. If the slab supports a transverse wall at 7 ft from the left support line, determine the maximum wall load per transverse ft that will cause a collapse mechanism to occur.

18.2 Solve Example 18.7.1, but use $M_{nx} = 10$ ft-kips and $M_{ny} = 8$ ft-kips, each per foot width of slab.

18.3 For the regular yield pattern solution to Case 1 of Example 18.9.1, investigate the effect of a corner yield pattern in which the negative moment yield line intersects the edges at 5 ft and 4 ft from the corner along the 25 ft and 20 ft edges, respectively.

18.4 Same as Prob. 18.3, except for Case 2 of Example 18.9.1.

18.5 Same as Prob. 18.3, except for Case 3 of Example 18.9.1.

18.6 Verify the solution for Case 1 in Table 18.11.1.

18.7 Verify the solution for Case 2 in Table 18.11.1.

18.8 Verify the solution for Case 3 in Table 18.11.1.

18.9 Verify the solution for Case 4 in Table 18.11.1.

18.10 Verify the solution for Case 5 in Table 18.11.1.

18.11 Verify the solution for Case 6 in Table 18.11.1.

19

Torsion

19.1 GENERAL

Reinforced concrete members may be subjected to torsion, frequently in combination with bending and shear. The cantilever member in Fig. 19.1.1(a) is largely subjected to torsion, although some bending and shear also exist due to its own weight. The fixed-ended beam of Fig. 19.1.1(b) is subjected to substantial amounts of bending, shear, and torsion.

Spandrel beams at the edge of a building built integrally with the floor slab are subjected not only to transverse loads but also to a torsional moment per unit length equal to the restraining moment at the exterior end of the slab. Similarly, spandrel girders receive torsional moments from the exterior ends of the floor beams that frame into them.

Torsion on structural systems may be classified into two types: (1) *Statically determinate torsion* (sometimes called "equilibrium torsion"), for which the torsion *can* be determined from statics alone; and (2) *statically indeterminate torsion* (sometimes called "compatibility torsion"), for which the torsion *cannot* be determined from statics and a rotation (twist) is required for deformational compatibility between interconnecting elements, such as a spandrel beam, slab, or column. Both examples in Fig. 19.1.1 are cases of statically determinate torsion.

In cases of statically determinate torsion, as in Figs. 19.1.2(a) and (b), the amount of torsion that the member is required to resist is based on the requirement of statics and is independent of the stiffness of the member. Statically indeterminate torsion, as shown in Figs. 19.1.2(c) and (d), exists in some situations where there would be no torsion if the statical indeter-

Inclined cracks due to torsion; test by J. P. Klus at the University of Wisconsin, Madison.

minacy were eliminated. For instance, if the support at A is eliminated in Fig. 19.1.2(c), the torsion is eliminated. Similarly, in Fig. 19.1.2(d), if a flexural hinge is put at B, the torsion is eliminated. For such statically indeterminate torsion situations the amount of torsion in a member depends on the magnitude of the torsional stiffness of the member itself in relation to the stiffnesses of the interconnecting members.

For an excellent overview of the entire subject of torsional phenomena as it affects structures, with specific reference to reinforced concrete design, the reader is referred to Tamberg and Mikluchin [1]. An extensive bibliography is also included. The most extensive treatment of torsion for reinforced and prestressed concrete is the 1983 book by Hsu [83]. Collins and Mitchell [2] have provided a unified rational treatment of the function of reinforcement to resist shear and torsion, and have made design recommendations for reinforced and prestressed concrete. Warwaruk [3] has provided a recent state-of-the-art discussion of torsion in reinforced concrete.

In this chapter, brief treatment is given to the computation of torsional stress and torsional rigidity of homogeneous sections, the torsional strength of reinforced concrete sections, the development and background for the ACI Code requirements, design examples, and the effect of torsional stiffness in a continuity analysis.

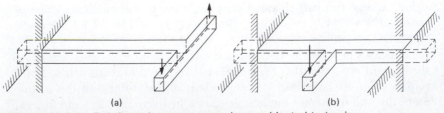

(a)   (b)

Figure 19.1.1 Reinforced concrete members subjected to torsion.

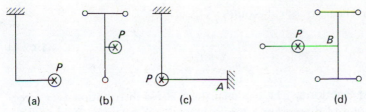

Figure 19.1.2 Comparison of statically determinate torsion (cases *a* and *b*) and statically indeterminate torsion (cases *c* and *d*). (Structures shown in plan view with load *P* at 90° to plane of frame).

19.2 TORSIONAL STRESS IN HOMOGENEOUS SECTIONS

A torsional moment T acting on a shaft of homogeneous material as shown in Fig. 19.2.1 causes shear stresses v over the cross section.

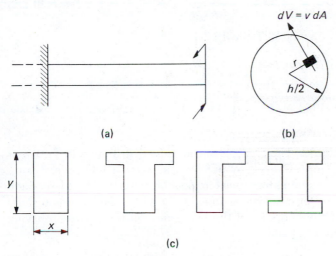

Figure 19.2.1 Torsional stress in homogeneous sections.

Circular Sections. For a circular section, a plane transverse section before twisting remains plane after twisting. Consequently, the resultant shear stress v at any point is proportional to its distance from the center and is in a direction perpendicular to the radius. Calling h the diameter of the circle [Fig. 19.2.1(b)] and v_t the maximum torsional shear stress at the perimeter,

$$T = \int_A rv \, dA = \int_A r\left(\frac{2v_t}{h} r\right) dA = \frac{2v_t}{h} \int_A r^2 \, dA = \frac{2C}{h} v_t$$

in which C, the polar moment of inertia, is

$$C = \int_A r^2 \, dA = \int_0^{h/2} r^2 2\pi r \, dr = \left(\frac{2\pi r^4}{4}\right)_0^{h/2} = \frac{\pi h^4}{32}$$

The torsional shear stress v_t becomes

$$v_t = \frac{16T}{\pi h^3}$$

(19.2.1)

Rectangular Sections. The torsional shear stress distribution over a rectangular section of dimension x by y cannot be as easily derived as for a circular section. Unlike the circular section where plane transverse sections remain plane after twisting, the noncircular cross section warps under torsion. If plane sections were maintained after twisting, the maximum shear stress would exist at a point farthest from the axis of twist. Such is not the case for rectangular sections. From the mathematical theory of elasticity [4], it has been found that the maximum torsional shear stress v_t occurs at the midpoint of the long side and parallel to it. The magnitude of v_t is a function of the ratio of y to x (long to short sides) [Fig. 19.2.1(c)]; and

$$v_t = \frac{T}{\alpha x^2 y}$$

(19.2.2)

The values for α are given in Table 19.2.1.

Table 19.2.1

y/x	1.0	1.2	1.5	2.0	2.5	3	5	∞
α	0.208	0.219	0.231	0.246	0.256	0.267	0.290	0.333

T-, L-, and I-Sections. The torsional shear stress distribution in T-, L-, or I-sections may be approximated by dividing the section into several component rectangles assuming that each component rectangle has a large ratio y/x so that the value for α may be assumed to be $\frac{1}{3}$ [4]. The maximum shear stress v_t occurs at the midpoint of the long side y of the rectangle having the greatest thickness x_m and

$$v_t = \frac{Tx_m}{\sum \frac{1}{3} x^3 y}$$

(19.2.3)

in which x and y are the thickness and side, respectively, of each component rectangle. Since the web of the sections considered is usually thicker than the flange, x_m will usually be the web thickness.

19.3 TORSIONAL STIFFNESS OF HOMOGENEOUS SECTIONS

The torsional stiffness K_t of a member is defined as the ratio of torsional moment T to the angle of twist θ in the length L. The torsional rigidity is usually represented by the symbol GC in which G is the modulus of elasticity in shear and C is the torsion constant. Thus if θ is the total angle of twist in a length L,

$$K_t = \frac{T}{\theta} = \frac{GC}{L} \tag{19.3.1}$$

Circular Sections. It has been shown in Sec. 19.2 (Circular Section) that the torsion constant C of a circular section of diameter h is the polar moment of inertia

$$C = \frac{\pi h^4}{32} \tag{19.3.2}$$

Rectangular Sections. The torsion constant C of a rectangular section of height y and width x may be expressed [4] as

$$C = \beta x^3 y \tag{19.3.3}$$

in which β is a function of y to x. The values for β are given in Table 19.3.1.

Table 19.3.1

y/x	1.0	1.2	1.5	2.0	2.5	3	4	5
β	0.141	0.166	0.196	0.229	0.249	0.263	0.281	0.291

T-, L-, and I-Sections. The torsion constant C of a T-, L-, or I-section may be approximated [4] by the expression

$$C = \Sigma \tfrac{1}{3} x^3 y \tag{19.3.4}$$

in which y and x are the side and thickness of each of the component rectangles into which the section may be divided.

The following more exact expression for the torsion constant, giving values closer to Table 19.3.1 for sections composed of rectangular elements having y/x less than about 10, has been derived by Timoshenko,*

$$C = \Sigma \tfrac{1}{3} x^3 y \left(1 - 0.63 \frac{x}{y} \right) \tag{19.3.5}$$

The reader may note that the ACI Code in its design provisions for torsion (ACI-Chap. 11) makes use of the simpler expression, Eq. (19.3.4) in computing C_t, as explained in Sec. 19.13. However, in the provisions for slabs as discussed in Chaps. 16 and 17 (ACI-13.7.5.3), Eq. (19.3.5) is preferred as it gives a lower estimate of the torsional restraint, and for the uncracked stiffness is more accurate. Although there is some inconsistency in using two different expressions for the torsion constant, the overestimate of stiffness using Eq. (19.3.4) is an acceptable simplification when determining the nominal torsional strength for design. A lower estimate of stiffness using Eq. (19.3.5) is desirable in structural analysis when determining the restraining effect of spandrel members in two-way floor systems.

*p. 278 of Ref. 4.

19.4 EFFECTS OF TORSIONAL STIFFNESS ON COMPATIBILITY TORSION

General Treatment of Torsion on Statically Indeterminate Systems. In order to perform a statically indeterminate structural analysis, it is necessary first to be able to determine the relative stiffnesses of interacting members. The so-called "compatibility torsion" discussed in Sec. 19.1 is involved. For example, if a spandrel member is *uncracked* and its torsional stiffness GC/L is computed as shown in Sec. 19.3, the torsional moment that the member will attempt to carry may be very large. As the member cracks, its torsional stiffness reduces drastically, the member will rotate, and the torsional moment carried is likewise reduced.

Postcracking stiffness and torsional moment have been studied by Lampert [5] and Collins and Lampert [6], who proposed an expression for the torsional rigidity of a cracked section. Since the stiffness is needed *before* the torsional moment can be determined, the cracked section stiffness is not available because it requires knowledge of the steel reinforcement.

Alternatively, Collins and Lampert [6] have indicated that in cases of compatibility torsion (where torsional moment depends on torsional stiffness) analysis on the basis of zero torsional stiffness resulted in a design as satisfactory as an analysis using uncracked stiffness. In fact, added steel may increase the torsional moment in the member but have little effect on the twist (rotation). Consequently, it may be more effective to design for a twist (rotation) than for a torque (moment). The purpose of the torsional reinforcement then is to provide ductility and distribute cracks caused by the twist. Such a procedure would be within the spirit of the ACI Code, where ACI-8.6.1 states, "Any reasonable assumptions may be adopted for computing relative flexural and torsional stiffnesses of columns, walls, floors, and roof systems."

Using the zero torsional stiffness assumption, the member resisting torsion (say, a spandrel beam) would be designed for flexure and shear, neglecting torsion; then the torsional stiffness based on cracked section might be computed. The structure may then be analyzed [7] to determine the torsional moments, and the section may be checked according to the ACI Code rules for torsion design.

ACI Code Procedure. The ACI Code (ACI-11.6.3) provides an optional simple procedure to reduce the design complexity for cases involving compatibility torsion. When a statically indeterminate situation involves torsion, and an internal redistribution of forces can occur as a result of cracking, the factored torsional moment to be used in design is reduced to a minimum value sufficient to provide the necessary rotation capacity (ductility). In other words, if the torsional restraint is omitted in determining the bending moments and shears on the structural elements, the design of those elements may be more conservative than otherwise, but the difficulty of determining the torsional moments has been eliminated. The torsional members must, however, have the ductility to *twist* the necessary amount.

To summarize, the ACI Code provides two options for the design of torsional members when the torsional moment is dependent on the relative stiffness of the interacting members.

1. Estimate the torsional and flexural stiffnesses of all interacting members making "any reasonable assumptions . . ." (ACI-8.6.1). Determine the moments, shears, and torsional moments by the statically indeterminate analysis using factored loads. Then apply the ACI Code provisions for torsion design.
2. Neglect torsional stiffness in the statically indeterminate structural analysis. Since no torsional moment will then be available, the torsional members must be designed for a strength based on a nominal torsional shear stress under factored load of $4\sqrt{f_c'}$. (The computation of strength based on this stress is discussed in detail in Sec. 19.11.)

Spandrel Beams and Girders. Consider the spandrel beam 2B4 (Fig. 19.4.1) in the typical slab-beam-girder floor of Example 8.3.1. This beam receives a vertical load and a torsional moment per unit length from the slab 2S1, which are equal, respectively, to the reaction and restraining moment at the exterior end of slab 2S1. In addition, the beam supports the weight of whatever walls or windows may rest directly on it. Thus the spandrel beam 2B4 is subjected to a torsional moment per unit length in addition to bending and shear. A similar condition exists in the spandrel girder 2G4; however, the torsional moments are applied only at the junction points with the beams.

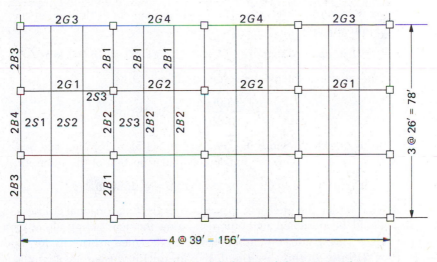

Figure 19.4.1   Floor plan of typical slab–beam–girder construction.

The torsional moments in the spandrel beams or girders cause torsional shear stresses, which are additive to the bending shear stresses at the *inside* face of the member. The usual approach in design is to provide for the sum of the torsional shear and flexural shear requirements. Since torsional shear stress goes around the member, closed stirrups (hoops) are necessary.

The magnitude of the torsional moment acting uniformly along a spandrel beam (ACI-11.6.3.2) such as 2B4 of Fig. 19.4.1 might be roughly approximated as equal to the restraining moment along the exterior edge of the slab, using a value such as $\frac{1}{24}wL^2$ as given by ACI-8.3.3. Alternatively, the torsional moment may be neglected if the slab is designed assuming there is no restraining moment along the spandrel beam. In such a case the spandrel beam must be designed for a minimum torsional strength corresponding to that which will provide adequate ductility to twist (see Example 19.15.3).

The design and behavior of spandrel beams has been the subject of many studies [8–15,72,73].

EXAMPLE 19.4.1 Estimate the maximum torsional shear stress in the spandrel beam 2B4 of Fig. 19.4.1 if the restraining moment at the exterior end of slab panel (4.5-in. slab and a clear span of 11.92 ft) 2S1 is $M = wL^2/24$. The service live and dead loads are 100 and 56 psf, respectively. Assume an 18×18 in. column and a $13 \times 22\frac{1}{2}$ in. overall size beam. Use $f'_c = 3000$ psi.

Solution: The restraining moment along the edge of the slab is approximately

$$w_u = 1.4(56) + 1.7(100) = 249 \text{ psf}$$

$$M_u = \tfrac{1}{24}(0.249)(11.92)^2 = 1.47 \text{ ft-kips/ft width}$$

The torsional moment is largest at the face of the column and decreases nearly linearly to zero at midspan. The torsional moment at the face of the column is approximately

$$T_u = (\tfrac{1}{2} \text{ clear span of spandrel})(M_u) = 12.25(1.47) = 18.0 \text{ ft-kips}$$

The maximum nominal torsional stress is (see Sec. 19.13)

$$v_{tn} = \frac{T_u}{\phi\frac{1}{3}\Sigma x^2 y} = \frac{18.0(12,000)}{0.85(1360)} = 187 \text{ psi}$$

where, considering that a width of slab equal to $3(4.5) = 13.5$ in. (see Sec. 19.11) is effective to act with the main rectangular portion of the member,

$$\tfrac{1}{3}\Sigma x^2 y = \tfrac{1}{3}[(13)^2(22.5) + (4.5)^2(13.5)] = 1360 \text{ in.}^3$$

Note that under the alternate procedure neglecting torsional stiffness, the slab would be designed without restraining moment acting along its exterior edge. The spandrel member would be designed arbitrarily for $v_{tn} = 4\sqrt{f'_c} = 219$ psi. In this case the design for ductility using $4\sqrt{f'_c}$ would be the more conservative approach, as well as the simpler.

19.5 STRENGTH OF PLAIN CONCRETE RECTANGULAR SECTIONS IN TORSION—SKEW BENDING THEORY

According to theory of elasticity, a plain concrete rectangular section would reach its torsional strength T_e when the maximum torsional stress v_t [Eq. (19.2.2)] equals the maximum principal tensile stress $f_t(\text{max})$, since pure torsion gives a pure shear stress condition. Equation (19.2.2) would then become

$$T_e = \alpha x^2 y [f_t(\text{max})] \qquad (19.5.1)$$

Experiments by Hsu [16,17], however, indicate that Eq. (19.5.1) underestimates the nominal strength by about 50%, partly due to the fact that the maximum stress indicated by Eq. (19.2.2) occurs only at the midpoint of the long side of the rectangle.

Studies by Hsu [16] have shown that a torsion failure of a rectangular section does not occur in a spiral form as may be expected from a circular shaft. Failure of a rectangular section in torsion occurs by *bending* about an axis parallel to the wider face of the section and inclined at about 45° to the axis of the beam, as shown in Fig. 19.5.1. In this skew bending theory Hsu showed that Eq. (19.5.1) can represent the nominal torsional strength of a plain concrete rectangular section when α is taken as $\frac{1}{3}$ and $f_t(\text{max})$ is taken at about $5\sqrt{f_c'}$.

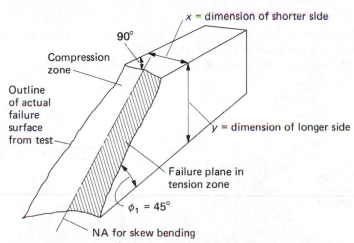

Figure 19.5.1 Skew bending of plain concrete rectangular section (according to Hsu, Ref. 16).

19.6 STRENGTH OF REINFORCED CONCRETE RECTANGULAR SECTIONS IN TORSION—SKEW BENDING THEORY

Once steel reinforcement, both longitudinal and transverse, is placed in a rectangular section, the behavioral pattern changes from that of plain concrete. The resisting action of transverse reinforcement in the form of closed hoops is similar to that of stirrups resisting flexural shear. Prior to cracking, the reinforcement participates little if at all; but after cracking, the reinforcement carries a large portion of the total torsional moment. The contribution of concrete is only about 40% of the torsional strength of an unreinforced section. The failure mode according to this theory, however, does continue to be one of skew bending such as described in Sec. 19.5.

The skew bending concept was proposed by Lessig [19] and extended by Goode and Helmy [20], Collins, Walsh, Archer, and Hall [21], and

Below, Rangan, and Hall [22], all of whom applied it to the case of combined bending and torsion. Hsu [18] has applied the concept to the case of torsion alone and has developed the expression that forms the basis for the ACI Code procedure. Hsu [17,23], Zia [24,25], and Warwaruk [3] have provided summaries of the theories relating to rectangular sections in torsion. McMullen and Rangan [74] have reviewed the research to clarify the contradictions between the skew bending and the space truss theories (see Sec. 19.7 for the space truss theory). The following development presents some of the ideas relating to the strength expression developed by Hsu [18].

Referring to Fig. 19.6.1, the failure section is assumed to be a plane that is perpendicular to the wider face of the member and inclined at 45° to the axis of the member. The failure plane may be as shown in Fig. 19.6.1, due to twisting moment in the direction indicated. Since a bending mode of failure is assumed, the compression zone is treated as in any beam analysis; that is, it has a depth a over which the compressive stress may be assumed uniform. On the tension side where the concrete cracks and is assumed not to resist tension, the reinforcing hoops have tensile forces P_v in them and the longitudinal bars resist shear across the cracked concrete via dowel action (see Sec. 5.4). The horizontal and vertical components of this dowel action are designated Q_x and Q_y. As long as the concrete is uncracked and the concrete itself transmits shear, no dowel action exists.

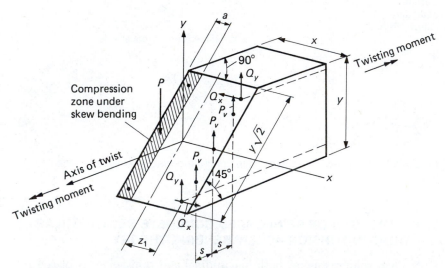

Figure 19.6.1 Forces acting on skew bending failure section.

On the compression side [Fig. 19.6.2(a)], the longitudinal bars contribute tensile force P_l; the concrete contributes shear resistance P_s in the failure plane and also compressive resistance P_c normal to the failure plane. The components of the resultant force P are shown in Fig. 19.6.2(b). The hoop reinforcement in compression is neglected because, as has been shown in Sec. 3.10, the nominal moment strength M_n is not significantly affected by the inclusion of compression reinforcement.

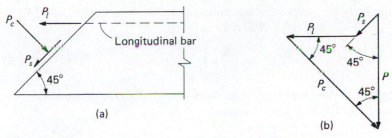

Figure 19.6.2 Components of resultant force P acting on compression zone of failure plane.

On the tension side, it is noted that no longitudinal force can exist. If such a force were to act, it would have to be counterbalanced by a component of resistance acting oppositely. Since only P_v, Q_x, and Q_y are assumed to be acting, and the resultant force must be directed upward (opposite to P on the compression side), no resultant tension or compression can exist in the longitudinal direction of the tension zone under skew bending.

The reader is reminded that unless the direction of the twisting moment is definitely established by analysis, in usual conditions of providing nominal torsional strength (ACI-11.6.3), a potential failure plane can exist opposite to that in Fig. 19.6.1, having the compression and tension sides interchanged. Thus the longitudinal forces, stirrup forces, and dowel forces must be resisted on each side of the section.

Strength Attributable to Concrete. The shear resistance P_s (Fig. 19.6.2) may be expressed as

$$P_s = v_{avg}(y\sqrt{2})a \tag{19.6.1}$$

where v_{avg} is the average shear stress acting over the compression zone; $y\sqrt{2}$ is the width of the compression zone; and a is the depth of the compression zone.

Alternatively, the shear strength P_s may be considered proportional to the effective area $xy\sqrt{2}$ and to $\sqrt{f_c'}$ [see Eq. (5.5.6) omitting effect of reinforcement]. Thus

$$P_s = k_1 xy\sqrt{f_c'} \tag{19.6.2}$$

where k_1 is a proportionality constant.

From Fig. 19.6.2(b), one may note that

$$P = \sqrt{2}P_s + P_l \tag{19.6.3}$$

The first term of Eq. (19.6.3) represents the portion attributable to concrete. Thus the torsional strength T_c attributable to concrete equals the force $\sqrt{2}P_s$ times the moment arm (say, $0.80x$),

$$T_c = \sqrt{2}P_s(\text{arm})$$
$$= \sqrt{2}k_1 xy\sqrt{f_c'}(0.80x)$$
$$= k_2 x^2 y\sqrt{f_c'} \tag{19.6.4}$$

Experimentally the proportionality constant k_2 has been determined [23] to be $2.4/\sqrt{x}$. Thus Eq. (19.6.4) may be written

$$T_c = \left(\frac{2.4}{\sqrt{x}}\right)x^2 y \sqrt{f_c'} \qquad (19.6.5)$$

Equation (19.6.5) represents the torsional moment strength available from the concrete in the compression zone.

Strength Attributable to Hoops and Longitudinal Reinforcement. Referring to Figs. 19.6.1 and 19.6.2, the forces P_v, Q_x, and Q_y on the tension side and P_l on the compression side have yet to be considered.

1. The contribution of the closed vertical stirrups (hoops) is

$$P_v = A_t f_y\left(\frac{y_1}{s}\right) \qquad (19.6.6)$$

where y_1/s (see Fig. 19.6.3) is the number of hoops intercepted by the 45° failure plane.

2. The tensile force P_l in the longitudinal bars intercepting the compression concrete zone is

$$P_l = \xi\left(\frac{A_l}{2}\right)f_y \qquad (19.6.7)$$

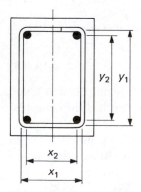

Figure 19.6.3 Cross-section dimensions.

where ξ is an efficiency factor to account for the longitudinal bars being located at two or more points within the compression zone, and A_l is the total area of all longitudinal bars (assumed to be $A_l/2$ in the compression zone). P_l contributes to the torsional strength since from Eq. (19.6.3) it is part of P.

3. The dowel forces Q_x and Q_y act after the concrete has cracked and these forces may be assumed proportional to the bar cross-sectional area and to the bar lateral displacement, which is proportional to the distance (either $0.5x_2$ or $0.5y_2$) from the center of twist to the bar. Thus

$$Q_x = k_3 A_l y_2 \tag{19.6.8}$$
$$Q_y = k_3 A_l x_2$$

where k_3 is a proportionality constant. Dowel action in beams subject to torsion has been reviewed by Youssef and Bishara [75].

Next let m equal the ratio of the volume of longitudinal bars to the volume of closed hoops such that

$$m = \frac{A_l s}{2A_t(x_1 + y_1)} \tag{19.6.9}$$

or

$$A_l = A_t \left[\frac{2m(x_1 + y_1)}{s} \right] \tag{19.6.10}$$

Substitution of Eq. (19.6.10) into Eq. (19.6.7) gives

$$P_l = \xi m \left(1 + \frac{y_1}{x_1} \right) \left(\frac{x_1 A_t f_y}{s} \right) \tag{19.6.11}$$

Substituting Eq. (19.6.10) into Eq. (19.6.8) gives

$$Q_x = k_3 y_2 \left(\frac{2m A_t(x_1 + y_1)}{s} \right)$$
$$= 2 \frac{k_3}{f_y} \left(\frac{y_2}{y_1} \right) m \left(1 + \frac{y_1}{x_1} \right) \left(\frac{x_1 y_1 A_t f_y}{s} \right) \tag{19.6.12}$$

Similarly,

$$Q_y = 2 \frac{k_3}{f_y} \left(\frac{x_2}{y_1} \right) m \left(1 + \frac{y_1}{x_1} \right) \left(\frac{x_1 y_1 A_t f_y}{s} \right) \tag{19.6.13}$$

The torsional resistance from reinforcement then is

$$T_s = P_v \left(\frac{x_1}{2} \right) + P_l \left(\frac{x_2}{2} \right) + 2Q_x \left(\frac{y_2}{2} \right) + 2Q_y \left(\frac{x_2}{2} \right) \tag{19.6.14}$$

Substitution of Eqs. (19.6.6), (19.6.11), (19.6.12), and (19.6.13) into Eq. (19.6.14) gives

$$T_s = \alpha_t \left(\frac{x_1 y_1 A_t f_y}{s} \right) \tag{19.6.15}$$

where

$$\alpha_t = \frac{1}{2} + \xi m \left(1 + \frac{y_1}{x_1} \right) \left(\frac{x_2}{2y_1} \right) + 2 \frac{k_3}{f_y} m \left(1 + \frac{y_1}{x_1} \right) (x_2^2 + y_2^2) \left(\frac{1}{y_1} \right) \tag{19.6.16}$$

Assuming $x_2 \approx x_1$ and $y_2 \approx y_1$, the quantity α_t is essentially a function of two parameters, m and y_1/x_1 and might be written

$$\alpha_t = C_1 + C_2 m + C_3 \left(\frac{y_1}{x_1} \right) \tag{19.6.17}$$

where the constants C_1, C_2, and C_3 may be experimentally determined.

Hsu [18] has shown that for equal volume of longitudinal bars to closed hoops (i.e., $m = 1$), α_t may be expressed as

$$\alpha_t = 0.66 + 0.33\left(\frac{y_1}{x_1}\right) \tag{19.6.18}$$

which is used by the ACI Code (ACI-11.6.9.1).

Thus the full nominal strength T_n of rectangular reinforced concrete sections may be written by combining Eqs. (19.6.5) and (19.6.15),

$$T_n = T_c + T_s = \left(\frac{2.4}{\sqrt{x}}\right)x^2 y \sqrt{f_c'} + \alpha_t\left(\frac{x_1 y_1 A_t f_y}{s}\right) \tag{19.6.19}$$

An alternate approach proposed by Victor et al. [26] is to consider the contribution of the dowel action separately rather than include it with T_s.

19.7 STRENGTH OF REINFORCED CONCRETE RECTANGULAR SECTIONS IN TORSION—SPACE TRUSS ANALOGY

The space truss analogy approach to the evaluation of torsional capacity was first suggested by Rausch [27] and developed by Lampert and Thürlimann [5,28]. It was then further explained by Lampert and Collins [29], and more recently treated by Müller [81] and Rabbat and Collins [82]. Recent advances in knowledge of reinforced concrete in torsion seem to lead to the space truss theory as the most rational way to give unified treatment to flexure and torsion. The treatment presented by Collins and Mitchell [2] suggests a rational design approach which may eventually supercede the ACI Code approach that is linked to the skew bending theory. The European Concrete Committee (CEB) [30] uses the space truss analogy theory as its basis for design.

In determining the torsional strength of reinforced concrete rectangular sections, it has been found that nearly all of the strength is derived from the reinforcement and the concrete that immediately surrounds the steel. Then one may consider a thin-walled box section, as shown in Fig. 19.7.1, as a space truss. The longitudinal bars in the corners contribute tensile forces while the concrete strips between cracks provide compressive resistance. The inclined compressive forces act in a spiral fashion around the box section, giving the compressive forces D_h on the vertical sides and D_b on the horizontal sides.

Consider equilibrium requirements on the section of Fig. 19.7.1. From force equilibrium in the z direction,

$$4Z = 2\frac{\tau t x_2}{\tan \phi_1} + 2\frac{\tau t y_2}{\tan \phi_1} \tag{19.7.1}$$

where Z is the tensile force in each longitudinal bar in the corner of the section, and τt is the shear flow (force per unit length) in the thin-walled section. Summation of moments of the shear flow forces about the z axis

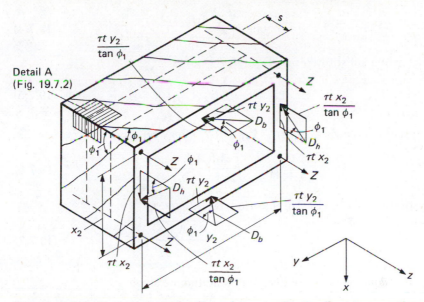

Figure 19.7.1 Forces on section in torsion according to space truss analogy (from Ref. 5).

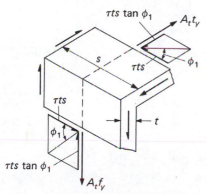

Figure 19.7.2 Free body of detail A (Fig. 19.7.1) on the hollow box section used for the space truss analogy.

must equal the applied torsional moment,

$$T_n = \tau t y_2 x_2 + \tau t x_2 y_2 \tag{19.7.2}$$

Next consider the detail A of Fig. 19.7.1 whose free body diagram is given in Fig. 19.7.2. Detail A represents the portion of concrete along the z axis of the member tributary to one hoop. Equilibrium in the direction of the hoop tensile force requires

$$A_t f_y = \tau t s \tan \phi_1 \tag{19.7.3}$$

From Eq. (19.7.3)

$$\tan \phi_1 = \frac{A_t f_y}{\tau t s} \tag{19.7.4}$$

which when substituted into Eq. (19.7.1) gives

$$2Z = \frac{(\tau t)^2 s}{A_t f_y}(x_2 + y_2) \tag{19.7.5}$$

Letting the force $Z = (A_l/4)f_y$, Eq. (19.7.5) may be solved for τt,

$$\tau t = \sqrt{\frac{A_l f_y A_t f_y}{2s(x_2 + y_2)}} \tag{19.7.6}$$

Substitution of Eq. (19.7.6) into Eq. (19.7.2) gives

$$T_n = 2x_2 y_2 \sqrt{\frac{A_l f_y A_t f_y}{2s(x_2 + y_2)}} \tag{19.7.7}$$

For equal volumes of hoop and longitudinal steel,

$$A_l = \frac{2A_t(x_1 + y_1)}{s} \tag{19.7.8}$$

Substitution of Eq. (19.7.8) into (19.7.7) gives

$$T_n = 2\left(\frac{A_t f_y}{s}\right)x_2 y_2 \sqrt{\frac{x_1 + y_1}{x_2 + y_2}} \tag{19.7.9}$$

If $x_1 \approx x_2$ and $y_1 \approx y_2$, Eq. (19.7.9) becomes

$$T_n = 2\left(\frac{A_t f_y}{s}\right)x_2 y_2 \tag{19.7.10}$$

Comparison of Eqs. (19.7.10) and (19.7.2) will show that

$$\tau t = \frac{A_t f_y}{s}$$

which if substituted into Eq. (19.7.4) will show that $\tan \phi_1 = 1$, or $\phi_1 = 45°$.

One of the features of the more recent treatment of the truss analogy theory, according to the Collins and Mitchell [2,31] approach of using inclined compression struts (so-called *compression field action*), is that ϕ_1 is *not* necessarily 45° but will take an optimum angle to permit the greatest strength. Collins and Mitchell [2] and the European approach [30] accept that $\tan \phi_1$ can be as low as $\frac{3}{5}$ and as high as $\frac{5}{3}$, and need not be 1.0.

One may also note that Eq. (19.7.10) is of the same form as T_s of Eq. (19.6.15), and becomes essentially identical if $\alpha_t = 2$. The simple space truss analogy neglects any torsional strength due to the concrete alone; that is, if either longitudinal *or* hoop steel is not present, this theory gives no torsional capacity.

19.8 STRENGTH OF SECTIONS IN COMBINED BENDING AND TORSION*

In most practical situations, torsion will occur simultaneously with flexure. Although there have been many studies of the interaction between bending and torsion [2,5,19–22,28,32–39,76], there does not seem to be agreement on what is the correct interaction criterion. Zia [24,25] has summarized much of the available information.

Both the skew bending theory [19,21,34] and the space truss analogy as developed by Lampert [5,28] are in general agreement on the interaction behavior. According to Collins and Lampert [29,40] both theories indicate that under positive bending, yielding of bottom reinforcement occurs when equal volumes of longitudinal and transverse steel are used and when equal amounts of longitudinal steel are used in the top and bottom faces (i.e., $A_s' = A_s$). The theories may be approximated by the following, as shown in Fig. 19.8.1,

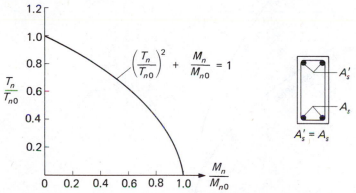

Figure 19.8.1 Bending-torsion interaction diagram for equal tension and compression longitudinal steel (positive moment yield).

$$\left(\frac{T_n}{T_{n0}}\right)^2 + \frac{M_n}{M_{n0}} = 1 \qquad (19.8.1)$$

where

T_n = nominal torsional strength in the presence of flexure
T_{n0} = nominal torsional strength when the member is subjected to torsion alone
M_n = nominal flexural strength in the presence of torsion
M_{n0} = nominal flexural strength when the member is subjected to flexure alone

---

*The discussion in this section is for the combined action of torsional moment and *positive* bending moment, which if acting alone would cause tension in the bottom steel A_s and compression in the top steel A_s'. In the event that the bending moment is negative, the reader should then interpret A_s' as being the bottom steel and A_s the top steel.

Unequal Top and Bottom Steel, $A_s' < A_s$. When torsion alone acts, the top and bottom longitudinal steel must resist *equal tensile forces*. When positive bending moment acts with torsion, the bottom steel will yield before the top steel if equal amounts of top and bottom steel are used. However, if there is less top steel than bottom steel, the application of a small amount of positive bending moment will result in an increase in the interactive torsional strength because the compressive force due to flexure tends to counteract the tensile force due to torsion. This mutually beneficial effect reaches its peak (Fig. 19.8.2) when both the top and bottom steel yield simultaneously.

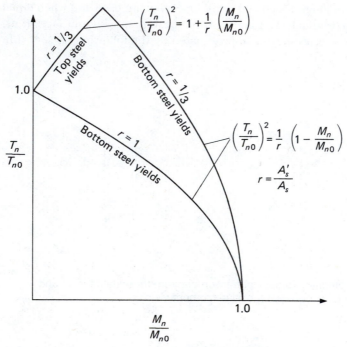

Figure 19.8.2 Bending-torsion interaction relationships (from Ref. 40).

In general, the strength interaction relationship may be stated [40], for the case of first yield in bottom reinforcement,

$$\left(\frac{T_n}{T_{n0}}\right)^2 = \frac{1}{r}\left(1 - \frac{M_n}{M_{n0}}\right) \tag{19.8.2}$$

where $r = (A_s' f_y')/(A_s f_y)$ and f_y' and f_y are the yield stresses of the top and bottom steel, respectively. For the case when the top steel yields first, the expression is

$$\left(\frac{T_n}{T_{n0}}\right)^2 = 1 + \frac{1}{r}\left(\frac{M_n}{M_{n0}}\right) \tag{19.8.3}$$

Equations (19.8.2) and (19.8.3) are shown graphically in Fig. 19.8.2.

Design Implications. In order to examine the implications of Eqs. (19.8.2) and (19.8.3), first consider the bottom steel yield failure mode. Expanding Eq. (19.8.2),

$$rT_n^2 M_{n0} = T_{n0}^2 (M_{n0} - M_n) \qquad (19.8.4)$$

The strength of a section under flexure alone is

$$M_{n0} = A_s f_y \left(d - \frac{a}{2} \right) \qquad (19.8.5)$$

and using Eq. (19.7.7) for the theoretical strength of a member subjected to torsion alone,

$$T_{n0} = 2x_2 y_2 \sqrt{\frac{A_l f_y A_t f_y}{2s(x_2 + y_2)}} \qquad (19.8.6)$$

where x_2 is defined in Fig. 19.6.3. Substitution of Eqs. (19.8.5) and (19.8.6) into Eq. (19.8.4) gives

$$rT_n^2 A_s f_y \left(d - \frac{a}{2} \right) = 4x_2^2 y_2^2 \left[\frac{A_l f_y A_t f_y}{2s(x_2 + y_2)} \right] \left[A_s f_y \left(d - \frac{a}{2} \right) - M_n \right] \qquad (19.8.7)$$

Let $A_l/2$ equal A_s' or A_s, whichever is smaller. Then if hoop steel is designed for torsion alone, with equal volumes of longitudinal and transverse steel, substituting Eq. (19.7.8) for A_l in Eq. (19.8.6) would give

$$T_n = 2 \left(\frac{A_t f_y}{s} \right) x_2 y_2$$

which gives the amount of hoop steel to be used as

$$\frac{A_t f_y}{s} = \frac{T_n}{2x_2 y_2} \qquad (19.8.8)$$

Substitution of Eq. (19.8.8) into Eq. (19.8.7) with $r = A_s'/A_s$ and $A_l/2 = A_s'$ gives

$$\frac{A_s'}{A_s} (T_n^2) A_s f_y \left(d - \frac{a}{2} \right) = 4x_2^2 y_2^2 \left[\frac{A_s' f_y}{(x_2 + y_2)} \right] \left(\frac{T_n}{2x_2 y_2} \right) \left[A_s f_y \left(d - \frac{a}{2} \right) - M_n \right]$$

$$T_n \left(d - \frac{a}{2} \right) = \frac{2x_2 y_2}{x_2 + y_2} \left[A_s f_y \left(d - \frac{a}{2} \right) - M_n \right] \qquad (19.8.9)$$

which upon solving for A_s gives

$$A_s = \frac{T_n(x_2 + y_2)}{2x_2 y_2 f_y} + \frac{M_n}{f_y(d - a/2)} \qquad (19.8.10)$$

Equation (19.8.10) shows that the bottom steel required is the sum of the requirements for flexure and torsion computed separately.

For the top steel yield failure mode, expand Eq. (19.8.3) and let $r = A_s'/A_s$,

$$A_s'T_n^2 = \left[A_s' + A_s\left(\frac{M_n}{M_{n0}}\right)\right]T_{n0}^2 \qquad (19.8.11)$$

Using Eqs. (19.8.5) and (19.8.6) for M_{n0} and T_{n0}, respectively, and letting $A_l/2 = A_s'$,

$$A_s'T_n^2 = \left[A_s' + A_s\frac{M_n}{A_s f_y(d - a/2)}\right]\left[4x_2^2 y_2^2\left(\frac{A_s' f_y}{x_2 + y_2}\right)\left(\frac{A_t f_y}{2}\right)\right] \qquad (19.8.12)$$

In the same manner as for the bottom steel yield case, assume the hoop steel is designed for torsion alone. Then substituting Eq. (19.8.8) into Eq. (19.8.12) gives

$$A_s' = \frac{T_n(x_2 + y_2)}{2x_2 y_2 f_y} - \frac{M_n}{f_y(d - a/2)} \qquad (19.8.13)$$

Thus the top steel requirement is the difference between the flexure and torsion requirements. Usually only minimal longitudinal steel A_s' is needed for combined flexure and torsion.

As will be discussed again in Sec. 19.12, the ACI Code does not explicitly consider combined bending and torsion. However, the flexural requirement is added to the requirement of longitudinal reinforcement for torsion, so that actually Eq. (19.8.10) is used.

19.9 STRENGTH OF SECTIONS IN COMBINED SHEAR AND TORSION

Rectangular and L-shaped sections have been studied under combined shear and torsion by a few investigators [41–43]. However, since shear usually accompanies flexure, it is the combination of flexure, shear, and torsion that has received the greatest attention [44–63]. Generally, though, flexural shear and torsional shear are of significance in those regions where bending moment is low. Thus for design purposes it is most necessary to know the strength interaction relationship between shear and torsion.

Test data have provided a wide range of points on the interaction relationship using torsion and shear coordinates. Because of the unknowns involved, some investigators have proposed a linear interaction equation [24,25,40,49] for design purposes. However, a number of studies at the University of Texas [41,43,44,48,58] on rectangular, L-shaped, and T-shaped beams have indicated that a quarter-circle interaction relationship is acceptable for members *without* web reinforcement. For members *with* web reinforcement, the interaction is curved but flatter than the quarter circle. The quarter-circle expression is

$$\left(\frac{T_n}{T_{n0}}\right)^2 + \left(\frac{V_n}{V_{n0}}\right)^2 = 1 \qquad (19.9.1)$$

where T_n and V_n are the nominal strengths in torsion and shear, respectively, acting simultaneously; T_{n0} is the nominal strength under torsion alone;

and V_{n0} is the nominal strength under shear alone. As will be shown in Sec. 19.11, Eq. (19.9.1) is used for the ACI Code expression for the strength in combined shear and torsion on sections *without* web reinforcement. However, for web reinforcement the separate requirements for shear and torsion are added together, rather than following Eq. (19.9.1).

19.10 STRENGTH INTERACTION SURFACE FOR COMBINED BENDING, SHEAR, AND TORSION

The effect of the simultaneous application of bending, shear, and torsion may be most easily examined by means of an interaction surface. Such a concept has been used in Sec. 13.21 for biaxial bending of compression members. Various investigators have proposed interaction surfaces; notably those of Hsu [46], Mirza and McCutcheon [47], Victor and Ferguson [48], Collins, Walsh, and Hall [49], and Rangan and Hall [77,78]. Combined bending, shear, and torsion on L-beams has been studied by Rajagopalan [79] and on prestressed box girders by Taylor and Warwaruk [80]. Some of these surfaces are shown in Figs. 19.10.1 and 19.10.2. The work of Collins

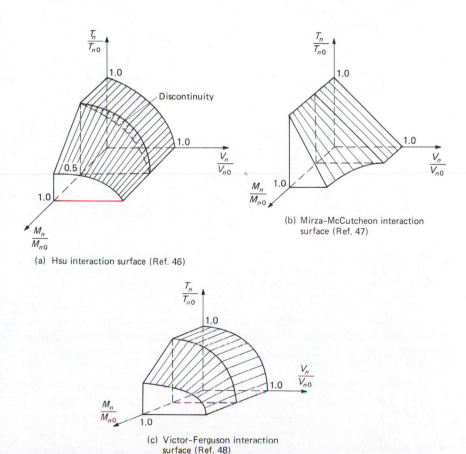

(a) Hsu interaction surface (Ref. 46)

(b) Mirza–McCutcheon interaction surface (Ref. 47)

(c) Victor–Ferguson interaction surface (Ref. 48)

Figure 19.10.1   Interaction surfaces for combined bending, shear, and torsion (for members *without* transverse reinforcement).

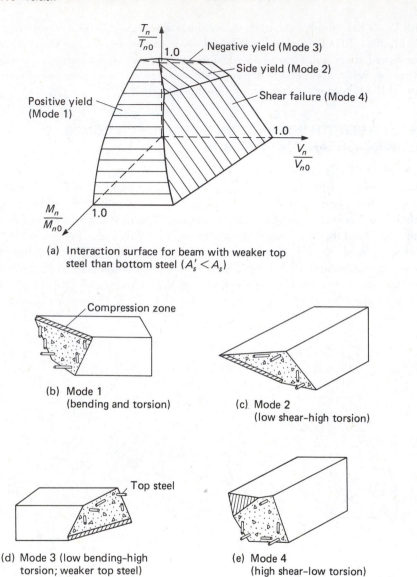

(a) Interaction surface for beam with weaker top
steel than bottom steel ($A_s' < A_s$)

(b) Mode 1
(bending and torsion)

(c) Mode 2
(low shear–high torsion)

(d) Mode 3 (low bending–high
torsion; weaker top steel)

(e) Mode 4
(high shear–low torsion)

Figure 19.10.2 Interaction surface and failure modes according to Collins et al. (from Ref. 49) (for members *with* transverse reinforcement).

et al. seems to provide the most complete rational approach by correlating the regions of the failure surface with modes of failure as shown in Fig. 19.10.2.

Since most members that are subjected to bending, shear, and torsion will have less longitudinal steel in the compression zone due to flexure alone (top steel in the positive moment zone) than in the tension zone (bottom steel in the positive moment zone), only that case is given in Fig. 19.10.2(a). When equal amounts of top and bottom longitudinal steel are used, the Mode 3 ("negative" yield or yield in top steel of positive moment zone) cannot occur; in such a case the surfaces for Modes 1 and 2 should extend upward until they intersect the T_n/T_{n0} axis.

The following conclusions may be drawn regarding the interaction of bending, shear, and torsion.

1. The interaction between torsion and shear may be represented for most situations (with $A_s' < A_s$) by a quarter circle, and is relatively unaffected by the simultaneous application of bending moment of a magnitude equal to one-third to one-half of the nominal bending strength when shear and torsion are absent.
2. When equal amounts of top and bottom longitudinal steel are used, the quarter-circle shear-torsion interaction still seems acceptable, but there is a reduction in strength when bending moment is also applied.
3. Based on most test results, the straightline shear-torsion interaction, while simple to use, appears overly conservative.

As will be noted in the next section, the use of a quarter-circle shear-torsion interaction, although slightly on the nonconservative side based on a number of experiments, is acceptable by ACI procedures because the ACI shear strength (in the absence of torsion) is taken as that causing an inclined crack. The actual shear strength in many cases is greater; particularly when the shear span to depth ratio is less than about 2.5.

19.11 TORSIONAL STRENGTH OF CONCRETE AND HOOP REINFORCEMENT—ACI CODE

The ACI procedure of design for torsion is based largely on the work of Hsu and the recommendations of ACI Committee 438 [64–68]. The general philosophy for designing to accommodate torsional shear is the same as that for flexural shear—that is, to consider that a part of the torsional moment may be carried by the concrete without web reinforcement and the remainder must be carried by closed hoop web reinforcement.

Thus

$$T_n = T_c + T_s \qquad (19.11.1)$$

in which

T_n = total nominal torsional strength
T_c = nominal torsional strength provided by concrete
T_s = nominal torsional strength provided by hoop steel

Whenever the required nominal torsional strength $T_n = T_u/\phi$ exceeds $0.5\sqrt{f_c'}(\Sigma x^2 y)$ (as shown in Sec. 19.13 this corresponds to a nominal ultimate torsional stress v_{tn} of $1.5\sqrt{f_c'}$ psi), the torsion effects must be included with shear and flexure. The strength reduction factor ϕ used for torsion is 0.85 (ACI-9.3.2.3).

Strength Provided by Concrete Subject to Torsion Alone. Essentially the expression developed by Hsu [18], Eq. (19.6.19), has been taken by the

ACI Code as the basis for the nominal torsional strength of reinforced concrete rectangular sections; or

$$T_n = T_c + T_s$$

$$= \frac{2.4}{\sqrt{x}} x^2 y \sqrt{f'_c} + \alpha_t \left(\frac{x_1 y_1 A_t f_y}{s} \right) \qquad \textbf{[19.6.19]}$$

The ACI Code (ACI-11.6.6.1) has taken $2.4/\sqrt{x}$ as 0.8, corresponding to an x of 9 in. Thus the torsional strength attributable to the concrete *when torsion alone is acting* is

$$T_c = 0.8\sqrt{f'_c}\, x^2 y \qquad (19.11.2)$$

for a rectangular section, and

$$T_c = 0.8\sqrt{f'_c}(\Sigma x^2 y) \qquad (19.11.3)$$

when the cross section consists of several rectangular segments, each having a short dimension x and a long dimension y.

Computation of $\Sigma x^2 y$. In order to compute the torsional property $\Sigma x^2 y$, the cross section is divided into rectangles in such a way that the largest rectangle obtainable is used as one of the components, and the closed hoop reinforcement should then follow its boundaries.

When a monolithic slab is acting as shown in Fig. 19.11.1(a), an effec-

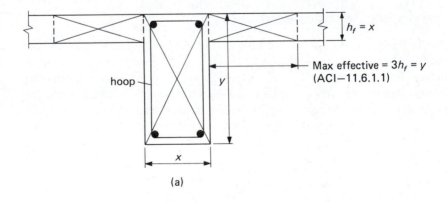

(a)

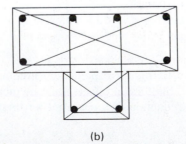

(b)

Figure 19.11.1 Computation of $\Sigma x^2 y$ for use in calculating nominal torsional stress.

tive overhanging width not exceeding 3 times the flange thickness (ACI-11.6.1.1) is to be used [64,69]. In this case the largest rectangle extends from top of flange to bottom of web. However, when the flange is thicker, the main rectangle should extend throughout the entire flange width, as shown in Fig. 19.11.1(b). The smaller rectangles of the configuration should also contain transverse reinforcement, frequently accomplished by using hairpin-type (U-shaped) stirrups anchored within the region of the closed hoop.

The hollow box section of Fig. 19.11.2 may be treated as a solid rectangle (ACI-11.6.1.2) when its wall thickness h is at least $x/4$. When the wall thickness is between $x/10$ and $x/4$, it may still be treated as a solid rectangle, except $\Sigma x^2 y$ must be multiplied by $4h/x$. When the wall thickness is less than $x/10$, the wall stiffness must be considered. The design of thin-walled box girders having h less than $x/10$ should generally be avoided because of the high flexibility and susceptibility to buckling of such walls, and the fact that few experimental results are available to guide the designer.

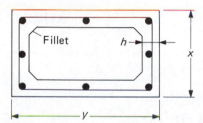

Figure 19.11.2   Hollow box sections.

Critical Section for Torsion. According to ACI-11.6.4 the critical section is to be taken at the effective depth d from the face of support, the same as is done for flexural shear alone. The same amount of transverse steel required at the critical section is to be used between the face of support and the critical section.

Hoop Reinforcement Required for Strength. The second term of Eq. (19.6.19) gives (ACI-11.6.9.1) ACI Formula (11-23),

$$T_s = \alpha_t \left(\frac{x_1 y_1 A_t f_y}{s} \right) \tag{19.11.4}$$

where $\alpha_t = 0.66 + 0.33(y_1/x_1)$ but not more than 1.50. For the design of transverse reinforcement, solving Eq. (19.11.4) for A_t/s, and using $T_s = $ required $T_n - T_c$,

$$\frac{A_t}{s} = \frac{\text{required } T_n - T_c}{\alpha_t x_1 y_1 f_y} \tag{19.11.5}$$

Note that A_t is the area of *one* leg of a closed hoop. The design yield stress shall not exceed 60,000 psi (ACI-11.6.7.4).

Longitudinal Reinforcement. The development by Hsu [18] assumed that the same volume of longitudinal steel would be provided in the distance s as occurs in one hoop of transverse reinforcement. In mathematical terms

$$A_l s = 2A_t(x_1 + y_1) \qquad (19.11.6)$$

or

$$A_l = 2\left(\frac{A_t}{s}\right)(x_1 + y_1) \qquad (19.11.7)$$

which is ACI Formula (11-24).

19.12 PROVISIONS FOR COMBINED TORSION WITH SHEAR OR BENDING—ACI CODE

Combined Shear and Torsion. As discussed in Sec. 19.9, the interaction between shear and torsion for members *without* web reinforcement is assumed to be a quarter circle, represented by Eq. (19.9.1),

$$\left(\frac{T_n}{T_{n0}}\right)^2 + \left(\frac{V_n}{V_{n0}}\right)^2 = 1 \qquad [19.9.1]$$

The shear strength V_n in the presence of torsion T_n is obtained by solving Eq. (19.9.1), after first dividing through by $(V_n/V_{n0})^2$,

$$V_n = \frac{V_{n0}}{\sqrt{1 + \left(\frac{V_{n0}}{T_{n0}}\right)^2\left(\frac{T_n}{V_n}\right)^2}} \qquad (19.12.1)$$

For beams *with* web reinforcement, it is assumed that the quarter-circle interaction relationship is also applicable to the concrete contribution to shear and torsional strengths. Consequently V_n can be replaced by V_c; also $V_{n0} = 2\sqrt{f_c'}b_w d$ the basic shear strength (ACI-11.3.1.1) of concrete according to the simplified method; and $T_{n0} = 0.8\sqrt{f_c'}(\Sigma x^2 y)$ the basic torsion strength from Eq. (19.11.3). Thus Eq. (19.12.1) becomes

$$V_c = \frac{2\sqrt{f_c'}b_w d}{\sqrt{1 + \left(\frac{V_{n0}}{T_{n0}}\frac{T_n}{V_n}\right)^2}} \qquad (19.12.2)$$

Note that

$$\frac{V_{n0}}{T_{n0}} = \frac{2\sqrt{f_c'}b_w d}{0.8\sqrt{f_c'}(\Sigma x^2 y)} = 2.5C_t \qquad (19.12.3)$$

where

$$C_t = \frac{b_w d}{\Sigma x^2 y}$$

Substitution of Eq. (19.12.3) into Eq. (19.12.2) by using $T_n/V_n = T_u/V_u$ gives

$$V_c = \frac{2\sqrt{f_c'}b_w d}{\sqrt{1 + \left(2.5C_t\dfrac{T_u}{V_u}\right)^2}} \tag{19.12.4}$$

which is ACI Formula (11-5).

The torsion strength T_n in the presence of shear V_n is obtained by solving Eq. (19.9.1), after first dividing through by $(T_n/T_{n0})^2$,

$$T_n = \frac{T_{n0}}{\sqrt{1 + \left(\dfrac{T_{n0}}{V_{n0}}\right)^2\left(\dfrac{V_n}{T_n}\right)^2}} \tag{19.12.5}$$

For beams *with* web reinforcement, T_n may be replaced by T_c; also $T_{n0} = 0.8\sqrt{f_c'}(\Sigma x^2 y)$, $V_{n0} = 2\sqrt{f_c'}b_w d$ and $V_n/T_n = V_u/T_u$; thus

$$T_c = \frac{0.8\sqrt{f_c'}(\Sigma x^2 y)}{\sqrt{1 + \left(\dfrac{0.4}{C_t}\dfrac{V_u}{T_u}\right)^2}} \tag{19.12.6}$$

which is ACI Formula (11-22). Since Eqs. (19.12.4) and (19.12.6) represent the strengths attributable to the concrete in members *with* web reinforcement, they underestimate the actual strengths of members *without* web reinforcement. As discussed in Sec. 19.14, the torsional strength of a concrete member without web reinforcement is considerably greater than the torsional strength T_c representing the concrete contribution to the strength of a member containing web reinforcement.

For the upper limits to strength carried by hoop reinforcement,

$$\max V_s = 8\sqrt{f_c'}b_w d$$

as given by ACI-11.5.6.8. Also

$$\max T_s = 4T_c$$

a limit recommended by ACI Committee 438 [66] and given by ACI-11.6.9.4.

Combined Bending and Torsion. The ACI Code does not explicitly consider this combination of loadings. For beams *without* web reinforcement bending was neglected since it can be taken care of by the conservatism inherent in Eqs. (19.12.4) and (19.12.6). For beams *with* web reinforcement, however, the procedure is to design for bending separately and add the longitudinal steel required for flexure to that required for torsion or combined shear and torsion (ACI-11.6.1). From the discussion presented in Sec. 19.8, it may be apparent that this is an acceptable procedure.

19.13 NOMINAL TORSIONAL STRESSES—ACI CODE

Even though strength is the basis for design, it may be convenient to compute nominal torsional stress in the same manner as for the elastic homogeneous section. Thus referring to Eqs. (19.2.1), (19.2.2), and (19.2.3), and noting that T_u corresponds to the factored torsional moment that may be carried when the torsional strength is achieved, in general one may write

$$v_{tn} = \frac{T_u x_m}{\phi C} \tag{19.13.1}$$

where C = torsion constant = $\frac{1}{3}\Sigma x^3 y$, and x_m is the smaller dimension of the thickest component rectangle.

For rectangular sections, b_w is usually less than h making $b_w = x$, and for T- and L-sections any web is usually thicker than the slab; therefore $x_m = x$ for the rectangle contributing the most significant part of $\frac{1}{3}\Sigma x^3 y$. For practical purposes the approximate expression may be used,

$$\frac{C}{x_m} = \frac{1}{3}\Sigma x^2 y \tag{19.13.2}$$

Thus the nominal torsional stress may be computed by

$$v_{tn} = \frac{3T_u}{\phi \Sigma x^2 y} \tag{19.13.3}$$

Note that $\phi = 0.85$, the strength reduction factor for torsion (ACI-9.3.2.3).

Torsion effects must be included whenever the nominal stress v_{tn} exceeds $1.5\sqrt{f_c'}$ psi, which corresponds to $T_u = \phi 0.5\sqrt{f_c'}\Sigma x^2 y$ of ACI-11.6.1.

Using nominal unit stresses, the torsional stress v_{tc} attributable to the concrete is obtained by substituting Eq. (19.11.3) for T_n in Eq. (19.13.3), giving

$$v_{tc} = 2.4\sqrt{f_c'} \tag{19.13.4}$$

Similarly, the torsional stress v_{ts} attributable to the hoop reinforcement is obtained by substituting Eq. (19.11.4) for T_u/ϕ in Eq. (19.13.3), giving

$$v_{ts} = \frac{T_u}{\phi\frac{1}{3}\Sigma x^2 y} = \frac{3\alpha_t x_1 y_1}{\Sigma x^2 y}\left(\frac{A_t f_y}{s}\right) \tag{19.13.5}$$

For design, the unit stress counterpart to the expression for A_t/s in Eq. (19.11.5) is

$$\frac{A_t}{s} = \frac{(v_{tn} - v_{tc})\Sigma x^2 y}{3\alpha_t x_1 y_1 f_y} \tag{19.13.6}$$

In a similar fashion, the combined stress equations Eqs. (19.12.4) and (19.12.6) may be expressed in terms of nominal unit stresses, $v_n = V_u/(\phi b_w d)$ for shear and $v_{tn} = T_u/(\phi\frac{1}{3}\Sigma x^2 y)$ for torsion. Thus

$$v_c = \frac{2\sqrt{f_c'}}{\sqrt{1 + \left(\dfrac{v_{tn}}{1.2v_n}\right)^2}} \tag{19.13.7}$$

corresponds to ACI Formula (11-5). Also,

$$v_{tc} = \frac{2.4 \sqrt{f_c'}}{\sqrt{1 + \left(1.2 \dfrac{v_n}{v_{tn}}\right)^2}} \tag{19.13.8}$$

corresponds to ACI Formula (11-22).

19.14 MINIMUM REQUIREMENTS FOR TORSIONAL REINFORCEMENT—ACI CODE

The minimum requirements for the transverse reinforcement A_t and longitudinal reinforcement A_l are to ensure that there is ductile behavior prior to failure. Hsu [23] and Collins [65] have found that the strength attributable to concrete when closed hoops *are* present is only about 40% of the torsional capacity of a plain concrete member. Thus, for ductile behavior, a beam *with* web reinforcement should have at least the capacity of one without web reinforcement. In other words it is desired that

$$T_n \text{ (with hoops)} \geq T_n \text{ (without hoops)} \tag{19.14.1}$$

or for a rectangular section,

$$\text{Eq. (19.6.19)} \geq T_n \text{ (without hoops)}$$

$$\underbrace{2.4\sqrt{f_c'}\left(\frac{x^2 y}{\sqrt{x}}\right)}_{\text{term } A} + \underbrace{\alpha_t\left(\frac{A_t}{s}\right)}_{\text{term } B} \underbrace{(x_1 y_1 f_y)}_{\text{term } C} \geq T_n \text{ (without hoops)} \tag{19.14.2}$$

Thus accepting the 40% value for term A, term B must represent 60% of the capacity of a member containing only a small amount of torsional reinforcement. The condition for the minimum amount of torsional reinforcement then becomes

$$\text{term } B \geq 1.5(\text{term } A)$$

or, letting $\sqrt{x}$ in term A equal to 3,

$$\alpha_t\left(\frac{A_t}{s}\right)x_1 y_1 f_y \geq 1.5\left(\frac{x^2 y}{3}\right)2.4\sqrt{f_c'} \tag{19.14.3}$$

Solving for A_t gives, for torsion acting alone,

$$A_t \geq \frac{1.2}{\alpha_t}\left(\frac{x^2 y s}{x_1 y_1}\right)\frac{\sqrt{f_c'}}{f_y} \tag{19.14.4}$$

This requirement could reasonably be reduced when transverse reinforcement (stirrups or hoops) is also used to provide for flexural shear. Transverse reinforcement for either purpose improves ductility. Thus when flexural shear requiring transverse reinforcement is present, the minimum transverse reinforcement for both torsion and shear ($A_t + A_v/2$) is

$$A_t + \frac{A_v}{2} \geq \frac{1.2}{\alpha_t}\left(\frac{x^2 y s}{x_1 y_1}\right)\frac{\sqrt{f_c'}}{f_y}\left(\frac{v_{tn}}{v_{tn} + v_n}\right) \tag{19.14.5}$$

where the term $[v_{tn}/(v_{tn} + v_n)]$ represents the assumption that the total shear strength of plain concrete is divided into resistance for torsional shear and flexural shear in the same ratio as the respective values of nominal stresses. Note that as defined in Chap. 5 on shear, A_v is the total area for two legs of a vertical stirrup provided to resist flexural shear.

To simplify Eq. (19.14.5), assume $\alpha_t = 1.2$, $x/x_1 = y/y_1 = 1.2$, and $f'_c = 5000$ psi,

$$A_t + \frac{A_v}{2} \geq \frac{1.2}{1.2}(1.2)(1.2)xs \frac{\sqrt{5000}}{f_y}\left(\frac{v_{tn}}{v_{tn} + v_n}\right) \geq 102 \frac{xs}{f_y}\left(\frac{v_{tn}}{v_{tn} + v_n}\right)$$

Rounding the constant off to 100 and multiplying through by 2 gives

$$2A_t + A_v \geq 200 \frac{xs}{f_y}\left(\frac{v_{tn}}{v_{tn} + v_n}\right) \tag{19.14.6}$$

For equal volumes of longitudinal and transverse steel,

$$A_l = 2A_t\left(\frac{x_1 + y_1}{s}\right) \tag{19.14.7}$$

Substitution of Eq. (19.14.6) into Eq. (19.14.7) gives

$$A_l \geq \left[200 \frac{xs}{f_y}\left(\frac{v_{tn}}{v_{tn} + v_n}\right) - A_v\right]\left(\frac{x_1 + y_1}{s}\right) \tag{19.14.8}$$

ACI Committee 438 [66] recommended that the minimum transverse reinforcement for combined shear and torsion should not be required to exceed that for shear alone; that is,

$$2A_t + A_v \geq \frac{50b_w s}{f_y} \tag{19.14.9}$$

as now stated in ACI-11.5.5.5.

Since Eq. (19.14.9) will usually give a lower requirement than Eq. (19.14.6), the recommendation of Committee 438 [66] was to increase the required amount of longitudinal reinforcement. There is no complete agreement on whether this is an acceptable procedure [65]. Thus the first term of Eq. (19.14.8) was doubled and A_v replaced by $2A_t$,

$$A_l \geq \left[400 \frac{xs}{f_y}\left(\frac{v_{tn}}{v_{tn} + v_n}\right) - 2A_t\right]\left(\frac{x_1 + y_1}{s}\right) \tag{19.14.10}$$

Expressing v_{tn} and v_n in terms of forces T_u and V_u gives

$$A_l \geq \left[400 \frac{xs}{f_y}\left(\frac{T_u}{T_u + V_u/3C_t}\right) - 2A_t\right]\left(\frac{x_1 + y_1}{s}\right) \tag{19.14.11}$$

which is ACI Formula (11-25) with $C_t = b_w d/(\Sigma x^2 y)$. In the use of Eqs. (19.14.10) or (19.14.11) the value for $2A_t$ is not to be taken less than $50b_w s/f_y$.

Spacing Limitations. The spacing of closed hoops may not exceed $(x_1 + y_1)/4$ nor 12 in., whichever is smaller (ACI-11.6.8.1).

Longitudinal bars must be at least #3 in size and be distributed around the perimeter not farther than 12 in. apart and with one bar in each corner (ACI-11.6.8.2).

These limits (ACI Commentary-11.6.8.1) are intended to "insure the development of the ultimate torsional strength of the beam, prevent excessive loss of torsional stiffness after cracking, and control crack widths."

Termination of Torsion Reinforcement. Closed hoops may be terminated at a location equal to the sum of the effective depth d plus the width b_t of that part of the cross section containing the closed stirrups resisting torsion, beyond the point where the required nominal torsional shear strength T_n (i.e., T_u/ϕ) is $0.5\sqrt{f_c'}(\Sigma x^2 y)$—that is, a nominal unit stress v_{tn} equal to $1.5\sqrt{f_c'}$ psi (ACI-11.6.7.6).

19.15 EXAMPLES

Several examples are presented to illustrate use of the ACI Code procedures. Table 19.15.1 summarizes most of the ACI provisions on torsion; it is prepared to help the reader in reviewing the computations in the examples.

EXAMPLE 19.15.1 A reinforced concrete spandrel beam has overall dimensions of 10×18 in. and is joined integrally with a 6-in. slab (based on Example of Ref. 46) as shown in Fig. 19.15.1(a). The section shown is that at the critical location a distance d from the face of support. At this section the factored loads are negative bending moment $M_u = 75$ ft-kips, shear force $V_u = 9$ kips, and torsional moment $T_u = 8$ ft-kips. Assume that the torsional stiffness was estimated and used in a structural analysis to obtain these design loadings. Check the adequacy of this section, and select the transverse hoop steel required, if any, according to the ACI Code using $f_c' = 4000$ psi and $f_y = 40,000$ psi.

Solution: (a) Flexural strength. The required coefficient of resistance R_n is

$$\text{required } R_n = \frac{\text{required } M_n}{bd^2} = \frac{M_u}{\phi bd^2}$$

$$= \frac{75(12,000)}{0.90(10)(16)^2} = 390 \text{ psi}$$

From Fig. 3.8.1, the required reinforcement ratio ρ is

$$\text{required } \rho = 0.011$$
$$\text{required } A_s = 0.011(10)(16) = 1.76 \text{ sq in.}$$
$$\text{provided } A_s = 3(0.79) = 2.37 \text{ sq in.} > 1.76 \qquad \text{OK}$$

Available steel area for longitudinal torsion reinforcement,

$$\text{available } A_s = 2.37 - 1.76 = 0.61 \text{ sq in.}$$
$$\text{available } A_s' = 2(0.60) = 1.20 \text{ sq in.}$$

Table 19.15.1 ACI CODE PROVISIONS ON TORSION

---

1. $T_u = 0$ if computed $T_u \leq \phi(0.5\sqrt{f_c'}\Sigma x^2 y)$ $\qquad v_{tn} \leq 1.5\sqrt{f_c'}$ $\qquad$ 11.6.1

$\quad T_u = \phi(4\sqrt{f_c'}\Sigma\frac{1}{3}x^2 y)$, with redistribution $\qquad v_{tn} \leq 4\sqrt{f_c'}$ $\qquad$ 11.6.3

---

2. $V_c = \dfrac{2\sqrt{f_c'}b_w d}{\sqrt{1 + \left(2.5C_t\dfrac{T_u}{V_u}\right)^2}}$ $\qquad v_c = \dfrac{2\sqrt{f_c'}}{\sqrt{1 + \left(\dfrac{v_{tn}}{1.2v_n}\right)^2}}$ $\qquad$ 11.3.1.4; Formula (11-5)

$\quad T_u > \phi(0.5\sqrt{f_c'}\Sigma x^2 y)$ $\qquad v_{tn} = \dfrac{T_u}{\phi\Sigma\frac{1}{3}x^2 y}$

$\quad C_t = \dfrac{b_w d}{\Sigma x^2 y}$ $\qquad v_n = \dfrac{V_u}{\phi b_w d}$

$\quad N_u$ = axial tension (negative):

$\quad V_c = 0$; or multiply above V_c by $\left(1 + \dfrac{N_u}{500A_g}\right)$ $\qquad$ 11.6.6.2

---

3. $\phi T_n \geq T_u$ $\qquad v_{tn} = \dfrac{T_u}{\phi\Sigma\frac{1}{3}x^2 y}$ $\qquad$ 11.6.5; Formula (11-20)

$\quad T_n = T_c + T_s$ $\qquad v_{tn} = v_{tc} + v_{ts}$ $\qquad$ 11.6.5; Formula (11-21)

---

4. $T_c = \dfrac{0.8\sqrt{f_c'}\Sigma x^2 y}{\sqrt{1 + \left(\dfrac{0.4V_u}{C_t T_u}\right)^2}}$ $\qquad v_{tc} = \dfrac{2.4\sqrt{f_c'}}{\sqrt{1 + \left(\dfrac{1.2v_n}{v_{tn}}\right)^2}}$ $\qquad$ 11.6.6.1; Formula (11-22)

$\quad T_u > \phi(0.5\sqrt{f_c'}\Sigma x^2 y)$ $\qquad v_{tn} = \dfrac{T_u}{\phi\Sigma\frac{1}{3}x^2 y}$

$\quad C_t = \dfrac{b_w d}{\Sigma x^2 y}$ $\qquad v_n = \dfrac{V_u}{\phi b_w d}$

$\quad N_u$ = axial tension (negative):

$\quad T_c = 0$; or multiply above T_c by $\left(1 + \dfrac{N_u}{500A_g}\right)$ $\qquad$ 11.6.6.2

---

5. $T_s = \dfrac{A_t \alpha_t x_1 y_1 f_y}{s}$ $\qquad v_{ts} = \dfrac{A_t \alpha_t x_1 y_1 f_y}{s\Sigma\frac{1}{3}x^2 y}$ $\qquad$ 11.6.7.4

$\quad f_y \leq 60{,}000$ psi $\qquad \alpha_t = (0.66 + 0.33y_1/x_1) < 1.50$ $\qquad$ 11.6.9.1; Formula (11-23)

$\quad$ max $T_s = 4T_c$ $\qquad$ max $v_{ts} = 4v_{tc}$ $\qquad$ 11.6.9.4

$\quad A_l = 2A_t\left(\dfrac{x_1 + y_1}{s}\right)$ $\qquad$ 11.6.9.3; Formula (11-24)

$\quad A_l = \left[\dfrac{400xs}{f_y}\left(\dfrac{T_u}{T_u + \dfrac{V_u}{3C_t}} \text{ or } \dfrac{v_{tn}}{v_{tn} + v_n}\right) - 2A_t\right]\dfrac{x_1 + y_1}{s}$; $2A_t \geq \dfrac{50b_w s}{f_y}$ $\qquad$ 11.6.9.3; Formula (11-25)

Table 19.15.1   *(Continued)*

6. min $A_v = \dfrac{50 b_w s}{f_y}$

	11.5.5.3; Formula (11-14)

for $T_u \leq \phi(0.5\sqrt{f'_c}\Sigma x^2 y)$ and $V_u > \frac{1}{2}\phi V_c$

min A_t to satisfy $A_v + 2A_t = \dfrac{50 b_w s}{f_y}$

	11.5.5.5; Formula (11-16)

for $T_u > \phi(0.5\sqrt{f'_c}\Sigma x^2 y)$ and $V_u > \frac{1}{2}\phi V_c$

---

7. min s (closed stirrups) $\leq \dfrac{x_1 + y_1}{4} \leq 12$ in.   11.6.8.1

torsion reinforcement to terminate
at $(d + b_t)$ beyond theoretical point   11.6.7.6

min s $\left(\begin{array}{l}\text{longitudinal bars, \#3 or}\\ \text{larger, one in each corner}\end{array}\right) \leq 12$ in.   11.6.8.2

(b) Torsional moment strength. The maximum factored torsional moment T_u that may be neglected (ACI-11.6.1) is

$$\text{limit } T_u = \phi[0.5\sqrt{f'_c}(\Sigma x^2 y)]$$

$$\Sigma x^2 y = (10)^2(18) + (6)^2(18) = 2450 \text{ in.}^3$$

Note that the maximum effective width of slab is 3 times its thickness (ACI-11.6.1.1). Thus,

$$\text{limit } T_u = 0.85 \, [0.5\sqrt{4000} \, (2450)] \, \tfrac{1}{12,000} = 5.5 \text{ ft-kips}$$

Since T_u of 8 ft-kips exceeds limit T_u, torsion must be included in design.

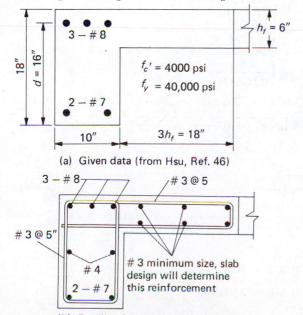

(a) Given data (from Hsu, Ref. 46)

(b) Details of torsion reinforcement

Figure 19.15.1   Spandrel beam of Example 19.15.1.

(c) Shear strength. The strength attributable to the concrete is given by Eq. (19.12.4) or ACI Formula (11-5),

$$V_c = \frac{2\sqrt{f'_c}b_w d}{\sqrt{1 + \left(2.5C_t \dfrac{T_u}{V_u}\right)^2}}$$

$$C_t = \frac{b_w d}{\Sigma x^2 y} = \frac{10(16)}{2450} = 0.0653 \text{ per in.}$$

$$V_c = \frac{2\sqrt{4000}(10)(16)/1000}{\sqrt{1 + \left(2.5(0.0653)\dfrac{8(12)}{9}\right)^2}} = 10.1 \text{ kips}$$

$$\phi V_c = 0.85(10.1) = 8.6 \text{ kips}$$

Since $V_u > \phi V_c/2$ (ACI-11.5.5.1), the requirement for transverse reinforcement based on strength is

$$\frac{A_v}{s} = \frac{V_u - \phi V_c}{\phi d f_y} = \frac{9 - 8.6}{0.85(16)(40)} = 0.00074$$

This amount will be combined with the torsion requirement, and the total $A_v + 2A_t$ must be at least $50b_w s/f_y$ (ACI-11.5.5.5).

(d) Transverse torsional reinforcement. From Eq. (19.12.6) or ACI Formula (11-22),

$$T_c = \frac{0.8\sqrt{f'_c}(\Sigma x^2 y)}{\sqrt{1 + \left(\dfrac{0.4 V_u}{C_t T_u}\right)^2}}$$

$$T_c = \frac{0.8\sqrt{4000}(2450)/12,000}{\sqrt{1 + \left(\dfrac{0.4}{0.0653}\dfrac{9}{8(12)}\right)^2}} = 8.95 \text{ ft-kips}$$

$$\phi T_c = 0.85(8.95) = 7.6 \text{ ft-kips}$$

The requirement of closed hoops for strength is, from Eq. (19.11.5) or ACI Formula (11-23),

$$\frac{A_t}{s} = \frac{T_u - \phi T_c}{\phi \alpha_t x_1 y_1 f_y}$$

$$= \frac{(8.0 - 7.6)12}{0.85(1.39)(6.625)(14.625)40} = 0.00105$$

where

$$x_1 = 10 - 2(1.5) - 0.375 = 6.625 \text{ in.}$$

$$y_1 = 18 - 2(1.5) - 0.375 = 14.625 \text{ in.}$$

$$\alpha_t = 0.66 + 0.33\left(\frac{y_1}{x_1}\right) = 0.66 + 0.33\left(\frac{14.625}{6.625}\right) = 1.39 < 1.50 \quad \text{OK}$$

The total transverse reinforcement required for strength is

$$\frac{A_v}{s} + \frac{2A_t}{s} = 0.00074 + 0.00210 = 0.00284$$

$$\min\left(\frac{A_v}{s} + \frac{2A_t}{s}\right) = \frac{50b_w}{f_y} = \frac{50(10)}{40,000} = 0.0125 \text{ controls}$$

For #3 closed hoops,

$$\max s = \frac{2(0.11)}{0.0125} = 17.6 \text{ in.}$$

Since this exceeds the spacing limitations of ACI-11.6.8.1,

$$\max s = \frac{x_1 + y_1}{4} = \frac{6.625 + 14.625}{4} = 5.3 \text{ in.}$$

Since this is less than the 12-in. upper limit, the 5.3-in. limit controls.

Use #3 hoops at 5 in. spacing.

(e) Longitudinal torsional reinforcement. According to ACI Formula (11-24) with the strength requirement $2A_t/s = 0.00210$

$$A_l = \frac{2A_t}{s}(x_1 + y_1)$$

$$= 0.00210(6.625 + 14.625) = 0.045 \text{ sq in.}$$

or as a minimum,

$$A_l = \left[\frac{400xs}{f_y}\left(\frac{T_u}{T_u + \dfrac{V_u}{3C_t}}\right) - 2A_t\right]\left(\frac{x_1 + y_1}{s}\right)$$

$$= \left[\frac{400(10)}{40,000}\left(\frac{8(12)}{8(12) + 9/(0.196)}\right) - \frac{2A_t}{s}\right](6.625 + 14.625)$$

$$= \left(0.0676 - \frac{2A_t}{s}\right)(21.25)$$

Since the required $2A_t/s$ of 0.00210 for strength is less than $50b_w/f_y$ of 0.0125, the latter is to be used for $2A_t/s$ in the above equation

$$\text{required } A_l = (0.0676 - 0.0125)(21.25) = 1.17 \text{ sq in.}$$

This is to be distributed around the perimeter of the section at a spacing not to exceed 12 in. In this case, bars must be placed at mid-depth

$$\frac{A_l}{3} = \frac{1.17}{3} = 0.39 \text{ sq in.}$$

Use 2-#4 bars at mid-depth.

The longitudinal steel at the top and bottom in excess of that required for flexure is more than adequate for the torsional requirement. Also #3 transverse reinforcement should be placed in the effective flange portion of the slab and it should be anchored within the main rectangle resisting torsion. Though the ACI Code requires only standard hooks at the location where the hoop closes, Mitchell and Collins [70] have recommended the use of 105° hooks (the 105° is the amount of bend from the initial straight bar). Furthermore, when longitudinal steel is to carry torsion at the face of support, the steel should be embedded an amount L_d into the support [70]. The reinforced section is shown in Fig. 19.15.1(b).

EXAMPLE 19.15.2 For the continuous spandrel beam shown in Fig. 19.15.2, design the longitudinal and transverse reinforcement for the factored moment, factored flexural shear, and factored torsional moment, also given in Fig. 19.15.2. Assume that these design loads were obtained after estimating the torsional stiffness and performing a structural analysis. Use $f'_c = 4000$ psi and $f_y = 60,000$ psi.

Solution: (a) Flexural strength. The effective depth d is approximately 17.5 in. for one layer of reinforcement. At midspan, neglecting T-beam effect,

$$\text{required } R_n = \frac{M_u}{\phi bd^2} = \frac{83.3(12,000)}{0.90(12)(17.5)^2} = 302 \text{ psi}$$

From Fig. 3.8.1, the required $\rho = 0.0055$ which gives

$$\text{required } A_s = 0.0055(12)(17.5) = 1.15 \text{ sq in.}$$

$$\min A_s = \frac{200}{f_y}bd = \frac{200(12)(17.5)}{60,000} = 0.70 \text{ sq in.}$$

At the support,

$$\text{required } A_s = 1.15(61.7/83.3) = 0.85 \text{ sq in.}$$

The reader may note that less reinforcement than 0.70 sq in. may be used if the amount used is one-third more than required for strength (ACI-10.5.2). Thus the sloping straight line of Figs. 19.15.3(a) and (b) represents approximately $\frac{4}{3}$ of the required A_s in the region between the controlling minimum requirements of ACI-10.5.1 and ACI-12.11.1 (or ACI-12.12.3).

(b) Nominal stresses for shear and torsion at critical section. In this example, the unit stress format is illustrated, as presented in Sec. 19.13. At d from the face of support,

$$v_n = \frac{V_u}{\phi bd} = \frac{19,400}{0.85(12)(17.5)} = 109 \text{ psi}$$

$$v_{tn} = \frac{T_u}{\phi\frac{1}{3}\Sigma x^2 y} = \frac{47.7(12,000)}{0.85(1180)} = 571 \text{ psi}$$

where

$$\tfrac{1}{3}\Sigma x^2 y = \tfrac{1}{3}[(12)^2(20) + (6)^2(18)] = 1180 \text{ in.}^3$$

In computing $\Sigma x^2 y$, the effective width of flange may not exceed 3 times the flange thickness (ACI-11.6.1.1).

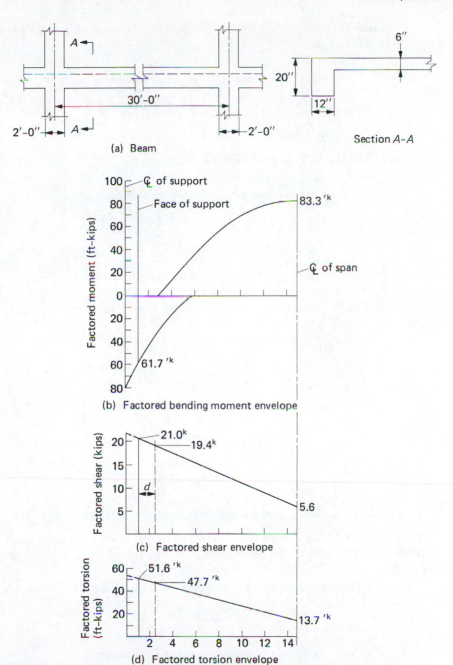

Figure 19.15.2 Spandrel beam including design loadings, for Example 19.15.2 (from Mattock, Ref. 68).

(c) Nominal stress carried by the concrete. For torsion, Eq. (19.13.8) or ACI-11.6.6.1 gives

$$v_{tc} = \frac{2.4\sqrt{f_c'}}{\sqrt{1 + \left(\dfrac{1.2v_n}{v_{tn}}\right)^2}} = \frac{2.4\sqrt{f_c'}}{\sqrt{1 + \left(\dfrac{1.2(109)}{571}\right)^2}}$$

$$= 2.34\sqrt{f_c'} = 148 \text{ psi}$$

For shear, Eq. (19.13.7) or ACI-11.3.1.4 gives

$$v_c = \frac{2\sqrt{f_c'}}{\sqrt{1 + \left(\dfrac{v_{tn}}{1.2v_n}\right)^2}} = \frac{2\sqrt{f_c'}}{\sqrt{1 + \left[\dfrac{571}{1.2(109)}\right]^2}}$$

$$= 0.44\sqrt{f_c'} = 28 \text{ psi}$$

For this example v_{tc} and v_c are constant along the member since the ratio of V_u to T_u is constant along the member.

(d) Maximum nominal stresses permitted. For torsion, ACI-11.6.9.4 gives

$$\max T_s = 4T_c$$

Since

$$T = T_c + T_s$$

$$\max T_n = T_c + 4T_c = 5T_c$$

Then,

$$\max v_{tn} = 5v_{tc}$$

For this problem,

$$\max v_{tn} = 5(2.34)\sqrt{f_c'} = 11.7\sqrt{f_c'}$$

$$= 740 \text{ psi} > 571 \text{ psi} \qquad \text{OK}$$

For shear, ACI-11.5.4.3 allows $(v_n - v_c) = 4\sqrt{f_c'}$ with maximum stirrup spacings up to $d/2$. In this case $v_n = 109$ psi $< 4\sqrt{f_c'}$, which is acceptable.

(e) Transverse reinforcement. For shear, ACI-11.5.6.2 gives

$$\frac{A_v}{s} = \frac{V_s}{f_y d} = \frac{\text{required } V_n - V_c}{f_y d}$$

or in terms of unit stresses and evaluated *at the critical section*

$$\frac{A_v}{s} = \frac{(v_n - v_c)b_w}{f_y} = \frac{(109 - 28)12}{60,000} = 0.0162$$

For torsion, Eq. (19.13.6) derived from ACI-11.6.9.1 gives *at the critical section*

$$\frac{A_t}{s} = \frac{(v_{tn} - v_{tc})\frac{1}{3}\Sigma x^2 y}{\alpha_t x_1 y_1 f_y} = \frac{(571 - 148)1180}{1.30(8.5)(16.5)60,000} = 0.0455$$

where

$$x_1 = 12 - 2(1.5) - 0.50 = 8.5 \text{ in.}$$
$$y_1 = 20 - 2(1.5) - 0.50 = 16.5 \text{ in.}$$
$$\alpha_t = 0.66 + 0.33 y_1/x_1 = 0.66 + 0.33(16.5/8.5) = 1.30 < 1.50 \quad \text{OK}$$

The minimum transverse reinforcement, Eq. (19.14.9) or ACI-11.5.5.5 gives

$$\frac{2A_t}{s} + \frac{A_v}{s} \geq \frac{50b_w}{f_y} = \frac{50(12)}{60,000} = 0.01$$

For strength, the total requirement is

$$\frac{2A_t}{s} + \frac{A_v}{s} = 2(0.0455) + 0.0162 = 0.107 > 0.01$$

Try #4 closed hoops, $A_v = 2(0.20) = 0.40$ sq in. The maximum spacing of hoops between the face of support and the critical section is

$$\max s = \frac{0.40}{0.107} = 3.74 \text{ in.}$$

The upper limit for spacing of closed hoops anywhere is (ACI-11.6.8.1),

$$\max s = \frac{x_1 + y_1}{4} = \frac{8.5 + 16.5}{4} = 6.25 \text{ in.}$$

and not in any case to exceed 12 in.

(f) Longitudinal reinforcement for torsion at the critical section. Using Eq. (19.11.7) or ACI Formula (11-24),

$$A_l = \frac{2A_t}{s}(x_1 + y_1) = 0.091(8.5 + 16.5) = 2.27 \text{ sq in.}$$

The minimum value is, from using Eq. (19.14.10) the unit stress counterpart of ACI Formula (11-25)

$$A_l = \left[\frac{400x}{f_y}\left(\frac{v_{tn}}{v_{tn} + v_n}\right) - \frac{2A_t}{s}\right](x_1 + y_1)$$

$$= \left[\frac{400(12)}{60,000}\right]\left(\frac{571}{571 + 109}\right) - 0.091\right](8.5 + 16.5)$$

$$= \text{negative} \quad \text{(does not control)}$$

ACI-11.6.8.2 requires the longitudinal bars to be distributed around the perimeter with a spacing not to exceed 12 in. Since the member depth exceeds 12 in., a layer intermediate between top and bottom is necessary. Thus use at mid-depth

$$\frac{A_l}{3} = \frac{2.27}{3} = 0.76 \text{ sq in.}$$

Use 2-#6 bars, one at each side of the member.

(g) Total longitudinal reinforcement for flexure and torsion. At the face of support, top steel = 0.85 (flexure) + 0.76 (torsion) = 1.61 sq in.

bottom steel = 0.29 (flexure) + 0.76 (torsion) = 1.05 sq in.

At d from the face of support,

top steel = 0.70 (flexure) + 0.76 (torsion) = 1.46 sq in.

bottom steel = 0.29 (flexure) + 0.76 (torsion) = 1.05 sq in.

(h) Transverse steel requirement along the span. Using the same procedure as at the critical section, the requirements are computed as shown in Table 19.15.1.

Table 19.15.1 TRANSVERSE STEEL REQUIREMENT ALONG THE SPAN (EXAMPLE 19.15.2)

LOCATION FROM CENTER OF SUPPORT	V_u	v_n	$\dfrac{A_v}{s}$	T_u	v_{tn}	$\dfrac{A_t}{s}$	$\dfrac{50b_w}{f_y}$	REQUIRED $\dfrac{2A_t}{s} + \dfrac{A_v}{s}$	MAX s FOR #4 HOOPS
	(kips)	(psi)	(in.)	(ft-kips)	(psi)	(in.)	(in.)	(in.)	(in.)
4	17.7	99.0	0.0142	43.5	522	0.0402	0.010	0.095	4.20
6	15.5	86.8	0.0118	38.1	457	0.0332	0.010	0.078	5.13
8	13.3	74.4	0.0092	32.7	392	0.0262	0.010	0.062	6.25[b]
10	11.1	62.1	0.0064	27.2	326	0.0191	0.010	0.045	6.25[b]
12	8.9	49.8	0.0044	21.8	262	0.0123	0.010	0.029	6.25[b]
15	5.6	31.4	0.0006	13.7	164	0.0017	0.010	0.010[a]	6.25[b]

[a] min $50b_w/f_y$ controls
[b] max $s = (x_1 + y_1)/4$ controls

The maximum spacing curve and selected spacings are shown in Fig. 19.15.3(c). Since the torsional stress v_{tn} never is less than $1.5\sqrt{f'_c} = 95$ psi, closed stirrups are required along the entire span.

(i) Longitudinal steel requirements along the span. The flexural requirements have already been shown in Figs. 19.15.3(a) and (b). The torsional requirements are computed in Table 19.15.2. Because the member

Table 19.15.2 LONGITUDINAL STEEL REQUIREMENT FOR TORSION ALONG THE SPAN (EXAMPLE 19.15.2)

LOCATION FROM CENTER OF SUPPORT (ft)	REQUIRED $\dfrac{2A_t^*}{s}$ (in.)	ACI FORMULA (11-24)[†] (sq in.)	$\dfrac{400x}{f_y}\left(\dfrac{v_{tn}}{v_{tn}+v_n}\right)$	ACI FORMULA (11-25)[‡] (sq in.)	REQUIRED A_l (sq in.)
4	0.0804	2.01	0.067	negative	2.01
6	0.0664	1.66	0.067	negative	1.66
8	0.0524	1.31	0.067	0.38	1.31
10	0.0382	0.96	0.067	0.73	0.96
12	0.0246	0.62	0.067	1.06	1.06
15	0.0100	0.25	0.067	1.43	1.43

[*] $2A_t/s$ required for torsion, obtained from Table 19.15.1, but need not be less than $50b_w/f_y = 0.010$ for computing longitudinal reinforcement (ACI-11.6.9.3).
[†] $A_l = (2A_t/s)(x_1 + y_1)$
[‡] min $A_l = \left[\dfrac{400x}{f_y}\left(\dfrac{v_{tn}}{v_{tn}+v_n}\right) - \dfrac{2A_t}{s}\right](x_1 + y_1)$, in terms of unit stress.

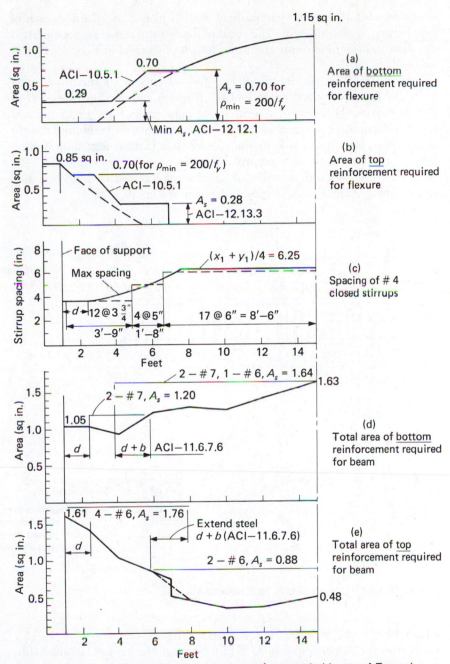

Figure 19.15.3 Reinforcement requirements for spandrel beam of Example 19.15.2.

depth exceeds 12 in., the longitudinal steel is placed one-third at each of top, bottom, and midheight. The sums of the flexural requirement and $A_t/3$ are shown as the total requirement for longitudinal steel in Figs. 19.15.3(d) and (e).

In determining the length of longitudinal bars, a conservative interpretation has been made of ACI-11.6.7.6 which requires an extension of $d + b_t$ beyond the theoretical termination point, as in Figs. 19.15.3(d) and (e). The theoretical termination point is taken as that for combined flexure and torsion. It is noted that in some cases ACI-12.10.5 regarding cutting bars in a tension zone may control. The crack control provisions of ACI-10.6.4 must be checked, as well as the deflection if excessive deflection may cause damage. The design details are shown in Fig. 19.15.4.

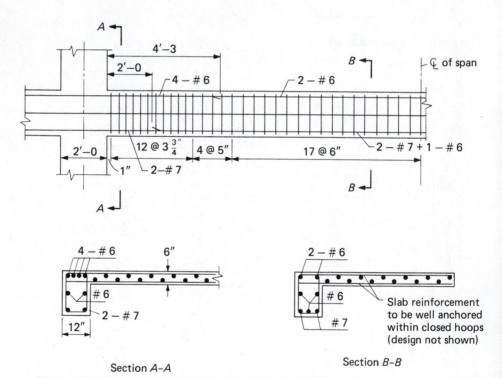

Section *A-A*

Section *B-B*

Figure 19.15.4 Design details for Example 19.15.2.

EXAMPLE 19.15.3 Redesign the spandrel beam shown in Fig. 19.15.2 taking the option permitted in ACI-11.6.3 of considering the torsional stiffness to be zero in performing the structural analysis to obtain the design loads. The torsional member is then designed for a nominal torsional stress of $4\sqrt{f'_c}$.

Solution: (a) Flexural strength. Assume that the factored bending moment and shear are not significantly affected by a change in the assumption for torsional stiffness; that is, use Figs. 19.15.2(b) and (c). Note that most structural analysis methods traditionally do not include the effect of torsional stiffness; thus this approach is simple. After completing the structural analysis disregarding torsional stiffness, any member that needs to undergo torsional deformation is then arbitrarily designed to include a torsional stress v_{tn} of $4\sqrt{f_c'}$.

For this example, the requirements for flexure are as in Example 19.15.2 and are shown in Figs. 19.15.3(a) and (b).

(b) Factored shear V_u and torsion T_u forces for which strength must be provided. At d from face of support,

$$V_u = 19.4 \text{ kips} \quad [\text{see Fig. 19.15.2(c)}]$$

$$T_u = \phi(4\sqrt{f_c'})(\tfrac{1}{3}\Sigma x^2 y) = 0.85(4\sqrt{4000})(1180)\tfrac{1}{12,000} = 21.1 \text{ ft-kips}$$

Note that the $4\sqrt{f_c'} = 253$ psi compares with the nominal unit stress of 571 psi that was used in Example 19.15.2.

(c) Nominal strengths T_c and V_c attributable to the concrete. For torsion, Eq. (19.12.6) or ACI Formula (11-22) gives at d from the face of support

$$T_c = \frac{0.8\sqrt{f_c'}(\Sigma x^2 y)}{\sqrt{1 + \left(\dfrac{0.4}{C_t}\dfrac{V_u}{T_u}\right)^2}} = \frac{0.8\sqrt{4000}(3540)/12,000}{\sqrt{1 + \left(\dfrac{0.4}{0.0593}\dfrac{19.4}{(21.1)(12)}\right)^2}} = 13.3 \text{ ft-kips}$$

where

$$C_t = \frac{b_w d}{\Sigma x^2 y} = \frac{12(17.5)}{3540} = 0.0593 \text{ per in.}$$

For shear, Eq. (19.12.4) or ACI Formula (11-5) gives at d from the face of support

$$V_c = \frac{2\sqrt{f_c'}b_w d}{\sqrt{1 + \left(2.5C_t \dfrac{T_u}{V_u}\right)^2}} = \frac{2\sqrt{4000}(12)17.5/1000}{\sqrt{1 + \left(2.5(0.0593)\dfrac{21.1(12)}{19.4}\right)^2}} = 12.2 \text{ kips}$$

Though the factored shear force V_u decreases farther from the support, the code-specified factored torsional force T_u, based on a stress of $4\sqrt{f_c'}$ psi used to provide ductility, is taken constant along the length of the beam.

At midspan,

$$V_u = 5.6 \text{ kips}; \quad T_u = 21.1 \text{ ft-kips}$$

$$T_c = \frac{0.8\sqrt{4000}(3540)/12,000}{\sqrt{1 + \left(\dfrac{0.4}{0.0593}\dfrac{5.6}{21.1(12)}\right)^2}} = 14.8 \text{ ft-kips}$$

$$V_c = \frac{2\sqrt{4000}(12)(17.5)/1000}{\sqrt{1 + \left(2.5(0.0593)\dfrac{21.1(12)}{5.6}\right)^2}} = 3.91 \text{ kips}$$

(d) Transverse reinforcement. For shear, using ACI Formula (11-17) from ACI-11.5.6.2 *at the critical section*

$$\frac{A_v}{s} = \frac{V_u - \phi V_c}{\phi f_y d} = \frac{19.4 - 0.85(12.2)}{0.85(60)(17.5)} = 0.0101$$

For torsion, using Eq. (19.11.5) or ACI Formula (11-23) *at the critical section*

$$\frac{A_t}{s} = \frac{T_u - \phi T_c}{\phi \alpha_t x_1 y_1 f_y} = \frac{[21.1 - 0.85(13.3)]12}{0.85(1.30)(8.5)(16.5)60} = 0.0126$$

where $\alpha_t = 1.30$, $x_1 = 8.5$ in., and $y_1 = 16.5$ in. as computed in Example 19.15.2. The total requirement for strength at d from face of support is

$$\frac{2A_t}{s} + \frac{A_v}{s} = 2(0.0126) + 0.0101 = 0.0353$$

This exceeds the minimum of $50 b_w / f_y$ of 0.01.
    The requirement at midspan is

$$\frac{A_v}{s} = \frac{5.6 - 0.85(3.91)}{0.85(60)(17.5)} = 0.0026$$

$$\frac{A_t}{s} = \frac{[21.1 - 0.85(14.8)]12}{0.85(1.30)(8.5)(16.5)60} = 0.0110$$

$$\frac{2A_t}{s} + \frac{A_v}{s} = 2(0.0110) + 0.0026 = 0.0246 > 0.01 \text{ min} \qquad \text{OK}$$

Try #3 hoops, $A_v = 2(0.11) = 0.22$ sq in. The maximum spacing of hoops between the face of support and the critical section is

$$\text{max } s = \frac{0.22}{0.0353} = 6.2 \text{ in.}$$

The upper limit for spacing of closed hoops is (ACI-11.6.8.1),

$$\text{max } s = \frac{x_1 + y_1}{4} = \frac{8.625 + 16.625}{4} = 6.3 \text{ in.}$$

Thus for this beam a practical spacing of 6 in. may be used for the entire beam. Note that the distances x_1 and y_1 have been corrected to agree with the use of #3 hoops.
    (e) Longitudinal reinforcement for torsion. Using Eq. (19.11.7) or ACI Formula (11-24), at the critical section,

$$A_l = \frac{2A_t}{s}(x_1 + y_1) = 2(0.0126)(8.625 + 16.625) = 0.64 \text{ sq in.}$$

where A_t / s has been corrected to agree with the values of x_1 and y_1 using #3 hoops. The minimum value at the critical section is, from Eq. (19.14.11) or ACI Formula (11-25),

$$A_l = \left[\frac{400x}{f_y} \left(\frac{T_u}{T_u + \dfrac{V_u}{3C_t}} \right) - \frac{2A_t}{s} \right](x_1 + y_1)$$

$$= \left[\frac{400(12)}{60,000} \left(\frac{21.1}{21.1 + \dfrac{19.4}{3(0.0593)12}} \right) - 0.0252 \right](8.625 + 16.625)$$

$$= 0.0307(25.25) = 0.78 \text{ sq in.} \qquad\qquad \text{Controls}$$

Note that the amount A_l will be divided into three equal parts, top of beam, bottom of beam, and midheight in accordance with ACI-11.6.8.2. The longitudinal requirement for torsion is added to the requirement for flexure [Figs. 19.15.3(a) and (b)] and the total requirements for the top and bottom of the beam are shown in Fig. 19.15.5. These requirements can be compared with Figs. 19.15.3(d) and (e) that were obtained when the torsional moments were computed after making stiffness assumptions. Note that in this alternate method, fewer and smaller hoops are required (i.e., #3 @ 6 throughout) and less longitudinal reinforcement likewise is needed.

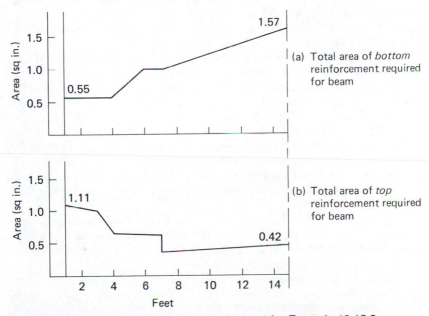

(a) Total area of *bottom* reinforcement required for beam

(b) Total area of *top* reinforcement required for beam

Figure 19.15.5 Longitudinal steel requirement for Example 19.15.3.

SELECTED REFERENCES

1. K. G. Tamberg and P. T. Mikluchin. "Torsional Phenomena Analysis and Concrete Structure Design," *Analysis of Structural Systems for Torsion* (SP-35). Detroit: American Concrete Institute, 1973 (pp. 1–102).
2. Michael P. Collins and Denis Mitchell. "Shear and Torsion Design of Prestressed and Non-Prestressed Concrete Beams," *PCI Journal*, **25**, September/October 1980, 32–100. Disc., **26**, November/December 1981, 96–118.

3. J. Warwaruk. "Torsion in Reinforced Concrete," *Significant Developments in Engineering Practice and Research* (SP-72). Detroit: American Concrete Institute, 1981 (pp. 247–277).

4. S. Timoshenko and J. N. Goodier. *Theory of Elasticity.* New York: McGraw-Hill, 1951 (pp. 275–288).

5. Paul Lampert. "Postcracking Stiffness of Reinforced Concrete Beams in Torsion and Bending," *Analysis of Structural Systems for Torsion* (SP-35). Detroit: American Concrete Institute, 1973 (pp. 385–433). (Also presented at 1971 Annual Convention, American Concrete Institute, Denver, March 1971.)

6. Michael P. Collins and Paul Lampert. "Redistribution of Moments at Cracking—The Key to Simpler Torsion Design?" *Analysis of Structural Systems for Torsion* (SP-35). Detroit: American Concrete Institute, 1973 (pp. 343–383).

7. Chu-Kia Wang. *Matrix Methods of Structural Analysis* (2nd ed.). Madison, Wisconsin: American Publishing Co., 1970 (Chaps. 12 and 15).

8. Robert A. Shoolbred and Eugene P. Holland. "Investigation of Slab Restraint on Torsional Moments in Fixed-Ended Spandrel Girders," *Torsion of Structural Concrete* (SP-18). Detroit: American Concrete Institute, 1968 (pp. 69–88).

9. Kolbjorn Saether and N. M. Prachand. "Torsion in Spandrel Beams," *ACI Journal, Proceedings,* **66,** January 1969, 24–30.

10. James O. Jirsa, John L. Baumgartner, and Nathan C. Mogbo. "Torsional Strength and Behavior of Spandrel Beams," *ACI Journal, Proceedings,* **66,** November 1969, 926–932. Disc., **67,** 434–435.

11. Mario G. Salvadori. "Spandrel-Slab Interaction," *Journal of the Structural Division,* ASCE, **96,** January 1970 (ST1), 89–106.

12. James O. Jirsa. "Torsion in Floor Slab Structures," *Analysis of Structural Systems for Torsion* (SP-35). Detroit: American Concrete Institute, 1973 (pp. 265–292).

13. Ugur Ersoy. "Distribution of Torsional and Bending Moments in Beam-Slab Systems," *Analysis of Structural Systems for Torsion* (SP-35). Detroit: American Concrete Institute, 1973 (pp. 293–324).

14. E. L. Kemp and W. J. Wilhelm. "Influence of Spandrel Beam Torsion on Slab Capacity Based on Yield Line Criteria," *Analysis of Structural Systems for Torsion* (SP-35). Detroit: American Concrete Institute, 1973 (pp. 325–341).

15. Thomas T. C. Hsu and Kenneth T. Burton. "Design of Reinforced Concrete Spandrel Beams," *Journal of the Structural Division,* ASCE, **100,** January 1974 (ST1), 209–229.

16. Thomas T. C. Hsu. "Torsion of Structural Concrete—Plain Concrete Rectangular Sections," *Torsion of Structural Concrete* (SP-18). Detroit: American Concrete Institute, 1968 (pp. 203–238). (Also Portland Cement Association Development Department Bulletin D134.)

17. Thomas T. C. Hsu. "Torsion of Structural Concrete—A Summary of Pure Torsion," *Torsion of Structural Concrete* (SP-18). Detroit: American Concrete Institute, 1968 (pp. 165–178). (Also Portland Cement Association. Development Department Bulletin D133.)

18. Thomas T. C. Hsu. "Ultimate Torque of Reinforced Rectangular Beams," *Journal of the Structural Division,* ASCE, **94,** February 1968 (ST2), 485–510. (Also Portland Cement Association Development Department Bulletin D127.)

19. N. N. Lessig. "Determination of Load Carrying Capacity of Reinforced Concrete Element with Rectangular Cross-Section Subjected to Flexure with Torsion," Work No. 5. Moscow: Institut Betona i Zhelezobetona (Concrete and Reinforced Concrete Institute), 1959 (pp. 5–28). (Also available as Foreign Literature Study No. 371, PCA Research and Development Labs., Skokie, Illinois.)

20. C. D. Goode and M. A. Helmy. "Ultimate Strength of Reinforced Concrete

Beams in Combined Bending and Torsion," *Torsion of Structural Concrete* (SP-18). Detroit: American Concrete Institute, 1968 (pp. 357–377).

21. M. P. Collins, P. F. Walsh, F. E. Archer, and A. S. Hall. "Ultimate Strength of Reinforced Concrete Beams Subjected to Combined Torsion and Bending," *Torsion of Structural Concrete* (SP-18). Detroit: American Concrete Institute, 1968 (pp. 379–402).

22. Kevin D. Below, B. Vijaya Rangan, and A. Stanley Hall. "Theory for Concrete Beams in Torsion and Bending," *Journal of the Structural Division*, ASCE, August 1975 (ST8), 1645–1660.

23. Thomas T. C. Hsu. "Torsion of Structural Concrete—Behavior of Reinforced Concrete Rectangular Members," *Torsion of Structural Concrete* (SP-18). Detroit: American Concrete Institute, 1968 (pp. 261–306). (Also Portland Cement Association Development Department Bulletin D135.)

24. Paul Zia. "Torsion Theories for Concrete Members," *Torsion of Structural Concrete* (SP-18). Detroit: American Concrete Institute, 1968 (pp. 103–132).

25. Paul Zia. "What Do We Know About Torsion in Concrete Members?" *Journal of the Structural Division*, ASCE, **96**, June 1970 (ST6), 1185–1199.

26. David J. Victor, Narayanan Lakshmanan, and Narayanan Rajagopalan. "Ultimate Torque of Reinforced Concrete Beams," *Journal of the Structural Division*, ASCE, **102**, July 1976 (ST7), 1337–1352.

27. E. Rausch. *Berechnung des Eisenbetons gegen Verdrehung und Abscheren (Design of Reinforced Concrete for Torsion and Shear)*. Berlin: Springer Verlag, 1929.

28. Paul Lampert and Bruno Thürlimann. "Ultimate Strength and Design of Reinforced Concrete Beams in Torsion and Bending," *Publications*, International Association for Bridge and Structural Engineering, 31-I, 1971, 107–131.

29. Paul Lampert and Michael P. Collins. "Torsion, Bending and Confusion—An Attempt to Establish the Facts," *ACI Journal, Proceedings*, **69**, August 1972, 500–504.

30. CEB-FIP. *Model Code for Concrete Structures, CEB-FIP International Recommendations* (3rd ed.). Paris: Commite Euro-International du Beton, 1978. 348 pp.

31. Denis Mitchell and Michael P. Collins. "Diagonal Compression Field Theory—A Rational Model for Structural Concrete in Pure Torsion," *ACI Journal, Proceedings*, **71**, August 1974, 396–408.

32. Hans Gesund, Frederick J. Schuette, George R. Buchanan, and George A. Gray. "Ultimate Strength in Combined Bending and Torsion of Concrete Beams Containing Both Longitudinal and Transverse Reinforcement," *ACI Journal, Proceedings*, **61**, December 1964, 1509–1522.

33. John P. Klus and C. K. Wang. "Torsion in Grid Frames," *Torsion of Structural Concrete* (SP-18). Detroit: American Concrete Institute, 1968 (pp. 89–101).

34. G. S. Pandit and Joseph Warwaruk. "Reinforced Concrete Beams in Combined Bending and Torsion," *Torsion of Structural Concrete* (SP-18). Detroit: American Concrete Institute, 1968 (pp. 133–163).

35. A. A. Gvozdez, N. N. Lessig, and L. K. Rulle. "Research on Reinforced Concrete Beams Under Combined Bending and Torsion in the Soviet Union," *Torsion of Structural Concrete* (SP-18). Detroit: American Concrete Institute, 1968 (pp. 307–336).

36. David J. Victor and Phil M. Ferguson. "Reinforced Concrete T-Beams Without Stirrups Under Combined Moment and Torsion," *ACI Journal, Proceedings*, **65**, January 1968, 29–36. Disc., **65**, 560–566.

37. V. Ramakrishnan and Y. Ananthanarayana. "Tests to Failure in Bending and

Torsion of Reinforced Concrete," *ACI Journal, Proceedings,* **66,** May 1969, 428–431. Disc., 943–944.

38. D. W. Kirk and S. D. Lash. "T-Beams Subject to Combined Bending and Torsion," *ACI Journal, Proceedings,* **68,** February 1971, 150–159.

39. Hota V. S. GangaRao and Paul Zia. "Rectangular Prestressed Beams in Torsion and Bending," *Journal of the Structural Division,* ASCE, **99,** January 1973 (ST1), 183–198.

40. Michael P. Collins and Paul Lampert. "Designing for Torsion," *Structural Concrete Symposium.* Toronto: University of Toronto Civil Engineering Department, May 1971 (pp. 38–79).

41. Ugur Ersoy and Phil M. Ferguson. "Behavior and Strength of Concrete L-Beams Under Combined Torsion and Shear," *ACI Journal, Proceedings,* **64,** December 1967, 797–801. Disc., **65,** 477–479.

42. John P. Klus. "Ultimate Strength of Reinforced Concrete Beams in Combined Torsion and Shear," *ACI Journal, Proceedings,* **65,** March 1968, 210–215. Disc., 786–791.

43. Huey Ming Liao and Phil M. Ferguson. "Combined Torsion in Reinforced Concrete L-Beams with Stirrups," *ACI Journal, Proceedings,* **66,** December 1969, 986–993. Disc., **67,** 475–478.

44. Larry E. Farmer and Phil M. Ferguson. "T-Beams Under Combined Bending, Shear and Torsion," *ACI Journal, Proceedings,* **64,** November 1967, 757–766. Disc., **65,** 417–421.

45. E. L. Kemp. "Behavior of Concrete Members Subject to Torsion and to Combined Torsion, Bending, and Shear," *Torsion of Structural Concrete* (SP-18). Detroit: American Concrete Institute, 1968 (pp. 179–201).

46. Thomas T. C. Hsu. "Torsion of Structural Concrete—Interaction Surface for Combined Torsion, Shear, and Bending in Beams Without Stirrups," *ACI Journal, Proceedings,* **65,** January 1968, 51–60, Disc., 566–572.

47. M. S. Mirza and J. O. McCutcheon. Discussion of "Torsion of Structural Concrete—Interaction Surface for Combined Torsion, Shear, and Bending in Beams Without Stirrups," *ACI Journal, Proceedings,* **65,** July 1968, 567–570.

48. David J. Victor and Phil M. Ferguson. "Beams Under Distributed Load Creating Moment, Shear, and Torsion," *ACI Journal, Proceedings,* **65,** April 1968, 295–308. Disc., 892–894.

49. Michael P. Collins, Paul F. Walsh, and A. S. Hall. Discussion of "Ultimate Strength of Reinforced Concrete Beams in Combined Torsion and Shear," *ACI Journal, Proceedings,* **65,** September 1968, 786–788.

50. D. L. Osburn, B. Mayoglou, and Alan H. Mattock. "Strength of Reinforced Concrete Beams With Web Reinforcement in Combined Torsion, Shear, and Bending," *ACI Journal, Proceedings,* **66,** January 1969, 31–41. Disc., 593–595.

51. M. S. Mirza and J. O. McCutcheon. "Behavior of Reinforced Concrete Beams Under Combined Bending, Shear, and Torsion," *ACI Journal, Proceedings,* **66,** May 1969, 421–427. Disc., 940–942.

52. Alfred Bishara. "Prestressed Concrete Beams Under Combined Torsion, Bending, and Shear," *ACI Journal, Proceedings,* **66,** July 1969, 525–538. Disc., **67,** 61–63.

53. Arthur E. McMullen and Joseph Warwaruk. "Concrete Beams in Bending, Torsion, and Shear," *Journal of the Structural Division,* ASCE, **96,** May 1970 (ST5), 885–903.

54. Umakanta Behera and Phil M. Ferguson. "Torsion, Shear, and Bending on Stirruped L-Beams," *Journal of the Structural Division,* ASCE, **96,** July 1970 (ST7), 1271–1286.

55. Einar Gausel. "Ultimate Strength of Prestressed I-Beams Under Combined Torsion, Bending, and Shear," *ACI Journal, Proceedings*, **67**, September 1970, 675–678.

56. Priya R. Mukherjee and Joseph Warwaruk. "Torsion, Bending, and Shear in Prestressed Concrete," *Journal of the Structural Division*, ASCE, **97**, April 1971 (ST4), 1063–1079.

57. P. K. Syamal, M. S. Mirza, and D. P. Ray. "Plain and Reinforced Concrete L-Beams Under Combined Flexure, Shear, and Torsion," *ACI Journal, Proceedings*, **68**, November 1971, 848–860.

58. K. S. Rajagopalan and Phil M. Ferguson. "Distributed Loads Creating Combined Torsion, Bending, and Shear on L-Beams with Stirrups," *ACI Journal, Proceedings*, **69**, January 1972, 46–54.

59. K. S. Rajagopalan, Umakanta Behera, and Phil M. Ferguson. "Total Interaction Method for Torsion Design," *Journal of the Structural Division*, ASCE, **98**, September 1972 (ST9), 2097–2117.

60. Thomas G. Barton and D. Wayne Kirk. "Concrete T-Beams Subject to Combined Loading," *Journal of the Structural Division*, ASCE, **99**, April 1973 (ST4), 687–700.

61. Robert L. Henry and Paul Zia. "Prestressed Beams in Torsion, Bending, and Shear," *Journal of the Structural Division*, ASCE, **100**, May 1974 (ST5), 933–952.

62. Lennart Elfgren, Inge Karlsson, and Anders Losberg. "Torsion-Bending-Shear Interaction for Concrete Beams," *Journal of the Structural Division*, ASCE, **100**, August 1974 (ST8), 1657–1676.

63. D. Wayne Kirk and David G. McIntosh. "Concrete L-Beams Subject to Combined Torsional Loads," *Journal of the Structural Division*, ASCE, **101**, January 1975 (ST1), 269–282.

64. ACI Committee 438. "Tentative Recommendations for the Design of Reinforced Concrete Members to Resist Torsion," *ACI Journal, Proceedings*, **66**, January 1969, 1–8. Disc., **66**, 576–588.

65. Michael P. Collins. Discussion of "Tentative Recommendations for the Design of Reinforced Concrete Members to Resist Torsion," *ACI Journal, Proceedings*, **66**, July 1969, 577–579.

66. ACI Committee 438. Discussion of "Proposed Revision of ACI 318-63: Building Code Requirements for Reinforced Concrete," *ACI Journal, Proceedings*, **67**, September 1970, 686–689.

67. Thomas T. C. Hsu and E. L. Kemp. "Background and Practical Application of Tentative Design Criteria for Torsion," *ACI Journal, Proceedings*, **66**, January 1969, 12–23. Disc., 591–593.

68. Alan H. Mattock. "How to Design for Torsion," *Torsion of Structural Concrete* (SP-18). Detroit: American Concrete Institute, 1968 (pp. 469–495).

69. David J. Victor. "Effective Flange Width in Torsion," *ACI Journal, Proceedings*, **68**, January 1971, 42–46.

70. Denis Mitchell and Michael P. Collins. "Detailing for Torsion," *ACI Journal, Proceedings*, **73**, September 1976, 506–511.

71. M. A. Gouda. "Distribution of Torsion and Bending Moments in Connected Beams and Slabs," *ACI Journal, Proceedings*, **56**, February 1960, 757–774. Disc., 1425–1446.

72. Thomas T. C. Hsu and Ching-Sheng Hwang. "Torsional Limit Design of Spandrel Beams," *ACI Journal, Proceedings*, **74**, February 1977, 71–79.

73. Alfred G. Bishara, Larry Londot, Peter Au, and Majety V. Sastry. "Flexural Rotational Capacity of Spandrel Beams," *Journal of the Structural Division*, ASCE, **105**, January 1979 (ST1), 147–161.

74. Arthur E. McMullen and B. Vijaya Rangan. "Pure Torsion in Rectangular Sections—A Re-Examination," *ACI Journal, Proceedings*, **75**, October 1978, 511–519.

75. Mahmoud A. Reda Youssef and Alfred G. Bishara. "Dowel Action in Concrete Beams Subject to Torsion," *Journal of the Structural Division*, ASCE, **106**, June 1980 (ST6), 1263–1277.

76. David J. Victor and P. K. Aravindan. "Prestressed and Reinforced Concrete T-Beams Under Combined Bending and Torsion," *ACI Journal, Proceedings*, **75**, October 1978, 526–532.

77. B. Vijaya Rangan and A. S. Hall. "Strength of Rectangular Prestressed Concrete Beams in Combined Torsion, Bending and Shear," *ACI Journal, Proceedings*, **70**, April 1973, 270–278.

78. B. Vijaya Rangan and A. S. Hall. "Strength of Prestressed Concrete I-Beams in Combined Torsion and Bending," *ACI Journal, Proceedings*, **75**, November 1978, 612–618.

79. K. S. Rajagopalan. "Combined Torsion, Bending, and Shear on L-Beams," *Journal of the Structural Division*, ASCE, **106**, December 1980 (ST12), 2475–2492.

80. G. Taylor and J. Warwaruk. "Combined Bending, Torsion, and Shear of Prestressed Concrete Box Girders," *ACI Journal, Proceedings*, **78**, September–October 1981, 335–340.

81. P. Müller. "Failure Mechanisms for Reinforced Concrete Beams in Torsion and Bending," *Publications*, International Association for Bridge and Structural Engineering, 36-II, 1976, 146–163.

82. B. G. Rabbat and M. P. Collins. "A Variable Angle Space Truss Model for Structural Concrete Members Subjected to Complex Loading," *Douglas McHenry International Symposium on Concrete and Concrete Structures* (SP-55). Detroit: American Concrete Institute, 1978 (pp. 547–587).

83. Thomas T. C. Hsu. *Torsion of Reinforced Concrete*. New York: Van Nostrand Reinhold, 1983.

PROBLEMS

All problems are to be worked in accordance with the strength method of the ACI Code, and all loads given are service loads, unless otherwise indicated.

19.1 Determine the reinforcement required on a 12 × 22 in. overall size member to carry a torsional moment of 10 ft-kips. Use f'_c = 4000 psi and f_y = 40,000 psi.

19.2 Determine the reinforcement required for the member in the accompanying figure to carry a torsional moment of 20 ft-kips. Use f'_c = 4000 psi and f_y = 50,000 psi.

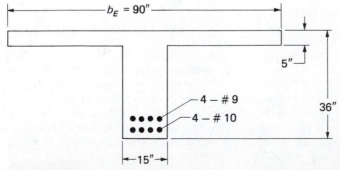

Problem 19.2

19.3   For the beam of the accompanying figure, assume that 4-#9 bars are used in the bottom for the main flexural reinforcement at midspan, with 2-#9 bars extended into the support and properly anchored. Further assume that 2-#7 are used in the top at the supports. What is the nominal torsional strength T_n of the section according to the ACI Code, assuming no simultaneous transverse shear? What is the factored negative bending moment M_u that might be permitted to act at the supports simultaneously when $T_u = \phi T_n$, according to the logic of Sec. 19.8?

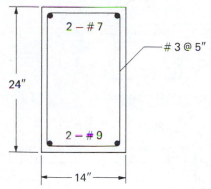

Problem 19.3

19.4   Determine the reinforcement required (including torsion) at midspan and at the supports for the beam of Example 19.4.1 (2B4 of Fig. 19.4.1). Assume the shear is the same as that of a simply supported beam carrying a uniform dead load of 0.65 kips/ft and uniform live load of 0.65 kips/ft. Assume that flexure alone requires longitudinal steel areas of 1.60 sq in. and 2.30 sq in., for positive and negative moment regions, respectively. Use $f'_c = 3000$ psi and $f_y = 40,000$ psi. Show sketches of the cross sections.

19.5   Design the reinforcement to include torsion on the spandrel girder of Example 19.4.2 (2G4 of Fig. 19.4.1). Assume simple beam shears for this span which is continuous at both ends. Assume the reactions to girder 2G4 from beams 2B1 are concentrated loads of 12.3 kips dead load and 15.9 kips live load, and also that the girder has uniform dead load (including its own weight) of 1 kip/ft. Use $f'_c = 3000$ psi and $f_y = 40,000$ psi. Show design sketch.

19.6   Redesign the transverse reinforcement for the beam of Prob. 5.1, except consider the total loading to be acting along a line at a distance of 2 in. from the midwidth of the beam. Use the rough approximation that equilibrium torsion is acting and that the torsional moment per unit length equals the uniform loading times 2 in.

19.7   Redesign the transverse reinforcement for the beam of Prob. 5.9 except consider the line uniform loading to be acting at 3 in. from the centerline of the beam cross section. Assume equilibrium torsion and use the same approximation as in Prob. 19.6.

19.8   Design the reinforcement for the edge beam which is continuous at both ends shown in the accompanying figure. Assume the torsional moment is 76 ft-kips (50% live load) at the face of column and that the torsional moment varies proportionally with the flexural shear (this problem is similar to that of Hsu and Kemp, Ref. 67).

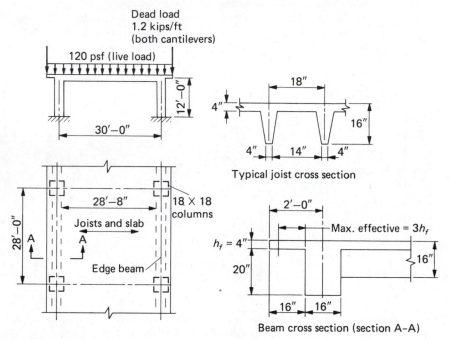

Problem 19.8

19.9 Design the parking garage spandrel beam having a simply supported span of 25 ft center-to-center of bearing (from Ref. 2). The floor system, consisting of double tees spanning 60 ft, carries a live load of 50 psf. The double tees are supported on the ledge of the spandrel beam. The contribution to R_1 from the dead load of the floor system is 10.7 kips, and from the 50 psf live load is 6.0 kips. Consider the given cross section as a preliminary trial section, with the 8 in. eccentricity of R_1 with respect to R_2 held as constant. The spandrel member will have an attachment to the supporting column, providing lateral support at each vertical support of the spandrel girder.

Use $f'_c = 4000$ psi and $f_y = 60,000$ psi and the ACI Code. (Note that the preliminary section is from Ref. 2 and is for a prestressed concrete member. The final design for a nonprestressed member may require changes from the preliminary given dimensions; those given dimensions may serve as guides for arriving at reasonable proportions.)

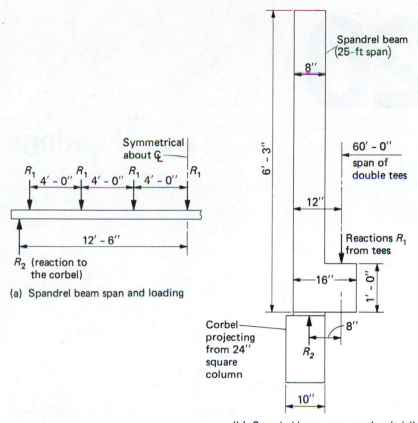

Symmetrical
about ℄

R_1 R_1 R_1 R_1

4' – 0''   4' – 0''   4' – 0''

12' – 6''

R_2 (reaction to
the corbel)

(a) Spandrel beam span and loading

Spandrel beam
(25–ft span)

8''

6' – 3''

12''

60' – 0''

span of
double tees

Reactions R_1
from tees

16''

1' – 0''

Corbel
projecting
from 24''
square
column

R_2

8''

10''

(b) Spandrel beam cross section (trial)

Problems 19.9 and 19.10

19.10 Design a spandrel beam of similar L-shape to that of Prob. 19.9 for loads R_1
of 5.5 kips dead load and 4.0 kips live load, plus the spandrel girder weight.
The span of the spandrel and the locations of the R_1 loads are the same as in
Prob. 19.9. The eccentricity e is 8 in. (same as for Prob. 19.9). Consider the
8 in. used in Prob. 19.9 as the minimum thickness; however, the other
dimensions should be appropriate for the loads given in the problem. Use
$f'_c = 4000$ psi and $f_y = 60,000$ psi.

20

Footings

20.1 PURPOSE OF FOOTINGS

Footings are structural elements that transmit to the soil column loads, wall loads, or lateral loads from retained earth. If these loads are to be properly transmitted, footings must be designed to prevent excessive settlement or rotation, to minimize differential settlement, and to provide adequate safety against sliding and overturning.

20.2 BEARING CAPACITY OF SOIL

There must be reliable information on the safe bearing capacity of the soil prior to the design of a footing. It is not within the scope of this text to discuss the details of arriving at the bearing capacity of soil. The allowable bearing capacity of soil is usually determined by the ruling building code, by comparison with existing footings and with related information in the area, by close examination of the soil and study of logs of test borings, by the application of the science of soil mechanics, by load test, or by combinations of the various sources and methods mentioned here.

The following are some of the most common reasons for the many uncertainties concerning soil behavior under a footing.

1. There may be wide variations in soil types, which depend on their geological source, mode of transportation, and sedimentation mechanism.
2. The physical properties and probable behavior under load are unknown and may require extensive testing.

Wall and square spread footings; Engineering Library, University of Wisconsin, Madison, Wis. (Photo by C. G. Salmon.)

3. Frost action may cause heaving or subsidence.
4. Vibration may cause consolidation of granular material which results in nonuniform settlement.
5. Man-made hazards may exist below the earth surface, such as rock heaps, old sewers, and questionable fill.

Soil pressure is usually utilized in design under a working stress philosophy. The soil mechanics/foundations specialist (geotechnical engineer) will establish the ultimate bearing capacity, applying the appropriate margin for safety, and specify a service load bearing capacity (allowable bearing capacity) to be used in design.

In general, rock is considered the best foundation material; graded sand and gravel are good materials; fine particles of sand and silt are generally

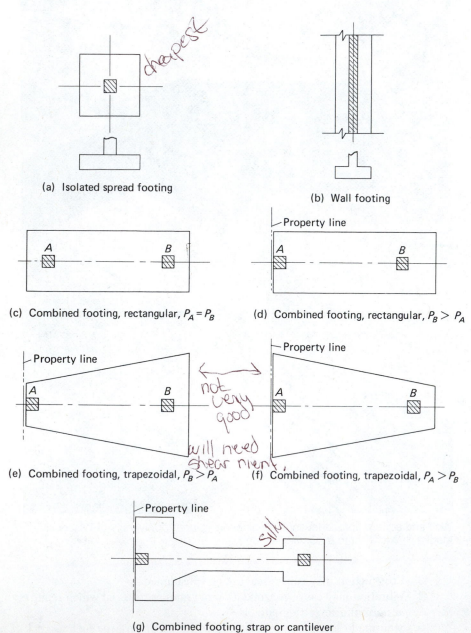

(a) Isolated spread footing

(b) Wall footing

(c) Combined footing, rectangular, $P_A = P_B$

(d) Combined footing, rectangular, $P_B > P_A$

(e) Combined footing, trapezoidal, $P_B > P_A$

(f) Combined footing, trapezoidal, $P_A > P_B$

(g) Combined footing, strap or cantilever

Figure 20.3.1 Types of footings.

questionable; and clay should be studied carefully. The allowable bearing capacity used for design may range from 12,000 psf (575 kN/m^2) or higher for rock to 2000 psf (96 kN/m^2) for soft clay or silty clay. Soils unable to carry 2000 psf generally require piling.

20.3 TYPES OF FOOTINGS

Most building footings may be classified as one of the following types (Fig. 20.3.1):

1. Isolated spread footings under individual columns. These may be square, rectangular, or occasionally circular in plan.
2. Wall footings, either flat or stepped, which support bearing walls.
3. Combined footings supporting two or more column loads. These may be continuous with a rectangular or trapezoidal plan or they may be isolated footings joined by a beam. The latter case is referred to as a strap, or cantilever, footing.
4. A mat foundation, which is one large continuous footing supporting all the columns of the structure. This is used when soil conditions are poor but piles are not used.
5. Pile caps, structural elements that tie a group of piles together. These may support bearing walls, isolated columns, or groups of several columns.

20.4 TYPES OF FAILURE OF FOOTINGS

The procedures used for the design of footings in the United States are based primarily on the work of Talbot in 1907 [1], Richart in 1946 [2], and Moe in 1957–1959 [3].

Moe defines the several types of failure that may occur in a slab acted on by concentrated loads. These failure modes are related to the shear span to depth (a/d) ratio, that is, $M_u/V_u d$, and are similar to those described for beams in Sec. 5.4 (Fig. 5.4.4). The failure mechanisms may be reviewed as follows:

1. Shear-compression failure [Fig. 20.4.1(a)]. Typical with deep sections of short span (low a/d ratios), inclined cracks form that do not cause failure but do extend into the compression zone, thus reducing its size until finally the compression zone fails under the combined compressive and shear stresses.
2. Flexure failure *after* inclined cracks form. Also typical with low a/d ratios, inclined cracks that form first do not cause failure or prevent the development of the theoretical ultimate bending moment. If embedment of tension steel is adequate, and no failure in the compression zone occurs, the tension steel may reach its yield strength.
3. Diagonal-tension failure [Fig. 20.4.1(b)]. Sometimes called *punch-*

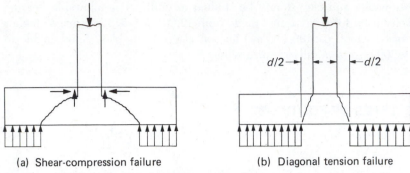

(a) Shear-compression failure   (b) Diagonal tension failure

Figure 20.4.1 Shear-related failure mechanisms.

ing shear, this is typical with medium-span average depth sections (intermediate values of a/d); the slab fails on formation of the inclined cracks at all four sides of the concentrated load. Test results indicate that the critical section can reasonably be considered at $d/2$ from the periphery of a column.

4. Flexure failure before inclined cracks form. Typical with large values of a/d, no inclined cracks form before flexural strength is reached.

In the design of a footing as well as of a beam, a shear failure should not occur prior to reaching the member flexural strength.

20.5 SHEAR STRENGTH OF FOOTINGS

The shear strength of footings is essentially the shear strength of two-way slab systems, as discussed in Secs. 16.15 and 16.18. An excellent summary of the current status is presented by ASCE-ACI Task Committee 426 under Chairman N. M. Hawkins [4] and confirmed by a 1979 ACI-ASCE Committee 426 Report [5]. The six principal variables involved in the strength of slabs without shear reinforcement are (a) the concrete strength, f'_c; (b) the ratio of the side length c of the loaded area to the effective depth d of the slab; (c) the relationship (V/M) between shear and moment near the critical section; (d) the column shape in terms of the ratio β_c of the long side to the short side of the rectangular column; (e) lateral restraints such as may be provided by stiff beams along the boundaries of a slab (no such restraints would normally be acting in the case of footings); and (f) the rate of loading.

Assuming that the design objective is to achieve a shear strength high enough so that any possible failure would be in a flexural mode, under the conditions of no lateral restraints and static loading rather than dynamic, the shear strength may be expressed [4]

$$V_n = \left(2.5 + \frac{3.0}{\beta_c} \right) \sqrt{f'_c} b_0 d \leq 4\sqrt{f'_c} b_0 d \qquad (20.5.1)$$

where

V_n = nominal shear strength

b_0 = periphery around critical section taken at $d/2$ from the loaded area

d = effective depth of slab

β_c = ratio of long side to short side of rectangular loaded area

Though the critical section at which inclined cracking occurs in the two-way shear action was found to follow the perimeter of the loaded area, Moe [3] showed that the variable c/d could be eliminated from the shear strength expression if a pseudo critical section at $d/2$ from the loaded area is used for computing this strength; thus the variable c/d does not appear in Eq. (20.5.1). One might consider that taking the critical section at $d/2$ from the loaded area in two-way shear action computation is for the same reason as taking the critical section at d from the face of support in one-way

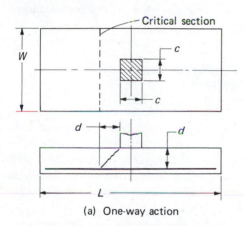

(a) One-way action

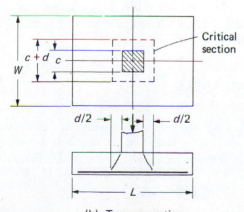

(b) Two-way action

Figure 20.5.1   Critical section for shear (inclined cracking) in footings.

shear action computation, the reason being that statistical correlation between computation and actual behavior is best when those critical sections are used. In both cases inclined cracks tend to begin *at* the face of support (i.e., edge of loaded area for a footing).

ACI-11.11.2 uses a modified version of Eq. (20.5.1) for strength in two-way shear action for sections without shear reinforcement ($V_n = V_c$), as follows:

$$V_c = v_c b_0 d$$

$$V_c = \left(2 + \frac{4}{\beta_c}\right)\sqrt{f_c'} b_0 d \leq 4\sqrt{f_c'} b_0 d \qquad (20.5.2)*$$

Note that v_c approaches $2\sqrt{f_c'}$, the value used for one-way action on beams, when β_c becomes large.

For footings in which the bending action is primarily in one direction, the procedure used for beams should be applied as described in Chap. 5. The critical section for such an investigation is to be taken at a distance d from the column face. Figure 20.5.1 summarizes the critical sections to be investigated in regard to shear.

20.6 MOMENT STRENGTH OF FOOTINGS AND DEVELOPMENT OF REINFORCEMENT

Research has shown that critical sections for moment and development of reinforcement occur at the face of a reinforced concrete column or wall. The bending in each direction should be considered separately. Richart's tests [2] showed that, under service loads, the moment is greater on a strip under the column load than it is out near the corners. However, failure in flexure does not occur until all of the steel has reached yield—that is, after some redistribution of load has taken place.

The critical sections for moment and development of reinforcement, however, are to be taken (ACI-15.4.2) at halfway between the middle and the edge of the wall for footings under masonry walls, and at halfway between the face of the column and the edge of the metallic base for footings under steel bases.

20.7 PROPORTIONING FOOTING AREAS FOR EQUAL SETTLEMENT

Differential settlement between footings should be eliminated as much as possible because it may adversely affect the strength of the structure as well as interfere with the fitting of such items as partitions, doors, and ceilings.

---

*For SI, with f_c' in MPa and b_0 and d in mm, ACI 318-83M gives

$$V_c = \frac{1}{6}\left(1 + \frac{2}{\beta_c}\right)\sqrt{f_c'} b_0 d \leq \frac{1}{3}\sqrt{f_c'} b_0 d \qquad (20.5.2)$$

It is generally assumed that there will be equal settlement if the unit soil pressures due to service loads are equal under all footings. However, the service loads in the columns consist of dead and live loads. The dead load is always there, but in a tall building there is little chance for a column to receive maximum live load from all floors. Therefore most building codes allow a reduction of live loads in columns. Typically, the total live load would have to be carried by columns supporting the roof and one floor. As the number of floors carried by a column increases, the percentage of the total live load that the column must be designed to carry may be reduced from 100% progressively. This percentage might be a minimum of 50% when the number of floors carried by the column is at least eight or nine. The maximum soil pressure under a footing is, then, due to the sum of the dead load in the column, the maximum reduced live load in the column, and the weight of the footing itself.

Soil is a substance that may be relatively elastic such as granular material like sand, or it may be a relatively plastic substance such as clay exhibiting time-dependent deformation under sustained load. Often, for design purposes, equal settlement of footings is presumed when the soil pressure due to sustained loads is the same under all footings. The sustained load may be taken as the sum of the dead load in the column, the weight of the footing itself, and a certain percentage of the maximum reduced live load in the column. In such cases, the relative areas of the footings should be so proportioned that the unit soil pressures under sustained loads would be the same under all footings. This requirement is additional to the requirement that the soil pressure under maximum possible load must not exceed the allowable bearing capacity at each footing.

20.8 INVESTIGATION OF SQUARE SPREAD FOOTINGS

The investigation of square spread footings can best be treated by an illustrative example in which the items considered are (1) soil pressure under the footing, (2) shear (inclined cracking), (3) bending moment, (4) development of reinforcement, and (5) load transfer from column to footing.

EXAMPLE 20.8.1 Check the adequacy of the square footing of Fig. 20.8.1, according to the strength method of the ACI Code. The column axial load is 300 kips dead load and 160 kips live load. The concrete for both the column and the footing has $f'_c = 3000$ psi and $f_y = 40,000$ psi. The allowable soil pressure is 5000 psf. There is a 2-ft earth overburden having a unit weight of 100 pcf.

Solution: (a) Soil pressure. The action of soil on the footing is taken to be uniform for isolated footings under concentric loads. The base of the footing must have an area large enough so that the allowable soil pressure will not be exceeded under the action of the column service load, footing weight, and weight of overburden.

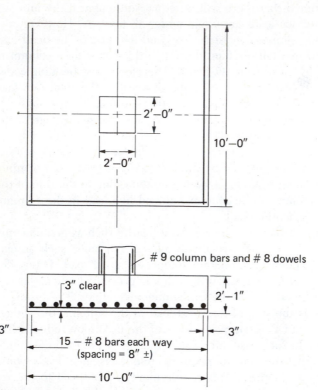

Figure 20.8.1 Square spread footing for Example 20.8.1.

$$
\begin{array}{lr}
\text{column load } (300 + 160) & = 460 \text{ kips} \\
\text{footing weight } 10(10)(2.08)(0.150) & = \ \ 31 \\
\text{earth } (100 - 4)2(0.100) & = \ \ \underline{19} \\
\text{Total weight on soil} & = \overline{510 \text{ kips}}
\end{array}
$$

$$
\text{soil pressure } p = \frac{510}{(10)(10)} = 5.1 \text{ ksf} \approx 5 \text{ ksf} \qquad \text{OK}
$$

(b) Shear (inclined cracking) strength. Two possible critical sections must be investigated: one-way action as a beam and two-way action as a slab, as shown in Fig. 20.5.1. The shear to be used is the upward soil pressure less the downward overburden and the footing weight acting outside of the critical section. Since the footing weight and the overburden are usually uniform, the forces acting on the footing may be obtained by using the net upward pressure, which is caused by the column load only. Since the strength method is to be used, the overload factors U must be applied.

Since the action of a square concentrically loaded footing is symmetrical about both axes, the reinforcement in each direction is presumed to do the same work. However, the effective depth cannot be the same for both directions since the bars must cross each other. The average depth d is commonly used except for very shallow footings (say, less than 15 in. deep) where the more conservative value should probably be used. Here the average d is

$$d = 25 - 3(\text{cover}) - 1(\text{bar diameter}) = 21 \text{ in.}$$

For one-way diagonal tension action, the net earth pressure acting upward due to factored loads is

$$p_{\text{net}} = \frac{300(1.4) + 160(1.7)}{100} = 6.92 \text{ ksf}$$

Using the loaded area shown in Fig. 20.8.2(a) the factored shear is

$$V_u = (p_{\text{net}})(\text{effective area}) = 6.92(2.25)10 = 156 \text{ kips}$$

When no shear reinforcement is used, $v_c = 2\sqrt{f_c'}$ unless the more detailed Vd/M procedure is used; thus

$$V_n = V_c = 2\sqrt{f_c'}b_w d = 2\sqrt{3000}(120)(21)\tfrac{1}{1000} = 276 \text{ kips}$$

$$[\phi V_n = 0.85(276) = 235 \text{ kips}] > V_u = 156 \text{ kips} \qquad \text{OK}$$

For two-way diagonal tension action, the factored shear [Fig. 20.8.2(b)] is

$$V_u = (p_{\text{net}})(\text{effective area})$$
$$= 6.92[100 - 3.75(3.75)] = 595 \text{ kips}$$

When no shear reinforcement is used, the strength is based on $v_c = 4\sqrt{f_c'}$ when $\beta_c = 1.0$ (see ACI-11.11.2); thus

$$V_n = V_c = 4\sqrt{f_c'}b_0 d = 4\sqrt{3000}(4)(3.75)(12)(21)\tfrac{1}{1000} = 828 \text{ kips}$$

$$[\phi V_n = 0.85(828) = 704 \text{ kips}] > V_u = 595 \text{ kips} \qquad \text{OK}$$

Since isolated footings are rarely designed with shear reinforcement, V_c will control the thickness.

(c) Bending moment strength. The critical section and loaded area are as shown in Fig. 20.8.2(a). The factored bending moment is

$$M_u = \frac{p_{\text{net}}bl^2}{2} = \frac{6.92(10)(4.0)^2}{2} = 554 \text{ ft-kips}$$

$$\text{required } R_n = \frac{M_u}{\phi bd^2} = \frac{554(12,000)}{0.90(10)(12)(21)^2} = 140 \text{ psi}$$

Referring to Sec. 3.8,

$$\text{required } \rho = \frac{1}{m}\left(1 - \sqrt{1 - \frac{2mR_n}{f_y}}\right)$$

$$= \frac{1}{15.7}\left(1 - \sqrt{1 - \frac{2(15.7)140}{40,000}}\right) = 0.0036$$

For spread footings, the minimum reinforcement (min $\rho_g = 0.002$) "for structural slabs of uniform thickness" as stated in ACI-10.5.3 applies; thus

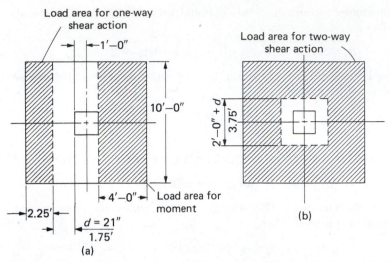

Figure 20.8.2 Critical sections and loaded area for square spread footing.

$$\text{required } A_s = 0.0036(10)(12)21 = 9.1 \text{ sq in.}$$

$$\text{min required } A_s = 0.002(10)(12)25 = 6.00 \text{ sq in.}$$

$$\text{provided } A_s = 15(0.79) = 11.9 \text{ sq in.}$$

Using 12-#8 would satisfy the requirement of 9.1 sq in.; however, with the low reinforcement ratio used, the 15 bars are not excessive.

(d) Development of reinforcement. The reinforcement must be embedded from the face of the column a distance equal to the development length L_d for #8 bars. By using the formulas of ACI-12.2, or from Table 6.7.1,

$$L_d(\#8) = 23.1 \text{ in.}$$

Allowing an inch or so cover on the end of the #8 bars, the embedment provided is

$$\text{actual embedment} = 48 - 2 = 46 \text{ in.} > L_d \qquad \text{OK}$$

(e) Load transfer from column to footing. ACI-15.8.1 requires all forces acting at the column base to be transferred into the footing. Tensile forces, if any, must be transferred by developed reinforcement, such as bar reinforcement, dowels, or mechanical connectors; however, compressive forces may be transmitted directly by bearing.

The nominal ultimate bearing stress f_b that the base of the column can withstand is $0.85f_c'$ (ACI-10.15.1). The nominal strength P_n in compression based on the *column* concrete strength f_c' is

$$P_n = 0.85f_c'A_g = 0.85(3)(576) = 1470 \text{ kips}$$

$$P_u = 1.4(300) + 1.7(160) = 692 \text{ kips}$$

$$[\phi P_n = 0.70(1470) = 1029 \text{ kips}] > P_u = 692 \text{ kips} \qquad \text{OK}$$

Regarding the bearing on the footing concrete, the capacity is increased because the footing area is much larger than the column area, thus permitting a distribution of the concentrated load. The bearing strength for the *footing* concrete is the regular value based on $0.85f_c'$ increased by the multiplier α_b that varies between 1 and 2, as follows:

$$\alpha_b = \sqrt{\frac{A_2}{A_1}} \le 2$$

where A_1 is the load area, the column area in this example, 576 sq in.; A_2 is the maximum area of the portion of the supporting surface that is geometrically similar to and concentric with the loaded area, the entire footing area in this case, 14,400 sq in.

The bearing stress *on the footing* may control when columns of high-strength concrete rest on footings of low-strength concrete. Since both the column and footing contain the same strength concrete in this example, only the check on bearing in the column was necessary. Bearing in the bottom of the column is important unless the longitudinal reinforcement is developed by extension into the footing or by embedding dowels lapped to the column bars. In this case the load can be carried without using developed reinforcement.

When the transfer is made by bearing, as in this case, ACI-15.8.2.2 still requires a minimum amount of reinforcement across the joint between the column and footing. Extended longitudinal reinforcement or dowels at least equal to 0.005 times the gross cross-sectional area of the supported member must be provided. Additionally, when dowels are used, their diameter shall not exceed the diameter of longitudinal bars by more than 0.15 in. The minimum area of developed reinforcement is

$$\text{required } A_s = 0.005(576) = 2.88 \text{ sq in.}$$

Using a practical minimum of 4 bars, one in each corner,

$$\text{required } A_s \text{ per bar} = \frac{2.88}{4} = 0.72 \text{ sq in.}$$

Thus the #8 dowels shown are adequate. They must be embedded into the footing a distance equal to the development length L_d for #8 bars.

Using the basic development length formulas of ACI-12.3 for compression bars,

$$L_{db} = \frac{0.02f_y d_b}{\sqrt{f_c'}} \qquad\qquad [6.11.1]$$

$$= \frac{0.02(40,000)1.0}{\sqrt{3000}} = 14.6 \text{ in.}$$

but not less than

$$L_{db} = 0.0003f_y d_b \qquad\qquad [6.11.2]$$

$$= 0.003(40,000)(1.0) = 12.0 \text{ in.}$$

and the final L_d cannot be less than 8 in. Thus *really only .003*

$$L_d(\#8) = L_{db} = 14.6 \text{ in.} < 25 \text{ in. available}$$

If the available footing thickness is inadequate, a greater number of smaller diameter bars should be used.

Thus the footing of Fig. 20.8.1 satisfies all ACI requirements.

20.9 DESIGN OF SQUARE SPREAD FOOTINGS

The design of square spread footings involves the determination of the size and depth of the footing and the amount of main reinforcement and dowels so that all of the requirements of the preceding section are fulfilled. The design procedures are illustrated by the following example.

EXAMPLE 20.9.1 Design a square spread footing to carry a column dead load of 197 kips and a live load of 160 kips from a 16-in square tied column containing #11 bars as the principal column steel. The allowable soil pressure is 4.5 ksf. Consider there is a 2-ft overburden weighing 100 pcf. Use $f'_c = 3000$ psi, $f_y = 40,000$ psi, and the strength method of the ACI Code.

Solution: (a) Estimate the footing weight and determine the plan of the footing. The total weight of the footing, plus any overburden, may be estimated and added to the column load or, as an alternative, the effect of these items in terms of the unit soil pressure may be estimated. In this case, the footing is estimated to be about 2 ft thick, that is, 300 psf, frequently the minimum used by designers. This leaves the net allowable soil pressure that must carry the column load as

$$p_{net} = 4500 - 200 - 300 = 4000 \text{ psf}$$

$$\text{required } A = \frac{357}{4.0} = 89.5 \text{ sq ft}$$

Try 9 ft–6 in. square, $A = 90.3$ sq ft. Note that ACI-15.2.2 requires the base area of a footing to be determined using service loads (unfactored loads) with the allowable soil pressure. This is reasonable since the allowable soil pressure should be determined using principles of soil mechanics and may incorporate varying factors of safety depending on the type of soil and condition of loading.

For the design of the reinforced concrete member, factored loads must be used. Applying overload factors to the column load,

$$P_u = 1.4(197) + 1.7(160) = 548 \text{ kips}$$

$$p_{net} = \frac{548}{90.3} = 6.07 \text{ ksf}$$

The strength reduction factor ϕ will be applied later in the calculations.

(b) Determine depth based on shear (inclined cracking) strength. In most cases the depth necessary for shear without using stirrups controls the footing thickness.

For two-way action [Fig. 20.9.1(a)], assuming a thickness of 24 in.,

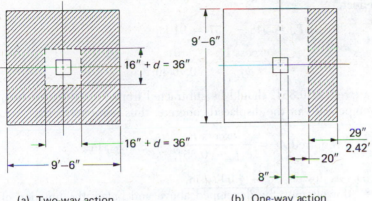

(a) Two-way action   (b) One-way action

Figure 20.9.1 Critical sections for shear in square footing design.

average $d = 24 - 3$ (cover) $- 1$ (bar diameter) ≈ 20 in.

$$V_u = (p_{net})(area) = 6.07\ [9.5(9.5) - 3.0(3.0)] = 493 \text{ kips}$$

According to ACI-11.11.2 for $\beta_c = 1.0$,

$$V_c = 4\sqrt{f'_c}b_0 d = 4\sqrt{3000}\ [4(16 + 20)](20)\tfrac{1}{1000} = 631 \text{ kips}$$

$$\phi V_c = 0.85(631) = 536 \text{ kips} > [V_u = 493 \text{ kips}] \qquad\qquad \text{OK}$$

No shear reinforcement is required.

For one-way action [Fig. 20.9.1(b)],

$$V_u = 6.07(2.42)(9.5) = 140 \text{ kips}$$

According to ACI-11.11.1.1 and 11.3.1.1,

$$V_c = 2\sqrt{f'_c}b_w d = 2\sqrt{3000}(9.5)(12)\tfrac{1}{1000} = 250 \text{ kips}$$

$$\phi V_c = 0.85(250) = 212 \text{ kips} > [V_u = 140 \text{ kips}] \qquad\qquad \text{OK}$$

No shear reinforcement is required. Note that V_u is computed from the simplified procedure which neglects the effect of $\rho Vd/M$. The 24-in. thickness is satisfactory for shear.

(c) Check transfer of load at the base of column (ACI-15.8). The compressive design strength ϕP_n based on the nominal ultimate bearing stress $0.85f'_c$ *in the column* is

$$\phi P_n = \phi(0.85f'_c)A_g = 0.70(0.85)(3)(256) = 457 \text{ kips}$$

$$\phi P_n < [P_u = 548 \text{ kips}] \qquad\qquad \text{NG}$$

Thus the column load cannot be transferred by bearing alone. It may well be that the minimum dowels required by ACI-15.8.2.1 will be adequate to transfer the excess load.

$$\text{min required } A_s = 0.005(256) = 1.28 \text{ sq in.} \qquad \text{(ACI-15.8.2.1)}$$

The excess load to be carried by the dowels is, neglecting the displaced concrete effect,

$$\text{excess } P_u = 548 - 457 = 91 \text{ kips}$$

$$\text{required } A_s = \frac{\text{excess } P_u}{\phi f_y} = \frac{91}{0.70(40)} = 3.25 \text{ sq in.}$$

Logically a stress of $0.85f'_c$ should be subtracted from f_y in the dowels in order to compensate for the displaced concrete; thus more correctly,

$$\text{required } A_s = \frac{\text{excess } P_u}{\phi(f_y - 0.85f'_c)} = 3.47 \text{ sq in.}$$

Use 4-#9 bars as dowels, $A_s = 4.00$ sq in.

The #9 dowels must be developed above and below the junction of column and footing. The development length L_d required in compression according to ACI-12.3 is

$$L_{db} = 0.02f_y d_b / \sqrt{f'_c}$$

$$= 0.02(40,000)(1.128)/\sqrt{3000} = 16.5 \text{ in.} \qquad \text{Controls}$$

but not less than

$$L_{db} = 0.0003f_y d_b = 0.0003(40,000)(1.128) = 13.5 \text{ in.}$$

and final L_d cannot be less than 8 in. Thus, $L_d = L_{db} = 16.5$ in. The 24-in. thick footing is adequate for straight dowels.

Hooks or bending of bars should not be considered effective in adding to the compressive resistance of bars (ACI-12.5.3). Very often engineers will specify bending of the dowels as shown in Fig. 20.9.2 to prevent their being pushed through the footing during construction and thus reducing the effective embedment distance L_2. In such cases of bending of the dowels, full development of the compressive force is required over the distance L_1.

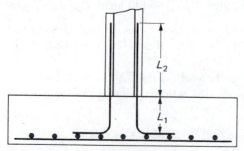

Figure 20.9.2 Dowel anchorage.

For this design, if such a bend is used, the available length L_1 is

$$L_1 = 24 - 3 \text{ (cover)} - 2(1) \text{ (footing bars)} - 1.128 \text{ (dowels)} = 17.9 \text{ in.}$$

This exceeds the 16.5 in. required and is therefore acceptable. If it were unacceptable, alternatives would include a thicker footing, a larger number of smaller-sized dowels, or the use of a pedestal.

(d) Design for bending moment strength. The critical section for moment is at the face of the column (Fig. 20.9.3).

$$M_u = \tfrac{1}{2}(6.07)(9.5)(4.08)^2 = 480 \text{ ft-kips}$$

$$\text{required } R_n = \frac{M_u}{\phi bd^2} = \frac{480(12,000)}{0.90(114)(20)^2} = 140 \text{ psi}$$

Referring to Sec. 3.8,

$$\rho = \frac{1}{m}\left(1 - \sqrt{1 - \frac{2mR_n}{f_y}}\right)$$

$$\rho = \frac{1}{15.7}\left[1 - \sqrt{1 - \frac{2(15.7)(140)}{40,000}}\right] = 0.00360$$

$$\text{required } A_s = \rho bd = 0.00360(114)(20) = 8.21 \text{ sq in.}$$

$$\text{min required } A_s = 0.002(114)(22) = 5.00 \text{ sq in.} \qquad \text{(ACI-10.5.3)}$$

Try 19-#6, $A_s = 8.36$ sq in.; $d = 24 - 3 - 0.75 = 20.3$ in.

$$C = 0.85f_c'ba = 0.85(3)(9.5)(12)a = 291a$$

$$T = A_sf_y = 8.36(40) = 334 \text{ kips}$$

$$C = T; \quad a = 1.15 \text{ in.}$$

$$\phi M_n = 0.90(334)[20.3 - 0.5(1.15)]\tfrac{1}{12} = 494 \text{ ft-kips}$$

$$\phi M_n > [M_u = 480 \text{ ft-kips}] \qquad\qquad \text{OK}$$

Use 19-#6 bars ($A_s = 8.36$ sq in.) each way.

(e) Check development of reinforcement. The required embedment measured from the face of the column equals the development length L_d. For #6 bars, using the formulas of ACI-12.2, or Table 6.7.1,

$$L_d(\#6) = L_{db} = 12.8 \text{ in.} < 48 \text{ in. available} \qquad \text{OK}$$

(f) Design sketch. A design sketch as shown in Fig. 20.9.4 is necessary to convey the designer's decision properly.

For practical design of square footings, design aids are available [6,7].

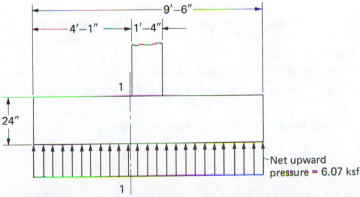

Figure 20.9.3 Critical section for bending moment and development of reinforcement.

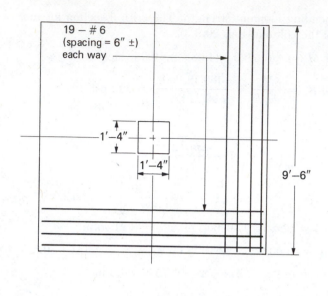

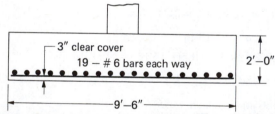

Figure 20.9.4 Design sketch for spread footing of Example 20.9.1.

20.10 DESIGN OF RECTANGULAR FOOTINGS

Rectangular footings may be used in locations where space is restricted to prevent the use of a square footing. The procedure for their design is essentially identical with that of square footings, except that one-way shear action and bending moment must be considered in both principal directions.

EXAMPLE 20.10.1 Design a rectangular spread footing to carry 235 kips dead load and 115 kips live load from an 18-in. square tied column that contains #9 bars. One dimension of the footing is limited to a maximum of 7 ft. The allowable soil pressure is 5500 psf. Neglect the effect of overburden. Use $f'_c = 3000$ psi, $f_y = 40,000$ psi, and the strength method of the ACI Code.

Solution: (a) Determine plan of footing. Assume a footing depth of 2 ft, or 300 psf.

$$\text{net soil pressure} = 5500 - 300 = 5200 \text{ psf}$$

$$\text{required } A = \frac{350}{5.2} = 67.3 \text{ sq ft}$$

Space limitation prevents one dimension from exceeding 7 ft; thus

$$\text{length} = \frac{67.3}{7.0} = 9.6 \text{ ft}$$

Try 7 ft × 9 ft–8 in. (area = 67.7 sq ft).

(b) Determine depth required for shear strength. This footing may be long enough for one-way beam action to govern. In such a case, a direct solution for the effective depth d is reasonably practical. The factored column load is

$$P_u = 235(1.4) + 115(1.7) = 525 \text{ kips}$$

The net upward pressure under factored load condition is

$$p_{\text{net}} = \frac{525,000}{67.7} = 7750 \text{ psf}$$

Using section *A-A* in Fig. 20.10.1, and making the nominal shear strength $V_n = V_c$, so that shear reinforcement is not required, which means

$$V_u = \phi V_c = \phi(2\sqrt{f_c'})b_w d$$

$$7750(7.0)(4.08 - d) = 0.85(2\sqrt{f_c'})(7)(d)144$$

$$d = 1.49 \text{ ft (17.9 in.)}$$

Total depth = 17.9 + 3 (cover) + 1 (estimated bar diameter) = 21.9 in. Try 22 in. for total depth. Check this depth for two-way shear action as a slab, using critical section *B-B-B-B* shown in Fig. 20.10.1 with $d = 18$ in.

$$V_u = 7750[7.0(9.67) - 3.0(3.0)]\tfrac{1}{1000} = 455 \text{ kips}$$

For $\beta_c = 9.67/7.0 = 1.38 < 2.0$,

$$V_c = 4\sqrt{f_c'}b_o d = 4\sqrt{3000}[4(18 + 18)](18)\tfrac{1}{1000} = 568 \text{ kips}$$

$$\phi V_c = 0.85(568) = 483 \text{ kips} > [V_u = 455 \text{ kips}] \qquad\qquad \text{OK}$$

(c) Check transfer of load at base of column. Assuming transfer without aid of dowels, the design strength is

$$\phi P_n = \phi(0.85f_c')A_g = 0.70(0.85)(3)(324) = 578 \text{ kips}$$

$$\phi P_n > [P_u = 525 \text{ kips}] \qquad\qquad \text{OK}$$

Only the minimum dowels required by ACI-15.8.2.1 are needed.

$$\text{min required } A_s = 0.005(324) = 1.62 \text{ sq in.}$$

Use 4-#6 bars as dowels ($A_s = 1.76$ sq in.).

Minimum embedment length L_1 (Fig. 20.10.1) required,

$$\text{min } L_1 = L_d(\#6) = 11.0 \text{ in. (ACI-12.3)}$$

$$\text{available } L_1 = 22 - 3 - 1.5 - 0.75 = 16.75 \text{ in.} > 11.0 \text{ in.} \qquad \text{OK}$$

(d) Design for bending moment strength. The reinforcement in the long direction is distributed uniformly across the 7-ft width, while that in

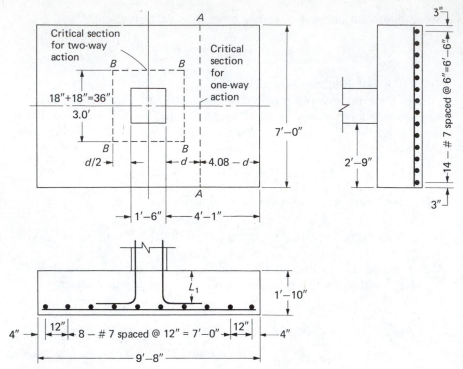

Figure 20.10.1 Rectangular footing for Example 20.10.1.

the short direction is concentrated more heavily under the column in a band equal to the footing width and less heavily near the ends. It is prescribed in ACI-15.4.4 that the portion $2/(\beta + 1)$ of the total transverse reinforcement should be placed in the central band (of a width equal to the short side of the footing) wherein β is the ratio of the long side to the short side of the footing. The ratio $2/(\beta + 1)$ may be derived on the basis that the intensity of reinforcement in the central band is twice that of the outer portions.

In the long direction,

$$M_u = 7.75(7.0)\frac{(4.08)^2}{2} = 452 \text{ ft-kips}$$

$$\text{required } R_n = \frac{M_u}{\phi b d^2} = \frac{452(12,000)}{0.90(7)(12)(18.5)^2} = 209 \text{ psi}$$

Since the longitudinal bars are placed below the transverse bars, d has been taken as 18.5 in. in the above calculation.

Using the trial moment arm method, rather than the formula, Eq. (3.8.5), for ρ, assume the moment arm as $0.95d = 17.6$ in. since the value of required R_n is very low.

$$\text{required } A_s = \frac{M_u}{\phi f_y(\text{arm})} = \frac{452(12)}{0.90(40)(17.6)} = 8.55 \text{ sq in.}$$

Check:

$$C = 0.85f_c'ba = 0.85(3)(84)a = 214a$$

$$T = A_sf_y = 8.55(40) = 342 \text{ kips}$$

$$a = \frac{342}{214} = 1.60 \text{ in.}$$

$$\text{arm} = 18.5 - 0.80 = 17.70 \text{ in.} \approx 17.6 \text{ in. assumed}$$

$$\text{revised required } A_s = 8.55\left(\frac{17.6}{17.7}\right) = 8.5 \text{ sq in.}$$

Use 14-#7, $A_s = 8.4$ sq in. (close enough).

In the short direction,

$$M_u = 7.75(9.67)\frac{(2.75)^2}{2} = 283 \text{ ft-kips}$$

$$\text{assume arm} = 0.95d = 0.95(17.5) = 16.6 \text{ in.}$$

$$\text{required } A_s = \frac{M_u}{\phi f_y(\text{arm})} = \frac{283(12)}{0.90(40)(16.6)} = 5.7 \text{ sq in.}$$

Check:

$$C = 0.85(3)(116)a = 296a$$

$$T = 5.7(40) = 228$$

$$a = \frac{228}{296} = 0.77 \text{ in.}$$

$$\text{arm} = 17.5 - 0.39 = 17.11 \text{ in.}$$

$$\text{revised required } A_s = 5.7\left(\frac{16.6}{17.11}\right) = 5.5 \text{ sq in.}$$

For minimum reinforcement, ACI-10.5.3 requires $\rho_g = 0.002$. For the short direction,

$$\text{min required } A_s = 0.002(116)(22) = 5.1 \text{ sq in.}$$

In this case, the strength requirement controls.
Try 9-#7, $A_s = 5.6$ sq in.

$$\frac{\text{reinforcement in band width}}{\text{total reinforcement}} = \frac{2}{\beta + 1} = \frac{2}{9.67/7.0 + 1} = 0.84$$

Number of bars in the 7-ft band = $9(0.84) = 7.6$, say 8. If one bar is placed on each side outside the 7-ft band, a total of 10 bars would be required.

Use 10-#7 bars.

(e) Check development of reinforcement. Using the formulas of ACI-12.2, or Table 6.7.1,

$$L_d(\#7) = 17.5 \text{ in.}$$

The minimum available embedment is in the short direction and equals 33 in. less an inch or two of cover; this is more than adequate.

The complete design is shown in Fig. 20.10.1.

20.11 DESIGN OF PLAIN AND REINFORCED CONCRETE WALL FOOTINGS

Wall footings carrying direct concentric loads may be of either plain or reinforced concrete. Those that are required to carry moment, such as for the cantilever retaining wall, are treated in Chap. 12. Since the wall footing has bending in only one direction, it may be designed or investigated by considering a typical 12-in. strip along the wall. Many typical walls carry relatively light loads, and the supporting footings are proportioned by using arbitrary minimums. Footings carrying light loads on good soil are often made of plain concrete and may be designed in accordance with *Building Code Requirements for Structural Plain Concrete* [8].

EXAMPLE 20.11.1 Determine the adequacy of the plain concrete wall footing of Fig. 20.11.1 to carry a load of 20 kips/linear ft dead load including the wall weight and 8 kips/linear ft live load. Use $f'_c = 3000$ psi, an allowable soil pressure of 6 ksf, and the ACI Code.

Solution:

$$\text{total service load} = 28 + 6(2.5)(0.145) = 30.2 \text{ kips/ft}$$

$$\text{maximum soil pressure} = \frac{30.2}{6.0} = 5.0 \text{ ksf} < 6 \text{ ksf} \qquad \text{OK}$$

The factored bending moment is computed on the critical section at the face of the wall (ACI-15.4.2).

$$w_u = 20(1.4) + 8(1.7) = 41.6 \text{ kips/ft}$$

$$\text{net soil pressure under factored load} = 41.6/6 = 6.95 \text{ ksf}$$

$$M_u = 6.95 \frac{(2.5)^2}{2} = 21.7 \text{ ft-kips/ft}$$

For computing the moment of inertia of the section, the bottom 2 in. of concrete placed against the ground are assumed to be of poor quality and required [8, Sec. 6.3.5] to be neglected. Neglecting the bottom 2 in.,

$$I_g = \frac{12(28)^3}{12} = 21,950 \text{ in.}^4$$

For bending of plain concrete, the design strength is based on a linear stress-strain relationship for both tension and compression [8, Sec. 6.3.1] and a conservatively low value of $5\sqrt{f'_c}$ for the modulus of rupture [8, Sec. 6.2.1]. A strength reduction factor of 0.65 is also prescribed [8, Sec. 6.2.2]. Thus,

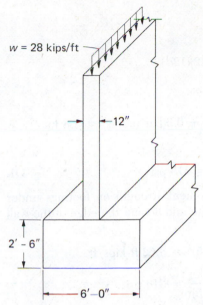

Figure 20.11.1   Wall footing for Example 20.11.1.

$$\phi M_n = \phi f_t I_g/(h/2)$$

$$= \frac{0.65(5\sqrt{3000})(21,950)}{14(12,000)} = 23.3 \text{ ft-kips/ft}$$

$$\phi M_n > [M_u = 21.7 \text{ ft-kips/ft}] \qquad\qquad \text{OK}$$

Shear strength is adequate since the critical section at a distance d from the face of wall falls near the edge of the footing; ACI-15.5.2, 11.11.1.1, and 11.1.3.1 apply here.

EXAMPLE 20.11.2 Design a reinforced concrete footing for a 12-in. masonry wall carrying 10 kips/linear ft dead load including the wall weight and 5 kips/ft live load. Use $f'_c = 3000$ psi, $f_y = 40,000$ psi, an allowable soil pressure of 4000 psf, and the strength method of the ACI Code.

Solution: Assume the footing depth to be 10 in. at 125 psf. Allowable net soil pressure $= 4000 - 125 = 3875$ psf. Footing width $= 15/3.875 = 3.87$ ft. Use 4 ft. It is probable that the thickness will be governed by shear (inclined cracking) which is taken to be critical at a distance d from the face of the wall (Fig. 20.11.2).

Applying the overload factors,

$$w_u = 10(1.4) + 5(1.7) = 22.5 \text{ kips/ft}$$

Net soil pressure under factored load $= 22.5/4 = 5.63$ ksf. When no shear reinforcement is used, the nominal strength for one-way action, using the simplified procedure, is

$$V_n = V_c = 2\sqrt{f'_c}b_w d$$

which requires that

$$V_u = \phi V_c$$

$$5630(1.5 - d) = 0.85(2\sqrt{3000})(12)(12d)$$

$$2.38d = 1.5 - d$$

$$d = 0.44 \text{ ft } (5.3 \text{ in.})$$

$$\text{total thickness} = 5.3 + 3(\text{cover}) + 0.5(\text{bar radius}) = 8.8 \text{ in.}$$

Use 9 in. thickness.

$$\text{check weight} = 113 \text{ psf} \qquad \text{OK}$$

The critical section for bending moment strength on footings under masonry walls occurs halfway between the middle and the edge of the wall (ACI-15.4.2b)

$$M_u = \tfrac{1}{2}(5.63)(1.75)^2 = 8.62 \text{ ft-kips/ft}$$

$$\text{required } R_n = \frac{8.62(12,000)}{0.90(12)(5.5)^2} = 316 \text{ psi}$$

The steel area may be obtained by formula, Eq. (3.8.5), by trial as in Example 20.9.1, or from Fig. 3.8.1. Using the last,

$$\rho \approx 0.0085$$

$$\text{required } A_s = \rho bd = 0.0085(12)(5.5) = 0.56 \text{ sq in.}$$

Try #5 @ $6\frac{1}{2}$ in. ($A_s = 0.57$ sq in./ft). Check strength.

$$C = 0.85f'_c ba = 0.85(3)(12)a = 30.6a$$

$$T = 0.57(40) = 22.8 \text{ kips}$$

$$a = \frac{22.8}{30.6} = 0.75 \text{ in.}$$

$$\phi M_n = 0.90(22.8)[5.5 - 0.5(0.75)]\tfrac{1}{12} = 8.75 \text{ ft-kips} > M_u \qquad \text{OK}$$

For development of reinforcement, the embedment required is

$$L_d(\#5) = 12 \text{ in. minimum} \qquad (\text{see Table } 6.7.1)$$

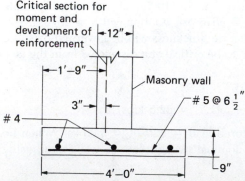

Figure 20.11.2 Wall footing for Example 20.11.2.

Since 21 in., less an inch or two for cover, are available for embedment of main reinforcement, this is acceptable.

See Fig. 20.11.2 for the details of the complete design. Some longi-- tudinal reinforcement for shrinkage should probably be provided; say 3-#4 bars.

20.12 COMBINED FOOTINGS

A combined footing is one that usually supports two columns. These may be two interior columns [Fig. 20.12.1(a)] which are so close to each other that isolated footing areas would overlap. If a property line exists at or near the edge of an exterior column, a rectangular [Fig. 20.12.1(b)] or a trape- zoidal [Fig. 20.12.1(c)] combined footing may be used to support the exterior column and its adjacent interior column. The area of the combined footing may be proportioned for uniform settlement by making its centroid coincide with the resultant of the respective portions of the two column loads that are sustained for long duration. It may be noted that for footings of constant thickness the centroid of the bearing area always coincides with the resultant of the weight of the footing itself.

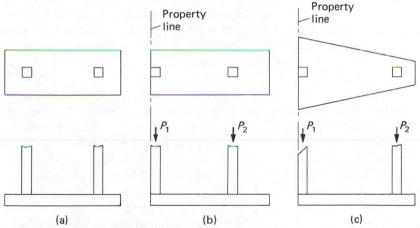

Figure 20.12.1   Combined footings.

Referring to the frequently occurring situation in Fig. 20.12.1(b), the load P_1 is close to the property line; however, there is adequate space to the right of P_2. Whenever P_2 exceeds P_1, a rectangular combined footing can be used, since it may be made long enough to make the load resultant and the footing centroid coincide. It may be shown that if $\frac{1}{2} < P_2/P_1 < 1$ approximately, a trapezoidal footing could be used. If $P_2/P_1 < \frac{1}{2}$ approxi- mately, then either a strap (Fig. 20.12.2) or a T-shaped spread footing would have to be used.

In the strength computation for a combined footing, maximum loads in columns (full dead load plus reduced live load as discussed in Sec. 20.7) should be used. Since the resultant of the maximum column loads does not

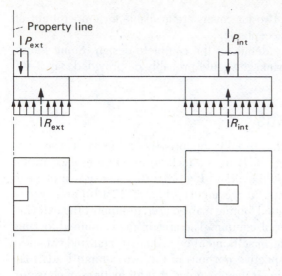

Figure 20.12.2 Cantilever, or strap, combined footing.

necessarily coincide with that of the sustained column loads producing uniform settlement, under the former loading condition the distribution of the net soil pressure along the footing is not uniform. The deviation from uniformity is usually so small that certain approximate short-cut procedures may be used in determining the shear and moment diagrams in the longitudinal direction.

The ACI Code (see ACI-15.10) does not provide full recommendations for combined footings. Kramrisch [9] has provided a detailed treatment of footing design, with particular emphasis on combined footings. However, ACI Committee 436 [10] has given design procedures for combined footings and mats, and additional suggestions have been given by Kramrisch and Rogers [11], Szava-Kovats [12], and Davies and Mayfield [13]. Basically, transverse steel under each column tends to distribute the column load in the transverse direction. This being considered accomplished, the combined footing itself becomes a beam in the longitudinal direction. Thus it is suggested that the provisions for isolated footings be applied to the transverse direction, and those for beams to the longitudinal direction.

The cantilever or strap footing shown in Fig. 20.12.2 is an alternate design to prevent overturning of an exterior footing placed eccentrically under an exterior column, the edge of which is at or close to a property line. Overturning of the exterior footing is prevented by connecting it with the adjacent interior footing by a strap beam. This strap beam is subjected to a constant shearing force and a linearly varying negative bending moment. Thus it behaves like a cantilever beam; hence the name "cantilever footing."

In the strength computation for a cantilever footing (Fig. 20.12.2), the weight of the strap, the exterior footing, and the interior footing is each assumed to be balanced by the soil pressure caused by it and thus such weight causes no shears and moments in any part of the structure. In the longitudinal direction, the column loads P_{ext} and P_{int} are in equilibrium with the total "net" upward soil pressures R_{ext} and R_{int} under the exterior and interior footings; the upward soil pressure is assumed to be uniform over the entire area of each footing. The exterior footing may be considered as under one-way transverse bending although some reinforcement in the longitudinal direction is desirable, and the interior footing is under two-way bending as in isolated footings.

Use of a cantilever footing may be justifiable under conditions where the distance between columns is large and a large area of excavation must be avoided. It is usual practice that the bottom surfaces of the exterior footing, the strap, and the interior footing be at the same elevation, but the total thickness of each element may be different, depending on the strength requirements. Certainly it is desirable, unless there are good reasons to the contrary, to make all three elements of constant thickness.

20.13 DESIGN OF COMBINED FOOTINGS

The design of two common types of combined footings will be shown. The first is a rectangular footing, and the second is a strap or cantilever footing.

EXAMPLE 20.13.1 Design a combined footing to support two columns as shown in Fig. 20.13.1(a): P_A = 350 kips (40% live load); P_B = 400 kips (40% live load); column A is centered 1 ft–3 in. from property line, and column B is centered 19 ft–3 in. from property line. Use f'_c = 3000 psi, f_y = 40,000 psi, maximum soil pressure = 5000 psf, and the strength method of the ACI Code. Assume that the ratio of maximum column loads as given is equal to that of long duration loads in the exterior and interior columns.

Solution: (a) Length and width of footing. ACI-15.2.2 indicates base area of footings is to be determined using service loads and allowable soil pressure.

$$\bar{x} \text{ from property line} = \frac{350(1.25) + 400(19.25)}{750} = 10.85 \text{ ft}$$

$$\text{length of footing, } L = 10.85(2) = 21.70 \text{ ft}$$

Use 21 ft–9 in.

Since the design for strength of the footing involves factored loads, there will be an eccentricity no matter how "exact" is the length determination. In the design for shear and bending moment, the soil pressure under factored loads might be taken as linearly varying to account for the eccentric loading, but in the case of a small eccentricity it is probably sufficient to assume a uniform soil pressure as is done in this example.

Assume the footing thickness to be about 2 ft–6 in., or 375 psf.

$$\text{net soil pressure} = 5000 - 375 = 4625 \text{ psf}$$

$$\text{footing width} = \frac{750,000}{4625(21.75)} = 7.46 \text{ ft}$$

Try 7 ft–6 in.

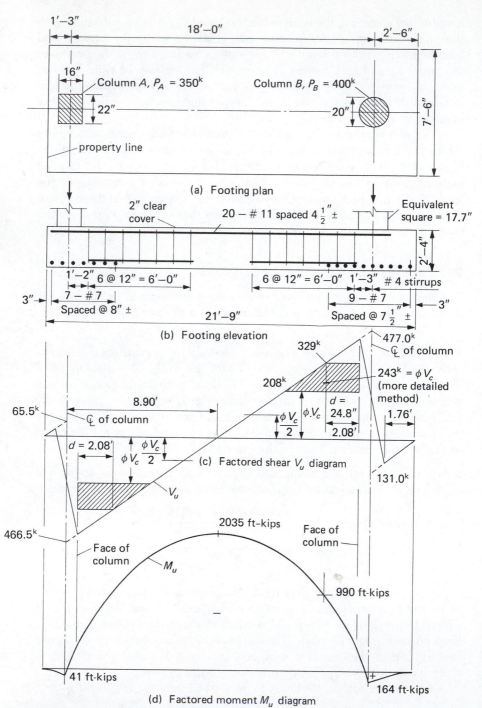

Figure 20.13.1 Rectangular combined footing for Example 20.13.1.

(b) Longitudinal factored shears and factored moments. For gravity loading,

$$\text{column } A, P_u = 210(1.4) + 140(1.7) = 532 \text{ kips}$$

$$\text{column } B, P_u = 240(1.4) + 160(1.7) = 608 \text{ kips}$$

$$\text{net soil pressure under factored load } = \frac{1,140,000}{21.75(7.5)} = 6990 \text{ psf}$$

The factored shear V_u diagram is computed as for a beam and given in Fig. 20.13.1(c). For simplicity, the column loads are taken to be acting along the column centerlines, thus producing the dashed portions within the column widths on the shear diagram. In the computations, the 20-in. diameter column is treated as an equivalent square of side 17.7 in. in accordance with ACI-15.3.

$$\text{net upward uniform pressure } = 7.5(6.99) = 52.4 \text{ kips/ft}$$

$$V_u \text{ at centerline of column } A = +52.4(1.25) = +65.5 \text{ kips}$$

$$+65.5 - 532 = -466.5 \text{ kips}$$

$$V_u \text{ at centerline of column } B = -52.4(2.5) = -131.0 \text{ kips}$$

$$-131.0 + 608 = +477.0 \text{ kips}$$

$$\text{point of zero shear } = 18\left(\frac{466.5}{466.5 + 477.0}\right)$$

$$= 8.90 \text{ ft from centerline of column } A$$

The factored moment M_u diagram as computed for a beam is given in Fig. 20.13.1(d). Note that the numerical values on the two small end portions of the factored moment diagram are based on assuming all of the column loads to be concentrated at the column centerlines.

$$\text{max } M_u \text{ (computed from left side) } = \frac{52.4(10.15)^2}{2} - 532(8.90)$$

$$= -2035 \text{ ft-kips}$$

$$\text{max } M_u \text{ (computed from right side) } = \frac{52.4(11.60)^2}{2} - 608(9.10)$$

$$= -2006 \text{ ft-kips}$$

The moments as computed from both sides are not exactly the same because a footing length of 21 ft–9 in. is used instead of the computed 20.70 ft and because the distance 8.90 ft to the point of zero shear contains only three significant figures. Use $M_u = 2035$ ft-kips.

(c) Thickness of slab. For moment, the thickness may be based on a desired reinforcement ratio ρ. The maximum permitted by the ACI Code is from Table 3.8.1,

$$\text{max } \rho = 0.75\rho_b = 0.0278$$

For reasonable deflection control, select $\rho = 0.014$, that is, approximately one-half of the maximum permitted. For this value of ρ, using Eq. (3.8.4),

$$R_n = \rho f_y (1 - \tfrac{1}{2}\rho m)$$

$$m = \frac{f_y}{0.85 f_c'} = \frac{40}{0.85(3)} = 15.7$$

$$R_n = 0.014(40,000)[1 - 0.5(0.014)(15.7)] = 498 \text{ psi}$$

$$\text{required } d = \sqrt{\frac{M_u}{\phi R_n b}} = \sqrt{\frac{2035(12,000)}{0.90(498)(7.5)(12)}} = 24.6 \text{ in.}$$

The footing is considered a beam in shear computations. One-way action is assumed to control at the distance d from the face of the columns. The shear at a distance d from the face of the 17.7-in. equivalent square column is

$$V_u = 477.0 - \left(\frac{8.85 + d}{12}\right)(52.4) = 438.3 - 4.37d$$

The nominal shear strength when no shear reinforcement is to be used is

$$V_n = V_c = 2\sqrt{f_c'}\,b_w d$$

unless the more detailed expression involving $\rho V_u d / M_u$ is used. Then,

$$V_u = \phi V_c$$

$$438.3 - 4.37d = 0.85(2\sqrt{3000})(7.5)(12)d/1000$$

$$438.3 - 4.37d = 8.38d$$

$$d = 34.4 \text{ in.}$$

It seems desirable to make the footing deep enough at the desirable reinforcement ratio for the bending moment, but not deep enough to give the extra 10 in. that would be required to eliminate stirrups.

total depth $= 24.6 + 2(\text{cover}) + 0.5(\text{stirrup}) + 0.6(\text{bar radius}) = 27.7$ in.

Use 28 in. Check the weight, $28(150)/12 = 350$ psf.

$$\text{max soil pressure} = \frac{750,000}{21.75(7.5)} + 350 = 4950 \text{ psf} < 5000 \text{ psf} \quad \text{OK}$$

(d) Main longitudinal reinforcement. At the middle of the span,

$$\text{required } R_n = \frac{M_u}{\phi b d^2} = \frac{2035(12,000)}{0.90(7.5)(12)(24.9)^2} = 484 \text{ psi}$$

$$\text{required } A_s \approx 0.014\left(\frac{484}{498}\right)(7.5)(12)(24.9) = 30.5 \text{ sq in.}$$

Try 20-#11 (approximate $4\frac{1}{2}$ in. spacing); $A_s = 31.2$ sq in.

The anchorage required from the maximum moment point equals the development length L_d for #11 top bars, which from Tables 6.7.1 and 6.8.1 equals

$$L_d(\#11) = 1.4 L_{db} = 1.4(45.5) = 64 \text{ in.}$$

The 1.4 is the modification factor for bars cast with more than 12 in. of concrete beneath them. If all 20 of the #11 bars are run into the centerlines of the columns, the development requirement is more than adequate, since an anchorage distance of 108 in. is provided on both sides of the point of maximum moment.

Use 20-#11 bars at 18 ft long (spaced approximately at $4\frac{1}{2}$ in.).

Check development length requirement at the points of inflection according to ACI-12.11.3. The Code provision is checked here even though the reinforcement is negative moment reinforcement rather than the positive moment reinforcement as prescribed in ACI-12.11.3. The footing may be visualized for this purpose as an inverted beam with the soil pressure as loading and the columns as supports. The situation is similar to the positive moment requirement in the sense that the required embedment is into the support (column) rather than out into the span as for the negative moment requirement in an ordinary continuous beam. Furthermore, since the inflection points are inside the faces of the columns, the 1.3 factor might be used; however, since the column width (22 in.) is only a small fraction of the footing width (7.5 ft), the authors do *not* recommend using the 1.3 factor.

Since all 20-#11 bars extend through the inflection points,

$$M_n = \frac{2035}{0.90}\left(\frac{31.2}{30.5}\right) = 2300 \text{ ft-kips}$$

$$V_u = 466.5 \text{ kips (at column } A)$$

$$V_u = 477.0 \text{ kips (at column } B)$$

$$L_a = 15 \text{ in. approx (near both columns } A \text{ and } B)$$

In this case the actual distance L_a from the inflection point to the end of the bars is less than the maximum L_a limits of effective depth d or 12 bar diameters; thus the actual distance of 15 in. controls. Since M_u and L_a are the same at both inflection points, the one near column B having the larger shear V_u controls. At column B inflection point,

$$\frac{M_n}{V_u} + L_a = \frac{2300(12)}{477.0} + 15 = 73 \text{ in.} > L_d = 64 \text{ in.} \qquad \text{OK}$$

(e) Alternate design of the main longitudinal reinforcement using cutoff. If it seems desirable to cut off some of the tension bars, say 6-#11, and extend the remaining ones into the supports, the general anchorage requirements of ACI-12.10 must be applied. The theoretical point at which the 6-#11 bars are no longer required is indicated in Fig. 20.13.2, corresponding to the moment capacity ϕM_n of 1500 ft-kips. The provision in ACI-12.10.3 requires an extension of at least the effective depth d or 12 bar diameters, whichever is greater, beyond the theoretical cutoff point. In this case the effective depth of 24.8 in. (or 2.1 ft) controls over 12 diameters = 16.9 in. (or 1.4 ft).

In addition, for such a cutoff in the tension bars to be permitted. ACI-12.10.5 must be satisfied. It is assumed here that stirrups will be provided sufficient to satisfy ACI-12.10.5.1. The continuing bars must be embedded

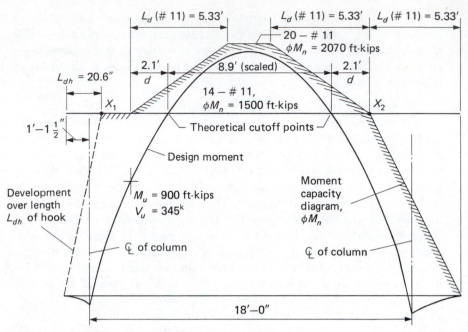

Figure 20.13.2 Bar cutoff alternate for Example 20.13.1.

L_d beyond the cutoff of the 6-#11 bars according to ACI-12.10.4. At the end under column A there is not adequate distance available for embedment (≈ 3.5 ft available $< L_d = 5.33$ ft required). Hooks would need to be provided at the end under column A so the full capacity of the continuing bars would be available at the cutoff for the 6 bars. At the end under column B adequate embedment length is available.

When hooks are provided as anchorage beyond an inflection point, ACI-12.11.3 need not be satisfied. Referring to Table 6.11.2 and Fig. 6.11.2 for #11 hooked bars,

$$L_{dh} = 20.6 \text{ in.}$$

Allowing 1.5-in. cover from the property line, the #11 hooked bars will develop their full strength at $20.6 + 1.5 = 22.1$ in. from the property line (point X_1 on Fig. 20.13.2).

At column B inflection point, no hooks are required if straight embedment L_a is available. Solving ACI Formula (12-1) for L_a,

$$\frac{M_n}{V_u} + L_a = \frac{1670(12)}{477.0} + L_a = L_d = 64 \text{ in.}$$

$$\text{required } L_a = 22 \text{ in.}$$

Thus for the alternate design, 14-#11 bars would be required for 20 ft–9 in. (i.e., $1.25 + 8.90 + 8.90 + 22/12 - 1.5/12$) including the hook at the end under column A. This would not, however, provide an embedment L_d beyond point X_2 (Fig. 20.13.2) where the 6-#11 are terminated (ACI-12.10.4).

If it may be assumed ACI-12.10.4 refers to the *theoretical* cutoff point, then the 20 ft–9 in. could be used, otherwise the 14 bars would have to run full length (21 ft–9 in. less cover on ends; say 21 ft–6 in.). The 6-#11 cut bars would have to be 13 ft–3 in. long. Further, stirrups near the cutoffs would have to be provided so that ACI-12.10.5.1 is satisfied (this calculation is not shown).

(f) Longitudinal reinforcement at bottom of footing beyond column centers. The bending moment at the face of column B is

$$M_u = \tfrac{1}{2}(52.4)(1.76)^2 = 81.2 \text{ ft-kips}$$

Though certainly not always the case, the moment here appears small enough to require no reinforcement. The strength of the unreinforced section in flexure is computed by using $\phi = 0.65$ according to Sec. 6.2.2 of *Building Code Requirements for Structural Plain Concrete* [8]. Neglecting the bottom 2 in. of thickness,

$$I_g = \tfrac{1}{12}(7.5)(12)(26)^3 = 132{,}000 \text{ in.}^4$$

$$\phi M_n = \phi\left(\frac{f_t I_g}{h/2}\right) = 0.65\left[\frac{5\sqrt{3000}(132{,}000)}{13(12{,}000)}\right] = 151 \text{ ft-kips}$$

$$\phi M_n > [M_u = 81.2 \text{ ft-kips}] \qquad\qquad \text{OK}$$

No flexural reinforcement in the longitudinal direction is required for strength at the bottom of either cantilever.

(g) Transverse reinforcement. Bending in the transverse direction may be treated in a manner similar to isolated spread footings. The 1940 Joint Committee [14] recommended that the transverse reinforcement at each column should be placed uniformly within a band having a width not greater than the width of the column plus twice the effective depth of the footing.

The procedure seems to be reasonable. Certainly the behavior of the footing depends on the overall length-to-width ratio as well as the spacing of the columns. In this design example, the large spacing of the columns and the relatively narrow footing mean that most of the footing between the columns will be subjected only to longitudinal curvature while locally, in the vicinity of the concentrated loads, curvature in both directions will result. Thus the transverse reinforcement is put into bands as shown in Fig. 20.13.3.

$$\text{column } A \text{ band width, } W_A = 1.25 + \frac{8 + 24.8}{12} = 4.0 \text{ ft}$$

$$\text{net factored load pressure in transverse direction} = \frac{532}{7.5} = 71.0 \text{ kips/ft}$$

$$M_u = \tfrac{1}{2}(71.0)(2.83)^2 = 284 \text{ ft-kips}$$
$$d = 28 - 3 \text{ (cover at bottom)} - 0.5(\text{bar radius}) = 24.5 \text{ in.}$$
$$\text{required } R_n = \frac{M_u}{\phi b d^2} = \frac{284(12{,}000)}{0.90(4.0)(12)(24.5)^2} = 132 \text{ psi}$$

From Fig. 3.8.1, $\rho \approx 0.0035$, which exceeds the minimum reinforcement ratio required by ACI-10.5.3 (0.002 from ACI-7.12).

$$\text{required } A_s = 0.0035(4.0)(12)(24.5) = 4.1 \text{ sq in.}$$

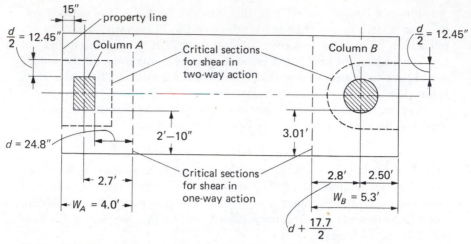

Figure 20.13.3 Band width for transverse reinforcement.

Try 7-#7 bars ($A_s = 4.20$ sq in.). Check strength.

$$C = 0.85f'_cba = 0.85(3)(4.0)(12)a = 122.4a$$

$$T = 4.20(40) = 168 \text{ kips}$$

$$a = \frac{168}{122.4} = 1.37 \text{ in.}$$

$$\phi M_n = 0.90(168)\,[24.5 - 0.5(1.37)]\,\tfrac{1}{12} = 300 \text{ ft-kips} > 284 \text{ ft-kips} \quad \text{OK}$$

Since 2 ft–10 in. is available from face of column, embedment can be provided exceeding the required development length L_d for #7 bars ($L_d = 17.5$ in. from Table 6.7.1).

Use 7-#7 bars (approximate 8 in. spacing).

$$\text{column } B \text{ band width } W_B = 2.50 + \frac{8.85 + 24.8}{12} = 5.3 \text{ ft}$$

$$\text{net factored load pressure in transverse direction} = \frac{608}{7.5} = 81.0 \text{ kips/ft}$$

$$M_n = \tfrac{1}{2}(81.0)(3.01)^2 = 367 \text{ ft-kips}$$

$$\text{required } R_n = \frac{M_u}{\phi bd^2} = \frac{367(12,000)}{0.90(5.3)(12)(24.5)^2} = 128 \text{ psi}$$

From Fig. 3.8.1, $\rho \approx 0.0035$, approximately the same as for column A.

$$\text{required } A_s = 0.0035(5.3)(12)(24.5) = 5.5 \text{ sq in.}$$

Try 9-#7 ($A_s = 5.40$ sq in.). Check strength.

$$C = 0.85f'_cba = 0.85(3)(5.3)(12)a = 162a$$

$$T = 5.40(40) = 216 \text{ kips}$$

$$a = \frac{216}{162} = 1.33 \text{ in.}$$

$$\phi M_n = 0.90(216)[24.5 - 0.5(1.33)]\tfrac{1}{12} = 386 \text{ ft-kips} > 367 \text{ ft-kips} \quad \text{OK}$$

Anchorage of #7 bars is adequate since available embedment of 3.01 ft exceeds L_d of 17.5 in.

Use 9-#7 bars (approximately $7\tfrac{1}{2}$ in. spacing).

(h) Shear reinforcement. The usual approach is to consider the footing as a beam and to provide shear reinforcement on the assumption that the shear (inclined cracking) effect is uniform across the width. This approach seems appropriate in this case with the large distance between columns and the relatively narrow footing width.

The maximum shear to be provided for is at the critical section a distance d from the face of the column. From Fig. 20.13.1, $V_u = 329$ kips. The design shear strength ϕV_c of a beam without shear reinforcement, using the simplified procedure, is

$$\phi V_c = \phi(2\sqrt{f_c'})b_w d$$
$$= 0.85(2\sqrt{3000})(90)(24.8)\tfrac{1}{1000} = 208 \text{ kips}$$

$$\text{required } \phi V_s = V_u - \phi V_c = 329 - 208 = 121 \text{ kips}$$

When shear reinforcement is required, there must be at least the minimum specified by ACI-11.5.5.3, as follows:

$$\text{min } \phi V_s = \phi(50)b_w d = 0.85(50)(90)(24.8)\tfrac{1}{1000} = 95 \text{ kips} > 121 \text{ kips} \quad \text{OK}$$

Thus the 121 kips controls the closest spacing for shear reinforcement. Even though this is a footing, the rules for beams are believed appropriate here. For design of shear reinforcement,

$$\frac{A_v}{s} = \frac{\phi V_s}{\phi f_y d} = \frac{121}{0.85(40)24.8} = 0.144 = \frac{A_s N}{s}$$

In the above expression, A_s is the cross section of the stirrup bar, and N is the number of times the multiple-loop stirrup crosses the neutral axis of the beam. Use a #4 stirrup, with $N = 8$ (see Fig. 20.13.4), spaced at 12 in.

Figure 20.13.4 Multiple-loop stirrup ($N = 8$).

This gives

$$\text{provided } \frac{A_v}{s} = \frac{0.20(8)}{12} = 0.133$$

which is about 8% too low.

This may be a situation where computation using the more detailed expression for V_c is justified to verify using the above arrangement.

$$\phi V_c = \phi \left[1.9\sqrt{f_c'} + 2500\frac{\rho V_u d}{M_u} \right] b_w d$$

$$\rho = \frac{A_s}{bd} = \frac{20(1.56)}{7.5(12)(24.8)} = 0.0140$$

$$\frac{V_u d}{M_u} = \frac{329{,}000(24.8)}{990(12{,}000)} = 0.69 < 1.0 \qquad \text{OK}$$

$$[1.9\sqrt{3000} + 2500(0.0140)(0.69) = 2.34\sqrt{f_c'}] < [3.5\sqrt{f_c'}](\text{max}) \quad \text{OK}$$

$$\phi V_c = 0.85(2.34\sqrt{3000})(90)(24.8)\tfrac{1}{1000} = 243 \text{ kips}$$

$$\phi V_s = V_u - \phi V_c = 329 - 243 = 86 \text{ kips} < [\text{min } \phi V_s = 95 \text{ kips}]$$

The minimum requirement now controls, giving

$$\text{required } \frac{A_v}{s} = \frac{95}{0.85(40)24.8} = 0.113 < 0.133$$

Use #4 stirrups @ 12 in. with $N = 8$, as in Fig. 20.13.4. The 12-in. spacing is the maximum permitted (ACI-11.5.4.3) when v_s does not exceed $4\sqrt{f_c'}$, which corresponds to $\phi V_s = 415$ kips.

The first stirrup is placed at $s/2 = 6$ in. from the face of the column, and the last one should be within $s/2 = 6$ in. of the location where $V_u = \phi V_c/2$, at which shear reinforcement is no longer theoretically required for beams (ACI-11.5.5.1). The stirrup arrangement will be made identical at each column. A few longitudinal bars are placed in the bottom of the footing in order to space and hold in position the stirrups and the transverse reinforcement.

(i) Check shear strength based on two-way action. At the exterior column A, the calculation based on a perimeter at $d/2$ (12.45 in.) from the column face on all four sides would be unrealistic on the side nearest the property line. Instead a three-sided perimeter is used as at column A in Fig. 20.13.3,

$$b_0 = 2(15 + 16 + 12.45) + 22 + 2(12.45) = 133.8 \text{ in.}$$

Actually the resisting section for shear is not symmetrical about the column; therefore, the shear distribution is nonuniform and eccentricity of shear should be considered in accordance with ACI-11.12.2. In this example, where there is no moment assumed at the base of column, and the moment in the slab at the face of column is small, it may be reasonable to neglect the eccentric effect. Thus, neglecting any eccentricity of shear,

$$V_u = 532 - 6.99 \left[\frac{43.45(46.90)}{144} \right] = 433 \text{ kips}$$

$$\phi V_c = 0.85 \left[4\sqrt{3000}\,(133.8)(24.9) \right]\tfrac{1}{1000} = 620 \text{ kips}$$

Neglecting any possible contribution of the stirrups used for ϕV_s in one-way shear action,

$$\phi V_n = \phi V_c = 620 \text{ kips} > [V_u = 433 \text{ kips}] \qquad \text{OK}$$

A similar calculation for two-way shear action at column B will show the thickness to be adequate without shear reinforcement. Again, it is prudent to consider the failure section (see Fig. 20.13.3) to extend to the end of the footing, rather than merely the circular path at $d/2$ from the column.

(j) Design sketch. The complete design details are given in Fig. 20.13.1.

EXAMPLE 20.13.2 Design a cantilever or strap footing for the situation shown in Fig. 20.13.5, where the property line is at the exterior edge of the exterior column. Column data are given in Table 20.13.1. The distance between column centers is 18 ft. Equal settlement is assumed for DL plus $\frac{1}{2}$LL condition at a uniform pressure of 3.34 ksf. Use $f'_c = 3000$ psi for footing, $f'_c = 3750$ psi for columns, $f_y = 40,000$ psi, and the strength method of the ACI Code.

Solution: (a) Size of exterior and interior footings. According to ACI-15.2.2 footing size is always determined using service loads, rather than factored loads. The size of exterior and interior footings is a function of the assumed width of the exterior footing (which affects the cantilever action and therefore the reactions required of each footing) and the assumed total thickness of each footing (which affects the available net soil bearing capacity). Should these assumed values be revised in the subsequent computation, the sizes of the footings must be revised accordingly. Assume that the total thickness of the entire footing is 24 in. The net uniform soil pressure for both exterior and interior footings is $3.34 - 2(0.150) = 3.04$ ksf. Referring to Fig. 20.13.5, the equal settlement condition, taking moments about the interior column, gives

$$R_{ext} = \frac{(55 + 35)18}{16.25} = 99.7 \text{ kips}$$

and, taking moments about the exterior column gives

$$R_{int} = \frac{(80 + 65)16.25 - (55 + 35)(2.25 - 0.50)}{16.25} = 135.3 \text{ kips}$$

$$\text{area of exterior footing required} = \frac{99.7}{3.04} = 32.8 \text{ sq ft}$$

Assuming a footing width of 4 ft–6 in.,

$$\text{length of exterior footing} = \frac{32.8}{4.50} = 7.29 \text{ ft} \qquad \text{(try 7 ft–4 in.)}$$

$$\text{area of interior footing required} = \frac{135.3}{3.04} = 44.5 \text{ sq ft}$$

$$\text{side of square interior footing} = \sqrt{44.5} = 6.67 \text{ ft} \qquad \text{(try 6 ft–8 in.)}$$

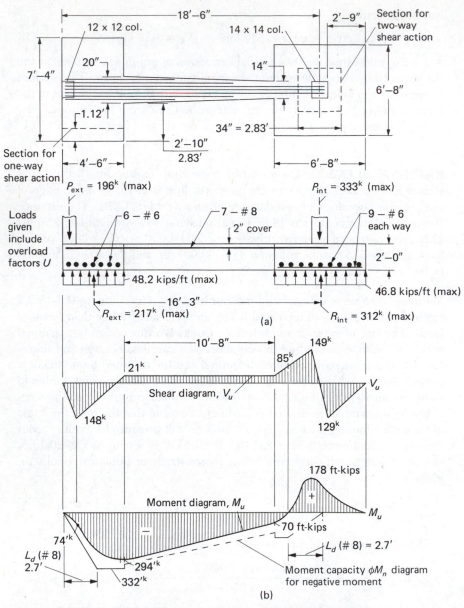

Figure 20.13.5 Cantilever, or strap, footing of Example 20.13.2.

Table 20.13.1

COLUMN	SIZE	REINFORCEMENT	LL	DL
Exterior	12 × 12 in.	4-#7	70 kips	55 kips
Interior	14 × 14 in.	8-#8	130 kips	80 kips

(b) Factored shear and factored moment diagram for strap. Referring to Fig. 20.13.5, applying overload factors U to the maximum load condition gives

$$R_{ext} = \frac{[55(1.4) + 70(1.7)]18}{16.25} = \frac{196(18)}{16.25} = 217 \text{ kips}$$

$$R_{int} = [80(1.4) + 130(1.7)] - \frac{196(1.75)}{16.25} = 333 - 21 = 312 \text{ kips}$$

$$V_u \text{ in strap} = -196 + 217 = +21 \text{ kips}$$

$$M_u \text{ at right end of strap} = 21(3.33) = 70.0 \text{ ft-kips}$$

$$M_u \text{ at left end of strap} = 21(14) = 294 \text{ ft-kips}$$

(c) Design of strap. For the shear requirement, assuming no shear reinforcement is to be used and the simplified expression for strength is used,

$$\phi V_c = \phi(2\sqrt{f_c'})b_w d$$

$$\text{width } b_w \text{ of strap required} = \frac{21,000}{0.85(2\sqrt{3000})21.5} = 10.5 \text{ in.}$$

Assume desirable reinforcement,

$$\rho \approx 0.5\rho_{max} = 0.014$$

for which

$$R_n = 498 \text{ psi} \quad \text{(see Example 20.13.1, part c)}$$

At the junction with the interior footing,

$$\text{width of strap required} = \frac{M_u}{\phi R_n d^2} = \frac{70(12,000)}{0.90(498)(21.5)^2} = 4.1 \text{ in.}$$

At the junction with the exterior footing,

$$\text{width of strap by moment requirement} = \frac{294(12,000)}{0.90(498)(21.5)^2} = 17.1 \text{ in.}$$

Adopt a strap width to vary from 14 in. to 20 in.

$$\text{required } R_n \text{ at wide end} = \frac{M_u}{\phi b d^2} = \frac{294(12,000)}{0.90(20)(21.5)^2} = 424 \text{ psi}$$

$$\text{required } R_n \text{ at narrow end} = \frac{70(12,000)}{0.90(14)(21.5)^2} = 144 \text{ psi}$$

Approximately,

$$\text{required } A_s \text{ at wide end} \approx 0.014\left(\frac{424}{498}\right)(20)(21.5) = 5.2 \text{ sq in.}$$

$$\text{required } A_s \text{ at narrow end} \approx 0.014\left(\frac{144}{498}\right)(14)(21.5) = 1.2 \text{ sq in.}$$

The minimum requirement of ACI-10.5.1 or 10.5.2 applies here because the strap is a beam.

$$\min \rho = \frac{200}{f_y} = 0.005$$

$$\min A_s = 0.005(14)(21.5) = \underline{1.51 \text{ sq in. controls}}$$

or

$$\min A_s = \tfrac{4}{3}(\text{required } A_s) = \tfrac{4}{3}(1.2) = 1.60 \text{ sq in.}$$

Use 7-#8 (A_s = 5.53 sq in.) at the wide end. Extend 3-#8 through narrow end. Two bars would satisfy min $A_s \geq 1.51$ sq in. but extend three bars to maintain symmetry.

$$L_d(\#8) = 1.4L_{db} = 1.4(23.1) = 32.4 \text{ in.}$$

The maximum available embedment length into the footing from end of strap is 4 ft–6 in. which is more than adequate.

(d) Investigate one-way shear at the distance $d(21.5$ in.) from the face of exterior column,

$$V_u = 148 - \frac{21.5}{12}(48.2) = 62 \text{ kips}$$

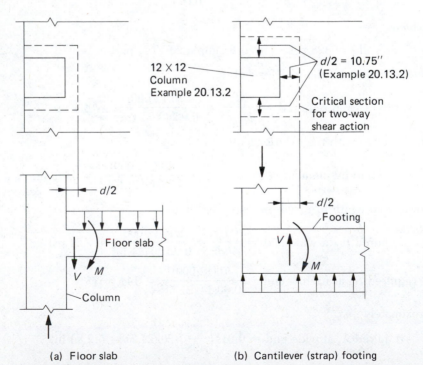

Figure 20.13.6 Comparison of forces to be transferred between floor slab and column to those between exterior footing and column in cantilever (strap) footing.

For no shear reinforcement and the simplified expression for strength,

$$\phi V_c = \phi(2\sqrt{f_c'})b_w d$$

$$= 0.85(2\sqrt{3000})(7.33)(12)(21.5)\tfrac{1}{1000} = 176 \text{ kips} > 62 \text{ kips} \quad \text{OK}$$

(e) Investigate two-way shear action at the critical section $d/2$ from the face of the exterior column. At this location there is an unsymmetrical three-sided perimeter, and there is a bending moment on the section across the 7 ft–4 in. width at the column face. ACI-11.12.2 applies. This shear transfer check is an empirical check of stresses under factored load similar to that for a flat plate at an exterior column. Referring to Fig. 20.13.6, the forces to be considered in analysis of the floor slab and the footing are compared. The floor slab treatment has been discussed in Sec. 16.18, where the details of the computation are shown. In the floor slab the direction of the shear is the opposite of that in the strap footing; however, the moment direction is the same. Thus, the maximum factored shear stress occurs at the interior side of the critical section for the floor slab, but at the exterior edge of the footing slab.

At the face of the column, the factored moment* is

$$M_u = 74 \text{ ft-kips (at face of column)}$$

ACI-11.12.2.3 requires the following fraction of unbalanced moment to be transferred by eccentricity of shear:

$$1 - \frac{1}{1 + \frac{2}{3}\sqrt{\dfrac{c_1 + d/2}{c_2 + d}}} = 1 - \frac{1}{1 + \frac{2}{3}\sqrt{\dfrac{12 + 21.5/2}{12 + 21.5}}} = 0.355$$

Thus,

$$M_v = 0.355 M_u = 0.355(74) = 26.3 \text{ ft-kips}$$

Using the symbols of Fig. 16.18.1, for this example $a = 22.75$ in. and $b = 33.5$ in. The factored shear that must be carried is

$$V_u = \frac{217}{4.5(7.33)}\left[4.5(7.33) - \frac{22.75(33.5)}{144}\right] = 182 \text{ kips}$$

The section properties for the three-sided critical section for shear transfer, as shown in Fig. 20.13.6(b), are

$$A = 21.5[2(22.75) + 33.5] = 1699 \text{ sq in.}$$

$$x_2 = \frac{2(22.75)(11.375)}{2(22.75) + 33.5} = 6.55 \text{ in.}$$

$$J_c = 21.5\left[\frac{2(22.75)^3}{3} - (45.5 + 33.5)(6.55)^2\right] + \frac{22.75(21.5)^3}{6}$$

$$= 133{,}600 \text{ in.}^4$$

*This longitudinal moment is taken from the factored moment diagram of Fig. 20.13.5(c). Where the strap joins the exterior footing there is a sudden discontinuity which may cause an indeterminate redistribution of longitudinal and transverse moments.

At the exterior side of column,

$$v_1 = \frac{182,000}{0.85(1699)} + \frac{26.3(12,000)(22.75 - 6.55)}{0.85(133,600)} = 126 + 45 = 171 \text{ psi}$$

At the interior side of column,

$$v_2 = \frac{182,000}{0.85(1699)} - \frac{26.3(12,000)(6.55)}{0.85(133,600)} = 126 - 18 = 108 \text{ psi}$$

For β_c not exceeding 2.0, the factored load stress is limited to $4\sqrt{f_c'}$ as a maximum; for this example, 219 psi. The computed maximum stress of 171 psi is below $4\sqrt{f_c'} = 219$ psi, and therefore acceptable.

(f) Investigate development of reinforcement at the inflection point (ACI-12.11.3). The Code provision is checked here even though the reinforcement is negative moment reinforcement rather than the positive moment reinforcement as prescribed in ACI-12.11.3 because the footing may be visualized as an inverted beam as discussed in Example 20.13.1, part (d).

Tension reinforcement is 3-#8 at the narrow end of the strap near the inflection point.

$$C = 0.85f_c'ba = 0.85(3)(14)a = 35.7a$$

$$T = 3(0.79)40 = 95 \text{ kips}$$

$$a = \frac{95}{35.7} = 2.66$$

$$M_n = 95(21.5 - 1.33)\tfrac{1}{12} = 160 \text{ ft-kips}$$

$$V_u = 85 \text{ kips}$$

The #8 bars are proposed to be terminated at the interior side of the interior column, giving about 2.5 ft of embedment from the inflection point.

$$L_a = 2.5 \text{ ft}, \quad \text{or } d = 1.79 \text{ ft}, \quad \text{or } 12d_b = 1.0 \text{ ft}$$

$$\frac{M_n}{V_u} + L_a = \left(\frac{160}{85} + 1.79\right)12 = 44 \text{ in.} > [L_d = 32.4 \text{ in.}] \qquad \text{OK}$$

(g) Investigate development of reinforcement in the top of the exterior column footing. Tension reinforcement is 7-#8.

$$C = 0.85f_c'ba = 0.85(3)(7.33)(12)a = 224a$$

$$T = 7(0.79)40 = 221 \text{ kips}$$

$$a = \frac{221}{224} = 0.99 \text{ in.}$$

$$M_n = 221[21.5 - 0.5(0.99)]\tfrac{1}{12} = 387 \text{ ft-kips}$$

$$V_u = 148 \text{ kips}$$

$$1.30\frac{M_n}{V_u} + L_a = 1.30\frac{387(12)}{148} + 4 = 45 \text{ in.} > [L_d = 32.4 \text{ in.}] \qquad \text{OK}$$

Note that ACI-12.11.3 allows a 30% increase to reflect the ends of the reinforcement compressed by a reaction; also L_a is the embedment beyond the *centerline of the support*, which is 4 in. in this case.

It is observed that the assumption of zero earth pressure under the strap has been made; hence in construction the region below the strap should be disturbed and the strap should be formed on the bottom. Furthermore, liberal anchorage lengths (perhaps even hooks) should be provided into the exterior footing [as shown in Fig. 21.13.5(a)] to accommodate fully the tensile force in the top of the footing across to the exterior column. Often some of the steel extending into the exterior footing is flared to distribute better the load from the 20-in. width to the 88-in. width.

(h) Design of exterior footing. In the transverse direction,

$$d = 24 - 3(\text{cover}) - 0.5(\text{est. bar radius}) = 20.5 \text{ in.}$$

$$M_u \text{ at edge of strap} = \left(\frac{217}{7.33}\right)\frac{(2.83)^2}{2} = 118 \text{ ft-kips}$$

$$\text{required } R_n = \frac{M_u}{\phi bd^2} = \frac{118(12,000)}{0.90(54)(20.5)^2} = 70 \text{ psi}$$

From Fig. 3.8.1, required $\rho = 0.002 < [\text{min } \rho_g = 0.002 \text{ (ACI-10.5.3)}]$,

$$\text{required } A_s = 0.002(54)(24) = 2.59 \text{ sq in.}$$

The beam shear (one-way action) at d from the face of the column in the 7.33-ft direction of the footing is

$$V_u = \frac{217}{7.33}(1.12) = 33 \text{ kips}$$

$$\phi V_c = \phi(2\sqrt{f_c'})b_w d$$

$$= 0.85(2\sqrt{3000})(54)(20.5)\tfrac{1}{1000} = 103 \text{ kips} > 33 \text{ kips} \qquad \text{OK}$$

Use 6-#6 bars ($A_s = 2.64$ sq in.). Since bars are to extend the full 7 ft–4 in. length (less minimum cover) of the exterior footing, development of reinforcement for #6 bars is automatically provided; that is,

$$\text{available embedment} = 36 \text{ in.} > [L_d(\#6) = 12.8 \text{ in.}] \qquad \text{OK}$$

(i) Design of interior footing.

$$\text{net soil pressure under overload} = \frac{312}{(6.67)^2} = 7.0 \text{ ksf}$$

$$M_u = 7.0(6.67)\frac{(2.75)^2}{2} = 177 \text{ ft-kips}$$

$$\text{required } R_n = \frac{M_u}{\phi bd^2} = \frac{177(12,000)}{0.90(80)(20)^2} = 74 \text{ psi}$$

Check two-way action for shear. Using nominal stress,

$$v_n = \frac{V_u}{\phi b_0 d} = \frac{7000[(6.67)^2 - (2.83)^2]}{0.85(4)(34)(20)} = 110 \text{ psi} < 4\sqrt{f_c'} \qquad \text{OK}$$

From Fig. 3.8.1, required $\rho \approx 0.0025$,

$$\text{required } A_s = 0.0025(80)(20) = 4.0 \text{ sq in.}$$

Use 9-#6 bars each way ($A_s = 3.96$ sq in.). The 2 ft–9 in. from face of column to edge of footing provides adequate embedment to develop the #6 bars ($L_d = 12.8$ in.).

(j) Design sketch. The final details of the design are shown in Fig. 20.13.5(a). Note that the moment capacity ϕM_n diagram in Fig. 20.13.5(b) for the strap portion is approximate; the #8 bars are extended as far as possible toward the narrow end as the width narrows from 20 in. to 14 in.

20.14 PILE FOOTINGS

The principles and methods to be used in the design of pile caps are little different from those of spread footings. The following, however, may be noted.

1. Computations for moments and shears may be based on the assumption that the reaction from any pile is concentrated at the center of the pile (ACI-15.2.3).
2. In computing the external shear on any section through a footing supported on piles, the portion of the pile reaction to be assumed as producing shear on the section shall be based on straightline interpolation between full value when the pile center is at one-half the pile diameter outside the section and zero value when the pile center is at one-half the pile diameter inside the section (ACI-15.5.3).
3. In reinforced concrete pile footings the thickness above the reinforcement at the edge shall not be less than 12 in. (ACI-15.7).

Note that pile caps frequently must be designed for shear considering the member as a deep beam. In other words, when piles are located inside the critical sections d (for one-way action) or $d/2$ (for two-way action) from face of column, the shear cannot be neglected. ACI Commentary-15.5.3 suggests using ACI-11.8 for designing such pile caps. The problem requires the most attention when large loads are carried by a few piles, such as the two-pile cap having 100-ton piles. There is no agreement about the proper procedure to use. For guidance, the reader is referred to the *CRSI Handbook* [15], Rice and Hoffman [16,17], and Gogate and Sabnis [18].

SELECTED REFERENCES

1. A. N. Talbot. *Reinforced Concrete Wall Footings and Column Footings*. Urbana: Eng. Experiment Station Bulletin No. 67, University of Illinois, March 1913.
2. F. E. Richart. "Reinforced Concrete Wall and Column Footings," *ACI Journal, Proceedings*, **45**, October–November 1948, 97–127, 237–260.
3. Johannes Moe. *Shearing Strength of Reinforced Concrete Slabs and Footings Under Concentrated Loads* (Bulletin No. D47). Chicago: Portland Cement Association Research and Development Laboratories, April 1961.
4. Neil M. Hawkins, (Chmn.). "The Shear Strength of Reinforced Concrete Members–Slabs," by the Joint ASCE-ACI Task Committee 426 on Shear and Diag-

onal Tension of the Committee on Masonry and Reinforced Concrete of the Structural Division, *Journal of the Structural Division*, ASCE, **100**, August 1974 (ST8), 1543–1591.

5. ACI-ASCE Committee 426. *Suggested Revisions to Shear Provisions for Building Codes*. Detroit, Michigan: American Concrete Institute, 1979. 82 pp.

6. *Design Handbook—In Accordance with the Strength Design Method of ACI 318-77* Vol. 1, (3d ed.) (SP-17). Detroit: American Concrete Institute, 1980.

7. Richard W. Furlong. "Design Aids for Square Footings," *ACI Journal, Proceedings*, **62**, March 1965, 363–371.

8. ACI Committee 318. *Building Code Requirements for Structural Plain Concrete* (ACI 318.1). Detroit: American Concrete Institute, 1983.

9. Fritz Kramrisch. "Footings," *Handbook of Concrete Engineering* (Mark Fintel, ed.). New York: Van Nostrand Reinhold Company, 1974 (Chapter 5, pp. 111–140).

10. ACI Committee 436. "Suggested Design Procedures for Combined Footings and Mats," *ACI Journal, Proceedings*, **63**, October 1966, 1041–1057. Disc., 1537–1544.

11. Fritz Kramrisch and Paul Rogers. "Simplified Design of Combined Footings," *Journal of Soil Mechanics and Foundations Division*, ASCE, **87**, October 1961 (SM5), 19–44.

12. Leslie J. Szava-Kovats. "Design of Combined Footings Using Support Reaction and Moment Influence Lines of Continuous Beam on Elastic Supports," *ACI Journal, Proceedings*, **64**, June 1967, 312–319.

13. Gwynne Davies and Brian Mayfield. "Choosing Plan Dimensions for an Eccentrically Loaded Footing Slab," *ACI Journal, Proceedings*, **69**, May 1972, 285–290.

14. *Report of the Joint Committee of Standard Specifications for Concrete and Reinforced Concrete*. Detroit: American Concrete Institute, 1940.

15. *CRSI Handbook* (3rd ed.). Chicago: Concrete Reinforcing Steel Institute, 1978, (pp. 13–21 to 13–41).

16. Paul F. Rice and Edward S. Hoffman. "Pile Caps—Theory, Code, and Practice Gaps," CRSI Professional Members' Structural Bulletin No. 2, February 1978.

17. Paul F. Rice and Edward S. Hoffman. *Structural Design Guide to the ACI Building Code* (2nd ed.). New York: Van Nostrand Reinhold Company, 1979, (pp. 368–372).

18. Anand B. Gogate and Gajanan M. Sabnis. "Design of Thick Pile Caps," *ACI Journal, Proceedings*, **77**, January–February 1980, 18–22.

PROBLEMS

All problems* are to be done in accordance with the strength method of the ACI Code, and all loads given are *service* loads, unless otherwise indicated.

20.1 Design a square spread footing to support a 14-in square tied column carrying a dead load of 120 kips and a live load of 90 kips. The column reinforcement consists of #8 bars. Use f'_c = 4000 psi for the column, f'_c = 3000 psi for the footing, and f_y = 60,000 psi. Use a 6 ft–9 in. square footing. (Column, 360 mm square; DL = 530 kN; LL = 400 kN; column bars, #25M ; f'_c = 30 MPa (column); f'_c = 20 MPa (footing); f_y = 400 MPa; footing, 2 m square.)

---

*Many problems may be solved as problems stated in Inch-Pound units, or as problems in SI units using quantities in parenthesis at the end of the statement. The SI conversions are approximate to avoid implying higher precision for the given information in metric units than that for Inch-Pound units.

20.2 Design a square spread footing to support a 20-in square tied column carrying a dead load of 400 kips and a live load of 264 kips. The column reinforcement consists of #11 bars. Use $f'_c = 3000$ psi for both the column and footing, $f_y = 60,000$ psi, and allowable soil pressure = 5 ksf. Include a design sketch (Column, 500 mm square; DL = 1800 kN; #40M bars; $f'_c = 20$ MPa; $f_y = 400$ MPa; soil pressure = 240 kN/m².)

20.3 Investigate the transfer of stress from column to footing for the two conditions of the accompanying figure. The column is 20 in. square ($f'_c = 4000$ psi) containing 12-#10 bars ($f_y = 40,000$ psi) spirally reinforced. The footing is 10 ft square and 28 in. thick ($f'_c = 3000$ psi) and is adequately reinforced. Utilize the provisions of ACI-15.8 and 15.11.

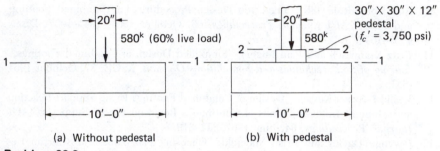

(a) Without pedestal   (b) With pedestal

Problem 20.3

20.4 Design a spread footing to carry a load from an 18×32 in. tied column. The dead load and live load are each 230 kips. Because of the closeness of the property line (see the accompanying figure), the footing cannot exceed 8 ft. perpendicular to that line. Use $f'_c = 3000$ psi, $f_y = 60,000$ psi, and allowable soil pressure = 5 ksf. (460 mm $\times$ 810 mm column; DL = LL = 1000 kN; $f'_c = 20$ MPa; $f_y = 400$ MPa; allowable soil pressure = 240 kN/m².)

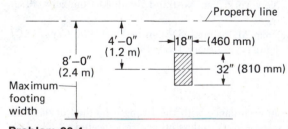

Problem 20.4

20.5 Design a plain concrete footing to carry a long 12-in. concrete block wall which must transmit 6 kips/ft (60% live load) to the footing. Use $f'_c = 3000$ psi and allowable soil pressure = 4 ksf. (300 mm wall; loading = 90 kN/m; $f'_c = 20$ MPa; soil $p = 190$ kN/m².)

20.6 Design a reinforced concrete footing to carry a 12-in. concrete wall to carry 20 kips/ft (60% live load). Use $f'_c = 3000$ psi, allowable soil pressure = 3 ksf and $f_y = 40,000$ psi. (300 mm wall; 300 kN/m; $f'_c = 20$ MPa; soil $p = 140$ kN/m²; $f_y = 300$ MPa.)

20.7 Redesign for the conditions of Example 20.13.2 a rectangular combined footing.

20.8 Design a rectangular combined footing for the situation shown in the accompanying figure. Column data are in the following tabulation:

COLUMN	SIZE	REINFORCEMENT	LL	DL
Exterior	12 in. square	#9 bars	45 kips	120 kips
Interior	12 in. square	#9 bars	90 kips	155 kips

Equal settlement is taken for DL plus LL conditions at a uniform soil pressure of 3 ksf. Use $f'_c = 4000$ psi and $f_y = 60,000$ psi.

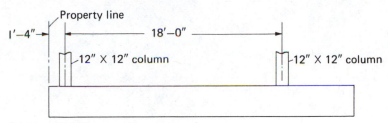

Problem 20.8

20.9 Design a rectangular combined footing for the conditions of the accompanying figure. Assume that the system is within a building basement and is to have a 4-in. concrete slab over the footing. Use $f'_c = 3000$ psi, $f_y = 60,000$ psi, and allowable soil pressure for the given loads $= 5$ ksf. ($f'_c = 20$ MPa; $f_y = 400$ MPa; soil $p = 240$ kN/m².)

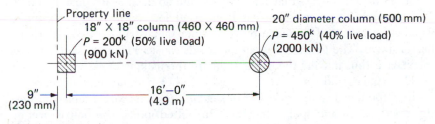

Problem 20.9

20.10 Rework Prob. 20.8 using a strap, or cantilever, footing.

20.11 Rework Prob. 20.9 except assume that the exterior column load is 450 kips and the interior column load is 250 kips with 50% and 40% live load, respectively. In this assignment, however, only the plan size and the shear and moment diagrams are required. (Exterior column, 2000 kN; interior column, 1100 kN.)

20.12 Rework Prob. 20.8, except assume that, in addition, the space to the right of the interior column is restricted to 1 ft–6 in. In this assignment, however, only the plan size and the shear and moment diagrams are required.

21

Introduction
to Prestressed Concrete

21.1 PRESTRESS

Prestress means a stress that acts even though no dead or live load is acting. The principle of prestressing has been used for centuries. For example, wooden barrels may be made by tightening metal bands or ropes around barrel staves. The tensile stress in the bands causes a compression between the staves, thus making the barrel tight. In the making of early wheels, the wooden spokes and rim were first held together by a hot metal tire. Upon cooling, the tensile stress due to shrinkage in the metal would then compress the wooden rim and spokes together. In bolted joints, the bolt is pretensioned by tightening, which in turn precompresses the elements being joined.

The primary application of prestressing on a large scale today is in concrete construction. In general, prestress involves the imposition of stresses opposite in sign to those which are caused by the subsequent application of service loads. For example, prestressing wires placed eccentrically in a simple beam as shown in Fig. 21.1.1 produce in the concrete an axial compression as well as a negative bending moment. Thus it is possible to keep the entire section in compression when service loads are added. This is a great advantage since concrete is weak in tension but strong in compression. Of course, steel is used to impose the prestress though less is required for prestressed concrete than in ordinary reinforced concrete. In general, it may be said that prestress provides a means for the most efficient use of material—that is, steel in tension and concrete in compression.

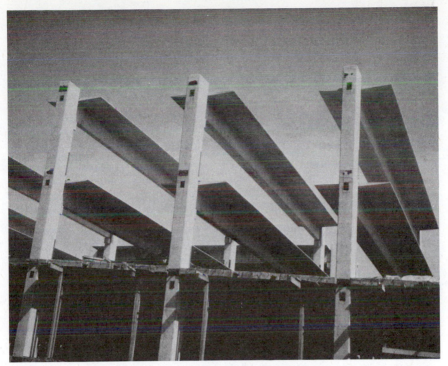

Prestressed concrete T-girders erected on precast columns for parking garage.
(Photo by C. G. Salmon.)

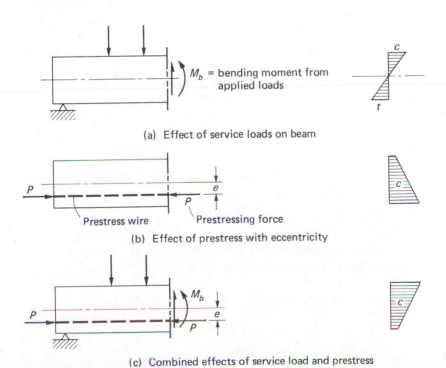

(a) Effect of service loads on beam

(b) Effect of prestress with eccentricity

(c) Combined effects of service load and prestress

Figure 21.1.1 Opposite effects of service load and prestress on simple beam.

21.2 HISTORICAL BACKGROUND

The general concepts of prestressed concrete were first formulated in the period 1885–1890 by C. F. W. Doehring in Germany and P. H. Jackson in the United States. These early applications were handicapped by the low steel strengths obtainable. Steel stressed to low tension levels will not pre-compress concrete adequately to maintain its compression after shrinkage and creep take place.

The theory of prestressed concrete was first propounded by J. Mandl of Germany in 1896. It was further advanced by M. Koenen of Germany in 1907 (first recognition of losses in prestress force from elastic shortening of concrete), and by G. R. Steiner in the United States in 1908. Steiner recognized shrinkage losses and suggested retensioning after shrinkage had occurred.

In practical uses, R. E. Dill of the United States in 1928 produced prestressed planks and fence posts. Circular prestressing of storage tanks began about 1935, but no significant linear prestressing (beams, slabs, planks, etc.) was done until about 1950. The Walnut Lane Bridge in Philadelphia, built in 1949–1950, was the first major use of linear prestressing in the United States.

In Europe, however, linear prestressing began about 1928 and advanced rapidly with the work of F. Dischinger, E. Freyssinet, E. Hoyer, G. Magnel, Y. Guyon, P. Abeles, and F. Leonhardt. With the publication of Magnel's work on the loss of stress in work-hardened steels in 1944, the basic theory of prestressed concrete was sufficiently complete for successful economical applications. In the United States, T. Y. Lin has been a leading proponent and practitioner.

The use of prestress is now widespread in nearly every type of simple structural element, as well as in many statically indeterminate structures. Zollman [1] has presented his interesting "Reflections" on the beginnings of prestressed concrete in America. Methods of inducing prestress are ingenious and unique. Many textbooks are available that describe the methods and applicable theories [2–8].

21.3 ADVANTAGES AND DISADVANTAGES OF PRESTRESSED CONCRETE CONSTRUCTION

The original concept of prestressed concrete was that it was crack free under service loads. Especially when a structure is exposed to the weather, elimination of cracks prevents corrosion. Also a crack free prestressed member has greater stiffness under service loads because its entire section is effective.

Prestressed concrete in several respects is more predictable than ordinary reinforced concrete. It permits accommodation of both shrinkage and creep reasonably well. High-strength concrete may be more efficiently utilized by merely adjusting the prestress force.

Precompression of the concrete reduces the tendency for inclined cracking, and the use of curved tendons provides a vertical component to aid in carrying the shear. Shear strength is more consistent than in ordinary reinforced concrete.

Other features of prestressed concrete are its high ability to absorb energy (impact resistance), its high fatigue resistance due especially to the low steel stress variation resulting from the high initial pretension, and its high live-load capacity arising from the ability of the prestressing tendon to support the dead load. Use of prestressed concrete also permits partial testing of both steel and concrete through application of prestress.

Some of the disadvantages of prestressed concrete construction are as follows: (1) the stronger materials used have a higher unit cost; (2) more complicated formwork may be necessitated; (3) end anchorages and bearing plates are usually required; (4) labor costs are greater; and (5) more conditions must be checked in design and closer control of every phase of construction is required.

Short-span members and single-unit applications of any kind are likely to be uneconomical in prestressed concrete. However, economy is usually achieved when units can be standardized and the same unit repeated many times. For many situations, the desirability of achieving a certain advantage is sufficient to justify a higher initial cost.

Currently (1984), prestressing need not create a crack free structure at service load. In fact, prestressing to various levels of stress can provide the whole range of results, from "fully" prestressed, where at service load tension does not occur, to nonprestressed as discussed in preceding chapters. The level of prestressing can be used that will accomplish the desired crack control or stiffness objective. So-called "partial" prestressing has become common in construction. In general, this chapter focuses on prestressed concrete where the section is *uncracked* at service load.

21.4 PRETENSIONED AND POSTTENSIONED BEAM BEHAVIOR

Since discussion in this chapter is limited to beams, it is well to consider at this stage the behavior of such a member as it relates to the method of inducing the prestress. The most commonly used procedure is to put a specified tensile force into the wires by stretching them between two anchorages prior to placing the concrete. The concrete is then placed and the wires become bonded to the concrete throughout their length. After the concrete has cured, the wires are cut at the anchorages. The immediate shortening of the wires transfers through bond a compressive stress to the concrete. Such a process is called *pretensioning*.

The behavior associated with *pretensioning* will be described step by step. In addition the terminology and allowable values will be given in accordance with Chap. 18 of the ACI Code.

Step 1, as shown schematically in Fig. 21.4.1(a), is to stretch the wires between two anchorages in the casting yard sufficiently to introduce a tensile stress f_{si} into the wires, which according to ACI-18.5.1 may not exceed the smaller of 85% of the specified tensile strength f_{pu} or 94% of the specified yield strength f_{py} of the steel. The quality controlled concrete is then placed in the forms and frequently is steam cured. Concrete strength must be

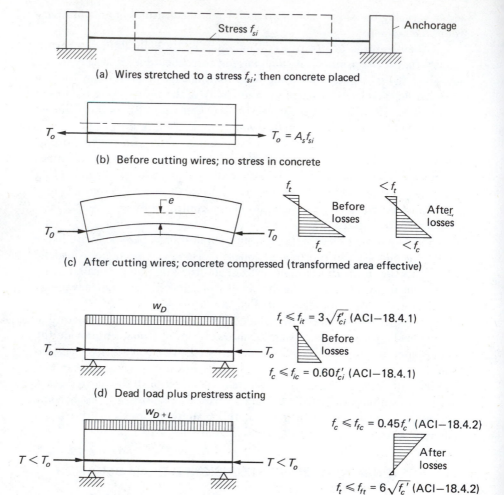

(a) Wires stretched to a stress f_{si}; then concrete placed

(b) Before cutting wires; no stress in concrete

(c) After cutting wires; concrete compressed (transformed area effective)

(d) Dead load plus prestress acting

(e) Dead load, prestress, and live load after all prestress losses

Figure 21.4.1 Stages of behavior up to working load—*pretensioned* beam.

adequately developed by the time the compression is to be introduced; thus high early-strength cement is usually used. Generally, the concrete strength f'_{ci} at transfer is specified by designers to be 4000 to 4500 psi.

Step 2 is to cut the wires. Acting through bond, the force T_0 in the wires acts as a compressive force on the entire effective (transformed) section. The stress in the concrete goes from zero [Fig. 21.4.1(b)] before the wires are cut to that shown in Fig. 21.4.1(c) after they have been cut. Once the prestress has been introduced, certain losses of prestress begin to occur. Loss of prestress may arise from slip at the anchorage, elastic shortening of the concrete member, creep and shrinkage of the concrete, relaxation of steel stress, and frictional losses due to intended or unintended curvature in the tendons. It is true that some small portion of such losses may occur prior to the transfer of stress to the concrete; however, it is practical and conservative to assume that the losses occur after the transfer.

Dead load of the flexural member will, of course, be acting simultaneously with the prestressing force once the wires have been cut and the transfer of stress is accomplished. Dead load combined with prestress is shown in Fig. 21.4.1(d), where the most critical stress situation occurs immediately after transfer and before most losses have taken place. Limiting values (ACI-18.4.1) for this temporary situation are a tensile stress at the top of the beam of $3\sqrt{f'_{ci}}$ (approximately 40% of the cracking strength) and a compression stress at the bottom equal to 60% of the concrete strength f'_{ci} which has been developed at the time of transfer. One reason for holding the temporary tensile stress to such a low value is to prevent any possibility of an upward buckling of the beam resulting from sudden cracking at the top. Frequently no reinforcement (nonprestressed) exists to restrain such cracking.

Step 3 is the service condition of dead load, live load, and prestress where, after losses, ACI-18.4.2 permits a net tensile stress at the bottom not to exceed $6\sqrt{f'_c}$, along with a compressive stress at the top not to exceed $0.45f'_c$. Since tendons are usually placed near the bottom surface, there will be little danger of sudden cracking and buckling. For this reason the allowable tension stress is only slightly below $7.5\sqrt{f'_c}$, which is the generally accepted value of the modulus of rupture for normal-weight concrete.

Previous to the 1971 ACI Code, no tensile stress was permitted if the member was exposed to freezing temperatures or to a corrosive environment. ACI Commentary-18.4.2(b) suggests using reduced tensile stress to eliminate possible cracking under adverse corrosive atmosphere conditions, but does not suggest specific reduced values. Corrosive atmosphere is defined as an atmosphere in which chemical attack may occur from such sources as seawater, corrosive industrial atmosphere, or sewer gas. The ACI Code limit of $6\sqrt{f'_c}$ implies cover according to ACI-7.7.3.1; with corrosive atmosphere conditions the cover should be increased above the minimum.

Further, the tensile stress in the precompressed zone may exceed $6\sqrt{f'_c}$ when special calculations show that the deflection requirements of ACI-9.5 are satisfied.

The alternative to pretensioning is *posttensioning*. In a posttensioned beam, the concrete is first cast either with a hollow tube enclosed or with the unstressed tendons coated with grease or mastic to prevent bond with the concrete, as shown in Fig. 21.4.2. An end plate or anchorage is placed against each end of the member; then, once the concrete is sufficiently cured, the wires are pulled by jacking against the end plates. During the tensioning process, elastic shortening occurs, frictional losses take place, and the dead load moment becomes partially active due to the induced curvature. Thus the jacking force must account for these losses. Losses that

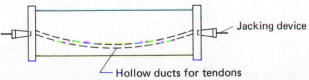

Jacking device

Hollow ducts for tendons

Figure 21.4.2   Section for posttensioning.

occur after tensioning will thus be less in this case than in pretensioning. For posttensioning, the stresses induced and the allowable values at the different stages are essentially the same as those described in detail for pretensioning. However, since the tendons are not bonded to the concrete, the posttension force acts only on the plain concrete, at least until reaching the situation of Fig. 21.4.1(d). Prior to the imposition of live load, the tendons are usually grouted (space in the ducts is filled). If such grouting is properly done, the live load stresses may be computed on the transformed section, the same as for pretensioning.

21.5 SERVICE LOAD STRESSES ON FLEXURAL MEMBERS— TENDONS HAVING VARYING AMOUNTS OF ECCENTRICITY

In order to demonstrate some of the attributes of prestressed concrete, the first example is one with the prestressing elements placed at the centroid of the section giving a uniform precompression of the concrete.

EXAMPLE 21.5.1 For the section shown in Fig. 21.5.1 assume that the member is pretensioned by 2.30 sq in. of steel wire having a maximum acceptable initial tensile stress of 175,000 psi. The prestress wires are centered at the centroid of the section. The concrete has $f'_c = 5000$ psi ($n = 6$), and it is to be assumed that the concrete has attained a strength of $f'_{ci} = 4000$ psi at the time of transfer. Determine (a) the stresses due to prestress immediately after transfer; (b) the temporary stresses when the member is used on a 40-ft simple span; and (c) the service live load moment capacity according to the ACI Code, allowing for a 20% loss of prestress due to creep, shrinkage, and other sources.

Solution: (a) Stress due to prestress immediately after transfer.

$$T_0 = f_{si}A_s = 175(2.30) = 402.5 \text{ kips}$$

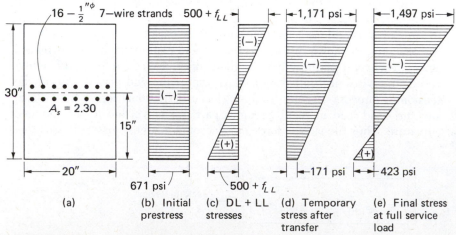

Figure 21.5.1 Section and stresses for Example 21.5.1.

This force T_0 acts as a compressive force on the transformed section immediately after cutting the wires. Thus

$$f_c = \frac{T_0}{A_c + nA_s} = \frac{T_0}{A_g + (n-1)A_s} = \frac{402,500}{600 + 5(2.3)} = \frac{402,500}{611.5} = 658 \text{ psi}$$

The decrease in steel stress is

$$\Delta f_s = nf_c = 6(658) = 3950 \text{ psi}$$

Thus elastic shortening may be considered to have caused a loss of tensile stress, so that the remaining tensile stress in the wires is $175,000 - 3950 = 171,050$ psi. The loss in this case is $2\frac{1}{4}\%$, but it could be as high as 5%.

It should further be noted that, although it may be theoretically correct to use the transformed section, in ordinary practice it is common and sufficiently accurate in most cases to use the gross section. Since the prestressing force is applied at the centroid of the gross section in the present problem, f_c is uniform over the entire section, or

$$f_c = \frac{402,500}{600} = 671 \text{ psi}$$

which is little different from the value of 658 psi determined above by using the transformed section.

(b) Temporary stress—prestress plus dead load.

$$w_D = \frac{20(30)}{144}(150) = 625 \text{ plf}$$

$$M_D = \tfrac{1}{8}(0.625)(40)^2 = 125 \text{ ft-kips}$$

Using the approximate method with gross moment of inertia I_g and neglecting the transformed area of reinforcement,

$$I_g = \tfrac{1}{12}(20)(30)^3 = 45,000 \text{ in.}^4$$

$$f(\text{initial prestress} + \text{DL}) = -\frac{402,500}{600} \mp \frac{125(12,000)(15)}{45,000}$$

$$= -671 \mp 500$$

$$= -1171 \text{ psi} \qquad \text{(compression, top)}$$

$$= -171 \text{ psi} \qquad \text{(compression, bottom)}$$

These stresses are acceptable based on temporary stress restrictions immediately after transfer and before losses,

$$f_c(\text{max}) = 0.6f'_{ci} = 2400 \text{ psi}$$

$$f_t(\text{max}) = 3\sqrt{f'_{ci}} = 190 \text{ psi}$$

(c) Service live load moment capacity.

$$f(\text{prestress} - \text{losses} + \text{DL}) = -0.8(671) \mp 500 = -537 \mp 500$$

$$= -1037 \text{ psi} \qquad \text{(compression, top)}$$

$$= -37 \text{ psi} \qquad \text{(compression, bottom)}$$

Based on stress at service load, after allowance for all prestress losses,

$$f_c(\max) = 0.45f'_c = 2250 \text{ psi}$$
$$f_t(\max) = 6\sqrt{f'_c} = 423 \text{ psi}$$

The stress available for live load may then be computed.

$$f(\text{prestress} - \text{losses} + DL + LL) = -1037 + f_{LL} = -2250 \text{ psi} \quad \text{(top)}$$
$$= -37 + f_{LL} = +423 \text{ psi} \quad \text{(bottom)}$$
$$f_{LL}(\max)(\text{top}) = -1213 \text{ psi}$$
$$f_{LL}(\max)(\text{bottom}) = +460 \text{ psi} \qquad \text{(Controls)}$$

Thus

$$\frac{M_L(15)}{45,000} = 460 \text{ psi}$$

$$M_L = \frac{45,000(460)}{15(12,000)} = 115 \text{ ft-kips}$$

The wide divergence between the maximum acceptable live load stresses of 1213 psi compression and 460 psi tension indicates the need for an unsymmetrical section or some arrangement to equalize them better. A study of Fig. 21.5.1 will show that the most economical arrangement would be for the initial prestress variation to offset the pattern of final stress under full dead plus live load.

EXAMPLE 21.5.2 Repeat the solution of Example 21.5.1, except locate the tendons 5 in. from the bottom of the section (Fig. 21.5.2).

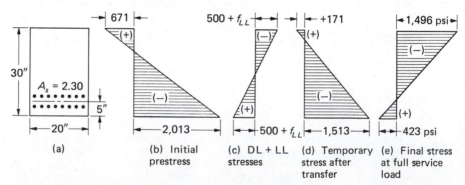

Figure 21.5.2 Section and stresses for Example 21.5.2.

Solution: (a) Temporary stress (prestress + DL immediately after transfer). Using properties of the gross section as for the preceding example, the prestressing force $T_0 = 402.5$ kips applied with an eccentricity of 10 in. gives

$$f(\text{initial prestress} + \text{DL}) = -671 \pm \frac{402,500(10)(15)}{45,000} \mp 500$$

$$= -671 \pm 1342 \mp 500$$

$$= +171 \text{ psi} \qquad (\text{tension, top})$$

$$= -1513 \text{ psi} \qquad (\text{compression, bottom})$$

It is to be noted that the temporary tensile stress of 171 psi at the top is nearly equal to the allowable value of $3\sqrt{f'_{ci}} = 190$ psi for such stress. Any significantly greater eccentricity, therefore, would require a reduction in the prestressing force.

(b) Final stress (prestress + DL), after allowance for 20% prestress loss, is

$$f(\text{prestress} + \text{DL} - \text{losses}) = +171 - 0.2(-671 + 1342)$$

$$= +37 \text{ psi} \qquad (\text{top})$$

$$= -1513 - 0.2(-671 - 1342)$$

$$= -1110 \text{ psi} \qquad (\text{bottom})$$

(c) Service live load moment capacity (see Example 21.5.1 for allowables). Based on final dead load plus live load conditions,

$$+37 + f_{LL} = -2250 \text{ psi} \qquad (\text{top}), \qquad f_{LL} = -2287 \text{ psi}$$

$$-1110 + f_{LL} = +423 \text{ psi} \qquad (\text{bottom}), \qquad f_{LL} = +1533 \text{ psi}$$

Since the neutral axis for live load resistance is assumed to be at middepth, $f_{LL} = +1533$ psi controls.

$$M_L = \frac{45,000(1533)}{15(12,000)} = 383 \text{ ft-kips}$$

Thus increasing the eccentricity of the prestressing force increases the live load capacity until the limit is reached when the temporary stress at transfer reaches its maximum permissible value, either at the top or at the bottom of the section.

It is to be noted that the magnitude of the prestress over the concrete section is constant for the entire span when the tendons are straight, whereas the magnitude of dead and live load stresses is a maximum at only one point. For straight tendons, the complete stress situation near the supports on simple spans approaches that of Fig. 21.5.2(b), less losses, because the superimposed dead and live load stresses vanish. Because of this difficulty, tendons frequently are placed so as to have an eccentricity that varies from zero at points of low external bending moment to a maximum in the region of high external bending moment. This variation in eccentricity may be accomplished in pretensioning by holding down the stressed tendons at midspan, or at other locations such as at the one-third points. In posttensioning, the ducts or greased tendons are simply draped (held at the ends and permitted to take a natural deflected shape) such that desired eccentricities at the ends and at midspan are achieved; the points in between will lie on a curved path.

21.6 THREE BASIC CONCEPTS OF PRESTRESSED CONCRETE

When considering the stresses in prestressed concrete under service load conditions, there are three general patterns of thought that may be applied.

Homogeneous Beam Concept. The homogeneous beam concept is used in Sec. 21.5 wherein the prestressing effectively eliminates cracking and the combined stress formula, $P/A \pm Mc/I$, may be used to investigate the section. Two examples appear in Sec. 21.5.

Internal Force Concept. The internal force approach uses the equilibrium of internal forces; steel takes the tension and concrete the compression as shown in Fig. 21.6.1. This approach is analogous to the internal-couple method used for nonprestressed reinforced concrete. At service load in reinforced concrete the *points of action* of the forces C and T ($C = T$) are *independent* of the magnitude of applied bending moment, depending only on the cross-sectional dimensions and the modulus of elasticity ratio n; thus the magnitude of the forces is directly proportional to the applied bending moment. In prestressed concrete, the *magnitude* of internal forces is *independent* of applied bending moment, depending only on the prestress and the percentage of losses. In this case the location of the force C must vary with the applied loading. The approach may be summarized by the following steps:

1. A known prestress force put into the steel defines T.
2. An applied moment M is put on the beam.
3. For equilibrium, the moment arm $= M/T$ and $C = T$.
4. Knowing the magnitude and point of action of the force C, the stress in the concrete may be computed as

$$f = \frac{C}{A} \pm \frac{Cey}{I} \qquad (21.6.1)$$

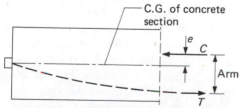

Figure 21.6.1 Internal force concept of prestressing.

EXAMPLE 21.6.1 Apply the dead load moment of 125 ft-kips and live load moment of 383 ft-kips (as computed in Example 21.5.2) to the rectangular beam of Fig. 21.5.2, using the internal force concept. Determine the service load stresses at transfer and under final conditions.

Solution: (a) At transfer, the prestress force T_0 is 402.5 kips.

$$C = T_0 = 402.5 \text{ kips}$$

When the applied moment is 125 ft-kips, the moment arm of the internal forces must be

$$\text{arm} = \frac{M_D}{C} = \frac{125(12)}{402.5} = 3.73 \text{ in.}$$

This means the compressive force C is eccentric to the middepth by an amount,

$$e = 15 - 5 \text{ (to steel)} - 3.73 \text{ (arm)}$$
$$= 6.27 \text{ in.} \qquad \text{(below middepth)}$$
$$f = -\frac{402.5}{600} \pm \frac{402.5(6.27)15}{45,000}$$
$$f(\text{top}) = -671 + 842 = +171 \text{ psi} \qquad \text{(tension)}$$
$$f(\text{bottom}) = -671 - 842 = -1513 \text{ psi} \qquad \text{(compression)}$$

exactly the same as in Example 21.5.2, Fig. 21.5.2(d).

(b) At the final condition, the prestress force T_e is $0.8(402.5) = 322$ kips after losses.

$$C = T_e = 322 \text{ kips}$$
$$\text{arm} = \frac{M_D + M_L}{C} = \frac{(125 + 383)12}{322} = 18.95 \text{ in.}$$
$$e = 15 - 5 - 18.95 = -8.95 \text{ in.} \qquad \text{(above middepth)}$$
$$f = -\frac{322}{600} \mp \frac{322(8.95)15}{45,000}$$
$$f(\text{top}) = -537 - 960 = -1497 \text{ psi} \qquad \text{(compression)}$$
$$f(\text{bottom}) = -537 + 960 = +423 \text{ psi} \qquad \text{(tension)}$$

exactly as shown in Fig. 21.5.2(e).

Load Balancing Concept.

The load balancing approach visualizes prestressing primarily as a process of balancing loads on the member. The prestressing tendons are placed so that the eccentricity of the prestressing force varies in the same manner as the moments from applied loads, which if exactly done would result in zero flexural stress. Only the axial stress P/A (P is the horizontal component of force in tendon) would act. Refer to Fig. 21.6.2(a) showing the parabolically draped prestressing tendon. Figure 21.6.2(b) shows the free body of forces acting on the concrete due to prestress alone. The prestressing effect may be considered as an upward uniform load if the tendon is parabolically draped. The maximum prestress moment of $Te_{\max}$ at midspan can be equated to an equivalent uniformly loaded beam moment, $w_p L^2/8$; thus

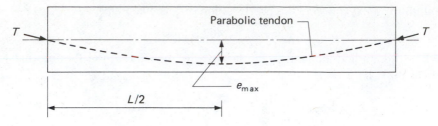

(a) Member having parabolically draped tendon

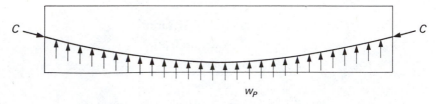

(b) Forces acting on concrete from prestress alone

Figure 21.6.2 Load balancing concept of prestressing.

$$w_p = \frac{8Te_{max}}{L^2} = \text{equivalent uniform load (acting upward)} \quad \textbf{(21.6.2)}$$

Let

$$w_{net} = w \text{ (actual downward load)} - w_p$$

then

$$M_{net} = \frac{w_{net}L^2}{8} \quad \textbf{(21.6.3)}$$

and

$$f = -\frac{C}{A} \mp \frac{M_{net}y}{I} \quad \textbf{(21.6.4)}$$

If the tendons are not parabolically draped, the actual net moment (applied load moment minus prestress moment) may be used for M_{net} in Eq. (21.6.4).

EXAMPLE 21.6.2 Compute the stresses on the beam of Fig. 21.5.2 using dead load and live load moments of 125 and 383 ft-kips, respectively, using the load balancing concept.

Solution: The maximum moment due to prestress at transfer is

$$M_{prestress} = 402.5\left(\frac{10}{12}\right) = 335 \text{ ft-kips}$$

$$M_{net} = M_D - M_{prestress} = +125 - 335 = -210 \text{ ft-kips}$$

(negative bending)

$$f = -\frac{402.5}{600} \mp \frac{(-210)(12)15}{45,000}$$

$$f \text{ (top)} = -671 + 842 = +171 \text{ psi} \qquad \text{(tension)}$$
$$f \text{ (bottom)} = -671 - 842 = -1513 \text{ psi} \qquad \text{(compression)}$$

At the final condition,

$$M_{\text{net}} = 125 + 383 - 322(10)/12 = 240 \text{ ft-kips (positive bending)}$$

$$f = -\frac{322}{600} \mp \frac{240(12)(15)}{45,000}$$

$$f \text{ (top)} = -537 - 960 = -1497 \text{ psi} \qquad \text{(compression)}$$
$$f \text{ (bottom)} = -537 + 960 = +423 \text{ psi} \qquad \text{(tension)}$$

The results agree with those previously obtained.

21.7 LOSS OF PRESTRESS

The amount of prestress actually existing in a prestressed concrete member is not easily measured. The total force in the tendons at the time of prestressing is all that may conveniently be determined. Various losses reduce the prestress to that lower value available to resist the load. The difference between the final available prestress and the initial value is referred to as the loss of prestress.

In practice the initial prestress is usually determined by a pressure gage on the jack and may be verified by a direct measurement of the tendon elongation. In the pretensioned members, the uniformity of initial prestress may be verified at several points. In certain posttensioning procedures, the prestressing force will diminish due to friction at points remote from the jacking source. Initial prestress is, however, generally known with good accuracy.

Elastic Shortening. The loss of prestress due to elastic shortening can be easily determined. For example, let T_0 be the prestressing force that is applied at the centroid of the concrete section in a pretensioned member. If T_f is the final tensile force in the tendons just after elastic shortening has occurred, the strain (unit shortening) in the concrete may be expressed as

$$\epsilon_c = \frac{f_c}{E_c} = \frac{T_f}{A_c E_c} \qquad (21.7.1)$$

where $A_c = A_g - A_s$. The change in strain in the tendons as a result of losses is

$$\Delta\epsilon_s = \frac{T_0 - T_f}{A_s E_s} \qquad (21.7.2)$$

Equating the expressions for ϵ_c and $\Delta\epsilon_s$ gives

$$\frac{T_0}{T_f} = \frac{A_c + nA_s}{A_c} = \frac{A_T}{A_c} \qquad (21.7.3)$$

The loss of prestress Δf_s is

$$\Delta f_s = \frac{T_0 - T_f}{A_s} = \frac{nT_f}{A_c} = \frac{nT_0}{A_T} \qquad (21.7.4)$$

As a practical matter, the loss in prestress Δf_s, regardless of whether or not the prestressing force is applied at the centroid of the gross section, may be taken approximately as

$$\Delta f_s = \frac{nT_0}{A_g} \qquad (21.7.5)$$

More correctly, the loss in prestress due to elastic shortening and bending of the section should be obtained as n times the computed compressive stress in the concrete adjacent to the tendons.

In the posttensioning case, usually the tendons are not stretched simultaneously. Further, the elastic shortening occurs gradually during the tensioning operation. The various methods of accounting for these gradual losses are adequately described elsewhere [2–8].

EXAMPLE 21.7.1 Determine the percent loss of prestress due to elastic shortening and bending in the pretensioned member of Fig. 21.5.2 $f'_c = 5000$ psi with $n = 7$; $f_{si} = 175,000$ psi.

Solution: (a) Loss due to elastic shortening, neglecting bending. "Exact" method [see also part (a) of Example 21.5.1],

$$\Delta f_s = \frac{nT_0}{A_T} = \frac{7(402.5)}{611} = 4.61 \text{ ksi}$$

$$A_T = 20(30) + (7 - 1)2.30 = 614 \text{ sq in.}$$

$$\text{percent loss} = \frac{4.61}{175} = 2.63\%$$

Approximate method,

$$\Delta f_s = \frac{nT_0}{A_g} = \frac{7(402.5)}{600} = 4.70 \text{ ksi}$$

$$\text{percent loss} = \frac{4.70}{175} = 2.69\%$$

There is no significant difference in the two results. Three percent is a typical value for loss due to elastic shortening in a pretensioned beam, whereas in a posttensioned beam it would be on the order of $1\frac{1}{2}\%$ average.

(b) Loss including bending due to dead load of 125 ft-kips. By the approximate method (using gross section instead of transformed section), the stress in *the concrete adjacent to the tendons* is (using internal-force concept),

$$\text{arm} = \frac{M_D}{T_0} = \frac{125(12)}{402.5} = 3.72 \text{ in.}$$

$$C = T_0 = 402.5 \text{ kips}$$

$$f_c = -\frac{C}{A_g} - \frac{C(15 - 5 - 3.72)10}{I_g}$$

$$= -\frac{402.5}{600} - \frac{402.5(6.28)10}{45,000}$$

$$= -671 - 562 = -1233 \text{ psi}$$

This is the stress in concrete 5 in. from the bottom of the beam. The *change* in steel stress is nf_c.

$$\Delta f_s = nf_c = 7(1.233) = 8.6 \text{ ksi}$$

$$\text{percent loss} = \frac{8.6}{175} = 4.9\%$$

Creep in Concrete. Creep is the time-dependent deformation that occurs in concrete under stress, and has already been discussed in Sec. 1.10 (Chap. 1) and Sec. 14.6 (Chap. 14). The strain due to creep will vary with the magnitude of stress and in general may be assumed to vary with the elastic strain from about 100% in humid atmosphere to about 300% in very dry atmosphere [9].

In Chap. 14, the creep coefficient C_t is defined as

$$C_t = \frac{\text{creep strain, } \epsilon_{cp}}{\text{initial elastic strain, } \epsilon_i} \tag{21.7.6}$$

The elastic strain in the concrete at the centroid of the section is ($f_c =$ stress at centroid)

$$\epsilon_i = \frac{f_c}{E_c}$$

$$\epsilon_{cp} = C_t \epsilon_i = C_t \left(\frac{f_c}{E_c} \right) \tag{21.7.7}$$

The strain in the concrete due to creep equals the decrease in strain in the steel; thus

$$\Delta \epsilon_s = \epsilon_{cp} = C_t \left(\frac{f_c}{E_c} \right) \tag{21.7.8}$$

also

$$\Delta \epsilon_s = \frac{\Delta f_s}{E_s} \tag{21.7.9}$$

Then, equating Eqs. (21.7.8) and (21.7.9),

$$\Delta f_s = C_t nf_c \tag{21.7.10}$$

The coefficient C_t may be determined using the general expressions given in Sec. 14.6 (Chap. 14), or as suggested by Zia, Preston, Scott, and Workman [10], use $C_t = 2.0$ for pretensioned members and $C_t = 1.6$ for post-tensioned members. The stress f_c in Eq. (21.7.10) is recommended [10] to be taken as $(f_{cir} - f_{cds})$, where f_{cir} (the subscript r means residual) is the net compressive stress in the concrete at center of gravity of tendons immediately after the prestress has been applied to the concrete, and f_{cds} (the subscript s means superimposed) is the stress in concrete at center of gravity of tendons due to all superimposed permanent dead loads that are applied to the member after it has been prestressed. For example, f_{cir} would correspond to the $f_c = 1233$ psi computed in Example 21.7.1(b), because the prestress and the girder dead load act simultaneously when the prestress is applied. Typical values for percentage loss of prestress due to creep are from 5 to 6%. Pretensioned beams will exhibit more creep than posttensioned beams because the prestress is imposed when the concrete is at an earlier age; age at loading is a major factor in determining the magnitude of creep.

Shrinkage in Concrete. Shrinkage is the volume change in concrete that occurs with time, as discussed in Sec. 1.10 (Chap. 1) and Sec. 14.7 (Chap. 14). The loss of prestress due to shrinkage may be expressed

$$\Delta f_s = \epsilon_{sh} E_s \qquad (21.7.11)$$

where ϵ_{sh} is the shrinkage strain in concrete (see Sec. 14.7). In Chap. 14 are given general expressions for evaluating shrinkage strain. Zia et al. [10] recommend computing ϵ_{sh} by starting with a value of 550×10^{-6} in./in. as the basic ultimate shrinkage strain, multiplying by $(1 - 0.06V/S)$ to correct for the volume V to surface S ratio, and then multiplying by $(1.5 - 0.015H)$ to correct for the relative humidity H. For posttensioned members, an additional reduction factor multiplier is used to account for the time between the end of moist curing and the application of prestress.

Relaxation of Steel Stress. Relaxation is taken to mean the loss of stress in steel under nearly constant strain at constant temperature. Loss due to relaxation varies widely for different steels, and such loss should be provided for in accordance with test data furnished by the steel manufacturers. This loss is generally assumed to be in the range of 2 to 3% of the initial steel stress [9]. Zia et al. [10] have provided a formula for this computation. The percentage loss of prestress relating to relaxation varies with the type of tendon and the ratio of initial prestress to tensile strength of tendon.

Friction Losses in Posttensioned Members. There will be frictional losses, which are generally small, in the jacking equipment as well as friction between the tendons and the surrounding material (either duct or actual concrete member), due to intended or unintended curvature in the tendons. The friction between tendons and surrounding material is not small and may be considered as partly a length effect and partly a curvature effect.

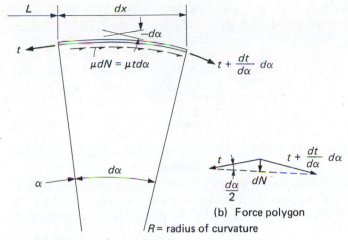

(a) Element of curved tendon

(b) Force polygon

R = radius of curvature

Figure 21.7.1 Friction losses in posttensioned member.

Referring to Fig. 21.7.1, let dx be a segment of a curved tendon. Assume that the tendon is being jacked from the left end by the force P_s which results in a force P_x at some distance to the right; these forces define the limits for the tension t. The full angle enclosed within the arc is α.

For equilibrium of the entire segment dx, refer to Fig. 21.7.1(b); the normal force dN is

$$dN = t\left(\frac{d\alpha}{2}\right) + \left(t + \frac{dt}{d\alpha}\,d\alpha\right)\frac{d\alpha}{2} \qquad (21.7.12)$$

and neglecting infinitesimals of higher order,

$$dN = 2t\left(\frac{d\alpha}{2}\right) = t\,d\alpha \qquad (21.7.13)$$

The friction force developed along the length dx is

$$\mu\,dN = \mu t\,d\alpha \qquad (21.7.14)$$

Summation of forces along the tendon gives

$$t - \mu t\,d\alpha - (t + dt) = 0$$

$$\frac{dt}{t} = -\mu\,d\alpha \qquad (21.7.15)$$

Integrating to obtain the total effect over the entire curved portion included within the angle α,

$$\int_{P_s}^{P_x} \frac{dt}{t} = \int_0^\alpha -\mu\,d\alpha \qquad (21.7.16)$$

$$\log_e P_x - \log_e P_s = -\mu\alpha \qquad (21.7.17)$$

$$\frac{P_x}{P_s} = e^{-\mu\alpha} \qquad (21.7.18)$$

Replacing the friction force term $\mu\alpha$ with the following expression, which contains a friction part due to curvature and a length effect (wobble effect), $\mu\alpha + KL$, thus

$$\frac{P_x}{P_s} = e^{-(\mu\alpha + KL)} \tag{21.7.19}$$

or

$$P_s = P_x e^{(\mu\alpha + KL)} \tag{21.7.20}$$

which is ACI Formula (18-1) in ACI-18.6.2.1. Note that $\alpha = L/R$, the length L of the curve divided by R, the radius of curvature.

As an approximation, when $P_s - P_x$ is small (such as not more than 15 to 20% of the jacking force P_s) the friction force may be assumed to be constant. If the friction force is assumed to be proportional to the force P_x, then (see Fig. 21.7.2)

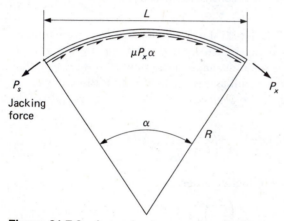

Figure 21.7.2 Approximate procedure for friction losses.

$$\mu N = \mu P_x \alpha \tag{21.7.21}$$

and assuming the wobble effect KL is also proportional to P_x, equilibrium requires

$$P_s = P_x + P_x(\mu\alpha + KL)$$

or

$$P_s = P_x(1 + \mu\alpha + KL) \tag{21.7.22}$$

which is ACI Formula (18-2), permitted for use when $(\mu\alpha + KL)$ does not exceed 0.3.

Practical Design Consideration—Total Losses. The total loss in prestress may be expressed in unit strains, total strains, unit stresses, or in percentage of initial prestress. Although it is difficult to generalize the amount of prestress loss, Lin and Burns [2] have suggested that for average steel and concrete properties and for average curing conditions the values in Table 21.7.1 may be taken as representative.

Zia et al. [10] have indicated that the upper limit for total loss in steel stress (not including friction loss) for stress-relieved strand in normal-weight concrete may be assumed as 50,000 psi.

Table 21.7.1   AVERAGE PERCENTAGES OF LOSS OF PRESTRESS [2]

	PRETENSIONING (PERCENT)	POSTTENSIONING (PERCENT)
Elastic shortening and bending of concrete member	4	1
Creep of concrete	6	5
Shrinkage of concrete	7	6
Relaxation (creep) in steel	8	8
Totals	25	20

More detailed treatment of losses of prestress is to be found in the textbooks by Lin and Burns [2], Naaman [3], and Nilson [4]; as well as in the PCI Committee Report [11], and Zia et al. [10].

EXAMPLE 21.7.2 The posttensioned beam of Fig. 21.7.3 containing a cable of 72 parallel wires, $A_s = 3.60$ sq in. is to be tensioned 2 wires at a time. The jacking stress is to be measured by a pressure gage. The wires are to be stressed from one end of the member to a value f_1 to overcome frictional loss, then released to a value f_2, so that immediately after anchoring an initial prestress of 144 ksi is obtained. Compute f_1 and f_2, as well as the final design stress after all losses, according to the ACI Code. Assumptions are as follows:

(a) Coefficient of friction $\mu = 0.50$
(b) Wobble coefficient, $K = 0.0008$ (per ft)
(c) Deformation at anchorages and slip of wires = 0.06 in.
(d) Shrinkage strain, $\epsilon_{sh} = 0.0002$
(e) Steel relaxation = 5% of initial prestress (144 ksi)
(f) $E_s = 29,000$ ksi; $E_c = 5000$ ksi

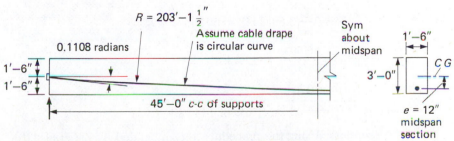

Figure 21.7.3   Posttensioned beam of Example 21.7.2.

Solution: (a) Loss due to friction. Using the "exact" method, Eq. (21.7.20),

$$P_s = P_x e^{(\mu\alpha + KL)}$$

$$\mu\alpha = 0.5\left(\frac{45}{203.125}\right) = 0.1108$$

$$KL = 0.0008(45) = \frac{0.0360}{0.1468}$$

$$P_s = P_x e^{(0.1468)} = 1.158 P_x$$

The above expression can also be used in terms of unit stress f; in this case if f_x is desired to be 144 ksi, then the initial stress f_1 to overcome frictional loss is

$$f_1 = 1.158(144) = 167 \text{ ksi}$$

Using the approximate expression, Eq. (21.7.22),

$$P_s = P_x(1 + 0.1468) = 1.1468 P_x$$

or

$$f_1 = 1.1468(144) = 165 \text{ ksi}$$

(b) Anchorage slip. For tensioning from one end,

$$\epsilon_s = \frac{0.06}{45(12)} = 0.00011$$

$$\Delta f_s = \epsilon_s E_s = 0.00011(29,000) = 3.2 \text{ ksi}$$

To allow for anchorage slip, tension to $f_1 = 167$ ksi then release to $f_2 = 144 + 3.2 = 147.2$ ksi. The minimum stress $f_{si} = 144$ ksi will then exist at both ends.

(c) Elastic shortening due to posttensioning two wires at a time. The wires tensioned first will have the greatest loss, having some additional loss as each succeeding pair of wires is tensioned. The pair tensioned last will have zero loss. Thus Δf_s in the first pair, given by Eq. (21.7.5), is

$$\Delta f_s = \frac{nT_0}{A_g}$$

$$n = \frac{29,000}{5000} = 5.8, \qquad \text{say 6 to nearest whole number}$$

$$T_0 = \frac{35}{36}(3.60)(144) = 504 \text{ kips}$$

$$\Delta f_s \text{ (first pair)} = \frac{6(504)}{18(36)} = 4.66 \text{ ksi}$$

$$\Delta f_s \text{ (last pair)} = 0 \text{ ksi}$$

average $\Delta f_s = 2.33$ ksi (1.6% of 144 ksi)

(d) Creep loss. Using the procedure recommended by Zia et al. [10], f_c in Eq. (21.7.10) is computed as $(f_{cir} - f_{cds})$. The *net* concrete stress at midspan at the centroid of the steel tendons is

$$f_{cir} = \frac{T_0}{A_g} + \frac{T_0 e^2}{I_g} - \frac{M_D e}{I_g}$$

where

$$T_0 = 3.60(144) = 518.4 \text{ kips}$$

$$M_D = \frac{1}{8}\frac{18(36)}{144}(0.15)(45)^2 = 170.9 \text{ ft-kips}$$

Then,

$$f_{cir} = \frac{518.4}{18(36)} + \frac{518.4(12)12}{18(36)^3/12} - \frac{170.9(12)12}{18(36)^3/12}$$

$$= 0.80 + 1.07 - 0.35 = 1.52 \text{ ksi}$$
$$\text{(at the midspan)}$$

In this case, if there is no additional superimposed permanent dead load, f_{cds} is zero. At the support where there is zero eccentricity of the tendons and zero dead load moment,

$$f_{cir} = \frac{T_0}{A_g} = 0.80 \text{ ksi}$$

The average value of $(f_{cir} - f_{cds})$ should be used as f_c in Eq. (21.7.10) for posttensioned beams,

$$f_c = \frac{1.52 + 0.80}{2} = 1.16 \text{ ksi}$$

Note again that in this example f_{cds} is taken as zero.

$$\epsilon_c = \text{elastic strain} = \frac{f_c}{E_c} = \frac{1.16}{5000} = 0.000232 \text{ in./in.}$$

and for a posttensioned beam using $C_t = 1.6$,

$$\epsilon_{cp} = C_t \epsilon_c = 1.6(0.000232) = 0.000371 \text{ in./in.}$$
$$\Delta f_s = \epsilon_{cp} E_s = 0.000371(29,000) = 10.8 \text{ ksi} \qquad (7.5\%)$$

(e) Shrinkage loss. Using the procedure of Zia et al. [10], the basic shrinkage strain is

$$\text{basic } \epsilon_{sh} = 550 \times 10^{-6} \text{ in./in.}$$

The correction factor (CF) for volume/surface (V/S) ratio is

$$(CF)_{V/S} = 1 - 0.06\frac{V}{S} = 1 - 0.06(6) = 0.64$$

where

$$\frac{V}{S} = \frac{18(36)}{2(18) + 2(36)} = 6$$

The correction factor (CF) for humidity H is

$$(CF)_h = 1.5 - 0.015H = 1.5 - 0.015(70) = 0.45$$

for 70% relative humidity. The adjusted shrinkage strain then becomes

$$\epsilon_{sh} = (\text{basic } \epsilon_{sh})(CF)_{V/S}(CF)_h = (550 \times 10^{-6})(0.64)(0.45)$$

$$= 0.000158 \text{ in./in.}$$

$$\Delta f_s = \epsilon_{sh}E_s = 0.000158(29,000) = 4.58 \text{ ksi} \qquad (3.2\%)$$

(f) Relaxation in steel.

$$\Delta f_s = 0.05(144) = 7.2 \text{ ksi} \qquad (5.0\%)$$

(g) Total losses

	LOSS OF STRESS	
	ksi	PERCENT
Elastic shortening	2.33	1.6
Creep in concrete	10.8	7.5
Shrinkage	4.6	3.2
Relaxation in steel	7.2	5.0
Total =	24.9 ksi	17.3%

The final design prestress under dead load plus live load, after losses, is

$$f_{se} = 144 - 24.9 = 119.1 \text{ ksi}$$

21.8 STRENGTH OF FLEXURAL MEMBERS—ACI CODE

Design must include consideration of all significant load stages. Primarily, these stages are (1) initial stage, including the period before and during prestressing, as well as the transfer of prestress to the concrete; (2) intermediate stage during transportation and erection; (3) final stage under service load, after losses; and (4) overload stage, where cracking and ultimate strength are important. Though the actual number of conditions to be investigated varies with the situation, ordinarily the initial service condition involves beam dead load plus prestress *before* losses; the final service condition involves full dead plus live load *after* losses; and, the strength must be adequate for possible overload.

The initial and final service load stages have been considered in Sec. 21.5. A beam may be properly prestressed to carry service loads with little deflection and generally without cracking but, because of a small moment arm to the centroid of the steel, may have a nominal moment strength M_n that gives an inadequate margin of safety. On the other hand, if only the strength M_n were considered, service loads might cause excessive camber or deflection.

Balanced Strain Condition. As first described in Sec. 3.5, the balanced strain condition occurs when the concrete strain ϵ_c at the extreme compression fiber is 0.003 at the instant the steel reaches its yield strain $\epsilon_y = f_y/E_s$.

In prestressed concrete members, the steel used does not exhibit the well-defined yield point that occurs with ordinary deformed bars, so that the concept of balanced failure is nebulous. The balanced condition was used with ordinary deformed bars as a frame of reference to ensure a ductile failure mode. ACI-18.8.1 uses the criterion $\omega_p = \rho_p f_{ps}/f'_c = 0.36\beta_1$ for prestressed concrete members to represent the equivalent of $0.75\rho_b$ used for ordinary reinforced concrete. The term ρ_p refers to the reinforcement ratio A_{ps}/bd for the prestressing steel, and f_{ps} is the stress in the prestressed reinforcement when the nominal moment strength M_n is reached.

Prior to 1983, 0.3 was used instead of $0.36\beta_1$; for $f'_c = 5000$ psi, $\beta_1 = 0.80$ giving $0.36\beta_1 = 0.288$. The recent use of concrete strengths above 5000 psi has required a lower limit on ω_p for satisfactory ductility. The criterion $\rho_p f_{ps} \leq 0.36\beta_1$ essentially limits the neutral axis location to be not farther than $0.423d$ from the extreme compression fiber when the strength M_n is reached, which is comparable to the limit of $0.75x_b$ as defined in Chap. 3. (This is further explained by Example 21.8.2.)

The constant 0.423 may be explained as follows. Starting with the ACI Code limitation,

$$\frac{\rho_p f_{ps}}{f'_c} \leq 0.36\beta_1$$

Then, multiplying both sides by $f'_c bd$ gives

$$\rho_p f_{ps} bd \leq 0.36\beta_1 f'_c bd$$

The left-hand side of this inequality is $T = A_{ps}f_{ps} = \rho_p f_{ps} bd$. The right-hand side is the compression force $C = 0.85 f'_c b\beta_1 x$. When $x = 0.423d$ the two sides of the above inequality are identical.

The following discussion in regard to under- and overreinforcement is restricted to pretensioned construction and posttensioned construction in which the tendons are grouted. The nominal moment strength M_n of post-tensioned members in which tendons are ungrouted is, in general, less than that of a bonded beam.

Most of the following development applies for members containing only prestressed reinforcement (so-called *fully* prestressed members); thus, any nonprestressed steel effect is omitted. Most of the 1983 changes to the ACI Code Chapter 18 relate to the inclusion of nonprestressed steel effects to provide for *partially* prestressed members.

Underreinforced Beams. For cases where $\rho_p f_{ps}/f'_c \leq 0.36\beta_1$, the nominal moment strength may be determined for rectangular sections, in accordance with the principles of Chap. 3 (Sec. 3.3), by

$$M_n = T\left(d - \frac{a}{2}\right) \qquad (21.8.1)$$

where $T = A_{ps}f_{ps}$, $C = 0.85 f'_c ba$, and from $C = T$,

$$a = \frac{A_{ps}f_{ps}}{0.85 f'_c b} = \frac{\rho_p bd f_{ps}}{0.85 f'_c b} = \frac{\rho_p f_{ps}}{0.85 f'_c} d \qquad (21.8.2)$$

and A_{ps}, ρ_p, and f_{ps} refer to the area, reinforcement ratio, and tensile stress for the prestressed reinforcement. Note that f_{ps} is the average stress in the prestressing steel *when nominal moment strength M_n is reached*; it is used instead of f_y because the steel usually exhibits no well-defined yield point (see Fig. 21.8.1).

Substitution of Eq. (21.8.2) into Eq. (21.8.1) gives

$$M_n = A_{ps}f_{ps}\left(d - \frac{\rho_p f_{ps}}{1.7f'_c}\,d\right) \tag{21.8.3}$$

Thus for any concrete cross section, the nominal moment strength depends on ρ_p and f_{ps}.

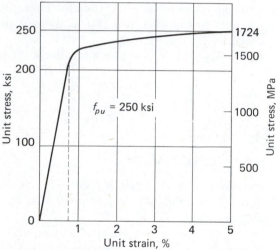

Figure 21.8.1 Typical stress-strain curve for high-tensile steel wire.

The actual stress in the prestressed reinforcement when the nominal strength is reached may not be easily determined, particularly when the specific stress-strain curve for the steel used is not available. Referring to Fig. 21.8.1, for low percentages of steel and therefore higher stress f_{ps}, the strain may be nearly 0.05; whereas for higher percentages of reinforcement, the strain may be closer to 0.01 (approximately corresponding to yield stress). Thus f_{ps} is not the same for all beams.

Since test results have indicated good agreement [12] with Eq. (21.8.3) up to $\rho_p f_{pu}/f'_c = 0.3$, the ACI-ASCE Joint Committee [9] proposed that f_{ps} be taken at $0.85f_{pu}$ (representing the sharp change in slope of the stress-strain curve) when $\rho_p f_{pu}/f'_c = 0.3$ and that it vary linearly from the aforementioned value to $f_{ps} = f_{pu}$ when $\rho_p f_{pu}/f'_c = 0$. Thus, in lieu of more accurate stress-strain data, the steel stress f_{ps} in members with bonded tendons could be taken as

$$f_{ps} = f_{pu}\left(1 - 0.5\rho_p\,\frac{f_{pu}}{f'_c}\right) \tag{21.8.4}$$

which was the 1977 ACI Code formula. Note that f_{pu} is the tensile strength of the prestressed reinforcement (see Fig. 21.8.1).

With the increasing use of *low relaxation prestressing strands* where the yield strength f_{py} equals $0.90f_{pu}$, instead of *stress relieved tendons* where f_{py} equals $0.85f_{pu}$, and the increasing use of f_c' greater than 5000 psi, where β_1 is less than 0.80, there was need to improve the estimate of f_{ps} previously based on Eq. (21.8.4). Thus, Eq. (21.8.4) has been modified in 1983 to become

$$f_{ps} = f_{pu}\left(1 - \frac{\gamma_p}{\beta_1}\frac{\rho_p f_{pu}}{f_c'}\right) \tag{21.8.5}$$

which is ACI Formula (18-3) omitting the terms for nonprestressed reinforcement. Note the values of γ_p/β_1 in Table 21.8.1. The value used for f_{ps} from Eq. (21.8.5) will be *lower than previously* for higher strength concrete with stress relieved tendons; however, with low relaxation strands the values used for f_{ps} will be higher than previously.

Table 21.8.1   VALUES FOR γ_p/β_1 FOR EQ. (21.8.5)

γ_p	CONCRETE STRENGTH, f_c' (psi)			
	5000	6000	7000	8000
0.40 (for $f_{py}/f_{pu} \geq 0.85$)	0.50	0.53	0.57	0.62
0.28 (for $f_{py}/f_{pu} \geq 0.90$)	0.35	0.37	0.40	0.43

Overreinforced Beams. For situations in which $\rho_p f_{pu}/f_c' > 0.36\beta_1$ (i.e., equivalent to $x > 0.423d$), corresponding roughly to ρ exceeding $0.75\rho_b$ for nonprestressed concrete beams, the higher prestressed reinforcement ratio is permitted but the strength M_n must be evaluated using no more than the amount corresponding to $\rho_p f_{pu}/f_c' = 0.36\beta_1$. When $\rho_p f_{pu}/f_c' > 0.36\beta_1$ for a member containing only prestressed reinforcement, ACI-18.8.2 states "design moment strength shall not exceed the moment strength based on the compression portion of the moment couple."

EXAMPLE 21.8.1 Determine the nominal moment strength M_n of the pretensioned bonded section investigated in Example 21.5.2. The concrete has $f_c' = 5000$ psi and the stress relieved prestressing strand has $f_{pu} = 250{,}000$ psi. Assume 20% prestress losses and an average stress-strain relationship for the steel, as given in Fig. 21.8.1.

Solution: The approximate stress when nominal strength is reached may be taken, according to ACI-18.7.2, or Eq. (21.8.5), as

$$f_{ps} = f_{pu}\left(1 - \frac{\gamma_p}{\beta_1}\rho_p\frac{f_{pu}}{f_c'}\right)$$

$$\rho_p = \frac{A_{ps}}{bd} = \frac{2.30}{20(25)} = 0.0046$$

For $f'_c = 5000$ psi, $\beta_1 = 0.80$; and for stress relieved strand $f_{py}/f_{pu} \geq 0.85$, $\gamma_p = 0.40$. Thus,

$$f_{ps} = 250\left[1 - \frac{0.40}{0.80}(0.0046)\frac{250}{5}\right] = 250(1 - 0.115) = 221 \text{ ksi}$$

Check $\rho_p f_{ps}/f'_c$ against $0.36\,\beta_1$.

$$\rho_p \frac{f_{ps}}{f'_c} = 0.0046\left(\frac{221}{5}\right) = 0.203 < [0.36\beta_1 = 0.288] \qquad\qquad \text{OK}$$

From Fig. 21.8.2(c),

$$C_u = 0.85f'_c ba = 0.85(5)(20)a = 85a$$

$$T_u = A_{ps}f_{ps} = 2.30(221) = 508 \text{ kips}$$

$$C_u = T_u$$

$$a = \frac{508}{85} = 5.98 \text{ in.}$$

$$x = \frac{a}{\beta_1} = \frac{5.98}{0.8} = 7.47 \text{ in.}$$

The additional strain ϵ_{s2} due to the ultimate moment is

$$\epsilon_{s2} = 0.003 \frac{17.53}{7.47} = 0.00704$$

which, when added to the strain due to prestress after losses, gives

$$\epsilon_s = \epsilon_{s1} + \epsilon_{s2} = \frac{140}{29,000} + 0.00704 = 0.00483 + 0.00704 = 0.0119$$

$$f_{si} \text{ (initial)} = 175,000 \text{ psi}$$

$$\Delta f_s \text{ (losses)} = 0.2(175,000) = 35,000 \text{ psi}$$

The strain ϵ_s of 0.0119 corresponds approximately to that for a stress f_{ps} of 225 ksi, using a typical stress-strain relationship for a steel with $f_{pu} = 250$ ksi, such as in Fig. 21.8.1. The value of f_{ps} agrees closely with the starting assumption of 221 ksi based on Eq. (21.8.4).

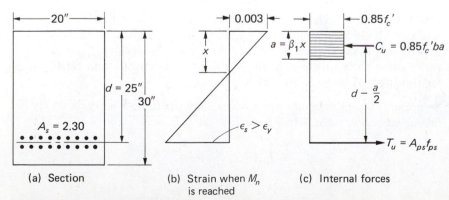

Figure 21.8.2 Nominal strength M_n of section for Example 21.8.1.

Thus, the nominal moment strength is

$$M_n = T_u\left(d - \frac{a}{2}\right) = \left[508\left(25 - \frac{5.98}{2}\right)\right]\frac{1}{12} = 932 \text{ ft-kips}$$

The required nominal moment strength based on the service loads of Example 21.5.2 is

$$\text{required } M_n = \frac{M_u}{\phi} = \frac{1.4(125) + 1.7(383)}{0.90} = 918 \text{ ft-kips} < 932 \text{ ft-kips}$$

OK

EXAMPLE 21.8.2 For a typical prestressed concrete beam, using high tensile strength steel wire having a stress-strain curve as in Fig. 21.8.1, show that the balanced strain condition at ultimate strength may be approximated by $\rho_p = 0.48\beta_1 f'_c/f_{ps}$.

Solution: (a) Consider that crushing strain for concrete is 0.003 as prescribed by the ACI Code. Referring to Fig. 21.8.1, the proportional limit is about $0.85f_{pu}$, above which value strain increases more rapidly than stress. Thus the approximate balanced strain condition occurs when $\epsilon_c = 0.003$ and $\Delta\epsilon_s$ (change in steel strain due to external loading) equals the strain ϵ_{ps} at $f_{ps} = 0.85f_{pu}$ less the initial prestress strain ϵ_i. For the balanced strain condition,

$$C_b = 0.85f'_c ba_b$$

$$T_b = A_{psb}f_{ps} = \rho_{pb}bdf_{ps}$$

$$C_b = T_b$$

$$a_b = \left(\frac{d}{0.85}\right)\rho_{pb}\frac{f_{ps}}{f'_c}$$

$$a_b = \beta_1 x_b$$

$$\frac{x_b}{d} = \frac{1}{0.85\beta_1}\rho_{pb}\frac{f_{ps}}{f'_c}$$

or

$$\rho_{pb} = 0.85\beta_1\frac{f'_c}{f_{ps}}\left(\frac{x_b}{d}\right)$$

where ρ_{pb} is the prestressed reinforcement ratio at the balanced stain condition.

At $f_{ps} = 0.85f_{pu}$, $\epsilon_{ps} = 0.0075$ from Fig. 21.8.1. Also, if the initial tension in the prestressing steel is $0.60f_{pu}$, then the initial tensile strain in the steel (representing a compressive strain in concrete) is

$$\epsilon_i = \frac{0.60f_{pu}}{E_s} = \frac{150}{30,000} \approx 0.005$$

Thus

$$\Delta\epsilon_s = \epsilon_{ps} - \epsilon_i = 0.0075 - 0.005 = 0.0025$$

Then, from Fig. 21.8.3,

$$\frac{x_b}{d} = \frac{0.003}{0.003 + 0.0025} = 0.545$$

Substituting into the ρ_{pb} expression gives

$$\rho_{pb} = 0.85\beta_1 \frac{f'_c}{f_{ps}}(0.545) = 0.46\beta_1 \frac{f'_c}{f_{ps}}$$

(b) Suppose that $\epsilon_c = 0.0035$ instead of 0.003. There will be no change in $\Delta\epsilon_s$ if it is still assumed $f_{ps} = 0.85f_{pu}$,

$$\frac{x_b}{d} = \frac{0.0035}{0.0035 + 0.0025} = 0.584$$

$$\rho_{pb} = 0.85\beta_1 \frac{f'_c}{f_{ps}}(0.584) = 0.50\beta_1 \frac{f'_c}{f_{ps}}$$

In effect the values of ρ_{pb} represent the "balanced" amount of reinforcement, which if exceeded would have the concrete reach crushing strain while the steel is still in the elastic range. Probably $0.50\beta_1 f'_c/f_{ps}$ represents a reasonable upper bound and $0.46\beta_1 f'_c/f_{ps}$ represents a typical value for the balanced condition. Thus the ACI Code implied use of $0.48\beta_1 f'_c/f_{ps}$ as the balanced reinforcement ratio seems reasonable; consequently, $0.36\beta_1 f'_c/f_{ps}$ reasonably represents 0.75 of the balanced reinforcement ratio.

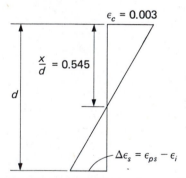

Figure 21.8.3 Strain diagram for the approximate balanced strain condition due to superimposed loads.

21.9 CRACKING MOMENT

One of the features of prestressed concrete is that under service load it is usually crack free. To be sure that an adequate reserve exists against cracking, ACI-18.8.3 requires the total amount of prestressed and nonprestressed reinforcement to be adequate to develop a factored load in flexure at least 1.2 times the cracking load calculated using a modulus of rupture equal to $7.5\sqrt{f'_c}$ for normal-weight concrete (see ACI-9.5.2.3 for lightweight concrete). The basic cracking moment requirement is a means of ensuring that cracking will occur *before* flexural strength is reached, and by a large enough margin so that significant deflection will occur to warn that the strength M_n

is being approached. The typical member will have a fairly large margin between cracking strength and flexural strength but the designer must be certain by checking it.

The 1983 ACI Code change to this section waives the above requirement when the flexural member has ϕV_n and ϕM_n at least twice V_u and M_u, respectively.

EXAMPLE 21.9.1 Compute the cracking moment and check its acceptability according to the ACI Code, for the beam of Example 21.8.1 (Fig. 21.8.2). Use $f'_c = 5000$ psi and assume the effective prestress, after losses, is 140 ksi.

Solution: (a) Use the homogeneous beam concept (Sec. 21.6). The effective prestress force is

$$T_e = A_{ps}f_{se} = 2.30(140) = 322 \text{ kips}$$

Using the following equation to find the cracking moment,

$$-\left(\begin{array}{c}\text{axial}\\\text{prestress}\end{array}\right) - \left(\begin{array}{c}\text{moment}\\\text{prestress}\end{array}\right) + \left(\begin{array}{c}\text{external}\\\text{loads}\end{array}\right) = \left(\begin{array}{c}\text{cracking}\\\text{stress}\end{array}\right)$$

$$-\frac{T_e}{A_g} - \frac{T_e e y_t}{I_g} + \frac{M_{cr} y_t}{I_g} = f_r = 7.5\sqrt{5000} = 530 \text{ psi}$$

$$-\frac{322}{600} - \frac{322(10)15}{45,000} + \frac{M_{cr}(15)}{45,000} = 0.530$$

$$M_{cr} = \left[\frac{0.530(45,000)}{15} + 321.5(10) + \frac{322(45,000)}{600(15)}\right]\frac{1}{12}$$

$$= 133 + 268 + 134 = 535 \text{ ft-kips}$$

(b) Use the internal force concept (Sec. 21.6). Consider that the cracking moment M_{cr} is comprised of two parts,

$$M_{cr} = M_1 + M_2$$

where M_1 is the superimposed moment necessary to give zero stress in the precompressed tension zone (see Fig. 21.9.1); and M_2 is the additional moment to cause cracking assuming that zero stress exists at the tension face when M_2 is applied (see Fig. 21.9.1).

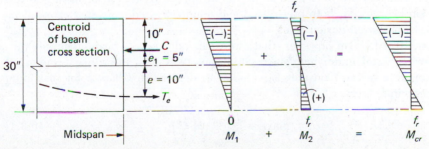

Figure 21.9.1 Computation of cracking moment.

$$C = T_e = 322 \text{ kips}$$

$$M_1 = T_e(e + e_1) = 322\left(\frac{10 + 5}{12}\right) = 403 \text{ ft-kips}$$

$$M_2 = \frac{f_r I_g}{y_t} = \frac{0.530(45,000)}{15(12)} = 133 \text{ ft-kips}$$

$$M_{cr} = M_1 + M_2 = 403 + 133 = 536 \text{ ft-kips}$$

(c) Check ACI-18.8.3. From Example 21.8.1, the nominal moment strength M_n provided is 932 ft-kips. This must exceed $1.2M_{cr}$ to satisfy ACI-18.8.8,

$$\phi M_n > 1.2 M_{cr}$$

$$0.90(932) > 1.2(536)$$

$$840 \text{ ft-kips} > 640 \text{ ft-kips} \qquad \text{OK}$$

Thus cracking will occur soon enough before reaching nominal moment strength so that large deflection will give a warning of impending failure.

In addition it is noted that the overload factor against cracking at mid-span is

$$\text{FS against cracking} = \frac{536}{125 + 383} = 1.05$$

If more reserve against cracking is desired, the tensile stress permitted under final service conditions should be reduced below $6\sqrt{f_c'}$. Cracking at service load may or may not be detrimental. Nonprestressed beams usually crack under service load.

21.10 SHEAR STRENGTH OF MEMBERS WITHOUT SHEAR REINFORCEMENT

In general the ideas presented in Chap. 5 regarding shear strength for nonprestressed beams are also applicable to prestressed beams. An excellent summary of background for the ACI Code expressions for shear strength of prestressed concrete beams is given by MacGregor and Hanson [13]. More information on shear strength of prestressed concrete is available in the ASCE-ACI Committee 426 Reports [14,15]. Collins and Mitchell [Chap. 19, Ref. 2] have presented an excellent unified treatment of shear and torsion behavior for both prestressed and nonprestressed concrete beams, along with proposed design rules.

It is known, however, that a prestressed concrete beam generally performs better under high shear conditions than does an ordinary reinforced concrete beam. Consider Eq. (5.3.1) as derived in Chap. 5 for maximum principal stress,

$$f_t(\text{max}) = \frac{f_t}{2} + \sqrt{\left(\frac{f_t}{2}\right)^2 + v^2} \qquad [5.3.1]$$

In nonprestressed concrete the normal stress f_t is a tensile stress on one side of the neutral axis. When the concrete is prestressed, f_t is a compressive stress throughout the member. Replacing f_t with $-f_t$ it is apparent that the magnitude of the principal tensile stress decreases,

$$f_t(\text{max}) = \frac{-f_t}{2} + \sqrt{\left(\frac{f_t}{2}\right)^2 + v^2} \qquad (21.10.1)$$

The angle α that the maximum principal tensile stress makes with the beam axis is greater for a prestressed concrete beam than for an ordinary reinforced concrete beam, as shown in Fig. 21.10.1. Inclined cracking will be less likely to occur and, if it occurs, will be more nearly horizontal than in nonprestressed members.

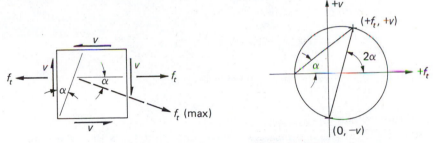

(a) Principal stress — reinforced concrete beam

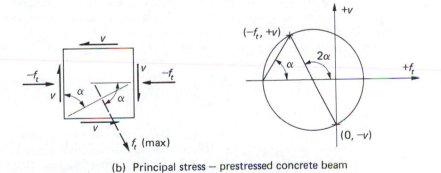

(b) Principal stress — prestressed concrete beam

Figure 21.10.1 Comparison of directions of principal tensile stress.

Two types of inclined cracks are possible in prestressed concrete beams: (1) the flexure-shear crack that occurs in a beam previously cracked due to flexure; and (2) the web-shear crack that occurs in the thin web of a previously uncracked beam (see Fig. 5.4.1). Whereas only the flexure-shear crack was common in nonprestressed beams, both types represent potential cracks in prestressed concrete beams.

Flexure-Shear Cracking Strength. The flexure-shear crack arises from high principal stress near the interior extremity of a flexural crack. Experimental studies have shown that the shear corresponding to the flexure-shear crack is that which causes flexural cracking at approximately a distance $d/2$ from the load point, in the direction of decreasing bending moment [16] (see Fig. 21.10.2). The 1963 ACI Code formula for flexure-shear cracking strength was the experimentally determined linear relationship of Fig. 21.10.3. Because the relationship between moment and shear (M/V) was not the same for both dead and live loads in the tests, the effects of dead load were not included in the linear relationship.

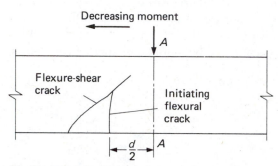

Figure 21.10.2   Flexure-shear type of inclined crack.

The ordinate of the plot (see Fig. 21.10.3) is thus in terms of $V_{ci} - V_d$, where V_d is the service dead load shear force and V_{ci} is the total nominal shear strength. The abscissa involves the net cracking moment M_{cr} due to service loads. The net cracking stress used for computing M_{cr} equals the modulus of rupture (assumed conservatively at $6\sqrt{f'_c}$ psi), plus the compressive stress f_{pe} provided by the prestressing force (after losses) occurring at the extreme fiber at which tensile stresses are caused by the applied loads, less the tensile stress f_d due to service dead load. Thus

$$M_{cr} = \frac{I}{y_t}(6\sqrt{f'_c} + f_{pe} - f_d) \qquad (21.10.2)*$$

which is ACI Formula (11-12).

The abscissa (Fig. 21.10.3) involves $M_{cr}/(M/V - d/2)$, which represents the applied load shear at $d/2$ from the cross section being investigated. A moment M and a shear V act on section $A\text{-}A$ of Fig. 21.10.4. Assume that M and V give rise to a moment crack at $d/2$ from section $A\text{-}A$. From the basic shear-moment relationship the change in moment between two points equals the area under the shear diagram between those points. Thus referring to Fig. 21.10.4,

$$M - M_{cr} = \frac{V + V_{cr}}{2}\left(\frac{d}{2}\right) \qquad (21.10.3)$$

Since the difference between V and V_{cr} will usually be small, assume

*For SI, ACI 318-83M, for f'_c, f_{pe}, and f_d in MPa, M_{cr} in N·mm, I in mm^4, and y_t in mm, gives

$$M_{cr} = \frac{I}{y_t}\left(\frac{1}{2}\sqrt{f'_c} + f_{pe} - f_d\right) \qquad (21.10.2)$$

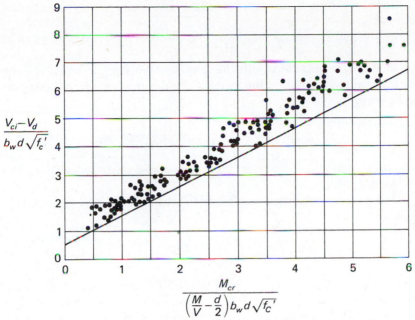

Figure 21.10.3 Comparison of the shear corresponding to flexure-shear cracking in prestressed concrete beams with the ratio of the flexural cracking moment to the shear span (both axes nondimensionalized by dividing by $b_w d \sqrt{f_c'}$) (from Sozen and Hawkins [16]).

$V_{cr} = V$; thus

$$M - M_{cr} = V\left(\frac{d}{2}\right) \tag{21.10.4}$$

Solving for M_{cr},

$$M_{cr} = M - V\left(\frac{d}{2}\right)$$

$$= V\left(\frac{M}{V} - \frac{d}{2}\right) \tag{21.10.5}$$

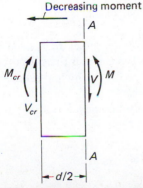

Figure 21.10.4 Shear-moment relationship.

Finally, the shear on the section investigated becomes

$$V = \frac{M_{cr}}{M/V - d/2} \qquad (21.10.6)$$

Thus the linear relationship for the flexure-shear cracking strength resulting from high principal stress in the vicinity of a flexural crack is, according to Fig. 21.10.3,

$$\frac{V_{ci} - V_d}{b_w d \sqrt{f'_c}} = 0.6 + \frac{M_{cr}}{(M/V - d/2)b_w d \sqrt{f'_c}} \qquad (21.10.7)$$

Since 1971, ACI-11.4.2 has simplified the expression by eliminating the subtracted $d/2$ term.

Thus the nominal flexure-shear cracking strength V_{ci}, ACI Formula (11-11), is

$$V_{ci} = 0.6\sqrt{f'_c}b_w d + \frac{V_i M_{cr}}{M_{max}} + V_d \geq 1.7\sqrt{f'_c}b_w d \qquad (21.10.8)*$$

In Eq. (21.10.8), M_{max} replaces M of Eq. (21.10.7) and represents the maximum moment that can occur at the section under consideration, due to externally applied factored loads (i.e., applied loads other than beam weight and prestress unless the prestress causes an external reaction). V_i replaces V of Eq. (21.10.7) and represents the shear force at the section due to the factored loading that caused maximum moment. In other words, one uses the moment *envelope* values for M_{max} along with the corresponding shears, rather than the shear envelope values which would be larger. Of course, where partial span loadings are not considered, the full-span loading gives V_i and M_{max} for each point along the span. When full-span uniform loading is used in a simply supported span, $V_i = w(L - 2x)/2$ and $M_{max} = wx(L - x)/2$, and

$$\frac{V_i}{M_{max}} = \frac{L - 2x}{x(L - x)} \qquad (21.10.9)$$

In effect, Eq. (21.10.8) gives the flexure-shear cracking strength as the sum of (1) the shear due to actual dead load (beam weight) without overload factor, (2) the superimposed factored live load which is sufficient to cause flexural cracking, and (3) the additional load ($0.6\sqrt{f'_c}b_w d$) which will cause the flexural crack to initiate the inclined flexure-shear crack.

Web-Shear Cracking Strength. The web-shear crack arises in a beam previously uncracked due to flexure. Such a crack is typical near the support of a thin-webbed section (see Fig. 5.4.1) as a result of high principal tensile stress.

For this development the beam may reasonably be considered as homogeneous. Thus Eq. (21.10.1) may be directly applied,

---

*For SI, ACI 318-83M, for f'_c in MPa, b_w and d in mm, and V in N, gives

$$V_{ci} = \frac{1}{20}\sqrt{f'_c}\,b_w d + \frac{V_i M_{cr}}{M_{max}} + V_d \geq \frac{1}{7}\sqrt{f'_c}b_w d \qquad (21.10.8)$$

$$f_t(\text{max}) = \frac{-f_t}{2} + \sqrt{\left(\frac{f_t}{2}\right)^2 + v^2} \qquad [21.10.1]$$

Since tests have demonstrated that cracks of this type usually originate near the centroid of the section, the stresses refer to that location. Solving Eq. (21.10.1) for v,

$$\left[f_t(\text{max}) + \frac{f_t}{2}\right]^2 = \left(\frac{f_t}{2}\right)^2 + v^2$$

$$[f_t(\text{max})]^2 + f_t(\text{max})f_t + \left(\frac{f_t}{2}\right)^2 = \left(\frac{f_t}{2}\right)^2 + v^2$$

$$v = f_t(\text{max})\sqrt{1 + \frac{f_t}{f_t(\text{max})}} \qquad (21.10.10)$$

where f_t is the compressive stress at the level of the centroid, and $f_t(\text{max})$ = principal tensile stress ≤ tensile strength of concrete if no cracks are to form.

Since Eq. (21.10.10) should agree with the criteria for shear strength of nonprestressed beams, $f_t(\text{max})$ is taken as $3.5\sqrt{f_c'}$. The equation then becomes

$$v = 3.5\sqrt{f_c'}\sqrt{1 + \frac{f_t}{3.5\sqrt{f_c'}}} \qquad (21.10.11)$$

A plot of Eq. (21.10.11) is given in Fig. 21.10.5, illustrating that this equation may be approximated by a straight line,

$$v = 3.5\sqrt{f_c'} + 0.3f_t \qquad (21.10.12)$$

or, in ACI Code terminology, multiplying by $b_w d$ to give nominal shear strength,

$$V_{cw} = (3.5\sqrt{f_c'} + 0.3f_{pc})b_w d \qquad (21.10.13)$$

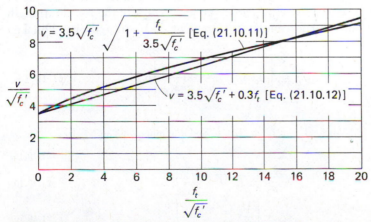

Figure 21.10.5 Comparison of theoretical maximum shear stress with straightline approximation.

where f_{pc} is defined as the compressive stress (psi) in the concrete, after losses, at the centroid of the section resisting the applied loads, or if the centroid lies within the flange on a T-section, f_{pc} is the stress at the junction of the flange and web.

If prestressing tendons are draped, a vertical component V_p arises which assists in carrying the shear. Therefore ACI-11.4.2 gives ACI Formula (11-13) as

$$V_{cw} = (3.5\sqrt{f'_c} + 0.3f_{pc})b_w d + V_p \qquad (21.10.14)*$$

Alternatively, the web-shear cracking strength may be determined (ACI-11.4.2.2) using the principal stress equation, Eq. (21.10.10), where the principal stress $f_t(\text{max})$ is limited to $4\sqrt{f'_c}$. This would give Eq. (21.10.11) with coefficients 4 instead of 3.5. Thus using f_{pc} for f_t, Eq. (21.10.11) multiplied by $b_w d$ with V_p added becomes

$$V_{cw} = 4\sqrt{f'_c}\sqrt{1 + \frac{f_{pc}}{4\sqrt{f'_c}}}b_w d + V_p \qquad (21.10.15)*$$

which is an acceptable alternate to using Eq. (21.10.14).

The nominal shear strength V_n at which inclined cracking is imminent is given by the lesser of V_{ci} and V_{cw}, determined from ACI Formulas (11-11) and (11-13) for normal-weight concrete.

When applying Eqs. (21.10.8) and (21.10.14) or (21.10.15) for V_{ci} and V_{cw}, the effective depth d is to be taken as the distance from the extreme compression fiber to the centroid of the prestressing tendons, or as 80% of the overall depth of the member, whichever is greater.

In computing the prestress effect f_{pc}, full prestress after losses may be used only when the prestressing tendons are embedded a distance exceeding their required development length from the section being investigated. The critical section for maximum shear is generally at $h/2$ from the face of support (ACI-11.1.2.2), so that the region between the face of support and $h/2$ therefrom must be designed for the shear at the critical section.

ACI Code Simplified Alternative for Shear Strength. When the member has an effective prestress f_{se} at least equal to 40% of the tensile strength f_{pu} of the flexural reinforcement, the nominal shear strength may be taken as

$$V_c = \left[0.6\sqrt{f'_c} + 700\frac{V_u d}{M_u}\right]b_w d \qquad (21.10.16)*$$

*For SI, ACI 318-83M, for f'_c and f_{pc} in MPa, b_w and d in mm, and V in N, gives

$$V_{cw} = 0.3(\sqrt{f'_c} + f_{pc})b_w d + V_p \qquad (21.10.14)$$

$$V_{cw} = \frac{1}{3}\sqrt{f'_c}\sqrt{1 + \frac{3f_{pc}}{\sqrt{f'_c}}}b_w d + V_p \qquad (21.10.15)$$

$$V_c = \left[\frac{1}{20}\sqrt{f'_c} + 5\frac{V_u d}{M_u}\right]b_w d \qquad (21.10.16)$$

which is ACI Formula (11-10). This equation may be considered a linear gradient between that using the minimum nominal unit stress $v_c = 2\sqrt{f_c'}$ and an upper bound of $v_c(\max) = 5\sqrt{f_c'}$. Supporting data for this alternate relationship are shown in Fig. 21.10.6 from MacGregor and Hanson [13].

When applying Eq. (21.10.16) from ACI-11.4.1, V_u is the maximum shear due to factored loading at the section and M_u is the simultaneously occurring moment; $V_u d/M_u$ is also limited to a maximum value of 1.0; and d is the actual effective depth to the centroid of prestressed reinforcement. Equation (21.10.16) represents essentially the flexure-shear cracking strength expressed in a manner similar to that for nonprestressed reinforced concrete. Since the percentage of reinforcement is low in prestressed concrete, a constant has been used instead of the variable ρ.

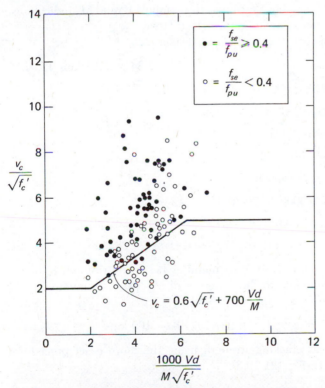

Figure 21.10.6 Alternate equation for computing v_c for prestressed beams (from MacGregor and Hanson [13]).

EXAMPLE 21.10.1 Determine the nominal shear strength $V_n = V_c$ at a section 4 ft from the supports of the 40 ft simple span shown in Fig. 21.10.7. The effective prestress force T_e (after losses) is 322 kips. The beam is to support a service live load of 1.38 kips/ft in addition to the beam weight of 0.625 kip/ft. The concrete has $f_c' = 5000$ psi.

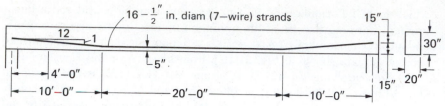

Figure 21.10.7 Section for Example 21.10.1.

Solution: (a) Simplified alternate procedure, using ACI Formula (11-10), Eq. (21.10.16).

$$V_c = \left(0.6\sqrt{f_c'} + 700\frac{V_u d}{M_u}\right)b_w d$$

In the application of the above formula, the symbols V_u and M_u indicate use of the shear and moment including overload factors.

Assume partial span loading is not considered.

$$w_u = 0.625(1.4) + 1.38(1.7) = 0.87 + 2.34 = 3.21 \text{ kips/ft}$$

$$V_u = 3.21(20 - 4) = 51.4 \text{ kips}$$

$$M_u = \tfrac{1}{2}(3.21)(4)(36) = 231 \text{ ft-kips}$$

$$\frac{V_u d}{M_u} = \frac{51.4(30 - 11)}{231(12)} = 0.35$$

The nominal unit stress capacity is

$$v_c = 0.6\sqrt{5000} + 700(0.35) = 42 + 245 = 287 \text{ psi}$$
$$v_c \text{ (upper limit)} = 5\sqrt{f_c'} = 354 \text{ psi} > 287 \text{ psi} \qquad \text{OK}$$
$$v_c \text{ (lower limit)} = 2\sqrt{f_c'} = 141 \text{ psi} < 287 \text{ psi} \qquad \text{OK}$$

Thus at this section the prestressed member has a contribution v_c to the shear strength attributable to the concrete of 287 psi. The nominal shear strength V_c at 4 ft from the support would be

$$V_c = v_c b_w d = 0.287(20)(19) = 109 \text{ kips}$$

(b) Flexure-shear cracking strength using the more exact procedure, ACI Formula (11-11); Eq. (21.10.8),

$$V_{ci} = 0.6\sqrt{f_c'}b_w d + \frac{V_i M_{cr}}{M_{max}} + V_d$$

using Eq. (21.10.2),

$$M_{cr} = \frac{I}{y_t}(6\sqrt{f_c'} + f_{pe} - f_d)$$

$$I = \tfrac{1}{12}(20)(30)^3 = 45{,}000 \text{ in.}^4 \qquad \text{(neglecting steel)}$$

$y_t = 15$ in.

f_{pe} = compressive stress due to prestressing force

$$= \frac{322,000}{20(30)} + \frac{322,000(4)(15)}{45,000} = 537 + 429 = 966 \text{ psi}$$

$e = 4$ in. at 4 ft–0 in. from center of support

$$f_d = \text{dead load stress} = \frac{My}{I} = \frac{625(4)(36)(12)(15)}{2(45,000)} = 180 \text{ psi}$$

$$M_{cr} = \frac{45,000}{15(12,000)}(6\sqrt{5000} + 966 - 180)$$

$$= 0.25(424 + 966 - 180) = 0.25(1210) = 302 \text{ ft-kips}$$

M_{max} = maximum moment due to externally applied factored loads except beam weight at section being investigated

$$= \tfrac{1}{2}(1.38)(4)(36)(1.7) = 169 \text{ ft-kips}$$

V_i = corresponding shear due to factored load at section investigated

$$= 1.38(20 - 4)(1.7) = 37.5 \text{ kips}$$

$$\frac{V_i M_{cr}}{M_{max}} = \frac{37.5(302)}{169} = 67.0 \text{ kips}$$

V_d = dead load shear = $0.625(20 - 4) = 10.0$ kips

$d = 30 - 11 = 19$ in.   or $0.80h = 0.8(30) = \underline{24 \text{ in.}}$

$V_{ci} = 0.6\sqrt{5000}(20)(24)\tfrac{1}{1000} + 67.0 + 10.0 = 97.4$ kips

The flexure-shear cracking strength is not to be taken less than

$$\min V_{ci} = 1.7\sqrt{f'_c}b_w d$$

$$= 1.7\sqrt{5000}(20)(24)\tfrac{1}{1000} = 57.7 \text{ kips} < 97.4 \text{ kips} \qquad \text{OK}$$

(c) Web-shear cracking strength using the more exact procedure, ACI Formula (11-13), Eq. (21.10.14),

$$V_{cw} = [3.5\sqrt{f'_c} + 0.3f_{pc}]b_w d + V_p$$

f_{pc} = compressive stress in concrete at centroid of section resisting live load, due to all applied loads (zero due to all except T_e/A_c in this case)

$$= \frac{322,000}{20(30)} = 537 \text{ psi}$$

$d = 19$ in.   or $0.80(30) = \underline{24 \text{ in.}}$ (use larger value)

$V_{cw} = [3.5\sqrt{5000} + 0.3(537)](20)(24)\tfrac{1}{1000} + \tfrac{1}{12}(322) = 223$ kips

Since $V_{ci} < V_{cw}$, $V_c = V_{ci} = 97.4$ kips at 4 ft from the support for the member without shear reinforcement. The simplified alternate equation gives a nominal shear strength 12% higher than obtained from the more exact procedure; 109 kips compared to 97.4 kips.

21.11 SHEAR REINFORCEMENT FOR PRESTRESSED CONCRETE BEAMS

The computations for shear reinforcement are essentially the same as for nonprestressed concrete as developed in Chap. 5. The total nominal shear strength V_n may be represented as the sum of the amount V_c attributable to the concrete and the amount V_s attributable to the shear reinforcement; thus Eq. (5.7.1) still applies,

$$V_n = V_c + V_s \qquad\qquad [5.7.1]$$

The nominal strength V_c attributable to the concrete may be determined by (1) Eq. (21.10.16) when the effective prestress is at least 40% of the tensile strength of the steel, or (2) the *smaller* of Eqs. (21.10.8) and (21.10.14) for V_{ci} and V_{cw}. The second and more accurate method using V_{ci} and V_{cw} may be used whatever the magnitude of prestress.

For the shear reinforcement contribution, Eq. (5.10.7) is used,

$$V_s = \frac{A_v f_y d}{s} \qquad\qquad [5.10.7]$$

where

A_v = effective area of shear reinforcement at any section
s = spacing of the shear reinforcement

Just as for nonprestressed reinforced concrete beams, prestressed concrete beams must have at least a minimum amount of shear reinforcement whenever $V_u > \phi V_c/2$ (ACI-11.5.5.1). However, this requirement may be waived if tests are made showing that the required flexural and shear strengths can be developed when shear reinforcement is omitted. Since prestressed concrete members are usually of standardized shapes with many identical or similar members manufactured, manufacturers frequently have made tests demonstrating that no shear reinforcement is required.

When minimum shear reinforcement is required, ACI-11.5.5.3 gives for the minimum area A_v

$$A_v = 50\frac{b_w s}{f_y} \qquad\qquad (21.11.1)$$

which is also used for nonprestressed concrete.

Alternatively, for prestressed members only, where the effective prestress force equals at least 40% of the tensile strength of the steel, the minimum area may be taken as (ACI-11.5.5.4)

$$A_v = \frac{A_{ps}}{80}\left(\frac{f_{pu}}{f_y}\right)\left(\frac{s}{d}\right)\sqrt{\frac{d}{b_w}} \qquad\qquad (21.11.2)$$

where

A_{ps} = area of prestressed reinforcement
f_{pu} = tensile strength of prestressed reinforcement
f_y = specified yield strength of shear reinforcement

s = stirrup spacing, which may not exceed 0.75 of the depth of the member, nor 24 in., whichever is smaller

d = distance from the extreme compressive fiber to the centroid of the prestressed reinforcement (but need not be taken less than $0.8h$)

EXAMPLE 21.11.1 Determine the maximum spacing of #3 U stirrups at a point 4 ft from the support on the beam of Example 21.10.1 (Fig. 21.10.7). The prestressing steel has an area of 2.30 sq in. and a tensile strength f_{pu} = 250,000 psi, the yield strength of the stirrup reinforcement f_y = 40,000 psi, and the concrete has f'_c = 5000 psi. The service live load is 1.38 kips/ft, and the dead load is 0.625 kip/ft.

Solution: The inclined cracking strength $V_c = V_{ci}$ as determined in Example 21.10.1, is

$$V_{ci} \text{ (controls)} = 97.4 \text{ kips} \qquad \text{(more exact procedure)}$$

At 4 ft from centerline of support,

$$V_u = [1.4(0.625) + 1.7(1.38)](20 - 4) = 3.21(20 - 4) = 51.4 \text{ kips}$$

$$V_u = 51.4 \text{ kips} > \frac{\phi V_c}{2} = \frac{0.85(97.4)}{2} = 41 \text{ kips}$$

Thus a minimum amount of shear reinforcement must be provided, unless tests are made to justify its omission. Thus using Eq. (21.11.1)

$$\frac{A_v}{s} = 50\frac{b_w}{f_y} = 50\left(\frac{20}{40,000}\right) = 0.025$$

or, alternatively, using the more elaborate formula, Eq. (21.11.2),

$$\frac{A_v}{s} = \frac{A_{ps}}{80}\left(\frac{f_{pu}}{f_y}\right)\left(\frac{1}{d}\right)\sqrt{\frac{d}{b_w}}$$

$$= \frac{2.30}{80}\left(\frac{250}{40}\right)\left(\frac{1}{24}\right)\sqrt{\frac{24}{20}} = 0.0082$$

gives for minimum A_v/s (for #3 U stirrups),

$$s = \frac{2(0.11)}{0.0082} = 30.1 \text{ in.}$$

but, according to ACI-11.5.4.1, the spacing may not exceed $0.75h = 0.75(30)$ = 22.5 in. or 24 in., whichever is smaller. Thus #3 stirrups could be placed no farther apart than 22.5 in. unless tests are performed.

21.12 DEVELOPMENT OF REINFORCEMENT

Following the basic concepts established for nonprestressed reinforced concrete in Chap. 6, development of reinforcement must also be considered in prestressed members. The prestressing force must be transferred into the concrete by bond (interaction between concrete and steel strand) in the pretensioned beam, and the length required to accomplish this is called the "transfer length." This, of course, occurs in end regions and, in effect, anchors the tendons.

The mechanism of transfer is summarized by Zia and Mostafa [17]. The high stress in the pretensioned cable must, on cutting of the cable, be transferred to the concrete so that equilibrium is achieved. The situation in the transfer zone is quite different from that in the anchorage zones of non-prestressed concrete.

In nonprestressed concrete, the bars are stressed after being cast in the concrete. As they are stressed they decrease slightly in diameter and tend to draw away (at least create a slight tensile stress in the direction transverse to the bars) from the surrounding concrete. Thus a reasonably large length is necessary to transfer a stress f_y in the steel to the surrounding concrete.

In prestressed concrete, the wires or strands are pretensioned to a high level of stress, thus initially reducing the wire diameter prior to the placing of the concrete. Once the concrete is cured and the wires are cut at the ends, the wires tend to shorten and correspondingly increase in diameter. Thus a compression between the wires and concrete is produced. The stress in the cut wire must increase from zero at the free end to the prestress value at a certain distance from the free end. This distance is known as the "transfer length." During the accomplishment of the transfer, probably in the first few inches from the free end, it is solely the friction that is developed as the slipping wires compress against the concrete. Farther from the end, adhesive bond, that is, slip resistance, certainly contributes to the transfer.

The transfer length L_t typically is about 50 diameters for a strand, and about 110 to 120 diameters for a single wire [17]. These values assume a clean strand (or wire) surface, a gradual release of the prestressing force to the concrete, and a steel stress of 140 to 150 ksi after transfer. For strands with a slightly rusted surface, the transfer distance is certainly less, and for a sudden release of stress such as can occur by burning the strand, the transfer length may easily be 20% greater. The transfer length does not seem to be affected by variation in concrete strength [18].

The importance of the transfer length depends on the member under consideration. It seems to be of little importance except on short members and in situations where flexural cracking may occur in the transfer zone.

Development Length for Prestressing Strand. Exactly as for nonprestressed reinforcement, the tension in the prestressing steel necessary to achieve nominal flexural strength M_n must be developed by embedment or end anchorage or a combination thereof (ACI-12.1). The purpose is to prevent general slip prior to achieving the necessary moment strength M_n; thus the steel stress must increase from the effective prestress value f_{se} to the value f_{ps} used in computing the moment strength. The net anchorage length available to accommodate a change in tension force is the development length L_d to the end of the strand from the section in question, less the transfer length L_t.

The following empirical relationship was established by Hanson and Kaar [18],

$$(f_{ps} - f_{se}) \text{ in ksi} = \frac{L_d - L_t}{d_b} \tag{21.12.1}$$

where d_b is the nominal strand diameter.

The calculated stress f_{ps} in the prestressed reinforcement occurring when the nominal strength M_n is reached must not cause the reinforcement to slip relative to the surrounding concrete; thus

$$f_{ps}(\text{ksi}) \le f_{se}(\text{ksi}) + \frac{L_d - L_t}{d_b} \tag{21.12.2}$$

Next, it is necessary to obtain an expression for L_t/d_b. Using the relationship from Chap. 6 (Sec. 6.2) for embedment length,

$$A_s f_{se} = u\pi d_b L_t$$

$$\left(\frac{\pi d_b^2}{4}\right) f_{se} = u\pi d_b L_t \tag{21.12.3}$$

Since the actual strand (three- or seven-wire) properties differ from those based on the nominal diameter, a correction must be applied. The true circumference is taken as $\frac{4}{3}\pi d_b$; and the true area A_s as $0.725\pi d_b^2/4$. Thus

$$0.725\left(\frac{\pi d_b^2}{4}\right) f_{se} = u\left(\frac{4}{3}\right)\pi d_b L_t \tag{21.12.4}$$

$$\frac{L_t}{d_b} = \frac{2.175 f_{se}}{16u} \tag{21.12.5}$$

If the average bond stress in the transfer zone is taken as 400 psi for clean strands [18],

$$\frac{L_t}{d_b} = \frac{2.175}{16(0.4)} f_{se} = \frac{1}{2.94} f_{se} \quad \text{say } \frac{1}{3} f_{se} \tag{21.12.6}$$

where f_{se} is measured in ksi.

Substituting Eq. (21.12.6) in Eq. (21.12.2) gives

$$f_{ps} \le f_{se} + \frac{L_d}{d_b} - \frac{1}{3} f_{se}$$

and solving for L_d, the necessary development length to the end of the strand,

$$L_d = (f_{ps} - \tfrac{2}{3}f_{se})d_b \tag{21.12.7}$$

which is the formula given in ACI-12.9.1 for the three- or seven-wire pretensioning strands. Note that the expression in parenthesis uses stresses f_{ps} and f_{se} in ksi units, but the resulting expression $(f_{ps} - \tfrac{2}{3}f_{se})$ is treated as a dimensionless constant. Investigation may be restricted to those cross sections nearest each end of the member that are required to develop their nominal strengths M_n.

EXAMPLE 21.12.1 Investigate the development of reinforcement for the beam of Example 21.10.1 (Fig. 21.10.7). Assume $f_{ps} = 221$ ksi as computed in Example 21.8.1. The effective prestress, after losses, may be assumed to be $0.8(175) = 140$ ksi.

Solution: Since only at midspan is the maximum moment strength M_n required, the total embedment from midspan must exceed the development length L_d,

$$L_d = [221 - \tfrac{2}{3}(140)]d_b$$

$$L_d = (221 - 93.3)(0.5) = 64 \text{ in.} < 240 \text{ in.} \qquad \text{OK}$$

It is noted that f_{ps} used in this equation may be taken as that value required for the necessary strength at a section. In this case required nominal strength M_u/ϕ of 918 ft-kips was only slightly less than the provided M_n of 932 ft-kips. However, if the requirement was much below that provided, f_{ps} could have been taken as less than 221 ksi even at midspan. Development of reinforcement usually is not a problem on simply supported prestressed concrete beams.

21.13 PROPORTIONING OF CROSS SECTIONS FOR FLEXURE WHEN NO TENSION IS PERMITTED

In order to give further insight into the variables involved, the following discussion is presented on proportioning the cross section. All cross sections used here will consist of rectangular flanges and web, though in practical design 90° junctions between flanges and web are avoided because of forming difficulties.

In this brief treatment, the only case treated is when no tension is permitted, either at the initial condition (at transfer of prestress before losses) or at the final condition (full dead and live load after losses). Thus the desired stress distributions are shown in Fig. 21.13.1.

It can be shown that on any cross section having an axis of symmetry, there are two points on the axis of symmetry, known as the "kern points," that are located at distances k_b and k_t from the centroid, as shown in Fig. 21.13.1. Each kern point represents the farthest distance from the centroid at which a resultant force can act without inducing a stress of opposite sign at the extreme fiber in the opposite direction from the centroid. This means that the stress distribution varies from zero at the top to maximum at the bottom, when the resultant force acts on the lower kern point; or it varies from zero at the bottom to maximum at the top when the resultant force acts on the upper kern point. Referring to Fig. 21.13.1, and using Eq. (21.6.1) with the internal force concept

$$f = \frac{C}{A} - \frac{Ck_b y_t}{I} = 0 \qquad \textbf{(21.13.1a)}$$

and

$$f = \frac{C}{A} - \frac{Ck_t y_b}{I} = 0 \qquad \textbf{(21.13.1b)}$$

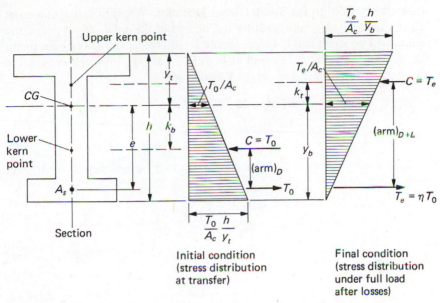

Figure 21.13.1 Stress distributions desired when no tension is permitted—*small* girder moment.

Solving Eqs. (21.13.1ab) for k_b and k_t,

$$k_b = \frac{I}{Ay_t} = \frac{r^2}{y_t} \tag{21.13.2a}$$

and

$$k_t = \frac{I}{Ay_b} = \frac{r^2}{y_b} \tag{21.13.2b}$$

in which r is the radius of gyration of the cross section.

The reader may recall from Sec. 21.6 that as the load applied to a prestressed concrete beam changes, the *position* of the internal force C must also change; that is, the moment arm (see Fig. 21.13.1) measured from the position of the prestressing steel must increase as the load increases. Thus when the moment M_D due to dead load of the girder is acting at the initial condition (immediately after transfer of prestress), the moment arm of the internal couple is $(\text{arm})_D$; and when moment due to dead plus live load is acting, the moment arm is $(\text{arm})_{D+L}$.

In order to achieve the triangular distribution at the initial condition, the prestressing steel must have its centroid below the kern point by the amount $(\text{arm})_D = M_D/T_0$ (see Fig. 21.13.1). If the girder moment M_D is small, it may be possible and practical to position the steel at $e = k_b + (\text{arm})_D$ below the centroid of the section. If the moment M_D is large, the distance $[k_b + (\text{arm})_D]$ may extend so far below the centroid of the section as to be too close to the bottom for proper cover or even outside the section. Thus for larger girder weight (large M_D), the optimum condition of triangular stress distribution at the initial condition may be impossible to achieve.

Preliminary Design for Small Girder Moment. When the girder moment M_D is small (say M_D representing 0.2 or less of the total, $M_D + M_L$), the $(\text{arm})_D$ will be small and C can probably be located at the lower kern point; that is, the steel will be located at $e = k_b + M_D/T_0$ from the centroid of the section. Observing Fig. 21.13.1 again,

$$(\text{arm})_{D+L} - (\text{arm})_D = k_t + k_b$$

$$\frac{M_D + M_L}{T_e} - \frac{M_D}{T_0} = k_t + k_b$$

If M_D is small, M_D/T_0 in the above equation may be assumed to be M_D/T_e, then

$$\frac{M_L}{T_e} \approx k_t + k_b$$

or

$$T_e \approx \frac{M_L}{k_t + k_b} \qquad (21.13.3)$$

The maximum stress in Fig. 21.13.1 may be obtained from the stress at the centroid (which is T/A_c without bending stress) by the linear relationship. Normally the ratio of the allowable stresses f_{ic} at the initial condition to f_{fc} at the final condition is approximately the same as the ratio of the initial prestress force T_0 to the final prestress force T_e. This means that the economical section should have $y_t \approx y_b$, or $k_t \approx k_b$, that is, be symmetrical. For a rectangular section $k_t = k_b = h/6$, thus $k_t + k_b = h/3$. For the I-shaped section that is more efficient to resist bending, $k_t + k_b > h/3$, say about $h/2$. Thus Eq. (21.13.3) may be approximated

$$T_e \approx \frac{M_L}{0.5h} \qquad (21.13.4)$$

If the section is symmetrical, the stress at the centroid will be half of the maximum; thus the required area A_c of the section would be, for the initial condition at transfer,

$$\text{required } A_c = \frac{T_0}{0.50 f_{ic}} \qquad (21.13.5)$$

and for the final condition of dead plus live load after losses,

$$\text{required } A_c = \frac{T_e}{0.50 f_{fc}} \qquad (21.13.6)$$

For the preliminary selection of cross section in case of small girder moment M_D, the following steps may be followed:

1. Select the overall depth h of the section. This is somewhat arbitrary but in the absence of other limitations, the guidelines of Lin and Burns [2] may be followed:

(a) Use 70% of the depth that would be used for nonprestressed reinforced concrete construction.

(b) For *slabs* having light loading, $h \geq 1/55$ of the span L.

(c) For *slabs* having heavy loading, $h \geq L/35$.

(d) On bridges, h ranges between $L/15$ and $L/25$.

(e) As a rule of thumb, h (inches) for *beams* can be approximated by the empirical formula, $1.5\sqrt{M}$(ft-kips) to $2.0\sqrt{M}$.

2. Compute the approximate T_e from Eq. (21.13.4).

3. Determine the approximate A_c from Eq. (21.13.6).

4. Proportion a symmetrical I-shaped section.

5. With this preliminary section, compute the section properties and locate the desired distance e of the steel centroid from the section centroid (CG),

$$e = k_b + (\text{arm})_D = k_b + \frac{M_D}{T_0} \qquad (21.13.7)$$

where $T_0 = T_e/\eta$ and η is the proportion of initial prestress remaining after losses.

6. If the steel can be located at the desired e, then T_e is more correctly determined,

$$T_e = \frac{M_D + M_L}{(\text{arm})_{D+L}} = \frac{M_D + M_L}{e + k_t} \qquad (21.13.8)$$

Then $T_0 = T_e/\eta$ and a new value of e is established; the iterative process is repeated until the desired accuracy is obtained.

7. Equations (21.13.5) and (21.13.6) are then used to determine the required A_c. When the equations give significantly different requirements, the section may be changed to become somewhat unsymmetrical with respect to the centroidal axis. In such a case the average stress is not half of the maximum. In general, from Fig. 21.13.1,

$$\text{required } A_c = \frac{T_0 h}{f_{ic} y_t} \qquad (21.13.9)$$

$$\text{required } A_c = \frac{T_e h}{f_{fc} y_b} \qquad (21.13.10)$$

The minimum area A_c is obtained when Eqs. (21.13.9) and (21.13.10) give the same result.

Preliminary Design for Large Girder Moment. When the girder moment M_D exceeds about 0.2 to 0.3 of the total moment $M_D + M_L$, the $(\text{arm})_D$ will be too large to permit the steel distance e to be at $k_b + M_D/T_0$ from the centroid of the section. Thus the initial stress distribution at transfer cannot be triangular but instead will be trapezoidal (see Fig. 21.13.2). The final condition, which can still give a triangular stress distribution, will probably govern. Thus

$$T_e = \frac{M_D + M_L}{e + k_t} \qquad (21.13.11)$$

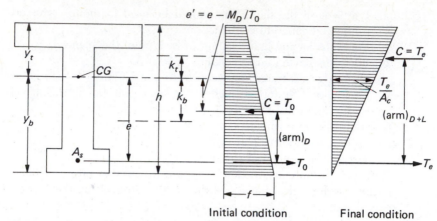

Initial condition   Final condition

Figure 21.13.2 Stress distributions when no tension is permitted—*large* girder moment.

When the final condition controls, more of the area A_c should be located at the top where the highest stress occurs. Thus an unsymmetrical section is indicated—for instance, a T-shaped section. As an approximation, Lin and Burns [2] suggests $(e+k_t) \approx 0.65h$ for use in Eq. (21.13.11). Generally $e+k_t$ will vary from $0.3h$ to $0.8h$, with the average about $0.65h$. For preliminary design, Eq. (21.13.11) then becomes

$$T_e \approx \frac{M_D + M_L}{0.65h} \tag{21.13.12}$$

The required area A_c would then be

$$\text{required } A_c = \frac{T_e}{\text{avg}f_c} = \frac{M_D + M_L}{0.65h(\text{avg}f_c)} \tag{21.13.13}$$

For the unsymmetrical section, y_b/h in Eq. (21.13.10) will be greater than 0.5, say 0.6. Equation (21.13.13) for design would then be

$$\text{required } A_c = \frac{M_D + M_L}{0.65h(0.6f_{fc})} = \frac{2.6(M_D + M_L)}{hf_{fc}} \tag{21.13.14}$$

For the preliminary selection of cross section in case of large girder moment, the following steps may be followed:

1. When $M_D/(M_D + M_L) > 0.2$ to 0.3, use Eq. (21.13.14) to estimate A_c after establishing the overall depth according to step 1 for small girder moment.
2. Proportion an unsymmetrical section; a T-section may be a practical choice.
3. With the preliminary section, compute the section properties and establish the distance e from the centroid of the section to the centroid of the prestressing steel. For large girder moment, one will find

$$e < k_b + \frac{M_D}{T_0}$$

If $e \geq k_b + M_D/T_0$, then the procedure for the small girder moment case is to be followed.

4. With the steel located, T_e can be determined using Eq. (21.13.8). From that, $T_0 = T_e/\eta$.
5. The required area A_c based on the final condition is then determined using Eq. (21.13.10).
6. Check the required area based on the initial condition. Referring to Fig. 21.13.2, a trapezoidal stress distribution should occur.

The maximum compressive stress f is, using Eq. (21.6.1) with the internal force concept,

$$f = \frac{C}{A} + \frac{Cey}{I} \tag{21.13.15}$$

Since $C = T_0$, $A = A_c$, $I = A_c r^2$, $e = e'$ (Fig. 21.13.2), and $y = y_b$,

$$f = \frac{T_0}{A_c} + \frac{T_0 e' y_b}{A_c r^2} \leq f_{ic}$$

$$= \frac{T_0}{A_c}\left[1 + \frac{e - M_D/T_0}{k_t}\right] \leq f_{ic} \tag{21.13.16}$$

The required area A_c based on the initial condition is

$$\text{required } A_c = \frac{T_0}{f_{ic}}\left[1 + \frac{e - M_D/T_0}{k_t}\right] \tag{21.13.17}$$

Again the minimum area A_c will be obtained when Eqs. (21.13.10) and (21.13.17) give the same result.

EXAMPLE 21.13.1 Design a cross section for a 30-in. deep girder whose girder moment $M_D = 45$ ft-kips. The live load moment to be carried is 300 ft-kips. The initial prestress $f_{si} = 175,000$ psi and assume 20% losses. The allowable service load stresses are $f_{it} = 0$, $f_{ic} = 2400$ psi, $f_{ft} = 0$, and $f_{fc} = 2250$ psi. Omit checks of flexural strength, cracking moment, shear strength, and development of reinforcement.

Solution: (a) Preliminary design. The girder moment as a percent of the total moment is

$$\frac{M_D}{M_D + M_L} = \frac{45}{45 + 300} = 0.13 < 0.2$$

Approach as a small girder moment design. Using Eqs. (21.13.4) and (21.13.6),

$$T_e \approx \frac{M_L}{0.5h} = \frac{300(12)}{0.5(30)} = 240 \text{ kips}$$

$$T_0 = \frac{T_e}{\eta} = \frac{240}{0.8} = 300 \text{ kips}$$

$$\text{required } A_c = \frac{T_0}{0.5f_{ic}} = \frac{300}{0.5(2.40)} = 250 \text{ sq in.}$$

Since $f_{fc}/f_{ic} = 2.25/2.4 = 0.94$ is greater than $T_e/T_0 = 0.8$, the equation based on the initial condition is controlling if the section is symmetrical. Though a slightly unsymmetrical section (with the centroid below the mid-depth in this case) would give the section of minimum area, the procedure is illustrated using a symmetrical one.

Try a 30-in. deep section with flanges 5×17 in. and a 4-in. thick web, $A_c = 250$ sq in. [Fig. 21.13.3(a)].

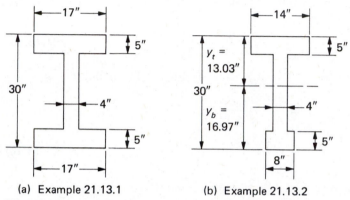

(a) Example 21.13.1   (b) Example 21.13.2

Figure 21.13.3 Sections for design examples.

(b) Determine the section properties.

	AREA, A_c (sq in.)	I (in.^{4})
17×30 rectangle	510	38,250
13×20 sides	-260	$-8,667$
	250	29,583

$$r^2 = \frac{I}{A_c} = 118.3 \text{ sq in.}$$

$$k_t = k_b = \frac{r^2}{y_b} = \frac{r^2}{y_t} = \frac{118.3}{15} = 7.89 \text{ in.}$$

(c) Locate centroid of prestressed reinforcement. Using approximate T_e, find approximate T_0

$$T_0 \approx \frac{T_e}{\eta} = \frac{249}{0.8} = 300 \text{ kips}$$

$$\text{desired } e = k_b + \frac{M_D}{T_0} = 7.89 + \frac{45(12)}{300} = 9.69 \text{ in.}$$

The available distance is 15 in., less appropriate cover. Thus $e = 9.69$ in.

would be acceptable. Recalculate T_e using Eq. (21.13.8),

$$T_e = \frac{M_D + M_L}{e + k_t} = \frac{(45 + 300)12}{9.69 + 7.89} = 235 \text{ kips}$$

$$\text{revised } T_0 = \frac{235}{0.8} = 294 \text{ kips}$$

$$\text{revised } e = 7.89 + \frac{45(12)}{294} = 9.73 \text{ in.}$$

This is in close agreement with the previous value of 9.69 in. If the first estimate had been farther off, additional iterations may have been needed.

(d) Check whether the area A_c is adequate. Using Eqs. (21.13.9) and (21.13.10),

$$\text{required } A_c \text{ (initial condition)} = \frac{T_0 h}{f_{ic} y_t} = \frac{294(30)}{2.4(15)} = 245 \text{ sq in.}$$

$$\text{required } A_c \text{ (final condition)} = \frac{T_e h}{f_{fc} y_b} = \frac{235(30)}{2.25(15)} = 209 \text{ sq in.}$$

The section is adequate.

Use 5 × 17 in. flanges with a 4-in. web (A_c = 250 sq in.).
If made slightly unsymmetrical, the area could be reduced to somewhere between 209 and 245 sq in. A final check of initial and final stresses should be made as done in Sec. 21.6, but the illustration is omitted here.

EXAMPLE 21.13.2 Redesign the cross section of Example 21.13.1 for M_D = 200 ft-kips and M_L = 145 ft-kips. Note the total $M_D + M_L$ = 345 ft-kips is the same as in Example 12.13.1.

Solution: (a) Preliminary design. The girder moment as a percent of the total moment is

$$\frac{M_D}{M_D + M_L} = \frac{200}{345} = 0.58 > 0.2 \text{ to } 0.3$$

Approach as a large girder moment design. Using Eq. (21.13.14),

$$\text{required } A_c = \frac{2.6(M_D + M_L)}{h f_{fc}} = \frac{2.6(345)12}{30(2.25)} = 159 \text{ sq in.}$$

An unsymmetrical shape is to be selected; try flanges 5 × 14 in. and 5 × 8 in. with a 4-in. web having A_c = 190 sq in. [Fig. 21.13.3(b)]. The larger the ratio of M_D to $(M_D + M_L)$ the larger may be the numerator coefficient (2.6); thus an area somewhat larger than indicated by the formula was used.

(b) Determine section properties. Referring to Fig. 21.13.3(b), first locate the centroid of the area measured from the top.

	AREA, A_c (sq in.)	ARM, y (in.)	$A_c y$ (in.^{3})	I (in.^{4})
5 × 10 top flange projection	50	2.5	125	313
I_0				104
5 × 4 bottom flange projection	20	27.5	550	15,125
I_0				42
4 × 30 web (full depth)	120	15	1800	27,000
I_0				9,000
	190		2475	51,584

$$y_t = \bar{y} = \frac{\Sigma Ay}{\Sigma A} = \frac{2475}{190} = 13.03 \text{ in.}$$

$$I_0 = I - A\bar{y}^2 = 51,584 - 190(13.03)^2 = 19,300 \text{ in.}^4$$

$$r^2 = \frac{I_0}{A_c} = 101.7 \text{ sq in.}$$

$$k_t = \frac{r^2}{y_b} = \frac{101.7}{16.97} = 5.99 \text{ in.}$$

$$k_b = \frac{r^2}{y_t} = \frac{101.7}{13.03} = 7.81 \text{ in.}$$

(c) Locate centroid of prestressed reinforcement. Assume with adequate cover the steel may be centered 4 in. from the bottom of the section. Then

$$e = y_b - 4 = 16.97 - 4 = 12.97 \text{ in.}$$

and using Eq. (21.13.8),

$$T_e = \frac{M_D + M_L}{e + k_t} = \frac{345(12)}{12.97 + 5.99} = 218 \text{ kips}$$

then

$$T_0 = \frac{218}{0.8} = 273 \text{ kips}$$

Check

$$k_b + \frac{M_D}{T_0} = 7.81 + \frac{200(12)}{273} = 7.81 + 8.79 = 16.60 \text{ in.} > e$$

This shows the tendons cannot be located far enough from the centroid of the section to give a triangular stress distribution at the initial condition.

(d) Check whether the area A_c is adequate. Using Eqs. (21.13.10) and (21.13.17),

$$\text{required } A_c \text{ (final condition)} = \frac{T_e h}{f_{fc} y_b} = \frac{218(30)}{2.25(16.97)} = 171 \text{ sq in.}$$

$$\text{required } A_c \text{ (initial condition)} = \frac{T_0}{f_{ic}}\left[1 + \frac{e - M_D/T_0}{k_t}\right]$$

$$= \frac{273}{2.4}\left[1 + \frac{12.97 - 8.79}{5.99}\right] = 193 \text{ sq in.}$$

This is close enough but shows that the initial condition governs in this case, the reason being that area has been shifted from the bottom of the section to the top where it is needed for the final condition. The minimum area section for this case would be slightly more symmetrical than the chosen one.

A final check of stresses (initial and final conditions) should be made (see Sec. 21.6) to verify the result.

The general line of reasoning presented here may also be used when tension is permitted at initial or final conditions, or both. Lin and Burns [2] provide a detailed treatment of design of sections when tension is permitted.

21.14 ADDITIONAL TOPICS

Many other topics have been omitted from this introductory treatment of prestressed concrete. Such topics as the practical design approaches, use of I-shaped and nonsymmetrical sections, prestressing of continuous members, stresses in end blocks, partial prestressing, deflections, composite construction, and other specific applications are adequately and extensively treated in recent textbooks devoted entirely to the subject [2–5].

SELECTED REFERENCES

1. Charles C. Zollman. "Reflections on the Beginnings of Prestressed Concrete in America," *PCI Journal*, Part 1, **23**, May/June 1978, 22–48; Part 2, **23**, July/August 1978, 29–63; Part 9, **25**, January/February 1980, 124–145, March/April 1980, 94–117, May/June 1980, 123–152.
2. T. Y. Lin and Ned H. Burns. *Design of Prestressed Concrete Structures* (3d ed.). New York: Wiley, 1981.
3. Antoine E. Naaman. *Prestressed Concrete Analysis and Design–Fundamentals.* New York: McGraw–Hill, 1982.
4. Arthur H. Nilson. *Design of Prestressed Concrete.* New York: Wiley, 1978.
5. James R. Libby. *Modern Prestressed Concrete* (2d ed.). Princeton, N.J.: Van Nostrand, 1977.
6. Narbey Khachaturian and German Gurfinkel. *Prestressed Concrete.* New York: McGraw–Hill, 1969.
7. Yves Guyon. *Prestressed Concrete*, Vols. 1 and 2. New York: Wiley, 1960.
8. Gustave Magnel. *Prestressed Concrete* (2d ed.). London: Concrete Publications, 1950.
9. ACI-ASCE Joint Committee 323. "Tentative Recommendations for Prestressed Concrete," *ACI Journal, Proceedings,* **54**, January 1958, 545–578.
10. Paul Zia, H. Kent Preston, Norman L. Scott, and Edwin B. Workman. "Estimating Prestress Losses," *Concrete International,* **1**, June 1979, 32–38.
11. PCI Committee on Prestress Losses. "Recommendations for Estimating Prestress Loss," *PCI Journal,* **20**, July–August 1975, 43–75.
12. J. R. Janney, E. Hognestad, and D. McHenry. "Ultimate Flexural Strength of Prestressed and Conventionally Reinforced Concrete Beams," *ACI Journal, Proceedings,* **52**, January 1956, 601–620.
13. James G. MacGregor and John M. Hanson. "Proposed Changes in Shear Provisions for Reinforced and Prestressed Concrete Beams," *ACI Journal, Proceedings,* **66**, April 1969, 276–288. Disc., 849–851.

14. ACI-ASCE Committee 426. "The Shear Strength of Reinforced Concrete Members—Chapters 1 to 4," *Journal of the Structural Division*, ASCE, **99**, June 1973 (ST6), 1091–1187.
15. ACI-ASCE Committee 426. *Suggested Revisions to Shear Provisions for Building Codes*. Detroit: American Concrete Institute, 1979 (pp. 10–15).
16. M. A. Sozen and N. M. Hawkins. Discussion of "Shear and Diagonal Tension Report," Report of ACI-ASCE Committee 326, *ACI Journal, Proceedings*, **59**, September 1962, 1341–1347.
17. Paul Zia and Talat Mostafa. "Development Length of Prestressing Strands," *PCI Journal*, **22**, September/October 1977, 54–65.
18. Norman W. Hanson and Paul H. Kaar. "Flexural Bond Tests of Pretensioned Prestressed Beams," *ACI Journal, Proceedings*, **55**, January 1955, 783–802.

PROBLEMS

All problems are to be worked in accordance with the ACI Code using the allowable tension stresses in ACI-18.4.1(b) and ACI-18.4.2(b), unless otherwise indicated.

21.1 The rectangular beam of the accompanying figure contains pretensioned steel with an initial tensile stress of 160 ksi (f_{pu} = 250 ksi). The concrete has f'_{ci} = f'_c = 5000 psi (n = 7). The beam is on a simple span of 35 ft.
(a) Determine the concrete stresses at top and bottom, and the steel stress, at transfer immediately after the wires are cut at the ends.
(b) Recompute the stresses in (a) immediately after a 20% loss in prestress. What is the maximum service live load that can be superimposed on the beam? Consider only the section of maximum bending moment, and *omit* consideration of flexural strength M_n.

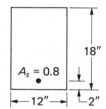

A_s = 0.8

— 12" — ⌐ 2"

Problem 21.1

21.2 Based on the midspan cross section of the figure for Prob. 21.1, investigate whether or not it is possible to increase the live load moment capacity by either or both of the following:
(a) Increase the initial prestress above 160 ksi.
(b) Decrease the eccentricity.
Assume there is a 20% loss of initial prestress. Determine the maximum service live load capacity possible by adjusting the prestress or the eccentricity or both but still not violating the ACI Code limitations. Omit consideration of flexural strength M_n.

21.3 The rectangular section of the accompanying figure has been pretensioned by a force of 300 kips after all losses. If f'_{ci} = f'_c = 5000 psi, what uniformly distributed live load may be safely carried on a 40-ft simple span? Omit consideration of flexural strength M_n.

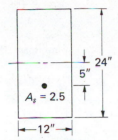

Problem 21.3

21.4 For the live load determined in Prob. 21.1, determine the maximum permissible eccentricity of the prestressing tendons at the $\frac{1}{4}$ point of the span to satisfy service load allowable stress limits. Omit consideration of flexural strength M_n.

21.5 For the live load determined in Prob. 21.3, determine the maximum permissible eccentricity of the prestressing tendons at the $\frac{1}{8}$ point of the span to satisfy service load allowable stress limits. Omit consideration of flexural strength M_n.

21.6 A straight pretensioned member 35 ft long is 18 in. square in cross section. It is concentrically prestressed with 2.24 sq in. of high tensile strength steel wire. The wires are stressed originally to 145 ksi and are anchored to end bulkheads. Calculate the loss and percent of loss of prestress in the wires due to elastic shortening of the concrete at transfer using both the approximate and the "exact" methods. Use $f'_c = 6000$ psi $(n = 6)$.

21.7 An 18-in. square concrete member is posttensioned by four cables each with an area of 0.56 sq in. These cables are stressed one after another, each to a stress of 145 ksi. Without taking any account of the eccentricity of the cables, compute the loss and percent of loss of prestress in each cable due to the elastic shortening of the concrete. Compute the average loss of prestress. Assume $n = 6$.

21.8 The symmetrical double cantilever beam shown is to be prestressed by a single cable $ABCDE$. The cable consists of 12 wires, each of 0.20 in. diameter, and is to be prestressed simultaneously from both ends of the member. It is desired that the minimum stress in the cable immediately after stressing and before any creep or shrinkage losses take place be 145 ksi. The cable is such that the friction constant $\mu = 0.50$ and the wobble effect $K = 0.0010$. Determine the steel stress at the jack, the percentage of friction losses, and the extension that will be required at each jack. Solve by the following methods; (a) Neglect variation in tension throughout the length, using ACI Formula (18-2), $P_s = P_x(1 + \mu\alpha + KL)$; (b) Neglect variation in tension at every point along the length of the curve but consider variation from segment to segment, using ACI Formula (18-2); and (c) Use "exact" expression, ACI Formula (18-1).

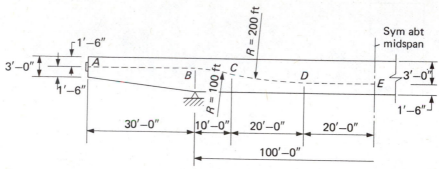

Problem 21.8

21.9 Determine the nominal moment strength M_n and the cracking moment M_{cr} for the pretensioned bonded section of the figure for Prob. 21.1. The concrete has $f'_c = 5000$ psi and the steel has $f_{pu} = 250$ ksi (stress-relieved strand). Assume the average stress-strain curve of Fig. 21.8.1 is to be used for the steel.

21.10 Assuming no special tests are to be made, determine the number and spacing for #3 U stirrups for the beam of Prob. 21.1 if a service load of 0.48 kips/ft is acting. Use the alternate procedure for V_c of ACI-11.4.1, as well as the more exact procedure of ACI-11.4.2.

21.11 Assuming no special tests are to be made, determine the number and spacing for #3 U stirrups for the beam of Prob. 21.3 if the maximum service load computed in that problem is acting. Use and compare both procedures of ACI-11.4.

21.12 Design of section for *small* dead load moment.
(a) Make a preliminary design (use rectangular flanges and a web) for a section of a prestressed beam to resist a total bending moment of 960 ft-kips assuming that the moment due to the girder weight is 70 ft-kips. The overall depth of the section is to be 42 in. and the effective prestress f_{se} in the steel is 136 ksi. In selecting a section assume the minimum thickness of components (flanges or web) is 5 in.
(b) Make the final design for the preliminary section you selected for part (a) making such changes as you find necessary, assuming a *minimum* cross-sectional area is desired. The selected cross-sectional area should not exceed 460 sq in.; however, the "best" design will be presumed to be the one having the least cross-sectional area.
(c) Make a check of stresses at initial (transfer) and final conditions. Use the following control stresses:

$$f_{ic} = 2400 \text{ psi} \qquad f_{fc} = 2250 \text{ psi} \qquad f_{si} = 160 \text{ ksi}$$
$$f_{it} = 0 \text{ psi} \qquad f_{ft} = 0 \text{ psi} \qquad f_{se} = 136 \text{ ksi}$$

21.13 Design of section for *large* dead load moment. The data are the same as Prob. 21.12 except that the moment due to the girder weight is 650 ft-kips, instead of 70 ft-kips; the minimum thickness for components (flanges or web) is 4 in.; and the cross-sectional area should not exceed 340 sq in.

22

Composite Construction

22.1 INTRODUCTION

Composite construction, as defined herein, is the use of a cast-in-place concrete slab placed upon and interconnected to a prefabricated beam so that the combined beam and slab will act together as a unit. The prefabricated beam may be a rolled or built-up steel shape, a precast reinforced concrete beam, a prestressed concrete beam, a timber beam, or even light-gage steel decking. The interconnection to obtain the single unit action is by combinations of mechanical shear connectors, friction, and shear keys.

In the early 1900s a type of composite construction was used where a steel I-shaped section was fully encased in concrete placed integrally with the slab. The use of encased beams is still permitted [2] but such use is rare. The composite beam and slab construction presently used began to appear in the 1930s. Since about 1940, nearly all usage has been with a slab attached to one flange of a prefabricated beam by means of mechanical connectors. This type of composite construction has been widespread in bridge design since the early 1950s and in buildings since about 1960. Present design methods are the result of extensive research into composite section behavior [4–13].

Throughout this chapter, emphasis is on the slab composite with precast reinforced concrete and prestressed concrete, as covered by the recommendations of the Joint ASCE-ACI Committee on Composite Construction [1], and by the ACI Code. The slab composite with a steel beam is covered by the AISC Specification [2], and detailed treatment is provided elsewhere [3,15,16,18,19].

22.2 COMPOSITE ACTION

Consider a concrete slab atop the flange of a steel or precast concrete beam as shown in Fig. 22.2.1. First, if the system of slab and beam is not acting compositely, only friction will provide interaction; thus little of the longitudinal action is carried by the slab. The static system with friction neglected is shown in Fig. 22.2.2(a), wherein the slab and the beam each carry separately a portion of the load. When the noncomposite system deforms under vertical load, the lower surface of the slab is in tension and elongates while the upper surface of the beam is in compression and shortens. Thus a discontinuity will occur at the plane of contact. Since friction is neglected, only vertical internal forces act between the slab and beam.

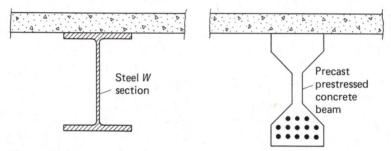

Figure 22.2.1   Concrete slab and prefabricated beam.

When a system acts compositely [Fig. 22.2.2(b)], no relative slippage occurs between the slab and the beam. Horizontal forces (shears) are developed which would shorten the lower surface of the slab and elongate the upper surface of the beam. Thus the discontinuity at the contact surface may be eliminated when sufficiently large horizontal shear resistance can develop. It is noted that the deflection of the composite system will be significantly less than that of the noncomposite system.

In an actual beam-slab system, the degree of composite action may vary over a wide range. For instance, a steel beam used with a concrete slab

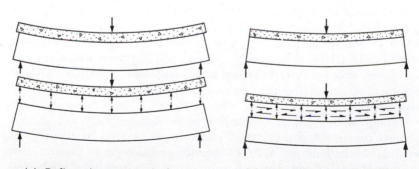

(a) Deflected noncomposite beam   (b) Deflected composite beam

Figure 22.2.2   Comparison of deflected beams with and without composite action.

without mechanical shear connectors will generally develop little composite action, while friction between a concrete beam and a slab may develop nearly the full composite action. Present methods entirely neglect friction (bond) between a steel beam and the concrete slab (unless beam is encased) but consider such bond under certain conditions when the beam is made of precast concrete.

22.3 ADVANTAGES AND DISADVANTAGES OF COMPOSITE CONSTRUCTION

The significant feature of a composite system is a stiffer and stronger structure than can be obtained from the same beam and slab acting noncompositely. In general, the advantages over noncomposite construction are (1) smaller and shallower beams may be used, (2) longer spans are possible without encountering deflection problems, (3) the toughness (impact capacity or energy absorption) is greatly increased, and (4) the overload capacity is substantially greater.

Some of the factors that tend to weigh against this construction are (1) the cost of the connectors which offsets some of the saving in beam material; (2) the cost of placing the mechanical shear connectors, particularly on nonencased steel beams where they are required without exception; and (3) the erection and construction difficulties encountered when the projecting connectors impede or prevent workmen from walking on the beams.

Most indications are, however, in favor of designing for composite interaction wherever a cast-in-place slab is used.

22.4 EFFECTIVE SLAB WIDTH

A slab acting compositely with a beam behaves the same as in an ordinary reinforced concrete T-section, as discussed in Chap. 9. Referring to Fig. 22.4.1, the variables that control the effective slab width are (1) the ratio of slab thickness to total beam depth, t/h; (2) the ratio of beam span to beam width, L/b_w; (3) the ratio of beam span to beam spacing, L/b_0; (4) the type of loading; and (5) Poisson's ratio. As for the T-section in Chap. 9, here also the effective width b_E is to be taken as the smallest of the following for interior beams: (1) one-fourth the beam span length, $L/4$; (2) center-to-

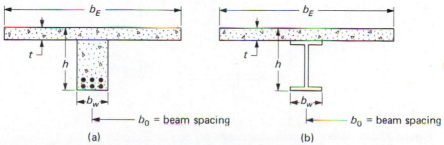

Figure 22.4.1 Variables controlling effective slab width.

center spacing of beams, b_0; (3) beam web width plus 16 times the slab thickness, $b_w + 16t$. These provisions, as well as those for isolated beams in which the T-shape is used and for beams having a flange on one side only, appear in ACI-8.10 for cases [Fig. 22.4.1(a)] where the prefabricated beam is concrete, and in Section 1.11 of the AISC Specification [2] for cases [Fig. 22.4.1(b)] where the prefabricated beam is steel.

22.5 COMPUTATION OF SECTION PROPERTIES

The elastic section properties (area, moment of inertia, section modulus) of the composite section are needed for computation of actual working stresses and deflections under service loads. For such section properties the transformed section concept (see Sec. 4.5) is used to convert all areas of the composite section into an equivalent homogeneous member. When a steel beam is used, the concrete slab is converted into equivalent steel by using a slab width equal to b_E/n, where $n = E_s/E_c$, the ratio of the modulus of elasticity of the steel beam to that of the concrete slab. When the prefabricated beam is either reinforced or prestressed concrete, the 28-day compressive strength f'_c is frequently different for the beam and slab; thus E_c is different. In that case the slab may be converted into equivalent beam material by using a slab width of

$$\text{equivalent } b_E \cdot = \frac{b_E E_c(\text{slab})}{E_c(\text{beam})} = \frac{b_E n_{\text{beam}}}{n_{\text{slab}}} \qquad (22.5.1)$$

EXAMPLE 22.5.1 Compute the elastic section modulus values for the composite steel-concrete section of Fig. 22.5.1. The W21 × 57 steel section has a depth of 21.06 in., flange width of 6.555 in., moment of inertia about its middepth of 1170 in.^{4}, and an area of 16.7 sq in. The yield strength of steel is 36,000 psi. The slab is of concrete with $f'_c = 3000$ psi $(n = 9)$.

Solution: The properties for the composite section are shown in Table 22.5.1. The distance y is measured from the centroid (axis x-x of Fig. 22.5.1) of the steel section.

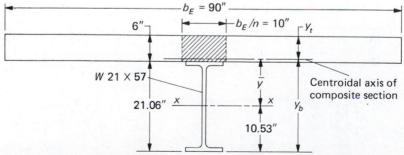

Figure 22.5.1 Steel-concrete composite section for Example 22.5.1.

Table 22.5.1 PROPERTIES OF COMPOSITE SECTION IN EXAMPLE 22.5.1

	EFFECTIVE AREA, A	ARM y	Ay	Ay^2	I_0
Slab	60.0	13.53	811.8	10,984	180
W section	16.7	0	0	0	1170
Totals	76.7		811.8	10,984	1350

$$I_x = I_0 + Ay^2 = 1350 + 10,984 = 12,334 \text{ in.}^4$$

$$\bar{y} = \frac{811.8}{76.7} = 10.58 \text{ in.}$$

$$I = 12,334 - 76.7(10.58)^2 = 3748 \text{ in.}^4$$

$$y_t = 10.53 + 6.00 - 10.58 = 5.95 \text{ in.}$$

$$y_b = 10.53 + 10.58 = 21.11 \text{ in.}$$

$$S_t = \frac{I}{y_t} = \frac{3748}{5.95} = 630 \text{ in.}^3$$

$$S_b = \frac{I}{y_b} = \frac{3748}{21.11} = 177 \text{ in.}^3$$

It is to be noted that the neutral axis of the composite section falls slightly in the concrete slab (0.05 in.). Usually the concrete on the tension side is entirely neglected, but here no correction is made since the amount of concrete in tension is negligible. For cases where the tension concrete is to be considered inactive, the neutral axis under service load is located as for ordinary beams (see Chap. 4).

EXAMPLE 22.5.2 Compute the elastic section modulus values for the composite precast concrete beam and concrete slab system of Fig. 22.5.2. The slab concrete has $f'_c = 3000$ psi ($n = 9$), while the precast beam has $f'_c = 6000$ psi ($n = 6$).

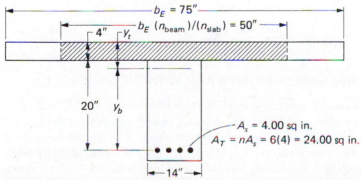

Figure 22.5.2 Precast beam and cast-in-place slab composite section for Example 22.5.2.

Solution: The cross section may be converted into a homogeneous beam of material having the same E_c as concrete with $f'_c = 6000$ psi. Using Eq. (22.5.1), the transformation factor η for the slab is either

$$\eta = \frac{E_c \text{ (slab)}}{E_c \text{ (beam)}} = \frac{57,000\sqrt{3000}}{57,000\sqrt{6000}} = 0.707$$

or

$$\eta = \frac{n_{\text{beam}}}{n_{\text{slab}}} = \frac{6}{9} = 0.666$$

In this case, $\eta = 0.666$ is used.

The conversion into equivalent beam concrete ($f'_c = 6000$ psi) is not actually necessary except in cases either where the precast beam is prestressed and therefore its gross section is fully effective or where the concrete area in the web and above the neutral axis is significant enough to be considered. In this example, the computations of Table 22.5.2 have neglected compression in the web.

Table 22.5.2 PROPERTIES OF COMPOSITE SECTION IN EXAMPLE 22.5.2

	EFFECTIVE AREA, A	ARM y	Ay	Ay^2	I_0
Slab	200	2.0	400	800	267
Steel in precast beam	24	24.0	576	13,824	—
Totals	224		976	14,624	267

$$I_{\text{top}} = I_0 + Ay^2 = 267 + 14,624 = 14,891 \text{ in.}^4$$

$$\bar{y} = y_t = \frac{976}{224} = 4.36 \text{ in.}$$

$$I_{cr} = 14,891 - 224(4.36)^2 = 10,630 \text{ in.}^4$$

$$y_b = 24.0 - 4.36 = 19.64 \text{ in.}$$

$$S_t = \frac{I_{cr}}{y_t} = \frac{10,630}{4.36} = 2440 \text{ in.}^3$$

$$S_b = \frac{I_{cr}}{y_b} = \frac{10,630}{19.64} = 541 \text{ in.}^3$$

22.6 WORKING STRESSES WITH AND WITHOUT SHORING

When no temporary falsework or shoring is used to prevent deflection of the precast member while the slab is being placed and cured, the precast member must support alone its own weight plus the weight of the freshly placed slab. The composite section then resists the live load and any additional superimposed dead load, together with long-time effects from creep and shrinkage. On the other hand, if temporary supports are used to carry the precast beam and the slab concrete until such concrete has achieved about 75% of its 28-day compressive strength f'_c, then the composite section will carry the entire load. Thus working stresses under service load may be computed as follows:

Without shoring,

$$f = \frac{M_D}{S_p} + \frac{M_L}{S_c} \qquad (22.6.1)$$

where M_D is the moment due to dead load, produced prior to the time at which the cast-in-place concrete attains 75% of its specified 28-day compressive strength; M_L is the moment due to live load and superimposed dead load; S_p is the effective section modulus of the precast or prefabricated beam; and S_c is the effective section modulus of the composite section.

With shoring,

$$f = \frac{M_D + M_L}{S_c} \qquad (22.6.2)$$

EXAMPLE 22.6.1 For the section of Example 22.5.2, compute the stresses due to a service dead load moment of 70 ft-kips and a service live load moment of 105 ft-kips. Reinforcement has $f_y = 60,000$ psi. Consider the case (a) without shoring and (b) with shoring.

Solution: (a) Without shoring. Since the precast beam must carry the dead load prior to curing of the slab, its neutral axis location is required; thus

$$\tfrac{1}{2}(14)x^2 = 24(20 - x)$$

$$x = 6.74 \text{ in.}$$

$$\text{arm} = 20 - \frac{x}{3} = 20 - 2.25 = 17.75 \text{ in.}$$

$$f \text{ (tension, steel)} = \frac{M_D}{A_s \text{ (arm)}} = \frac{70(12)}{4.0(17.75)} = 11.8 \text{ ksi}$$

$$f \text{ (compression, concrete)} = \frac{11.8}{6}\left(\frac{6.74}{13.26}\right) = 1.00 \text{ ksi}$$

The additional stresses due to the live load acting on the composite section are

$$f \text{ (tension, steel)} = \frac{nM_L}{S_b} = \frac{6(105)(12)}{541} = 14.0 \text{ ksi}$$

$$f \text{ (compression, } n{=}6 \text{ concrete)} = \frac{M_L}{S_t} = \frac{105(12)}{2440} = 0.52 \text{ ksi}$$

$$f \text{ (actual compression, } n{=}9 \text{ concrete)} = 0.52(6)/9 = 0.35 \text{ ksi}$$

The maximum stress in the reinforcement is $11.8 + 14.0 = 25.8$ ksi, which exceeds the ACI allowable value of 24 ksi. The maximum stress in the concrete at the top fiber of the precast beam is

$$f = 1.00 + \frac{105(12)(0.36)}{10,630} = 1.00 + 0.04 = 1.04 \text{ ksi}$$

The stress distribution without shoring is given in Fig. 22.6.1(b).
(b) With shoring.

$$f \text{ (tension, steel)} = \frac{n(M_D + M_L)}{S_b} = \frac{6(175)(12)}{541} = 23.3 \text{ ksi}$$

$$f \text{ (compression, concrete)} = \frac{(M_D + M_L)(n_{\text{beam}})}{S_t(n_{\text{slab}})} = \frac{175(12)(6)}{2440(9)} = 0.57 \text{ ksi}$$

The corresponding stress distribution on a homogeneous section where $n = 6$ is given in Fig. 22.6.1(c).

Thus it would appear that the reinforcement is overstressed in the system without shoring, whereas it is within the allowable value when shoring is used.

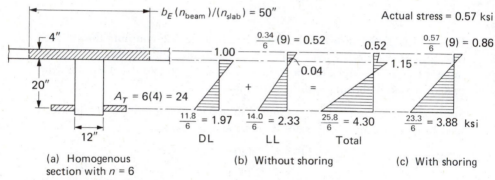

(a) Homogenous section with $n = 6$ (b) Without shoring   (c) With shoring

Figure 22.6.1 Service load stresses in precast concrete-concrete composite section with and without shoring.

22.7 STRENGTH OF COMPOSITE SECTIONS

Moment strength M_n of a composite section entirely of reinforced concrete is computed as explained for T-sections in Sec. 9.4. When the composite section includes a steel shape or a prestressed concrete beam, the basic principles are only slightly modified. Since the moment strength M_n is unrelated to the sequence of loading and the relative amounts of live and dead load, it is independent of whether or not shoring is used.

EXAMPLE 22.7.1 Determine the nominal moment strength M_n of the steel-concrete composite section of Example 22.5.1, using basic statics. The steel section has $f_y = 36,000$ psi and the concrete has $f'_c = 3000$ psi.

Solution: Determine whether the neutral axis lies above or below the bottom of the slab. If the neutral axis is at the bottom of the slab (i.e., $a = 0.85t$),

$$C_{\max} = 0.85f'_c b_E a = 0.85(3)(90)(0.85)(6) = 1170 \text{ kips}$$

$$T_{\max} = A_s f_y = 16.7(36) = 601 \text{ kips}$$

It will be observed from Fig. 22.7.1(c) that the value of T_{max} as computed above is an overestimate if the neutral axis lies at the bottom of the slab. However, by comparing C_{max} to T_{max} it is also obvious that the distance to the neutral axis from the top of the slab is less than 6 in. Thus as in an ordinary T-section where the neutral axis falls in the flange,

$$C = 0.85f_c'ab = 0.85(3)(a)(90) = 230a$$

$$T = A_s f_y = 16.7(36) = 601 \text{ kips}$$

$$a = \frac{601}{230} = 2.62 \text{ in.}$$

$$x = \frac{2.62}{0.85} = 3.08 \text{ in.}$$

For this case the strain at the top of the steel section is

$$\epsilon_s' = \frac{0.003}{3.08}(6.0 - 3.08) \approx 0.0028 > \left[\epsilon_y = \frac{36}{29,000} = 0.00124 \right]$$

which means the strain on the entire steel section is at least yield strain ϵ_y. Thus

$$\text{arm} = \frac{d}{2} + t - \frac{a}{2} = \frac{21.06}{2} + 6.0 - \frac{2.62}{2} = 15.22 \text{ in.}$$

$$M_n = T(\text{arm}) = 601(15.22)\tfrac{1}{12} = 762 \text{ ft-kips}$$

Thus the nominal moment strength M_n is 762 ft-kips whether or not shoring is used.

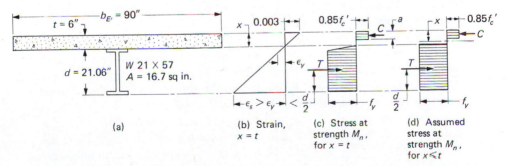

(a)

(b) Strain,
$x = t$

(c) Stress at
strength M_n,
for $x = t$

(d) Assumed
stress at
strength M_n,
for $x \leqslant t$

Figure 22.7.1   Section for Example 22.7.1.

EXAMPLE 22.7.2 Determine the nominal moment strength M_n of the concrete composite section of Fig. 22.7.2. The slab has $f_c' = 3000$ psi while the precast beam has $f_c' = 6000$ psi, and the reinforcement has $f_y = 60,000$ psi.

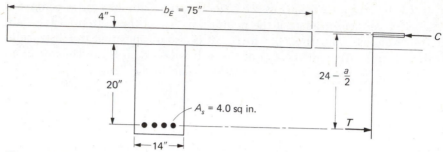

Figure 22.7.2 Section for Example 22.7.2.

Solution:

$$C_{max} = 0.85f'_c b_E t = 0.85(3)(75)(4) = 765 \text{ kips}$$
$$T_{max} = A_s f_y = 4(60) = 240 \text{ kips}$$

Since $C_{max} > T_{max}$, $a < t$. Thus

$$a = \frac{240}{0.85(3)(75)} = 1.25 \text{ in.}$$

$$M_n = 240[24.0 - 0.5(1.25)]\tfrac{1}{12} = 468 \text{ ft-kips}$$

This most typical situation with $a < t$ is treated exactly as in Chap. 9. If $a > t$, some contribution from the 6000 psi concrete would be included in the compressive force.

22.8 SHEAR CONNECTION

Working Stress Concept. As discussed in Sec. 22.2, in order for the slab to act together with the prefabricated beam, the horizontal shear forces must be developed between the slab and beam. Consider the uniformly loaded beam of Fig. 22.8.1 along with the shear stress distribution across a typical section of the beam. It is the shear stress v_1 that must be developed by the connection between slab and beam. Under the service load (working stress) condition, it is seen from Fig. 22.8.1 that the shear stress v_1 varies from zero at midspan to a maximum at the support. Consider an elemental slice of the beam, as in Fig. 22.8.2. The shear force per unit distance along the span is $dC/dz = v_1 b_E = VQ/I$. Thus if a given connector has an allowable capacity of q kips, the maximum spacing s to provide the required capacity is

$$s \leq \frac{q}{VQ/I} \tag{22.8.1}$$

where V is the total shear at the section, Q is the statical moment of the effective slab area above the neutral axis with respect to the neutral axis of

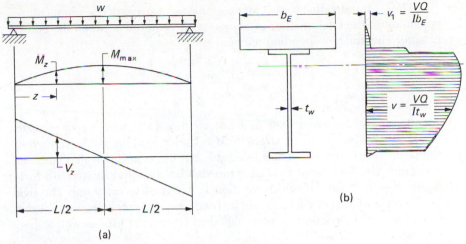

Figure 22.8.1 Shear stress distribution across a steel-concrete composite section.

the composite section, and I is the moment of inertia of the transformed composite section neglecting area of concrete in tension.

Strength Concept. If one uses the strength concept, the shear connectors share equally in carrying the total compressive force developed in the concrete slab as the nominal moment strength M_n is approached. This means, referring to Fig. 22.8.1(a), that shear connection is required to transfer the compressive force developed at midspan to the prefabricated beam in the distance $L/2$, since no compressive force can exist in the concrete slab at the end of the span where zero moment exists. The compressive force at strength M_n to be accommodated could not exceed that which the concrete can carry,

$$C_{max} = 0.85 f'_c b_E t \quad \text{(upper bound)} \quad (22.8.2)$$

or, if the tensile force below the bottom of the slab at strength M_n is less than C_{max},

$$T_{max} = A_s f_y \quad (22.8.3)$$

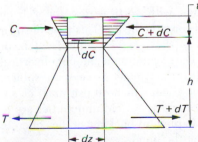

Figure 22.8.2 Force required of shear connectors—working stress method.

Thus, for individual connectors each having a strength q_{ult} when failure is imminent, the total number of connectors N required between the points of maximum and zero bending moment is

$$N = \frac{C_{max}}{q_{ult}} \quad \text{or} \quad \frac{T_{max}}{q_{ult}} \tag{22.8.4}$$

whichever is smaller.

It can also be noted that the connection and the beam must provide the same nominal moment strength M_n. Under working loads, however, the beam resists dead load and live load, but the connection may need to resist only the load coming on after the slab has acquired its strength (primarily the live load). If the connection is designed to carry only the live load, a higher factor of safety should be used. Approximately the same result is achieved if the connection is designed to carry dead load as well as live load with the usual safety provisions.

Several types of connectors are as follows:

1. Stud shear connector, straight [Fig. 22.8.3(a)] and L-shaped [Fig. 22.8.3(b)], welded to the steel beam in concrete-steel construction.
2. Flexible channel shear connector [Fig. 22.8.3(c)], welded to the steel beam in concrete-steel construction.
3. Spiral shear connector [Fig. 22.8.3(d)], welded to the steel beam in concrete-steel construction.

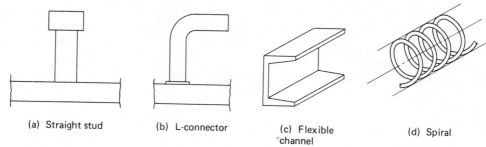

(a) Straight stud   (b) L-connector   (c) Flexible channel   (d) Spiral

Figure 22.8.3 Shear connectors for concrete slab steel beam construction.

4. Reinforcing bar stirrups from the precast beam, fully anchored into the slab.
5. Friction, or bond, in combination with vertical ties, for slab on precast reinforced or prestressed concrete beam. This type of shear connection is adequate for most of these cases. While friction, or bond, alone may be sufficient, at least a minimum amount of vertical ties must be used (ACI-17.6) unless the contact surfaces are subject to a nominal stress $V_u/(\phi b_v d)$ not greater than 80 psi, *and* "clean, free of laitance, and intentionally roughened." Minimum ties are to provide a sort of clamping action to prevent buckling of the concrete slab, which would suddenly break the bond.
6. Shear keys, for all cases of concrete-to-concrete composite action where friction, or bond, is inadequate. Since these keys are acting

very nearly in pure shear rather than in diagonal tension, determination of the strength of such a key according to the general principles in Chap. 5 will be unduly conservative. The shear-friction concept of ACI-11.7 should be used for the design of the keys, as discussed in Sec. 5.15.

22.9 DEFLECTIONS

Since, in general, one of the advantages of using composite construction is to obtain shallower members, the calculation of deflection is important. For deflections arising from live load, or dead load in shored construction, properties of the composite section may be used in accordance with the principles of Chap. 14. For the time-dependent effect, the work of Branson [14, 17–19] may be used. Specifically, for a slab on precast reinforced or prestressed concrete the provisions of ACI-9.5.5 should be followed. For a slab on a steel beam, the AISC [2] suggests using the ordinary value of $n = E_s/E_c$ (short-time loading) when computing either immediate live load or sustained load deflections, while the Joint ACI-ASCE Committee [1] recommended using $2n$ for sustained load deflections.

22.10 SLAB ON PRECAST REINFORCED CONCRETE BEAM—STRENGTH DESIGN

Essentially, the principles of Chap. 3 are used, with full realization that no distinction is made between shored and unshored members. To ensure that the stresses at working load level will not exceed 75% of the yield strength specified for the reinforcement, the 1963 ACI Code required that the effective depth d_c of the composite section as used for computing the nominal moment strength M_n should not exceed

$$d_c \le \left(1.15 + 0.24 \frac{M_L}{M_D}\right) d_p \qquad (22.10.1)$$

where d_p is the effective depth of the precast beam.

Equation (22.10.1) may be developed as follows. Two requirements must be satisfied; first, the working stress when no shores are used must not exceed $0.75f_y$ in the steel,

$$\frac{M_D}{A_s j_p d_p} + \frac{M_L}{A_s j_c d_c} \le 0.75f_y \qquad (22.10.2)$$

where $j_p d_p$ is the moment arm of internal resisting couple for precast beam assuming a linear stress distribution, and $j_c d_c$ is the moment arm of internal resisting couple for composite beam assuming also a linear stress distribution. The second requirement is that the nominal moment strength M_n of the composite section must be adequate.

$$\phi M_n \ge M_u \qquad (22.10.3)$$

$$0.90(A_s f_y)\left(d_c - \frac{a}{2}\right) \ge 1.4M_D + 1.7M_L \qquad (22.10.4)$$

The ACI Code has eliminated the requirement on d_c, in all likelihood for two reasons. First, it is not the stresses under service load that need to be limited, but factors that affect serviceability such as deflection and cracking. Second, Eq. (22.10.1) specifically applies to dead load and live load using particular overload factors, and composite design is not limited to those classes of loads, though certainly they are the most common.

Instead of using several specific rules aimed at providing proper serviceability *indirectly*, ACI-17.2 simply states general requirements, including ACI-17.2.7 which requires composite members to meet the deflection control requirements of ACI-9.5.5. This means that whenever excessive deflection may cause damage, deflections must be computed.

To assure the composite action, the horizontal shear must be transferred across the contact surface. For ordinary design, ACI-17.5.2 uses a horizontal shear nominal strength V_{nh} [instead of Eq. (22.8.1)] computed as

$$V_{nh} = v_{nh}b_v d_c \qquad (22.10.5)$$

where

> v_{nh} = nominal ultimate unit stress capable of being transmitted on contact surface (values as obtained from ACI-17.5.2 are described below)
>
> b_v = width of cross section being investigated for horizontal shear
>
> d_c = distance from extreme compression fiber to centroid of tension reinforcement, for the entire composite section

The maximum values of v_{nh} from ACI-17.5.2 are as follows:

1. When the contact surface is intentionally roughened, clean and free of laitance, with no vertical ties used, max v_{nh} = 80 psi.
2. When the contact surface is clean but *not* intentionally roughened, and when vertical ties having a minimum area of $A_v = 50b_w s/f_y$ are spaced not more than 4 times the slab thickness (i.e., least dimension of the supported element), nor 24 in., max v_{nh} = 80 psi.
3. When the contact surface is intentionally roughened,* clean, free of laitance, and minimum ties as in (2) are used, max v_{nh} = 350 psi.
4. When the nominal ultimate stress exceeds 350 psi, design for horizontal shear must be made using the shear-friction provisions of ACI-11.7, as explained in Sec. 5.15.

Thus the design requirement of ACI-17.5.2 may be stated as

$$\phi V_{nh} \geq V_u \qquad (22.10.6)$$

where

> V_u = total shear force at section due to factored loads
>
> ϕ = 0.85 for shear (ACI-9.3)
>
> V_{nh} = nominal strength computed according to Eq. (22.10.5)

---

*ACI-17.5.2.3 indicates that the interface must have a full amplitude of $\frac{1}{4}$ in. of roughness to satisfy the requirement of intentional roughness (based on Ref. 6) to use v_{nh} = 350 psi.

As an alternative to the above procedure of ACI-17.5.2, ACI-17.5.3 provides that the actual compressive or tensile force in any segment may be computed, and then provision is made to transfer that force as horizontal shear to the supporting element. Since the term "any segment" could mean anything from an elemental segment of span to the full distance between the maximum moment point and a point of contraflexure, the ACI statement would seem to suggest any procedure from Eq. (22.8.1) relating to an elemental segment to Eq. (22.8.4) for a longer finite length of span. In other words, the total horizontal shear to be transferred must be accommodated by some rational process.

For the strength method, it seems Eq. (22.10.6) should be used when designing for a distributed load transfer, such as by friction with or without ties, and Eq. (22.8.4) should be used with individual mechanical connectors, such as extended and anchored stirrups or ties and steel studs.

For precast prestressed concrete beams, the minimum tie area may be taken as ACI Formula (11-15), Eq. (21.11.4), if the effective prestress force is at least equal to 40% of the tensile strength of the flexural reinforcement.

EXAMPLE 22.10.1 Design a composite slab on a simply supported precast reinforced concrete beam span of 24 ft. The spacing of beams is 8 ft center to center. The cast-in-place slab is 4 in. thick, and the live load to be carried is 200 psf. Use f'_c (slab) = 3000 psi, f'_c (precast beam) = 4000 psi, f_y = 40,000 psi, and the strength method of the ACI Code.

Solution: (a) Loads and solution procedure. As a preliminary to the actual solution, it is to be noted that the use of precast members speeds construction and the use of composite action reduces the required depth of the beam.

To design a composite beam without using temporary shoring, the precast beam is first designed to carry its own weight plus the weight of freshly placed concrete. Of course, the noncomposite system must also carry temporary load due to workmen, equipment, runways, and impact, plus the dead weight of forms. The loads are:

Loads on the noncomposite precast beam,

$$\begin{array}{r} \text{4-in. slab, } (4/12)(0.15)(8) = 0.4 \text{ kip/ft} \\ \underline{\text{estimated beam weight} = 0.2 \text{ kip/ft}} \\ \text{dead load} = 0.6 \text{ kip/ft} \\ \text{temporary load, } 0.050(8) = 0.4 \text{ kip/ft} \end{array}$$

Load on the composite section,

$$\text{live load, } 0.200(8) = 1.6 \text{ kips/ft}$$

Temporary live and dead construction loads frequently are not included in the design of the precast noncomposite section, but rather the overload that may occur is accepted as a short-duration reduction in the factor of safety.

When deflection is to be investigated, service load moments are needed; so they could be computed first and then overload factors are applied. For permanent loads on the noncomposite section,

$$M_D = \tfrac{1}{8}(0.6)(24)^2 = 43.2 \text{ ft-kips}$$

For the live load on the composite section,

$$M_L = \tfrac{1}{8}(1.6)(24)^2 = 115 \text{ ft-kips}$$

Several factors may control the size of the precast beam: (1) dead load moment requirement for a rectangular precast beam, (2) total load moment requirement acting on the T-shaped composite section, and (3) total load shear on the T-shaped section. Items (2) and (3) are treated similarly to the procedure discussed in Secs. 9.6 and 10.2 for T-sections.

(b) Moment on precast noncomposite section. Assume a desirable reinforcement ratio ρ about one-half the maximum permitted, say 0.018 (see Table 3.6.1 for ACI Code maximum). Then the desired R_n is

$$R_n = \rho f_y(1 - \tfrac{1}{2}\rho m)$$

$$= 0.018(40,000)[1 - 0.5(0.018)(11.8)] = 644 \text{ psi}$$

$$m = \frac{f_y}{0.85 f_c'} = \frac{40,000}{0.85(4000)} = 11.8$$

$$M_u = 1.4(43.2) = 60.5 \text{ ft-kips}$$

$$\text{required } bd^2 = \frac{M_u}{\phi R_n} = \frac{60.5(12,000)}{0.90(644)} = 1250 \text{ in.}^3$$

$$\text{min } h = \frac{L}{16}(0.8) = \frac{24(12)}{16}(0.8) = 14.4 \text{ in.}$$

The precast beam must be at least 14.4 in. deep (ACI-Table 9.5a) unless deflection is computed even if the member is not supporting or attached to construction likely to be damaged by excessive deflection. If $h = 15$ in., $d \approx 12.5$ in.; then

$$\text{required } b = \frac{1250}{(12.5)^2} = 8 \text{ in.}$$

$$\text{required } A_s \approx 0.018(8)(12.5) = 1.8 \text{ sq in.}$$

There is no problem to fit the steel required for the beam *before* the live load is applied. However, the greater reinforcement requirement for the *total* load will probably make it desirable to use a width exceeding 8 in.

(c) Determine reinforcement for composite section.

$$M_u = 1.4M_D + 1.7M_L = 1.4(43.2) + 1.7(115) = 256 \text{ ft-kips}$$

$$\text{required } M_n = \frac{M_u}{\phi} = \frac{256}{0.90} = 284 \text{ ft-kips}$$

$$\text{effective width } b_E = \tfrac{1}{4}(24)(12) \quad \text{or } 12 + 16(4) \quad \text{or } 8(12)$$
$$= 72 \text{ in.} \qquad\qquad \text{or } 76 \text{ in.} \qquad \text{or } 96 \text{ in.}$$
$$= 72 \text{ in.}$$

Assuming neutral axis to be in the slab when nominal strength M_n is reached,

$$C = 0.85f'_c b_E a = 0.85(3)(72)a = 184a$$

$$T = f_y A_s = 40A_s$$

$$C = T$$

$$A_s = \frac{184a}{40} = 4.60a$$

If a is typically somewhat less than $t/2$, say 1 to 2 in., two layers of steel will be required even if the beam width is increased. Try b = 12 in. and h = 15 in. for precast beam (Fig. 22.10.1). Then d_c = 15 + 4 − (≈3.5) = 15.5 in.

$$M_n = C(d_c - 0.5a)$$

$$284(12) = 184a(15.5 - 0.5a)$$

$$a^2 - 31a = -37.04$$

$$a = 1.24 \text{ in.} < 4.0 \text{ in.} \qquad \text{OK}$$

$$\text{required } A_s = 4.60a = 4.60(1.24) = 5.70 \text{ sq in.}$$

Try 6-#9 bars in two layers (A_s = 6.00 sq in.).

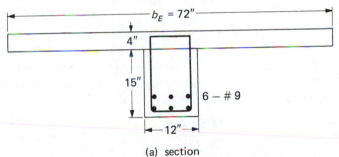

(a) section

Figure 22.10.1   Section for Example 22.10.1.

A check by basic statics may be made as follows:

$$T = 40(6.0) = 240 \text{ kips}$$

$$a = \frac{240}{184} = 1.30 \text{ in.}$$

$$M_n = 240(15.5 - 0.65)\tfrac{1}{12} = 297 \text{ ft-kips} > 284 \text{ ft-kips} \qquad \text{OK}$$

(d) Check d_c by Eq. (22.10.1). Though not required by the ACI Code, the check on maximum d_c using Eq. (22.10.1) may still provide a serviceability check. Maximum effective depth d_c of the composite section should not exceed

$$\left(1.15 + 0.24\frac{M_L}{M_D}\right)d_p = \left[1.15 + 0.24\left(\frac{115}{43.2}\right)\right](11.5)$$

$$= 20.6 \text{ in.} > 15.5 \text{ in. (actual)} \qquad \text{OK}$$

(e) Investigate construction loads.

$$M_{\text{temp}} = \tfrac{1}{8}(0.6 + 0.4)(24)^2 = 72 \text{ ft-kips} \qquad \text{(service load)}$$

For the precast section,

$$C = 0.85f_c'b_wa = 0.85(4)(12)a = 40.8a$$

$$T = f_yA_s = 40(6.0) = 240 \text{ kips}$$

$$a = \frac{240}{40.8} = 5.88 \text{ in.}$$

$$M_n = 240(11.5 - 2.94)\tfrac{1}{12} = 171 \text{ ft-kips}$$

Though no special safety requirements are given for temporary loads, it would be conservative to use the regular overload factors.

$$\text{required } M_n = \frac{1.4(72)}{0.90} = 112 \text{ ft-kips} < 171 \text{ ft-kips} \qquad \text{OK}$$

If deflection is important, it should be checked.

(f) Investigate shear transfer.

$$V_u = \tfrac{1}{2}[1.4(0.6) + 1.7(1.6)]24 = 42.7 \text{ kips}$$

$$v_{nh} = \frac{V_{nh}}{b_vd_c} = \frac{V_u}{\phi b_vd_c} = \frac{42,700}{0.85(12)(15.5)} = 270 \text{ psi}$$

Since $v_{nh} > 80$ psi, friction alone may not be relied on to transmit horizontal shear; however, since $v_{nh} < 350$ psi, the surface is to be intentionally roughened and minimum ties will be used. For #3 ties,

$$\text{max } s = \frac{A_vf_y}{50b_w} = \frac{0.22(40,000)}{50(12)} = 14.6 \text{ in.} \qquad \text{(Controls)}$$

or

$$\text{max } s = 4t = 4(4) = 16 \text{ in.}$$

but not greater than 24 in. in any case.

Use #3 ties @ 12 in. In general, these should be stirrups which project out at the top of the precast beam. This projecting portion of the stirrup is then cast into the slab as shown in Fig. 22.10.1. Even when stirrups in excess of the minimum percentage are required for shear, all of these should be extended into the slab to serve as ties.

22.11 SLAB ON STEEL BEAM

The design requirements for the slab on a steel beam are covered by the AISC Specification [2]. Working stress method is used with ultimate strength concepts so that for strength no distinction is made between shored and unshored construction so long as certain conditions are fulfilled.

To ensure that stresses at the working load level do not exceed approximately 80 to 90% of the yield stress, the section modulus of the composite section S_c is not permitted to exceed

$$S_c \leq \left(1.35 + 0.35 \frac{M_L}{M_D}\right) S_p \qquad (22.11.1)$$

which is Formula (1.11-2) of the AISC Specification.

In addition, the stress on the steel beam acting alone prior to the development of the concrete strength must not exceed the allowable bending stress.

Thus any section must have enough composite section modulus S_c to carry the total load; must have adequate steel section S_p to carry the dead load; and must satisfy Eq. (22.11.1) so as not to exceed a stress of about 80 to 90% of the yield stress under working stresses when no shores are used. Design under the AISC Specification permits the option of using shoring for cases when Eq. (22.11.1) is not satisfied.

To ensure the composite action, mechanical connectors are required for all cases except where beams are totally encased. Connectors are designed using ultimate strength principles and may be spaced uniformly between sections of maximum and zero moment.

Examples of design are not presented because steel section properties are necessary, and this subject is treated in detail by Salmon and Johnson [3].

22.12 COMPOSITE COLUMNS

The composite column was first discussed in Chap. 13 where the two major types of such columns are shown in Fig. 13.2.1. The general approach to the short column is the same as for regular reinforced concrete columns described in Chap. 13. The specific ACI Code rules for both short and long composite columns are in ACI-10.14. The work of Furlong [20–23] provides the basis for the ACI Code design of steel-encased concrete columns, with supporting data from the work of Roderick and Rogers [24] and Knowles and Park [25].

Basically, every composite column, whether concrete encased steel sections or steel encased concrete, must be specifically designed to have shear transfer between concrete and steel. The so-called combination column where concrete merely fills a pipe column, does not fit this category.

ACI-10.14.3 requires that any axial load strength assigned to be carried by concrete must be transferred to concrete by "members or brackets in direct bearing" on the concrete. Connectors such as lugs, plates, or reinforcing bars welded to the structural shape before the concrete is cast are suitable to transfer by direct bearing the force in the concrete. If the force assigned to the concrete is

$$C_c = 0.85 f_c' A_c \qquad (22.12.1)$$

and the connectors each have a capacity q_{ult}, the total number N of connectors required is

$$N = \frac{C_c}{q_{ult}} \qquad (22.12.2)$$

Another modification for composite column design is a modified expression for radius of gyration given by ACI-10.14.5

$$r = \sqrt{\frac{0.2E_cI_g + E_sI_t}{0.2E_cA_g + E_sA_t}} \qquad (22.12.3)$$

where I_t and A_t represent the moment of inertia and area, respectively, of structural steel or tubing in a composite section.

Further, in computing the moment magnification factor (see Chap. 15) the effective EI may not exceed

$$\max EI = \frac{0.2E_cI_g}{1 + \beta_d} + E_sI_s \qquad (22.12.4)$$

Though Eqs. (22.12.3) and (22.12.4) are mentioned above for completeness, their use relates directly to length effects on columns dealt with in Chap. 15. Symbols not defined herein are standard ACI symbols and are used throughout Chap. 15.

SELECTED REFERENCES

1. ACI-ASCE Committee 333. "Tentative Recommendations for Design of Composite Beams and Girders for Buildings," *ACI Journal, Proceedings,* **57,** December 1960, 609–628; also *Journal of the Structural Division,* ASCE, **86,** December 1960 (ST12), 73–92.
2. *Specification for the Design, Fabrication and Erection of Structural Steel for Buildings* (Effective November 1, 1978). Chicago: American Institute of Steel Construction, 1978 (Section 1.11).
3. Charles G. Salmon and John E. Johnson. *Steel Structures: Design and Behavior* (2d ed.). New York: Harper & Row, 1980 (Chap. 16).
4. Ivan M. Viest. "Review of Research on Composite Steel-Concrete Beams," *Journal of the Structural Division,* ASCE, **86,** June 1960 (ST6), 1–21.
5. B. Grossfield and C. Birnstiel. "Tests of T-Beams with Precast Webs and Cast-in-Place Flanges," *ACI Journal, Proceedings,* **59,** June 1962, 843–851.
6. J. C. Saemann and George W. Washa. "Horizontal Shear Connections Between Precast Beams and Cast-in-Place Slabs," *ACI Journal, Proceedings,* **61,** November 1964, 1383–1409.
7. N. W. Hanson. "Precast-Prestressed Concrete Bridges: (2) Horizontal Shear Connections," *Journal,* PCA Research and Development Labs., **2,** No. 2 (May 1960), 38–58. (Also PCA Development Department Bulletin D35.)
8. Peter R. Barnard. "A Series of Tests on Simply Supported Composite Beams," *ACI Journal, Proceedings,* **62,** April 1965, 443–456. Disc., 1629–1631.
9. William R. Spillers. "On Composite Beams," *Journal of the Structural Division,* ASCE, **91,** August 1965 (ST4), 17–21.
10. John C. Badoux and C. L. Hulsbos. "Horizontal Shear Connection in Composite Concrete Beams Under Repeated Loads," *ACI Journal, Proceedings,* **64,** December 1967, 811–819.
11. Alan H. Mattock and Sterling B. Johnston. "Behavior Under Load of Composite

Box-Girder Bridges," *Journal of the Structural Division,* ASCE, **94,** October 1968 (ST10), 2351–2370.

12. R. Paul Johnson. "Research on Steel-Concrete Composite Beams," *Journal of the Structural Division,* ASCE, **96,** March 1970 (ST3), 445–459.

13. R. Paul Johnson. "Longitudinal Shear Strength of Composite Beams," *ACI Journal, Proceedings,* **67,** June 1970, 464–466.

14. Dan E. Branson. "Time-Dependent Effects in Composite Concrete Beams," *ACI Journal, Proceedings,* **61,** February 1964, 213–230. Disc., 1207–1209.

15. Charles G. Salmon and James M. Fisher. "Composite Steel-Concrete Construction," *Handbook of Composite Construction Engineering* (ed. by Gajanan Sabnis). New York: Van Nostrand Reinhold, 1979 (Chapter 2).

16. John P. Cook. *Composite Construction Methods.* New York: Wiley, 1977.

17. Dan E. Branson. *Deformation of Concrete Structures.* New York: McGraw-Hill, Inc., 1977 (pp. 226–249).

18. Dan E. Branson. "Reinforced Concrete Composite Flexural Members," *Handbook of Composite Construction Engineering* (ed. by Gajanan Sabnis). New York: Van Nostrand Reinhold, 1979 (Chapter 4).

19. Dan E. Branson. "Prestressed Concrete Composite Flexural Members," *Handbook of Composite Construction Engineering* (ed. by Gajanan Sabnis). New York: Van Nostrand Reinhold, 1979 (Chapter 5).

20. Richard W. Furlong. "Strength of Steel-Encased Concrete Beam-Columns," *Journal of the Structural Division,* ASCE, **93,** October 1967 (ST10), 113–124.

21. Richard W. Furlong. "Design of Steel-Encased Concrete Beam-Columns," *Journal of the Structural Division,* ASCE, **94,** January 1968 (ST1), 267–281.

22. Richard W. Furlong. "Design Tables for Composite Columns," Preprint 1531, ASCE Annual Meeting, October 18–22, 1971, St. Louis, Mo.

23. Richard W. Furlong. "Steel-Concrete Composite Columns," *Handbook of Composite Construction Engineering* (ed. by Gajanan Sabnis). New York: Van Nostrand Reinhold, 1979 (Chapter 6).

24. J. W. Roderick and D. F. Rogers. "Load-Carrying Capacity of Simple Composite Columns," *Journal of the Structural Division,* ASCE, **95,** February 1969 (ST2), 209–228.

25. Robert B. Knowles and Robert Park. "Axial Load Design for Concrete Filled Steel Tubes," *Journal of the Structural Division,* ASCE, **96,** October 1970 (ST10), 2125–2153.

PROBLEMS

All problems are to be worked in accordance with the strength method of the ACI Code unless otherwise indicated.

22.1 Determine the nominal moment strength M_n for live load plus superimposed dead load applied to the composite section for the section of the accompa-

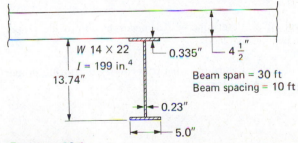

Problem 22.1

nying figure. The concrete has $f'_c = 4500$ psi $(n = 7.5)$, and the steel is A36 with $f_y = 36,000$ psi.

22.2   Determine the service live load moment capacity available for superimposed load on the composite slab on the precast beam of the accompanying figure. Use $f'_c(\text{slab}) = 3000$ psi $(n = 9)$, f'_c (beam) $= 6000$ psi $(n = 6)$, and $f_y = 40,000$ psi.

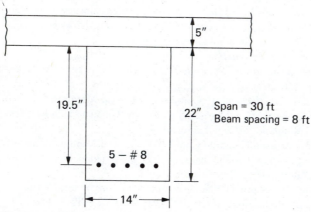

Span = 30 ft
Beam spacing = 8 ft

Problem 22.2

22.3   For the composite beam of the accompanying figure, determine the service live load moment capacity. Use $f'_c(\text{slab}) = 3000$ psi $(n = 9)$, $f'_c(\text{beam}) = 4500$ psi $(n = 7.5)$, and $f_y = 50,000$ psi.

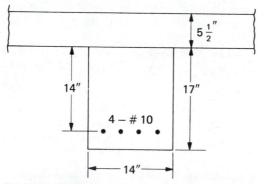

Problem 22.3

22.4   For the beam of Prob. 22.2, determine what is necessary to provide proper shear transfer for maximum live load.

22.5   For the beam of Prob. 22.3, determine what is necessary to provide proper shear transfer for maximum live load.

22.6 Determine the depth and reinforcement required for a 12-in. wide by 20-in. deep precast reinforced concrete beam on a span of 36 ft. The supported slab is 4 in. thick, and the beam spacing is 8 ft. The live load is 125 psf. Use f'_c(slab) = 3500 psi (n = 8.5), f'_c(beam) = 5000 psi (n = 7), and f_y = 60,000 psi. Assume that the construction is to be made without shores.

22.7 Investigate the economics of the composite concrete-concrete beam for the data of Prob. 22.6, except use a different beam spacing (4 ft, 6 ft, 7 ft, 9 ft, 10 ft, as assigned by instructor). As a consequence of different beam spacing, the required slab thickness may change. Consider slab concrete at $85.00/cu yd (including forms) and steel at $0.25/lb.

22.8 Repeat Example 22.10.1, except use a simply supported span of 22 ft, beam spacing 7 ft–6 in., with a 5 in. thick cast-in-place slab. The live load is 250 psf, and f_y = 60,000 psi. Other data are the same as the example.

Index

-32-33 Ø values (beams)
   Flexure with or without axial tension .90
   Axial tension   .90
   Shear & torsion   .85
-p 21-22' steel bars info
-Ø values for colum's & beams
.70 -tied colomn
.90 beam in bending
.85 shear
.75 Spiral column

f_y	$E_y - \dfrac{f_y}{E}$	$\min \rho = \dfrac{200}{f_y}$ ks.
40	.001379	.00500
50	.001724	.00400
60	.002069	.00333
75	.002586	.00267
80	.002759	.00250

f'_c	$.85\, f'_c$ ks.
3000	2.55
4000	3.40
5000	4.25
6000	5.10
3500	2.975

Quadradic Eq.

$$x = \frac{-B \pm \sqrt{B^2 - 4AC}}{2a}$$

142-143 shear bending only
161 shear - bending plus axial load
p 210 Basic develop· length
p 212 development length factors

	column and slab at a joint (Secs. 16.18 and 20.13)
n	modulus of elasticity ratio, E_s/E_c
N_n	nominal axial strength, taken positive for compression and negative for tension (Sec. 5.13)
N_{nc}	nominal strength in axial tension (Sec. 5.16)
N_u	factored axial force (taken positive for compression) (Sec. 11.4)
N_{uc}	factored tensile force on bracket or corbel acting simultaneously with V_u (taken positive for tension) (Secs. 5.15 and 5.16)
P_b	nominal axial strength at the balanced strain condition
P_c	Euler buckling load, $\pi^2 E_t I/(kL_u)^2$
P_i	approximation of nominal compressive strength in biaxial bending and compression obtained from the reciprocal load method (Sec. 13.21)
P_n	nominal axial load strength
$P_{n(max)}$	maximum nominal compression strength permitted for use by the ACI Code; $0.80P_0$ for tied column and $0.85P_0$ for spirally reinforced column
P_0	nominal strength P_n for an axially loaded column (i.e., for $e = 0$)
P_u	factored axial force
P_x	nominal compressive strength in combined compression and uniaxial bending about the x axis (Sec. 13.21)
P_y	nominal compressive strength in combined compression and uniaxial bending about the y axis (Sec. 13.21)
q	$\rho_g f_y/f_c'$ (Fig. 13.21.11)
Q_u	$\Sigma P_u \Delta_{1u}/(H_u h_s)$, stability factor (Chap. 15)
r	radius of gyration, $\sqrt{I/A}$
R	coefficient of resistance for working stress method, $M_{uc}/(bd^2)$ (Chap. 4)
	clockwise rotation of element axis (Sec. 16.20 and Chap. 17)
R_n	coefficient of resistance for strength design, $M_n/(bd^2)$
s	spacing of shear reinforcement measured along the axis of the member
	pitch of spiral reinforcement in a column measured center-to-center of spiral bar (Fig. 13.9.1)
s_2	vertical spacing of longitudinal shear reinforcement (Sec. 5.14)
S_{ii}	stiffness at end i of a flexural member ij; moment per unit angle at end i (Sec. 16.20; Chap. 17)
S'_{ii}	$(S_{ii} + F_{tj}T_1)/T_2$ (Sec. 17.4); $S_{ii}/(1 + S_{ii}/K_{ti})$ (Sec. 17.5)
S_{ij}	cross stiffness, moment at end i caused by unit angle at end j of member ij (Sec. 16.20; Chap. 17)
S'_{ij}	S_{ij}/T_2 (Sec. 17.4)
S_{ji}	cross stiffness; moment at end j caused by unit angle at end i of member ij (Sec. 16.20; Chap. 17)
S_{jj}	stiffness at end j; moment per unit angle at end j of member ij (Sec. 16.20; Chap. 17)
S'_{jj}	$(S_{jj} + F_{ti}T_1)/T_2$ (Sec. 17.4)
S_p	effective section modulus of precast or prefabricated beam
t	thickness of slab (Chaps. 9, 16, and 17)
	time; days after loading (Sec. 14.6)
t_a	age at loading, days (Sec. 14.6)
T	tensile force
	torsional force (Chap. 19)
T_0	force in concrete due to prestress before losses have occurred (Chap. 21)
T_1	tensile force in steel to balance C_c (Sec. 3.10)
	$S_{ii}S_{jj} - (S_{ij})^2$ (Sec. 17.4)
T_2	tensile force in steel to balance C_s (Sec. 3.10)
	$1 + F_{ti}S_{ii} + F_{tj}S_{jj} + F_{ti}F_{tj}T_1$ (Sec. 17.4); $1 + F_{ti}S_{ii}$ (Sec. 17.5)
T_b	tensile force in steel for the balanced strain condition
T_c	nominal torsional strength attributable to concrete
T_e	effect prestress force after all prestress losses (Chap. 21)
T_f	final tensile force in the tendons after elastic shortening has occurred (Chap. 21)
T_n	nominal torsional strength
T_{n0}	nominal torsional strength when member is subject to torsion alone (Chap. 19)
T_s	nominal torsional strength attributable to steel reinforcement, $\alpha_t x_1 y_1 A_t f_y/s$ (Chap. 19)
T_u	force in prestressed reinforcement when nominal strength M_n is reached, $A_{ps}f_{ps}$ (Chap. 21)
	factored torsional moment

LIST OF SYMBOLS

U	ACI Code factors for safety related to overload; factored load, factored moment, factored shear, factored axial force (Sec. 2.6)
v_c	nominal shear stress attributable to concrete, $V_c/(b_w d)$
v_n	nominal shear stress, $V_u/(\phi b_w d)$
v_{nh}	nominal ultimate unit stress capable of being transmitted on contact surface (values from ACI-17.5.4) (Chap. 22)
v_s	nominal shear stress attributable to the reinforcement, $V_s/(b_w d)$
v_{tc}	nominal torsional stress, $T_c/(\Sigma x^2 y)$ (Chap. 19)
v_{tn}	nominal torsional stress, $T_u/(\phi \frac{1}{3} \Sigma x^2 y)$ (Chap. 19)
v_{ts}	nominal torsional stress, $T_s/(\Sigma x^2 y)$ (Chap. 19)
V_c	nominal shear strength attributable to concrete
V_{ci}	nominal flexure-shear cracking strength (Chap. 21)
V_{cw}	nominal web-shear cracking strength (Chap. 21)
V_n	nominal shear strength, $V_c + V_s$
V_{n0}	nominal shear strength in the absence of torsion (Chap. 19)
V_s	nominal shear strength attributable to shear reinforcement
V_u	factored shear force
w	crack width at tension face of a section (Sec. 4.12)
w_c	density of concrete (for computing E_c)
w_D	service dead load per unit length or area
w_L	service live load per unit length or area
w_u	factored load per unit length or area
x	distance from compression face of section to neutral axis
	M_n/V_u (Sec. 6.15)
	short dimension of a rectangular element for use in computing torsional properties (Chaps. 16 and 19)
x_1	narrow dimension of core of a section subject to torsion (Fig. 19.6.3)
x_2	narrow dimension center-to-center of longitudinal bars of a section subject to torsion (Fig. 19.6.3)
x_b	neutral axis distance for the balanced strain condition
y	long dimension of a rectangular element for use in computing torsional properties (Chaps. 16 and 19)
y_1	long dimension of core of a section subject to torsion (Fig. 19.6.3)
y_2	long dimension center-to-center of longitudinal bars of a section subject to torsion (Fig. 19.6.3)
y_t	distance from neutral axis to extreme fiber of concrete in tension
z	factor of ACI Code relating to crack width, $f_s \sqrt[3]{d_c A}$ (Sec. 4.12)
α	angle of inclination of shear reinforcement measured from axis of member
	$(E_{cb}I_b/E_{cs}I_s)$; ratio of flexural stiffness of beam section to the flexural stiffness of a width of slab bounded laterally by the centerline of adjacent panel, if any, on each side of the beam (Chap. 16)
	constant used to determine the load contour curve for biaxial bending at a constant compression strength P_n (Sec. 13.21)
	$PL^2/(\pi^2 EI)$ (Chap. 15)
α_1	α in the direction of L_1 (Chap. 16)
α_2	α in the direction of L_2 (Chap. 16)
α_c	$\Sigma K_c/\Sigma(K_s + K_b)$, ratio of flexural stiffness of the columns above and below the slab to the combined flexural stiffness of the slab and beams at a joint taken in the direction moments are being determined
α_f	angle between shear-friction reinforcement and the shear plane
α_m	average α for all beams along the edges of a panel (Chap. 16)
α_{min}	minimum α_c to avoid using the positive moment multiplier (Chap. 16, Table 16.9.1)
α_t	dimensionless factor relating to torsional strength, $0.66 + 0.33 y_1/x_1$
β	ratio of long side to short side of a rectangular spread footing (Chap. 20)
	M_{nx}/M_{0x} or M_{ny}/M_{0y} (Sec. 13.21)
β_a	ratio of service dead load per unit area to service live load per unit area (Chap. 16)
β_1	ratio a/x; depth of rectangular stress distribution to the depth to the neutral axis
β_c	ratio of long side to short side of a concentrated load or reaction area
β_d	proportion of factored load moment that is sustained (Chap. 15)
β_s	ratio of length of continuous edges to total perimeter of a slab panel (Chap. 16)
β_t	$E_{cb}C/(2E_{cs}I_s)$ (Chap. 16)
γ	ratio $(d - d')/h$ (Fig. 13.16.4)
γ_p	dimensionless constant used in predicting f_{ps} (Sec. 21.8)
δ	magnification factor